FISH DISEASES AND DISORDERS
Volume 1
Protozoan and Metazoan Infections

Edited by

P.T.K. Woo

Department of Zoology,
University of Guelph,
Canada

CAB INTERNATIONAL

CAB INTERNATIONAL
Wallingford
Oxon OX10 8DE
UK

Tel: +44 (0)1491 832111
Telex: 847964 (COMAGG G)
E-mail: cabi@cabi.org
Fax: +44 (0)1491 833508

A catalogue entry for this book is available from the British Library.

ISBN 0 85198 823 7

Typeset in 10/12 pt English Times by Colset Pte Ltd, Singapore
Printed and bound in the UK at the University Press, Cambridge

Contents

v

Fish Diseases and Disorders: Volume 1 (Edited by P.T.K. Woo)
©CAB INTERNATIONAL 1995
Please note that the figure captions on the following pages are
incorrect.

p.483 should read
Fig. 13.3. *Sarcotaces verrucosus*, female, ventral view, alongside
host capsule (from *Mora moro*, Tasmania). Scale: units 1 mm.
p.484 should read
Fig. 13.4. *Sarcotaces verrucosus*, male (from *Epinephelus
undulostriatus*, Queensland). Scale bar = 0.5 mm.
p.485 should read
Fig. 13.5. *Sarcotaces verrucosus*. Section through collagenous
capsule wall (arrow) showing muscle bundles (m) and
vascularization (vc). (From Bullock *et al.*, 1986.)

Contributors

S.M. Bower, *Department of Fisheries and Oceans, Biological Sciences Branch, Pacific Biological Station, Nanaimo, British Columbia, Canada V9R 5K6.*

E.M. Burreson, *School of Marine Science, Virginia Institute of Marine Science, The College of William and Mary, Gloucester Point, Virginia 23062, USA.*

A. Choudhury, *Department of Zoology, The University of Manitoba, Winnipeg, Manitoba, Canada R3T 2N2.*

D.K. Cone, *Department of Biology, St Mary's University, Halifax, Nova Scotia, Canada B3H 3C3.*

D.L. Dawe, *Department of Medical Microbiology, University of Georgia, Athens, Georgia 30602, USA.*

T.A. Dick, *Department of Zoology, The University of Manitoba, Winnipeg, Manitoba, Canada R3T 2N2.*

H.W. Dickerson, *Department of Medical Microbiology, University of Georgia, Athens, Georgia 30602, USA.*

I. Dyková, *Institute of Parasitology, Czech Academy of Sciences, 370 05 České Budějovice, Branišovská 31, Czech Republic.*

R.C. Ko, *Department of Zoology, University of Hong Kong, Hong Kong.*

R.J.G. Lester, *Department of Parasitology, The University of Queensland, Brisbane, Queensland 4072, Australia.*

M.G. Levy, *Department of Microbiology, Pathology and Parasitology, College of Veterinary Medicine, North Carolina State University, Raleigh, North Carolina 27606, USA.*

J. Lom, *Institute of Parasitology, Czech Academy of Sciences, 370 05 České Budějovice, Branišovská 31, Czech Republic.*

K. Molnár, *Veterinary Medical Research Institute, Hungarian Academy of Sciences, H-1581 Budapest, Hungary.*

B.B. Nickol, *School of Biological Sciences, University of Nebraska-Lincoln, Lincoln, Nebraska 68588-0118, USA.*

E.J. Noga, *Department of Companion Animal and Special Species Medicine,*

College of Veterinary Medicine, North Carolina State University, Raleigh, North Carolina 27606, USA.

I. Paperna, *Department of Animal Sciences, Faculty of Agriculture, Hebrew University of Jerusalem, Rehovot 76–100, Israel.*

S.L. Poynton, *Division of Comparative Medicine, School of Medicine, The Johns Hopkins University, Baltimore, Maryland 21205, USA.*

F.R. Roubal, *Department of Parasitology, The University of Queensland, Brisbane, Queensland 4072, Australia.*

P.T. Thomas, *Department of Zoology, U.C. College PO, Aluva 683 102, Kerala, India.*

W.H. van Muiswinkel, *Department of Experimental Animal Morphology and Cell Biology, Wageningen Agricultural University, PO Box 338, 6700 AH Wageningen, The Netherlands.*

P.T.K. Woo, *Department of Zoology, University of Guelph, Guelph, Ontario, Canada N1G 2W1.*

Preface

Fin fish is the primary source of protein for humans in many parts of the world and this is especially true in most developing countries. The catch-fish industry has declined significantly and the decline is due to a series of factors which include over fishing, loss of fish habitats and environmental pollution. In the past few decades numerous international agencies and national governments have encouraged and continue to encourage private industries to be involved in aquaculture or have themselves gone into intensive fish culture, usually under artificial and/or semi-artificial conditions. Disease outbreaks (infectious and noninfectious) with resulting high mortalities occur more often when fish are held under relatively crowded and confined conditions. Also, mass mortality of healthy fish may occur even under good environmental conditions when an infectious agent is accidentally introduced into the culture system.

This volume is the first of three proposed volumes on diseases and disorders of freshwater and marine fishes (fin and shellfish). It is on parasitic infections while the second and third volumes are on microbial and noninfectious diseases/disorders. No single author can hope to write these volumes with any authority, hence I have chosen well qualified, internationally recognized experts to write the chapters. Each chapter deals with a specific disease/disorder or a group of closely related diseases. The primary purpose is to produce comprehensive and authoritative reviews by experts who are actively working in the area or have contributed greatly to our understanding of the disease/disorder.

The principal audience for the books is research scientists in the aquaculture industry and universities, fish health consultants, and managers of private and government fish health laboratories. The books are also appropriate for graduate and senior undergraduate students who are studying diseases of aquatic organisms. The series may also serve as reference textbooks for undergraduate and graduate courses in general parasitology, microbiology, environmental studies and for courses on impacts of diseases in aquaculture.

The secondary audience is research scientists with expertise in related disciplines (e.g. immunology, molecular biology) who wish to know more about specific important fish disease(s) so that they may be able to initiate research programmes in their areas of expertise. I expect this secondary audience to increase as it becomes evident that fish disease is an important component of the aquaculture industry and that fish health can be used as an indicator of problems in the aquatic ecosystem.

P.T.K. Woo

1

Dinoflagellida
(Phylum Sarcomastigophora)

E.J. Noga[1] and M.G. Levy[2]

[1]Department of Companion Animal and Special Species Medicine;
[2]Department of Microbiology, Pathology and Parasitology, North Carolina State University College of Veterinary Medicine, 4700 Hillsborough Street, Raleigh, North Carolina 27606, USA.

INTRODUCTION

Dinoflagellates are protozoa that are commonly found in aquatic ecosystems. They are important primary producers and consumers, as well as endosymbionts in many invertebrates (Taylor, 1987). Many dinoflagellates produce ichthyotoxins, which have caused mass mortalities in wild and cultured fishes (Steidinger and Baden, 1984). About 140 of the approximately 2000 known living species are parasites. They are mainly parasites of invertebrates (Drebes, 1984). Five genera have been reported as parasites of fishes: *Amyloodinium*, *Piscinoodinium*, *Crepidoodinium*, *Ichthyodinium*, and *Oodinioides*.

ECONOMIC IMPORTANCE

Amyloodinium

Amyloodinium ocellatum is the most common and important dinoflagellate parasitizing fish (Becker, 1977; Lawler, 1980; Lauckner, 1984). It causes morbidity and mortality in brackish and marine food fishes at aquaculture facilities throughout the world (Table 1.1) and is often considered the most consequential pathogen of marine fishes cultivated in warm water. (Paperna *et al.*, 1981). One outbreak in Mississippi, United States, resulted in the loss of over 300,000 juvenile striped bass (75–80% of the stock) in less than one week (McIlwain, 1976). *Amyloodinium ocellatum* is also a major problem in aquarium fishes (Nigrelli, 1936; Brown and Hovasse, 1946; Hojgaard, 1962; Dempster, 1972; Lawler, 1977b).

Amyloodinium ocellatum is one of the few fish parasites that infect both

1

elasmobranchs and teleosts (Lawler, 1980) and most fishes that live within its ecological range are susceptible to infections. As warm-water mariculture develops, there has been a concomitant increase in the number of susceptible species, which presently is over 100. Even freshwater fishes, such as centrarchids or tilapia, are susceptible to infection when they are in brackish water (Lawler, 1980; Table 1.1).

Piscinoodinium

Piscinoodinium is closely related to *Amyloodinium*. It is a problem in tropical freshwater fishes. Most reports of the parasite have been on aquarium fishes (Jacobs, 1946; Reichenbache-Klinke, 1955; Geus, 1960; van Duijn, 1973; Lom and Schubert, 1983).

Piscinoodinium limneticum has been identified in North American aquarium fishes, and *P. pillulare* occurs on European aquarium fishes and Malaysian food fishes (Shaharom-Harrison *et al.*, 1990). Many tropical fishes are susceptible to *Piscinoodinium*, with anabantids (Siamese fighting fish, gouramis), cyprinids (carp, goldfish, barbs) and cyprinodontids (killifish) frequently affected. Temperate species (e.g. *Cyprinus carpio*, *Tinca tinca*) and larval amphibians (*Ambystoma mexicanum*, *Rana temporaria* and *R. arvalis*) are also susceptible (Geus, 1960).

Crepidoodinium

Crepidoodinium, which is closely related to *Amyloodinium*, has been reported on some estuarine cyprinodontid fishes in the Gulf of Mexico and the Atlantic Ocean near Virginia, USA. Susceptible host species include *Fundulus majalis*, *F. heteroclitus*, *F. similis*, *F. luciae*, *Lucania parva* and *Cyprinodon variegatus* (Lawler, 1967, 1968; Williams, 1972). The parasite causes relatively minor damage to its host and is not fatal.

Ichthyodinium

Ichthyodinium chabelardi parasitizes sardines (*Sardina pilchardus*) and other marine clupeoid fishes in the Mediterranean Sea (Hollande and Cachon, 1952, 1953). Its impact is uncertain, but high prevalences have been reported in fish populations. Sardine eggs are infected mainly in winter, with prevalence in eggs ranging from 3.59% in April/May to 33.67% in January (Meneses and Re, 1988).

Oodinioides

This parasite was reported on freshwater and marine aquarium fishes. Little is known about its importance, because it has not been reported since the original description (Reichenbache-Klinke, 1970).

Table 1.1. Amyloodiniosis infestations reported in food fishes[1,2].

Geographic location	Hosts	References
Gulf of Mexico (Mississippi, Texas, Louisiana, USA)	Striped bass (*Morone saxatilis*) Redfish (*Sciaenops ocellata*) Mullet (*Mugil cephalus*) Pompano (*Trachinotus carolinus*)	Lawler (1980)
Atlantic Ocean (North Carolina, USA)	Hybrid striped bass (*Morone chrysops* × *Morone saxatilis*)	Noga *et al.* (1991)
Atlantic Ocean (Florida Keys, USA)	Snapper (*Lutjanus* sp.)	S. Citino (Miami Metrozoo, personal communication)
Caribbean Sea (Martinique, French West Indies)	Sea bass (*Dicentrarchus labrax*)	Gallet de Saint-Aurin (1987)
Caribbean Sea (Isla de Margarita, Venezuela)	Pompano (*Trachinotus goodei*, *Trachinotus carolinus*)	Gaspar (1987)
Mediterranean Sea (Italy)	Sea bass (*Dicentrarchus labrax*)	Ghittino *et al.* (1980); Giavenni (1988)
Mediterranean Sea (Sete, France)	Sea bream (*Sparus aurata*)	Paperna and Baudin Laurencin (1979)
Mediterranean Sea (Israel)	Sea bass (*Dicentrarchus labrax*) Sea bream (*Sparus aurata*)	A. Colorni (Israel Oceanographic and Limnological Research Ltd, unpublished data)
Mediterranean Sea (Sicily, Italy)	Yellowtail (*Seriola dumerili*)	Aiello and D'Alba (1986)
Adriatic Sea (Italy)	Sea bass (*Dicentrarchus labrax*)	Ghittino (personal communication cited in Paperna, 1980)
Aegean Sea (Turkey)	Sea bream (*Sparus aurata*) Sea bass (*Dicentrarchus labrax*)	Tareen (unpublished data cited in Tareen, 1986)
Red Sea (Eilat, Israel)	Sea bream (*Sparus aurata*) Sea bass (*Dicentrarchus labrax*) Mullet (Mugilidae)	Paperna *et al.* (1981)
Red Sea (Eilat, Israel)	Tilapia (*Oreochromis mossambicus*)	A. Colorni (unpublished data)
Persian Gulf (Kuwait)	*Acanthopagrus cuvieri*	Tareen (1986)
Pacific Ocean (Iloilo, Philippines)	Grey mullet (*Mugil cephalus*)	Baticados and Quinitio (1984)
Pacific Ocean (Thailand)	Sea bass (*Lates calcarifer*)	Chonchuenchob *et al.* (1987)

[1] This list only includes reported infestations where the contagion was identified in a local source or was introduced into local waters with the fish. It does not include numerous other reports from captive aquarium fishes worldwide.
[2] Note that taxonomic confirmation of isolates as *A. ocellatum* was not done in all cases.

TAXONOMY/SYSTEMATICS

Having characteristics of both plants and animals, dinoflagellates are classified in both taxa. They are members of the zoological Phylum Sarcomastigophora, Subphylum Mastigophora (flagellates), Class Phytomastigophorea (phytoflagellates), order Dinoflagellida (Levine *et al.*, 1980).

Two groups within the Dinoflagellida have fish-parasitic species: the zoological Tribe Blastodinida (botanical Order Blastodiniales), Family Oodinidae, and the Tribe Syndinida (botanical Order Syndiniales), Family Syndinidae. Parasitic dinoflagellates may be a polyphyletic group, which are evolutionarily derived from different ancestors (Cachon and Cachon, 1987). The reader should refer to Cachon and Cachon (1987) for details of parasitic dinoflagellate taxonomy.

Oodinid dinoflagellates

Oodinids, which are entirely extracellular parasites, include the ectoparasites *Amyloodinium* Brown and Hovasse, 1946 (*A. ocellatum* (Brown, 1931) Brown and Hovasse, 1946), *Piscinoodinium* Lom, 1981 (*P. pillulare* (Schaperclaus, 1954) Lom, 1981, *P. limneticum* (Jacobs, 1946) Lom, 1981) and *Crepidoodinium* Lom and Lawler, 1981 (*C. virginicum* Lom and Lawler, 1981).

Differentiation of the fish-parasitic oodinids is based primarily upon the morphology of the trophont, especially the type of host attachment and the mode of nutrition (Lom, 1981) (see pp. 6–18 and Table 1.2). All fish-pathogenic oodinids were previously classified in the genus *Oodinium*; however, there are important differences which justify their separation into different genera (Lom, 1981).

Many undescribed oodinid dinoflagellates parasitize fishes (Overstreet, 1968; Lom and Lawler, 1973; Paperna and Zwerner, 1976). Apparent differences in environmental tolerances, as well as morphological differences (dinospore morphology, presence/absence of a stigma, etc.) among different isolates of *Amyloodinium* and *Piscinoodinium* suggest that undescribed species or strains exist. Many reported amyloodiniosis outbreaks (Table 1.1) did not include taxonomic confirmation of the organism.

Syndinid dinoflagellates

The syndinids include intracellular parasites, and those that inhabit cavities, such as *Ichthyodinium* Hollande and Cachon, 1953 (*I. chabelardi* Hollande and Cachon, 1953).

Uncertain taxonomy

Oodinioides Reichenbache-Klinke, 1970 (*O. vastator* Reichenbache-Klinke, 1970) (Reichenbache-Klinke, 1970) is of questionable taxonomic validity (Lom, 1981).

Table 1.2. Key differentiating criteria of oodinid dinoflagllates infesting fish.

Genus	Host cytopathology	Penetration of holdfasts into cytoplasm	Stomopode	Clove-like bodies	Stigma	Chloroplasts	Food vacuoles	Lytic bodies	Mucocysts	Starch granules
Amyloodinium	Severe	Deep	Y	Y	Y	N	Y	Y	Y	Y
Piscinoodinium	Moderate	Moderate	N	N	N?[1]	Y	N	N	Y	Y
Crepidoodinium	Mild	None	N	N	N	Y	N	N	Y	Y

[1] *P. pillulare* dinospores have been reported by Hirschmann and Partsch (1953) as having a stigma. However, others have not observed a stigma in either *P. pillulare* (Lom, 1981) or *P. limneticum* (Jacobs, 1946).

PARASITE MORPHOLOGY/LIFE CYCLE

All fish-parasitic dinoflagellates are highly adapted to a parasitic mode of existence, and the trophozoite stage bears little resemblance to free-living dinoflagellates. It is only during the disseminative (dinospore) stage that typical dinoflagellate morphology is apparent (Fig. 1.1).

Oodinids

All oodinids have a similar life cycle (Fig. 1.1). They feed as sessile, sac-like trophozoites (trophonts) on the skin or gill epithelium (Figs 1.2, 1.3). Trophonts are bounded by a cell wall with a thin, resistant, pellicular envelope. Beneath the cell membrane is a layer of flat amphiesmal vesicles, as is found in free-living dinoflagellates (Dodge and Crawford, 1970). Various organelles, including chloroplasts, mitochondria, golgi apparatus, trichocysts, and mucocysts may be present.

Trophonts have a prominent stalk (Fig. 1.4), a cytoplasmic evagination with holdfasts that anchor the parasite to the host and in some cases absorb

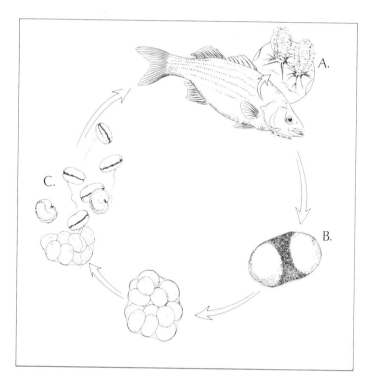

Fig. 1.1. Life cycle of blastodinid parasites of fish. A. Parasitic trophont that feeds on the skin and gill epithelium. B. Encysted tomont stage that divides to produce free-swimming infective dinospores, C. (From Noga, 1987; courtesy of *Science.*)

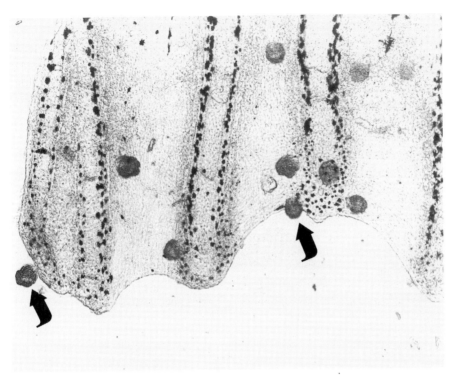

Fig. 1.2. Photomicrograph of *Amyloodinium* trophonts (arrows) on a damselfish (*Dacyllus* sp.) fin.

nutrients. The stalk is a characteristic of oodinid (and blastodinid) dino-flagellates. After feeding, the trophont detaches and withdraws its stalk, forms a reproductive cyst (tomont) and divides asexually several times. After the last asexual division, motile forms (dinospores; also referred to as gymnospores) are released, and they can infect a new host. This type of reproduction is termed palintomy. Dinospores have a girdle, sulcus and two flagella (Fig. 1.3). Dinospores are fairly uniform in size and the number of dinospores produced per tomont is roughly proportional to the size of the trophont.

Amyloodinium

LIFE CYCLE AND MORPHOLOGY

Details of the life cycle have been reported (Brown, 1934; Nigrelli, 1936; Brown and Hovasse, 1946). The unpigmented trophont is pear-shaped to ovoid and up to 350 μm long (Figs 1.2, 1.3). Trophonts are reported to have a cellulose wall consisting of a theca with amphiesmal plates. Thecal armour is complete except for a gap at the base of the trophont. An osmophilic ring encircles the basal region, and an attachment plate bearing numerous filiform rhizoids (Fig. 1.4) exits through the break in the theca. The rhizoids are enlarged at their bases, which serves to anchor them in the host cell. A 30 μm

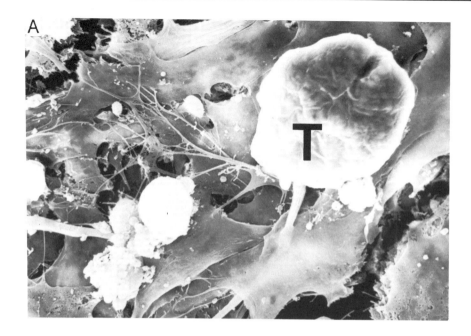

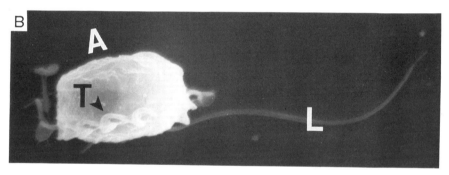

Fig. 1.3. Scanning electron micrographs of *Amyloodinium ocellatum.* A. The attachment of a trophont (T) to G1B cells in cell culture. B. A dinospore from a North Carolina isolate, showing longitudinal flagellum (L) and transverse flagellum (T). Note the antero-posteriorly flattened appearance. A = anterior end of dinospore. (From Noga *et al.*, 1991; courtesy of *Journal of Aquatic Animal Health.*)

stomopode is associated with the attachment plate. It passes from deep within the trophont and exits proximal to the sulcal fold. This organelle, which probably functions in feeding, contains two types of inclusions: 'lytic bodies' and 'clove-like structures', both believed to aid digestion. Starch granules and food vacuoles containing host cell cytoplasm are scattered throughout the cytoplasm, and chloroplasts are absent (Lom and Lawler, 1973).

After feeding, the trophont detaches from the host, retracts its rhizoids, and becomes a tomont. Divisions within a common cyst wall produce up to 256 dinospores (Brown, 1934; Nigrelli, 1936). Dinospores are 8–13.5 μm long

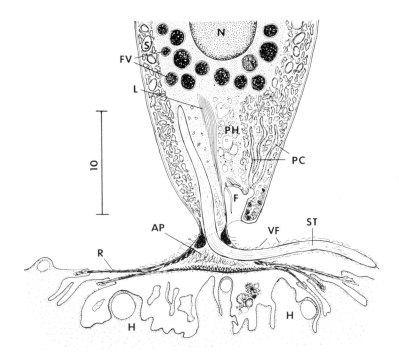

Fig. 1.4. Host attachment mechanism of *Amyloodinium*. Scale is in microns. H = host cell; AP = attachment plate; R = rhizoid; ST = stomopode tube; VF = velum-like pellicular folds; F = flagellum; PC = pusular canal; L = fibrillar ledge; FV = food vacuole; N = nucleus; PH = phagoplasm; S = starch grains. (From Lom and Lawler, 1973; courtesy of *Protistologica*.)

by 10–12.5 µm wide. Brown (1934) and Nigrelli (1936) described *A. ocellatum* as having gymnodinoid (barrel-shaped) dinospores, but recent studies have described isolates with antero-posteriorly compressed dinospores (Lawler, 1980, Noga *et al.*, 1991; Fig. 1.3). The taxonomic significance of dinospore morphology is uncertain. A red stigma, which accounts for the name 'ocellatum', is present throughout the life cycle.

ECOLOGICAL FACTORS AFFECTING GROWTH

Amyloodiniosis is limited to warm waters. In Red Sea isolates, the optimal temperature for tomont division and sporulation ranges from 23–27°C. Completion of tomont division is limited to 16–30°C (Paperna, 1984). Infections do not occur at less than 17°C (A. Colorni, personal communication). At 15°C, strain DC-1 tomonts stop dividing after four divisions; however, after at least 17 weeks exposure to 15°C, these tomonts can produce dinospores when returned to 25°C (C.E. Bower, personal communication).

Temperature also affects salinity tolerance, which narrows as the deviation from optimal temperature increases. Effective tomont division and sporulation of Red Sea isolates occurred at up to 50 ppt salinity. The minimum effective salinity varied from 12 to 20 ppt, depending upon the isolate

(Paperna, 1984). Hojgaard (1962) reported that isolates in Danish public aquaria were inhibited by 10 ppt salinity. In the Gulf of Mexico, however, outbreaks in oligohaline waters are common (McIlwain, 1976) with epidemics reported in salinities of 2.8–45 ppt (Lawler, 1977b). The reason for such differences in salinity tolerance are unknown. The salinity of the Red Sea usually fluctuates very little (around 40 ppt), whereas near-shore waters in the Gulf of Mexico may vary greatly; this depends upon freshwater runoff.

Bacteria and protozoa inhibit the development of *Amyloodinium* (Brown, 1934; Bower *et al.*, 1987; Noga and Bower, 1987) by delaying dinospore emergence and killing sporulating tomonts before dinospores emerge (Brown, 1934; Bower *et al.*, 1987). It is uncertain whether this inhibition is due to a direct toxic effect or environmental changes (e.g. hypoxia) caused by these contaminants.

Piscinoodinium

LIFE CYCLE AND MORPHOLOGY

The trophont is a yellow-green, pyriform or sac-like cell and is up to $12 \times 96\,\mu m$. It is almost round when mature; it appears to possess a rudimentary sulcus (Lom and Schubert, 1983). A short stalk with an attachment disc extends from its base (Fig. 1.5) and thin nail-like holdfasts (rhizocysts) radiate from the stalk. Head-parts of the rhizocysts are inverted in separate compartments (rhizothecas) in the sole of the disc while their shafts are firmly embedded in host cell cytoplasm (Fig. 1.5).

The theca covers the entire cell except for the area of the attachment disc. The periphery of the cell contains disc-shaped chloroplasts and numerous starch granules. Up to 256 dinospores (10–19 μm long $\times$ 8–15 μm wide in *P. limneticum*) are produced from each tomont.

ECOLOGICAL FACTORS AFFECTING GROWTH

The life cycle may be completed in 10–14 days under optimal conditions. The optimal temperature for *P. pillulare* is 23–25°C, with sporulation requiring 50–70 h for an average-sized tomont. At 15–17°C, sporulation requires 11 days (van Duijn, 1973). Optimal conditions are probably similar for *P. limneticum*. Under crowded conditions or in stagnant water, sporulation is inhibited and smaller dinospores are produced. Lower temperature slows the life cycle and results in larger dinospores (Jacobs, 1946).

Crepidoodinium

LIFE CYCLE AND MORPHOLOGY

The trophonts measure up to 670 μm long by 130 μm wide (Lom and Lawler, 1973). There are well-developed chloroplasts, spongy cytoplasm and abundant starch granules. The holdfast organelle consists of numerous attachment projections and rhizoids (Fig. 1.6) formed by cytoplasmic stems evaginated through thecal openings. The finger-like rhizoids are attached only superficially to the host cell membrane, which is modified at the point of contact.

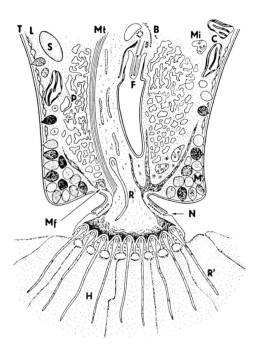

Fig. 1.5. Host attachment mechanism of *Piscinoodinium*. H = host cell; R = rhizocyst supposedly migrating into the attachment disc; R' = rhizocyst 'in position' within the rhizotheca, embedded into the host cell cytoplasm; Mf = microfibrillar strands converging to form a perinema-like ring around the 'neck' of the attachment disc; N = notch on the upper surface of the attachment disc; above it, the theca is subtended by a circular ribbon of microtubules and an electron-dense substance; C = chloroplast; S = starch grains; Mi = mitochondrion; Mt = microtubular ribbons along the zone of special cytoplasm extending from nucleus into the attachment disc; B = complex of basal bodies; F = flagellum; T = theca; L = subthecal lacunae; P = pusular system. (From Lom, 1981; courtesy of *Folia Parasitologica*.)

The attachment resembles the cell junction of epithelial cells. Penetration of host cells has not been observed. Up to 2048 dinospores are produced from a single tomont.

Syndinids

Ichthyodinium

LIFE CYCLE AND MORPHOLOGY

Ichthyodinium chabelardi is an endoparasite in the vitelline sac of sardines and some other teleosts in the Mediterranean Sea. The earliest parasitic stage in the egg yolk or larval yolk sac is a spherical, amoeboid, uninucleate trophont less than 8 µm in diameter. One to three trophonts may be seen and they grow

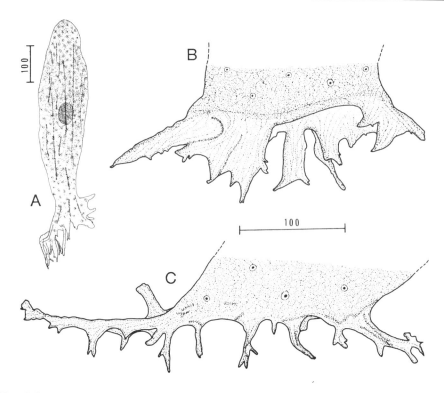

Fig. 1.6. Morphology of *Crepidoodinium*. Scales are in microns. A. Trophont with a rounded nucleus and the flattened holdfast organelle ramified into numerous lobose or finger-like projections from which the small attachment rhizoids arise. B and C. Enlarged holdfast portions of the trophonts; only the tips of small rhizoids emanating from these major and minor branches contact the host epithelial cells. (From Lom and Lawler, 1973; courtesy of *Protistologica*.)

rapidly to produce large (100–200 µm), multinucleated plasmodia or 'primary schizonts', via karyokinesis without plasmotomy. The 'primary schizonts' cleave into elongated 'secondary schizonts' (Fig. 1.7), which differentiate into subspherical units (15–20 µm), which resemble young trophonts. The host dies and the freed parasite divides once or twice more in the water. The resultant free-swimming dinospores survive several days in culture, but are not infective to sardine eggs by either exposure to eggs in water or direct inoculation of dinospores into the yolk mass (Hollande and Cachon, 1952, 1953). The mechanism responsible for transmission is unknown.

Uncertain taxonomy

Oodinioides

Oodinioides is primarily an ectoparasite on the skin and gills of fishes. It is also sometimes found in the viscera. The colourless trophont is spherical to

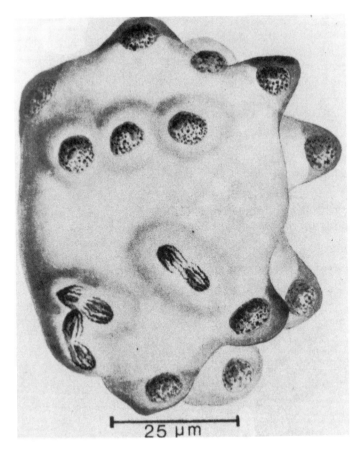

Fig. 1.7. *Ichthyodinium chabelardi.* A primary schizont in the process of differentiating into secondary schizonts (From Hollande and Cachon, 1952; courtesy of *Compte Rendu Hebdomadaire des Séances de l'Académie des Sciences, Paris.*)

pyriform and has refractile inclusions, which supposedly produce small microspores or larger flagellated macrospores. These differentiate into dinospores (Reichenbache-Klinke, 1970). The absence of proof that this is a dinoflagellate and the lack of further reports since its initial description, make this taxon highly questionable (Lom, 1981).

HOST–PARASITE RELATIONSHIPS

Amyloodinium

Amyloodinium causes amyloodiniosis, marine velvet disease, marine *Oodinium* disease and oodiniosis.

CLINICAL SIGNS AND PATHOLOGY

Clinical signs of amyloodiniosis include anorexia, depression, dyspnoea (swimming near the water surface with laboured breathing) and pruritis (Brown, 1934; Brown and Hovasse, 1946; Lawler, 1977a,b). The gills are usually the primary site of infection. Heavy infections may also involve the skin, fins and eyes. Paperna (1980) found that the skin was the primary site of infestation in *Sparus aurata* larvae. Heavily infested skin may have a dusty appearance ('velvet disease'), but this is not a common finding and fish often die without obvious gross skin lesions.

Trophonts may also occur on the pseudobranch, branchial cavity and nasal passages (Lawler, 1980). Although tomonts have been reported in the oesophagus, stomach and intestines (Hojgaard, 1962; Brown, 1934), they probably developed elsewhere and were subsequently swallowed. Cheung *et al.* (1981) reported on a case of presumptive amyloodiniosis in the viscera of a porkfish (*Anisostremus virginicus*). However, only tomont-like structures were seen and there was no strong morphological evidence that this organism was *A. ocellatum*.

A single trophont can feed on multiple epithelial cells simultaneously (Paperna, 1980; Fig. 1.3). This is at least partly responsible for the extensive damage caused by the parasite. Damage may be exacerbated by the hydrodynamic forces exerted on the trophont, which projects into the water, when the fish swims. Trophonts can also be seen moving slowly back and forth while attached (Noga, 1987). This may facilitate host cell fragmentation.

Mild infestations (e.g. 1–2 trophonts per gill filament) cause little pathology. However, heavy infestations (up to 200 trophonts per gill filament) cause serious gill hyperplasia, inflammation, haemorrhage and necrosis. Death is usually attributed to anoxia and occurs within 12 hours in heavy infestations (Lawler, 1980). However, acute mortalities that are sometimes associated with apparently mild infestations suggests that hypoxia may not always be the cause of death. Osmoregulatory impairment and secondary microbial infections due to severe epithelial damage may also be important.

Paperna (1984) felt that the serious gill pathology associated with many *A. ocellatum* infestations was more severe than the parasitosis would indicate. He speculated that *A. ocellatum*, like some other dinoflagellates, may produce an exotoxin responsible for some of the hyperplasia and necrosis. This remains to be proven, but in cell culture, *A. ocellatum* trophonts produce a very localized area of cell necrosis and this appears to be associated with the feeding activity of the parasite (Noga, 1987).

INNATE IMMUNITY

Although most fish species in *Amyloodinium*-infested waters are susceptible, certain factors clearly limit the ability of *Amyloodinium* to proliferate, as feral fishes usually have very few parasites (Lawler, 1980). Epidemics have only been reported in the stressful confines of aquaculture (Lawler, 1977b, 1980; Lauckner, 1984), even though there is often very high prevalence in feral fish (Lawler, 1980).

Host factors must play an important role in host–parasite interactions.

Oodinids feed exclusively on or within the epithelial tissues of the skin or gills. Thus, all host–parasite interactions (i.e. host recognition, defensive mechanisms responsible for protecting against these pathogens, etc.) are located in the mucus, or on/in epithelial cells, and extracellular fluid of the epithelium.

Some fish species are naturally resistant to infection. They include *Fundulus grandis*, *Anguilla rostrata*, *Poecilia latipinna* and others belonging to seven families. Resistant species are generally those which produce thick mucus or can tolerate low oxygen levels (Lawler, 1977b). In some instances, older fish appear to be more resistant. Quantitative comparisons of susceptibility also showed that clownfish (*Amphiprion ocellaris*) are more susceptible than striped bass (*Morone saxatilis*), which are more susceptible than tilapia (*Oreochromis mossambicus*) (C.E. Bower, personal communication). Mucus and serum from tilapia or striped bass with no previous exposure to *Amyloodinium* possess anti-*Amyloodinium* activity. As little as 1.25% serum is parasiticidal (Landsberg *et al.*, 1992).

ACQUIRED IMMUNITY

There is some anecdotal evidence that fish recovered from amyloodiniosis are resistant to reinfestation (Lawler, 1977b, 1980; Paperna, 1980), and serum from fish immunized with dinospores agglutinates living dinospores and kills *Amyloodinium* in cell culture (Smith *et al.*, 1993).

Although *A. ocellatum* has a pantropical distribution, little is known about differences (e.g. antigenic) between geographic isolates. Using a recently developed ELISA (enzyme-linked immunosorbent assay), specific anti-*Amyloodinium ocellatum* antibody was detected in tilapia (*Oreochromis aureus*) and sea bream (*Sparus aurata*), immunized with the DC-1 strain and a Red Sea isolate respectively (Smith *et al.*, 1992). Anti-*A. ocellatum* antibody has also been identified in hybrid striped bass (*Morone saxatilis* × *M. chrysops*) that had recovered from an amyloodiniosis outbreak. Strain DC-1 antigens were used as the target antigen (Smith *et al.*, 1994). Although only three isolates have thus far been examined, it is evident that the three isolates possess common antigens.

Piscinoodinium

Piscinoodinium causes freshwater velvet, rust disease, gold dust disease, pillularis disease and freshwater *Oodinium* disease.

CLINICAL SIGNS AND PATHOLOGY

Piscinoodinium infests the skin and gills. Clinical signs (Fig. 1.8) are similar to amyloodiniosis, except that fish can withstand much heavier infestations. The parasite is most pathogenic in young fish. They may die within 1–2 weeks, while older fish may live for months. Heavy infestations produce a yellow or rusty sheen to the skin. There is excess mucus, darkening of the skin, dyspnoea, anorexia and/or depression (Jacobs, 1946; van Duijn, 1973; Shaharom-Harrison *et al.*, 1990). Skin ulcers (Shaharom-Harrison *et al.*, 1990)

Fig. 1.8. *Piscinoodinium* infestation of the dorsal fin of a barb. Each refractile white focus is a parasite.

and tattered, sloughing, epithelium (Schaperclaus, 1951) have been seen in some cases.

Histopathology ranges from separation of the respiratory epithelium to severe hyperplasia of the entire gill filament. Filament degeneration and necrosis may occur. Some parasites may become almost entirely covered by hyperplastic epithelium (van Duijn, 1973; Shaharom-Harrison *et al.*, 1990), probably due to the chronic irritation caused by infestation. Some 'encysted' parasites may even sporulate (Geus, 1960).

Deeply penetrating rhizocysts anchor the trophont to its host. They damage the epithelium and this may be exacerbated by the hydrodynamic forces exerted on the trophont projecting into the water moving past the fish. Cellular damage is thought to attract parasites because trophonts typically cluster in damaged areas.

Outbreaks may be precipitated by stress (i.e. poor water quality or unfavourable temperatures). There is some evidence that recovered fish may be resistant to reinfestation (Jacobs, 1946).

Crepidoodinium

CLINICAL SIGNS AND PATHOLOGY

This photosynthetic dinoflagellate of marine and estuarine fish is attached only superficially to host epithelium and is not considered to be pathogenic.

IN VITRO CULTURE

Amyloodinium

Amyloodinium can be maintained *in vivo* by serial exposure of infective dinospores to susceptible hosts (Lawler, 1980; Bower *et al.*, 1987). This method is very labour-intensive, and requires considerable space for holding fish. Noga and Bower (1987) developed a method of passaging the parasite on gnotobiotic fish larvae. Guppies (*Poecilia reticulata*) were successfully infected with germ-free dinospores. The parasites completed at least one full life cycle on the host fish and were serially passaged up to six times.

In some cases, the parasites on gnotobiotic larvae continued to grow and reproduce long after the host had died. This suggests that factors other than death of the host *per se*, are responsible for the parasite normally leaving the host. Rapid exodus from a dead host is common among other fish ecto-parasites (Hoffmann, 1967). *Amyloodinium* survives for several days after its host dies so long as the tissues that support its growth are nourished (Noga and Bower, 1987).

Organ cultures of fish larvae also supported parasite growth (Noga and Bower, 1987). *A. ocellatum* can grow in a fish gill cell culture maintained in a 1 : 1 mixture of cell culture medium and artificial seawater (Noga, 1987, 1992). One isolate (DC-1) has been continuously propagated *in vitro* for over 9 years and 400 cell culture passages (Noga, unpublished data). Cell-cultured parasites are similar morphologically and developmentally to those produced *in vivo* (Noga, 1987). Isolates from the Gulf and Atlantic coasts of the United States, the Red Sea, and the Mediterranean Sea, have also been cultured *in vitro* (Noga *et al.*, 1991; Colorni, Diamant, Noga and Levy, unpublished data).

The parasite can be propagated in cell culture medium consisting of a 1 : 1 mixture of Hank's Balanced Salt Solution and artificial seawater. Serum or other additives are not needed (Noga, 1989).

Media of less than 300 mOsm osmolarity does not support parasite survival. The effects of mineral composition on parasite survival indicate that ionic strength is only one determinant of survival (see p. 18).

This cell culture system can be used to quantitatively evaluate the infectivity of *Amyloodinium* dinospores. Exposure to parasite toxins or immune fish serum produces a reproducible, dose-dependent inhibition of infectivity (Noga, 1987; Smith *et al.*, 1993). It is a much more sensitive indicator of parasite survival than measuring inhibition of motility (Noga, 1987) and allows the quantitative evaluation of infectivity in a way which is analogous to colony counts of bacteria.

PARASITE NUTRITION AND PHYSIOLOGY

Amyloodinium

The mechanism by which *A. ocellatum* feeds on host cells is unclear. A single trophont can damage and kill several host cells (Lom and Lawler, 1973; Noga, 1987). This probably accounts for the severe injuries inflicted on the host by trophonts. Rhizoids anchor the parasite to the host cells, but probably do not absorb nutrients (Lom and Lawler, 1973). The stomopode may be a source of digestive enzymes that are injected into host cells or it may serve as a feeding tentacle that gathers fragments of cells severed by the pulling motion of the rhizoids. This is supported by the observation of the constant twisting motion of the trophont (Noga, 1987), which may facilitate the severance of the host cell fragments.

There is *in vitro* evidence that *A. ocellatum* derives probably all its nutrition from host cells. It grows equally well in a medium with only mineral salts, phosphate and glucose as in a medium with trace minerals, vitamins and amino acids (Noga, 1989). It is likely that absorption of nutrients from the water, as with free-living dinoflagellates, is of minimal importance. Further studies are needed, however, to substantiate this hypothesis.

Mineral composition has an important influence on the growth and survival of *A. ocellatum in vitro* (Noga, 1989). Mineral composition had different effects on various life stages. Media with K^+, Mg^{2+}, Ca^{2+} and SO_4^{2-} content were best for continued growth. Media which enhanced dinospore differentiation to trophonts would not support continued survival. Certain mineral ions (possibly K^+, Mg^{2+}, Ca^{2+} and SO_4^{2-}) inhibit dinospore differentiation while stimulating development of other stages (Noga, 1989).

The effect of mineral composition on parasite growth and development suggests that mineral fluxes occur even while the parasite is attached to the host. The intracellular fluid of fishes differs considerably from that of seawater with respect to concentrations of ions such as K^+ and Na^+. Hence, the parasite may need to regulate by ionic exchange with the environment (Noga, 1989).

Other dinoflagellates

The presence of chloroplasts and lack of food vacuoles suggests that nutrition in *Piscinoodinium* is derived from photosynthesis. The extensive ramifying nature of the attachment organelle implies that nutrition is also obtained from the host, probably by osmotrophy through the rhizocysts.

In *Crepidoodinium virginicum*, the minute rhizoids contact but do not penetrate enough of an area of the host cell membrane for osmotrophy to supply most of the nutrition (Lom and Lawler, 1973). Thus, *C. virginicum* is more appropriately termed a symphoriont ectocommensal than a parasite. It is an autotroph that probably uses the fish primarily as an attachment site.

DIAGNOSIS OF INFECTIONS

Gross skin infestations of oodinids are most easily seen on dark coloured fish. Skin parasites are best observed using indirect illumination, such as by shining a flashlight on top of the fish in a darkened room. Observing fish against a dark background also helps. While presumptive diagnosis of oodinid infestation may sometimes be made from the gross clinical appearance (e.g. 'velvet'), microscopic identification of trophonts or tomonts is required for definitive diagnosis. If fish are small, they are restrained in a dish of water, and eyes, skin, and fins are examined under a dissecting microscope. Lifting the operculum allows examination of the gills. Trophonts are removed by brushing the fish gently, followed by microscopic examination of the sediment, which contains detached parasites. Snips of gill are also removed from living or recently dead fish for examination (Lawler, 1977b; Lawler, 1980; Noga, 1988).

Freshwater dips dislodge marine dinoflagellates and are especially useful for small fish. Fish are placed in a beaker of fresh water for 1–3 minutes. After 15–20 minutes, tomonts settle to the bottom of the beaker. They are detected using a dissecting or inverted microscope (Bower *et al.* 1987). Fish are best examined while still living or soon after death, as parasites detach shortly after host death.

Other techniques, such as antigen or antibody detection systems (ELISA, agglutination assays, etc.) or DNA probes, have not been developed for diagnosing parasitic dinoflagellate infestations.

PREVENTION AND CONTROL

Amyloodinium

The economic importance of warm-water mariculture has created an effort toward development of methods for the prophylaxis and treatment of amyloodiniosis. *Amyloodinium* has a very rapid reproductive rate and can complete its life cycle in less than one week under optimal conditions; thus, prompt treatment is imperative to prevent it from quickly overwhelming a susceptible fish population. The free-swimming dinospore is susceptible to chemotherapy (Lawler, 1980; Paperna, 1984), but trophonts and tomonts are relatively resistant, making eradication difficult. For example, tomonts tolerate copper concentrations that are over ten times the levels that are toxic to dinospores (Paperna, 1984). Even when tomonts are inhibited from dividing, they can often resume dividing when returned to untreated water (Paperna *et al.*, 1981). Thus, periodic examination for re-infestation after treatment is advisable.

Copper is most widely used for treatment (Bower, 1983; Cardeilhac and Whitaker, 1988). The free copper ion is the active component and free copper must be maintained at 0.12–$0.15 \, \text{mg} \, \text{l}^{-1}$ for 10–14 days to control epidemics. Higher concentration of free copper should be avoided because it is toxic to fish. Copper levels needed to treat amyloodiniosis are also acutely toxic to most invertebrates and algae. Free copper ion is unstable in seawater and thus copper levels should be monitored closely, and adjusted as needed. Chelating agents (e.g citrate or EDTA) increase the stability of copper in water but still carry concerns about efficacy and safety (Noga, 1994).

Bower (personal communication) discovered that the antimalarial chloroquine diphosphate is very safe and effective in treating amyloodiniosis. Experimentally infested clownfish (*Amphiprion ocellaris*) were free of *A. ocellatum* infestation after a 10-day exposure to a single treatment of 5–10 mg l^{-1} chloroquine diphosphate. Chloroquine has no effect on tomont division, but kills dinospores immediately upon their excystment. The concentration is nontoxic to fish, but is highly toxic to micro- and macroalgae and to various invertebrates (Bower, personal communication). Chloroquine is very expensive. Lewis *et al.* (1988) examined the pharmacokinetics of water-borne chloroquine in cultured red drum, *Sciaenops ocellatus*.

Many other agents have limited succcess against amyloodiniosis. Flush treatment with 100–200 mg l^{-1} formalin for 6–9 hours caused trophonts to detach from the fish, but tomonts resumed division after removal of formalin (Paperna, 1984). Nitrofurazone (50 mg l^{-1}), Furanace® (2.5 mg l^{-1}), acriflavin (6 mg l^{-1}) or malachite green (0.5 mg l^{-1}) also halted tomont division or killed this stage over a 4 to 17 day treatment period, but parasite development often resumed after the therapeutant was removed from the water.

Long-term baths with chlortetracycline, tetracycline, acetic acid, and potassium permanganate were ineffective (Lawler, 1977a). Johnson (1984) found that the herbicides simazine, endothall, or diuron did not control amyloodiniosis. Benzalkonium chloride was reportedly curative (Johnson, 1984). The efficacy of this chemical is most encouraging and more work needs to be done.

Amyloodinium tolerates wide temperature and salinity ranges, making environmental control difficult (Paperna, 1984). Lowering the temperature to 15°C arrests the disease process, but this is almost never feasible. Lowering salinity delays but does not prevent infestations (Babaro and Francescon, 1985), unless fish are placed in fresh water. A short freshwater bath of up to 5 minutes dislodges most but not all trophonts (Kingsford, 1975; Lawler, 1977b).

The risk of introducing infectious dinospores into an aquaculture system may be reduced by ultraviolet irradiation of incoming water (Lawler, 1977b). Aging of water beyond the survival time of dinospores, and quarantine of new fish for at least 20 days, are additional measures which may reduce, but not eliminate, the risk of parasite introduction. Dinospores remain infective for at least 6 days at 26°C (Bower *et al.*, 1987).

Other parasites

Heating water to 33–34°C reportedly controls *Piscinoodinium* infestations (Untergasser, 1989), but some aquarium fish cannot tolerate high temperatures. The temperature should be raised no more than 1°C per hour. Immersion for 10 days in methylene blue ($3\,\text{mg}\,\text{l}^{-1}$) or acriflavine ($10\,\text{mg}\,\text{l}^{-1}$) is also reportedly curative (Jacobs, 1946; van Duijn, 1973). Higher dosages of methylene blue are recommended in organically rich water. However, methylene blue is toxic to nitrifying bacteria (Collins *et al.*, 1975) and many aquatic plants (van Duijn, 1973). While *Piscinoodinium* is susceptible to copper (van Duijn, 1973), copper is unsafe for fish when used in waters having less than $50\,\text{mg}\,\text{l}^{-1}$ alkalinity. A $35\,\text{g}\,\text{l}^{-1}$ salt (NaCl) dip for 1–3 minutes dislodges trophonts, and immersion for 3–5 days in $7\,\text{g}\,\text{l}^{-1}$ salt combined with $40\,\text{mg}\,\text{l}^{-1}$ potassium permanganate is reportedly curative (van Duijn, 1973). However, some tropical aquarium fish cannot tolerate such high salt concentrations and levels of potassium permanganate that exceed $2\,\text{mg}\,\text{l}^{-1}$ of active ingredient are not considered safe for fish (Plumb, 1979). Chloroquine deserves investigation. The temperature should be 24–27°C during treatment. Reducing lighting to inhibit autotrophy has also been advocated during drug treatment (van Duijn, 1973). Dinospores remain infective for up to 48 hours (Jacobs, 1946; van Duijn, 1973). Attempts to treat other parasitic dinoflagellates have not been reported.

SUMMARY AND RECOMMENDED RESEARCH

More effective methods to control amyloodiniosis in aquaculture systems are needed. The most promising approach is environmental manipulation. This includes a better understanding of environmental conditions that affect parasite growth and survival, as well as development of novel approaches to manipulate the environment to not only cure but also prevent parasitosis. Determining factors that inhibit trophont growth and development *in vivo* may provide methods of treating the parasites while on the host. This would be especially useful for eliminating latent amyloodiniosis, which appears quite common (Lawler, 1980). It is very difficult to eliminate the infestation and, with the increasing regulations on the use of drugs in aquaculture, it is necessary to try other approaches, some of which are increasingly attractive.

There is also a need for basic research to determine the strains and species of oodinids that infect fish. Other areas of interest include mechanisms of feeding and possibility of toxin production by these parasites.

Immunological potentiation via vaccination or stimulation of non-specific immunity also holds promise. Immunodiagnosis of infections in fish for certification and preventive health management would also be useful, as would the use of nucleic acid probes for identifying latent infestations. In a related area, there is a need to determine differences in ecological tolerances. The introduction of exotic parasites should be avoided. Thus, immunological and other methods of screening are needed, especially when some infections are asymptomatic.

ACKNOWLEDGEMENTS

Our research reported in this review has been supported by grants from the University of North Carolina Sea Grant College and the US-Israel Binational Agricultural Research and Development Program. We thank A. Colorni, A. Diamant and C. Bower for reviewing a draft of the manuscript and A. Lawler for helpful discussions.

REFERENCES

Aiello, P. and D'Alba, A. (1986) *Amyloodinium ocellatum* infestation in yellowtail, *Seriola dumerili*, intensively reared in Sicily, Italy. *Bulletin of the European Association of Fish Pathologists* 6, 110–111.

Barbaro, A. and Francescon, A. (1985) Parassitosi da *Amyloodinium ocellatum* (Dinophyceae) su larve di *Sparus aurata* allevate in un impianto di riproduzione artificialer. *Oebalia* 11, 745–752.

Baticados, M.C. and Quinitio, G.F. (1984) Occurrence and pathology of an *Amyloodinium*-like protozoan parasite on the gills of grey mullet, *Mugil cephalus*. *Helgoländer Meeresuntersuchungen* 37, 595–601.

Becker, C.D. (1977) Flagellate parasites of fishes. In: Krier, J.P. (ed.) *Parasitic Protozoa*. Academic Press, New York, pp. 357–416.

Bower, C.E. (1983) *The Basic Marine Aquarium*. C.H. Thomas, Springfield, Illinois, 269 pp.

Bower, C.E., Turner, D.T. and Biever, R.C. (1987) A standardized method of propagating the marine fish parasite, *Amyloodinium ocellatum*. *Journal of Parasitology* 73, 85–88.

Brown, E.M. (1934) On *Oodinium ocellatum* Brown, a parasitic dinoflagellate causing epidemic disease in marine fish. *Proceedings of the Zoological Society of London* Part 3, 583–607.

Brown, E.M. and Hovasse, R. (1946) *Amyloodinium ocellatum* (Brown), a peridinian parasitic on marine fishes. *Proceedings of the Zoological Society of London* 116, 33–46.

Cachon, J. and Cachon, M. (1987) Parasitic dinoflagellates. In: Taylor, F.J.R. (ed.) *The Biology of Dinoflagellates*. Blackwell, Oxford, pp. 571–610.

Cardeilhac, P. and Whitaker, B. (1988) Copper treatments: Uses and precautions. *Veterinary Clinics of North America (Small Animal Practice)* 18, 435–448.

Cheung, P.J., Nigrelli, R.F. and Ruggieri, G.D. (1981) *Oodinium ocellatum* (Brown, 1931) (Dinoflagellata) in the kidney and other internal tissues of pork fish, *Anisostremus virginicus* (L.) *Journal of Fish Diseases* 4, 523–525.

Chonchuenchob, P., Sumpawapol, S. and Mearoh, A. (1987) Diseases of cage-cultured sea bass (*Lates calcarifer*) in southwestern Thailand. *Australian Centre for International Agricultural Research Proceedings Series* 20, 194–197.

Collins, M.T., Gratzek, J.B., Dawe, D.L. and Nemetz, T.G. (1975) Effects of parasiticides on nitrification. *Journal of the Fisheries Research Board of Canada* 32, 2033–2037.

Dempster, R.P. (1972) A description of the use of copper sulfate as a cure for gill disease in marine tropical fish tanks. *Anchor* 6, 450–452.

Dodge, J.D. and Crawford, R.M. (1970) A survey of thecal fine structure in the Dinophyceae. *Botanical Journal of the Linnean Society* 63, 53–67.

Drebes, G. (1984) Life cycle and host specificity of marine parasitic dinophytes. *Helgoländer Meeresuntersuchungen* 37, 603–622.

Gallet de Saint-Aurin, D. (1987) Diseases of the sea bass *Dicentrarchus labrax* in intensive rearing programs in Martinique French West Indies. *Proceedings of the Gulf and Caribbean Fisheries Institute* 38, 144–163.

Gaspar, A.G. (1987) Algunas enfermedades de Pampanas cultivados experimentale en Venezuela. *Revista Latinoamericana de Acuicultura, Lima, Peru* No. 33-27-44.

Geus, A. (1960) Nachtragliche Bemerkungen 24r Biologie des Fischpathogenen Dinoflagellater *Oodinium pillularis* Schaperclaus. *Aquarien Terrarien Zoologica* 13, 305–306

Ghittino, P.S., Bignami, I.S., Annibali, A. and Boni, L. (1980) First record of serious oodiniasis in seabass (*Dicentrarchus labrax*) intensively reared in brackish water. *Rivista Italiana Piscicoltura e Ittiopatologica* 15, 122–127.

Giavenni, R. (1988) Some parasitic and other diseases occurring in sea-bass (*Dicentrarchus labrax* L.) broodstock in Italy. *Bulletin of the European Association of Fish Pathologists* 8, 45–46.

Hirschmann, H. and Partsch, K. (1953) Der 'Colisaparasit' – ein Dinoflagellat aus der Oodiniumgruppe. *Aquarien Terrarien Zoologica* 6, 229–234.

Hoffmann, G.L. (1967) *Parasites of North American Freshwater Fishes*. University of California Press, Berkeley, California, 486 pp.

Hojgaard, M. (1962) Experiences made in Denmarks Akvarium concerning the treatment of *Oodinium ocellatum*. *Bulletin de l'Institut Oceanographigue (Monaco)* Numero Special 1A, 77–79.

Hollande, A. and Cachon, J. (1952) Un parasite des oeufs de sardine: l'*Ichthyodinium chabellardi*, nov. gen., nov. sp. (peridinian parasite). *Compte Rendu Hebdomadaire des Séances de l'Académie des Sciences, Paris* (Ser. D) 235, 976–977.

Hollande, A. and Cachon, J. (1953) Morphologie et evolutium d'un peridinien parasite des oeufs de sardine (*Ichthyodinium chabelardi*). *Bulletin des Travaux publies par la Station d'Aquiculture et de Peche de Castiglione (Alger)* No. 4, 1–17.

Jacobs, D.L. (1946) A new parasitic dinoflagellate from freshwater fish. *Transactions of the American Microscopical Society* 65, 1–17.

Johnson, S.K. (1984) Evaluation of several chemicals for control of *Amyloodinium ocellatum*, a parasite of marine fishes. *Texas A & M* FDDL-M5, 4 pp.

Kingsford, E. (1975) *Treatment of Exotic Marine Fish Diseases*. Palmetto Publishing Co., St Petersburg, Florida, 90 pp.

Landsberg, J.H., Smith, S.A., Noga, E.J. and Richards, S.A. (1992) Effect of serum and mucus of blue tilapia, *Oreochromis aureus* on infectivity of the parasitic dinoflagellate, *Amyloodinium ocellatum* in cell culture. *Fish Pathology* 27, 163–169.

Lauckner, G. (1984) Diseases caused by protophytans (algae). In: Kinne, O. (ed.) *Diseases of Marine Animals*. Vol. IV, Part 1, Biologische Anstalt Helgoland, Hamburg, pp. 169–180.

Lawler, A.R. (1967) *Oodinium cyprinodontum* n.sp., a parasitic dinoflagellate on gills of Cyprinodontidae of Virginia. *Chesapeake Science* 8, 67–68.

Lawler, A.R. (1968) Occurrence of the parasitic dinoflagellate *Oodinium cyprinodontum* Lawler, 1967, in North Carolina. *Virginia Journal of Science* 19, 240 (Abstract).

Lawler, A.R. (1977a) The parasitic dinoflagellate *Amyloodinium ocellatum* in marine aquaria. *Drum and Croaker* 17, 17–20.

Lawler, A.R. (1977b) Dinoflagellate (*Amylooainium*) infestation of pompano. In: Sindermann, C.J. (ed.) *Disease Diagnosis and Control in North American Marine Aquaculture*. Elsevier, Amsterdam, pp. 257–264.

Lawler, A.R. (1980) Studies on *Amyloodinium ocellatum* (Dinoflagellata) in Mississippi Sound: natural and experimental hosts. *Gulf Research Reports* 6, 403–413.

Levine, N.D., Corliss, J.O., Cox, F.E.G., Deroux, G., Grain, J., Honigberg, B.M., Leedale, G.F., Loeblich, A.R., Lom, J., Lynn, D., Merinfeld, E.G., Page, F.C., Polyansky, Y., Sprague, V., Vavra, J. and Wallace, F.G. (1980) A newly revised classification of the Protozoa. *Journal of Protozoology* 27, 37–58.

Lewis, D.H., Wenxing W., Ayers, A. and Arnold, C.R. (1988) Preliminary studies on the use of chloroquine as a systemic chemotherapeutic agent for amyloodinosis in red drum (*Sciaenops ocellatus*). *Contributions in Marine Science* Suppl. to Vol. 30, 183–189.

Lom, J. (1981) Fish invading dinoflagellates: A synopsis of existing and newly proposed genera. *Folia Parasitologica* 28, 3–11.

Lom, J. and Lawler, A.R. (1973) An ultrastructural study on the mode of attachment in dinoflagellates invading the gills of Cyprinodontidae. *Protistologica* IX, 293–309.

Lom, J. and Schubert, G. (1983) Ultrastructural study of *Piscinoodinium pillulare* (Schaperclaus, 1954) Lom, 1981 with special emphasis on its attachment to the fish host. *Journal of Fish Diseases* 6, 411–428.

McIlwain, T.D. (1976) Striped bass rearing and stocking program – Mississippi. Completion Report AFCS-5. National Marine Fisheries Service, 121 pp.

Meneses, I. and Re, P. (1988) Infection of sardine eggs by a parasitic dinoflagellate (*Ichthyodinium chabelardi*) off Portugal. *Rapports et Proces-Verbaux des Reunions Ciem* 191, 442 (Abstract).

Nigrelli, R.F. (1936) The morphology, cytology, and life history of *Oodinium ocellatum* Brown, a dinoflagellate parasitic on marine fishes. *Zoologica* 21, 129–164.

Noga, E.J. (1987) Propagation in cell culture of the dinoflagellate *Amyloodinium*, an ectoparasite of marine fishes. *Science* 236, 1302–1304.

Noga, E.J. (1988) Biopsy and rapid postmortem techniques for diagnosing diseases of fish. *Veterinary Clinics of North America (Small Animal Practice)* 18, 401–426.

Noga, E.J. (1989) Culture conditions affecting the in vitro propagation of *Amyloodinium ocellatum*. *Diseases of Aquatic Organisms* 6, 137–143.

Noga, E.J. (1992) Immune response to ectoparasitic protozoa: The infectivity assay. In: Stolen, J.S., Fletcher, T.C., Anderson, D.P., Kaattari, S.L. and Rowley, A.F. (eds) *Techniques in Fish Immunology*, Vol. 2. SOS Publications, Fair Haven, New Jersey, pp. 167–176.

Noga, E.J. (1995) *Fish Disease: Diagnosis and Treatment*. Mosby-Year Book, Inc., St Louis (in press).

Noga, E.J. and Bower, C.E. (1987) Propagation of the marine dinoflagellate *Amyloodinium ocellatum* under germ-free conditions. *Journal of Parasitology* 73, 924–928.

Noga, E.J., Landsberg, J.H. and Smith, S.A. (1991) Amyloodiniosis in cultured hybrid striped bass (*Morone saxatilis* × *M. chrysops*) in North Carolina. *Journal of Aquatic Animal Health* 3, 294–297.

Overstreet, R.M. (1963) Parasites of the inshore lizardfish, *Synodus foetens*, from South Florida, including a description of a new genus of Cestoda. *Bulletin of Marine Science* 18, 444–470.

Paperna, I. (1980) *Amyloodinium ocellatum* (Brown 1931) (Dinoflagellida) infestations in cultured marine fish at Eilat, Red Sea: epizootiology and pathology. *Journal of Fish Diseases* 3, 363–372.

Paperna, I. (1984) Reproduction cycle and tolerance to temperature and salinity of *Amyloodinium ocellatum* (Brown 1931) (Dinoflagellida). *Annales de Parasitologie Humaine et Comparée* 59, 7–30.

Paperna, I. and Baudin Laurencin, F. (1979) Parasitic infections of sea bass, *Dicen-*

trarchus labrax and gilt head sea bream, *Sparus aurata*, in mariculture facilities in France. *Aquaculture* 16, 173–175.

Paperna, I. and Zwerner, D. (1976) Parasites and diseases of striped bass, *Morone saxatilis* (Walbaum), from the lower Chesapeake Bay. *Journal of Fish Biology* 9, 267–287.

Paperna, I., Colorni, A., Ross, B. and Colorni, B. (1981) Diseases of marine fish cultured in Eilat mariculture project based at the Gulf of Aqaba, Red Sea. *European Mariculture Society Special Publication* 6, 81–91.

Plumb, J.A. (1979) *Principal Diseases of Farm-raised Catfish*. Southern Cooperative Series No. 225. Auburn University, Alabama, USA, 92 pp.

Reichenbache-Klinke, H.H. (1955) Die Atzubehorigkeit in Mitteleuropa Vorkommenden *Oodinium*- Aut und Beobachtongen uber ihr parasitares stadium (Dinoflagellata, Gymnodiniidae) *Giornale di Microbiologia*, 106–111.

Reichenbache-Klinke, H.H. (1970) Vorlaufige Mitteilung und Neubeschreibung einer parasitaren Blastodinidae (Dinoflagellata) bei Susswasserfischen. *Zeitscher Fischerei Neue Forschongen* 18, 289–297.

Reichenbache-Klinke, H.H. (1973) *Fish Pathology*. T.F.H. Publications, Neptune City, New Jersey, 380 pp.

Schaperclaus, W. (1951) Der Colisa-Parasit, ein neuer Krankheitserreger bei Aquarienfischen. *Die Aquarien- und Terrarienzeitschrift* 4, 169–171.

Shaharom-Harrison, F.M., Anderson, I.G., Siti, A.G., Shazili, N.A.M., Ang, K.J. and Azmi, T.I. (1990) Epizootics of Malaysian cultured freshwater pond fishes by *Piscinoodinium pillulare* (Schaperclaus, 1954) Lom, 1981. *Aquaculture* 86, 127–138.

Smith, S.A., Levy, M.G. and Noga, E.J. (1992) Development of an enzyme-linked immunosorbent assay (ELISA) for the detection of antibody to the parasitic dinoflagellate *Amyloodinium ocellatum* in *Oreochromis aureus*. *Veterinary Parasitology* 42, 145–155.

Smith, S.A. Noga, E.J., Levy, M.G. and Gerig, T.M. (1993) Effect of serum from tilapia *Oreochromis aureus*, immunized with dinospores of *Amyloodinium ocellatum* on the motility, infectivity and growth of the parasite in cell culture. *Diseases of Aquatic Organisms* 15, 73–80.

Smith, S.A., Levy, M.G. and Noga, E.J. (1994) Detection of anti-*Amyloodinium ocellatum* antibody from cultured hybrid striped bass (*Morone saxatilis* × *Morone chrysops*) during an epizootic of amyloodiniosis. *Journal of Aquatic Animal Health* 6, 79–81.

Steindinger, K.A. and Baden, D.G. (1984) Toxic marine dinoflagellates. In: Spector, D.L. (ed.) *Dinoflagellates*. Academic Press, New York, pp. 201–261.

Tareen, I.U. (1986) Parasitic infestations on cultured marine fish *Acanthopagrus cuvieri* (Sparidae), incidence and control. *Special Publication of the European Mariculture Society* 9, 85–90.

Taylor, F.J.R. (1987) General group characteristics; special features of interest; short history of dinoflagellate study. In: Taylor, F.J.R. (ed.) *The Biology of Dinoflagellates*. Blackwell, Oxford, pp. 1–23.

Untergasser, D. (1989) *Handbook of Fish Diseases*. Tropical Fish Hobbyist Publications, Neptune City, New Jersey. 160 pp.

van Duijn, C., Jr (1973) *Diseases of Fishes*. Charles H. Thomas, Springfield, Illinois, 3rd edn, 274 pp.

Williams, E.H., Jr (1972) *Oodinium cyprinodontum* Lawler (Dinoflagellida) on *Fundulus similis* (Baird and Girard) and *Cyprinodon variegatus* Lacepede from the Gulf of Mexico. *Alabama Marine Resources Bulletin* 8, 32–33.

2

Diplomonadida, Kinetoplastida and Amoebida (Phylum Sarcomastigophora)

P.T.K. Woo[1] and S.L. Poynton[2]

[1]*Department of Zoology, University of Guelph, Guelph, Ontario Canada, N1G 2W1;* [2]*Division of Comparative Medicine, School of Medicine, The Johns Hopkins University, Baltimore, Maryland 21205, USA.*

INTRODUCTION

Diplomonads (Order Diplomonadida), kinetoplastids (Order Kinetoplastida) and amoebae (Order Amoebida) are parasitic protozoans (Phylum Sarcomastigophora) that are normally found on the body surface, in the digestive tract, and body fluids/blood of vertebrates (fishes, amphibians, reptiles, birds and mammals).

Diplomonads typically have two karyomastigonts which are characterized by a two-fold axial symmetry. Each karyomastigont has one to four flagella of which one is recurrent, and an axial structure of microtubules and accessory filaments (Lee, 1985). They are extracellular parasites and are normally found in the intestinal tract of vertebrates. The mature mobile trophozoite stage multiplies by binary fission, and forms cysts. Parasites are passed into the environment with faeces, transmission is direct, and infection normally occurs orally (a monogenetic life cycle).

Kinetoplastids have one or two flagella, an axoneme and a paraxial rod which arise from a flagellar pit. The single mitochondrion usually extends the length of the body and the kinetoplast, which is part of the mitochondrial system, stains with Feulgen, Janus green or Romanovsky stains. The mitochondrion is believed to be concerned with respiratory adaptations to different environmental conditions (Vickerman, 1976b). Most species are extracellular endoparasites in the intestinal tract, internal organs, blood and/or tissue fluids of their vertebrate hosts. The ectoparasitic flagellates and those associated with the intestinal tract have direct transmissions while the haematozoic parasites are usually transmitted by blood-sucking invertebrates.

Amoebae are usually uninucleated organisms with broad pseudopodia and reproduce by asexual binary or multiple fission. They do not have a theca, are usually free-living and are found in moist soil, in fresh water and in sea

27

water. These amoebae commonly form cysts and do not have a flagellated stage (Bovee, 1985). A few free-living species (e.g. *Thecamoeba hoffmani*, *Paramoeba pemaguidensis*) may, under certain environmental conditions, become ectoparasitic and multiply on gills of fish. They cause amoebic gill disease which may lead to mortality in heavily infected fish. Also, there are free-living amoebae (e.g. *Acanthamoeba*) that may invade internal organs to cause systemic amoebosis and this may lead to fish mortality. In general, studies on piscine amoebosis, especially systemic amoebosis, are very sketchy and the disease is in need of more intensive studies.

The majority of diplomonads and kinetoplastids are not known to cause disease in their vertebrate hosts. Some pathogenic species (e.g. *Hexamita meleagridis* of turkey, *Trypanosoma cruzi* of man, *Trypanosoma vivax* of cattle) are very well studied because of their medical and/or economic importance. Relatively less is known about those that cause disease in fish, amphibians and reptiles. This review is on diplomonads (*Hexamita* and *Spironucleus*), kinetoplastids (*Ichthyobodo*, *Cryptobia* and *Trypanosoma*) and amoebae (*Thecamoeba*, *Paramoeba* and *Acanthamoeba*) that are known to cause disease in freshwater and marine fishes.

HEXAMITOSIS AND SPIRONUCLEOSIS

Introduction

The flagellates most frequently reported from the intestinal lumen, and less frequently as systemic invaders, are *Hexamita* and *Spironucleus*. These closely related organisms have direct life cycles, and are collectively called diplomonads or hexamitids. They are sometimes associated with morbidity and mortality (hexamitosis, spironucleosis, and possibly 'hole in the head' disease), particularly in juvenile salmonids, in cyprinids, and in ornamental tropical aquarium fishes. Their pathogenicity is poorly understood and subclinical infections are common. The recent outbreaks of fatal systemic hexamitid infections in salt-water reared Atlantic salmon in Norway (Mo *et al.*, 1990; Poppe *et al.*, 1992), and in chinook salmon in Canada (Kent, 1992; Kent *et al.*, 1992) indicate new problems for the fish farming industry. The taxonomy of hexamitids is complicated and *Hexamita* and *Spironucleus* are considered together here unless stated otherwise. A third hexamitid, *Trepomonas*, occasionally reported from the intestine of a marine fish, is mentioned briefly.

Geographical distribution and host range

Hexamitids commonly infect wild, farmed and aquarium fishes in cold, temperate and warm waters of diverse salinities, and in many parts of the world. Prevalence may vary markedly between adjacent geographic localities (Arthur *et al.*, 1976). *Hexamita salmonis* occurs in Europe, North America, and Asia, typically in freshwater, less commonly in marine fishes (Becker, 1977; Mo *et al.*, 1990). *Spironucleus anusmirabilisvortens* infects freshwater tropical aquarium fish in the southern United States (Poynton *et al.*, 1994), *S. elegans* is reported from freshwater tropical aquarium fish in Europe (Kulda and Lom, 1964a,b) and *S. torosa* occurs in fishes in the Canadian Atlantic (Poynton and Morrison, 1990). The hexamitid in chinook salmon is transmitted in salt and fresh water, although trophozoites survive poorly in fresh water (Kent, 1992; Kent *et al.*, 1992).

Hexamitids are common in the intestinal tract, where infections may reach high numbers (Becker, 1977) and they infect cold and temperate water fishes (Acipenseridae, Anguillidae, Catostomidae, Centrarchidae, Cyprinidae, Cyprinodontidae, Gadidae, Gasterosteidae, Mugilidae, Percichthyidae, Percidae, Salmonidae, Siganidae, and Sparidae). *Hexamita salmonis* (Moore) usually infects salmonids, particularly rainbow trout (*Oncorhynchus mykiss*) and Atlantic salmon (*Salmo salar*) (see Mo *et al.*, 1990). Disease due to *H. salmonis* is most commonly encountered in fingerling salmonids in aquaculture, although yearlings and smolts may be affected. The disease is usually sporadic in aquaculture and rare in wild fish (Allison, 1963; Uzmann and Hayduk, 1963; Becker, 1977; Mo *et al.*, 1990). *Spironucleus torosa* infects wild marine gadids, namely Atlantic cod (*Gadus morhua*) and haddock (*Melanogrammus aeglefinus*) (see Poynton and Morrison, 1990).

Warm water hosts of hexamitids include Acanthuridae, Anabantidae, Aspreninidae, Belontiidae, Cichlidae, Doradidae, Poeciliidae and Pomacentridae (Becker, 1977; Ferguson and Moccia, 1980; Bassleer, 1983; Post, 1987; Andrews *et al.*, 1988; Gratzek, 1988). *Spironucleus anusmirabilisvortens* infects angel fish (*Pterophyllum scalare*) in Florida (Poynton *et al.*, 1994). *Spironucleus elegans* Lavier, has been reported from angel fish (*Pterophyllum scalare*), and amphibians in Europe (Kulda and Lom, 1964a,b; Becker, 1977).

Hexamitids also have been found in digenean trematodes parasitizing fish (Hunninen and Wichterman, 1938; Overstreet, 1976). *Trepomonas agilis* is reported from the marine fish *Box salpa*, as well as amphibians and dead copepods; it can also live freely in fresh water (Grassé, 1952).

Systematics and taxonomic position

According to Lee (1985) and Vickerman (1990), members of the Order Diplomonadida having a double set of organelles (the diplozoic forms) belong to the Suborder Diplomonadina, Family Hexamitidae. *Trepomonas*, *Hexamita* and *Spironucleus* infect many hosts including fishes, and *Octomitus* and *Giardia* only infect higher vertebrates. Synonyms of *Hexamita* and *Spironucleus* include *Octomitus* and *Urophagus* (see Alexeieff, 1910; Becker, 1977). The name 'discus parasite' is sometimes used for infections in tropical fish because of frequent infections in discus fish (*Symphysodon discus*) (see Gratzek, 1988).

For many years the generic identity of most hexamitids from fish has been uncertain because of difficulties in distinguishing between the small ($<20\,\mu$m long) trophozoites using only light microscopy. It was believed that a spherical body and spherical nuclei characterized *Hexamita*, and a pyriform body and elongate nuclei characterized *Spironucleus* (see Becker, 1977; Lee, 1985). However, the shape of the body and the nuclei may be very variable in light microscope preparations, and this technique is therefore unreliable for distinguishing between the genera (Poynton and Morrison, 1990). An unambiguous ultrastructural feature, the position of the anterior kinetosomes (flagellar bases), is now the principal characteristic used to distinguish *Hexamita* from *Spironucleus* (see Lee, 1985 and below).

The taxonomy of hexamitids needs further studies because most descriptions are only based on light microscope studies, which do not provide adequate resolutions for species distinctions. Approximately 20 species of *Hexamita* and ten species of *Spironucleus* have been described (Vickerman, 1990), and many of those infecting fish are assumed to be *H. salmonis*. Becker (1977) suggested that hexamitids infecting marine fish are conspecific with *H. salmonis*, but this has recently been disproved (Poynton and Morrison, 1990). Only three hexamitids from fishes have been comprehensively studied, including by ultrastructural investigation, namely *Hexamita salmonis* (Figs 2.1, 2.2; Ferguson, 1979), *Spironucleus anusmirabilisvortens* (Poynton *et al.*, 1994) and *Spironucleus torosa* (Figs 2.3–2.8; Poynton and Morrison, 1990). A fourth hexamitid from fish, believed to be *S. elegans*, has been studied extensively using light microscopy (Kulda and Lom, 1964a,b); its ultrastructure (TEM only) is known only from specimens isolated from amphibians (Brugerolle *et al.*, 1973).

ORIGIN AND EVOLUTION OF THE PARASITE

The ancestor of diplozoic flagellates such as *Hexamita* and *Spironucleus* (Suborder Diplomonadina) may be *Enteromonas*, a monozoic flagellate (Suborder Enteromonadina) (see Brugerolle, 1975; Siddall *et al.*, 1992). This genus, one of the two most basal in the Order Diplomonadida (Siddall *et al.*, 1992) is a simple organism, with a single karyomastigont. Evolution towards parasitism within the Diplomonadina is believed to have proceeded from *Trepomonas* to *Hexamita*, *Spironucleus*, *Octomitus* and culminated in *Giardia* (see Brugerolle, 1975; Siddall *et al.*, 1992). *Trepomonas* and *Hexamita* may live freely in organically rich waters, or as intestinal parasites, while the remaining genera are parasitic. As evolution proceeds, the cytostome, or flagellar pocket, becomes reduced. It is present in *Trepomonas*, *Hexamita* and *Spironucleus* (Subfamily Hexamitinae), but absent in *Octomitus* and *Giardia* (Subfamily Giardiinae). A ventral disc, for attachment to the host's intestine, is present only in *Giardia*.

Parasite morphology and life cycle

MORPHOLOGY

Live *Hexamita* and *Spironucleus* trophozoites vary from nearly spherical to elongate, with considerable variations between and within individuals of one species (Figs 2.3, 2.4, 2.7; Poynton and Morrison, 1990). Pyriform trophozoites become more spherical before they divide by binary fission. The length of live organisms may reach 20 μm. Each cell is bilaterally symmetrical or diplozoic, with each side containing, at the anterior end, a nucleus (Figs 2.1, 2.5, 2.6), and four flagella (Fig. 2.2), together comprising the karyomastigont. The kinetosomes of each karyomastigont are arranged in two pairs (a characteristic of the Order), the most anterior being designated K1 and K2, the most posterior being designated K3 and R (recurrent); those of each pair are usually at right angles to one another. Three locomotory flagella extend anteriorly, while the fourth is recurrent and passes posteriorly through the body, and emerges as a free flagellum (Lee, 1985). Flagella are about one and a half times the length of the body. Some hexamitids such as *S. torosa* have a caudal projection at the posterior end of their body (Fig. 2.7).

The length and widths of trophozoites are as follows: *H. salmonis* 8–14 × 6–10 µm (fixed) (Kulda and Lom, 1964a); *S. anusmirabilisvortens* 13–21 × 5–11 µm (live) (Poynton *et al.*, 1994); *S. elegans* 5–11 × 2–5 µm (fixed) (Kulda and Lom, 1964a); *S. torosa* 10–18 × 3–13 µm (live) (Poynton and Morrison, 1990). Cysts measure 7 × 10 µm, and are oval or round, containing a flagellate that divides once (Becker, 1977). *Trepomonas* trophozoites have on each side a pyriform nucleus and four flagella. The longest flagellum is for locomotion, while the three shorter oral flagella beat within the long oval oral groove (Lee, 1985).

In *Hexamita* the cytostome openings are caudal, at the flattened posterior end of the body (Fig. 2.2), and in *Spironucleus* they are posterolateral and the body may taper posteriorly to a caudal projection (Fig. 2.7) (Table 2.1) (Brugerolle *et al.*, 1973; Becker, 1977; Kulda and Nohynkova, 1978; Lee, 1985). The oval or spherical nuclei of *Hexamita* are apposed at their flattened median portions, whereas in *Spironucleus*, the pyriform and S-shaped nuclei are adjacent only at their narrow anterior ends (Brugerolle *et al.*, 1973; Brugerolle, 1974; Kulda and Nohynkova, 1978). There are two opposed kinetosomal complexes, which are situated anterior-lateral near the free face of each nucleus in *Hexamita*, and anterior-medial near the extreme anterior in *Spironucleus* with the recurrent flagellum passing on the internal side of the nuclei (Brugerolle *et al.*, 1973; Brugerolle, 1974; Lee, 1985). The kinetosomal pocket in the nucleus is shallow in *Hexamita*, while it is deep in *Spironucleus* (see Siddall *et al.*, 1992).

Microtubular ribbons comprise supranuclear microtubules (M1), infra-

Table 2.1. Some ultrastructural features used to distinguish trophozoites of *Hexamita* and *Spironucleus*. (After Brugerolle *et al.*, 1973; Brugerolle, 1974; Becker, 1977; Kulda and Nohynkova, 1978; Lee, 1985; Siddall *et al.*, 1992.)

	Hexamita	*Spironucleus*
External morphology		
Cytostomal opening	Caudal	Postero-lateral
Posterior end of body	Flattened	Caudal projection may be present
Internal features		
Nuclei – shape	Oval/spherical	S-shaped
Nuclei – where apposed	Flattened medial portions	Narrow anterior ends
Kinetosomal complexes	Anterior-lateral	Anterior-medial
Depth of kinetosomal pocket in nucleus	Shallow	Deep

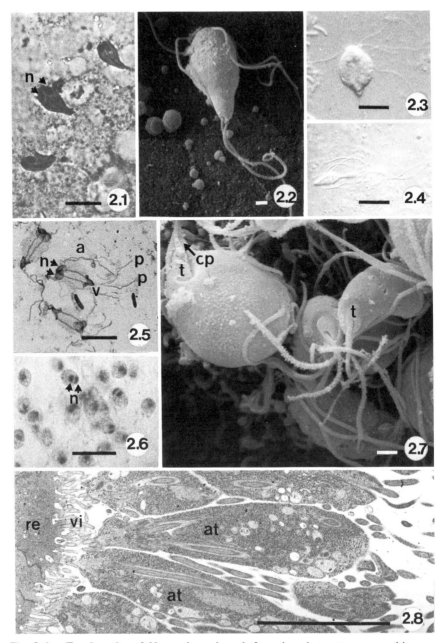

Fig. 2.1. Trophozoite of *Hexamita salmonis* from brook trout, protargol impregnation smear; note the paired nuclei (n), and blunt posterior end. Bar = 10 μm.
(From Poynton and Morrison, 1990; courtesy of *Journal of Eukaryotic Microbiology*, formerly *Journal of Protozoology*.) **Fig. 2.2.** Scanning electron micrograph of typical pyriform trophozoite of *Hexamita salmonis* from trout. Bar =
1 μm. (From Poynton and Morrison, 1990; courtesy of *Journal of Eukaryotic
Microbiology*, formerly *Journal of Protozoology*.) **Fig. 2.3.** Free swimming,
typically pyriform, trophozoite of *Spironucleus torosa* from a gadid, fresh

nuclear microtubules (M2), and funis (M3), with one band of each type present in each mastigont (Kulda and Nohynkova, 1978). The recurrent flagellum is accompanied by infranuclear microtubules and the funis (which together may be considered analogous to the axostyle), and a modified striated lamella (or rootlet fibril) and a tube-like invagination of the cell membrane. These together form a hollow tube, the flagellar pocket or cytostome that opens posteriorly (Kulda and Nohynkova, 1978). Rough endoplasmic reticulum, digestive vacuoles, and rosettes of glycogen granules are present. Bacteria are sometimes seen; however, mitochondria, Golgi apparatus, and microbodies are absent (Kulda and Nohynkova, 1978).

There are other ultrastructural differences between trophozoites of different species and these include lateral compound ridges, posterior counter-crossing ridges, papillae, surface tori and supporting cytoskeleton, and micro-filaments (Brugerolle *et al.*, 1973; Kulda and Nohynkova, 1978; Ferguson, 1979; Poynton and Morrison, 1990; Poynton *et al.*, 1994).

Diplomonad cysts are oval, and contain four nuclei, a flagellar sheath, partially disassembled microtubular ribbons, a striated rootlet fibre, and glycogen rosettes (Brugerolle *et al.*, 1973; Januschka *et al.*, 1988; Vickerman, 1990). The fibrillar walls are 0.3–0.5 µm thick.

LIFE CYCLE

Reproduction is by longitudinal binary fission of trophozoites (as described by Vickerman, 1990). In addition, there is a quadrinucleate encysted stage (Vickerman, 1990).

smear Normarski illumination. Bar = 10 µm. (From Poynton and Morrison, 1990; courtesy of *Journal of Eukaryotic Microbiology*, formerly *Journal of Protozoology*.) **Fig. 2.4.** Free swimming, elongate, trophozoite of *Spironucleus torosa* from a gadid, fresh smear Normarski illumination. Bar = 10 µm. (From Poynton and Morrison, 1990; courtesy of *Journal of Eukaryotic Microbiology*, formerly *Journal of Protozoology*.) **Fig. 2.5.** Free swimming trophozoites of *Spironucleus torosa* from a gadid, protargol impregnated smear; note the paired nuclei (n), tapering posterior end of the body with associated characteristic darkly staining v-shaped area (v), anterior flagella (a), and posterior flagella (p). Bar = 10 µm. (From Poynton and Morrison, 1990; courtesy of *Journal of Eukaryotic Microbiology*, formerly *Journal of Protozoology*.) **Fig. 2.6.** Free swimming trophozoites of *Spironucleus torosa* from a gadid, histological section stained by the Feulgen reaction; note the paired nuclei (n). Bar = 10 µm. (From Poynton and Morrison, 1990; courtesy of *Journal of Eukaryotic Microbiology*, formerly *Journal of Protozoology*.) **Fig. 2.7.** Scanning electron micrograph of free swimming trophozoites of *Spironucleus torosa* from a gadid; note variation in size and shape of the body, the torus (t) around the emergence of the recurrent flagella, and the caudal projection (cp). Bar = 1 µm. (From Poynton and Morrison, 1990; courtesy of *Journal of Eukaryotic Microbiology*, formerly *Journal of Protozoology*.) **Fig. 2.8.** Transmission electron micrograph of elongate trophozoites of *Spironucleus torosa* intimately associated with the rectal epithelium of a gadid; note that the flagellates (at) appear to be attached to the villi (vi) of the rectal epithelium (re). Bar = 5 µm. (From Poynton and Morrison, 1990; courtesy of *Journal of Eukaryotic Microbiology*, formerly *Journal of Protozoology*.)

The life cycle of hexamitids is direct (Becker, 1977). Infection is believed to be primarily oral, via ingestion of cysts or trophozoites, although rectal infection is also possible (Moore, 1922; Kulda and Lom, 1964a,b; McElwin and Post, 1968; Becker, 1977; Poynton and Morrison, 1990). Cysts and trophozoites are voided with faeces. Kent *et al.* (1992) did not find cysts and they suggested that transmission in chinook salmon was as trophozoites in the mucus or faecal material which provided protection from severe osmotic stress. Poppe *et al.* (1992) also suggested that skin lesions on heavily infected Atlantic salmon may be a source of infection in salt water.

Aquaculture conditions, particularly with high host population density for extended periods, facilitate transmission of most protozoa, and increase the probability of disease (Moore, 1922; Becker, 1977). The systemic disease in chinook salmon was thought to be transmitted directly from infected to uninfected fish during cohabitation or by exposure (through water or injection) of uninfected fish to ascitic fluid and tissues of infected fish (Kent *et al.*, 1992). Hexamitids may pass between wild and farmed trout (Poynton, 1986). Little is known of host specificity due to poor identification of most hexamitids. However, Kulda and Lom (1964a,b) found that *S. elegans* from a cichlid did not readily establish in salmonids. Also, Kent *et al.* (1992) showed that the systemic hexamitid from chinook did not infect Atlantic salmon. Amphibians may be a source of infection for fish since angel fish can acquire *S. elegans* from newt rectal fluid (Kulda and Lom, 1964a).

Host–parasite relationships
There are only a few rigorous experimental investigations on the host–parasite relationships (Uzmann *et al.*, 1965; Ferguson, 1979; Poynton and Morrison, 1990), and the disease is not well understood.

DIGESTIVE TRACT INFECTIONS
The most frequently encountered stage is the trophozoite which is usually free in the intestinal lumen. Cysts are less frequently observed in the intestine. The preferred locations of trophozoites of *H. salmonis* are the upper intestine and pyloric region, and less commonly in the gall bladder and rectum (Moore, 1922; Becker, 1977). *Spironucleus anusmirabilisvortens* most commonly occurs in the middle region of the intestine of angel fish (Poynton *et al.*, 1994). *Spironucleus elegans* prefers the lower intestine and rectum (Kulda and Lom, 1964a; Becker, 1977) while *S. torosa* is in the rectum (Poynton and Morrison, 1990). Organisms similar to *S. torosa* have been found in the rectum of marine codfish in Europe (Alexeieff, 1910). Abnormally slender *S. torosa* trophozoites also align themselves along the rectal microvilla (Fig. 2.8; Poynton and Morrison, 1990).

The suitability of regions of the intestine for parasite establishment may depend on the anatomy and physiology of the tissues, and the presence of suitable intestinal bacteria, which may in turn depend on host diet (Kulda and Lom, 1964b; Becker, 1977; Kulda and Nohynkova, 1978). Location of the parasite may also be influenced by portals of entry and barriers. The cloaca may permit *S. torosa* to enter and leave the rectum, while the ileorectal valve prevents anterior migration of the parasite (Poynton and Morrison, 1990).

An increase in the intestinal protozoa population, and the development of clinical signs, may be due to another aetiology. Certain hexamitid infections in trout can only get established if the intestinal tissue is not physiologically normal (Kulda and Lom, 1964b). Subclinical infections in tropical aquarium fishes may become pathogenic and more easily transmissible due to: (i) a decrease in the fishes' immune status; (ii) an increase in parasite numbers; (iii) development of systemic infections (see below); or (iv) secondary bacterial infections (Bassleer, 1983). Factors associated with problematic hexamitid infections include a change of diet or water temperature, stress (e.g. overcrowding low dissolved oxygen, considerable size variation between fish, poor nutrition (especially inadequate vitamins), improper handling and shipping, and concurrent nematode infections (Becker, 1977; Bassleer, 1983; Andrews *et al.*, 1988; Ferguson, 1989). Normal intestinal bacteria are considered important in hexamitid pathogenesis in mice (Boorman *et al.*, 1973).

SYSTEMIC INFECTIONS

Systemic hexamitid infections have been reported in cyprinids (Molnar, 1974), eels (Einszporn-Orecka, 1979), salmonids (Mo *et al.*, 1990; Poppe, *et al.*, 1992; Kent, 1992; Kent *et al.*, 1992), and tropical freshwater fish (Becker, 1977; Ferguson and Moccia, 1980; Gratzek, 1988), with trophozoites in the blood, gall bladder, heart, kidney, liver, spleen, eye, brain, muscles, mesentery and abdominal cavity. There are also reports of hexamitids in cranial skeletal tissues (Ferguson, 1989) and pustules on the skin of cichlids (Nigrelli and Hafter, 1947).

A recent review of parasitic diplomonads (Siddall *et al.*, 1992) reports that attachment to intestinal epithelium, and invasion of mucosa and localization in other tissues – possibly linked phenomena – are only seen in *Spironucleus*. Siddall *et al.* (1992) report that *Hexamita* species exist freely in the intestinal lumen. Thus reports of systemic invasions by *Hexamita* may be erroneous.

The factors prompting development of systemic infections are poorly understood. Invasion of the intestinal epithelium, and dissemination of the infection, may occur after the host's resistance has been lowered. The necrotic changes which accompany septicaemia, and lesions in the intestine, such as those caused by acanthocephalans, provide access, and the blood also serves to disseminate the infections (Molnar, 1974; Becker, 1977; Einszporn-Orecka, 1979; Ferguson and Moccia, 1980). Invasion of the intestinal epithelium is not believed to be the route of systemic infection in Atlantic salmon (Poppe *et al.*, 1992). It was suggested that infection was acquired in freshwater and the disease was triggered by stress (e.g. transportation, smoltification). Several authors have noted that extraintestinal infections may be accompanied by secondary infections, and the latter may be more pathogenic (Bassleer, 1983).

Systemic infections in tropical aquarium fishes are thought (Andrews *et al.*, 1988) to result in lesions on the skin (especially the head), at the base of the fins, and near the lateral line, i.e. 'hole in the head disease'. However, this is disputed by other authors (Bassleer, 1983; Gratzek, 1988). For a discussion of hole in the head disease and the role of bacterial infections of the sensory canal system, poor nutrition, and environmental factors, see Bassleer (1983), Gratzek (1988) and Ferguson (1989).

SUSCEPTIBILITY

Families and species of fishes differ in their susceptibility to hexamitids. Of numerous French marine fish, only codfishes (Gadidae) and porgies (Sparidae) are naturally infected with hexamitids (Alexeieff, 1910). Brook trout (*Salvelinus fontinalis*) is the most susceptible of five trout species in hatcheries to *H. salmonis* (Moore, 1922). The parasite is apparently innocuous to coho salmon (*Oncorhynchus kisutch*), whereas steelhead trout (*Oncorhynchus mykiss*) suffer low mortality (Uzmann *et al.*, 1965). In some freshwater ornamental fish, there are no clinical signs of hexamitid infections whereas in angel fish and discus (cichlids) and gouramis (belontiids), clinical disease is evident (Gratzek, 1988).

Acquired immunity allows adult salmonids to carry large numbers of *H. salmonis* with no clinical disease while younger fish may suffer from the disease (Becker, 1977). This is similar in hexamitid infection in tropical aquarium fishes (Bassleer, 1983). Clinical disease may occur if usually non-pathogenic flagellates infect an unusual but susceptible host (Kulda and Nohynkova, 1978). In rodents, susceptibility to *Spironucleus* is dependent on host genotype (species and strains), age and immunological status (Barthold, 1985).

CLINICAL SIGNS

Hexamitid infections in the intestine are usually chronic. In salmonids and tropical aquarium fishes the following signs may be associated with the infection: anorexia, emaciation, weakness, whirling, listlessness, pale gills, abdominal distention, faecal pseudocasts, pale stringy faeces, a red vent, dark coloration, and exophthalmia (Moore, 1922; Allison, 1963; Becker, 1977; Ferguson and Moccia, 1980; Post, 1987; Andrews *et al.*, 1988; Gratzek, 1988; Mo *et al.*, 1990). Mortalities may be moderate to severe in some salmonids (Ferguson, 1979). There may be no clinical signs in infected *S. fontinalis* infected with *H. salmonis* (see M'Gonigle, 1940), nor in gadoids infected with *S. torosa*. Hexamitids in adult breeding angelfish are reported to cause lower hatchability of eggs and/or death of young fry after hatching (Gratzek, 1988).

Atlantic salmon with systemic hexamitiosis may generally be in good condition, but are significantly smaller than healthy fish, behave abnormally and suffer increased morbidity and mortality (Mo *et al.*, 1990), while in chinook salmon the fish may appear normal or have abdominal distensions and pale gills (Kent, 1992). However, systemically infected tropical aquarium fish may be in poor condition and suffer mortality (Ferguson and Moccia, 1980). Fish with hole-in-the-head disease have small hole-like lesions on the body, which produce a yellow, cheesy string of mucus (Bassleer, 1983; Andrews *et al.*, 1988). Bilaterally symmetrical ulcers arise from large erosions in the cranial cartilage in *Spironucleus* infections (Ferguson, 1989).

NECROPSY FINDINGS AND HISTOPATHOLOGY

Intestinal *H. salmonis* infection may be accompanied by anaemia, ascites, enteritis, and intestinal contents that are yellow and watery or jelly-like, and contain excess mucus. Reports of pathology associated with the parasite vary, from no effects (Uzmann *et al.*, 1965), haemorrhage in the intestine (Roberts and Shepherd, 1974), catarrhal enteritis (Sano, 1970) and hepatocellular necrosis (Ferguson, 1979). Atlantic salmon with systemic hexamitiosis have dermatitis with whitish discoloration of peduncle and caudal fins. Infected post-smolts have whitish granulomatus nodules which contain many flagellates in the pale liver and kidneys (Poppe *et al.*, 1992). The most severe lesions are in large fish (4–5 kg) with additional brown abscesses in the caudal muscles. Livers and kidneys have multifocal necrosis with oedema, congestions, haemorrhages and fibrosis. Fish with advanced hexamitiosis have severe muscular degeneration, cholangiohepatitis and perihepatitis, encephalitis and meningitis.

Systemic infections in chinook salmon are characterized by hypertrophic livers which may be mottled with petechial haemorrhages and whitish, friable areas (Kent, 1992). There are also serosanguineous ascites, blood clots in the visceral cavity, enlarged spleen and kidney, and petechiae throughout the skeletal muscles. The blood vessels, especially those in the liver and lower intestine, have large numbers of parasites (Figs 2.11–2.13). The liver shows oedema, congestion and inflammation while the renal interstitium is hyperplastic (Kent, 1992; Kent *et al.*, 1992).

Cyprinids with intestinal hexamitosis may have enteritis, liver necrosis, and serous exudate in the abdominal cavity (Molnar, 1974).

Infected tropical fish may have yellow mucus in their intestines, enteritis, and enlarged and inflammed gall bladders (Molnar, 1974; Becker, 1977; Bassleer, 1983; Ferguson, 1989). The histopathology includes gastritis, perforation with serosal granuloma in the stomach (Ferguson, 1989). In systemic infections there may be perforation of the stomach, and chronic inflammation of many internal organs, atrophy of the renal tubules, and necrosis of the kidney (Amlacher, 1970; Ferguson and Moccia, 1980). Flagellates are sometimes in renal peritoneal exudates in kidneys.

In histological sections, hexamitid trophozoites are typically located extracellularly in the lumen of the intestine, less frequently they may be intimately associated with or invade the mucosal epithelium, or be in internal organs and surface lesions (Nigrelli and Hafter, 1947; Paperna and Overstreet, 1981; Mo *et al.*, 1990; Poynton and Morrison, 1990; Kent, 1992; Kent *et al.*, 1992).

Pathogenicity

Although intestinal hexamitids are often associated with disease and mortalities, their pathogenicity is disputed and poorly understood (Uzmann and Hayduk, 1963; Becker, 1977; Ferguson, 1979, 1989; Sommerville, 1981). Some authors believe that hexamitids may become serious pathogens when the host is adversely affected by other factors (Uzmann and Hayduk, 1963; Molnar, 1974; Becker, 1977; Lom, 1986; Post, 1987). Clinical signs and pathology may also depend, at least in part, on the hexamitid (Kiskaroly and Tafro, 1987). In contrast, some of the gross and pathological changes in systemic infection in chinook salmon are similar to those caused by blood flagellates (see pp. 58–74). Kent *et al.* (1992) suggested that the anaemia in infected chinook was caused partly by haemodilution and haemorrhage.

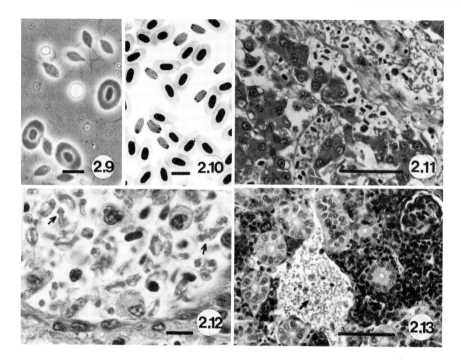

Fig. 2.9. Systemic hexamitid infection in chinook salmon; there are numerous flagellates in the blood (wet mount, phase contrast). Bar = 10 μm. (From Kent *et al.*, 1992; courtesy of Dr M. Kent and *Diseases of Aquatic Organisms*.) **Fig. 2.10.** Systemic hexamitid infection in chinook salmon; note the paired nuclei and flagellar pockets, Leishman's Giemsa stain. Bar = 10 μm. (From Kent *et al.*, 1992; courtesy of Dr M. Kent and *Diseases of Aquatic Organisms*.) **Fig. 2.11.** Systemic hexamitid infection in chinook salmon, showing a heavy infection in the liver; histological section. Bar = 25 μm. (From Kent *et al.*, 1992; courtesy of Dr M. Kent and *Diseases of Aquatic Organisms*.) **Fig. 2.12.** Systemic hexamitid infection in chinook salmon, flagellates can be seen in the blood vessel of the liver; histological section. Bar = 10 μm. (From Kent *et al.*, 1992; courtesy of Dr M. Kent and *Diseases of Aquatic Organisms*.) **Fig. 2.13.** Systemic hexamitid infection in chinook salmon, kidney with interstitial hyperplasia and proliferative glomerulonephritis, there are numerous flagellates in the blood vessels. Bar = 50 μm. (From Kent *et al.*, 1992; courtesy of Dr M. Kent and *Diseases of Aquatic Organisms*.)

The relationships between prevalence and intensity of infection, and disease are not well understood. Infections in the intestine of salmonids and tropical aquarium fishes are frequently considered pathogenic when parasite numbers are high (Post, 1987; Andrews *et al.*, 1988). Gratzek (1988) suggested that flagellates interfered with nutrition by competing for essential nutrients and/ or by damaging the intestinal epithelium. Yet apparently healthy wild marine fish have a high prevalence and intensity of *S. torosa* (Poynton and Morrison, 1990). This suggests that hexamitids in wild marine fish are not pathogenic, and this is consistent with the conclusions of Lavier (1936), Becker (1977) and Kulda and Nohynkova (1978).

The intimate association of trophozoites with the intestinal epithelium observed in *S. torosa* in cod fish (Fig. 2.8), may indicate that the parasite has the potential to be detrimental to the host as the trophozoites form an obstruction between the lumen and the epithelium. Also, there may be functional and/or structural damage to the microvilli (Poynton and Morrison, 1990). Trophozoites intimately associated with the microvilli appear to share cytoplasm with host cells. They may be able to evade the host's immune response by having an outside surface that mimics the glycocalyx covering of the microvilli (Poynton and Morrison, 1990).

In vitro culture and propagation of *Hexamita* and *Spironucleus*

Trophozoites of *H. salmonis* have been cultured axenically in a balanced organic medium, fortified with antibiotics (penicillin, streptopmycin and mycostatin), supersaturated with nitrogen and kept under a nitrogen atmosphere (Uzmann and Hayduk, 1963). The culture medium is medium 199, human cord serum, and lactalbumin hydrolysate. The medium is dispensed aseptically into screw cap culture tubes and saturated with nitrogen. The innoculum is overlaid with fresh nitrogen and incubated in darkness at 10°C for 48 hours.

The serendipitous growth of trophozoites of *S. anusmirabilisvortens* in a culture system with lip tumour tissue allowed the subsequent axenic cultivation of the flagellate in a medium originally developed for trichomonads. The flagellate is now routinely maintained in an axenic medium originally developed for the cultivation of *Entamoeba* and is stabilized in the cryopreserved state at the American Type Culture Collection (Poynton *et al.*, 1994).

Parasite nutrition

Endocytosis (e.g. of bacteria) and exocytosis occur at the anterior end of the cytostome (Brugerolle, 1974; Kulda and Nohynkova, 1978; Lee, 1985; Vickerman, 1990). In *Hexamita*, the cytostomes are capable of considerable dilation, allowing large bacteria to be ingested. In *Spironucleus*, reduced dilation allows bacteria and small particles to be ingested (Vickerman, 1990). Bacteria present in the cytoplasm may be food items or commensals (Poynton and Morrison, 1990). Certain hexamitids may also feed directly on the cytoplasm of host cells. Slender *S. torosa* trophozoites, intimately associated with the rectal epithelium, have anterior cytoplasm similar in appearance to that in the microvilli, and have less rough endoplasmic reticulum than free swimming trophozoites, which sugggests reduced polypeptide synthesis (Poynton and Morrison, 1990).

The growth and reproduction of intestinal flagellates are determined in part by intestinal bacteria, and host diet (Peterson, 1960; Feely and Erlansen, 1978).

Little is known about hexamitid metabolism, biochemistry or genetics (Becker, 1977). The absence of mitochondria, golgi apparatus and microbodies indicates a primitive metabolism (Kulda and Nohynkova, 1978). Glycogen is the main carbohydrate storage product (Honigberg *et al.*, 1981; Vickerman, 1990). Intestinal hexamitids are anaerobes (Kulda and Nohynkova, 1978) or, in the case of *S. anusmirabilisvortens*, an aerotolerant anaerobe (Poynton *et al.*, 1994).

Diagnosis of infection

All samples must be obtained within minutes from freshly killed fish, smears should be made from various sites along the gastrointestinal tract, and of the blood, and imprints made of internal organs.

To confirm the number and location of flagella and body shape which are consistent with hexamitids, living organisms should be observed in wet mounts. These are examined under a cover glass using bright field, phase contrast, or Normarski illumination at 400× and 1000× (Figs 2.3, 2.4, 2.9) (Poynton, 1994). The addition of viscous Protoslo, methyl cellulose or Polyox resin WSR 301 will slow or immobilize the organisms (Lee *et al.*, 1985). Also, trophozoites become less active at the margins of the cover glass as the preparation begins to dry. Detection of the organisms may be improved if the preparation is diluted with tap or distilled water, or 0.9% NaCl. Organisms are preserved in Bouins fixative for subsequent detailed examination. Fibrillar structures of the flagellates such as flagella and microtubules are viewed in protargol (silver protein) impregnation of Bouin's fixed wet mounts or suspensions of intestinal contents (Figs 2.1, 2.5) (Lee *et al.*, 1985; Montagnes and Lynn, 1987; Lynn, 1992; Poynton *et al.*, 1994). Smears and organ imprints may be stained with Leishman's or Giemsa's stain and the flagellates appear as dark staining oval bodies with two clear bands (flagellar pockets) and two nuclei (Fig. 2.10) (Kent, 1992).

In histological sections, the organisms are usually in the intestinal lumen, but may also be aligned along the rectal epithelium in the case of *S. torosa*. The protozoa in the lumen are usually rounded, with the two nuclei clearly seen in haematoxylin and eosin (H&E) preparations or those stained using the Feulgen reaction (Fig. 2.6). In *S. torosa* the caudal projection can also be seen in H&E preparations. Unstained imprints or smears or stained histological sections from internal organs should also be examined for systemic infections (Figs 12.11–2.13) (Mo *et al.*, 1990).

Samples from heavily infected fish may be fixed in glutaraldehyde or a combination of formaldehyde/glutaraldehyde fixative. For scanning electron microscopy (SEM) protozoa may be in suspension (Watson *et al.*, 1980) or in tissues (Figs 2.2, 2.7; Ferguson, 1979; Poynton and Morrison, 1990; Poynton *et al.*, 1994). The shape of the posterior end of the body, and the exit of the recurrent flagella can best be visualized under scanning electron microscopy (Table 2.1). Transmission electron microscopy is useful to confirm the genus and to distinguish species.

Caution should be exercised in interpreting the presence of flagellates in moribund fish, since they may merely accompany, rather than cause, observed pathologies (Becker, 1977).

Prevention and control

Elimination of hexamitids from fish-rearing facilites may be impractical (Becker, 1977); however, their introduction can be minimized by quarantine of new stock (14–21 days) and cleaning of filters and gravel, since accumulation of organic material can promote hexamitids (Gratzek, 1988). Good husbandry and nutrition are essential for reducing the number of outbreaks and severity of outbreaks. Amphibians, which can harbour *S. elegans*, may be a source of infection for fish.

Post (1987) recommends treatment of salmonids when the intensity of infection reaches an average of 15–30 tryphozoites per field of view (total magnification 100×). Emtryl (dimetridazole) and Flagyl (metronidazole) may be given in the feed, less commonly as bath treatments; neither is specifically licensed for use with fish for human consumption (Gratzek, 1988; Stoskopf, 1988). An antibacterial treatment (furazolidone), usually added to the feed, has been used successfully to control hexamitid parasites and reduce mortalities (Ferguson, 1979; Andrews *et al.*, 1988). Repeated treatments may be needed (Andrews *et al.*, 1988).

ICHTHYOBODOOSIS

Ichthyobodoosis in freshwater fish

Introduction

Ichthyobodoosis in freshwater fishes is caused by the ectoparasite *Ichthyobodo necator*. The disease has been known for a long time as costiasis after the previous name of the parasite, *Costia necatrix*. The organism has a free living stage in water (Fig. 2.14) and a parasitic stage (Fig. 2.16) which is usually attached on dorsal fins and tips of secondary gill lamellae (Fig. 2.17) of infected fish (Fish, 1940; Tavolga and Nigrelli, 1947). The free living stage has a long and a short flagellum (Fig. 2.14), however these flagella are not seen when the organism attaches to epithelial cells (Fig. 2.16).

Geographical distribution and host range

The parasite has worldwide distribution and is not host specific. The parasite is an important pathogen of young salmonids and cyprinids in hatcheries where fish are cultured in high numbers (e.g. Bohl, 1975; Ellis and Wootten, 1978; Bullock and Robertson, 1982; Awakura *et al.*, 1984; Rosengarten, 1985; Broderud and Poppe, 1986).

Systematic and taxonomic position

I. necator (Family Bodonidae) was first described by Henneguy as *Bodo necator* and was later transferred to the genus *Costia* by Leclerq. The name *Costia* was subsequently found to be preoccupied and according to Joyon and Lom (1969), Pinto replaced it with *Ichthyobodo* and renamed the parasite *I. necator*.

Two species (*I. necator* and *I. pyriformis*) have been reported from freshwater fishes. *I. pyriformis* was described from salmonids in the United States (Davis, 1943). Body measurements of the two parasites overlap and there are no morphological characters visible under the light microscope to distinguish them. Hence, *I. pyriformis* is generally accepted as a small form of *I. necator* and is a junior synonym (e.g. Tavolga and Nigrelli, 1947; Vickerman, 1976a; Becker, 1977).

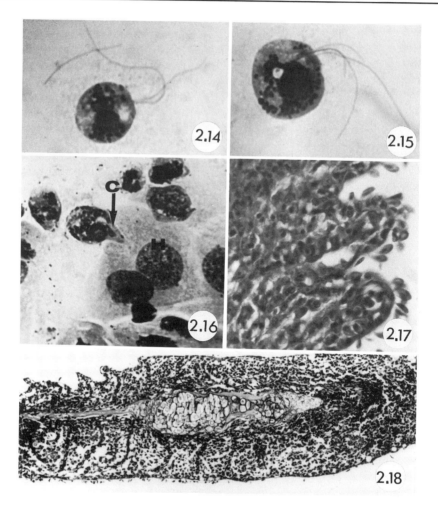

Fig. 2.14. Free living of *Ichthyobodo necator*; note the two free flagella (×1000). (From Miyazaki *et al.*, 1986; courtesy of Dr T. Miyazaki.) **Fig. 2.15.** Free living pre-division form of *Ichthyobodo necator*; note it is larger and has more than two free flagella (×1000). (From Miyazaki *et al.*, 1986; courtesy of Dr T. Miyazaki.) **Fig. 2.16.** A parasitic form of *Ichthyobodo necator* attached to a host cell; note the cytostome (C). (From Miyazaki *et al.*, 1986; courtesy of Dr T. Miyazaki.) **Fig. 2.17.** Large numbers of *Ichthyobodo necator* attached to gill filaments; note gill filaments show hyperplasia of epithelial cells, fusion of lamellae (×480). (From Miyazaki *et al.*, 1986; courtesy of Dr T. Miyazaki.) **Fig. 2.18.** Clubbing of gill filament due to heavy *Ichthyobodo necator* infection. (From Miyazaki *et al.*, 1986; courtesy of Dr T. Miyazaki.)

Morphology and life cycle

The free swimming form is ovoid to spherical and measures 5–18 μm by 3–8 μm (Fig. 2.14). It divides by binary fission and its predivision stage has three and four flagella and is larger than stages with two flagella (Fig. 2.15). The nucleus is round to oval, and basophilic granules are often seen in the cytoplasm. There is a large basophilic body at the base of the flagella (Fish, 1940; Tavolga and Nigrelli, 1947; Vickerman, 1976b). The parasitic stage attaches to a host cell by its pointed anterior end (Fig. 2.16). Tavolga and Nigrelli (1947) suggested that the organism also has a saprophagous stage which attaches to and feeds on detached cells and scales. Other workers (e.g. Bauer, 1959) had not seen such a form and were generally sceptical about its existence.

The cell membrane consists of two membranous layers with a fibrillar layer in between them (Schubert, 1966; Joyon and Lom, 1969). The nine peripheral fibrils of the flagella continue to form the kinetoplast–mitochondrion complex. The nuclear membrane also has two layers and there is a single nucleolus.

Infection occurs when free swimming forms attach to a fish. The flagella on the free swimming stage are presumed to be used for site selection, and the ventral flat disc for attachment in the parasitic form (Schubert, 1966, 1978). Both free swimming and parasitic stages multiply by longitudinal binary fission (Tavolga and Nigrelli, 1947; Bauer, 1959). Free living individuals with three to four flagella (predivision stage) are often seen in the aquatic environment (Fig. 2.15) and they occur in large numbers when conditions are favourable and fish become heavily infested (Fig. 2.17) 1–2 weeks after infection (e.g. Andai, 1933). The free swimming and parasitic forms are presumed to encyst under adverse environmental conditions and the cysts in the water become an additional source of infection. They may also be involved in the spread of the organism to new areas.

The organism has been found in the roe of spawning carp and it is presumed that larval fish become infected after hatching (Hlond, 1963). Other sources of infection in hatcheries are water supplies and feral fishes that come in with the water. Amphibian larvae (salamander, frog) are also considered potential sources of infection (Becker, 1977).

Host–parasite relationships

HOST SUSCEPTIBILITY

The parasite is not host specific. Malnourished and/or young fish are more severely affected than healthy adults (Robertson, 1985; Rosengarten, 1985) and high mortality (40–73%) is associated with heavy infections (e.g. Robertson, 1979; Poppe and Hastein, 1982; Awakura *et al.*, 1984; Rosengarten, 1985; Urawa, 1987). Outbreaks and infestations on cyprinids become more severe when infected fish from outdoors are transferred to indoor tanks (e.g. Bauer, 1959; Bauer *et al.*, 1969). The important factors include stress on the fish due to the transfer and the higher indoor temperature that promotes more rapid parasite multiplication.

Swordtails (*Xiphophorus hellerii*) are very susceptible and die 8–14 days after infection (Tavolga and Nigrelli, 1947) while platyfish (*Platypoecilus maculatus*) are less susceptible. *Tilapia (Tilapia macrocephala)* and guppies (*Poecilia reticulata*) usually have light infections or the infections are confined to the caudal region.

It is generally assumed that some protective immunity is acquired on recovery. Rainbow trout (*Oncorhynchus mykiss*) immunized with either whole *Tetrahymena thermophila* antigen or their cilia were more resistant than naive fish (Wolf and Markiw, 1982).

COURSE OF INFECTION AND FOCI OF INFESTATION

Parasitemias on salmonids peak at about 4 weeks after the fish larvae start feeding and mortality is highest at 4–8 weeks (Robertson, 1979).

Foci of infestations on the fish include the cuff of skin sheltered by the operculum, pectoral and pelvic fins and the area adjacent to the dorsal fin (Robertson *et al.*, 1981; Robertson, 1985). Parasites are not normally found on the head anterior to the operculum.

CLINICAL SIGNS AND PATHOLOGY

Fish with relatively light body infections may roll in the water (flashing) and rub against immersed objects or sides of the tank. These activities are apparently responses to skin irritations caused by the parasite. Fish with heavy infestations are often listless and anorexic. Spots appear on the body and these fuse to form greyish films on fins and body surface. Also, the gills are usually swollen (Fish, 1940; Miyazaki *et al.*, 1986) and other clinical signs include increased mucus secretions and frayed or destroyed fins. Some infected fish may not be able to maintain an upright position or to swim to the water surface (Savage, 1935; Bauer, 1959). Fish with only gill infestations are listless and anorexic but do not flash nor have excessive body mucus.

Goblet cells are often not seen in the epidermis and there is hyperplasia of Malphighian cells in salmonids with heavy body infections. Spongiosis, vacuoloation, loss of cytoplasmic and nuclear details in the suprabasal layers may occur in the epidermis. Oedema is followed by degeneration and sloughing of the epidermis (Ellis and Wootton, 1978; Robertson *et al.*, 1981; Robertson, 1985; Awakura *et al.*, 1984). It has been suggested that mortality is in part due to osmoregulatory problems with resultant haemodilution (Robertson *et al.*, 1981; Robertson, 1985). The histopathologies are similar in infected plaice and flounders that are maintained in sea water (Bullock and Robertson, 1982; Cone and Wiles, 1984).

Acute hyperplasia and fusion of secondary gill lamellae are obvious in heavily infected channel catfish (Miyazaki *et al.*, 1986). Hyperplasia of epithelial cells in the interlamellar spaces is extensive with fusion of the lamellae and marked clubbing of gill filaments (Fig. 2.18). Epithelial cells with parasites become necrotic and blood vessels collapse with proliferation of mucus cells in clubbed filaments.

Diagnosis of infection

Clinical signs of the disease are usually used for preliminary diagnosis of an outbreak. The infection is confirmed by examination of mucus from gills and body surface for flagellates under the light microscope. The parasite is quite fragile and often ruptures during the staining. Carefully prepared smears (e.g. from the body surface) are fixed in Shaudinn's fixative, transferred to ethyl alcohol before being stained (e.g. in iron haematoxylin, carmine, periodic acid-Schiff) (Becker, 1977). At present there are no serological techniques that can be used to either detect the infection or determine the immunological state of fish.

Control

Flush treatment with formalin (166 ppm) for 1 hour or a formalin bath (1 : 4000) for 15–30 min is well tolerated by salmon fry (Fish, 1940; Imamovic, 1986; Skrudland, 1987). Also, treatment for 1 hour in a formalin bath (1 : 6000) is also very effective. Formalin treatment should not be used if young fishes have bacterial gill disease. It is also important to have good aeration during formalin treatment because formalin (a reducing agent) combines with the dissolved oxygen in the water to form formic acid (Helms, 1967). Dipping infected salmonids in dilute acetic acid (2000 ppm) for 1 min is also recommended (Hora and Pillay, 1962). Other treatments include exposing fingerlings to a weak solution of malachite green (1 : 300,000 to 1 : 400,000) for 40–60 min (Becker, 1977). Modifications of these treatments (in amounts and treatment periods) are in Hoffman and Meyer (1974).

Cyprinids lose their infections when dipped daily in copper sulphate (500 ppm) for 1–2 min or in sodium chloride (10,000 ppm) for 15–30 min (Osborn, 1966; Schaperclaus, 1986). More concentrated salt solutions (5%) are recommended for larger fish. A 1 hour flush treatment with pyridylmercuric acetate (2 ppm) also eliminates parasites.

Ponds with infected fish are drained and quicklime or chloride of lime is added before restocking with new or treated fish. Other useful chemicals include methylene blue, potassium permanganate, aureomycin, globucid, lysol and quinine hydrochloride (Amlacher, 1970; Schaperclaus, 1986).

Bithionol (25 mg l^{-1} for 3 h or 2 consecutive days) is very effective in eliminating the parasite from rainbow trout (Tojo *et al.*, 1994). The drug is not toxic at the recommended dosage.

The parasite multiplies rapidly between 10 and 25°C. It encysts at about 8°C and dies at above 30°C (van Duijn, 1973). Hence raising the water temperature to 32°C for 5 days is also effective against temperature-sensitive strains.

Ichthyobodoosis in marine fish

Geographical distribution and host range

Ichthyobodo has been found on salmonid smolts in the sea (Ellis and Wootten, 1978; Roubal *et al.*, 1987; Bruno, 1992) and on wild plaice (*Pleuronectes platessa*) from Scotland (Bullock and Robertson, 1982; Bullock, 1985), winter flounders (*Pseudopleuronectes americanus*) from Newfoundland (Cone and Wiles, 1984) and Japanese flounders (*Paralichthys olivaceus*) at the Hokkaido Institute of Mariculture (Kusakari *et al.*, 1985).

Wood (1979) indicated that freshwater *Ichthyobodo* survived on fish following their transfer to the marine environment. Hence, it was assumed that the parasite on marine fish was acquired when fish were in fresh water and it survived when infected fish migrated or were transferred to sea water. However, Morrison and Cone (1986) described an *Ichthyobodo* from haddock (*Melanogrammus aeglefinus*) caught 120 km from Nova Scotia and Diamant (1987) found the parasite on common dabs (*Limanda limanda*) from the North Sea, thus suggesting the existence of purely marine species of *Ichthyobodo*.

Evidence for a marine *Ichthyobodo*
Bruno (1992) showed that mean width and length of *Ichthyobodo* from salmonid fry in fresh water were significantly greater than those of the parasite from fish in sea water. However, Bruno's measurements of the marine parasite were lower than those from other marine fishes (Morrison and Cone, 1986; Diamant, 1987). In a comparative ultrastructural study Roubal and Bullock (1987) found differences in the attachment discs and cytostome processes between freshwater and marine *Ichthyobodo* from salmonids. According to Lamas and Bruno (1992) the cytostome process of marine *Ichthyobodo* on Atlantic salmon is smooth along its entire length while that of *I. necator* has ridge-like projections.

I. necator from freshwater fish survived and multiplied on chums (*Oncorhynchus keta*) when infected salmon were transferred to sea water (Urawa and Kusakari, 1990). However, there were minor morphological changes (e.g. reductions in body width and in contractile vacuoles) when the parasite was in sea water. The parasite was also on chum frys in estuaries and *I. necator* from chums was not infective to Japanese flounders while the parasite from flounders was not very infective to chums.

Clinical signs and pathology
The parasite caused high mortality of juvenile Japanese flounders in a hatchery in Japan (Urawa *et al.*, 1991). The fish were maintained in canvas tanks supplied with sea water at 17°C and about 40% of 43,000 fish were infected. Large numbers of parasites were on the entire dorsal surface of infected fish including the head. The clinical signs were loss of appetite, greyish slime on the body surface and erosion of fins.

The histopathologies are similar to those in freshwater salmonids (Robertson *et al.*, 1981; Jones *et al.*, 1993) and they include extensive hyperplasia of epidermis with almost total depletion of goblet cells, nuclear and cytoplasmic degeneration of epithelial cells, and vacuolation and severe spongiosis in most of the epidermal layer. Also epidermal cells lose their interdigtating cell membranes (Urawa *et al.*, 1991).

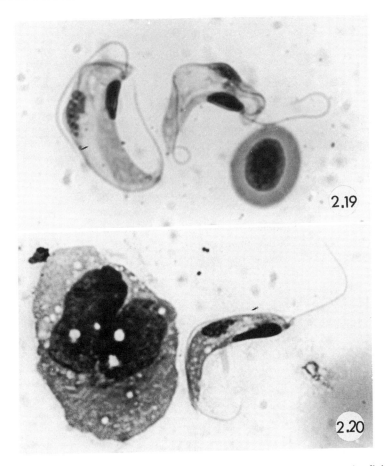

Fig. 2.19. *Cryptobia.* (*T.*) *salmositica* with red cell from anaemic fish; the anaemia is microcytic and hypochromic: note the red cell is not oval and there is reduced haemoglobin (×1150). (From Woo 1987a; courtesy of *Advances in Parasitology*.) **Fig. 2.20.** *C.(T.) salmositica* in mucus from body surface of a fish 6 weeks after infection; slender form next to epithelial cell (×1150). (From Woo and Wehnert, 1983; courtesy of *Journal of Protozoology*.)

CRYPTOBIOSIS

Introduction

Cryptobia (Fig. 2.19) infects and causes disease in many species of marine and freshwater food fishes (e.g. salmonids and flatfishes). The parasite has a worldwide distribution and there are at least 52 nominal species (Lom, 1979; Woo, 1987a). The ectoparasites (five species) are on the body surface (Fig. 2.20) or attached to the gills (Figs 2.21, 2.22) and body surface, while the endoparasites are either in the blood (Fig. 2.19; 40 species) or in the digestive tract (Figs 2.23–2.25; seven species). There are also numerous unidentified

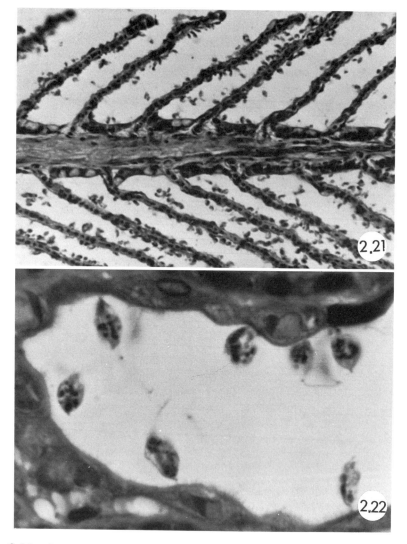

Fig. 2.21. Large numbers of *Cryptobia* sp. attached to gill filaments of a naturally infected summer flounder; note the absence of any obvious effect on the gill filaments. (From Burreson and Sypek, 1981; courtesy of Dr E.M. Burreson.) **Fig. 2.22.** *Cryptobia* sp. attached to gill filaments of naturally infected summer flounder (×1000). (From Burreson and Sypek, 1981; courtesy of Dr E.M. Burreson.)

Cryptobia sp. It is likely that many of the described species are not valid; however, their taxonomic status can only be settled after careful experimental studies. Many of these flagellates, e.g. *Cryptobia catostomi* of white sucker (*Catostomus commersoni*), are not known to cause disease (Bower and Woo, 1977; Thomas and Woo, 1992a) and such nonpathogenic species are not included in the present discussion.

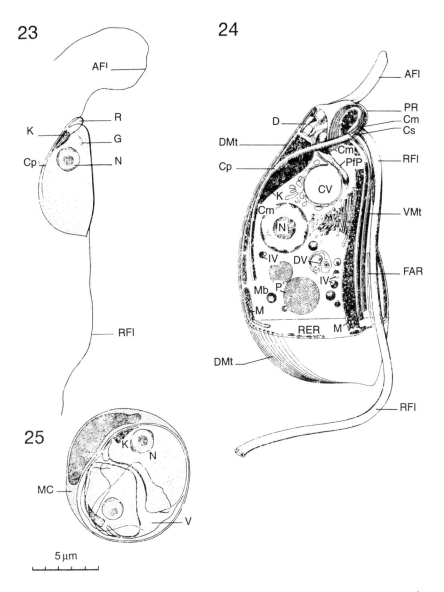

23

24

25

5 μm

Fig. 2.23. Free swimming *Cryptobia* (*C.*) *iubilans*; note the short prominent rostrum (R), slender cytopharynx (Cp), Golgi complex (G), anterior flagellum (AFI), recurrent flagellum (RFI), kinetoplast (K). (From Nohynkova, 1984; courtesy of *Protistologica*.) **Fig. 2.24.** Schematic drawing of *C.* (*C.*) *iubilans* to show principal structures at the electron microscopic level; note the microtubules (Cm), the mitochondrion (M) extending the length of the flagellate, the subpellicular microtubules (DMt), anterior flagellum (AFI), preoral ridge (PR), cytostome (Cs), recurrent flagellum (RFI), ventral subpellicular microtubules (VMt), flagellum associated reticulum (FAR), cytopharynx (Cp), dense lamina (D), postflagellar pit (PfP), contractile vacuole (CV), kinetoplast (K), nucleus (N), Golgi complex (G), digestive vacuole (DV), inclusion vacuole (IV), polysaccharide (P), microtubules (Mb), and rough endoplasmic reticulum (RER). (From Nohynkova, 1984; courtesy of *Protistologica*.) **Fig. 2.25.** Two intracellular parasites within the vacuole (V) of host macrophage-like cell (MC). (From Nohynkova, 1984; courtesy of *Protistologica*.)

The parasite (Figs 2.19, 2.20, 2.22) is an elongated (oval to ribbon-like) flagellate with two flagella which originate from the anterior end. The shorter anterior flagellum is free while the recurrent flagellum is attached to the body and it ends as a free flagellum at the posterior end. The oval to elongated kinetoplast is very prominent and is located at the anterior end in close proximity to the nucleus. The parasite multiplies by longitudinal binary fission under both *in vivo* and *in vitro* conditions (Woo, 1987a).

Systematics and taxonomy

The genus *Cryptobia* (Family Bodonidae) was initially proposed for biflagellated organisms in the reproductive systems of snails (Leidy, 1846). The genus was later expanded to include other morphologically similar flagellates infecting other invertebrates and vertebrates (e.g. ectoparasites on gills/body surface and endoparasites in blood/digestive tract of fishes). *Trypanoplasma* was proposed by Laveran and Mesnil (1901) for a biflagellated haemoflagellate in freshwater fish. Crawley (1909) synonymized *Trypanoplasma* with *Cryptobia* because of their morphological similarities. However, some workers (e.g. Lom, 1979) retained *Trypanoplasma* for haematozoic species and *Cryptobia* for non-haematozoic species because they believed the differences in life cycles (haematozoic species are transmitted indirectly by leeches, while non-haematozoic species have direct transmissions) and the presence of a contractile vacuole (in ectoparasitic species) were sufficient for their separation.

Cryptobia salmositica and *C. bullocki* are haematozoic parasites, however an ectoparasitic phase in the mucus on the body surface has been shown in both species (Woo and Wehnert, 1983; E.M. Burreson, personal communication). Ectoparasitic forms of *C. salmositica* (Fig. 2.20) are morphologically similar to blood forms and they are infective when inoculated into fish (Woo and Wehnert, 1983). The parasite is transmitted from infected to uninfected fish when fish are allowed to mix freely in a tank or are separated by a wire screen to eliminate bodily contacts (Woo and Wehnert, 1983) or when infected and uninfected fish are brought together briefly in a dip net (Bower and Margolis, 1983). A contractile vacuole was seen in electron micrographs of blood forms (Paterson and Woo, 1983). This organelle is essential for the ectoparasitic form to survive in the mucus on the body surface. Hence, Crawley was correct when he synonymized the two genera (Woo and Wehnert, 1983; Bower and Margolis, 1983).

Woo (1994) recently divided the *Cryptobia* into two subgenera. The haematozoic and normally digenetic species (i.e. indirect life cycle) belong to the subgenus *Trypanoplasma*, while the subgenus *Cryptobia* contains the non-haematozoic monogenetic (i.e. one host) parasites. This proposal still indicates the very close phylogenetic relationships between the groups but shows that there are biological differences between them. He further suggested that non-haematozoic species (ectoparasitic on gills/body surface and endoparasitic in intestine) may warrant division into additional subgenera because of the distinct physiological conditions under which they live. This, however should only be done after careful experimental studies.

Five species of *Cryptobia* are known to cause disease in fishes (Woo,

1987a). *Cryptobia* (*Cryptobia*) *branchialis* is an ectoparasite and the remaining species are endoparasites. *Cryptobia* (*C.*) *iubilans* is an intestinal parasite while *Cryptobia* (*Trypanoplasma*) *salmositica*, *C.* (*T.*) *bullocki* and *C.* (*T.*) *borreli* (Syn. *Trypanoplasma cyprini*) are haematozoic species.

Two hypotheses were proposed on the origin of the haematozoic parasites (subgenus *Trypanoplasma*) in fish. These were discussed more fully in an earlier review (Woo, 1987a). Briefly, Woo and Wehnert (1983) suggested that members of the subgenus *Trypanoplasma* arose from free living flagellates (e.g. genus *Procryptobia*) via the ectoparasites (subgenus *Cryptobia*) on the body surface of fish. In the second hypothesis (Nohynkova, 1984a) it was proposed that the haematozoic species were linked to the intestinal flagellates (subgenus *Cryptobia*).

Cryptobiosis in freshwater fish

C. (*C.*) *branchialis*, *C.* (*C.*) *iubilans*, *C.* (*T.*) *salmositica* and *C.* (*T.*) *borreli* are pathogenic to freshwater fishes and hence are of economic importance. *C.* (*C.*) *branchialis* is an ectoparasite while *C.* (*C.*) *iubilans* is in the digestive tract and associated organs. Both pathogens are always transmitted directly between fish. The haematozoic *C.* (*T.*) *salmositica* is not only transmitted indirectly by blood-sucking leeches (digenetic life cycle), it also can be transmitted in the absence of leeches (monogenetic life cycle). The haematozoic *C.* (*T.*) *borreli* requires blood-sucking leeches for indirect transmission.

Geographical distribution of parasite and host range

C. (*C.*) *branchialis* is ectoparasitic on the gills of a large number of freshwater fishes and has been implicated in mortality of cultured carp, goldfish and catfish in Asia, eastern Europe and North America (e.g. Chen, 1956; Naumova, 1969; Bauer *et al.*, 1969; Hoffman, 1978).

C. (*C.*) *iubilans* occurs extracellularly (Fig. 2.23) in the stomach, intestine, gall bladder, spleen, ovary and liver and intracellularly in macrophage-like cells (Fig. 2.25). It infects four species of freshwater aquarium cichlids (*Herichthys cyanoguttatum*, *Cichlasoma meeki*, *C. nigrofasciatum*, *C. octofasciatum*) in the former Czechoslovakia (Nohynkova, 1984a).

C. (*T.*) *salmositica* (Fig. 2.18) was initially described from coho salmon (*Oncorhynchus kisutch*) in Washington, USA (Katz, 1951). It is not host specific and has since been found in all species of Pacific *Oncorhynchus* spp., *Salmo trutta*, *Prosopium williamsoni*, seven species of sculpins (*Cottus* spp.), a cyprinid (*Rhinichthys cataractae*), a sucker (*Catostomus snyderi*) and a stickleback (*Gasterosteus aculeatus*) (Woo, 1987a). These fishes are in freshwater streams from California to British Columbia, and in southwestern Alaska (Wales and Wolf, 1955; Becker and Katz, 1965b; Katz *et al.*, 1966; Bower and Margolis, 1984b). It is likely that the parasites in the cyprinid and the sucker are not *C.* (*T.*) *salmositica* because other species of cyprinids and suckers cannot be experimentally infected with the parasite (Wehnert and Woo, 1980). Also, cyprinids and suckers have their own species of *Cryptobia* and they are not infective to salmonids (Woo, 1987a). Lom (1979) suggested that the poorly described *Cryptobia* (*T.*) *makeevi* from *Oncorhynchus gorbuscha* and *O. keta* from the Amur River in East Asia may be *C.* (*T.*) *salmositica*. The suggestion indicates the need for more careful comparative morphological, developmental (e.g. in the leech vector), antigenic (e.g. using monoclonal antibodies) and molecular (e.g. using DNA probes) studies. Using a 1 kb DNA fragment of *C.* (*T.*) *salmositica*, Li and Woo (unpublished) recently developed a species specific DNA probe which distinguishes *C.* (*T.*) *salmositica* from *C.* (*T.*) *bullocki*, *C.* (*T.*) *borreli* and *C.* (*T.*) *catostomi*. It is a practical technique because the parasite DNA can be extracted from infected blood dried on filter paper. Cross experimental transmissions and virulence (e.g. clinical signs of the disease) studies may also be useful in deciding the taxonomic status of the two organisms.

   C. (*T.*) *borreli* was described by Laveran and Mesnil (1901) from red-eye (*Leuciscus erythrophthalmus*) caught in ponds in Garches, France. The parasite has since been found in the blood and tissues of several freshwater cyprinids in Europe. It is also assumed to sometimes cause heavy losses in cultured carp, tench and crucian carp (Lom, 1979). The parasite is not host specific and causes mortality in experimentally infected goldfish (*Carassius auratus*) and the common carp (*Cyprinus carpio*) (see Dykova and Lom, 1979; Lom, 1979). After extensive morphological and cross transmission studies, Lom (1979) concluded that the parasites from carp, goldfish and crucian carp in Europe are all *C.* (*T.*) *borreli*, and that the pathogenic *Trypanoplasma cyprini* described by Plehn (1903) from carp in Germany is a junior synonym of *C.* (*T.*) *borreli*. Comparative antigenic and biochemical studies of various parasite isolates from cyprinds throughout Europe may help to confirm Lom's conclusion.

Geographical distribution and biology of leech vectors

Piscicola salmositica is the only known vector of *C.* (*T.*) *salmositica*. The freshwater leech is not host specific and is in cold fast flowing rivers and streams from British Columbia to northern California (Becker and Katz, 1965a, 1965c).

   Cocoons and adult *P. salmositica* are susceptible to drying and freezing, and draining areas where cocoons are deposited is a method to control leeches in hatcheries (Bower and Thompson, 1987). Adult leeches are susceptible to chlorine (Bower *et al.*, 1985). Hence it can be considered for controlling leeches.

   C. (*T.*) *borreli* develops in two freshwater leeches (*Hemiclepsis marginata* and *Piscicola geometra*). *H. marginata* is in small ponds and slow moving streams throughout Europe and Asia (Elliott and Mann, 1979; Sawyer, 1986; Shanavas *et al.*, 1989). *Piscicola geometra* prefers cool water and occurs in fast flowing streams and lakes throughout Eurasia (Mann, 1961; Malecha, 1984; Sawyer, 1986).

Parasite morphology

C. (*C.*) *branchialis*: body length (living specimens) 14.0–23.0 μm, (fixed specimens) 9.0–18.0 μm; body width (living specimens) 3.5–6.0 μm, (fixed specimens) 2.2–4.8 μm; anterior flagellum 7.7–11.0 μm; posterior flagellum 10–15 μm; nucleus round to oval (1.2 × 1–1.7 μm). Its cytoplasm contains refractile granules and the parasite attaches to host epithelium by its posterior flagellum (Chen, 1956).

The extracellular stage of *C. (C.) iubilans* (Fig. 2.23) is oval to elongated and measurements are based on fixed specimens: body length 5.5–12.5 μm; body width 3.5–5.5 μm. The anterior flagellum is 1.5–2× its body length and the posterior free flagellum is longer than the anterior flagellum. Its triangular kinetoplast is located at the anterior end of the body and is close to its round nucleus. Its morphology was also studied in detail under the electron microscope (Fig. 2.24; Nohynkova, 1984a).

C. (T.) salmositica (Fig. 2.19) is an elongated organism and its body measurements are based on air-dried blood smears: body length 14.9 (6.0–25.0) μm; body width 2.5 (1.3–4.0) μm; anterior flagellum 16.1 (6.5–27.0) μm; posterior free flagellum 9.0 (4.0–17.0) μm; kinetoplast length 2.0–9.0 μm; kinetoplast width 0.5–2.0 μm; nucleus length 1.5–3.5 μm; nucleus width 1.0–2.5 μm, ratio of anterior flagellum to body length 1.07 (0.40–1.95); ratio of posterior flagellum to body length 0.61 (0.25–1.15) and ratio of anterior flagellum to posterior free flagellum 1.97 (0.6–3.7) (Katz, 1951).

The ultrastructure of the parasite is generally similar to other *Cryptobia* spp. However, the blood form of *C. (T.) salmositica* has a functional contractile vacuole (with systole and diastole stages) and its lumen has electron dense filamentous materials (Paterson and Woo, 1983). Contractile vacuoles have not been found in other haematozoic species (Vickerman, 1971; Brugerolle *et al.*, 1979). In *C. (T.) salmositica*, the vacuole is at the base of the flagellar pocket and is associated with the postflagellar pit (Paterson and Woo, 1983). A similar contractile vacuole occurs in *C. (C.) iubilans* and it is also located just posterior to the postflagellar pit and close to the Golgi complex (Fig. 2.24; Nohynkova, 1984a).

C. (T.) borreli is an elongated parasite with considerable variations in size and shape. Small and slender forms predominate during acute infections while large stumpy parasites are the principal forms in chronic infections. Measurements (Kruse *et al.*, 1989b) are based on fixed and stained specimens: body length 24.4 (±3.0) μm; body width 4.9 (±1.2) μm; anterior flagellum 14.6 (±3.6) μm; posterior free flagellum 7.9 (±2.7) μm; kinetoplast length 7.6 (±1.6) μm; nucleus length 4.1 (±0.86) μm; nucleus width 2.1 (±0.47) μm; the oval nucleus is in the anterior part (near the convex margin) of the parasite.

The ultrastructure of *C. (T.) borreli* is similar to that of other *Cryptobia*. It was described in detail by Brugerolle *et al.* (1979); however, they did not find a contractile vacuole in the parasite (cf. *C. (T.) salmositica*).

Transmission of *Cryptobia*

DIRECT TRANSMISSION

The ectoparasitic *C. (C.) branchialis* presumably dislodge from gills of infected fish and become free in the water column. These are brought into the gill chamber via the mouth and the organisms attach to the gill filaments.

C. (C.) iubilans survives off the host for at least 4 hours in aquarium water at 20°C, and cichlids (*C. nigrofasciatum*) are infected by ingesting organs of infected fish. Nohynkova (1984a) suggested that ingestion of parasites in the water (e.g. via regurgitation and defecation) may also be involved in transmission.

C. (*T.*) *salmositica* is not only transmitted by blood-sucking leeches (see below) it is also transmitted directly between fish in the absence of a vector. Parasites in infected fish are released to the body surface through a blister caused by the dissociation of connective tissues near the abdominal pore (Bower and Margolis, 1983). The parasite is in the mucus on the body surface of *Oncorhynchus mykiss* at 6 weeks after experimental infection. Round and slender ectoparasitic forms (Fig. 2.20) are infective to fish and some are morphologically similar to those in the blood (Woo and Wehnert, 1983).

Woo and Wehnert (1983) showed that 67–80% of uninfected trout became infected in 27 weeks if they were allowed to mix freely with infected trout in the same tank. The infection was reduced to 50% if the two groups of fish were separated by a wire screen. They assumed that parasites were carried through the water in mucus strands and entered recipient fish either through lesions on the body surface or actively penetrated the epithelia of gills or the oral cavity.

Transmission was more efficient and rapid if heavily infected and uninfected juvenile sockeye (*Oncorhynchus nerka*) were held together in a dip net for a brief period out of water. Mortality of initially uninfected fish occurred in 27 days and prevalence of infection was 64–89% in fish in fresh water and was 94% in sea water (Bower and Margolis, 1983). There was evidence to suggest that direct transmission was enhanced during the transfer of fish because of direct contact between infected and uninfected fish. Bower and Margolis (1983) found that the prevalence in cohos rose from 29–74% after transfer, while the prevalence in non-transferred fish remained the same. They indicated the increase in prevalence might also be due to stress associated with the transfer. According to Becker and Katz (1965a) there were outbreaks of cryptobiosis in hatcheries where no leeches were found. Direct transmissions of C. (*T.*) *salmositica*, especially in hatcheries, is clearly an important area that needs more studies and further careful documentations.

INDIRECT TRANSMISSION

Haematozoic *Cryptobia* multiplies in the crop of blood-sucking leeches and the parasite from the crop is infective to fish. Slender forms accumulate in large numbers in the crop or in the proboscis sheath and these are transmitted to fish when the leech feeds.

Becker and Katz (1965a) showed that the leech *P. salmositica* became infected with C. (*T.*) *salmositica* when it fed on infected fish and transmitted the parasite when it fed again. Large numbers of parasites were in the crop of leeches 7–8 days after infective blood meals, however they gave no indications that flagellates were in the proboscis sheaths of infected leeches. Li and Woo (unpublished) found numerous slender *Cryptobia* in the proboscis sheaths of *P. salmositica* that were removed from naturally infected salmon. The *Cryptobia* from proboscis sheaths were morphologically similar to the slender ectoparasitic form of C. (*T.*) *salmositica* (Fig. 2.20) and some isolates were infective and caused disease when inoculated into trout. The development of *C. (T.) salmositica* in its leech vector needs more careful studies using laboratory-raised leeches.

The development of *C. (T.) borreli* was first described in the leech *H. marginata* by M. Robertson (1911). More recently, Kruse *et al.* (1989a) described its development in laboratory raised *P. geometra*. Parasites ingested with the blood meal multiplied in the crop. Long slender flagellates were in large numbers in the crop 8–10 days later. All stages in the leech were infective to fish. The parasite was transmitted experimentally to the common carp by allowing infected leeches to feed on them. There was no evidence of transovarian infection in the leech. The development of the parasite, its mode of transmission and the lack of transovarian transmission in the leech vector confirm the earlier study of Becker and Katz (1965a) on *C. (T.) salmositica* in *P. salmositica* (see above).

In vitro **culture and propagation of** *Cryptobia*

C. (T.) salmositica from fish was initially cultured and subcultured in Hanks's solution with 10% heat-inactivated fetal calf serum at 5 and 10°C (Woo, 1979). The parasite multiplies by longitudinal binary fission as blood forms and newly divided forms are slender (morphologically similar to the newly divided parasites in the blood of fish or on the body surface of infected fish). Cultures are always infective when inoculated into fish. *C. (T.) salmositica* and *C. (T.) bullocki* also multiply in minimum essential medium (MEM) with Hanks's salts (pH 7.2–7.3), 25 mM Hepes buffer, 25% heat-inactivated fetal bovine serum (FBS), L-glutamine and 0.33% dextrose (Woo and Li, 1990; E.M. Burreson, personal communication). Minimum essential medium is the medium of choice because *C. (T.) salmositica* multiplies rapidly and rosette colonies are present in about 5 weeks. Formation of colonies indicates a healthy culture and the number of parasites is 6–7 million parasites ml^{-1} of medium in about 6 weeks (Li and Woo, 1991a). A strain that has been in culture for over 4 years is still infective to fish; it however does not produce disease in fish but confers protection when it is used as a live vaccine (Woo and Li, 1990). The vaccine strain is similar to the pathogenic strain in that it secretes haemolysin and immune-complex forming antigens although the amounts secreted are significantly reduced (Woo and Thomas, 1992).

C. (T.) borreli from carp can be cultured at 25°C in SNB-9 diphasic blood agar supplemented with vitamins and antibiotics (Nohynkova, 1984b). It divided by binary fission and the parasite lost its infectivity to fish after six subcultures. A Hungarian strain cultured in SNB-9 blood at 20°C (Hajdu and Matskasi, 1984) was also not infective to fish (Jones *et al.*, 1993). Peckova and Lom (1990) were also unsuccessful in maintaining infective cultures *C. (T.) borreli* in supplemented SNB-9 blood agar. The cultures lost their infectivity to fish after 10–14 days in the primary culture. Li and Woo (unpublished) successfully cultured and subcultured the parasite continuously in a cell free medium (TDL 15 with 10% FBS, 1% goldfish serum and 17 mM Hepes). The parasite multiplied readily and was infective to goldfish and carp.

Parasite nutrition and physiology

C. (T.) salmositica multiplies rapidly by binary fission in fish. Parasitaemias are significantly higher in fish with high (e.g. $3.5\,g\,dl^{-1}$) than low (e.g. $1.0\,g\,dl^{-1}$) plasma proteins (Thomas and Woo, 1990b). Also, trout on an ascorbic acid deficient diet have lower parasitaemias than fish on an ascorbic acid supplemental diet (Li *et al.*, 1994). Serum supplement is an important component under *in vitro* conditions. The parasite does not multiply in minimum essential medium unless FBS is present. It multiplies less readily with 10 and 15% FBS than with 25% and 30% FBS (Li and Woo, 1991a). Glucose is depleted from the medium, and its depletion increases with increased numbers of parasites.

  C. (T.) salmositica and *C. (T.) bullocki* consume oxygen under *in vitro* conditions. Motilities and oxygen utilizations of both species are reduced significantly in sodium azide (Thomas *et al.*, 1992). However, both motilities and oxygen consumptions return to preazide treatment levels when the azide is removed even after the parasites have been exposed to azide for 24 hours.

Diagnosis of infection

Clinical signs (e.g. anorexia, exophthalmia, abdominal distension with ascites, see pp. 57–59) can be used for preliminary diagnosis. Parasites are easily detected using the wet mount technique in acute infections. Fresh samples (of excised gills and body mucus for ectoparasitic species; of internal organs and contents from the digestive tract for intestinal species; of blood/ascitic fluid for haematozoic species) are collected from living fish or within minutes of death and examined fresh under cover slips using bright field or phase contrast microscopy (at $160\times$ and $400\times$). For confirmation on the identification, air-dried smears of parasites are first fixed in 100% ethanol and then in buffered formalin, stained in Giemsa's stain and examined under a light microscope using an oil immersion lens (Woo, 1969).

  The clotting technique (Strout, 1962) and the haematocrit centrifuge technique (Woo, 1969) are used to concentrate and detect low chronic infections of haematozoic species. For parasites of cold water fishes, the haematocrit tubes with blood are centrifuged at 4–$10\,^{\circ}C$ for 5 min at 13,000 g (Woo and Wehnert, 1983; Bower and Margolis, 1984a). *C. (T.) salmositica* become very sluggish and die at about $21\,^{\circ}C$; hence the blood is kept cold at all times. Other species (e.g. *Cryptobia (T.) catostomi*) can be centrifuged at about $20\,^{\circ}C$ (Bower and Woo, 1977). After centrifugation, the junction of the plasma and packed red cells is examined under the microscope ($\times 160$) (Woo, 1969). The sensitivity of the technique is high, detecting *C. (T.) salmositica* infections when there are only about 75 organisms per ml of blood (Bower and Margolis, 1984a). Its sensitivity is increased if more than one capillary tube of centrifuged blood is examined, as in mammalian trypanosomosis (Woo and Rogers, 1974).

  The microscopic immuno-substrate-enzyme technique (MISET) and immunofluorescent antibody technique (IFAT) are equally sensitive in detecting specific antibodies to *C. (T.) salmositica* (Woo, 1990). MISET is preferred because it does not require a flourescent microscope. An ELISA (enzyme-linked immunosorbent assay) is also available to detect specific antibodies against *C. (T.) borreli* infections in carp (Jones *et al.*, 1993) and antibodies against *C. (T.) salmositica* in juvenile rainbow trout (Sitja-Bobadilla and Woo, 1994).

Verity and Woo (1994) developed an antigen-ELISA to detect *Cryptobia* antigen in the blood of infected fish. Preliminary results are most encouraging. They were able to detect parasite antigen in all experimentally infected rainbow trout (inoculated with either the virulent or avirulent strains of *C. (T.) salmositica*) as early as one week after infection. The test was negative for all fish prior to inoculation of the parasite but was consistently positive after experimental infection. They used a monoclonal antibody (Verity and Woo, 1993) against a major polypeptide (47 kDa) of *C. (T.) salmositica* in the test.

Fish–*C. (C.) branchialis* relationships

CLINICAL SIGNS, PATHOLOGY AND MORTALITY

Gills of infected fish with large numbers of *C. (C.) branchialis* are abnormally red. The body is usually covered with copious amount of mucus and the body generally darkens before the fish dies. Infected fish are anorexic and are usually close to the water surface (Bauer *et al.*, 1969).

The parasite destroys the epithelium of gill filaments of susceptible fish in heavy infestations and this leads to formation of thrombi and eventually death (Chen, 1956). The parasite has been associated with mortality of cultured carp, goldfish and catfish.

Lom (1980) was unable to show, at the ultrastructural level, that *C. (C.) branchialis* caused lesions to the gills of infected carp (*Cyprinus carpio*) from the former Czechoslovakia, and fantail darter (*Etheostoma flabellare*) from the United States. He concluded the parasite was not pathogenic and that mortality in infected fish was probably caused by an unknown pathogen. The disease is usually more severe in young grass carp (*Ctenopharyngdon idella*) than in older fish and in other species of carp (Chen, 1956). Hence, Woo (1987a) suggested that the study be repeated with experimentally infected young grass carp to help resolve the confusion.

Fish–*C. (C.) iubilans* relationships

CLINICAL SIGNS AND MORTALITY

C. (C.) iubilans infected aquarium cichlids are lethargic and anorexic. Most infected fish died within 3 months after the onset of clinical disease (Nohynkova, 1984a).

DEVELOPMENT IN FISH

Nohynkova (1984a) did not indicate finding obvious perforation of the gut in infected fish. She suggested that the parasite spread from the stomach to other organs including the spleen and ovary. It would be very challenging to document the route and mechanism of spread of the parasite to organs which are not continuous with the stomach. Perhaps the infected macrophage-like cells are responsible for transporting the parasite to other organs? This would indicate the significance of intracellular development in this parasite.

Another related area that needs experimental studies relates to the intracellular occurrence of the parasite. Nohynkova found active organisms in macrophage-like cells in infected organs (e.g. stomach, spleen, ovary) and concluded that they were developing parasites. However, it is also possible that the intracellular parasites were ingested by macrophage-like cells (part of the host's defence process). *C.* (*T.*) *salmositica* and *C.* (*T.*) *bullocki* were found in mononuclear macrophages under *in vivo* (Woo, 1979; Sypek and Burreson, 1983) and *in vitro* (Li and Woo, 1994) conditions. The ingestion is part of the mechanism (acquired cell-mediated immunity; pp. 61–62) the host uses to control the parasitaemia. Woo (1987a) had earlier suggested *in vitro* culture studies of *C.* (*C.*) *iubilans* with macrophage-like cells to confirm the intracellular development of the parasite.

Fish–*C.* (*T.*) *salmositica* relationships

COURSE OF INFECTION AND CLINICAL SIGNS

The number of *C.* (*T.*) *salmositica* action in the blood of rainbow trout fluctuates and this is particularly evident during the acute phase of the infection (Woo, 1979; Thomas and Woo, 1990b). The parasite multiplies more rapidly and with higher parasitaemias in fish with high plasma protein levels, as occurs in fish that are fed a high protein diet (Li and Woo, 1991b; Thomas and Woo, 1990b, 1992b).

The parasite multiples rapidly at 10°C in salmonids and in tissue culture medium (Woo, 1979; Woo and Li, 1990). However, it multiplies less readily under *in vivo* (Woo *et al.*, 1983) and *in vitro* (Woo and Thomas, 1992) conditions at 5, 15 or 20°C. The first indication of division is production of two new flagella. This is followed by duplications of the kinetoplast and nucleus. Cytokinesis begins at the posterior end and division is unequal and symmetrogenic. The parasite retains its elongated form during the division process (Woo, 1978).

The clinical signs in *Oncorhynchus* spp. include exophthalmia (Fig. 2.26), splenomegaly (Fig. 2.27), general oedema, abdominal distension with ascites (2.28), hepatomegaly, anaemia, positive antiglobulin reactions of red cells, and anorexia (Woo, 1979; Thomas and Woo, 1988, 1990a, 1992b; Li and Woo, 1991b). Some fish are obviously quite emaciated. The immune system is depressed (Jones *et al.*, 1986; Sin and Woo, 1993) and the haemolytic activity of complement is low (Thomas and Woo, 1989b). Also plasma thyroxine (T3 and T4), protein and glucose are lowered along with depletion of liver glycogen (Laidley *et al.*, 1988).

The parasite is not known to cause disease in sculpins and they are considered the principal natural reservoir hosts of the parasite (Becker and Katz, 1965a, 1966). Little is known about the sculpin–parasite relationships and we suggest that this should be one area of future studies. Laboratory raised *Cryptobia*-susceptible brook charr, *Salvelinus fontinalis*, do not suffer from cryptobiosis. None of the clinical signs (see above) were seen in infected fish,

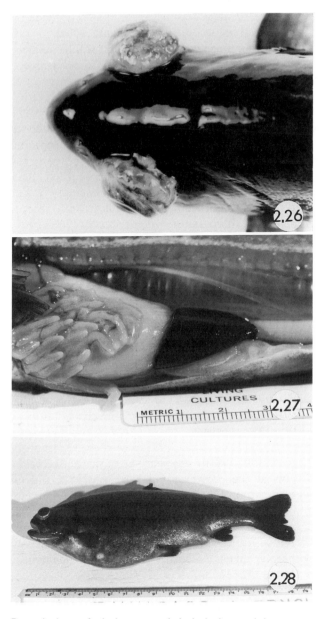

Fig. 2.26. Dorsal view of obvious exophthalmia in a rainbow trout experimentally infected with *Cryptobia (T.) salmositica* (original). Fig. 2.27. Splenomegaly of a trout with acute experimental cryptobiosis. (From Woo, 1979; courtesy of *Experimental Parasitology*.) Fig. 2.28. General oedema with abdominal distension and ascites during acute experimental cryptobiosis in trout. (From Woo, 1979; courtesy of *Experimental Parasitology*.)

hence Ardelli *et al.* (1994) suggested that charr would serve as reservoir host for *C. salmositica* in areas on the Pacific coast where brook charr and *Oncorhynchus* spp. occur together.

PATHOLOGY AND MORTALITY

The anaemia is microcytic and hypochromic and its severity is directly related to parasitaemias (Woo, 1979). There are obvious lesions in haemopoietic tissues during the acute phase of the disease in juvenile rainbow trout (Bahmanrokh and Woo, 1994). The severity of lesions is related to parasitaemias in the blood and extravascular localization of parasites. Blood components return to preinfection values in about 15 weeks after infection (Li and Woo, 1991b). Haemodilution, splenomegaly and haemolysis are also contributing factors to the anaemia (Woo, 1979; Laidley *et al.*, 1988; Thomas and Woo, 1988). Arcuri and Woo (unpublished) showed that trout had significantly lower blood osmolarities from 2 weeks after infection and this was related to high parasitaemias and anaemia. Haemolysis is caused by secretions of living parasites or by antigens liberated when parasites are disrupted due to complement fixing antibodies in infected fish. The lytic antigen (haemolysin) lyses red cells directly while the immune complex-forming antigen forms immune complexes on red cells with specific antibodies which result in intravascular/extravascular haemolysis (Thomas and Woo, 1988). Red cells from parasitaemic trout give positive anti-globulin reactions and some of these red cells are lysed when they are incubated with trout complement. Both antigens are also secreted by parasites in cultures (Thomas and Woo, 1989a; Woo and Thomas, 1992).

The severity of the disease, appearance of the clinical signs and mortality are related to size of the inoculum, water temperature, diets and the genetics of the fish (Woo, 1979; Woo *et al.*, 1983; Bower and Margolis, 1985; Jones *et al.*, 1986; Li and Woo, 1991b). Anorexia contributes to immunodepression (Thomas and Woo, 1992b); however, it is also beneficial to the host because it decreases plasma protein by reducing protein intake (Li and Woo, 1991b). Reduction in plasma protein lowers parasite multiplication and hence the parasitaemia. Lower parasitaemias decrease severity of the disease and the associated mortality. Future studies should be directed at identifying and characterizing the virulence factors (e.g. proteases and haemolysins) and their role(s) in the disease process and fish mortality.

The anaemia and large numbers of parasites occluding small blood vessels may combine to reduce oxygen delivery to tissues and vital organs. Part of this is manifested as an increase in susceptibility of infected fish to environmental hypoxia. This may be an important contributing factor to fish mortality under conditions when dissolved oxygen is reduced as a result of crowding, or algal blooms (Woo and Wehnert, 1986).

According to Putz (1972), *C. (T.) salmositica* was more pathogenic to cohos (*O. kisutch*; 100% mortality) than to chinooks (*O. tshawytscha*). However, Bower and Margolis (1985) found 100% mortality in chinooks while cohos from some stocks did not die (0% mortality). The differences are likely due to the genetics of the fish. For example sockeye salmon (*O. nerka*) from the Fulton River stock (in British Columbia) suffered high mortality when injected with about 100 parasites per fish, while the Weaver Creek stock of sockeye (also in British Columbia) had light mortality even when injected with about 10^6 parasites per fish (Bower and Margolis, 1984a). Mortality of infected sockeye is consistent within the same fish stock and to different parasite isolates (Bower and Margolis, 1985).

The histopathology includes focal haemorrhages, congestion of blood vessels, occlusion of capillaries with parasites and oedematous changes in kidney glomeruli (Putz, 1972).

Bahmanrokh and Woo (1994) further showed that the histopathology was a generalized inflammatory reaction and lesions were in connective tissues and the reticulo-endothelial systems in juvenile rainbow trout. Lesions were seen first in the liver, gills and spleen at 1–2 weeks after infection. Endovasculitis and mononuclear infiltration occurred at 3 weeks and these were followed by tissue necrosis and extravascular localization of parasites at 4 weeks post-infection. Extensive necrosis of tissues was related directly to high parasitaemias and extravascular localization of parasites. Necrosis in the liver and kidney, depletion of haematopoietic tissues, and anaemia were probably responsible for mortality of fish during the acute phase of the disease. The regeneration and replacement of necrotic tissues with normal structures were noticeable in haematopoietic and reticular tissues at 7–9 weeks after infection and these were associated with reduced parasitaemias in the blood and extravascular localizations of parasites (recovery phase).

PROTECTIVE IMMUNITY

The piscine immune system is well developed (Woo and Jones, 1989); both innate and acquired immunity are involved in protecting fish from parasitic diseases (Woo, 1992). The mechanism of innate resistance in refractory fishes (e.g. goldfish, suckers) to *C. (T.) salmositica* is the alternative pathway of complement activation (Wehnert and Woo, 1980). The undiluted plasma of refractory fish lyse the parasite in 30–60 min at 4°C and the lytic titres range from 1 : 4 to 1 : 8.

Brook charr raised in the laboratory are susceptible to *C. (T.) salmositica* infection (see pp. 58–60), however there are some individuals that are resistant to infection (Forward *et al.*, 1993). The plasma of resistant charr lyse the parasite (lytic titre 1 : 2 to 1 : 4) via the alternative pathway of complement activation. This innate resistance is inherited by progenies and seems to be controlled by a single dominant gene. Studies on innate resistance to *C. (T.) salmositica*, especially in *Oncorhynchus* spp., should continue because it is an excellent strategy to protect against the parasite (Woo, 1992).

Acquired humoral and cell-mediated responses in trout are detectable about 2 weeks after infection (Thomas and Woo, 1990a; Woo, 1990). Recovered fish are resistant to challenge. Complement fixing and neutralizing antibodies are present in infected or recovered fish (Wehnert and Woo, 1981; Jones and Woo, 1987; Li and Woo, 1994). Infected trout injected with cortisol had little or no detectable antibody. They also had shorter prepatent periods and their parasitaemias and mortality were higher than noncortisol injected fish (Woo *et al.*, 1987).

C. (T.) salmositica was phagocytized by mononuclear cells under *in vivo*

(Woo, 1979) and *in vitro* conditions (Li and Woo, 1994). Passive transfer of lymphocytes and plasma from immune fish conferred partial protection in naive trout (Jones and Woo, 1987). Cell-mediated response was demonstrated in intact (Thomas and Woo, 1990a) and thymectomized (Feng and Woo, 1993) infected trout. Humoral and cell-mediated immune responses were lowered significantly in infected fish on either pantothenic acid-deficient or low protein diets (Thomas and Woo, 1990a,b). However, neither dietary ascorbic acid (Li *et al.*, 1993) nor thymectomy (Feng and Woo, 1993) reduce protection and production of complement fixing antibody in infected or vaccinated trout (see below).

Infected fish lost their infections or there was no mortality when the water temperature was raised to 20°C (Woo *et al.*, 1983; Bower and Margolis, 1985). Woo (1987a) suggested that modification(s) of this approach might be useful for protecting fish. A recent study confirmed that infected juvenile sockeye salmon acclimated to 20°C survived while all infected fish maintained at 10°C died from the disease. Also, 60 of temperature acclimated infected fish survived a parasite challenge at 10°C while 95 of infected naive fish died (Bower and Evelyn, 1988).

Vaccine and vaccination

C. (T.) salmositica attenuated under *in vitro* conditions produces low parasitaemia and does not cause disease when it is inoculated into trout (Woo and Li, 1990). Vaccinated adult rainbow trout and brook charr were protected from disease when they were challenged with the pathogen (Woo and Li, 1990; Ardelli and Woo, 1993). Vaccinated adult trout were protected for at least 24 months after vaccination (Li and Woo, 1994). The titres of complement-fixing antibodies in vaccinated fish rose significantly after parasite challenge. This is a typical classical secondary response and it shows immunological memory in vaccinated fish. Also, under *in vitro* conditions, activated macrophages from head kidneys of vaccinated fish in the presence of antisera were very efficient in engulfing living parasites. They were significantly more efficient than activated macrophages with sera from naive fish or macrophages from naive fish with antisera (Li and Woo, 1994).

The efficacy of the vaccine and its protective period in pre-smolt and post-smolt salmon are unknown. However, juvenile rainbow trout were protected by the vaccine (Sitja-Bobadilla and Woo, 1994) and adult rainbow trout vaccinated in fresh water and transferred to sea water were protected from cryptobiosis (Li and Woo, unpublished). Future studies should include vaccination of pre-smolts in fresh water, transferring them to sea water for 3 years and then challenge (via the leech vector) the adults with the pathogen after they are transferred back to fresh water (see pp. 62–63).

The vaccine strain has lost five polypeptide bands and five of the remaining bands are antigenically different (Woo and Thomas, 1991). It is not known if the loss in virulence is related to the loss and/or changes in its proteins. The vaccine strain also multiplies much more readily than the pathogenic strain in tissue culture medium; however, it produces significantly less haemolysins and immune-complex forming antigens than the pathogenic strain (Woo and

Thomas, 1992). The identification and characterization of the haemolysin(s) and protective antigen(s) are areas that should be explored in future studies.

IMMUNODEPRESSION

C. (T.) salmositica is one of a few parasitic infections that are known to depress the piscine immune system (Woo, 1987b, 1992). Production of antibodies to sheep red cells and *Yersinia ruckeri* were depressed in *C. (T.) salmositica*-infected trout (Wehnert and Woo, 1981; Jones *et al.*, 1986; Sin and Woo, 1993). Jones *et al.* (1986) showed that when *Cryptobia*-infected fish were exposed to *Y. ruckeri* they suffered higher mortality than those infected with either pathogen alone. Also, when *Cryptobia*-infected fish that were earlier exposed to *Yersinia* were re-exposed to the bacteria, they were as susceptible as fish that had no prior exposure to *Yersinia*.

Anorexia contributes to humoral immunodepression in infected trout (Thomas and Woo, 1992b). The haemolytic levels of complement in infected trout are about 20% of preinfected levels and the low levels persist throughout the infection (Thomas and Woo, 1989b). Low complement decreases phagocytic activity and antigen presentation by macrophages and hence may contribute to immunodepression. Further studies should include the mechanism in immunodepression and factors that contribute to the immunodepression in cryptobiosis.

Effects of salmonid cryptobiosis on fish production

There have been a few documented serious outbreaks of the disease in young salmon in hatcheries in Washington State, USA. These include three outbreaks in chinook salmon in three localities in 1972 and 1973 (Wood, 1979). The fish had massive numbers of *Cryptobia* in their blood and they were anaemic, lethargic and some had abdominal distension and generalized oedema. Wood found parasites on the body surface and in the abdominal fluid in fish with acute disease. He found that the infection in juvenile chinook resulted in substantial losses while the same age coho salmon in the same or adjacent ponds were unaffected. More recently, P.C. Chapman (personal communication, 1993) described another outbreak of cryptobiosis in chinook in hatcheries in Washington State which began in December 1992 and peaked in February 1993. Infected fish had the typical clinical signs (anaemia, splenomegaly, ascites) of the disease and 65,000 fish were involved with peak mortality of 0.1% per day in February. In one hatchery the mortality of chinook broodstock was about 50%. He noted that cryptobiosis had been recorded in the same hatcheries in the past but they were not as severe. Bower and Thompson (1987) had earlier concluded that the parasite is a lethal pathogen of salmon in semi-natural and intensive salmon culture facilities on the west coast of North America.

Some young Pacific salmon become infected before they migrate to sea. The prevalence of the parasite ranged from 3 to 21% (Becker and Katz, 1966). Infected pre-smolt salmon retained the infection when transferred to salt water and mortality was not reduced in sea water (Bower and Margolis, 1985). Hence future studies should include vaccination (see pp. 62–63) of pre-smolts in fresh water and transferring them to salt water. The immune response to the vaccine and behaviour (e.g. multiplication of the flagellate, clinical disease, if any) of the vaccine in Pacific salmon in sea water should be documented. Also, the protection of vaccinated fish 3 years later when they are challenged with the pathogen after they are returned to fresh water should be investigated.

Cryptobiosis causes mortality in sexually mature salmon. Significant mortalities were associated with post-spawning rainbow trout and pre-spawning chinook salmon in North America (Wales and Wolf, 1955; P.C. Chapman, personal communication) and pre-spawning pink salmon (*Oncorhynchus gorbuscha*) in the former Soviet Union (Makeyeva, 1956). Wales and Wolf (1955) suggested that many post-spawning trout in a hatchery in northern California died from a combined effect of *Cryptobia* and the ectoparasitic fungus, *Saprolegnia parasitica*. According to Makeyeva (1956) heavy *Cryptobia* infection was the cause of pre-spawning mortality of pink salmon in a tributary of the Amur River. Makeyeva found lower numbers of parasites in fish that had spawned and assumed that the fish were infected just before spawning and hence were able to spawn successfully. Carefully controlled experimental studies (during the acute and chronic phases of the disease) should be conducted to better understand the conditions (e.g. crowding, water temperature, susceptibility to secondary infections) that contribute to mortality of infected fish during their reproductive state.

The indirect effect(s) cryptobiosis on reproduction is not known. Some adult Pacific salmon had detectable infections as early as 5 days after they returned to fresh water (in November) and parasitaemias in many fish were very high when they spawned (Bower and Margolis, 1984b). Since the severity of the diseases is related to parasitaemia (pp. 60–61), acute cryptobiosis may affect the reproductive capacity of salmon by causing lesions in the gonads and associated organs. Studies on effects (e.g. levels of reproductive and related hormones, quantity and quality of sperms and eggs) on the reproductive capacity of acutely and chronically infected salmon would prove rewarding.

Epizootiology of salmonid cryptobiosis

The prevalence of C. (*T.*) *salmositica* in salmon returning to fresh water to spawn is low in September but increases to about 100% in December and January. Also, returning salmon have detectable infections within 5 days of returning to fresh water and the longer fish are in fresh water the higher are their parasitaemias (Bower and Margolis, 1984b). These increases are related to increased numbers of leeches in November (Becker and Katz, 1965b, 1966; Bower and Margolis, 1984b).

Becker and Katz (1966) found that prevalence of the parasite in young salmon varied greatly (3–21%) between years. Fingerlings had detectable infections in the autumn and winter. The parasite was also in downstream migrants (normally 1 year old fish) in early spring. Wood (1979) had found *Cryptobia* in coho and chinook salmon less than 60 days old.

The torrent sculpin (*Cottus rhotheus*) is considered the principal reservoir host. This will have to be confirmed using a recently developed DNA probe which is specific to *Cryptobia salmositica* (Li, 1994). Becker and Katz (1966) found the parasite in 60% of sculpins between April 1961 and October 1962. The prevalence (27%) was lowest in small fish (<65 mm). However, the parasitaemia was highest in the small fish and it decreased in larger fish. The prevalence in sculpins (*Cottus aleuticus*) ranged from 8 to 95% between September 1981 and September 1982 (Bower and Margolis, 1984b). In small sculpins (<40 mm) the prevalence was low between August and November and was highest in April. This seasonal trend was seen in large sculpins (>40 mm) and their prevalence was higher than that in small fish.

Fish–C. (T.) borreli relationships

ROUTE AND COURSE OF INFECTION

Keysselitz (1906) indicated that post-spawning cyprinids seemed to be more susceptible to the parasite than sexually mature fish. This was also noted in trout infected with C. (T.) *salmositica* (see pp. 63–64). The effects of acute and chronic C. (T.) *borreli* infections on pre- and post-spawning cyprinds needs to be examined more closely as was suggested in salmoned cryptobiosis.

The parasite multiplies rapidly after infection and this is followed by their elimination from peripheral blood (Steinhagen *et al.*, 1989a; Jones *et al.*, 1993). Water temperatures affected the length of prepatent periods, height of parasitaemias and duration of infections (Steinhagen *et al.*, 1989a) as was also shown in salmonid cryptobiosis (Woo *et al.*, 1983). Also, the C. (T.) *borreli* infection lasted 20 weeks in fish kept at 20°C and it was shortened to 12 weeks at 30°C (Steinhagen *et al.*, 1989a).

PATHOLOGY AND MORTALITY

Experimentally infected fish are anaemic (Lom, 1979; Steinhagen *et al.*, 1990; Jones *et al.*, 1993). The anaemia is associated with high parasitaemia and is absent in recovered fish. The kinetics of the anaemia is not known but is presumed to be similar to that in salmonid cryptobiosis (pp. 60–61). The number of leucocytes (especially granuloblasts and granulocytes) in carp increased after infection with C. (T.) *borreli* and peaked at 44 days (Steinhagen *et al.*, 1990).

Parasites are located extra- and intravascularly in infected goldfish. The intercellular spaces in fat cells are dilated and are full of flagellates. Diffused degenerative changes and glomerulitis and tubulonephrosis are in the kidneys. Also, foci of necrotic tissues are obvious in livers (Dykova and Lom, 1979).

C. (T.) *borreli* killed 56–80% of experimentally infected goldfish (Lom, 1979). As in C. (T.) *salmositica* infections, mortality due to C. (T.) *borreli* is also related to the size of the inoculum and water temperature. Age seems to be another important factor. Lom (1979) indicated that 40–80% mortality in first year carp and this was reduced significantly in second year carp. This study should be repeated because experimental details (e.g. size of inoculum, water temperature) were not provided.

IMMUNITY

Infected goldfish that survived heavy experimental infections were protected (Lom, 1979). Experimentally infected carp rapidly produced antibodies in the first 4 weeks of the infection (Jones *et al.*, 1993). Peak antiboby production coincided with the decline in parasitaemia and most fish recovered 8–12 weeks after infection. Recovered fish were protected from challenge. Immunosuppressive agents significantly increased parasitaemias in carp which resulted in high fish mortalities (Steinhagen *et al.*, 1989b).

Cryptobiosis in marine fish

C. (T.) bullocki causes disease and mortality in marine fish. The parasite is transmitted by blood-sucking leeches. It is cultured in minimum essential medium (p. 55) at 15°C and culture forms are infective to fish (E.M. Burreson, personal communication). The diagnostic procedures for detection of *C. (T.) salmositica* (pp. 56–57) can be used for *C. (T.) bullocki*.

Geographical distribution of parasite and host range

The parasite, *C. (T.) bullocki*, was first described from the blood of winter flounders (*Pseudopleuronectes americanus*) in New Hampshire (Strout, 1965). It is found in other estuarine and inshore fishes along the Atlantic coast (from New Brunswick to Georgia) and also in the northern Gulf of Mexico (Strout, 1965; Laird and Bullock, 1969; Daily, 1978; Newman, 1978; Becker and Overstreet, 1979; Burreson and Zwerner, 1982). The parasite is common in flounders (70–100%) that use estuaries as nursery grounds. These include winter and summer flounders (*P. americanus* and *Paralichthys dentatus*) along the Atlantic coast, and southern flounder (*Paralichthys lethostigma*), hogchoker (*Trinectes maculatus*) and croaker (*Micropogonias undulatus*) from the southern Atlantic coast to the northern Gulf of Mexico. Also, smooth flounder (*Liopsetta putnami*) in New England and hogchoker and summer flounder in the Chesapeake Bay are commonly infected with the parasite.

Parasite morphology

Air-dried specimens are longer and more slender than those that are fixed in osmic vapour (Strout, 1965). The following description is based on stained air-dried specimens: body length 17.6 (12.5–23.1) µm; body width 2.7 (1.2–4.5) µm; anterior flagellum 13.1 (8.3–19.1) µm; posterior free flagellum 8.5 (4.4–15.7) µm; length of kinetoplast 3.3 (1.7–5.5) µm; width of kineoplast 1.1 (0.6–2.0) µm, length of nucleus 3.4 (1.8–6.8) µm; anterior of nucleus to anterior end 4.2 (0.9–6.3) µm; and anterior of kinetoplast to anterior end 1.5 (0.1–4.6) µm. The cytoplasm is alveolar and darkly stained chromatin granules are often seen in the posterior part of the body.

Geographical distribution of the leech vector

The marine leech, *Calliobdella vivida* is the vector. The leech is not host specific and occurs in estuaries or near shore from Newfoundland to northern

Gulf of Mexico and is present in the autumn, winter and spring (Sawyer *et al.*, 1975; Appy and Dadswell, 1981; Madill, 1988).

Life cycle and transmission of *C. (T.) bullocki*

C. (T.) bullocki rounded up at 10°C and started to divide within 24 hours in the crop of *C. vivida*. Slender forms appeared 4 days later, concentrated at the anterior part of the crop and appeared in the proboscis sheath a day later. At 20°C flagellates were in the proboscis sheath in 24 hours while their appearance was delayed to 10 days at 5°C. Infected leeches still had flagellates in their proboscis sheaths even after three consecutive feedings on uninfected fish (Burreson, 1982a).

Fish–*C. (T.) bullocki* relationships

CLINICAL SIGNS AND PATHOLOGY

The disease develops slowly in summer flounders and clinical signs include anaemia, splenomegaly, exophthalmia, abdominal distention with ascites (Fig. 2.29) and sluggishness (Burreson, 1982b; Burreson and Zwerner, 1984). Clinical disease is evident in experimentally infected juvenile summer flounders in 5 weeks and the fish are dead in 11 weeks after infection.

Ascites and haemorrhaging in the ventral musculature are evident in about 5 weeks (Burreson, 1982b; Burreson and Zwerner, 1984). In acute infections, the parasite occurs extravascularly and is in haemopoietic organs and interstitial adipose tissues. Necrotic foci are in the liver with diffuse necrosis throughout the spleen. Ulcerative and haemorrhagic lesions in the abdominal cavity, oedema, haemorrhage and necrosis of the intestine and oedema of the stomach may also be present. The parasite is also in the submucosa of the gut and liver. The glomeruli are congested and there are focal glomerular lesions in the kidneys (Newman, 1978).

IMMUNITY

Complement fixing antibody against the parasite was detected in experimentally and naturally infected summer flounders kept at 24 and 10–22°C respectively (Sypek and Burreson, 1983). Cell-mediated immunity seemed more important than antibodies below 12°C. It was suggested that antibodies were responsible for the recovery in fish and the decrease in prevalence of the disease in the spring. Sypek and Howe (1985), after studying more naturally infected fish, suggested that temperature would not be the only regulating factor.

A naturally infected fish that had recovered from the infection (no detectable parasitaemia) for at least 1 year was immune to challenge. Its antibody titre rose rapidly after challenge (Burreson and Frizzell, 1986). This study suggests immunological memory and should be repeated (Woo, 1987a).

Epizootiology of flatfish cryptobiosis

According to Burreson (see Chapter 14, this volume) the prevalence of *C. (T.) bullocki* is low in migrating fishes in Chesapeake Bay because they are exposed to *C. vivida* for only short periods. They migrate out of the Bay as leeches

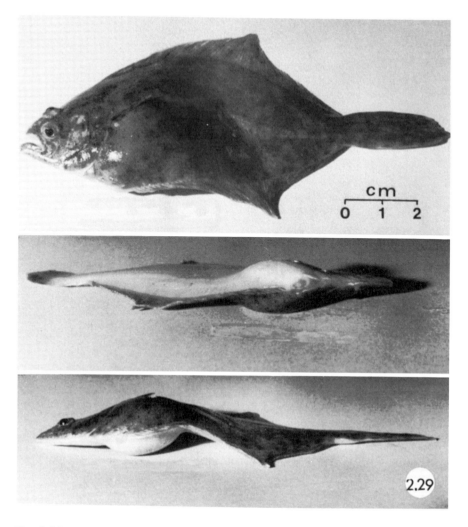

Fig. 2.29. Severe ascites in juvenile flounder experimentally infected with *Cryptobia* (*T.*) *bullocki*; three views of the same infected fish. (From Burreson and Zwerner, 1984; courtesy of Dr E.M. Burreson.)

hatch and re-enter the Bay as adult leeches are dying (p. 618). However, resident fishes (hogchoker, oyster toadfish and juvenile summer flounders) have high prevalence because they are exposed to leeches throughout winter, and mortality of juvenile summer flounders is highest when water temperature is low (0.5–1.5°C).

Infected fish with ascites brought into the laboratory died within 2–6 weeks at 5°C. However, if they were maintained at 10 and 13°C the ascites disappeared and fish survived the disease (Burreson and Zwerner, 1984).

Burreson (1982b) found the parasite in about 100% of juvenile summer flounders in the autumn in the lower Chesapeake Bay. Dead flounders (with clinical signs) were seen and also the number of juvenile flounders caught decreased rapidly through the winter. Infected juvenile flounders collected from warmer waters (8°C) did not have abdominal distension with ascites. Hence low water temperature seems to be important in the development of the disease and fish mortality. The higher temperatures may enhance immunological response in infected fish and this helps to control the infection (p. 67). The precise role(s) temperature plays in morbidity and mortality in infected fish should be studied more closely.

TRYPANOSOMOSIS

Introduction

Trypanosomes are haemoflagellates and they usually have a free flagellum at the anterior end (Fig. 2.30). About 190 species of piscine trypanosomes have been described and host specificity of the vast majority of species are not known. Some species are monomorphic (Fig. 2.30) while others are pleomorphic (Fig. 2.31). It is likely that many of the species are invalid, however this can only be confirmed after careful experimental studies (Woo, 1994). Piscine trypanosomes are always transmitted by blood sucking leeches and most species are not known to cause disease and/or mortality in the host, hence they will not be discussed in the present review.

Trypanosomosis in freshwater fish

Geographical distribution and host range

Trypanosoma danilewskyi was first described from the blood of the common carp (*Cyprinus carpio*) in Europe (Laveran and Mesnil, 1904). It has since been found in the common carp, goldfish (*Carassius auratus*), tench (*Tinca tinca*) and eel (*Anguilla* sp.) in Europe (Thompson, 1908; Pavlovskii, 1964; Lom, 1973) and in *Saccobranchus fossilis* in India (Qadri, 1962). The parasite is not host specific and is infective when inoculated experimentally into *Barbus conhus*, *Danio malabaricus*, *Catostomus commersoni*, *Notropis cornutus*, *Etheostoma caeruleum*, and *Ictalurus nebulosus* (Woo and Black, 1984).

Morphology and reproduction

Trypanosoma danilewskyi (Fig. 2.30) is a slender monomorphic trypanosome and the following measurements are based on specimens from a 4 week experimental infection in goldfish (Woo, 1981a). The strain was initially isolated from crucian carp (*Carassius auratus gibelio*) in the former Czechoslovakia by Dr J. Lom, Institute of Parasitology, Czech Republic. Morphometrics are based on 50 well stained specimens: distance from posterior tip to kinetoplast (PK) = 1.7 µm ± 0.61 (0.6–2.5); distance from kinetoplast to middle of nucleus (KN) = 10.7 µm ± 2.69 (7.8–15.0); length of body (PA) = 21.2 µm ± 3.64 (15.6–24.9); length of free flagellum (AF) = 14.3 µm ± 2.30

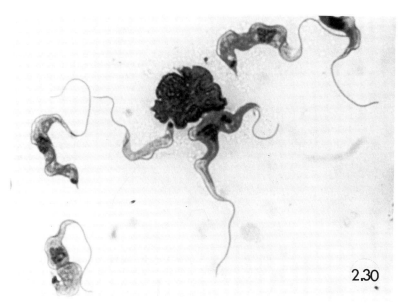

Fig. 2.30. *Trypanosoma danilewskyi* in the blood of an experimentally infected goldfish; note the monomorphic trypomastigotes, a non-nucleated abnormal form, and a pre-division stage with two kinetoplasts, two free flagella and a nucleus (×1150). (From Woo, 1981a; courtesy of *Journal of Parasitology*.)

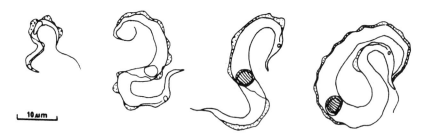

Fig. 2.31. Line drawings of small, intermediate and large forms of *Trypanosoma murmanensis* from the blood of experimentally infected cod. (From Khan, 1976; courtesy of *Canadian Journal of Zoology*.)

(9.2–18.2); maximum body width excluding undulating membrane (BW) 2.3 µm ± 0.40 (1.6–3.1).

Woo and Black (1984) showed that the host fish (see p. 69) significantly affects the morphometrics of the trypanosome. The PK, AF and area of the nucleus are similar among samples from the same species of fish host but differ significantly between samples from five different host species.

The ultrastructure of piscine trypanosomes is similar to that of mammalian trypanosomes. The surface coat of *T. danilewskyi* is fibrillar and thin and is morphologically similar to the stercorarian trypanosomes of mammals. The parasite has a well-developed cytostome which is located near the nucleus. Membrane-bound dense granules are in juxtaposition to the cytopharyngeal tube (Lom *et al.*, 1980; Paulin *et al.*, 1980) and the prominent nucleolus and chromatin patches are attached to the inner nuclear membrane (Paterson and Woo, 1984).

The parasite multiplies rapidly in the blood of goldfish (Woo, 1981a). The first division stage in the division process is the production of a new flagellum and this is followed by division of the kinetoplast. The new flagellum flips posteriorly, and along it new cytoplasm is produced or accumulated. The nucleus divides and one daughter nucleus migrates posteriorly past the two kinetoplasts. Body division is by transverse constriction at a point between the kinetoplasts. This division process is different from the traditional longitudinal binary division of most other trypanosomes.

Development in the leech vector

T. danilewskyi, after ingestion with the blood meal, develops in the leech *Hemiclepsis marginata* (see M. Robertson, 1911; Qadri, 1962). The biology of the leech vector is described in Chapter 14. Unequal division of trypanosome trypomastigotes in the crop of the leech results in tadpole-like epimastigotes. Sphaeromastigotes are also present but their significance is not known. Epimastigotes divide repeatedly by binary fission with eventual appearance of slender metatrypanosomes. These migrate and accumulate in the proboscis in about 10 days and may be present even several months after the infective blood meal. Transmission is presumed to occur when infected leeches feed.

In vitro culture of the parasite

T. danilewskyi from cyprinids was cultured and subcultured at 25°C in diphasic blood agar (Lom and Suchankova, 1974). Infectivity of cultures, which contained both epimastigotes and trypomastigotes, to goldfish decreased with subcultures. The exception was the Ma strain which was isolated from goldfish. It multiplied as trypomastigotes in diphasic medium and retained its infectivity after a considerable number of *in vitro* subcultures.

The parasite was also cultured at 25°C with fish cells in Eagle's minimum essential medium with 10% bovine fetal or calf serum supplement (Smolikova *et al.*, 1977). It grew better with the addition of haemin (2 mg per 100 ml of medium) and did not grow in the medium without living cells. However, another strain of *T. danilewskyi* was cultured and subcultured as trypomastigotes in MEM (minimum essential medium with Hank's salts and L-glutamine supplemented with 10% fetal bovine serum) at 20°C without fish cell (Islam and Woo, 1992). The strain also multiplied rapidly (a two- to fivefold increase in about 7 days) as infective trypomastigotes in a serum-free medium (Wang and Belosevic, 1994). However, it did not grow in four other media including diphasic blood agar medium (SNB9) supplemented with various vitamins (Islam and Woo, 1992). This combination supported growth of another strain (Skarlato *et al.*, 1987). The *in vitro* division process of the parasite is similar to that in fish (see above) and parasites in MEM are infective to goldfish (Islam and Woo, 1992).

The *in vitro* studies clearly indicate nutritional/cultural differences between various isolates. It is suggested here that there may be more than one species and future studies should include comparative developmental (e.g. in the vector and fish), antigenic (e.g. using monoclonal antibodies), molecular (e.g. using DNA probes) and virulence (e.g. using genetically known carp stocks) studies to establish the taxonomic status of the isolates. Cross transmission studies are also useful in this regard.

Host–parasite relationships

COURSE OF INFECTION AND CLINICAL SIGNS

The parasite multiplies rapidly in goldfish when they are kept at 20°C and slower at either 30 or 10°C (Lom, 1973; Woo *et al.*, 1983; Islam and Woo, 1992). The parasitaemia increases after infection and in fish maintained at 20°C it starts to decline at about 28 days after infection. These fish eventually recover from the infection and are immune to homologous challenge (Woo, 1981b; Islam and Woo, 1991a).

The parasite causes anorexia in experimentally infected goldfish (Islam and Woo, 1991b) and this is most evident during high parasitaemia. Fish that survive the disease return to normal feeding. *T. danilewskyi* also causes anaemia in goldfish (M. Robertson, 1911; Dykova and Lom, 1979; Islam and Woo, 1991c).

PATHOLOGY AND MORTALITY

The severity of the anaemia is directly related to the number of trypanosomes in the blood, and is partly caused by haemolysin and haemodilution (Islam and Woo, 1991c). The haemolysin(s) is secreted by living parasites and it lyses red cells in the absence of specific antibodies.

Mortality of infected goldfish is related to size of inoculum and small inoculum produces low mortality (Woo, 1981b; Islam and Woo, 1991a). Lom (1979) indicated that the infectivity and virulence of one parasite strain (isolated from carp) increased on repeated transfer in goldfish. However, the virulence of two other strains of *T. danilewskyi* (isolated from eels and tench) decreased on repeated transfer in goldfish. No explanations were provided for the difference and since no experimental details were given, the study should be carefully repeated as suggested above.

IMMUNITY

Goldfish that have recovered from a previous infection are protected on subsequent homologous challenge (Lom, 1973; Woo, 1981b; Islam and Woo, 1991a) and plasma from immune fish contain neutralizing antibodies (Woo, 1981b). The immunity is nonsterile as low numbers of parasites are in the blood of recovered fish. Intraperitoneal injection of corticosteroid into recovered goldfish significantly increases the number of trypanosomes in peripheral blood of immune fish (Islam and Woo, 1991a).

Diagnosis of infection

During the acute phase of the disease, parasites are readily detected by examination of a drop of freshly collected blood under the light microscope. Taxonomic identification of the parasite is best done with alcohol–formalin-fixed blood smears stained in Giemsa's stain (Woo, 1969). The haematocrit centrifuge technique (Woo, 1969) will detect infections which are not detectable using the wet mount technique. There are no serological techniques for the detection of piscine trypanosomosis; however the serodiagnostic techniques developed for cryptobiosis (see pp. 56–57) can easily be adapted for trypanosome infections.

Trypanosomosis in marine fish

Geographical distribution and host range

Trypanosoma murmanensis (Fig. 2.31) is a pleomorphic trypanosome. It was initially described from cod (*Gadus callarias*) from the Barents Sea, Murmansk in the former USSR (Nikitin, 1927). It is commonly found in Atlantic cod (*Gadus morhua*) from the North West Atlantic in coastal Newfoundland (Khan, 1972; Khan *et al.*, 1980a; Khan, 1986). Natural infections have also been detected in Pleuronectiformes and Perciformes (Khan, 1977a; Khan, *et al.*, 1980b).

The trypanosome is not host specific (Khan, 1977a; Khan *et al.*, 1980b). Experimentally infected fishes include the American eel (*Anguilla rostrata*), Atlantic cod (*G. morhua*), tomcod (*Microgadus tomcod*), cunner (*Tautogolabrus adspersus*), striped wolffish (*Anarhichas lupus*), Vahl's eelpout (*Lycodes vahlii*), Artic eelpout (*Lycodes reticulatus*), oceanpout (*Macrozoarces americanus*), longhorn sculpin (*Myoxocephalus octodecemspinosus*), shorthorn sculpin (*Myoxocephalus scorpius*), American plaice (*H. platessoides*), yellowtail flounder (*L. ferruginea*), and winter flounder (*P. americanus*).

Morphology and life cycle of parasite

MORPHOLOGY

Trypanosoma murmanensis is a pleomorphic trypanosome with small, intermediate and large forms (Fig. 2.31; Khan, 1972, 1977a). The following description is based on 50 stained trypanosomes from naturally infected cod (Khan, 1972). The nucleus and kinetoplast are oval and 6–8 longitudinal striations are present in large specimens, PK = 9.1 µm (6.0–14.0); KN = 32.9 µm (25.0–38.0); NA = 34.6 µm (26.0–40.0); PA = 76.7 µm (57.0–92.0); AF = 6.9 µm (6.0–8.0). The measurements of the trypanosome in other naturally infected fishes are somewhat different (Khan, 1977a).

LIFE CYCLE

The parasite multiplies as amastigotes and sphaeromastigotes in the digestive tract of the leech *Johanssonia arctica*. The epimastigotes, which are presumed to be derived from sphaeromastigotes, migrate to the proboscis of the leech where they transform into infective metatrypanosomes. These are inoculated into fish when the leech feeds (Khan, 1976). The development is completed in 62 days at 0–1°C and in 42 days at 4–6°C.

Small slender trypomastigotes appear in the blood of cod 3 days after infection. They grow into large forms in 29–55 days (Fig. 2.31). The pleomorphic nature of the parasite is more apparent when the parasitaemia starts to decline at about 29 days after infection (Khan, 1976).

Geographical distribution of leech vector

The vector of *T. murmanensis* is the marine leech *Johanssonia arctica*. It is commonly found at −1 to 2°C at depths of 165 m or more in the Arctic seas and the northwestern Atlantic Ocean (Epshtein, 1962; Meyer and Khan, 1979; Khan, 1982, 1991; Sawyer, 1986).

Fish–parasite relationships

COURSE OF INFECTION

The parasite is not known to multiply in the fish. Khan (1976) suggested that the slight increase in number of trypanosomes between 3 and 10 days after infection was due to late entry of parasites into the blood.

CLINICAL SIGNS AND PATHOLOGY

Infected fish are anaemic and lethargic, with emaciation and splenomegaly in some fish (Khan, 1985). The severity of the anaemia, and decreases in plasma proteins, are associated with parasitaemias (Khan, 1977b; Khan *et al.*, 1980c). The anaemia is associated with an inactive rather than an impaired haemopoietic system. The condition factor, somatic indices of liver, spleen and heart are altered in four species of experimentally infected fishes (*G. morhua, P. americanus, Myoxocephalus scorpius* and *M. octodecemspinosus*). Also, spleens of infected juvenile cod are congested with blood and melanomacrophage centres are more common in infected than in uninfected juvenile winter flounders (Khan, 1985).

MORTALITY

The parasite caused high mortality in experimentally infected cod and winter flounders kept at 0 to 1°C. Mortality ranged from about 7% in 3+ year to 65% in 0+ year cod. The mortality figures were similar in immature winter flounders where it ranged from 17–56%. Larger fish (presumably older) were less susceptible and adult flounders did not die from the infection (Khan, 1985).

Evidence of trypanosomosis in naturally infected fishes

There are several other species of trypanosomes that are presumed to cause lesions in organs and or changes in the blood of infected fish. However, these studies were on naturally infected fish and they did not have proper controls.

Fish from the field are often infected with a variety of organisms (viruses, fungi, bacteria, protozoans, and metazoans) and may have been subjected to numerous stress factors (e.g. toxic pollutants, nutritional deficiencies). Hence the abnormalities cannot/should not be ascribed to any one particular cause. At best, some of these field studies can only be considered as preliminary and are briefly included so that they will stimulate more carefully controlled studies in the future.

Neumann (1909) found inflammation of the brain, fatty degeneration in organs, anaemia and eosinophilia in skates (*Raja punctata*) that had high numbers of *Trypanosoma variabile*. According to Smirnova (1970) the number of red cells, haemogloblin contents and serum protein levels of *Trypanosoma lotae*-infected burbot (*Lota lota*) were lowered while the number of phagocytic white cells was elevated.

Fish infected with *Trypanosoma vittati* and *Trypanosoma maguri* had low numbers of red cells and haemogloblin (Tandon and Joshi, 1973). However, the numbers of white cells and immature or abnormal red cells were higher in the infected fish. *Clarias batrachus* and *Channa punctatus* infected with *Trypanosoma batrachi* and *Trypanosoma aligaricus* had lower red cell counts and lower haemogloblin (Gupta and Gupta, 1985). Also, the numbers of agranulocytes and granulocytes were higher in infected fishes except for neutrophils which were significantly lower. Also blood glucose levels were lower in *C. punctatus*, *C. batrachus*, *Heteropneustes fossilis* and *Mystus seenghala* infected with a trypanosome(s) (Tandon and Joshi, 1974). Similarly, serum alkaline phosphatase levels and cholesterol levels were lowered in five (*Cirrhina mrigala*, *C. batrachus*, *M. seenghala*, *Mastacembelus armatus*, *Wallago attu*) and six species of fishes (*C. mrigala*, *C. batrachus*, *H. fossilis*, *M. armatus*, *M. seenghala*, *W. attu*) infected with trypanosomes (Tandon and Chandra, 1977a,b).

We suggest that parasites from naturally infected fish should first be cloned to ensure that the parasite is a single species. The cloning is done using susceptible laboratory-raised fish (e.g. Woo, 1979) or an appropriate culture medium (e.g. Jones and Woo, 1991). The cloned parasite is then carefully described, identified and cultured. To determine its pathogenicity all three strains (field, cultured and cloned strains) should be used to experimentally infect hatchery-raised fish. This approach would satisfy Koch's postulates for determining the aetiological agent for a disease (Woo, 1979).

AMOEBOSIS

Amoebic gill disease

Introduction

Amoebic gill disease is not common in fishes but it has been reported in Europe, North America and Australia. Factors that trigger free-living amoebae to become pathogenic parasites are not known. However, some fish with amoebic gill infections (e.g. in Chatton, 1909; Kubota and Kamata, 1971;

Sawyer *et al.*, 1975) also had bacterial infections and it was presumed that the amoeba was a secondary invader.

Thecamoeba hoffmani

Thecamoeba hoffmani was described (Sawyer *et al.*, 1974) from the gills of fingerling rainbow trout (*O. mykiss*), coho salmon (*O. kisutch*), and chinook salmon (*O. tshawytscha*). Infected fingerlings were lethargic or moribund and the diseased fish were reported from four hatcheries in the United States (Washington, Oregon and Michigan States).

PARASITE MORPHOLOGY

The organism is a small ovoid amoeba which moves slowly with only slight changes in shape. It has broad hyaline pseudopodia, a large dense nucleolus, thick nuclear membrane and thick folded or wrinkled ectoplasm. The organism measured $31.0\,\mu m$ (21.6–$40.8\,\mu m$) $\times$ $23.1\,\mu m$ (16.8–$28.8\,\mu m$) in histological sections. Tissues were preserved in 10% formalin and tissue sections were stained in either Heidenhain iron haematoxylin and eosin or Feulgen reagent with fast-green counterstain (Sawyer *et al.*, 1974).

HOST-PARASITE RELATIONSHIPS AND MORTALITY

In rainbow trout and coho salmon, the amoeba was only seen between gill lamellae (Sawyer *et al.*, 1975). Some lamellae were short and abnormally wide while others had cellular hyperplasia at the junction of the lamella and the gill filament. Fish mortality was assumed to be due to respiratory impairment and asphyxiation because the gills were covered with amoebae.

In one infected chinook salmon, the amoeba was both on the gills and in gill filaments. The infected tissue was necrotic, vacuolated and hyperplastic. Miracidia of *Sanguinicola* (a parasitic blood digenean) were in the gills and numerous rod-shaped bacteria were in the mucus between lamellae. Sawyer *et al.* (1975) suggested that the bacteria and *Sanguinicola* predisposed the fish to *T. hoffmani* infection.

Paramoeba pemaquidensis

Paramoeba pemaquidensis was found on gills of coho salmon reared in sea water cages in British Columbia, Canada (Kent *et al.*, 1989). The organism (Fig. 2.32) on Atlantic salmon (*Salmo salar*) and rainbow trout in Tasmania was presumed to be the same parasite (Munday *et al.*, 1990; Roubal *et al.*, 1990).

PARASITE MORPHOLOGY AND *IN VITRO* CULTURE

The organism on gills is vacuolated and has one or two Feulgen positive parasomes adjacent to its nucleus (Fig. 2.33). Free floating and transitional forms (20–$30\,\mu m$ in diameter) may have up to 50 digitate pseudopodia (Kent *et al.*, 1989; Munday *et al.*, 1990). Amoeba that are attached to gills are larger than those in cultures (Roubal *et al.*, 1990).

The plasmalemma has surface filaments and the nucleus, with a central nucleolus, is closely apposed to the parasomes which contain a fibrous central

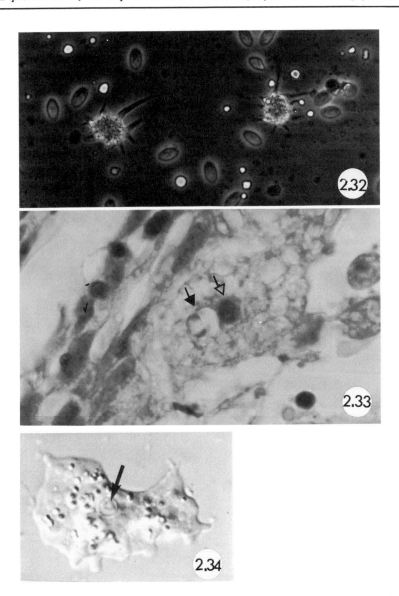

Fig. 2.32. *Paramoeba* sp. washed from the gills of an infected Atlantic salmon; note the large and small digitate pseudopodia and granular cytoplasm. (Courtesy of Dr F.R. Roubal.) **Fig. 2.33.** *Paramoeba* on the gills of an infected Atlantic salmon; note the nucleus (open arrow) and the parasome with polar condensations of chromatin (solid arrow) (×1000). (From Munday *et al.*, 1990; courtesy of Dr B. Munday.) **Fig. 2.34.** *Paramoeba* in culture; note the cytoplasmic granules and broad pseudopodia. (From Roubal *et al.*, 1989; courtesy of Dr F. Roubal.)

region. The ultrastructural features of the isolate from coho salmon in Canada are similar to those from Atlantic salmon in Tasmania and the isolates are assumed to be the same organism (Kent *et al.*, 1989; Munday *et al.*, 1990; Roubal *et al.*, 1990).

P. pemaquidensis grows well in malt-yeast extract in sea water supplemented with *Klebsiella* bacteria (Fig. 2.34; Kent *et al.*, 1989; Munday *et al.*, 1990; Roubal *et al.*, 1990). The disease was not reproduced under laboratory conditions in coho salmon exposed to cultures of the amoeba. The organism survived on fish and aquarium detritis for at least 4 weeks (Kent *et al.*, 1989).

CLINICAL SIGNS, PATHOLOGY AND MORTALITY

Moribund fish are sluggish, often congregate at the surface or corners of sea cages and swim with open opercula. Some fish are anorexic and may have elevated blood sodium (Kent *et al.*, 1989; Munday *et al.*, 1990).

The organism elicits epithelial hyperplasia on gill filaments (Kent *et al.*, 1989; Munday *et al.*, 1990; Roubal *et al.*, 1990) and the hyperplasia may lead to clubbing and fusion of secondary lamellae. The parasite is associated with necrosis of epithelial cells and the cytoplasmic processes of the organism are inserted into and between surface cells of the gill epithelium. The number of mucus-secreting cells increases during infection with corresponding reduction in the number of chloride cells.

Mortality of fish may reach 50% in the absence of treatment. Young fish are more susceptible than older fish (Munday *et al.*, 1990).

EPIDEMIOLOGY AND CONTROL

P. pemaquidensis is an opportunistic free living organism that multiplies rapidly on gills of fish and, under certain conditions, causes disease and fish mortality. In Canada, it caused 25% mortality in coho salmon in sea cages in 1985 and the disease subsequently recurred in 1986 and 1987 (Kent *et al.*, 1989). In Tasmania, although the amoeba was on gills of Atlantic salmon in the winter and spring (about 10°C), the disease occurred only in the summer when water temperature was high (15–20°C) and salinity of 35‰. The organism has been found on clean nets and in many native marine fishes (Munday *et al.*, 1990).

There is no chemotherapy and the parasite is controlled by exposing infected fish to low salinity sea water or to fresh water (Kent *et al.*, 1989; Munday *et al.*, 1990).

Systemic amoebosis

Introduction

Several amoebae have been reported in internal organs in fishes in Europe and North America. Most reports on systemic amoebosis are not well documented. As in amoebic gill disease, the factors that promote systemic disease are not known and the disease has not been reproduced experimentally in fish under laboratory conditions.

Schizamoeba salmonis

Davis (1926, 1946) described the parasite from the gastric lumen and mucosa of salmonid fingerlings in the United States. Cysts were also found in the stomach and intestine; however, no heavy loss of fish was associated with the infection.

The amoeba measures 10.0–25.0 µm. The nucleus of its trophozoite is vesicular, without a nucleolus (endosome) but with large discoid granules along the nuclear membrane. The organism has one to many nuclei and the nucleus in the cystic stage has a nucleolus.

Acanthamoeba spp.

Taylor (1977) found high numbers of *Acanthamoeba polyphaga* in peritoneal fluid, intestinal mucosa and gills and believed that it was responsible for severe losses of *Tilapia aureus*. The trophozoite measures 20.0–25.0 µm (average 22.0 µm) while the cyst is 10.0–12.5 µm (average 11.7 µm). Inoculation of a culture (isolated from the spleen of an infected *Micropterus coosae*) into fishes resulted in systemic infections but no mortality.

According to Nash *et al.* (1988), an *Acanthamoeba*-like amoeba was associated with severe proliferative gill disease and systemic granulomatosis in cultured European catfish (*Silurus glanis*) in Germany. About 30% of the fish died and it was suggested that the mortality was due to combined effects of poor water quality and the amoeba and bacterial infections. The water temperature was 24–26°C and water in the system was partially recirculated.

CLINICAL SIGNS AND PATHOLOGY

Acutely infected fish are emaciated and have marked ascites and generalized haemorrhages at the bases of fins and posterior dorsal and ventral surfaces.

Gills are severely swollen with confluence and fusion of secondary and primary gill lamellae. There are marked changes in colour and consistency of the liver, spleen and kidney of infected fish. Haemorrhages are sometimes seen in the swim bladder (Nash *et al.*, 1988). Other parasites found on/in the clinically sick catfish were the ciliates *Trichodina* and *Trichodinella* and the monogenean *Dactylogyrus* on the gills and massive numbers of the digenean *Allocreadium isosporum* in the gut.

Amoebae of unknown taxonomic status

Schaperclaus (1986) indicated that Plehn (1924) found an amoeba in the kidneys of rainbow trout (*O. mykiss*) and brook charr (*S. fontinalis*) in Europe and suggested that the organism was responsible for high fish mortality.

The small parasite (about 20 µm) causes inflamation in the kidneys. The clinical signs include cachexia, hydrops, cloudy cornea and lens, retinal detachment and blindness.

Voelker *et al.* (1977) found a pathogenic amoeba (Family Hartmannellidae) in *Carassius auratus*. There were large granulomas in the coelomic cavity, liver, skeletal muscles, swim bladder and cerebral meninges in infected fish. The clinical signs included lethargy, anorexia and abdominal distension. Mortality was sometimes associated with the disease.

The taxonomy and systematics of amoebae which cause disease need further careful studies. These should include light and electron microscopic descriptions along with biochemical characterizations of the organisms. Other priority areas of research are to elucidate factors (e.g. concurrent microbial infections, environmental effects) which encourage free-living amoebae to cause clinical disease and the host specificity, if any, of these organisms. We should also carefully examine the immune response of fish to these parasites, develop reliable diagnostic techniques and control measures especially for the systemic amoebosis. Better protective and preventive strategies against this group of parasites may emerge as we know more about predisposing factors, the host immune response and the biology of the organisms.

CONCLUSIONS AND NEW MAJOR AREAS OF STUDIES

Hexamitids, nonhaematozoic *Cryptobia* and amoebae are associated with morbidity and mortality of some economically important fishes. However, our knowledge of these monogenetic parasites and their host–parasite relationships are very limited compared to what we know about the digenetic haematozoic flagellates. These obvious gaps in our knowledge are of concern because outbreaks (in numbers and severity) are likely to increase as we venture into more intensive fish culture (e.g. in hatcheries, sea cages). Some of the reasons for the expected increase are that these parasites are transmitted directly in the water and they are generally not host specific.

Quarantine of new fish to a hatchery is a partial solution; however, these parasites can be introduced into a hatchery via its water supply. Chemical controls (prophylactic and therapeutic) may not be good long-term strategies. Fish for human consumption can only be treated with specific chemicals in accordance with national regulations and these may vary widely between countries and with time (Woo, 1987b). Also, some of these chemicals when discharged into the aquatic environment for prolonged periods may be toxic to other aquatic animals or they may accumulate to undesirable levels in the fish to affect their suitability for human consumption (Woo and Shariff, 1990). Hence, future areas of high research priority should be to better understand the disease process and to protect fish from disease outbreaks with minimal use of chemicals.

Strategies would include studies on: (i) the parasite (e.g. ultrastructure, culture, nutritional requirements, metabolic pathway(s), types of proteases excreted/secreted, identification and characterization of protective antigen); (ii) its pathogensis (e.g. environmental conditions that predispose fish to infection and disease, genetic factors in fish that affect susceptibility and severity of disease, elucidation of the mechanism of disease); and (iii) immune response of fish to parasites (e.g. mechanisms of innate and acquired immunity, heredity of innate immunity in resistant fish, vaccine development, immunization and duration of protective immunity in susceptible fish).

We know a little about immunodepression (e.g. contributing factors, its effects on susceptibility of fish to a second pathogen) in salmonid cryptobiosis (due to *C. (T.) salmositica*); however, nothing is known about immunomodulation in fish due to the other parasites. We hope that this will be rectified since immunodepression predisposes fish to concomitant secondary infections and increases mortality. It also decreases the efficacy of vaccination programmes against other diseases for which vaccines are available (e.g. against viral and bacterial diseases).

Other major areas for studies should include interactions between water quality (e.g. water pH, dissolved oxygen/organic materials), environmental contaminations (e.g. heavy metals, petroleum aromatic hydrocarbons, pesticides) and dietary factors (e.g. ascorbic acids, essential amino acids) with function of the piscine immune system and outbreaks and severity of disease. Also, the interactions of multiple infections (e.g. microbial and protozoan infections, protozoan and metazoan infections) on each other and their combined effects on survival, growth and reproduction of fish are neglected areas and they need careful research in the future.

REFERENCES

Alexeieff, A. (1910) Sur les flagellés intestinaux des poissons marins. *Archives de Zoologie Expérimentale et Generale.* Séries 5, Vol. VI, Notes et Revue, No. 1, pp. 1–20.

Allison, L.N. (1963) An unusual case of *Hexamita (Octomitus)* among yearling rainbow trout. *Progresive Fish Culturist* 25(4), 220.

Amlacher, E. (1970) *Textbook of Fish Diseases* (English Translation). TFH Publications, Neptune, New Jersey.

Andai, G. (1933) Uber *Costia necatrix. Archiv für Protistenkunde* 79, pt. 2.

Andrews, C., Exell, A. and Carrington, N. (1988) *Manual of Fish Health.* Salamander Books, London.

Appy, R.G. and Dadswell, M.J. (1981) Marine and estuarine piscicolid leeches (Hirudinea) of the Bay of Fundy and adjacent waters with a key to species. *Canadian Journal of Zoology* 59, 183–192.

Ardelli, B.F. and Woo, P.T.K. (1993) Protective immunity against cryptobiosis in *Oncorhynchus mykiss* and *Salvelinus fontinalis* inoculated with an attenuated strain of *Cryptobia salmositica. Bulletin, Canadian Society of Zoologists* 24(2), 30–31. (Abstract).

Ardelli, B.F., Forward, G. and Woo, P.T.K. (1994) Brook charr (*Salvelinus fontinalis*) and cryptobiosis: A potential salmonid reservoir host for *Cryptobia salmositica* Katz, 1951. *Journal of Fish Diseases* 17, 567–577.

Arthur, J.R., Margolis, L. and Arai, H.P. (1976) Parasites of fishes of Aishihik and Stevens Lakes, Yukon Territory, and potential consequences of their interlake transfer through a proposed water diversion from hydroelectric purposes. *Journal of the Fisheries Research Board of Canada* 33, 2489–2499.

Awakura, T., Kojima, H. and Tanaka, H. (1984) [Studies on parasites of masu salmon, *Oncorhynchus masou*. VIII. Costiasis of pond-reared masu salmon fry]. *Scientific Report of the Hokkaido Fish Hatchery* 39, 89–96.

Bahmanrokh, M. and Woo, P.T.K. (1994) The histopathology of cryptobiosis in juvenile *Oncorhynchus mykiss* (Walbaum). *VIII International Congress of Parasitology* 2, 434 (Abstract).

Barthold, S.W. (1985) *Spironucleus muris* infection, intestine, mouse, rat, hamster. In: Jones, T.C., Mohr, V. and Hunt, R.D. (eds) *Monographs on Pathology of Laboratory Animals. Digestive System.* Springer-Verlag, New York, pp. 356–358.

Bassleer, G. (1983) Disease prevention and control. *Spironucleus/Hexamita* infection, hole-in-the-head disease. *Freshwater Marine Aquarium* 6, 38–41, 58–60.

Bauer, O.N. (1959) Parasites of freshwater fishes and the biological basis for their control. *Bulletin of the State Scientific Research, Institute Lake and River Fish* 49, 236 pp. Israel Program for Scientific Translation, Jerusalem (1962).

Bauer, O.N., Mussellius, V.A. and Strelikov, Yu. A. (1969) *Diseases of Pond Fishes.* Israel Program for Scientific Translation, Jerusalem (1973); US Department of the Interior and the National Science Foundation, Washington, DC.

Becker, C.D. (1970) Haematozoa of fishes, with emphasis on North American records. In: Sniezsko, S.F. (ed.) *A Symposium on Diseases of Fishes and Shellfish.* Special Publication No. 5, American Fisheries Society, Washington. pp. 82–100.

Becker, C.D. (1977) Flagellate parasites of fish. In: Kreier, J.P. (ed.) *Parasite Protozoa* Volume I. Academic Press, New York. pp. 357–416.

Becker, C.D. and Katz, M. (1965a) Transmission of the haemoflagellate *Cryptobia salmositica* Katz 1951, by a rhynchobdellid leech vector. *Journal of Parasitology* 51, 95–99.

Becker, C.D. and Katz, M. (1965b) Infections of the haemoflagellate *Cryptobia salmositica* Katz 1951, in freshwater teleosts of the Pacific coast. *Transactions of the American Fisheries Society* 94, 327–333.

Becker, C.D. and Katz, M. (1965c) Distribution, ecology and biology of the salmonid leech, *Piscicola salmositica* Meyer 1946 (Rhynchobdella, Piscicolidae). *Journal of the Fisheries Research Board of Canada* 22, 1175–1195.

Becker, C.D. and Katz, M. (1966) Host relationships of *Cryptobia salmositica* (Protozoa, Mastigophora) in Washington hatchery stream. *Transactions of the American Fisheries Society* 95, 196–202.

Becker, C.D. and Overstreet, R.M. (1979) Haematozoa of marine fishes from the northern Gulf of Mexico. *Journal of Fish Diseases* 2, 469–479.

Bohl, M. (1975) [The adherent contamination of skin and gills (ichthyobodiasis = costiasis), a widespread parasitosis in the breeding of pond fish.] *Fisch Umwelt* 1, 25–33.

Boorman, G.A., Van Hooft, J.I.M., Van der Waaij, D. and Van Noord, H.J. (1973) Synergistic role of intestinal flagellates and normal intestinal bacteria in a postweaning mortality of mice. *Laboratory Animal Science* 23, 187–193.

Bovee, E.C. (1985) Class Lobosea Carpenter, 1861. In: Lee, J.J., Hunter, S.H. and Bovee, E.C. (eds) *An Illustrated Guide to the Protozoa.* Society of Protozoologists, Lawrence, Kansas, pp. 158–211.

Bower, S.M. and Evelyn, T.P.T. (1988) Acquired and innate resistance to the haemoflagellate *Cryptobia salmositica* in sockeye salmon (*Oncorhynchus nerka*). *Developmental and Comparative Immunology* 12, 749–760.

Bower, S.M. and Margolis, L. (1983) Direct transmission of the haemoflagellate *Cryptobia salmositica* among Pacific salmon (*Oncorhynchus* spp.). *Canadian Journal of Zoology* 61, 1242–1250.

Bower, S.M. and Margolis, L. (1984a) Detection of infection and susceptibility of different Pacific salmon stocks (*Oncorhynchus* spp.) to the haemoflagellate *Cryptobia salmositica. Journal of Parasitology* 70, 273–278.

Bower, S.M. and Margolis L. (1984b) Distribution of *Cryptobia salmositica*, a haemoflagellate of fishes in British Columbia and the seasonal pattern of infection in a coastal river. *Canadian Journal of Zoology* 62, 2512–2518.

Bower, S.M. and Margolis, L. (1985) Effects of temperature and salinity on the course of infection with the haemoflagellate *Cryptobia salmositica* in juvenile Pacific salmon (*Oncorhynchus* spp.). *Journal of Fish Diseases* 8, 25–33.

Bower, S.M. and Thompson, A.B. (1987) Hatching of the Pacific salmon leech (*Piscicola salmositica*) from cocoons exposed to various treatments. *Aquaculture* 66, 1–8.

Bower, S.M. and Woo, P.T.K. (1977) Morphology and host specificity of *Cryptobia catostomi* n.sp. (Protozoa: Kinetoplastida) from white sucker (*Catostomus commersoni*) in southern Ontario. *Canadian Journal of Zoology* 55, 1082–1092.

Bower, S.M., Margolis, L. and MacKay, R.J. (1985) Potential usefulness of chlorine for controlling Pacific salmon leeches, *Piscicola salmositica* in hatcheries. *Canadian Journal of Fisheries and Aquatic Sciences* 42, 1986–1993.

Broderud, A.E. and Poppe, T.T. (1986) [Costiasis (*Ichthyobodo necator* infection) in farmed turbot (*Psetta maxima* L.).] *Norsk Veterinaertidsskrift* 98, 883–884.

Brugerolle, C. (1974) Contribution a l'étude cytologique at phylétique des diplozoaires (Zoomastigophorea, Diplozoa, Dangeard 1910). III. Étude ultrastructurale du genre *Hexamita* (Dujardin, 1836) *Protistologica* 10, 83–90.

Brugerolle, G. (1975) Contribution a l'étude cytologique at phyletique des diplozoaires (Zoomastigophorea, Diplozoa, Dangeard 1910). VI. Caractères généraux des diplozoaires. *Protistologica* 11, 111–118.

Brugerolle, G., Joyon, L. and Oktem, N. (1973) Contribution a l'étude cytologique et phylétique des diplozoaires (Zoomastigophorea, Diplozoa, Dangeard 1910). II. Étude ultrastructurale du genre *Spironucleus* (Lavier, 1936). *Protistologica* 9, 495–502.

Brugerolle, C., Lom, J., Nohynkova, E. and Joyon, L. (1979) Comparison et evolution des structures cellulaires chez plusieurs especes de bodonides et cryptobiides appartenant aux genres *Bodo*, *Cryptobia* et *Trypanoplasma* (Kinetoplastida, Mastigophora). *Protistologica* 15, 197–221.

Bruno, D.W. (1992) *Ichthyobodo* sp. on farmed Atlantic salmon, *Salmo salar* L., reared in the marine environment. *Journal of Fish Diseases* 15, 349–351.

Bullock, A.M. (1985) The effect of ultraviolet-B radiation upon the skin of the plaice, *Pleuronectes platessa* L., infested with the bodonid ectoparasite, *Ichthyobodo necator* (Henneguy 1883). *Journal of Fish Diseases* 8, 547–550.

Bullock, A.M. and Robertson, D.A. (1982) A note on the occurrence of *Ichthyobodo necator* (Henneguy, 1883) in a wild population of juvenile plaice, *Pleuronectes platessa* L. *Journal of Fish Diseases* 5, 531–533.

Burreson, E.M. (1982a) The life cycle of *Trypanoplasma bullocki* (Zoomastigophorea, Kinetoplastida). *Journal of Protozoology* 29, 72–77.

Burreson, E.M. (1982b) Trypanoplasmiasis in flounder along the Atlantic coast of the United States. In: Anderson, D.P., Dorson, M. and Dubourget, Ph. (eds) *Les antigenes des micro-organismes pathogenes des poissons*. Collection Fondation Marcel Merieux, France, pp. 251–260.

Burreson, E.M. and Frizzell, L.J. (1986) The seasonal antibody response in juvenile summer flounder (*Paralichthys dentatus*) to the hemoflagellate *Trypanoplasma bullocki*. *Veterinary Immunology and Immunopathology* 12, 395–402.

Burreson, E.M. and Sypek, J.P. (1981) *Cryptobia* sp. (Mastigophorea, Kinetoplastida) from the gills of marine fishes in Chesapeake Bay. *Journal of Fish Diseases* 4, 519–522.

Burreson, E.M. and Zwerner, D.E. (1982) The role of host biology, vector biology and temperature in the distribution of *Trypanoplasma bullocki* infections in the lower Chesapeake Bay. *Journal of Parasitology* 68, 306–313.

Burreson, E.M. and Zwerner, D.E. (1984) Juvenile summer flounder, *Paralichthys dentatus*, mortalities in western Atlantic Ocean caused by the haemoflagellate *Trypanoplasma bullocki*, evidence from field and experimental studies. *Heligoländer Meeresuntersuchungen* 37, 343–352.

Chatton, E. (1909) Une amibe, *Amoeba mucicola* n.sp., parasite des branchies des labres, a ociée à une trichodine. *Comptes Rendus des Séances de la Société de Biologie* 67, 690–692.

Chen, C.L. (1956) The protozoan parasites from four species of Chinese pond fishes, *Ctenopharyngodon idellus, Mylopharyngodon piceus, Aristichthys nobilis* and *Hypophthalmichthys molithrix*. I. The protozoan parasites of *Ctenopharyngodon idellus*. *Acta Hydrobiologica Sinica* 1, 123–164.

Cone, D.K. and Wiles, M. (1984) *Ichthyobodo necator* (Henneguy, 1883) from winter flounder, *Pseudopleuronectes americanus* (Walbaum) in northwest Atlantic Ocean. *Journal of Fish Diseases* 7, 87–89.

Corliss, J.O. and Lom, J. (1985) An annotated glossary of protozoological terms. In: Lee, J.J., Hutner, S.H. and Bovee, E.C. (eds) *An Illustrated Guide to the Protozoa*. Society of Protozoologists, Lawrence, Kansas. pp. 576–602.

Crawley, H. (1909) Studies on blood parasites, II. The priority of *Cryptobia* Leidy 1846 over *Trypanoplasma* Laveran and Mesnil, 1901. *Bulletin of the US Bureau of Animal Industry* 119, 16–20.

Daily, D.D. (1978) Marine fish hematozoa from Maine. *Journal of Parasitology* 64, 361–362.

Davis, H.S. (1926) *Schizamoeba salmonis*, a new ameba parasitic in salmonid fishes. *Bureau of Fisheries Document* (987), 1–8.

Davis, H.S. (1946) Care and diseases of trout. *Fish and Wildlife Research Report* no. 12, 1–98.

Davis, H.S. (1943) A new polymastigine flagellate, *Costia pyriformis*, parasitic on trout. *Journal of Parasitology* 29, 385–386.

Diamant, A. (1987) Ultrastructure and pathogensis of *Ichthyobodo* sp. from wild common dab, *Limanda limanda* L., in the North Sea. *Journal of Fish Diseases* 10, 241–247.

Dykova, I. and Lom, J. (1979) Histopathological changes in *Trypanosoma danilewskyi* Laveran and Mesnil, 1904 and *Trypanoplasma borelli* Laveran and Mesnil, 1902 infections of goldfish, *Carassius aurata* (L). *Journal of Fish Diseases* 2, 381–390.

Einszporn-Orecka, T. (1979) Flagellates *Spironucleus anguillae* sp.n. parasites of eels (*Anguilla anguilla* L.). *Acta Protozoologica* 18, 237–241.

Elliott, J.M. and Mann, K.H. (1979) *A Key to the British Freshwater Leeches with Notes on their Life Cycle and Ecology*. Freshwater Biological Association, Scientific Publication No. 40.

Ellis, A.E. and Wootten, R. (1978) Costiasis of Atlantic salmon, *Salmo salar* L. smolts in sea water. *Journal of Fish Diseases* 1, 389–393.

Epshtein, V.M. (1962) A survey of fish leeches (Hirudinea Piscicolidae) from the northern seas of the SSSR. *Doklady Akademii Nauk SSSR* 141, 1121–1124.

Feely, D.E and Erlandsen, S.L. (1978) A method for the isolation of *Giardia muris* from rat small intestine. *Anatomical Record* 190, 393.

Feng, S. and Woo, P.T.K. (1993) The role of the thymus in protective immunity against salmonid cryptobiosis in *Oncorhynchus mykiss* (Walbaum). *Bulletin, Canadian Society of Zoologists*, 24(2), 50–51 (Abstract).

Ferguson, H.W. (1979) Scanning and transmission electron microscopic observations on *Hexamita salmonis* (Moore, 1922) related to mortalities in rainbow trout fry *Salmo gairdneri* Richardson. *Journal of Fish Diseases* 2, 57–67.

Ferguson, H.W. (1989) *Systemic Pathology of Fish*. Iowa State University Press, Ames, Iowa.

Ferguson, H.W. and Moccia. R.D. (1980) Disseminated hexamitiasis in Siamese fighting fish. *Journal of the American Veterinary Medical Association* 177, 854–857.

Fish, F.F. (1940) Notes on *Costia necatrix*. *Transactions of the American Fisheries Society* 70, 441–445.

Forward, G.M., Ferguson, M.M. and Woo, P.T.K. (1993) Susceptibility of brook charr, *Salvelinus fontinalis* to *Cryptobia salmositica* and the mechanism of innate resistance. *Bulletin, Canadian Society of Zoologists*, 24(2), 51–52 (Abstract).

Grassé, P.-P. (1952) Ordre des Distomatines ou Diplozoaires. In: Grassé, P.-P. (ed.) *Traite de Zoologie*. Vol. 1, Fasc. 1, Maison et Cie, Paris, pp. 963–982.

Gratzek, J.B. (1988) Parasites associated with ornamental fish. *Veterinary Clinics of North America, Small Animal Practice* (*Tropical Fish Medicine*) 18, 375–399.

Gupta, N. and Gupta, D.K. (1985) Haematological changes due to *Trypanosoma batrachi* and *T. aligaricus* infection in two fresh water teleosts. *Angewandte Parasitologie* 26, 193–196.

Hajdu, E. and Matskasi, I. (1984) *In vitro* cultivation of *Trypanoplasma* strains isolated from pike and leech (preliminary report). *Acta Veterinaria Hungarica* 32, 79–81.

Helms, D.R. (1967) Use of formalin for selective control of tadpoles in the presence of fishes. *Progressive Fish Culturalist* 29, 43–47.

Hlond, S. (1963) Occurrence of *Costia necatrix* Henneguy on the roe of carp. *Wiadomosci Parazytologiczne* 9, 249–251.

Hoffman, G.L. (1978) *Bodomonas concava*, a cryptic cryptogram for crippling *Cryptobia branchialis*. *American Fisheries Society, Fish Health Section Newsletter* 6, 9.

Hoffman, G.L. and Meyer, F.P. (1974) *Parasites of Freshwater Fishes, a Review of their Control and Treatment*. TFH Publications, Neptune City, New Jersey, 224 pp.

Honigberg, B.M., Vickerman, K., Kulda, J. and Brugerolle, G. (1981) Cytology and taxonomy of parasitic flagellates. Review of Advances in Parasitology. *Proceedings of the Fourth International Congress of Parasitology*. Polish Scientific Publishers, Warsaw, Poland, pp. 205–227.

Hora, S.L. and Pillay, T.V.R. (1962) Handbook of fish culture in the Indo-Pacific Region. *Food and Agriculture Organization Fish Biology Technical Paper* 14, 1–204.

Hunninen, A.V. and Wichterman, R. (1938) Hyperparasitism, a species of *Hexamita* (Protozoa, Mastigophora) found in the reproductive systems of *Deropristis inflata* (Trematoda) from marine eels. *Journal of Parasitology* 24, 95–101.

Imamovic, V. (1986) [Parasites and parasitoses in fish in some salmonid hatcheries in Bosnia-Hercegovina. I. *Ichthyobodo* and *Hexamita* infections.] *Veterinaria Yugoslavia* 35, 47–66.

Islam, A.K.M.N. and Woo, P.T.K. (1991a) *Trypanosoma danilewskyi* in *Carassius auratus*, the nature of protective immunity in recovered goldfish. *Journal of Parasitology* 77, 258–262.

Islam, A.K.M.N. and Woo, P.T.K. (1991b) Anorexia in goldfish *Carassius auratus* infected with *Trypanosoma danilewskyi*. *Diseases of Aquatic Organisms* 11, 45–48.

Islam, A.K.M.N. and Woo, P.T.K. (1991c) Anemia and its mechanism in goldfish *Carassius auratus* infected with *Trypanosoma danilewskyi*. *Diseases of Aquatic Organisms* 11, 37–43.

Islam, A.K.M.N. and Woo, P.T.K. (1992) Effects of temperature on the *in vivo* and *in vitro* multiplication of *Trypanosoma danilewskyi* Laveran and Mesnil. *Folia Parasitologica* 39, 1–12.

Januschka, M.M., Erlandsen, S.L., Bemrick, W.J., Schupp, D.G. and Freely, D.E. (1988) A comparison of *Giardia microti* and *Spironucleus muris* cysts in the vole: an immunocytochemical, light and electron microscope study. *Journal of Parasitology* 74, 452–458.

Jones, S.R.M. and Woo, P.T.K. (1987) The immune response of rainbow trout *Salmo gairdneri* Richardson to the haemoflagellate *Cryptobia salmositica* Katz 1951. *Journal of Fish Diseases* 10, 395–402.

Jones, S.R.M. and Woo, P.T.K. (1991) Culture characteristics of *Trypanosoma catostomi* and *Trypanosoma phaleri* from North American freshwater fishes. *Parasitology* 103, 237–243.

Jones, S.R.M., Woo, P.T.K. and Stevenson, R.M.W. (1986) Immunosuppression in *Salmo gairdneri* Richardson caused by the haemoflagellate, *Cryptobia salmositica* (Katz, 1951). *Journal of Fish Diseases* 9, 431–438.

Jones, S.R.M., Palmen, M. and van Muiswinkel, W.B. (1993) Effects of inoculum route and dose on the immune response of common carp, *Cyprinus carpio* to the blood parasite, *Trypanoplasma borreli. Veterinary Immunology and Immuno-pathology* 36, 369–378.

Joyon, L. and Lom, J. (1969) Etude cytologique, systematique et pathologique d'*Ichthyobodo necator* (Henneguy, 1883), Pinto, 1928 (Zooflagelle). *Journal of Protozoology* 16, 703–719.

Katz, M. (1951) Two new hemoflagellates (genus *Cryptobia*) from some western Washington teleosts. *Journal of Parasitology* 37, 245–250.

Katz, M., Woodey, J.C., Becker, C.D., Woo, P.T.K. and Adams, J.R. (1966) Records of *Cryptobia salmositica* from sockeye salmon from the Fraser River drainage and from the State of Washington. *Journal of the Fisheries Research Board of Canada* 23, 1965–1966.

Kent, M.L. (1992) Diseases of seawater netpen-reared salmonid fishes in the Pacific Northwest. *Canadian Special Publication of Fisheries and Aquatic Sciences*, No. 116, 76 pp.

Kent, M.L., Sawyer, T.K. and Hedrick, R.O. (1989) *Paramoeba pemaquidensis* (Sarcomastigophora: Paramoebidae) infestation of the gills of coho salmon, *Oncorhynchus kisutch* reared in sea water. *Diseases of Aquatic Organisms* 5, 163–169.

Kent, M.L., Ellis, J., Fournie, J.W., Dawe, S.C., Bagshaw, J.W. and Whitaker, D.J. (1992) Systemic hexamitid (Protozoa: Diplomonadida) infection in seawater pen-reared chinook salmon *Oncorhynchus tshawytscha. Diseases of Aquatic Organisms* 14, 81–89.

Keysselitz, C. (1906) Generations-und Wirtswechsel von *Trypanoplasma borreli*, Laveran und Mesnil. *Archiv für Protistenkunde* 7, 1–74.

Khan, R.A. (1972) On a trypanosome from Atlantic cod, *Gadus morhua* L. *Canadian Journal of Zoology* 50, 1051–1054.

Khan, R.A. (1976) The life cycle of *Trypanosoma murmanensis* Nikitin. *Canadian Journal of Zoology* 54, 1840–1849.

Khan, R.A. (1977a) Susceptibility of marine fish to trypanosomes. *Canadian Journal of Zoology* 55, 1235–1241.

Khan, R.A. (1977b) Blood changes in Atlantic cod (*Gadus morhua*) infected with *Trypanosoma murmanensis. Journal of the Fisheries Research Board of Canada* 34, 2193–2196.

Khan, R.A. (1982) Biology of the marine piscicolid leech, *Johanssonia arctica* (Johansson) from Newfoundland. *Proceedings of the Helminthological Society of Washington* 49, 266–278.

Khan, R.A. (1985) Pathogenesis of *Trypanosoma murmanensis* in marine fish of the northwestern Atlantic following experimental transmission. *Canadian Journal of Zoology* 63, 2141–2144.

Khan, R.A. (1986) Haematozoa of marine fish from Ungava Bay and adjacent northwestern Atlantic Ocean. *Canadian Journal of Zoology* 64, 153–159.

Khan, R.A. (1991) Trypanosome occurrence and prevalence in the marine leech *Johanssonia arctica* and its host preferences in the northwestern Atlantic Ocean. *Canadian Journal of Zoology* 69, 2374–2380.

Khan, R.A., Murphy, J. and Taylor, D. (1980a) Prevalence of a trypanosome in Atlantic cod (*Gadus morhua*) especially in relation to stocks in the Newfoundland area. *Canadian Journal of Fisheries and Aquatic Sciences* 37, 1467–1475.

Khan, R.A., Barrett, M. and Murphy, J. (1980b) Blood parasites of fish from northern Atlantic Ocean. *Canadian Journal of Zoology* 58, 770–781.

Khan, R.A., Campbell, J. and Barrett, M. (1980c) *Trypanosoma murmanensis*, its effect on the longhorn sculpin, *Myoxocephalus octodecemspinosus*. *Journal of Wildlife Diseases* 16, 359–361.

Kiškaroly, M. and Tafro, A. (1987) Symptoms, pathological picture and diagnosis of the most frequent parasitic diseases of fishes in freshwater ponds II. Protozoa-caused diseases 1. [in Slovenian] *Veterinarski Glasnik* 41, 619–624.

Kruse, P., Steinhagen, D. and Korting, W. (1989a) Development of *Trypanoplasma borreli* (Mastigophora, Kinetoplastida) in the leech vector *Piscicola geometra* and its infectivity for the common carp, *Cyprinus carpio*. *Journal of Parasitology* 75, 527–530.

Kruse, P., Steinhagen, D., Korting, W. and Friedhoff, K.T. (1989b) Morphometrics and redescription of *Trypanoplasma borreli* Laveran and Mesnil, 1901 (Mastigophora, Kinetoplastida) from experimentally infected common carp (*Cyprinus carpio* L.). *Journal of Protozoology* 36, 408–411.

Kubota, S. and Kamata, T. (1971) An amoeba observed on the gill of amago. *Fish Pathology* (Japan) 5, 155.

Kulda, J. and Lom, J. (1964a) Remarks on the diplomastigine flagellates from the intestine of fishes. *Parasitology* 54, 753–762.

Kulda, J. and Lom, J. (1964b) *Spironucleus elegans* Lavier, parasite of fish. *Československá Parasitologie* XI, 187–192.

Kulda, J. and Nohynkova, E. (1978) Flagellates of the human intestine and of intestines of other species. In: Kreier, J.P. (ed.) *Parasitic Protozoa* Vol. II. Academic Press, New York, pp. 1–138.

Kusakari, M., Mori, Y. and Miura, H. (1985) Technical studies on artificial production of fish larvae. *Annual Report of the Hokkaido Institute of Mariculture (1984)*, pp. 15–61.

Laidley, C.W., Woo, P.T.K. and Leatherland, J.F. (1988) The stress-response of rainbow trout to experimental infection with the blood parasite, *Cryptobia salmositica* Katz, 1951. *Journal of Fish Biology* 32, 253–261.

Laird, M. and Bullock, W.H. (1969) Marine fish haematozoa from New Brunswick and New England. *Journal of the Fisheries Research Board of Canada* 26, 1075–1102.

Lamas, J. and Bruno, D.W. (1992) Observations in the ultrastructure of the attachment plate of *Ichthyobodo* sp., from Atlantic salmon, *Salmo salar* L., reared in the marine environment. *Bulletin of the European Association of Fish Pathologists* 12, 171–173.

Laveran, A. and Mesnil, F. (1901) Sur les flagelles a membrane ondulante des poissons (genres *Trypanosoma* Gruby et *Trypanoplasma* n.gen.). *Compte Rendu Hebdomadaires des Séances de l'Académie des Sciences, Paris* 133, 670–675.

Laveran, A. and Mesnil, F. (1904) *Trypanosomes and Trypanosomiases* (Translated by D. Nabarro). W.T. Keener and Co., Chicago, USA. 538 pp.

Lavier, C. (1936) Sur quelques flagellés intestinaux de poissons marins. *Annales de Parasitologie*, 14, 278–289.

Lee, J.J. (1985) Diplomonadida. In: Lee, J.J., Hutner, S.H. and Bovee, E.C. (eds) *An Illustrated Guide to the Protozoa*. Society of Protozoologists, Lawrence, Kansas, pp. 130–134.

Lee, J.J., Small, E.B., Lynn, D.H. and Bovee, E.C. (1985) Some techniques for collecting, cultivating and observing protozoa. In: Lee, J.J. Hutner, S.H. and Bovee, E.C. (eds) *An Illustrated Guide to the Protozoa*. Society of Protozoologists, Lawrence, Kansas, pp. 1–7.

Leidy, J. (1846) Description of a new genus and species of Entozoa. *Proceedings of the Academy of Natural Sciences of Philadelphia* 3, 100–101.

Li, S. and Woo, P.T.K. (1991a) *In vitro* cultivation of *Cryptobia salmositica*, effects of fetal bovine serum and glucose on multiplication. *Journal of Parasitology* 77, 151–155.

Li, S. and Woo, P.T.K. (1991b) Anorexia reduces the severity of cryptobiosis in *Onchorhynchus mykiss Journal of Parasitology* 77, 467–471.

Li, S. and Woo, P.T.K. (1994) Efficacy of a live *Cryptobia salmositica* vaccine, and the mechanism of protection in vaccinated *Oncorhynchus mykiss* (Walbaum) against cryptobiosis. *Veterinary Immunology and Immunopathology* (in press).

Li, S., Cowey, C.B. and Woo, P.T.K. (1993) The effects of dietary ascorbic acid on *Cryptobia salmositica* infection and on vaccination against cryptobiosis in *Oncorhynchus mykiss. Bulletin, Canadian Society of Zoologists* 24(2), 74–75 (Abstract).

Li, S., Verity, C.K. and Woo, P.T.K. (1994) Monoclonal DNA probes in salmonid cryptobiosis. *VIII International Congress of Parasitology* 1, 5 (Abstract).

Lom, J. (1973) Experimental infection of goldfish with blood flagellates. In: *Progress in Protozoology*. 4th International Congress in Protozoology, Clermont-Ferrand, France, p. 255 (Abstract).

Lom, J. (1979) Biology of trypanosomes and trypanoplasms of fish. In: Lumsden, W.H.R. and Evans, D.A. (eds) *Biology of the Kinetoplastida*, Vol. 2. Academic Press, London, pp. 269–337.

Lom, J. (1980) *Cryptobia branchialis* Nie from fish gills, ultrastructural evidence of ectocommensal function. *Journal of Fish Diseases* 3, 427–436.

Lom, J. (1986) Protozoan infection in fish. In: Vivares, C.P., Bonami, J.-R. and Jaspers, J. (eds) *Pathology in Marine Aquaculture (Pathologie En Aquaculture Marine)*, pp. 95–104. European Aquaculture Society, Special Publication No. 9, Bredene, Belgium.

Lom, J. and Suchankova, E. (1974) Comments on the life cycle of *Trypanosoma danilewskyi. Proceedings of the 3rd International Congress in Parasitology*, pp. 66–67 (Abstract).

Lom, J., Paulin, J.J. and Nohynkova, E. (1980) The fine structure of fish trypanosome, *Trypanosoma danilewskyi*. I – Presence of a cytopharyngeal complex in bloodstream trypomastigotes. *Protistologica* 16, 365–373.

Lynn, D.H. (1992) Protargol staining. In: Lee, J.J. and Saldo, A.T. (eds) *Protocols in Protozoology*. Society of Protozoologists, Lawrence, Kansas, pp. C4.1–C4.8.

Madill, J. (1988) New Canadian records of leeches (Annelida, Hirudinea) parasitic on fish. *Canadian Field Naturalist* 102, 685–688.

Makeyeva, A.P. (1956) On one of the factors of prespawning mortality of pink salmon in rivers. In: *Pacific Salmon*. Israel Program for Scientific Translations, Jerusalem, 1961; Office of Technical Service, US Department of Commerce, Washington, DC. pp. 18–21.

Malecha, J. (1984) Cycle biologique de l'hirudinee rhynchodelle *Piscicola geometra* L. *Hydrobiologia* 118, 237–243.

Mann, K.H. (1961) *Leeches (Hirudinea). Their Structure, Physiology, Ecology and Embryology*. Pergamon Press, New York.

McElwin, I.B. and Post, C. (1968) Efficacy of cyzine for trout hexamitiasis. *Progressive Fish Culturist* 30, 84–91.

Meyer, M.C. and Khan, R.A. (1979) Taxonomy, biology and occurrence of some marine leeches in Newfoundland waters. *Proceedings of the Helminthological Society of Washington* 46, 254–264.

M'Gonigle, R.H. (1940) Acute catarrhal enteritis of salmonid fingerlings. *Transactions of the American Fisheries Society* 70, 297–302.

Miyazaki, T., Rogers, W.A. and Plumb, J.A. (1986) Histopathological studies on parasitic protozoan diseases of the channel catfish in the United States. *Bulletin of the Faculty of Fisheries, Mie University* 13, 1–9.

Mo, T.A., Poppe, T.T. and Iversen, L. (1990) Systemic hexamitosis in salt-water reared Atlantic Salmon (*Salmo salar* L.). *Bulletin of the European Association of Fish Pathologists* 10(3), 69–70.

Molnar, K. (1974) Data on the 'Octomitosis' (Spironucleosis) of cyprinids and aquary fishes. *Acta Veterinaria Academiae Scientarum Hungaricae* 24, 99–106.

Montagnes, D.J.S. and Lynn, D.H. (1987) A quantitative protargol stain (QPS) for ciliates: method, description and test of its quantitative nature. *Marine Microbial Food Webs* 2, 83–93.

Moore, E. (1922) *Octomitus salmonis*, a new species of intestinal parasite in trout. *Transactions of the American Fisheries Society* 52, 74–97 + plates I–II.

Morrison, C.M. and Cone, D.K. (1986) A possible marine form of *Ichthyobodo* sp. on haddock, *Melanogrammus aeglefinus* (L) in the north-west Atlantic Ocean. *Journal of Fish Diseases* 9, 141–142.

Munday, B.L., Foster, C.K., Roubal, F.R. and Lester, R.J.G. (1990) Paramoebic gill infection and associated pathology of Atlantic salmon, *Salmo salar* and rainbow trout, *Salmo gairdneri*, in Tasmania. In: Perkins, F.O. and Cheng, T.C. (eds) *Pathology in Marine Science*. San Diego, pp. 215–222.

Nash, G., Nash, M. and Schlotfeldt, H.J. (1988) Severe gill proliferative gill disease and systemic infection in cultured catfish, *Silurus glanis* in Germany caused by an *Acanthamoeba*-like amoeba. *Journal of Fish Diseases* 11, 57–71.

Naumova, A.M. (1969) [Parasitism of *Cryptobia branchialis*]. In: *Rybovodstvo i Bolezni Ryb*. Kolos, Moscow, pp. 253–254.

Neumann, R.O. (1909) Studien uber protozoische Parasiten im Blut von Meeresfischen. *Zeitschrift für Hygiene und Infectionskrankheiten* 64, 1–112.

Newman, M.W. (1978) Pathology associated with *Cryptobia* infection in a summer flounder (*Paralichthys dentatus*). *Journal of Wildlife Diseases* 14, 299–304.

Nigrelli, R.F. and Hafter, E. (1947) A species of *Hexamita* from the skin of two cichlids. *Anatomical Record* 99, 683–684.

Nikitin, S.A. (1927) Blood parasites of some northern vertebrates (in Russian). *Russian Journal of Tropical Medicine, Medical and Veterinary Parasitology* 6, 350–356.

Nohynkova, E. (1984a) A new pathogenic *Cryptobia* from freshwater fishes, a light and electron microscope study. *Protistologica* 20, 181-195.

Nohynkova, E. (1984b) *In vitro* cultivation of the bodonid flagellate *Trypanoplasma borreli. Journal of Protozoology* 32, 52 (Abstract).

Osborn, P.E. (1966) Effective chemical control of some parasites of goldfish and other pondfish. *1966 Annual Meeting of the Wildlife Disease Association* (unpublished; quoted in Hoffman and Meyer, 1974).

Overstreet, R.M. (1976) A redescription of *Crassicutis archosargi*, a digenean exhibiting an unusual tegumental attachment. *Journal of Parasitology* 62, 702-708.

Paperna, I. and Overstreet, R.M. (1981) Parasites and diseases of mullets (Mugilidae). In: Oreu, O.H. (ed.) *Aquaculture of Grey Mullets*. Cambridge University Press, Cambridge, pp. 411-493.

Paterson, W.B. and Woo, P.T.K. (1983) Electron microscopic observations of the bloodstream form of *Cryptobia salmositica* Katz, 1951 (Kinetoplastida, Bodonina). *Journal of Protozoology* 39, 431-437.

Paterson, W.B. and Woo, P.T.K. (1984) Ultrastructural studies on mitosis in *Trypanosoma danilewskyi* (Mastigophora, Zoomastigophora). *Canadian Journal of Zoology* 62, 1167-1171.

Paulin, J.J., Lom, J. and Nohynkova, E. (1980) The fine structure of *Trypanosoma danilewskyi*. II Structure and cytochemical properities of the cell surface. *Protistologica* 16, 375-383.

Pavlovskii, E.N. (1964) *Key to Parasites of Freshwater Fish of the USSR*. Academy of Sciences of the USSR, Zoological Institute. Translated from Russian by Israel Program for Scientific Translations, Jerusalem, Israel, 919 pp.

Peckova, H. and Lom, J. (1990) Growth, morphology and division of flagellates of the genus *Trypanoplasma* (Protozoa, Kinetoplastida) *in vitro. Parasitology Research* 76, 553-558.

Peterson, W.J. (1960) Population changes in the cecal protozoa of rats and some factors influencing them. *Experimental Parasitology* 10, 293-312.

Plehn, M. (1903) *Trypanoplasma cyprini* nov. sp. *Archiv für Protistenkunde* 3, 175-180.

Plehn, M. (1924) Praktikum der Fischkrankheiten. In: Demoll-Maier (ed.) *Handbuch der Binnenfischerei Mitteleuropas*. Vol. I.E. Schweizerbartsche Verlagsbuchhandlung, Stuttgart.

Poppe, T.T. and Hastein, T. (1982) [Costiasis in salmon smolt (*Salmo salar* L.) in sea water.] *Norsk Veterinaertidsskrift* 94, 259-262.

Poppe, T.T., Mo, T.A. and Iversen, L. (1992) Disseminated hexamitosis in sea-caged Atlantic salmon, *Salmo salar. Diseases of Aquatic Organisms* 14, 91-97.

Post, G. (1987) *Textbook of Fish Health*. TFH Publications, Neptune City, New Jersey.

Poynton, S.L. (1986) Distribution of the flagellate *Hexamita salmonis* Moore, 1922 and the microsporidian *Loma salmonae* Putz, Hoffman and Dunbar, 1965 in brown trout, *Salmo trutta* L., and rainbow trout, *Salmo gairdneri* Richardson, in the River Itchen (UK) and three of its fish farms. *Journal of Fish Biology* 29, 417-429.

Poynton, S.L. (1994) Hexamitiasis (Octomitosis), and Spironucleosis. In: *Procedures for the Detection and Identification of Certain Fish Pathogens*. Fish Health Section, American Fisheries Society.

Poynton, S.L. and Morrison, C. (1990) Morphology of diplomonad flagellates, *Spironucleus torosa* n.sp. from Atlantic cod *Gadus morhua* L., and haddock *Melanogrammus aeglefinus* (L.) and *Hexamita salmonis* Moore from brook trout

Salvelinus fontinalis (Mitchill). *Journal of Protozoology* 37, 369–383.

Poynton, S.L., Fraser, W., Francis-Floyd, R., Rutledge, P., Reed, P. and Nerad, T.A. (1994) *Spironucleus anusmirabilisvortens* n.sp. from angel fish *Pterophyllum scalare*: morphology and culture. *Journal of Eukaryotic Microbiology* (in press).

Putz, R.E. (1972) Biological studies on the hemoflagellates *Cryptobia cataractae* and *Cryptobia salmositica*. *Technical Paper Bureau of Sport Fishery and Wildlife* 63, 3–25.

Qadri, S.S. (1962) An experimental study of the life cycle of *Trypanosoma danilewskyi* in the leech, *Hemiclepsis marginata*. *Journal of Protozoology* 9, 254–258.

Roberts, R.J. and Shepherd, C.J. (1974) *Handbook of Trout and Salmon Diseases*. Fishing News Books, Farnham, UK.

Robertson, D.A. (1979) Host–parasite interactions between *Ichthyobodo necator* (Henneguy, 1883) and farmed salmonids. *Journal of Fish Diseases* 2, 481–491.

Robertson, D.A. (1985) A review of *Ichthyobodo necator* (Henneguy, 1883) an important and damaging fish parasite. In: Muir, J.F. and Roberts, R.J. (eds) *Recent Advances in Aquaculture*. Croom Helm, London, pp. 1–30.

Robertson, D.A., Roberts, R.J. and Bullock, A.M. (1981) Parthogenesis and auto-radiographic studies on the epidermis of salmonids infested with *Ichthyobodo necator* (Henneguy, 1883). *Journal of Fish Diseases* 4, 113–125.

Robertson, M. (1911) Transmission of flagellates of certain freshwater fishes. *Philosophical Transactions of the Royal Society of London B* 202, 29–50.

Rosengarten, R. (1985) [Parasitological examination of *Salmo gairdneri* on a trout farm in western lower Saxony]. Inaugural Diss. Tierarzliche Hochschule, Hannover, FRG, 125 pp.

Roubal, F.R., and Bullock, A.M. (1987) Differences between the host–parasite inter-face of *Ichthyobodo necator* (Henneguy, 1883) on the skin and gills of salmonids. *Journal of Fish Diseases* 10, 237–240.

Roubal, F.R., Robertson, D.A. and Roberts, J.A. (1987) Ultrastructural aspects of infection by *Ichthyobodo necator* (Henneguy, 1883) on the skin and gills of the salmonids, *Salmo salar* L. and *Salmo gairdneri* Richardson. *Journal of Fish Diseases* 10, 181–192.

Roubal, F.R., Lester, R.J.G. and Foster, C.K. (1990) Studies on cultured and gill-attached *Paramoeba* sp. (Gymnamoebae: Paramoebidae) and the cytopathology of amoebic gill disease in Atlantic salmon, *Salmo salar* L., from Tasmania. *Journal of Fish Diseases* 12, 481–492.

Sano, T. (1970) Etiology and histopathology of hexamitiasis and an IPN-like disease of rainbow trout. *Journal of the Tokyo University of Fisheries* 56, 23–30.

Savage, A. (1935) Notes on costiasis. *Transactions of the American Fisheries Society* 65, 332–333.

Sawyer, R.T. (1986) *Leech Biology and Behaviour*, Vol. 2. *Feeding Biology, Ecology and Systematics*. Oxford Scientific Publications, Oxford.

Sawyer, R.T., Lawler, A.R. and Overstreet, R.M. (1975) Marine leeches of the eastern United States and the Gulf of Mexico with a key to the species. *Journal of Natural History* 9, 633–667.

Sawyer, T.K., Hnath, J.G. and Conrad, J.F. (1974) *Thecamoeba hoffmani* sp.n. (Amoebida: Thecamoebidae) from gills of fingerling salmonid fish. *Journal of Parasitology* 60, 677–682.

Sawyer, T.K., Hoffman, G.K., Hnath, J.G. and Conrad, J.F. (1975) Infection of salmonid fish gills by aquatic amoebas (Amoebida: Thecamoebidae). In: Ribelin, W.E. and Magaki, G. (eds) *The Pathology of Fishes*. The University of Wisconsin Press, Madison, pp. 143–150.

Schaperclaus, W. (1986) *Fish Diseases I and II*. Translated by M.S.R. Chari (1991); US Departrment of the Interior and the National Science Foundation, Washington, DC.

Schubert, G. (1966) Zur Ultracytologie von *Costia necatrix* Leclerq, unter besonderer Berucksichtigung des Kinetoplast-Mitochondrions. *Zeitschrift für Parasitenkunde* 27, 271–286.

Schubert, G. (1978) *Krankheiten der Fische*. Kosmos, Franckh'sche Verlagshandlung, Stuttgart.

Shanavas, K.R., Ramachandran, P. and Janardanan, K.P. (1989) *Trypanoplasma ompoki* sp.n. from freshwater fishes in Kerala, India, with observations on its vector-phase development and transmission. *Acta Protozoologica* 28, 293–302.

Siddall, M.E., Hong, H. and Desser, S.S. (1992) Phylogenetic analysis of the Diplomonadida (Wenyon, 1926) Brugerolle, 1975: evidence for heterochrony in protozoa and against *Giardia lamblia* as a 'missing link'. *Journal of Protozoology* 39, 361–367.

Sin, Y.M. and Woo, P.T.K. (1993) Immunosuppression in rainbow trout, *Oncorhynchus mykiss*, caused by *Cryptobia salmositica*. In: Phang, V.P.E., Sin, Y.M., Lim, T.M., Tan, C.H., Shima, A. and Lam, T.J. (eds) *Fish Biology: From Genes to Organism*. National University of Singapore, pp. 160–169.

Sitja-Bobadilla, A. and Woo, P.T.K. (1994) An enzyme-linked immunosorbent assay (ELISA) for the detection of antibodies against the pathogenic haemoflagellate, *Cryptobia salmositica* Katz, and protection against cryptobiosis in juvenile rainbow trout, *Oncorhynchus mykiss* (Walbaum) inoculated with a live vaccine. *Journal of Fish Diseases* 17, 399–408.

Skarlato, S.O., Lom, J. and Nohynkova, E. (1987) Fine structural morphology of the nucleus of *Trypanosoma danilewskyi* (Kinetoplastida, Trypanosomatina) during mitosis. *Archiv für Protistenkunde* 133, 3–14.

Skrudland, A. (1987) [An outbreak of *Ichthyobodo necator* in salmon fry]. *Norsk Veterinaertidsskrift* 99, 729–730.

Smirnova, T.L. (1970) *Trypanosoma* in the blood of *Lota lota* L. - *Trypanosoma lotae* sp.n. *Parasitologyia* 4, 296–297.

Smolikova, V., Suchankova, E. and Lom, J. (1977) Growth of the carp trypanosome, *T. danilewskyi*, in fish tissue culture. *Journal of Protozoology* 24, 54 (Abstract).

Sommerville, C. (1981) Parasites of ornamental fish. *Journal of Small Animal Practice* 22, 367–376.

Steinhagen, D., Kruse, P. and Korting, W. (1989a) The parasitemia of cloned *Trypanoplasma borreli* Laveran and Mesnil, 1901, in laboratory-infected common carp (*Cyprinus carpio* L.). *Journal of Parasitology* 75, 685–689.

Steinhagen, D., Kruse, P. and Korting, W. (1989b) Effects of immunosuppressive agents on common carp infected with the haemoflagellate *Trypanoplasma borreli*. *Diseases of Aquatic Organisms* 7, 67–69.

Steinhagen, D., Kruse, P. and Korting, W. (1990) Some haematological observations on carp, *Cyprinus carpio* L., experimentally infected with *Trypanoplasma borreli* Laveran and Mesnil, 1901 (Protozoa, Kinetoplastida). *Journal of Fish Diseases* 13, 157–162.

Stoskopf, M.K. (1988) Fish chemotherapeutics. *Veterinary Clinics of North America, Small Animal Practice. Tropical Fish Medicine* 18, 331–348.

Strout, R.G. (1962) A method for concentrating hemoflagellates. *Journal of Parasitology* 48, 110.

Strout, R.G. (1965) A new hemoflagellate (genus *Cryptobia*) from marine fishes of northern New England. *Journal of Parasitology* 51, 654–659.

Sypek, J.P. and Burreson, E.M. (1983) Influence of temperature on the immune response of juvenile summer flounder, *Paralichthys dentatus* and its role in the elimination of *Trypanoplasma bullocki* infections. *Developmental and Comparative Immunology* 7, 277–286.

Sypek, J.P. and Howe, A.B. (1985) *Trypanoplasma bullocki*, natural infections in winter flounder, *Pseudopleuronectes americanus*. *International Meeting of Fish Immunology. Sandy Hook*, New Jersey, p. P3 (Abstract).

Tandon, R.S. and Chandra, S. (1977a) Physiology of host parasite relationship, effects on serum alkaline phosphatase levels of fish hosts parasitized by trypanosomes. *Zeitschrift für Parasitenkunde* 52, 195–198.

Tandon, R.S. and Chandra, S. (1977b) Studies on ecophysiology of fish parasites, Effects of trypanosome infection on the serum cholesterol levels of fishes. *Zeitschrift für Parasitenkunde* 52, 199–202.

Tandon, R.S. and Joshi, B.D. (1973) Studies on the physiopathology of blood of freshwater fishes infected with two new forms of trypanosomes. *Zeitschrift für Wissenschaftliche Zoologie* 185, 207–221.

Tandon, R.S. and Joshi, B.D. (1974) Effect of trypanosome infection on blood glucose levels of some fresh water teleosts. *Journal of the Inland Fisheries Society of India* 6, 81–82.

Tavolga, W.N. and Nigrelli, R.F. (1947) Studies on *Costia necatrix* Henneguy. *Transactions of the American Microscopical Society* 66, 366–378.

Taylor, A.E.R. and Baker, J.R. (1968) *Cultivation of Parasites in vitro*. Blackwell Scientific, Oxford.

Taylor, P.W. (1977) Isolation and experimental infection of free-living amoebae in freshwater fishes. *Journal of Parasitology* 63, 232–237.

Thomas, P.T. and Woo, P.T.K. (1988) *Cryptobia salmositica, in vitro* and *in vivo* study on the mechanism of anaemia in infected rainbow trout, *Salmo gairdneri* Richardson. *Journal of Fish Diseases* 11, 425–431.

Thomas, P.T. and Woo, P.T.K. (1989a) An *in vitro* study on the haemolytic components of *Cryptobia salmositica*. *Journal of Fish Diseases* 12, 89–393.

Thomas, P.T. and Woo, P.T.K. (1989b) Complement activity in *Salmo gairdneri* Richardson infected with *Cryptobia salmositica* and its relationship to the anaemia in cryptobiosis. *Journal of Fish Diseases* 12, 395–397.

Thomas, P.T. and Woo, P.T.K. (1990a) *In vivo* and *in vitro* cell-mediated immune responses of *Oncorhynchus mykiss* (Walbaum) against *Cryptobia salmositica* Katz, 1951 (Sarcomastigophora, Kinetoplastida). *Journal of Fish Diseases* 13, 423–433.

Thomas, P.T. and Woo, P.T.K. (1990b) Dietary modulation of humoral immune response and anaemia in *Oncorhynchus mykiss* (Walbaum) infected with *Cryptobia salmositica* Katz, 1951. *Journal of Fish Diseases* 13, 435–446.

Thomas, P.T. and Woo, P.T.K. (1992a) *In vitro* culture and multiplication of *Cryptobia catostomi* and experimental infection of white sucker (*Catostomus commersoni*). *Canadian Journal of Zoology* 70, 201–204.

Thomas, P.T. and Woo, P.T.K. (1992b) Anorexia in *Oncorhynchus mykiss* (Walbaum) infected with *Cryptobia salmositica* (Sarcomastigophora, Kinetoplastida), its onset and contribution to the immunodepression. *Journal of Fish Diseases* 15, 443–447.

Thomas, P.T., Ballantyne, J.S. and Woo, P.T.K. (1992) *In vitro* oxygen consumption and motility of *Cryptobia salmositica, Cryptobia bullocki* and *Cryptobia catostomi*. *Journal of Parasitology* 78, 747–749.

Thompson, J.D. (1908) Cultivation of the trypanosome found in the blood of the goldfish. *Journal of Hygiene* 8, 75–82.

Tojo, J.L., Santamarina, M.T., Leiro, J., Ubeira, F.M. and Sanmartin, M.L. (1994) Pharmacological treatments against *Ichthyobodo necator* (Henneguy, 1883) in rainbow trout, *Oncorhynchus mykiss* (Walbaum). *Journal of Fish Diseases* 17, 135–143.

Urawa, S. (1987) Effects of environmental stress on the mortality of chum salmon fry infected with *Ichthyobodo necator*. In: *Actual Problems in Fish Parasitology*. Hungarian Academy of Sciences, Budapest, p. 99.

Urawa, S. and Kusakari, M. (1990) The survivability of the ectoparasitic flagellate *Ichthyobodo necator* on chum salmon fry (*Oncorhynchus keta*) in seawater and comparison to *Ichthyobodo* sp. on Japanese flounder (*Paralichthys olivaceus*). *Journal of Parasitology* 76, 33–40.

Urawa, S., Ueki, N., Nakai, T. and Yamasaki, H. (1991) High mortality of cultured juvenile Japanese flounder, *Paralichthys olivaceus* (Temminck and Schlegel), caused by the parasite flagellate, *Ichthyobodo* sp. *Journal of Fish Diseases* 14, 489–494.

Uzman J.R. and Hayduk, S.H. (1963) *In vitro* culture of the flagellate protozoan *Hexamita salmonis*. *Science* 140, 290–292.

Uzmann, J.R., Paulik, G.J. and Hayduk, S.H. (1965) Experimental hexamitiasis in juvenile coho salmon (*Oncorhynchus kisutch*) and steelhead trout (*Salmo gairdneri*). *Transactions of the American Fisheries Society* 94, 53–61.

van Duijin, C., Jr (1973) *Diseases of Fishes*, 3rd edn. Thomas, Springfield, Illinois.

Verity, C. and Woo, P.T.K. (1993) Antigenic characterization of *Cryptobia salmositica* using monoclonal antibodies and Western immunoblot. *Bulletin, Canadian Society of Zoologists* 24(2), 112–113 (Abstract).

Verity, C. and Woo, P.T.K. (1994) Characterization of a monoclonal antibody against *Cryptobia salmositica*. Use of the antibody in an antigen-ELISA for diagnosis of cryptobiosis. *Bulletin, Canadian Society of Zoologists* 25(2), 106 (Abstract).

Vickerman, K. (1971) Morphological and physiological considerations of extracellular blood protozoa. In: Fallis, A.M. (ed.) *Physiology and Ecology of Parasites*. Toronto University Press, Toronto, pp. 59–91.

Vickerman, K. (1976a) The diversity of the kinetoplastid flagellates. In: Lumsden, W.H.R. and Evans, D.A. (eds) *Biology of the Kinetoplastida*, Volume I. Academic Press, London, pp. 1–34.

Vickerman, K. (1976b) Comparative cell biology of the kinetoplastid flagellates. In: Lumsden, W.H.R. and Evans, D.A. (eds) *Biology of the Kinetoplastida*, Volume I. Academic Press, New York, pp. 35–100.

Vickerman, K. (1990) Phylum Zoomastigina Class Diplomonadida. In: Margulis, L., Corliss, J.O., Melkanian, M. and Chapman, D. (eds) *Handbook of Protoctista*. Jones and Bartlett, pp. 200–210.

Voelker, F.A., Anver, M.R., McKee, A.E., Casey, H.W. and Brenniman, G.R. (1977) Amoebiasis in goldfish. *Veterinary Pathology* 14, 247–255.

Wales, J.H. and Wolf, H. (1955) Three protozoan diseases of trout in California. *California Fish and Game* 41, 183–187.

Wang, R. and Belosevic, M. (1994) Cultivation of *Trypanosoma danilewskyi* (Laveran and Mesnil 1904) in serum-free medium and assessment of the course of infection in goldfish, *Carassius auratus*. *Journal of Fish Diseases* 17, 47–56.

Watson, L.P., McKee, A.E. and Merrell, B.R. (1980) Preparation of microbiological specimens for scanning electron microscopy. *Scanning Electron Microscopy* II, 45–56.

Wehnert, S.D. and Woo, P.T.K. (1980) *In vivo* and *in vitro* studies on the host specificity of *Trypanoplasma salmositica*. *Journal of Wildlife Diseases* 16, 183–187.

Wehnert, S.D. and Woo, P.T.K. (1981) The immune responses of *Salmo gairdneri* during *Trypanoplasma salmositica* infection. *Bulletin, Canadian Society of Zoologists* 11, 100 (Abstract).

Wolf, K. and Markiw, M.E. (1982) Ichthyophthiriasis, immersion immunization of rainbow trout (*Salmo gairdneri*) using *Tetrahymena thermophila* as a protective immunogen. *Canadian Journal of Fisheries and Aquatic Science* 39, 1722–1725.

Woo, P.T.K. (1969) The haematocrit centrifuge for the detection of trypanosomes in blood. *Canadian Journal of Zoology* 47, 921–923.

Woo, P.T.K. (1978) The division process of *Cryptobia salmositica* in experimentally infected rainbow trout (*Salmo gairdneri*). *Canadian Journal of Zoology* 56, 1514–1518.

Woo, P.T.K. (1979) *Trypanoplasma salmositica*, experimental infections in rainbow trout, *Salmo gairdneri*. *Experimental Parasitology* 47, 36–48.

Woo, P.T.K. (1981a) *Trypanosoma danilewskyi*, a new multiplication process for *Trypanosoma* (Protozoa, Kinetoplastida). *Journal of Parasitology* 67, 522–526.

Woo, P.T.K. (1981b) Acquired immunity against *Trypanosoma danilewskyi* in goldfish, *Carassius auratus*. *Parasitology* 83, 343–346.

Woo, P.T.K. (1987a) *Cryptobia* and cryptobiosis in fishes. *Advances in Parasitology* 26, 199–237.

Woo, P.T.K. (1987b) Immune response of fish to protozoan infections. *Parasitology Today* 3, 186–188.

Woo, P.T.K. (1990) MISET, An immunological technique for the serodiagnosis of *Cryptobia salmositica* (Sarcomastigophora, Kinetoplastida) infection in *Oncorhynchus mykiss*. *Journal of Parasitology* 76, 389–393.

Woo, P.T.K. (1992) Immunological responses of fish to parasitic organisms. In: Faisal, M. and Hetrick, F.M. (eds) *Annual Review of Fish Diseases*, Vol. 2. Pergamon Press, New York, pp. 339–366.

Woo, P.T.K. (1994) Flagellate parasites of fishes. In: Kreier, J.P. (ed.) *Parasitic Protozoa*, 2nd edn, Vol. VIII. Academic Press, London, pp. 1–80.

Woo, P.T.K. and Black, G.A. (1984) *Trypanosoma danikewskyi*, host specificity and host's effects on morphometrics. *Journal of Parasitology* 70, 788–793.

Woo, P.T.K. and Jones, S.R.M. (1989) The piscine immune systems and the effects of parasitic protozoans on the immune response. In: Ko, R. (ed.) *Concepts in Parasitology*. Hong Kong University Press, Hong Kong, pp. 47–64.

Woo, P.T.K. and Li, S. (1990) *In vitro* attenuation of *Cryptobia salmositica* and its use as a live vaccine against cryptobiosis in *Oncorhynchus mykiss*. *Journal of Parasitology* 76, 752–755.

Woo, P.T.K. and Rogers, D.J. (1974) A statistical study of the sensitivity of the haematocrit centrifuge technique in the detection of trypanosomes in blood. *Transactions of the Royal Society of Tropical Medicine and Hygiene* 68, 319–326.

Woo, P.T.K. and Shariff, M. (1990) *Lernaea cyprinacea* L. (Copepoda: Caligidea) in *Helostoma temmincki* Cuvier and Valenciennes: the dynamics of resistance in recovered and naive fish. *Journal of Fish Diseases* 13, 485–493.

Woo, P.T.K. and Thomas, P.T. (1991) Polypeptide and antigen profiles of *Cryptobia salmositica, C. bullocki* and *C. catostomi* (Kinetoplastida, Sarcomastigophora) isolated from fishes. *Diseases of Aquatic Organisms* 11, 201–205.

Woo, P.T.K. and Thomas, P.T. (1992) Comparative *in vitro* studies on virulent and avirulent strains of *Cryptobia salmositica* Katz, 1951 (Sarcomastigophora,

Kinetoplastida). *Journal of Fish Diseases* 15, 261–266.

Woo, P.T.K. and Wehnert, S.D. (1983) Direct transmission of a haemoflagellate, *Cryptobia salmositica* Katz, 1951 (Kinetoplastida, Bodonina) between rainbow trout under laboratory conditions. *Journal of Protozoology* 39, 334–337.

Woo, P.T.K. and Wehnert, S.D. (1986) *Cryptobia salmositica*, susceptibility of infected trout, *Salmo gairdneri*, to environmental hypoxia. *Journal of Parasitology* 72, 392–396.

Woo, P.T.K., Wehnert, S.D. and Rodgers, D. (1983) The susceptibility of fishes to haemoflagellates at different ambient temperatures. *Parasitology* 87, 385–392.

Woo, P.T.K., Leatherland, J.F. and Lee, M.S. (1987) *Cryptobia salmositica*: cortisol increases the susceptibility of *Salmo gairdneri* Richardson to experimental cryptobiosis. *Journal of Fish Diseases* 10, 75–83.

Wood, J.W. (1979) *Diseases of Pacific Salmon: Their Prevention and Treatment*. 3rd edn. State of Washington, Department of Fisheries, Olympia, Washington.

3

Myxosporea (Phylum Myxozoa)

J. Lom and I. Dyková

Institute of Parasitology, Czech Academy of Sciences, 370 05 České Budějovice, Branišovská 31, Czech Republic.

INTRODUCTION

The Myxosporea is a large group of parasitic organisms. It has a unique position in the protistan kingdom because the multicellular spores possess nematocyst-like polar capsules. The parasite causes heavy infections and diseases in cultured fishes. In feral fishes, Myxosporea may also cause extensive lesions and mortality. In marine fishes, parasites of the genera *Kudoa*, *Hexacapsula* or *Unicapsula* may make the flesh of infected fish unsightly and destroy its texture, so that large parts of the catch are rendered unmarketable. In fish cultures, whole stocks may succumb to infections as is the case with *Ceratomyxa shasta* in North American salmonids.

There has been renewed interest in this parasite group in the last two decades. New myxosporean diseases have been described, e.g. swimbladder inflammation and proliferative kidney disease. Both diseases are provoked by presporogonic developmental stages. *Myxobolus cerebralis*, which causes whirling disease in trout, was shown to undergo a unique life cycle which involves transformation into an actinosporean protozoan.

Thus the current studies on Myxosporea centre in two major areas. One is on their pathogenicity and their significance as pathogens in aquacultures and in capture fisheries. The second area is on the peculiar features of their biology and life cycle, and also, relation to other organisms.

DISTRIBUTION

The majority of more than 1250 described species of Myxosporea occur in tissues and organ cavities of freshwater, brackish and marine fishes. Some are found in elasmobranchs, rarely in lampreys and myxines. Only a few species occur in aquatic reptiles, amphibians and invertebrates.

Some species seem to be host specific (e.g. *Hoferellus cyprini* in common carp or various *Sphaerospora* species in the kidneys of different hosts) while others are definitely highly polyxenous (e.g. *Myxobolus cerebralis* in various salmonids). The true extent of the host range for most species is still not known. The geographical distribution depends on that of the host(s). Thus *Myxidium lieberkuehni* is holarctic in Esocidae and *Myxobolus cerebralis* is believed to have spread, along with cultured trout, throughout Europe, northern Asia, in the Americas, Australia and New Zealand. Some of the myxosporeans of the common carp originally had restricted areas of distribution (e.g. *Myxobolus koi* and *Thelohanellus nikolskii* to East Asia) but were introduced to Europe with transfers of the fish. Parasites of ornamental fishes were similarly disseminated throughout several continents.

The prevalence of myxosporean infection in an area can sometimes be extremely high (up to 100%), or low. In some natural habitats, notably in the marine environment, host populations may have significantly different prevalences of a given myxosporean species. These marked differences in prevalence were used as parasite tags in stock separation. Parasite tags have several advantages over conventional mark-and-recapture methods (MacKenzie, 1987). One of their useful applications is in deep sea fish, where conventional tagging is either difficult or impossible. Szuks (1980) used *Myxidium coryphaenoidium* to separate stocks of the deep sea *Macrourus rupestris* in the Labrador Sea. *Kudoa clupeidae* was used to separate herring stocks off the Maine coast in the United States (Sindermann, 1961), *Ceratomyxa arcuata* and *Myxidium sphaericum* were used to separate whiting (*Merlangius merlangus*) stocks in the North Sea (Kabata, 1963) and *Myxobolus neurobius* was widely used to identify Canadian and Alaskan stocks of salmon (Margolis, 1982; Wood *et al.*, 1989).

Myxosporean infections may be long-lasting in fishes. *Myxidium lieberkuehni* are in the urinary bladder of pike all year round. *Ceratomyxa shasta* and *Myxidium salvelini* infections may survive the oceanic phase of the life of anadromous fishes. *M. salvelini* is retained in the trophozoite stage and sporogenesis resumes after the salmon returns to spawn. Other Myxosporea in temperate fishes exhibit marked seasonal fluctuations. In the perch (*Perca perca*) the infection of secondary gill lamellae with *Henneguya psorospermica* is limited to winter and early spring.

TAXONOMY

Myxosporea has been considered until now as one of two classes (Myxosporea Bütschli, 1881) in the phylum Myxozoa Grasse, 1960. The classification is based exclusively on the structure of the spores. Characteristics of the vegetative stages and life cycle are difficult to categorize and hence have not been used. When the myxosporean–actinosporean transformation is confirmed and its extent in genera assessed, the classification of the phylum may change drastically. At present, Myxosporea is divided into two orders, the Bivalvulida

Shulman, 1959 (spores with two shell valves and one to four polar capsules) and the Multivalvulida Shulman, 1959 (spores with three to seven shell valves and one to seven polar capsules). The Bivalvulida are subdivided into three suborders with 12 families. The division is based on the position of polar capsules in relation to spore shape and to the line of dehiscence of the shell valves. The number of polar capsules and shell valves and the shape and structure of these are also important characters (Lom and Noble, 1984). At the species level, there are difficulties, especially in the genera *Myxobolus, Chloromyxum, Ceratomyxa* with closely similar species whose spores have only a few diagnostic features and which are highly variable. At present, there are 51 genera with about 1250 species. The genus *Myxobolus* alone has 450 species (Landsberg and Lom, 1991).

Quite recently, Kent *et al.* (1994), taking into account the evidence that a complete myxosporean life consists of both myxosporean and actinosporean stages, proposed to suppress the class Actinosporea. Actinosporean species should be assigned to separate myxosporean species as soon as their myxosporean stages are identified; until then, they are preserved as species inquirendae. Only the family Tetractinomyxidae is retained and is transferred to the order Multivalvulida.

MORPHOLOGY AND LIFE CYCLE

The spore

The spore, which is the infective stage, consists of several (4 to 16) cells which are transformed during sporogenesis into parts of the spore. Depending on the genus, there may be one to seven capsulogenic cells which form the polar capsule, two to seven valvogenic cells which make up the shell valves, and one to two cells as the infective sporoplasm. The shell valves adhere together along the lines of dehiscence (suture lines) and cover the polar capsule which is situated in the spore apex. The sporoplasm is in the posterior (or middle) part of the spore and may be a binucleate or two uninucleate, or two uninucleate sporoplasms one inside the other. The sporoplasm contains dense bodies (sporoplasmosomes) discernible in transmission electron micrographs. In some genera, it contains β-glycogen particles usually but not always aggregated as a round inclusion designated as 'iodinophilous' (stains with Lugol's reagent) or 'glycogen' vacuole.

Shell valves may be smooth or ridged (Fig. 3.1C,G,H,I) or may bear various projections (Fig. 3.2E,H) or be invested with a transient mucous envelope (Fig. 3.2G) which usually disappears in a few days after the spore is released. These structures are presumed to increase the buoyancy of the spore and so enhance the spread of spores in the aquatic environment.

Thick-walled polar capsules which are refractile when freshly collected (Figs 3.1, 3.2), contain coiled polar filaments (e.g., Figs 3.1D; 3.2B,D). The polar filament is a hollow tube which is spirally twisted along its length and is capable of rapid extrusion when it everts inside out. The everted filament is sticky and probably is used to attach the hatching spore to the host's intestinal surface. In the electron microscope, the wall of the capsule is seen

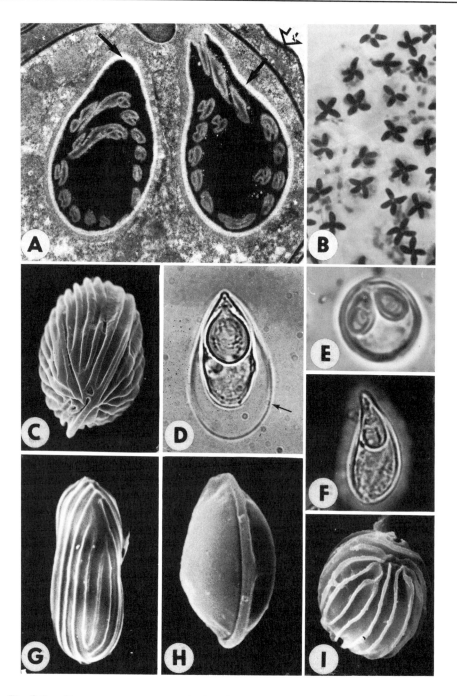

Fig. 3.1. Myxosporean spores. A. longitudinal section through almost mature polar capsules of *Chloromyxum leydigi* from *Raja clavata* (×15600); full arrows indicate the inner electron lucent layer of the capsule wall which continues in the polar filament wall; hollow arrow marks part of a valvogenic cell. Original. B. *Kudoa pericardialis* from the pericardial cavity of *Seriola quinqueradiata*,

to consist of two layers. The inner layer is electron lucent and is composed of chitin while the outer one is proteinaceous. Both layers continue into the polar filament wall (Fig. 3.1A). The arrangement of the polar filament is different in some genera. In *Sphaeromyxa* the filament is bent over several times in a zig-zag pattern (Lom, 1969), in some *Ceratomyxa* the filament has a straight basal part around which the rest of the filament coils (Desportes and Théodorides, 1982) . The mouth of the polar capsule is in contact with a cap-like structure which plugs the filament discharge canal in the shell valve.

The energy for the filament discharge may be supplied by pre-built pressure stored in the polar capsule, which is released after the cap-like structure has been removed by the host's digestive juices (Lom, 1990). Also, the discharge energy may be derived from contraction of Ca^{2+} dependent proteins in the polar capsule wall (Uspenskaya, 1982). The discharge of filaments can be elicited by various chemicals, and, in all cases, by saturated aqueous urea.

Spore size is usually between 8 to 20 µm. The largest spores (98 µm) are those of *Myxidium giganteum*.

DEVELOPMENTAL STAGES

The vegetative stages or trophozoites range from a few microns to large sporogonic plasmodia which are several mm in size (in *Sphaeromyxa miayai* up to 2 cm). Their cytoplasm contains most of the typical organelles, except for the missing centriole, and reserve materials (lipids, glycogen). Characteristic features include bundles of microtubules close to the nuclei of generative or sporogonic cells (see below) and free ribosomes which may sometimes occur in large numbers.

The vegetative stages assume a variety of forms in the course of their development in the fish.

Early infection stages
According to the actinosporean transformation theory (Wolf and Markiw, 1984) , the infectious germ is derived from the actinosporean sporoplasm. It

---

the polar capsules are heavily stained with Giemsa stain (×2000; original, courtesy of Prof. S. Egusa). C. *Myxidium giardi* from the gills of *Anguilla anguila*, (×4000) D. *Thelohanellus hovorkai* from the brain of *Cyprinus carpio*; arrow points at the membrane investing the spore (×2450; from Lom and Dyková, 1992, with permission of the publisher). E. *Myxobolus cerebralis* from head cartilage of rainbow trout (×2600). F. *T. pyriformis* from the gills of *Tinca tinca*; an India ink preparation to reveal the mucous envelope, (×1500). G. *Zschokkella nova* from the gall bladder of *Rutilus rutilus* (×5000). H. *M. neurobius* from the brain of *Salmo trutta* m. *fario*, in side (sutural) view (×6700). I. *Chloromyxum cristatum* from the gall bladder of *Barbus barbus* (×3200). A, transmission electron micrograph; C,G,H,I, scanning electron micrograph; D,E,F, light micrographs of fresh spores. C to I, originals.

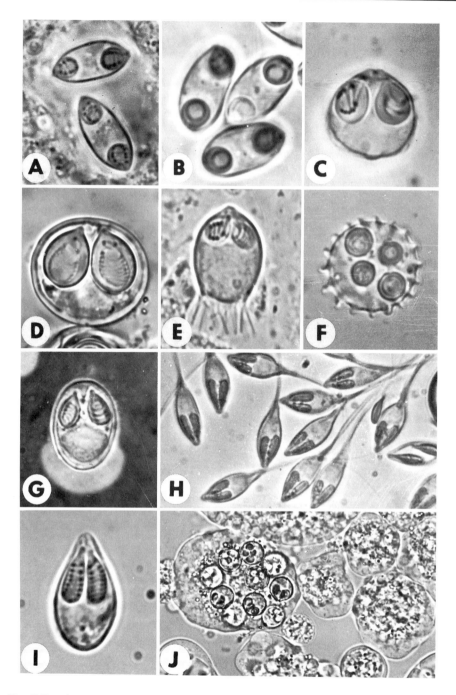

Fig. 3.2. A to I. Fresh myxosporean spores. A. *Myxidium rhodei* from the kidney of *Rutilus rutilus* (×2000). B. *Zschokkella nova* from the gall bladder of *R. rutilus* (×2000; see also Fig. 3.1G). C. *Sphaerospora molnari* from the gills of *Carassius carassius* (×2300). D. *Myxobolus magnasphaerus* from the kidney of *Lepomis gibbosus* (×1900). E. *Hoferellus carassii* from the kidney of *Carassius auratus*

enters the host via the digestive tract or through the skin of host larvae. Such stages were observed in the skin approximately 30 minutes after exposure to *Triactinomyxon* spores (Markiw, 1989, 1991) to initiate *Myxobolus cerebralis* infection. They were not found in the skin 8 hours later and their migration route in the host body is not known.

According to the theory on direct transmission of myxosporean spores, the sporoplasm escapes from the shell in the digestive tract as a small amoebula. It crosses the intestinal wall, reaches the bloodstream or lymphatic system and is carried to the target organ. It then transforms and divides into sporogonic plasmodia. This complies with the early ideas on myxosporean life cycles (summarized by Noble, 1944) but seems dubious because of recent evidence. Alternatively, in the target organ but prior to engaging in sporogonic plasmodia formation it may undergo a presporogonic, proliferative phase (Dyková and Lom, 1988a). Also, it may undergo one or more presporogonic cycles in an organ prior to reaching the target one.

Such a presporogonic proliferation may also take place in cases in which the infective germ is derived from an actinosporean sporoplasm. Since some phases of the presporogonic cycle were observed to persist parallel with the final sporogony phase, we prefer to call them extrasporogonic rather than presporogonic (Lom, 1987).

The amoebula starts its first cell division (not yet observed) via endogenous cleavage to produce an inner (secondary) cell. The amoebula becomes the primary cell and is homologous with a trophozoite. From this cell doublet we can deduce all subsequent developmental stages. The nucleus of the trophozoite, termed the vegetative nucleus, produces many vegetative nuclei. The inner cell, in fact a generative cell, also divides. It is the inner generative cell that carries on to the next generation. This 'cell-in-cell state' is typical of Myxosporea and regularly occurs during the life cycle (Lom, 1987).

The first division of the amoebula also exemplifies the separation of the somatic component (= primary cell which later becomes the plasmodium cell, spore shell and capsulogenic cells) and the generative component (=inner and later generative or sporogonic cell and the sporoplasm).

Extrasporogonic developmental cycles

They are quite common in Myxosporea and they produce large numbers of parasite stages which spread the infection throughout the fish and for the ensuing sporogonic cycle. This results in very heavy infections, which could otherwise be achieved by a massive ingestion of spores or by autoinfection.

(×2100). F. *Chloromyxum cristatum*, in apical view (×2600). G. *Myxobolus bramae* from the gills of *Abramis brama*, an India ink preparation revealing mucus coat on the posterior spore half (×1750). H. *Henneguya psorospermica* from the gills of *Perca fluviatilis* (×900). I. *Myxobolus aureatus* from the gills of *Hybognathus nuchalis* (×2400). J. plasmodia of *Chloromyxum thymalli* from the gall bladder of *Thymallus thymallus*; some of them contain mature spores (×820). A, E and F from Lom and Dyková, 1992, with permission of the publisher; all others originals.

Autoinfection is hatching of spores *in situ* or within the same host. This was claimed by many authors (e.g. Amandi *et al.*, 1985) but it has not been properly documented.

Extrasporogonic cycles have been studied mostly in *Sphaerospora*. There are two such cycles in *S. renicola*, in which sporogony takes place in the renal tubules of the common carp. One cycle takes place in the bloodstream (Csaba, 1976; Lom *et al.*, 1983b) and its stages are quite conspicuous in fresh mounts. The stages have a characteristic constant twitching motion. Up to eight cells may originate (see Fig. 3.13c below) within a primary cell, each of which has an inner (tertiary) cell of its own. Eventually, the primary cell disintegrates to release these cell doublets which start the cycle all over again. In other *Sphaerospora* species the stages may have a similar appearance (Fig. 3.13A,B, D below) or the primary cells may contain several decades of cell doublets (Baska and Molnar, 1989). A second cycle takes place intercellularly in tissues. This is primarily in the swimbladder wall and it causes inflammation (see below). These primary cells are immobile and may be very large. They produce up to 50 inner cell triplets. A triplet is one cell with two inner cells (Dyková *et al.*, 1990; Fig. 3.13G below). In *S. renicola* infections there is a third proliferative cycle in the epithelial cells of renal tubules. The primary cells may contain secondary, tertiary and quaternary cells (Lom and Dyková, 1985). There have been claims that these stages do not belong to *S. renicola* but are the sporogonic plasmodia of *Hoferellus cyprini*, another tubule-infecting myxosporean (Molnar and Kovacs-Gayer, 1986a).

Extrasporogonic cycles may be morphologically different in other species. In *S. elegans* they are in the rete mirabile of the eye of its stickleback host (Lom *et al.*, 1991b) In *Myxidium lieberkuehni* they are intracellular in glomerular cells in pike (*Esox lucius*) kidney where they elicit hypertrophic growth. The parasite turns the host cell into a large (up to 1.5 mm) xenoma similar to those produced by microsporidia (Weissenberg, 1921; Lom *et al.*, 1989a).

Sporogonic phase

The 'vegetative stages' or trophozoites of myxosporeans are the plasmodia of the sporogonic phase (Figs 3.2J; 3.3A,B,F; and see 3.12 below). They are primary cells and contain one to few or very numerous vegetative nuclei and enclose many secondary generative cells which generate spores. Vegetative nuclei are sometimes distinguished from the nuclei of the generative cells by their different size. According to Uspenskaya (1984) vegetative nuclei are tetraploid and generative nuclei are diploid.

There are three types of sporogonic trophozoites. The first is represented by small, monosporic or disporic trophozoites. They produce one or two spores and have only one vegetative nucleus and are thus called pseudoplasmodia. Spores originate by proliferation of generative cells which aggregate to produce sporoblasts. The early pseudoplasmodia, prior to sporogenesis, may divide by plasmotomy. Pseudoplasmodia are typical of coelozoic myxosporeans, which inhabit organ cavities, e.g. *Sphaerospora*. The second are large (up to several mm) polysporic plasmodia (Fig. 3.13H below), with many vegetative nuclei and generative cells (Grassé and Lavette, 1978). They produce many spores. A third category is intermediate between this and the preceding one and is found in some coelozoic myxosporeans. These mictosporic plasmodia produce one, several or many spores.

Coelozoic plasmodia may divide by plasmotomy and/or by budding. Buds with vegetative nuclei and a few generative cells may be formed on the surface (exogenously) or inside of the plasmodium (endogenously). The cell membrane is often raised in villosities to increase the surface area for nutrient absorption. Often there are cytoplasmic holdfast outgrowths for adhesion to the walls in organ cavities. Some plasmodia (e.g. in *Chloromyxum*) are mobile; they have pseudopodia. Sporogony is not synchronized and all stages may be found at the same time. Histozoic plasmodia are firmly wedged between cells in tissue, in a blood vessel or within a myocyte. They do not divide but often grow to enormous size. Their cell membrane forms numerous tiny invaginations or pinocytotic vesicles for endocytosis of nutrients from the host–parasite interface./Grown plasmodia are encased with a host fibroblast envelope and appear as large, often macroscopic structures usually referred to as myxosporidian cysts. In heavy infections the large number of cysts is the result of extrasporogonic proliferation rather than due to ingestion of a large number of spores. Sporogony is synchronized and spores mature at the same time. The plasmodium becomes an envelope which is packed full with spores.

Sporogenesis

This may proceed either within (Current and Janovy, 1977; Current, 1979; Current *et al.*, 1979) or without pansporoblasts (Weidner and Overstreet, 1979; Lom *et al.*, 1983a). Pansporoblasts are formed by the union of two generative cells (Lom and de Puytorac, 1965; Desser and Paterson, 1978). One is the pericyte and it envelops the other, which is the sporogonic cell. Prior to their union, the two cells are difficult to distinguish. The pericyte forms later than the pansporoblast envelope, and within it the sporogonic cell divides to produce one, but usually two spores. Valvogenic cells differentiate into the shell valves while capsulogenic cells produce the polar capsules and the sporoplasm cell. In the pseudoplasmodia of some genera (e.g. *Sphaerospora*) sporogony begins with the production of a number of cells which organize to form one or two spores. In species with large plasmodial trophozoites (*Kudoa*) spores are produced without sporoblast formation (Stehr and Whitaker, 1986; Lom and Dyková, 1988).

During sporogenesis, valvogenic cells spread themselves thinly around the sporoplasmic and capsulogenic cells. Their thick brims join at cell junctions and this is reminiscent of metazoan desmosomes. The cell membrane is thickened from beneath with nonkeratinous proteins and the cytoplasm shrinks into a dense mass. The nucleus becomes Feulgen-positive and later disappears completely.

Polar capsules in capsulogenic cells develop from a club-shaped structure which is externally lined with microtubules. It grows into a bulbous capsular

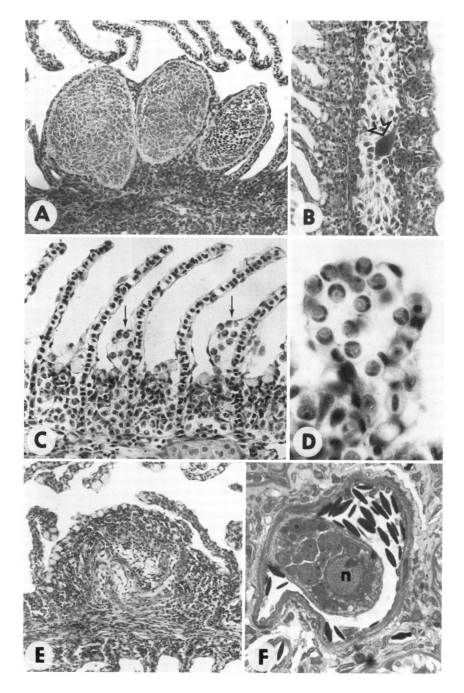

Fig. 3.3. A. Plasmodia of *Henneguya psorospermica* in the secondary gill lamellae of *Perca fluviatilis* (×200). B and F. *Thelohanellus pyriformis*, plasmodium within the axial blood vessel of a gill filament. B. Plasmodium with mature spores and a hypertrophic endothelial cell (hollow arrow) invaginated into it (×220). F. Young plasmodium next to host's erythrocytes; n = nucleus of the

primordium and passes into a narrow, long external tube. Later, the polar filament appears within this tube and seems to grow gradually into the primordium as the filament coil is laid down beneath the walls. Simultaneously, the external tube shortens and eventually disappears completely. Thus the filament is a transformed, inverted external tube. The primordium gradually assumes the form of the mature polar capsule (Lom and de Puytorac, 1965). The remainder of the capsulogenic cell nucleus can often be seen close to the bottom of the polar capsule.

Sexuality

The sexual process is autogamy. The nuclei of the sporoplasm fuse before or after hatching (this is the only uninucleate stage in the life cycle). The division products of the zygote nucleus differentiate later into vegetative nuclei and nuclei of the generative cells. According to Uspenskaya (1982, 1984) the vegetative nuclei are polyploid while nuclei of generative cells are diploid. The last cell divisions in the sporoblast are meiotic and sporoblast cells are haploid. Fusion of sporoplasm nuclei restores the diploid state.

Transmission

Myxosporean spores can retain their normal appearance in cold water for more than a year. The spores of *Myxobolus cerebralis* were viable and infective after 22 months of storage. This was verified in experimental infections (Hoffman *et al.*, 1962). Until 1984, a direct peroral transmission with spores was presumed to occur. The failures in most experimental infections using fresh spores was explained by a need for the spores to 'ripen' for several months in mud or water.

Wolf and Markiw (1984) showed that spores of *Myxobolus cerebralis* did not produce infection when ingested by salmonids. A tubificid worm (*Tubifex tubifex*) was needed as an intermediate host. In the worm the ingested spores developed to a stage which is morphologically similar to a species of the genus *Triactinomyxon*, a representative of the class Actinosporea. The *Myxobolus* stages in the oligochaete's intestine produced spores which, when ingested by the salmonid host, initiated the *M. cerebralis* infection. Such a life cycle is unparalleled among the protists, i.e. it has two types of asexual proliferation, two different sexual processes, two sporogonies which result in two completely

hypertrophic endothelial cell (×790). C and D. *Sphaerospora molnari* in the gills of *Cyprinus carpio* fingerlings; C, developmental stages in the epithelium at the base of the secondary lamellae, spores (arrows) in the lamellae, (×270). D, spores with two sporoplasms in the lamellae (×750). E. Tissue repair of a *Henneguya* lesion such as can be seen in A. (×220). B, F from Dyková and Lom, 1987, with permission of the publisher; all others are originals.

different types of infectious spores. The actinosporean theory has been further confirmed by the findings of Markiw (1989, 1991) and El-Matbouli and Hoffman (1989). The latter authors found this transformation in *Myxobolus cotti*, too. About twelve other recent findings also support the transformation theory. To quote just a few; El-Matbouli and Hoffman (1990) reported that *Myxobolus pavlovskii* formed *Hexactinomyxon* spores in *Tubifex*, Bartholomew *et al.* (1992) found some indication that *Ceratomyxa shasta* is transmitted via *Aurantiactinomyxon* spores formed in *Nais bretscheri*; Yokoyama *et al.* (1993) postulated that *Hoferellus carassii* is transmitted by *Neoactinomyxon* spores produced in *Branchiura sowerbyi*; Kent *et al.* (1993) suggested that *Myxobolus arcticus* transforms into a *Triactinomyxon* in *Stylodrilus heringianus* and Benajiba and Marques (1993) found that the source of *Myxidium giardi* infection is initiated by *Aurantiactinomyxon* spores formed in *Tubifex tubifex*. In addition, Markiw (1989) and Yokoyama (1993) demonstrated that infection in fish is initiated by penetration of the skin or gill epithelium by sporoplasms released from actinosporean spores that came into contact with the fish surface. Stages following this penetration have yet to be elucidated. Contradictory findings on direct transmission of *M. cerebralis* (Hoffman *et al.*, 1962; Uspenskaya 1984) and other species, e.g. *Sphaerospora* (Odening *et al.*, 1989) suggest that perhaps transmission via actinosporeans may alternate with direct transmission.

\PATHOGENICITY AND IMMUNE REACTIONS

Myxosporea are known from practically all organs and tissues in fish. However, only relatively few species are known to cause serious or fatal infections.

During their evolution the myxosporeans and their fish hosts have in most cases struck a balanced host–parasite equilibrium. /

Myxosporean spores have been shown to have no antigenicity to host fish (Pauley, 1974; McArthur and Sengupta, 1982) while a low degree of humoral response was reported by Griffin and Davis (1978). In heavy *Ceratomyxa shasta* infections in rainbow trout presenting various parasite stages, no humoral immune response was elicited (Bartholomew *et al.*, 1989b) and no such response was described in the proliferative kidney disease of salmonids. Furuta *et al.* (1993) observed, however, that developing stages of *Myxobolus artus* initiated antibody formation in the common carp. The spores have probably lost the antigenicity as evidenced by the absence of carp humoral response after experimental injection of spores.

The responses that are commonly observed are cell and tissue reactions, which are, however, usually not aimed at destruction of the parasite during its early development. Early histozoic plasmodia only elicit formation of a host cell envelope. This envelope is derived from the local population of connective tissue cells or from compressed cells of the neighbouring tissue. A granulomatous inflammation sets in only after the plasmodium is replete with mature spores. The host's granulation tissue replaces the lesion.

However, if the parasite develops in an atypical site, it readily provokes a vigorous tissue response. For example, if *Myxidium rhodei* plasmodia in *Rutilus rutilus* develop in kidney interstitium rather than in renal corpuscles, they are soon destroyed (Dyková *et al.*, 1987). If a myxosporean succeeds in infecting an atypical host, it may be able to multiply but not to complete sporogony because of the strong tissue reaction. This is the case of with PKX, the agent of salmonid proliferative kidney disease. The strong host tissue reaction is manifested by massive interstitial hyperplasia which is eventually deleterious for the host fish.

The extent of the damage to the tissue and organs depends on numerous factors, including the species of myxosporean and its life cycle stage, the intensity of infection and the host reaction. The broad range of lesions includes inocuous ones as well as lethal organ alterations. Histozoic plasmodia not only cause pressure atrophy of infected tissues but also alterations of neighbouring tissues. If the parasite is located in spinal cord nerves it will also elicit deformities of the spine.

The small plasmodia of *M. cerebralis* erode the trout cartilage and phagocytize the chondrocytes (Uspenskaya, 1984). In myocytes infected with some *Kudoa* species, the muscle bundles are completely replaced by parasite stages.

Small histozoic pseudoplasmodia pervading the tissue may elicit hyperplastic growth of the gill epithelium (e.g. *Sphaerospora molnari*) which render them non-functional (Dyková and Lom, 1988b) or cause hypertrophy of head kidney (*S. tincae*, Hermans and Körting, 1985) or of trunk kidney (PKX organism) (Kent and Hedrick, 1985). They can also simply completely destroy the tissue (*Ceratomyxa shasta* in salmonids).

Coelozoic plasmodia in the gall bladder may induce inflammation which results in jaundice (*Chloromyxum* or *Myxidium*) (Shulman, 1966) while plasmodia of *C. cristatum* may pervade the liver parenchyma in common carp and produce necrosis.

Muscle lesions due to several species of *Kudoa, Unicapsula*, and *Hexacapsula* result in dramatic changes in the muscles of heavily infected fishes. The muscle fibres are pervaded by small plasmodia or they are turned into pseudocysts replete with plasmodia. The muscles mottled with parasite foci become less elastic than usual. After death of the infected fish, the flesh rapidly softens or changes into a viscous mass or into a 'milky', jelly-like odourless ooze. Also, when heavily infected fish are frozen fresh and defrosted, the flesh is already jellified. This is probably because of proteolytic enzymes secreted by the myxosporeans, which, in living hosts, are continuously flushed by the bloodstream (Willis, 1949). These enzymes may also be confined within the pseudocysts (Patashnik *et al.*, 1982) and thus affect only the infected muscle fibre. After death of the host, they diffuse outwards and cause the autolysis of surrounding host tissue. Myxosporean infections may cause all types of regressive and progressive pathological changes. Also circulatory changes such as hyperaemia, oedema and haemorrhage have been recorded. Regressive changes include dystrophy which is common in the epithelial cell lining of urinary tracts infected with *Sphaerospora* (Dyková and Lom, 1982) or *Hoferellus* species. Atrophy is associated with development of large or abundant coelozoic or histozoic plasmodia. Necrosis of tissues is commonly encountered and is the result of intracellular localization of developmental stages.

Hypertrophy and hyperplasia are the most common progressive changes. Hypertrophy affects whole organs (kidney enlarged due to PKD, Kent and Hedrick, 1985) or their components (renal corpuscles infected with *Sphaerospora*). In myxosporean infections, hyperplasia of the epithelial lining is a common phenomenon. Epithelia are the only tissues which regenerate after a myxosporean infection.

The proliferative type of inflammation is the principal defence mechanism against myxosporean infections, and is initiated when plasmodia contain mature spores. Granulation tissue participates in reparation of lesions (Fig. 3.3E) left over following the destruction of large plasmodia or after an extensive necrosis. Encapsulation of plasmodia or of spore aggregates is also the result of proliferative inflammation.

Phagocytosis is pivotal in control of myxosporean infections. Melano-macrophages ingest mature spores released into tissue spaces and transport them to melanomacrophage centres in the kidney, spleen or liver, where they are encapsulated and gradually destroyed. Crumpled shell remains are the last to be digested in macrophage cytoplasm (Dyková, 1984).

A rather rare tissue reaction is melanization. Large plasmodia of some species of *Myxobolus* appear black due to melanin deposited extracellularly or based in melanocytes (Fig. 3.4B). Melanocytes may also be deposited around the muscle fibres filled by *Kudoa* stages. This gives a brown or black appearance (Kabata and Whitaker, 1989).

An especially serious pathogenic effect is exerted by extrasporogonic developmental stages of the myxosporeans as exemplified by *Sphaerospora renicola* and *Myxidium lieberkuehni* (see below).

MORPHOLOGICAL CHARACTERISTICS OF MYXOSPOREANS THAT CAUSE DISEASE

The full diversity of myxosporean genera and species are intriguing; however, they are not dealt with in detail in this review. We will only pay attention to species that are known to cause disease in their hosts. The pathogenic species belong to 15 genera and their taxonomic assignment and brief description are as follows:

Order Bivalvulida Shulman, 1959 (two shell valves)
  Suborder Variisporina Lom and Noble, 1984 (one to four polar capsules whose position in the spore is variable; if they are together at one pole of the spore, they are not located at the level of the suture of the two valves)
    Family Myxidiidae Thélohan, 1892 (spores are elongated, each with two polar capsules, with one at each end)
      Genus *Myxidium* Bütschli, 1882 (spores are spindle-shaped or fusiform); Genus *Zschokkella* Auerbach, 1910 (spores are ellipsoidal)
    Family Ortholineidae Lom and Noble, 1984 (two polar capsules at the anterior end lie in the sutural plane but are widely apart)

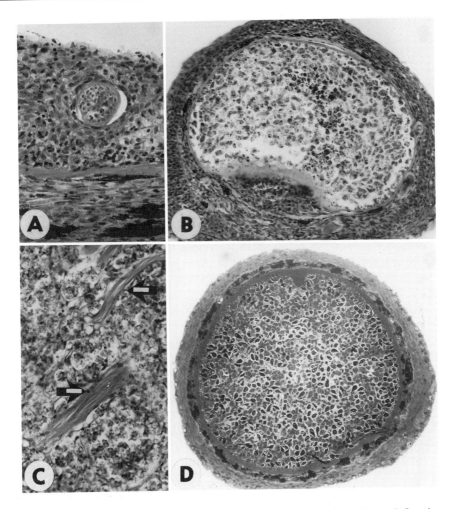

Fig. 3.4. A, B. *Thelohanellus nikolskyi* in the epidermis of the fins of *Cyprinus carpio* fingerlings. A. Small plasmodium in the hyperplastic epidermal layer (×300). B. Large plasmodium with several melanocytes on its periphery (×150). C. *Myxobolus lintoni* infection in *Cyprinodon variegatus*; the remnants of muscle tissue in the mass of plasmodia and spores are indicated by arrows (×200). D. *Myxobolus aureatus*, plasmodium in the skin of *Pimephales promelas*; in the enclosing host tissue melanophores and melanin containing macrophages are accumulated (×150). B from Dyková and Lom, 1988b, with permission of the publisher; all others are originals.

Genus *Triangula* Chen and Hsieh, 1984 (spores are rounded triangular and wide anteriorly)

Family Ceratomyxidae Doflein, 1899 (spores with extremely elongated valves which are perpendicular to the suture line; two polar capsules)

Genus *Ceratomyxa* Thélohan, 1892 (length of shell valves exceeds the axial diameter of the spore)

Family Sphaerosporidae Davis, 1917 (the two polar capsules at the anterior tip of the spore are in a plane perpendicular to the straight sutural line)

Genus *Sphaerospora* Thélohan, 1892 (spores are spherical or sub-spherical). Genus *Hoferellus* Berg, 1898 (syn. *Mitraspora* Fujita, 1912; spores with many filaments at the posterior end)

Family Chloromyxidae Thélohan, 1892 (four polar capsules in the apex of the spores)

Genus *Chloromyxum* Mingazzini, 1890 (spores are rounded or sub-spherical)

Family Parvicapsulidae Shulman, 1953 (asymmetrical spores with grossly unequal valves and with a curved suture, with two or four relatively small polar capsules in the apex)

Genus *Parvicapsula* Shulman, 1953 (the two polar capsules are in the spore apex)

Suborder Platysporina Kudo, 1919 (the polar capsules, usually two, are at the apex of the spore and in the sutural plane)

Family Myxobolidae Thélohan, 1892 (spores are flattened parallel to the straight sutural line)

Genus *Myxobolus* Bütschli, 1882 (two pyriform polar capsules, spores ellipsoidal or rounded). Genus *Henneguya* Thélohan, 1892 (two polar capsules, spores with caudal extensions of the shell valves). Genus *Thelohanellus* Kudo, 1933 (spores are pyriform with just one polar capsule in the apex)

Order Multivalvulida Shulman, 1959 (radially symmetrical spores with shell composed of three to seven valves; polar capsules are in the apex of the spore)

Family Trilosporidae Shulman, 1956 (spores with three shell valves)

Genus *Unicapsula* Davis, 1924 (spores with three unequal shell valves, a single spherical polar capsule)

Family Kudoidae Meglitsch, 1960 (spores with four shell valves, four polar capsules)

Genus *Kudoa* Meglitsch, 1947 (characters of the family)

Family Hexacapsulidae Shulman, 1959 (spores with six shell valves, six polar capsules)

Genus *Hexacapsula* Arai and Matsumoto, 1953 (spores with characters of the family)

Family Septemcapsulidae Hsieh and Chen, 1984 (seven spores with shell valves, seven polar capsules)

Genus *Septemcapsula* Hsieh and Chen, 1984 (characters of the family)

DISEASE-CAUSING MYXOSPOREANS

Gill infections

There are basically two types of gill infections. The tissue may be pervaded with small pseudoplasmodia (genus *Sphaerospora*) formerly designated as dif-fuse infiltration, or there may be small to large (up to several mm) plasmodia.

The large plasmodia appear as macroscopic whitish nodules located in the gill filaments. Growth impairment and/or mortalities are due to reduced respiratory capacity of the infected gills.

Sphaerospora molnari Lom, Dyková, Pavlásková and Grupcheva, 1983 (Figs 3.2C, 3.3C,D) infects the gills of the common carp, especially in fingerlings, and *Carassius carassius* in Europe. In some regions, about 20% of carp fingerling stocks may be affected. The parasite forms subspherical spores which average 10.5 µm in length. Monosporic pseudoplasmodia up to 17 µm in size may also occur free in the circulating blood. *S. molnari* infects the epithelium of gill filaments, the skin of the head and the lining of the gill cavity. It elicits epithelial hyperplasia. As the infection progresses, masses of parasites replace tissues and cause local circulatory disorders and dystrophic changes. Necrotic destruction of tissues follows the release of spores into the water. Massive infections cause heavy mortalities (Svobodová and Groch, 1986).

One of the major diseases of farmed alevins of channel catfish *Ictalurus punctatus* in southern USA is 'hamburger' (or proliferative) gill disease. It is caused by the extrasporogonic stages of *S. ictaluri* Hedrick, McDowell and Groff, 1990, which pervade gill tissue (Groff *et al.*, 1989; Hedrick *et al.*, 1990). The final phase of the infection (sporogony) is in renal tubules. Monosporic plasmodia produce mitre-shaped spores which are approximately 6 µm in length. The parasite causes a strong, diffuse perichondrial granulomatous inflammation. Multifocal degeneration and necrosis of gill ray cartilages is associated with hyperplasia of the gill epithelium (Duhamel *et al.*, 1986). There is a high prevalence of mortalities.

Species forming large plasmodia ('cysts') are very common. The damage to the host depends primarily on the number of plasmodia. In heavy infections most, if not all, secondary gill lamellae harbour a plasmodium. These lesions, together with the associated epithelial hyperplasia, render the gills non-functional. This is especially critical in young perch (*Perca fluviatilis*) in Europe which are infected with *Henneguya psorospermica* Thélohan, 1895 (Figs 3.2H, 3.3A). Infected gill lamellae may fuse together and this reduces the surface area for respiration to a minimum. Infected perch may die rapidly when the amount of dissolved oxygen in the water is lowered (Dyková and Lom, 1978).

H. exilis Kudo, 1929 is another serious pathogen of pond-reared channel catfish. It forms plasmodia up to 2 mm in diameter. The spindle-shaped spores are 65 µm long and this includes the tail appendage. If the plasmodia develop at the base of secondary lamellae, they produce the highly pathogenic 'interlamellar' disease. It is associated with a severe interlamellar tissue proliferation and this inflicts significant mortality of fingerlings and sometimes market-size fish. Infected fish show respiratory distress and they are also sensitive to stress. Plasmodia developing in other sites are benign (Current and Janovy, 1978).

Myxobolus muelleri Bütschli, 1882 is a widely distributed species. It forms large plasmodia (up to 4 mm) in the gills (and sometimes in other organs) of several tens of Eurasian fish species, especially cyprinids, without inflicting mortality. *M. exiguus* Thélohan, 1895 infects gills and other organs of cyprinids as well as other fishes in Eurasia. It caused massive mortality in *Mugil* in the Black and Azov seas (Shulman, 1966). *M. basilamellaris* Lom and Molnar, 1983 inflicts damage on the gill filaments of common carp in Europe by forming plasmodia close to the cartilaginous axis.

Thelohanellus pyriformis (Thélohan, 1892) (Figs 3.1F, 3.3B,F) is common in the freshwater *Tinca tinca* and several other hosts. Plasmodia grow in the blood vessels of gill filaments and elicit huge hypertrophy of single endothelial cells (Dyková and Lom, 1988b). *T. pyriformis* was assumed to form large plasmodia (3 cm and more) in subcutaneous tissue or muscles of various cyprinids in Europe and southeast Asia. Fatal epizootics were observed in the former USSR (Shulman 1966) and Indonesia.

Infections of the skin and subcutaneous tissue

Cyst-like plasmodia in the skin are easily seen. Species of *Henneguya* and *Myxobolus* are most common and are also found in other organs and tissues. *Myxobolus lintoni* Guerley, 1983 infects *Cyprinodon variegatus* along the Atlantic coast of the United States. It forms plasmodia about 0.12 mm in size, which aggregate to form large tumour-like masses (Fig. 3.4C) in the subcutaneous tissue and sometimes in the muscles.

M. aureatus Ward, 1919 (Fig. 3.4D) develops on the surface tissue of North American minnows. It forms spherical plasmodia (1.6 mm in diameter) in the skin of *Notropis argenteus* as golden cysts. In *Pimephales promelas* they are located in the fins and are black. In both cases, the colour is due to pigments deposited into the tissue encasing the plasmodium.

M. dermatobius (Ishii, 1915) forms small plasmodia in the subcutaneous tissue of eels, *Anguilla*, in Japan and Europe. Another species, *Thelohanellus nikolskii* Akhmerov, 1960 (syn. *T. cyprini* Hoshina and Hosoda, 1957) is a pathogen (Molnar, 1982) on the skin of cultured common carp fingerlings. It was introduced to Central Europe via the former USSR from the Far Eastern Amur region. Outbreaks in Europe are seasonal, in July it forms numerous, large, rounded, cyst-like plasmodia (about 2 mm), which resemble a bunch of grapes on the fins. They develop in close association with and erode the cartilaginous fin rays (Fig. 3.4A,B). Spores are large, 17 µm long and polar filaments are arranged in a double coil within the polar capsule.

Muscle infections

Some of the Myxosporea developing within the myocytes (or 'in the muscles') may cause disease and the macroscopic cysts render the fish unsightly and/or unmarketable.

Myxobolus cyprini Doflein, 1898 (syn. *M. pseudodispar* Gorbunova, 1936) (Fig. 3.5C) is widely distributed in Eurasian (and perhaps in American) cyprinid fishes. The spindle-shaped plasmodia, up to 300 µm long, within the myocytes produce spores that vary greatly in shape and size. Some spores

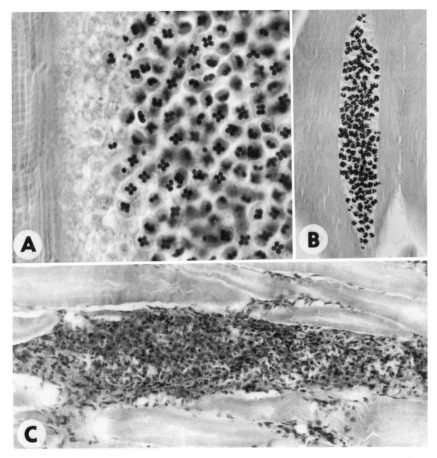

Fig. 3.5. Muscle-infecting Myxosporea. A. Lesion due to a *Kudoa* sp. within the muscle fibre of *Sebastes paucispinus*; polar capsules are densely stained with Giemsa's (×1100). B. A fusiform plasmodium of *Kudoa* sp. within the muscle of *Fundulus grandis* (×350). C. Spores of *Myxobolus cyprini* accumulated in the muscle tissue of *Cyprinus carpio* (×206). A, B are original, C is from Dyková and Lom, 1988b, with permission of the publisher.

often have at the posterior end a semicircular ledge or mucous cover. The two polar capsules are unequal in size, with loose oblique polar filament turns and may be situated apically, subapically or even laterally (Dyková and Lom, 1988b). Spores which eventually destroy the myocytes reach other organs via blood circulation. They accumulate in the kidney, spleen and liver melano-macrophage centres where they are gradually destroyed. Massive numbers of spores in the kidney were formerly incorrectly associated with 'pernicious anaemia of carp' (Shulman, 1966). Inflammatory tissue reaction towards the myocyte-based parasites is rare and pathogenicity may not be serious (Molnar and Kovacs-Gayer, 1985).

In the east and southeast Asian countries the muscles of common carp

may be infected with *M. artus* Akhmerov, 1960. The spores are wider
(10.5 μm) than long (7.5 μm). This species is quite common and often occurs
in heavy infections. It is a potential pathogen.

Salmonid fishes (*Oncorhynchus tschawytscha, O. kisutch, O. mykiss, O.
m. irideus, Salmo clarki*) in northwest America are very often infected with
M. insidiosus Wyatt and Pratt, 1963. The fusiform plasmodia are 140 μm long
and are between muscle bundles and/or within myocytes. They develop in
smolts within one month after exposure to water which contains spores.
Spores are released from dead fish after spawning in the autumn. The infection survives the period the salmon spend in the ocean (Amandi *et al.*, 1985).

Henneguya zschokkei Gurley, 1894 (the most important synonym of which
is *H. salminicola* Ward, 1919) is a common parasite in salmonids (genera
Coregonus, Salmo, Oncorhynchus). The parasite has large encapsulated plasmodia, or 'cysts' (2 mm to 2 cm in size) located in the myosepta. In American
salmonids, they are located mostly in the posterior part of the body. In some
coregonids the plasmodia occur in the connective tissue of the ventral cavity.
Spores are oval in front view, are approximately 11 × 9 μm, and have a long
tail. Salmon, infected as juveniles, retain the infection over their marine period
(Boyce *et al.*, 1985). Heavy infections with many cysts in the muscles render
the fish unmarketable. Since a milky fluid (spore suspension) oozes from cysts
cut during filleting, the infected fish are termed to be in 'milky condition'
(Awakura and Kimura, 1977). Numerous species (order Multivalvulidea)
infect muscles of marine fishes and elicit a variety of pathological changes,
e.g. in *Molva dipterygia* from the North Atlantic infected with *Trilospora
muscularis* Priebe, 1987 and in *Hippoglossus stenolepis* from the northeast
Pacific infected with *Unicapsula muscularis*. In the latter, the infected muscle
bundles are whitish and worm-like and hence the name 'wormy halibut'.
Trophozoites of *Kudoa clupeidae* (Hahn, 1917), which infect *Clupea harengus*
and other fish, form macroscopic spindle-shaped masses (0.5 cm long). Since
the parasite persisted for at least 2 years in the fish, they were used as biological tags (Sindermann, 1961). Differences in prevalence had been used to
separate different fish stocks. Spores are round, with a pointed apical projection; they are up to 7.5 μm wide. Numerous species of *Kudoa* infect muscles
of various hosts (Fig. 3.5A,B); however, only a few species infect other sites
such as the pericardial cavity (Fig. 3.1B).

Cultured *Seriola quinqueradiata* in Japan are infected with *K. amamiensis*
Egusa and Nakajima, 1980. The source of infection is said to be coral fishes
which serve as food or as disseminators of spores. Plasmodia up to 5 mm in
size massively pervade the musculature which becomes unsightly and the fish
lose all commercial value.

Enzymatic degradation of skeletal muscles is caused by many species of
Myxosporea. In some species the flesh of the fish disintegrates during cooking
or baking, often turning into a foul-smelling paste. This is due to the proteolytic enzymes which are relatively heat resistant. The peak activity of the
enzyme is at 55–60°C and this is a transitional temperature in the cooking process (Tsuyuki *et al.*, 1982) . This is the case with *Unicapsula seriolae* Lester,
1982 which infect *Seriola holandi* along the Australian coast. The parasites

form white streaks (up to 7 mm long) in the muscles. The Pacific hake, *Merluccius productus* along the British Columbia coast is commonly infected with *Kudoa paniformis* Kabata and Whitaker, 1981. The rounded rectangular spores are 7 µm wide and infections are usually heavy. In addition to melanization of the infected muscle fibres, the flesh is turned into a mushy mass during cooking (Kabata and Whitaker, 1981, 1986; Patashnik *et al.*, 1982).

In most multivalvulid infections, the post-mortem myoliquefaction occurs without high temperature. *Kudoa thyrsites* (Gilchrist, 1924) with stellate, large spores (width 16 µm) was reported to cause flesh liquefaction in *Coryphaena hippurus* off the Australian coast (Langdon, 1991). It is also in cage-reared *Salmo salar* smolts in Washington State and British Columbia (Harrel and Scott, 1985). Heavily infected fish were anaemic and had enlarged kidneys. Death was followed by rapid muscle deterioration. The parasite is recognized as a potential market problem in economically important fishes.

Spectacular postmortal flesh jellyfication in *Xiphias gladius* is associated with *K. musculoliquefaciens* (Matsumoto and Arai, 1954) when great mass of muscle turns into mash. *Hexacapsula neothunni* Arai and Matsumoto, 1953, produces petal-shaped spores (21 µm wide) and induces a similar condition in infected *Neothunnus macropterus* from the seas off Japan. Within 24–70 hours after the death of the host, almost all the body muscles appear paste-like.

Infection of the cartilage

Few species infect cartilages (Guilford, 1963; Hoffman *et al.*, 1965); the group includes the well studied pathogen, *Myxobolus cerebralis* Hofer, 1903 (syn. *Myxosoma cerebralis*) which causes 'whirling disease' in salmonid fry. It is thought to have been originally a mild pathogen of *Salmo trutta* in central Europe and of other salmonids in northeast Asia. After the rainbow trout, an unnatural but very susceptible host, was introduced in the area it contracted the disease. The virulence of the parasite increased so that we now have 80–90% mortalities even in *S. trutta*. *M. cerebralis* was introduced with rainbow trout to North and South America, Australia and New Zealand. In introduced areas the parasite is only rarely found outside hatcheries. It is known to occur in eight species of *Salmo*, six species of *Oncorhynchus*, four species of *Salvelinus, Thymallus thymallus* and *Hucho hucho*. The parasite destroys the cartilage in the head and vertebral column, the head and gill arch cartilage being the preferred site. Plasmodia have been detected in fish larvae 13 days after hatching. The plasmodia are about 40 µm in size, with tufts of finely branched pseudopodia. They divide and produce large foci of infection in the cartilage. These plasmodia may be separated by compartments (Fig. 3.6B) while the rest of the cartilage is destroyed by enzymatic lysis. The oval to circular spores (Fig. 3.1E) average 8.6 × 8.4 µm and are extremely variable in size. The mucous envelopes cover their posterior half (Lom and Hoffman, 1971).

The rate of development is temperature dependent. It takes 3 months at

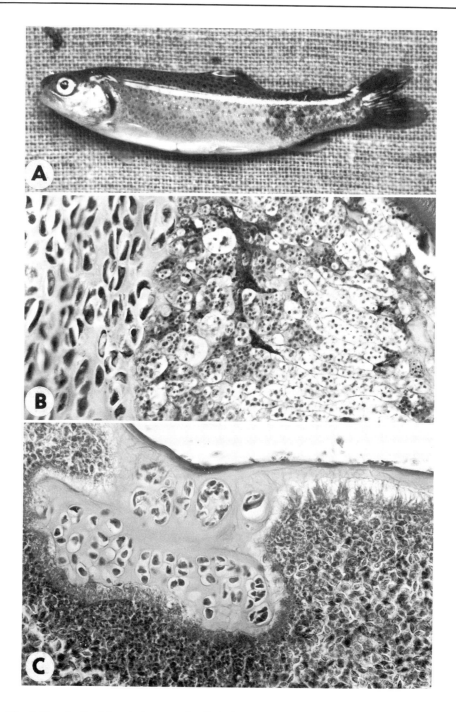

Fig. 3.6. A, B. *Myxobolus cerebralis* infection. A. An infected rainbow trout fingerling showing dark pigmentation of the posterior body part, natural size. B. A mass of young, small plasmodia eroding the cranial cartilage, the remains of which form the septae between the plasmodia (×450). C. *M. aeglefini*

12°C to produce spores and clinical signs appear 2 to 8 weeks after exposure. The first signs are dark pigmentation of the caudal peduncle and caudal fin (Fig. 3.6A) due to lesions in the cartilage of the vertebral column posterior to the 26th vertebra. The lesions exert pressure on the nerves which control the pigment cells. At 2–3 months after infection the abnormal tail-chasing behaviour (whirling) becomes noticable. The whirling is assumed to be caused because of the erosion of the cartilage around the auditory organ or pressure of the granulation tissue on the auditory capsule. Prolonged whirling may cause severe exhaustion, malnutrition and death. There is also deformation of the head and curvature of the spine. These clinical signs are more frequently seen at about 17°C than at other temperatures. Chronic infection may be manifested in the second year by cachexia, lethargy and head deformation.

In fish which are more than one year old the cartilage has fully ossified. Thus the infection in these fish develops only exceptionally in residual cartilage. In live fish, spores escape from destroyed cartilage into the intestine or directly to the outside. Survivors are thus the source of further infection. Spores are usually released from decaying dead fish.

Dyková (1989, unpublished) described the host reaction at the site of infection. There is hypertrophy of the cartilage and proliferation of connective tissue cells of the perichondrium and periost growing into the lesions. Inflammatory cells in the fragmented cartilage include mononuclear leucocytes and macrophages. Granulomas consist predominantly of epithelioid and mononuclear cells, fibroblasts and multinucleate giant cells.

Pathology of the disease and biology of its agent has been reviewed by Halliday (1976), Uspenskaya (1984) and Markiw (1986). Diagnosis by concentration methods and digestion is mentioned on p. 136.

Myxobolus aeglefini Auerbach, 1906 (Fig. 3.6C) is one of the few marine species. It infects the head cartilage of *Gadus morhua* and other gadid fishes as well as many other hosts from other taxonomic groups (e.g. flatfish), in the North Atlantic and the North and Baltic Seas. Spores are subspherical. Lesions in the cranial cartilage caused by the polysporic plasmodia are characterized by focal thickening with conspicuous difference in stainability. Later, focal erosions develop. Plasmodia have root-like pseudopodia, and cause pressure atrophy and lysis of the cartilage. *M. aeglefini* was used as a parasite tag for *Pleuronectes platessa* in the North Sea (van Banning *et al.*, 1978).

Infections of the central nervous system

Myxosporea may develop within the blood vessels in the brain and also within the axons (Ferguson *et al.*, 1985; Lom *et al.*, 1989b). Damage to the nervous

infection in *Cyclopterus lumpus*, a large plasmodium with thin pseudopodia inside the cranial cartilage (×360). A and C from Lom and Dyková, 1992, with permission of the publisher; B original.

tissue may not always result in disease. *Triangula percae* Langdon, 1987 was recorded in the brain of *Perca fluviatilis* in Victoria, Australia. It was in the white matter of the mesencephalon, diencephalon and medulla oblongata. Infected juveniles at 75 to 100 days of age developed S-shaped curvatures of the spine. The lower prevalence of deformities in older fish suggested considerable juvenile mortality (Langdon, 1987). Severe deformation of the vertebral column in juveniles of the same host in Scotland was caused by *Myxobolus sandrae* Reuss, 1906 (Fig. 3.7C), a species which was originally described from subcutaneous tissues (Lom *et al.*, 1991a).

Serious scoliosis of cultured marine *Seriola quinqueradiata* in Japan is due to *M. buri* Egusa, 1985. Trophozoites are up to 0.4 mm and are located in brain tissue and cerebral cavities (Egusa, 1985).

M. encephalicus (Mulsow, 1911) (Fig. 3.7A,B,E) is a widely distributed parasite of common carp fingerlings in Europe. The prevalence is up to 40% in some fish stocks. Elongated plasmodia in brain blood vessels may reach 120 µm in length and produce oval spores which may be variable in shape and size (7–12 × 6.3–10.5 µm). Plasmodia are coated with a layer of endothelial cells to prevent coagulation of blood. They impair blood circulation and cause oedema in their vicinity. Spores released from plasmodia may provoke heavy granulomatous inflammation in the brain tissue. Heavy infection in fry and fingerlings result in locomotory disorders, loss of equilibrium and circling motion. Emaciation and sunken eyes are associated clinical signs and heavy infections have poor prognosis.

M. cotti El-Matbouli and Hoffman, 1987 (Fig. 3.7D) develops in the brain tissue in the freshwater fish *Cottus gobio* in Europe. Smaller plasmodia are also found within nerve axons. There are several species of *Myxobolus* infecting the brain and spinal cord of salmonid fish; none of them is known to cause serious disease.

M. neurobius Schuberg and Schröder, 1905 is commonly found in Eurasian *Salmo trutta* m. *fario, Thymallus thymallus* and other salmonids. Plasmodia are up to 400 µm in size, and aggregations of spores cause atrophy of the white matter in the brain or spinal cord.

Cultured marine fishes, *Lateolabrax japonicus* and *Oplegnathus fasciatus*, in China and Japan are infected with *Septemcapsula yasunagai* Hsieh and Chen, 1984. Small plasmodia aggregate to form masses up to 1.2 cm in size in the brain. Spores are 12 µm wide. This parasite causes disease manifested by locomotory disorders and a swollen swimbladder.

Infections of the intestine

Numerous species of *Ceratomyxa* live as inocuous parasites in the gall and urinary bladder of marine fishes. *C. shasta* Noble, 1950 is an exception; it is a fresh water species and a pathogen. It causes serious losses in cultured and wild populations of anadromous salmonids on the west coast of North

America. *Oncorhynchus mykiss irideus* and *O. tschawytscha, O. keta* and *Salmo clarki* are highly susceptible to the parasite. Different strains of the same host species may vary in their susceptibility to the disease. Fish infected as fingerlings in fresh water keep the parasite as adults during their marine phase and may continue to die from ceratomyxosis. The surviving fish later spread the parasite to streams and waters. The disporic pseudoplasmodia are 19 μm in size and produce arched spores which average 7 × 15 μm.

The mode of infection is not clear. Feeding of mature spores to fish did not produce the infection, while fish exposed for as little as 15 to 120 minutes (Schafer, 1968) to contaminated bottom sediments or contaminated water became infected. The source of infection may actually be actinosporeans developing in oligochaetes (see p. 108).

The parasite first enters epithelium of the posterior intestine and elicits lymphocytic infiltration. At 12°C and approximately 30 days post-exposure it pervades all layers in the entire digestive tract and spreads to most other organs. The infection becomes systemic. On day 25 post exposure, the fish become anorexic, lethargic, have dark coloration, swollen abdomen with ascites and exophthalmia. Mature spores appear 30 to 35 days post exposure and mortalities in 38 to 42 days. Infected fish do not mount a detectable antibody response (Bartholomew *et al.*, 1989b). In the lesions, all the host tissues are destroyed and replaced with parasites. There is hyperplasia, leucocytic infiltration and necrosis. The infection is usually fatal to juvenile fish under culture conditions and severe losses occur when the temperature exceeds 10–12°C. Significant mortalities were also recorded in wild populations of salmonids, e.g. *O. tschawytscha* (Wales and Wolf, 1955; Zinn *et al.*, 1977; Ratliff, 1983; Bartholomew *et al.*, 1989a,b).

Relatively few myxosporeans are found only in the intestine. *Kudoa ciliatae* (Lom *et al.*, 1992) is found in *Sillago ciliata*, a marine fish off the eastern Australian coast. It develops in the lamina muscularis (Fig. 3.8A) of the intestine. *Zschokkella stettinensis* Wierzbicka, 1986 forms small plasmodia in the intestinal epithelium of eel, *Anguilla anguilla*. Most other myxosporeans also develop in other sites. These include several *Myxobolus* and *Henneguya* species and *Thelohanellus hovorkai* Akhmerov, 1960 (Fig. 3.8B) *T. hovorkai* forms plasmodia up to 3 mm in the intestine and other organs in common carp the (1+ and 2+ age class), in Europe and in the Far East. It is a potential pathogen (Molnar and Kovacs-Gayer, 1986c).

Infections of the kidney and urinary tract

This is the site of many coelozoic species. They infect the whole urinary system from the renal corpuscles to the urinary bladder. Histozoic species develop in the kidney interstitium. In addition, mature myxosporean spores released from plasmodia in other parts of the body eventually concentrate in the interstitium. They end up in melanomacrophages which are concentrated in melanomacrophage

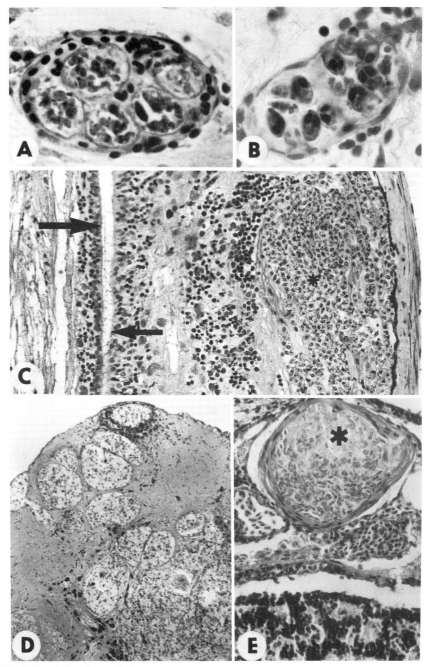

Fig. 3.7. Infections of the nervous tissue due to *Myxobolus* species. A, B, E. *M. encephalicus* infection in the brain of *Cyprinus carpio* fingerlings. A, B. Small plasmodia developing inside the blood vessels (×770 and ×900, respectively). E. A granuloma (asterisk) in the meningae (×280). C. A plasmodium of *M. sandrae* (asterisk) in the spinal cord of *Perca fluviatilis*; arrows delimit the central canal of the spinal cord (×250). D. A group of plasmodia of *M. cotti* in the brain of *Cottus gobio* (×80). All figures are originals.

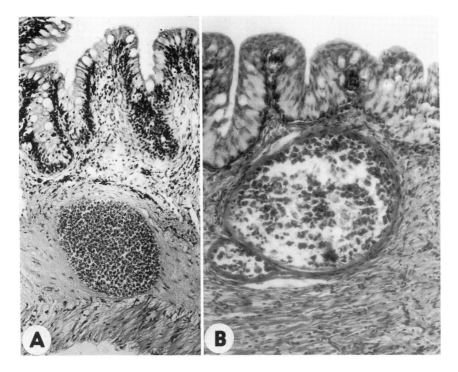

Fig. 3.8. A. Agglomeration of small plasmodia of *Kudoa ciliatae* in lamina muscularis of the intestine of *Silago ciliata* (×220). B. *Thelohanellus hovorkai* plasmodium in lamina muscularis of the oesophagus of *Cyprinus carpio* (×220). Both originals.

centres and encapsulated by fibroblasts where they are gradually destroyed (Dyková, 1984). Hence the spores of many species have been very often mistaken as kidney infection.

The species which lives in the urinary tract is *Myxidium lieberkuehni* Bütschli, 1882 (Fig. 3.9). It infects *Esox lucius* throughout its holarctic distribution and the prevalence is about 100%. Extrasporogonic stages are found in renal corpuscles (in about 20% of infected pike) transformed into whitish, cyst-like nodules up to 2 mm in diameter. They contain hypertrophic endothelial cells with huge hypertrophic nuclei and cytoplasm replete with parasites (Fig. 3.9C). The latter consist of a primary cell containing secondary and tertiary cells. Some of these stages may pass into the tubular lumen and transform into sporogonic plasmodia. Most nodules, however, are completely destroyed by the host tissue reaction (Lom *et al.*, 1989a). Other extrasporogonic stages are found intracellularly in hypertrophic epithelial cells of renal tubules. (Fig. 3.9E) and pass into the lumen to transform into sporogonic plasmodia. In the lumen the small plasmodia produce two or more spores. In the urinary bladder, polysporic plasmodia may reach up to 500 μm and their ends are

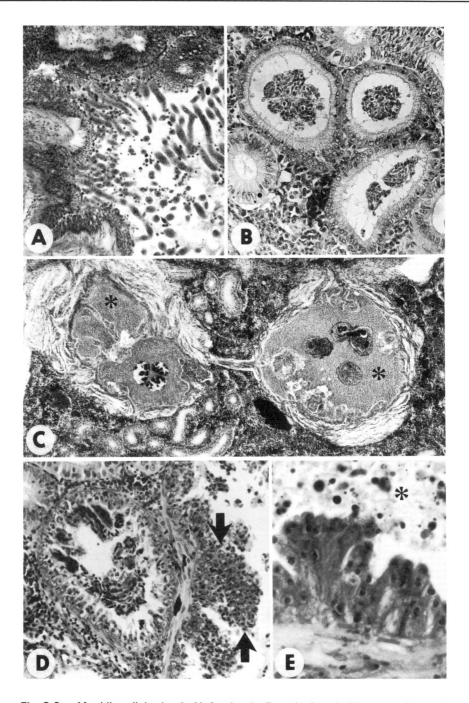

Fig. 3.9. *Myxidium lieberkuehni* infection in *Esox lucius*. A. Elongate plasmodia attached to the wall and free in the lumen of the urinary bladder (×85). B. Agglomerations of small plasmodia in the lumen of renal tubules (×240). C. Extensive alteration of the renal corpuscles; mass of extrasporogonic stages (asterisk) filling the hypertrophic endothelial cells of the renal corpuscles. At

wedged deep between epithelial cells (Fig. 3.9A). A bladder replete with plasmodia assumes a yellowish tinge. Spores have pointed ends and are fusiform in shape. They have longitudinal striation and average $20 \times 6 \,\mu m$. The epithelium of heavily infected tubules and collecting ducts shows dystrophic changes. Destruction of parasitic stages in renal corpuscles involves granulomatous inflammation. There are several other freshwater *Myxidium* species which cause disease in kidneys. *M. giardi* Cépede, 1906 (Fig. 3.1C) is a cosmopolitan species in eels, *Anguilla*. It is histozoic in the gills, skin, kidney and intestine and is often coelozoic in the gall bladder and urinary tract. The eels become infected after the 'glass eel' stage. Histopathological changes, usually a focal granulomatous inflammation, are caused by mature plasmodia or spores brought into organs by the blood. However, tubular and glomerular changes in the kidney are among the most serious. Elvers with tubular damage may develop dropsy and suffer massive mortality. Plasmodia of *M. rhodei* Léger, 1905 (Figs 3.10A,B) are commonly found in the renal corpuscles of cyprinid hosts in Eurasia and elicit hypertrophy of corpuscles. This is followed by atrophy of glomeruli and of the neighbouring renal parenchyma. If, however, the plasmodia develop in the renal interstitium, they provoke a inflammatory granulomatous reaction and are destroyed before spores (Fig. 3.2A) can reach maturity. *M. minteri* Yasutake and Wood, 1957 is common in salmonids *Oncorhynchus, Salvelinus* and *Prosopium* in the American Northwest. Trophozoites are up to $40 \,\mu m$ and cause renal tubule degeneration.

Another salmonid species, *M. truttae* Léger, 1930 (Fig. 3.10E) infects *Salmo salar, S. trutta, Oncorhynchus mykiss* and other salmonids throughout Eurasia. It causes serious inflammation of bile ducts and suppurative liver necrosis (Walliker, 1968).

An unidentified *Parvicapsula* caused epizootics in net pen-reared marine *Oncorhynchus kisutch* in Washington State, USA (Johnstone, 1985; Hoffman, 1981). The infected fish were lethargic and dark; mortalities reach 30% in some stocks. The trophozoites develop intracellularly in renal tubular epithelium and produce spores which are discharged via the urinary tract. The tubular epithelium is destroyed and the infection results in proliferative nephritis associated with renal hypertrophy.

their periphery, these cells are being attacked by host tissue reaction (×85). D, E. Intracellular development of *M. lieberkuehni* in the renal tubules accompanied with extensive hyperplasia of the tubular epithelium; D, full arrows indicate mass of developmental stages in the dystrophic epithelium, similar desquamated mass fills the lumen of the tubule (×260); E, polypoid type of epithelial hyperplasia, dystrophic developmental stages of *M. lieberkuehni* in the lumen (asterisk) (×580). All figures are originals.

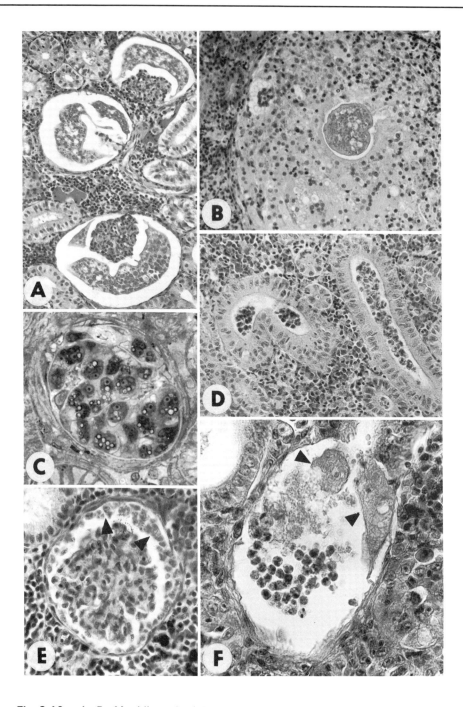

Fig. 3.10. A, B. *Myxidium rhodei* infection in the kidney of *Rutilus rutilus*. A. Coelozoic plasmodia within the renal corpuscles (×210). B. Histozoic plasmodium in the kidney interstitium, surrounded by inflammatory reaction (×280). C. *Sphaerospora cristata*, pseudoplasmodia developing in the renal corpuscle of *Lota lota* (×210). D, F. *S. renicola* infection in the kidney of *Cyprinus carpio*

Sphaerospora renicola Dyková and Lom, 1982 (Figs 3.10D,F, 3.13C,E) is widely distributed in intensive cultures of common carp throughout Europe and in Israel. In central Europe it is found in about 70% carp fingerling stocks; the prevalence may reach 100%. Disporic pseudoplasmodia (up to 20 μm) live in the lumen of renal tubules and produce subspherical spores (approximately 7.3 μm) in size. Direct transmission with spores or infected kidney tissue with or without spores was achieved (Odening *et al.*, 1989). In *S. renicola*, both sporogonic and extrasporogonic stages have serious pathogenic potential. Large masses of sporogonic stages in the renal tubules induce tubule dilatation (Fig. 3.10D,F), atrophy and result in necrosis of the epithelium. Dystrophic changes of tubule walls may be associated with histiocytary reaction in the interstitium with fibrosis affecting Bowman's capsules, and with formation of small granulomas. These changes may severely impair kidney function.

S. renicola has two cycles of extrasporogonic proliferation. The first takes place in the blood (Fig. 3.13C) (Csaba, 1976; Lom *et al.*, 1983b) and has already been mentioned. Kudryashova and Naumova (1978) postulated pathogenic effects of these bloodstream stages on the blood parameters. Some of the released parasites may pass between the epithelial cells of renal tubules into the tubular lumen and initiate sporogonic development (Molnar, 1988) or they may reach the swimbladder and cause swimbladder inflammation.

The abortive sequence, which takes place in the epithelial cells of renal tubules, degrades the cells into a huge syncytium with hypertrophic nuclei (Fig. 3.13E). Infected parts of the tubules appear as cyst-like nodules. Eventually, large numbers of extrasporogonic stages (up to 16 μm in size) degenerate. Some of the stages pass into the tubular lumen where they either gradually die or are discharged into the interstitium and undergo destruction through inflammatory granulomatous reaction (Grupcheva *et al.*, 1985). The extensive formation of granulomas with necrotic centres and massive layers of connective tissue may persist for several months in winter and impair excretory and haemopoietic function of the kidney.

A pathogen of European tench (*Tinca tinca*) is *Sphaerospora tincae* Plehn, 1925. It infects the pronephros and produces fatal epizootics in tench yearlings in France and Germany. The disporic pseudoplasmodia produce spores 8 μm long, with an anteriorly protruding sutural edge. Masses of parasites replace the parenchyma to provoke hypertrophy of the head kidney. This appears externally as conspicuous distension of the abdomen. Internal organs are damaged by the ensuing pressure; however, there is no obvious inflammation (Hermanns and Körting, 1985).

---

fingerlings. D. Pseudoplasmodia within the renal tubules (×220). F. Infected renal tubule with pronounced alteration of epithelium, its dystrophic remains are marked by the arrowheads (×460). E. Pseudoplasmodia of *S. truttae* (arrowheads) in the Bowman's space of the renal corpuscle of *Salmo trutta* (×330). All figures are originals.

S. truttae Fischer-Scherl, El-Matbouli and Hoffman, 1986 (Fig. 3.10E) infects renal tubules and Bowman's capsules of *Salmo trutta* m. *fario* in central Europe. Spores are broadly ellipsoid with fine oblique ridges. In heavy infections, renal corpuscles are hypertrophic and tubular epithelium undergoes atrophy. *S. cristata* Shulman, 1962 (Fig. 3.10C) infects renal tubules of *Lota lota* in Europe. Spores have marked ribs at the posterior end. This species also has bloodstream extrasporogonic stages. Sporogonic plasmodia may completely congest the tubules and inflict damage to the epithelium.

Hoferellus carassii Akhmerov, 1960 (Fig. 3.2E) (syn. *Mitraspora cyprini* Fujita, 1912) causes goldfish kidney enlargement disease in Asia, Europe and in North America where it is called 'kidney bloater'. The disease is quite common in goldfish farms and causes up to 20% mortality in Japan. The infection is seasonal, fish-of-the-year are infected in summer and spores mature in early spring when most of the heavily infected fish die. One of the clinical signs is abdominal swelling (often unilateral) due to enormous kidney hyperplasia (Ahmed, 1973, 1974; Hoffman, 1981). Extrasporogonic stages are intracellular and proliferate massively in epithelial cells of renal tubules. They cause papillary cystic hyperplasia and transform tubules into large cavernous lesions. Then the parasite stages (primary cells with secondary cells) pass into the lumen of the tubule and collecting duct where they grow into polysporic plasmodia (up to 130 μm) (Molnar *et al.*, 1989) and produce mitre-like spores (13 μm long).

Proliferative kidney disease (PKD) attracts much attention because of its high pathogenicity and also by its still mysterious identity. All that is known is that the parasite (PKX) is an extrasporogonic developmental stage of a myxosporean (Kent and Hedrick, 1985). It infects primarily kidney interstitium of fingerlings of *Oncorhynchus tschawytscha* and *O. kisutch* and steelhead in North America, of *Salmo salar*, *S. trutta* and *Thymallus thymallus* in Europe, and of cultured rainbow trout in both continents. The disease typically develops in spring (April to June) after the fingerling have come in contact with contaminated water. In rainbow trout the morbidity may be 100% while mortality is usually 30 to 50%. Mortalities seem to be temperature dependent, in susceptible salmon they are heaviest at 12–14°C (Ferguson and Needham, 1978; Clifton-Hadley *et al.*, 1984; Hedrick *et al.*, 1984; Smith *et al.*, 1984).

PKX resembles a proliferative extrasporogonic stage. It is a primary cell (up to 15 μm) with one or more secondary and tertiary cells (Figs 3.11C–G). The primary cell cytoplasm contains dense bodies which are similar to but different from sporoplasmosomes. PKX infiltrate kidney interstitium and develop both within and between host cells. They penetrate into the tubular lumen, and form sporoblasts which are reminiscent of those of *Sphaerospora* or *Parvicapsula* (Kent and Hedrick, 1986; Odening *et al.*, 1988a) which do not mature. This indicates that salmonids are not the normal hosts for the PKX agent (Kent and Hedrick, 1987). The possible sources of infection include stickleback (harbouring *Sphaerospora elegans*) and common carp (harbouring *Sphaerospora renicola*).

Gross clinical signs are bulging abdomen due to hypertrophic trunk kidney, exophthalmus and anaemia. PKX elicits an intense defence reaction of the host and this actually induces the disease. PKD is characterized by interstitial hyperplasia (Fig. 3.11A,B) associated with chronic, granulomatous interstitial nephritis and tubular atrophy (Clifton-Hadley *et al.*, 1987a; MacConnell *et al.*, 1989). There is an infiltration of macrophages and lympho-cytes and the severity of the inflammatory response can destroy large areas of kidney. This is unusual in myxosporean infections unless it is in an atypical site (Dyková *et al.*, 1987) or unless mature spores were produced (Dyková and Lom, 1978). These also suggest that salmonids are abnormal hosts. In heavy infections, PKX forms granulomatous changes in the spleen, intestine, gills, liver and muscles. PKX is spread to these organs via the blood and also causes necrotizing vasculitis.

Infected fish have a poor food conversion and lower stress tolerance. There is an associated immunosuppression (Angelidis *et al.*, 1987) and hence fish are susceptible to secondary infections. Kidney functions, e.g. fluid balance, waste excretion and blood cell production, become inexorably com-promised. Severe anaemia, either haemolytic (Hoffman and Lommel, 1984) or hypoplastic (Clifton-Hadley *et al.*, 1987b), contributes to fish mortality. The interstitial form of the PKX is eliminated in the autumn in fish that survive the disease (Kent and Hedrick, 1986; Foott and Hedrick, 1987). Interstitial hyperplasia subsides within 20 weeks and the kidney fully recovers. In the following year the recovered fish show strong resistance to reinfection or do not manifest the disease (Ferguson and Ball, 1979; Clifton-Hadley *et al.*, 1986). PKX can be syringe-passaged *in vivo* using infected blood and tissue.

The recommended control measure is to prevent salmonids from coming into contact with other fish during the critical season (April to June) and delay the use of river water in tanks with fingerlings until July. Increasing salinity (up to 0.8% with a further increase to 1.2%) helps to decrease mortality and development of the disease. A malachite green bath is useful in suppressing the PKD.

Infections of the gall bladder and liver

Gall bladder infection is always via the liver (Dyková and Lom, 1988a). Sporogonic plasmodia may also form in the biliary collecting ducts. Gall-bladder-infecting Myxosporea appear usually to be rather inocuous, yet there are clearly pathogenic species. A common lesion associated with the infection of biliary ducts is pericholangitis (Fig. 3.12A,D). It is detectable at early stages of plasmodium development (Fig. 3.12C).

Myxidium truttae Léger, 1930 (syn. *M. salmonis* Kulakovskaya, 1954) is in *Salmo trutta* and other salmonids throughout Eurasia (Fig. 3.12B). Walliker (1968) reported serious inflammation of bile ducts and suppurative liver necrosis in *Salmo salar*.

Unlike many nonpathogenic species of the genus *Zschokkella*, *Z. russelli* Tripathi, 1948 is commonly found in *Mustelus mustelus* and *Gaidropsarus mediterraneus* (prevalence is up to 89% and 75%, respectively) in the North

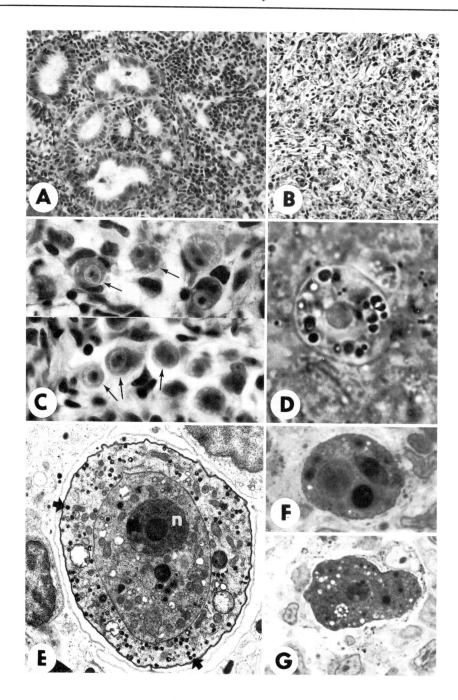

Fig. 3.11. The proliferative kidney disease of salmonids. A, B. Gradual disappearance of the renal tubules and their replacement by interstitial lymphoid tissue in the initial stage of infection (A, ×230) and by proliferating connective tissue in the later stage of infection (B, ×220). C. The agent (PKX) of proliferative kidney disease (PKD) in the sectioned kidney; the inner cells are clearly visible

Sea. The parasite has flat, round plasmodia up to 272 µm and ellipsoid spores 15 µm long. It elicits enlargement and thickening of hepatic ducts, degrades the epithelium and causes pericholangitis.

Several pathogenic species belong to the genus *Chloromyxum*. *C. cristatum* Léger, 1906 (syn. *C. cyprini* Fujita, 1927) (Fig. 3.2F) is rather common in cyprinids, especially common carp and *Ctenopharyngodon idella*, in Eurasia. Small plasmodia produce one or two spores (11 µm in diameter) with high ridges on the surface (Fig. 3.1I). It may pervade the liver parenchyma with a mass of sporogonic plasmodia and cause necrosis (Lom and Dyková, 1981). *C. truttae* Léger, 1906 may provoke a serious disease (Bauer *et al.*, 1981) in salmonids. In *S. salar* it infects cultures of yearlings, however, in rainbow trout, only producer fish at the time of spawning are affected. The clinical signs are anorexia, emaciation and yellow-tinged skin. The hypertrophic gall bladder is full of yellow-red fluid, the liver is discolored and there is inflammation in the intestine. The disease may last several months and be fatal.

C. trijugum Kudo, 1919 causes lesions in the gall bladder of several species of centrarchid fishes (genera *Lepomis* and *Pomoxis*) in the USA. Breakdown of the mucosal layer of the bladder may be due to irritation caused by plasmodia which are attached to the epithelial cells (Mitchell *et al.*, 1980).

Infections of the swimbladder

The swimbladder may be the site of sporogonic plasmodia as well as of extra-sporogonic proliferative stages of myxosporeans which produce spores in other organs. In common carp fingerlings, typically in their 3rd–4th month of age, the stages of *Sphaerospora renicola* (see p. 127) elicit swimbladder inflammation. Masses of extrasporogonic stages which are also found in other tissues pervade the swimbladder wall. They are located intra- and extravascularly and elicit an exudative-proliferative inflammation. Hyperplasia of the epithelium of lamina propria, proliferation of connective tissue and changes in serosa result in a greatly thickened swimbladder wall (Fig. 3.13F). The wall is full of haemorrhages. The inflammation is associated with kidney hypertrophy and peritonitis. Diseased fish display signs of distress along with locomotory disorders and swimming in circles. The damaged swimbladder is a poor prognosis for the infected fish. In addition to mortalities and reduced growth, there are additional losses (up to 15%) during the following winter (Körting, 1982; Kovács-Gayer *et. al.*, 1982; Csaba *et al.*, 1984; Dyková and Lom, 1988b).

(×950). D. The agent of PKD in the fresh sample of kidney tissue (fresh mount; ×1800). E. Electron micrograph of a secondary cell inside a primary one; arrows point at the dense bodies, n = nucleus (×5700). F, G. Semithin sections of primary cells containing secondary ones (×1600 and ×1200, respectively). A, B, E and G from Lom and Dyková, 1992, with permission of the publisher; other figures are originals.

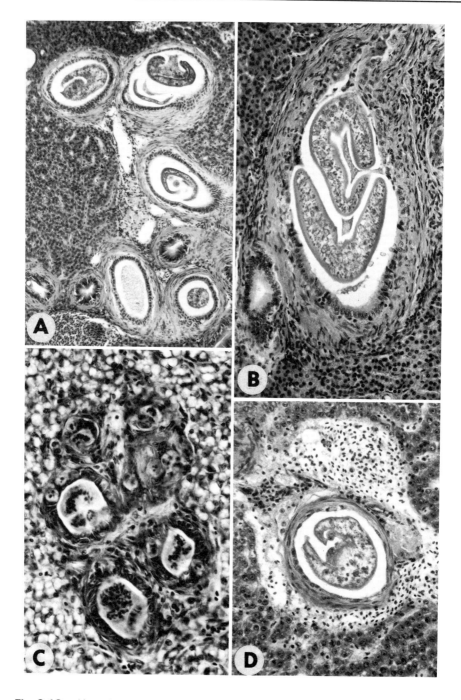

Fig. 3.12. Hepatic myxosporean infections. A. *Zschokkella costata*, polysporic plasmodia rolled up inside the collecting hepatic ducts of *Leuciscus cephalus* (×140). B. *Myxidium truttae*, plasmodia inside the intrahepatic bile duct of *Salmo trutta* m. *fario*; atrophy of the epithelium in the upper part of the sectioned duct lining (×210). C. *Chloromyxum reticulatum*; early plasmodial stages in the

Species which form large sporogonic plasmodia in the swimbladder usually inflict no serious injury. *Myxobolus macrocapsularis* Reuss, 1906 forms large plasmodia (up to 3 mm) in the swimbladder wall (and also in the intestinal wall) of *Leuciscus cephalus* and several other cyprinid hosts in Europe.

Infections of the gonads

Gonads are infected by very few species. Seminiferous tubules in the testis of the marine fish *Dicentrarchus labrax* in the Mediterranean may be infected with *Sphaerospora testicularis* Sitja-Bobadilla and Alvarez-Pellitero, 1990. The prevalence may be up to 6.25% in cultured fish. Disporic pseudoplasmodia form large spores, $12 \times 15 \, \mu m$ in size and damage to the testis greatly reduces male fecundity.

   H. oviperda (Cohn, 1895) is a rare parasite in oocytes of pike, *Esox lucius* in Europe. Infected ovaries are chalk white. The caudal appendages of the ellipsoid spores may be rudimentary or absent. The infection may cause atrophy of large numbers of oocytes and local circulatory disorders.

⸗ CULTURE

Little is known about *in vitro* cultivation of myxosporeans. Wolf and Markiw (1976) cultured trophozoites of *M. cerebralis* and achieved spore formation in an *in vitro* culture. The early development of *M. exiguus* trophozoites in tissue culture was observed by Siau (1977). Markiw (1986) passaged *Myxobolus cerebralis* in trout and in tubificids.

MODE OF NUTRITION

Developmental stages, especially of coelozoic species are assumed to feed by osmotrophy and by active transport of nutrients. Large histozoic plasmodia (e.g. of *Myxobolus*) have cell membranes with very active pinocytotic activity. Contact digestion followed by membrane transport is postulated for coelozoic plasmodia adhering to the epithelium (Uspenskaya, 1982). Extracellular digestion is normally associated with trophozoites infecting myocytes (*Kudoa*) and cartilage (*Myxobolus cerebralis*). Phagocytosis of chondrocytes was observed in *M. cerebralis*. Extrasporogonic stages of some *Sphaerospora* are known to phagocytize erythrocytes. The knowledge of myxosporean nutrition is limited to microscopic observations and no experimental study has been performed.

intrahepatic bile ducts of *Lota lota*, tiny plasmodial stages are also in the epithelial layer (×410). D. Plasmodium of *Ortholinea australis* in the hepatic duct of *Rhabdosargus sarga* (×290). All figures are originals.

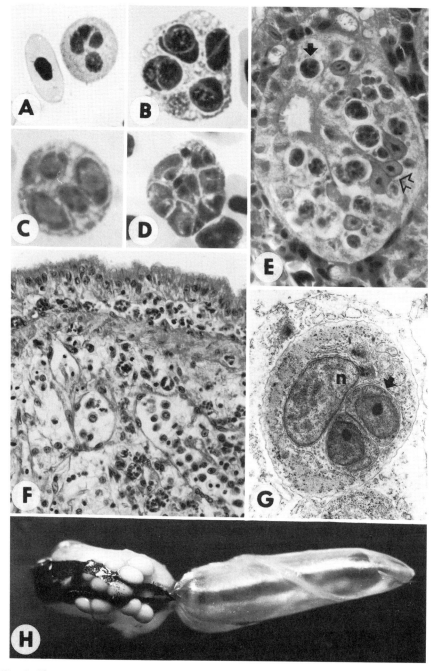

Fig. 3.13. A to D. Extrasporogonic bloodstream stages of the genus *Sphaerospora* as seen on stained blood smears; arrows point at the nucleus of the primary cell. A. *S. cristata* stage next to an erythrocyte (×1600). B. *Sphaerospora* sp. from *Leuciscus cephalus*; each of the four secondary cells contains a tertiary cell (×1700). C. *S. renicola* from the common carp – four simple secondary cells (×2300). D. *Sphaerospora* sp. from *Leuciscus leuciscus*; secondary

ORIGIN AND EVOLUTION

The phylum Myxozoa has no close affinities with other protistan phyla. The characters of the phylum include multicellularity with division of labour among the cells which are separated into somatic (e.g. plasmodium or spore shell cells) and germ cells (e.g. generative cells or sporoplasm), cell junctions, polar capsules and the cell-in-cell organization, when primary cells contain inner, secondary and tertiary cells. The latter feature is also found in another group, haplosporeans of the genera *Paramyxa* and *Marteilia*. The true affinities between the Myxosporea and the *Paramyxa* and *Marteilia* are not known. Some conjectures are that the amoebae are ancestors of Myxosporea (see e.g. Shulman, 1966). This can be dismissed as sheer speculation; it is based on ubiquitous amoeboid characters of myxosporean sporoplasms and plasmodia.

Structural similarities reaching the point of identity of polar capsules and coelenterate nematocysts led to the assumptions on a coelenterate ancestry for the myxozoans. These similarities are close enough to exclude mere evolutionary convergence, and there are parasitic coelenterates (e.g. *Polypodium hydriforme*) in fish. However, recent molecular biology studies (Smothers *et al.*, 1994; Schlegel and Lom, unpublished) indicate that myxosporeans are closely related to the bilateral animals, not to Radialia. Consequently, they should be considered metazoans, not protists.

The evolution within the class Myxosporea was discussed by Shulman (1966) and Donets (1979). They suggested that early myxosporeans lived in the gall bladder and later in the urinary bladder of marine bony fishes (subclass Actinopterygii) in the late Cretaceous period. In the Cenoman period, the Myxosporea diversified in perciform fishes; later they colonized freshwater environments with their hosts. From the urinary tract, they gradually spread to other organs and tissues. Platysporina are assumed to be associated with tissue parasitism in freshwater cyprinid fishes. Most of their genera seem closely related (e.g. *Myxobolus* giving rise to *Thelohanellus* by suppression of one capsulogenic cell, or by formation of various outgrowths to genera *Henneguya*, *Unicauda*, etc.). Multivalvulids are thought to originate from a lineage similar to *Ceratomyxa*, *Trilospora* and *Kudoa*. In the Variisporina, the taxonomic kinships of the genera are not clear. However, studies on the life cycles may contribute to more precise ideas on myxosporean evolution.

---

cells contain tertiary cells (×1600). E. Intracellular extrasporogonic stages (full arrow) of *S. renicola* in the renal tubule undergoing hypertrophy. Infected cells fuse to form a syncytium with hypertrophic nuclei (hollow arrow) (×780). F. A thickened wall in a case of swimbladder inflammation, extrasporogonic developmental stages are concentrated under the epithelium and in the oedematously changed connective tissue layer of the swimbladder (×350). G. The initial form of swimbladder extrasporogonic stages: a primary cell with a pair of secondary cells; n = nucleus. Electron micrograph (×8200). H. Swimbladder of *Leuciscus leuciscus* with cyst-like plasmodia of *Myxobolus macrocapsularis* (×3). F from Dyková and Lom, 1986, G from Dyková *et al.*, 1990, with permission of the publishers; all other figures are originals.

DIAGNOSIS

The proof of a myxosporean infection is finding the spores with polar capsules in fresh mounts. In histological sections, the spores can be detected more readily when stained with Giemsa. The polar capsules stain dark against a much lighter background. Identification to species and genera is by their morphological characteristics (Lom and Noble, 1984; Lom and Arthur, 1988; Shulman 1966, 1984).

In the absence of mature spores, the myxosporean nature can be reliably determined using electron microscope observations of the cell-in-cell organization, presence of microtubular bundles close to nuclei, abundant free ribosomes and absence of centrioles.

To avoid killing the fish, an intestinal lavage was developed to detect *Ceratomyxa shasta* infection (Coley *et al.*, 1983). Detection of light infections can be facilitated by concentration of spores. This was developed for the diagnosis of *Myxobolus cerebralis* but can also be used for other myxosporeans. Briefly, the tissue is homogenized, or minced or macerated, and then digested with trypsin, filtered, and centrifuged. The method of choice is differential centrifugation in dextrose solution (Markiw and Wolf, 1974a,b; Bailey *et al.*, 1989). Trypsinization may be preceded by decalcification if *M. cerebralis* is from head cartilages (Landolt, 1973). An aqueous polymer two-phase system may also be employed (Kozel *et al.*, 1980).

The use of serologic techniques is in its infancy. The fluorescent antibody technique was developed (Markiw and Wolf, 1978) to identify *Myxobolus cerebralis* spores. Monoclonal antibodies were prepared to identify prespore stages of *Ceratomyxa shasta* which do not cross-react with spores. Polyclonal antisera reacted with both spores and prespore stages (Bartholomew *et al.*, 1989a,b).

TREATMENT AND CONTROL

Myxosporean spores are extremely resistant and can tolerate extended freezing (Hoffman and Putz, 1969). Disinfectants (calcium hydroxide, calcium oxide, calcium cyanamide, potassium hydroxide or chlorine) have to be used to kill the spores at the bottom of ponds and tanks and to sanitize mud (Hoffman and Putz, 1969; Hoffman and Hoffman, 1972; Hoffman and Meyer, 1974; Bauer *et al.*, 1981). UV radiation is effective against spores of *Myxobolus cerebralis* (Hoffman 1974, 1975) or *Ceratomyxa shasta* (Bedell, 1971, Bower and Margolis, 1985) in the water feed.

Prevention is achieved by good fish husbandry. Delayed stocking of trout fingerlings can minimize proliferative kidney disease (Ferguson and Ball, 1979; Ferguson, 1981).

Chemotherapeutic trials have been promising though technically complicated and expensive. *M. cerebralis* spores were decreased in fish fed furazolidone (Taylor *et al.*, 1973). Proguanil in the diet also reduced spore production and alleviated lesions (Alderman, 1986). Fumagillin appeared to be effective in reducing infections with *Sphaerospora renicola* in carp (Molnar *et al.*, 1987) and with PKX in chinook salmon (Hedrick *et al.*, 1988). However, it was not effective with histozoic species of *Myxobolus* and *Thelohanellus* (Molnar *et al.*, 1987).

Medicated pellets containing 0.1% fumagillin fed to experimentally infected rainbow trout drastically reduced the infection rate and prevented clinical outbreaks of whirling disease (El-Matbouli and Hoffman, 1991). According to Mehlhorn *et al.* (1988) and Schmahl *et al.* (1989, 1991), the anticoccidial compound toltrazuril was effective against Myxosporea. It destroyed all the stages except for mature spores.

SUGGESTIONS FOR FUTURE STUDIES

The research on Myxosporea has now reached a point from where the solution of several most intriguing problems seems to be within easy reach. The key point in transmission is the *Myxobolus–Triactinomyxon* transformation. Does it alternate with or run parallel to simple direct transmission via the ingestion of myxosporean spores? Is this transformation widespread or limited to only a few Myxosporea? This in itself is quite fascinating and along with it are the practical epizootiological implications. Another basic question is the nature and role of extrasporogonic stages. Also, what are the stimuli that elicit their transformation into sporogonic plasmodia, and their persistence in various body tissues? Then there is the ultimate puzzle of the origin of Myxosporea and Myxozoa in general. Are Myxozoa and coelenterates phylogenetically related or is the close structural similarity of polar capsules and nematocysts merely an unbelievable case of convergence?

Careful work on the precise characterization of the common myxosporean species (genera such as *Myxobolus*, *Henneguya* and *Kudoa*) should continue. This would include well controlled studies on their host and tissue specificity.

The pathogenic mechanism(s), host response and antigenicity are equally interesting. Why is it that abnormal localization of e.g. *Myxidium rhodei* sporogonic plasmodium elicits such a strong host reaction? Why does the extrasporogonic stage of *Sphaerospora renicola* provoke such inflammation in the gall bladder, while the presporogonic stages of e.g. *Chloromyxum reticulatum* seem to pass unnoticed by the host? What is the nature of the PKX and the significance of the dense bodies in its primary cells? They do not seem to have counterparts in other Myxosporea.

Let us conclude that it is vitally important that we find a convenient, inexpensive, and effective way of controlling the parasite.

GLOSSARY OF TERMS

Autogamy – a sexual process taking place in the sporoplasm, in which two nuclei, products of division within a single cell, fuse together. If cytokinesis precedes the fusion, i.e. if two sporoplasm cells fuse, it is called **paedogamy**.

Capsulogenic cell – one of the cells constituting a sporoblast, it transforms into the polar capsule.

Cell-in-cell organization – myxosporean developmental stages in which a primary (mother) cell contains inner secondary (daughter) cells produced by endogenous cleavage. There may be also tertiary and quaternary cells included in secondary and tertiary cells, respectively. A single secondary cell within a primary cell is called a cell doublet. A primary cell with two secondary cells forms a cell triplet.

Coelozoic – infecting organ cavities.

Disporic – a small sporogonic plasmodium or pseudoplasmodium which produces two spores.

Extrasporogonic – a phase of the developmental cycle which occurs parallel to the sporogonic phase.

Generative cell – a secondary cell inside a plasmodium, destined to be engaged later in spore production.

Histozoic – infecting tissues.

Melanomacrophages – pigment-bearing macrophages which aggregate in the haemopoietic tissues to form melanomacrophage centres. They contain the pigments melanin, haemosiderin and lipid pigments (lipofuscin/ceroid). Melanomacrophages play an important part in the host's defence reaction. They are capable of attacking even large myxosporean spores and transporting them to melanomacrophage centres, where the parasite is destroyed. These centres are found in the spleen, kidney and, to a lesser extent, in the liver.

Mictosporic – a myxosporean parasite which forms plasmodia of various types – mono-, di- and polysporic.

Monosporic – a plasmodium or pseudoplasmodium which produces a single spore.

Nematocyst – a specialized cell in coelenterates which contains a capsule with extrudible polar filament.

Pansporoblast – a spore-producing formation within a polysporic plasmodium. It originates by the union of two generative cells, of which one, the pericyte, envelopes the other, the sporogonic cell. The former gives rise later to the pansporoblast envelope. The latter divides to produce the sporoblast cells. A pansporoblast usually produces two spores.

Pericyte – see pansporoblast.

Plasmotomy – division of large plasmodia by cleavage in two or more daughter parts.

Polysporic – a plasmodium which produces several spores.

Polar capsule – a thick-walled vesicle in myxozoan spores with a polar filament inverted inside it, forming a coil with two to many turns. Upon stimulation (in the digestive tract of the host) the filament quickly discharges, everting inside out.

Polar filament – see polar capsule.

Primary cell – see cell-in-cell organization.

Presporogonic – a sequence of development in the life cycle which precedes sporogony.

Pseudocyst – a cyst-like formation, i.e. a cavity-like lesion surrounded by a dense fibrous capsule. Unlike a true cyst, it does not possess an inner lining, the formation of a pseudocyst follows necrotic changes and the wall of a pseudocyst is formed by reparative inflammation.

Pseudoplasmodium – a myxosporean sporogony stage which produces one or two spores and has a single vegetative nucleus.

Pseudopodium – a temporary cytoplasmic projection of a pseudoplasmodium or plasmodium.

Secondary cell – see cell-in-cell organization.

Shell valve – one of the two or more parts of the myxosporean spore wall.

Spore – an infectious stage built up of several (four to many) specialized cells.

Sporoblast – a developmental stage preceding the spore. Its cells are not yet fully differentiated.

Sporogonic cell – a cell produced by division of the sporoblast cells.

Sporogony – a process of spore formation.

Sporogonic plasmodium – a multinuclear cell with many generative cells engaged in sporogony.

Sporoplasm – an amoeboid cell inside the spore, the actual infective germ.

Sporoplasmosome – electron dense inclusions in the sporoplasm.

Suture line – also the line of dehiscence, along which the shell valves adhere together.

Tertiary cell – see cell-in-cell organization.

Valvogenic cell – one of the cells constituting a sporoblast; it differentiates into a shell valve.

A more comprehensive Glossary can be found at the end of this book.

REFERENCES

Ahmed, A.T.A. (1973) Morphology and life history of *Mitraspora cyprini* Fujita parasitic in the kidney of goldfish. *Japanese Journal of Medical Science and Biology* 26, 87–101.

Ahmed, A.T.A. (1974) Kidney enlargement disease of goldfish in Japan. *Japanese Journal of Zoology* 17, 37–65.

Alderman, D.J. (1986) Whirling disease chemotherapy. *Bulletin of the European Association of Fish Pathologists* 6, 38–39.

Amandi, A., Holt, R.A. and Fryer, J.L. (1985) Observations on *Myxobolus insidiosus* (Myxozoa: Myxosporea) a parasite of salmonid fishes. *Fish Pathology* 20, 287–304.

Angelidis, P., Baudin-Laurencin, F., Quentel, C. and Youinou, P. (1987) Lower immune response induced by PKD. *Journal of Fish Biology* 31 (Suppl. A), 247–250.

Awakura, T. and Kimura, T. (1977) On the milky condition in smoked coho salmon (*Oncorhynchus kisutch*) caused by myxosporidian parasite (in Japanese). *Fish Pathology* 12, 179–184.

Bailey, R.E., Margolis, L. and Workman, G.D. (1989) Survival of certain naturally acquired freshwater parasites of juvenile sockeye salmon, *Oncorhynchus nerka* (Walbaum), in hosts held in fresh and sea water, and implications for their use as population tags. *Canadian Journal of Zoology* 67, 1757–1766.

Banning, P. van, Veen, J.F. and Leeuwen, P.J. van (1978) The myxosporidian parasite (*Myxobolus aeglefini* Auerbach, 1906) and its use as parasitological tag for plaice of the eastern North Sea. *International Council for Exploration of the Sea CM G* 48, 22 pp.

Bartholomew, J.L. , Rohovec, J.S. and Fryer, J.L. (1989a) Development, characterization, and use of monoclonal and polyclonal antibodies against the myxosporean, *Ceratomyxa shasta*. *Journal of Protozoology* 36, 397–401.

Bartholomew, J.L., Smith, C.E., Rohovec, J.S. and Fryer, J.L. (1989b) Characterization of a host response to the myxosporean parasite, *Ceratomyxa shasta* (Noble), by histology, scanning electron microscopy and immunological techniques. *Journal of Fish Diseases* 12, 509–522.

Bartholomew, J.L., Stevens, D.G., Rohovec, J.S. and Fryer, J.L. (1992) Ceratomyxosis is associated with an oligochaete and an actinosporean. *Abstracts of International Workshop on Myxosporea*, 6–8 October, České Budějovice.

Baska, F. and Molnar, K. (1989) Observation on the blood stages of *Sphaerospora* spp. (Myxosporea) in cyprinid fishes. *Diseases of Aquatic Organisms* 5, 23–28.

Bauer, O.N., Musselius, V.A. and Strelkov, Y.A. (1981) *Diseases of Pond Fishes*. (in Russian), 2nd edn. Legkaya i pishchevaya promyshlenost, Moskva, 319 pp.

Bedell, G.W. (1971) Eradicating *Ceratomyxa shasta* from infected water by chlorination and ultraviolet irradiation. *Progressive Fish Culturist* 33, 51–54.

Benajiba, M.H. and Marques, A. (1993) The alteration of actinomyxidian and myxosporidian sporal forms in the development of *Myxidium giardi* (parasite of *Anguilla anguilla*) through oligochaetes. *Bulletin of the European Association of Fish Pathologists* 13, 100–103.

Bower, S.M. and Margolis, L. (1985) Microfiltration and ultraviolet irradiation to eliminate *Ceratomyxa shasta* (Myxozoa: Myxosporea), a salmonid pathogen, from Fraser River water, British Columbia. *Canadian Technical Reports, Fisheries and Aquatic Science* 1364, p. 11.

Boyce, N.P., Kabata, Z. and Margolis, L. (1985) Investigations of the distribution, detection, and biology of *Henneguya salminicola* (Protozoa, Myxozoa), a parasite of the flesh of pacific salmon. *Canadian Technical Reports, Fisheries and Aquatic Science* No. 1405, 54 pp.

Clifton-Hadley, R.S., Richards, R.H. and Bucke, D. (1986) Proliferative kidney disease (PKD) in rainbow trout *Salmo gairdneri*: further observations on the effects of water temperature. *Aquaculture* 55, 165–171.

Clifton-Hadley, R.S., Bucke, D. and Richards, R.H. (1987a) A study of the sequential clinical and pathological changes during proliferative kidney disease in rainbow trout, *Salmo gairdneri* Richardson. *Journal of Fish Diseases* 10, 335–352.

Clifton-Hadley, R.S., Richards, R.H. and Bucke, D. (1987b) Further consideration of the haematology of proliferative kidney disease (PKD) in rainbow trout, *Salmo gairdneri* Richardson. *Journal of Fish Diseases* 10, 435–444.

Coley, T.C., Chacko, A.J. and Klontz, G.W. (1983) Development of a lavage technique for sampling *Ceratomyxa shasta* in adult salmonids. *Journal of Fish Diseases* 6, 317–319.

Csaba, G. (1976) An unidentifiable extracellular sporozoan parasite from the blood of the carp. *Parasitologica Hungarica* 9, 21–24.

Csaba, G., Kovacs-Gayer, E., Bekesi, L., Bucsek, M., Szakolczai, J. and Molnar, K. (1984) Studies into the possible protozoan aetiology of swimbladder inflammation in carp fry. *Journal of Fish Diseases* 7, 39–56.

Current, W.L. (1979) *Henneguya adiposa* Minchew (Myxosporida) in the channel catfish: ultrastructure of the plasmodium wall and sporogenesis. *Journal of Protozoology* 26, 209-217.

Current, W.L. and Janovy, J. Jr (1977) Sporogenesis in *Henneguya exilis* infecting the channel catfish: an ultrastructural study. *Protistologica* 13, 157-167.

Current, W.L. and Janovy, J. Jr (1978) Comparative study of ultrastructure of interlamellar and intralamellar types of *Henneguya exilis* Kudo from channel catfish. *Journal of Protozoology* 25, 56-65.

Current, W.L., Janovy, J. Jr and Knight, S.A. (1979) *Myxosoma funduli* Kudo (Myxosporida) in *Fundulus kansae*: ultrastructure of the plasmodium wall and of sporogenesis. *Journal of Protozoology* 26, 574-583.

Desportes, I. and Théodorides, J. (1982) Donneés ultrastructurales sur la sporogenese de deux myxosporidies rapportés aux genres *Leptotheca* et *Ceratomyxa* parasites de *Merluccius merluccius* (L.) (Téléostéen Merluciidae). *Protistologica* 18, 533-557.

Desser, S.S. and Paterson, W.B. (1978) Ultrastructural and cytochemical observations on sporogenesis of *Myxobolus* sp. (Myxosporida: Myxobolidae) from the common shiner *Notropis cornutus*. *Journal of Protozoology* 25, 314-326.

Donets, Z.S. (1979) The evolution of the myxosporidia. In: Bauer, O. (ed.) *Taxonomy and Ecology of the Sporozoa and Cnidosporidia*. Proceedings of the Zoological Institute of the Academy of Sciences of the USSR 87, 28-41.

Duhamel, G.E., Kent, M.L., Dybdal, N.O. and Hedrick, R.P. (1986) *Henneguya exilis* Kudo associated with granulomatous branchitis of channel catfish *Ictalurus punctatus* (Rafinesque). *Veterinary Pathology* 23, 354-361.

Dyková, I. (1984) The role of melanomacrophage centres in the tissue reaction to myxosporean infections of fishes. *Bulletin of the European Association of Fish Pathologists* 4, 65-67.

Dyková, I. and Lom, J. (1978) Histopathological changes in fish gills infected with myxosporidian parasites of the genus *Henneguya*. *Journal of Fish Biology* 12, 197-202.

Dyková, I. and Lom, J. (1982) *Sphaerospora renicola* n.sp., a myxosporean from carp kidney and its pathogenicity. *Zeitschrift für Parasitenkunde* 68, 259-268.

Dyková, I. and Lom, J. (1987) Host cell hypertrophy induced by contact with trophozoites of *Thelohanellus pyriformis* (Myxozoa: Myxosporea). *Archiv für Protistenkunde* 133, 285-293.

Dyková, I. and Lom, J. (1988a) *Chloromyxum reticulatum* (Myxozoa, Myxosporea) in the liver of burbot (*Lota lota* L.) and its migration to the final site of infection. *European Journal of Protistology* 23, 258-261.

Dyková, I. and Lom, J. (1988b) Review of pathogenic myxosporeans in intensive culture of carp (*Cyprinus carpio*) in Europe. *Folia Parasitologica* 35, 289-307.

Dyková, I., Lom, J. and Grupcheva, G. (1987) Pathogenicity and some structural features of *Myxidium rhodei* (Myxozoa: Myxosporea) from the kidney of the roach *Rutilus rutilus*. *Diseases of Aquatic Organisms* 2, 109-115.

Dyková, I., Lom, J. and Körting, W. (1990) Light and electron microscopic observations on the swimbladder stages of *Sphaerospora renicola*, the parasite of carp (*Cyprinus carpio*). *Parasitology Research* 76, 228-237.

Egusa, S. (1985) *Myxobolus buri* sp.n. (Myxosporea: Bivalvulida) parasitic in the brain of *Seriola quinqueradiata* Temminck et Schlegel. *Fish Pathology* 19, 239-244.

El-Matbouli, M. and Hoffmann, R.W. (1989) Experimental transmission of two *Myxobolus* spp. developing bisporogeny via tubificid worms. *Parasitology Research* 75, 461-464.

El-Matbouli, M. and Hoffmann, R.W. (1990) Transmissionsversuche mit *Myxobolus cerebralis* und *Myxobolus pavlovskii* und ihre Entwicklung in Tubificiden. *Zeitschrift für Fischerei* xx, 32–37.

El-Matbouli, M. and Hoffmann, R.W. (1991) Prevention of experimentally induced whirling disease in rainbow trout *Oncorhynchus mykiss* by fumagillin. *Diseases of Aquatic Organisms* 2, 109–113.

Ferguson, H.W. (1981) The effects of water temperature on the development of proliferative kidney disease in rainbow trout, *Salmo gairdneri* Richardson. *Journal of Fish Diseases* 4, 175–177.

Ferguson, H.W. and Ball, H.J. (1979) Epidemiological aspects of proliferative kidney disease amongst rainbow trout *Salmo gairdneri* Richardson in northern Ireland. *Journal of Fish Diseases* 2, 219–225.

Ferguson, H.W. and Needham, E.A. (1978) Proliferative kidney disease in rainbow trout *Salmo gairdneri* Richardson. *Journal of Fish Diseases* 1, 91–108.

Ferguson, H.W., Lom, J. and Smith, I. (1985) Intra-axonal parasites in the fish *Notropis cornutus* (Mitchill). *Veterinary Pathology* 22, 194–196.

Foott, J.S. and Hedrick, R.P. (1987) Seasonal occurrence of the infections of stage of proliferative kidney disease (PKD) and resistance of rainbow trout, *Salmo gairdneri* Richardson, to reinfection. *Journal of Fish Biology* 30, 477–484.

Furuta, T., Ogawa, K. and Wakabayashi, H. (1993) Humoral immune response of carp *Caprius carpio* to *Myxobolus artus* (Myxozoa: Myxobolidae) infection. *Journal of Fish Biology* 43, 441–450.

Grassé, P.P. and Lavette, A. (1978) La myxosporidie *Sphaeromyxa sabrazesi* et le nouvel embranchement des Myxozoaires (Myxozoa). Recherches sur l'état pluricellulaire primitif et considerations phylogénétiques. *Annales des Sciences Naturelles, Zoologie, Paris* 20, 193–285.

Griffin, B. and Davis, E. (1978) *Myxosoma cerebralis*: detection of circulating antibodies in infected rainbow trout (*Salmo gairdneri*). *Journal of the Fisheries Research Board of Canada* 35, 1186–1190.

Groff, J.M., McDowell, T. and Hedrick, R.P. (1989) Sphaerospores observed in the kidney of channel catfish (*Ictalurus punctatus*). *Fish Health Section/American Fisheries Society, Newsletter* 7, p. 5.

Grupcheva, G., Dyková, I. and Lom, J. (1985) Seasonal fluctuation in the prevalence of *Sphaerospora renicola* and myxosporean bloodstream stages in carp fingerlings in Bulgaria. *Folia Parasitologica* 32, 193–203.

Guilford, H.G. (1963) New species of Myxosporidia found in percid fishes from Green Bay (Lake Michigan). *Journal of Parasitology* 49, 474–478.

Halliday, M.M. (1976) The biology of *Myxosoma cerebralis*: the causative organism of whirling disease of salmonids. *Journal of Fish Biology* 9, 339–357.

Harrel, L.W. and Scott, T.M. (1985) *Kudoa thyrsitis* (Gilchrist) (Myxosporea: Multivalvulida) in Atlantic Salmon, *Salmo salar* L. *Journal of Fish Diseases* 8, 329–332.

Heckmann, R.A. and Jensen, L.A. (1978) The histopathology and prevalence of *Henneguya sebasta* and *Kudoa clupeidae* in the rockfish *Sebastes paucispinis* of southern California. *Journal of Wildlife Diseases* 14, 259–262.

Hedrick, R.P., Kent, M.L., Rosemark, R. and Manzer, D. (1984) Occurrence of proliferative kidney disease (PKD) among Pacific salmon and steelhead trout. *Bulletin of the European Association of Fish Pathologists* 4, 34–37.

Hedrick, R.P., Groff, J.M., Foley, P. and McDowell, T. (1988) Oral administration of fumagilin DCH protects chinook salmon *Oncorhynchus tschawytscha* from experimentally induced proliferative kidney disease. *Diseases of Aquatic Organisms* 4, 165–168.

Hedrick, R.P., McDowell, T. and Groff, J.M. (1990) *Sphaerospora ictaluri* n.sp. (Myxosporea: Sphaerosporidae) observed in the kidney of channel catfish, *Ictalurus punctatus* Rafinesque. *Journal of Protozoology* 37, 107–112.

Hermanns, W. and Körting, W. (1985) *Sphaerospora tincae* Plehn, 1925 in tench, *Tinca tinca*, fry. *Journal of Fish Diseases* 8, 281–288.

Hoffman, G.L. (1974) Disinfection of contaminated water by ultraviolet irradiation, with emphasis on whirling disease (*Myxosoma cerebralis*) and its effect on fish. *Transactions of the American Fisheries Society* 103, 541–550.

Hoffman, G.L. (1975) Whirling disease (*Myxosoma cerebralis*): control with ultraviolet irradiation and effect on fish. *Journal of Wildlife Diseases* 11, 505–507.

Hoffman, G.L. (1981) Two fish pathogens, *Parvicapsula* sp. and *Mitraspora cyprini* (Myxosporea) new to North America. In: Olah, J., Molnar, K. and Jeney, Z. (eds) *Fish Pathogens and Environment in European Polyculture.* Proc. Int. Seminar, Szarvas, Hungary, Fisheries Res. Inst., Szarvas, pp. 184–197.

Hoffman, G.L. Sr. and Hoffman, G.L. Jr (1972) Studies on the control of whirling disease (*Myxosoma cerebralis*). 1. The effects of chemicals on spores *in vitro*, and of calcium oxide as a disinfectant in simulated ponds. *Journal of Wildlife Diseases* 8, 49–53.

Hoffman, G.L. and Meyer, F.P. (1974) *Parasites of Freshwater Fishes.* TFH Publications Inc., New Jersey, 224 pp.

Hoffman, G.L. and Putz, R.E. (1969) Host susceptibility and the effect of aging, freezing, heat, and chemicals on spores of *Myxosoma cerebralis*. *Progressive Fish Culturist* 31, 35–37.

Hoffman, G.L., Dunbar, C.E. and Brandford, A. (1962) *Whirling Disease of Trouts caused by* Myxosoma cerebralis *in the United States.* US Department of Interior, Fish and Wildlife Service, Special Scientific Report No. 427.

Hoffman, G.L., Putz, R.E. and Dunbar, C.E. (1965) Studies on *Myxosoma cartilaginis* n.sp. (Protozoa: Myxosporidea) of centrarchid fish and a synopsis of the *Myxosoma* of North American freshwater fishes. *Journal of Protozoology* 12, 319–332.

Hoffmann, R. and Lommel, R. (1984) Haematological studies in proliferative kidney disease of rainbow trout, *Salmo gairdneri* Richardson. *Journal of Fish Diseases* 7, 323–326.

Johnstone, A.K. (1985) Pathogenesis and life cycle of the myxozoan *Parvicapsula* sp. infecting marine cultured coho salmon. PhD Thesis, University of Washington, Seattle, 70 pp.

Kabata, Z. (1963) Parasites as biological tags. *Special Publications of the International Commission of North-West Atlantic Fisheries* 4, 31–37.

Kabata, Z. and Whitaker, D.J. (1981) Two species of *Kudoa* (Myxosporea; Multivalvulida) parasitic in the flesh of *Merluccius productus* (Ayres, 1855) (Pisces: Teleostei) in the Canadian Pacific. *Canadian Journal of Zoology* 59, 2085–2091.

Kabata, Z. and Whitaker, D.J. (1986) Distribution of two species of *Kudoa* (Myxozoa: Multivalvulida) in the offshore population of the Pacific hake, *Merluccius productus* (Ayres, 1855). *Canadian Journal of Zoology* 64, 2103–2110.

Kabata, Z. and Whitaker, D.J. (1989) *Kudoa thyrsites* (Gilchrist, 1924) (Myxozoa) in the cardiac muscle of Pacific salmon (*Oncorhynchus* spp.) and steelhead trout (*Salmo gairdneri*). *Canadian Journal of Zoology* 67, 341–349.

Kent, M.L. and Hedrick, R.P. (1985) PKX, the causative agent of proliferative kidney disease (PKD) in Pacific salmonid fishes and its affinities with the Myxozoa. *Journal of Protozoology* 32, 254–260.

Kent, M.L. and Hedrick, R.P. (1986) Development of the PKX myxosporean in rainbow trout *Salmo gairdneri*. *Diseases of Aquatic Organisms* 1, 169–182.

Kent, M.L. and Hedrick, R.P. (1987) Effects of cortisol implants on the PKX myxosporean causing proliferative kidney disease in rainbow trout, *Salmo gairdneri*. *Journal of Parasitology* 73, 455–462.

Kent, M.L., Margolis, L. and Corliss, J.O. (1994) The demise of a class of protists: Taxonomic and nomenclatural revisions proposed for the protist phylum Myxozoa Grassé, 1970. *Canadian Journal of Zoology* 72, 932–937.

Kent, M.L., Whitaker, D.J. and Margolis, L. (1993) Transmission of *Myxobolus arcticus* Pugachev and Khokhlov, 1979, a myxosporean parasite of Pacific salmon, via a triactinomyxon from the aquatic oligochaete *Stylodrilus heringianus* (Lumbriculidae). *Canadian Journal of Zoology* 71, 1207–1211.

Körting, W. (1982) Protozoan parasites associated with swimbladder inflammation (SBI) in young carp. *Bulletin of the European Association of Fish Pathologists* 2, 25–28.

Kovacs-Gayer, E., Csaba, G., Békési, L., Bucsek, M., Szakolczai, J. and Molnar, K. (1982) Studies on the protozoan etiology of swimbladder inflammation in common carp fry. *Bulletin of the European Association of Fish Pathologists* 2, 22–24.

Kozel, T.R., Lott, M. and Taylor, R. (1980) Isolation of *Myxosoma cerebralis* (whirling disease) spores from infected fish by use of a physical separation technique. *Canadian Journal of Fisheries and Aquatic Science* 37, 1032–1035.

Kudryashova, I.V. and Naumova, A.M. (1978) Effect of *Haemogregarina* on the body of carp. (In Russian). *Veterinariya* 4, 76–78.

Landolt, M. (1973) *Myxosoma cerebralis*: isolation and concentration from fish skeletal elements. I. Trypsin digestion method. *Journal of the Fisheries Research Board of Canada* 30, 1713–1716.

Landsberg, J.H. and Lom, J. (1991) Taxonomy of the genus *Myxobolus* (Myxobolidae: Myxosporea): current listing of species and revision of synonyms. *Systematic Parasitology* 18, 165–186.

Langdon, J.S. (1987) Spinal curvatures and an encephalotropic Myxosporean *Triangula percae* sp.nov. (Myxozoa: Ortholineidae), enzootic in redfin perch, *Perca fluviatilis* L., in Australia. *Journal of Fish Diseases* 10, 425–434.

Langdon, J.S. (1991) Myoliquefaction post-mortem ('milky flesh') due to *Kudoa thyrsites* (Gilchrist) (Myxosporea: Multivalvulida) in mahi mahi, *Coryphaena hippurus* L. *Journal of Fish Diseases* 14, 45–54.

Lom, J. (1969) Notes on the ultrastructure and sporoblast development in fish-parasitizing myxosporidian of the genus *Sphaeromyxa*. *Zeitschrift für Zellforschung* 97, 416–437.

Lom, J. (1987) Myxosporea: a new look at long-known parasites of fish. *Parasitology Today* 3, 327–332.

Lom, J. (1990) Phylum Myxozoa. In: Margolis, L., Corliss, J.O., Melkonian, M. and Chapman, D.J. (eds) *Handbook of Protoctista*. Jones and Bartlett Publishers, Boston, pp. 36–52.

Lom, J. and Arthur, J.R. (1988) A guideline for the preparation of species description in Myxosporea. *Journal of Fish Diseases* 12, 151–156.

Lom, J. and Dyková, I. (1981) Pathogenicity of some protozoan parasites of cyprinid fishes. In: Olah, J., Molnar, K. and Jeney, Z. (eds) *Fish Pathogens and Environment in European Polyculture*. Proceedings of an International Seminar on Fish, Pathogens and Environment in European Polyculture, June 23–27, 1981. Szarvas, Hungary, Fisheries Research Institute, Szarvas, pp. 146–169.

Lom, J. and Dyková, I. (1985) *Hoferellus cyprini* Doflein, 1898 from carp kidney: a well established myxosporean species or a sequence in the developmental cycle of *Sphaerospora renicola* Dyková and Lom, 1982. *Protistologica* 21, 195–206.

Lom, J. and Dyková, I. (1988) Sporogenesis and spore structure in *Kudoa lunata* (Myxosporea, Multivalvulida). *Parasitology Research* 74, 521–530.

Lom, J. and Dyková, I. (1992) *Protozoan Parasites of Fishes*. Elsevier Science Publishers.

Lom, J. and Hoffman, G.L. (1971) Morphology of the spores of *Myxosoma cerebralis* (Hofer, 1903) and *M. cartilaginis* (Hoffman, Putz and Dunbar, 1965). *Journal of Parasitology* 57, 1302–1308.

Lom, J. and Noble, E.R. (1984) Revised classification of the Myxosporea Bütschli, 1881. *Folia Parasitologica* 31, 193–205.

Lom, J. and Puytorac, P. de (1965) Observations sur l'ultrastructure des trophozoites de myxosporidies. *Compte Rendu de l'Academie des Sciences (Paris)* 260, 2588–2590.

Lom, J., Dyková, I. and Lhotáková, Š. (1983a) Fine structure of *Sphaerospora renicola* Dyková and Lom, 1982 a myxosporean from carp kidney and comments on the origin of pansporoblast. *Protistologica* 18, 489–502.

Lom, J., Dyková, I. and Pavlásková, M. (1983b) 'Unidentified' mobile protozoans from the blood of carp and some unsolved problems of myxosporean life cycles. *Journal of Protozoology* 30, 497–508.

Lom, J., Dyková, I. and Feist, S. (1989a) Myxosporea-induced xenoma formation in pike (*Esox lucius* L.) renal corpuscles associated with *Myxidium lieberkuehni* infection. *European Journal of Prostistology* 24, 271–280.

Lom, J., Feist, S.W., Dyková, I. and Kepr, T. (1989b) Brain myxoboliasis of bullhead, *Cottus gobio* L., due to *Myxobolus jiroveci* sp. nov.: light and electron microscope observations. *Journal of Fish Diseases* 12, 15–17.

Lom, J., Pike, A.W. and Dyková, 1. (1991a) *Myxobolus sandrae* Reuss, 1906, the agent of vertebral column deformities of perch (*Perca fluviatilis*). *Diseases of Aquatic Organisms* 12, 49–53.

Lom, J., Pike, A.W. and Feist, S.W. (1991b) Myxosporean stages in rete mirabile in the eye of *Gasterosteus aculeatus* infected with *Myxobilatus gasterostei* and *Sphaerospora elegans*. *Diseases of Aquatic Organisms* 11, 67–72.

MacConnell, E., Smith, C.E., Hedrick, R.P. and Speer, C.A. (1989) Cellular inflammatory response of rainbow trout to the protozoan parasite that causes proliferative kidney disease. *Journal of Aquatic Animal Health* 1, 108–118.

MacKenzie, K. (1987) Parasites as indicators of host populations. *International Journal for Parasitology* 17, 345–352.

Margolis, L. (1982) Pacific salmon and their parasites: a century of study. The Eighth Invitational Wardle Lecture. *Bulletin of the Canadian Society of Zoology* 13, 7–11.

Markiw, M.E. (1986) Salmonid whirling disease: dynamics of experimental production of the infective stage – the triactinomyxon spore. *Canadian Journal of Fisheries and Aquatic Sciences* 43, 521–526.

Markiw, M.E. (1989) Portals of entry for salmonid whirling disease in rainbow trout. *Diseases of Aquatic Organisms* 6, 7–10.

Markiw, M.E. (1991) Whirling disease: earliest susceptible age of rainbow trout to the triactinomyxid of *Myxobolus cerebralis*. *Aquaculture* 1, 1–6.

Markiw, M.E. and Wolf, K. (1974a) *Myxosoma cerebralis*: comparative sensitivity of spore detection methods. *Journal of the Fisheries Research Board of Canada* 31, 1597–1600.

Markiw, M.E. and Wolf, K. (1974b) *Myxosoma cerebralis*: isolation and concentration from fish skeletal elements – sequential enzymatic digestions and purification by differential centrifugation. *Journal of the Fisheries Research Board of Canada* 31, 15–20.

Markiw, M.E. and Wolf, K. (1978) *Myxosoma cerebralis* fluorescent antibody techniques for antigen recognition. *Journal of the Fisheries Research Board of Canada* 35, 828–832.

McArthur, C.P. and Sengupta, S. (1982) Antigenic mimicry of eel tissues by a myxosporidian parasite. *Zeitschrift für Parasitenkunde* 66, 249–255.

Mehlhorn, H., Schmahl, G. and Haberkorn, A. (1988) Toltrazuril effective against a broad spectrum of protozoan parasites. *Parasitology Research* 75, 64–66.

Mitchell, L.G., Listebarger, J.K. and Bailey, W. (1980) Epizootology and histopathology of *Chloromyxum trijugum* (Myxospora: Myxosporidia) in centrarchid fishes from Iowa. *Journal of Wildlife Diseases* 16, 233–236.

Molnar, K. (1982) Biology and histopathology of *Thelohanellus nikolskii* Akhmerov, 1955 (Myxosporea, Myxozoa), a protozoan parasite of the common carp (*Cyprinus carpio*). *Zeitschrift für Parasitenkunde* 68, 269–277.

Molnar, K. (1988) Further evidence that C blood protozoa of the common carp are stages of *Sphaerospora renicola* Dyková et Lom, 1982. *Bulletin of the European Association of Fish Pathologists* 8, 3–4.

Molnar, K. and Kovacs-Gayer, E. (1985) The pathogenicity and development within the fish host of *Myxobolus cyprini* Doflein, 1898. *Parasitology* 90, 549–555.

Molnar, K. and Kovacs-Gayer, E. (1986a) Observation on the intracellular and coelozoic developmental stages of *Hoferellus cyprini* (Doflein, 1898) (Myxosporea, Myxozoa). *Parasitologica Hungarica* 19, 27–30.

Molnar, K. and Kovacs-Gayer, E. (1986b) Experimental induction of *Sphaerospora renicola* (Myxosporea) infection in common carp (*Cyprinus carpio*) by transmission of SB-protozoans. *Journal of Applied Ichthyology* 2, 86–94.

Molnar, K. and Kovacs-Gayer, E. (1986c): Biology and histopathology of *Thelohanellus hovorkai* Achmerov, 1960 (Myxosporea, Myxozoa), a protozoan parasite of the common carp (*Cyprinus carpio*). *Acta Veterinaria Hungarica* 34, 67–72.

Molnar, K., Baska, F. and Szekely, C. (1987) Fumagillin, an efficacious drug against renal sphaerosporosis of the common carp *Cyprinus carpio*. *Diseases of Aquatic Organisms* 2, 187–190.

Molnar, K., Fischer-Scherl, T., Baska, F. and Hoffmann, R.W. (1989) Hoferellosis in goldfish *Carassius auratus* and gibel carp *Carassius auratus gibelio*. *Diseases of Aquatic Organisms* 7, 89–95.

Noble, E.R. (1944) Life cycles in the Myxosporidia. *Quarterly Review of Biology* 19, 213–235.

Odening, K., Walter, G. and Bockhardt, I. (1988) Koinzidentes Auftreten von PKX und *Sphaerospora* sp. (Myxosporidia) in Bestanden von *Salmo gairdneri* (Osteichthyes). *Angewandte Parasitologie* 29, 137–148.

Odening, K., Walter, G. and Bockhardt, I. (1989) Zum Infektionsgeschehen bei *Sphaerospora renicola* (Myxosporidia). *Angewandte Parasitologie* 30, 131–140.

Patashnik, M., Groninger, H.S. Jr, Barnett, H., Kudo, G. and Koury, B. (1982) Pacific whiting, *Merluccius productus*; I. Abnormal muscle texture caused by myxosporidian-induced proteolysis. *Marine Fish Review* 44, 1–12.

Pauley, G. (1974) Fish sporozoa: Extraction of antigens from *Myxosoma cerebralis* which mimic tissue antigens of rainbow trout (*Salmo gairdneri* Rafinesque). *Journal of the Fisheries Research Board of Canada* 31, 1481–1484.

Ratliff, D.E. (1983) *Ceratomyxa shasta*: longevity, distribution, timing and abundance of the infective stages in central Oregon. *Canadian Journal of Fisheries and Aquatic Science* 40, 1622-1632.

Schafer, W.E. (1968) Studies on the epizootology of the myxosporidian *Ceratomyxa shasta* Noble. *California Fish and Game* 54, 90-99.

Schmahl, G., Mehlhorn, H. and Taraschewski, H. (1989) Treatment of fish-parasites. 7. Effects of sym. triazinone (toltrazuril) on developmental stages of *Myxobolus* sp. Bütschli, 1882, (Myxosporea, Myxozoa): a light and electron microscopic study. *European Journal of Protistology* 25, 26-33.

Schmahl, G., Sénaud, J. and Mehlhorn, H. (1991) Treatment of fish parasites. 8. Effects of sym. triazinone (toltrazuril) on developmental stages of *Henneguya* sp. (Myxosporea, Myxozoa): a light and electron microscopic study. *Archiv für Protistenkunde* 140, 83-94.

Shulman, S.S. (1966) *Myxosporidia of the Fauna of the USSR* (in Russian). Nauka, Moscow-Leningrad, 504 pp.

Shulman, S.S. (ed.) (1984) *Parasitic Protozoa*. (In Russian). Vol. 1. In: Bauer, O.N. (ed.) *Key to the Parasites of Freshwater Fauna of USSR, Vol. 140 of the Keys to the Fauna of the USSR*. Nauka, Leningrad, 428 pp.

Siau, Y. (1977) Premiers stades du développement experimental, en cultures, de spores de la Myxosporidie *Myxobolus exiguus Thélohan*, 1895. *Compte Rendu de l'Academie des Sciences (Paris)* 285 (D), 681-683.

Sindermann, C.J. (1961) Parasite tags for marine fish. *Journal of Wildlife Management* 25, 41-47.

Smith, C.E., Morrison, J.K., Ramsey, H.W. and Ferguson, H.W. (1984) Proliferative kidney disease: first reported outbreak in North America. *Journal of Fish Diseases* 7, 1-10.

Smothers, J.F., von Dohlen, C.D., Smith, L.H.Jr and Spall, R.D. (1994) Molecular evidence that the myxozoan protists are metazoans. *Science* 265, 1719-1721.

Stehr, C. and Whitaker, D.J. (1986) Host-parasite interaction of the myxosporeans *Kudoa paniformis* Kabata and Whitaker, 1981 and *Kudoa thyrsites* (Gilchrist, 1924) in the muscle of Pacific whiting, *Merluccius productus* (Ayers): an ultrastructural study. *Journal of Fish Diseases* 9, 505-517.

Svobodová, Z. and Groch, L. (1986) Žaberni onemocnĕni vyvolanĕ invazí *Sphaerospora molnari*. *Bulletin VÚRH Vodñany* 22, 13-17.

Szuks, H. (1980) Applicability of parasites in separating groups of grenadier fish *Macrourus rupestris*. *Angewandte Parasitologie* 21, 211-214.

Taylor, R.E.L., Coli, S.J. and Junell, D.R. (1973) Attempts to control whirling disease by continuous drug feeding. *Journal of Wildlife Diseases* 9, 302-305.

Tsuyuki, H., Williscroft, S.N., Kabata, Z. and Whitaker, D.J. (1982) The relationship between acid and neutral protease activities and the incidence of soft cooked texture in the muscle tissue of Pacific hake (*Merluccius productus*) infected with *Kudoa paniformis* and/or *K. thyrsitis*, held for varying times under different prefreeze chilled storage conditions. *Canadian Technical Reports, Fisheries and Aquatic Science* 1130, 39 pp.

Uspenskaya, A.V. (1982) New data on the life cycle and biology of Myxosporidia. *Archiv für Protistenkunde* 126, 309-338.

Uspenskaya, A.V. (1984) *Cytology of Myxosporidia* (in Russian). Nauka Publ. House, Leningrad, 122 pp.

Wales, J.H. and Wolf, H. (1955) Three protozoan diseases of trout in California. *California Fish and Game* 41, 183-187.

Walliker, D. (1968) Studies on *Myxidium oviforme*, a myxosporidian parasite of Irish salmon, *Salmo salar*. *Parasitology* 58, 839-844.

Weidner, E. and Overstreet, R.M. (1979) Sporogenesis of a myxosporidian with motile spores. *Cell and Tissue Research* 201, 331–342.

Weissenberg, R. (1921) Zur Wirtsgewebsableitung des Plasmakörpers der *Glugea anomala* Cysten. *Archiv für Protistenkunde* 42, 400–421.

Willis, A.G. (1949) On the vegetative forms and life history of *Chloromyxum thyrsites* Gilchrist and its doubtful systematic position. *Australian Journal of Scientific Research, Ser. B, Biological Science* 2, 379–398.

Wolf, K. and Markiw, M.E. (1976) *Myxosoma cerebralis*: *in vitro* sporulation of the myxosporidia of salmonid whirling disease. *Journal of Protozoology* 23, 425–427.

Wolf, K. and Markiw, M.E. (1984) Biology contravenes taxonomy in the Myxozoa: new discoveries show alternation of invertebrate and vertebrate hosts. *Science* 225, 1449–1452.

Wood, C.C., Rutherford, D.T. and McKinnell, S. (1989) Identification of sockeye salmon (*Oncorhynchus nerka*) stocks in mixed-stock fisheries in British Columbia and southeast Alaska using biological markers. *Canadian Journal of Fisheries and Aquatic Science* 46, 2108–2120.

Yokoyama, H. (1993) Studies on the life cycle of some myxosporeans. PhD. thesis, The University of Tokyo.

Yokoyama, H., Ogawa, K. and Wakabayashi, H. (1993) Involvement of *Branchiura sowerbyi* (Oligochaeta: Annelida) in the transmission of *Hoferellus carassii* (Myxosporea: Myxozoa), the causative agent of kidney enlargement disease (KED) in goldfish *Carassius auratus*. *Gyobyo Kenkyu (Fish Pathol.)* 28, 135–139.

Zinn, J.L., Johnson, K.A., Sanders, J.E. and Fryer, J.L. (1977) Susceptibility of salmonid species and hatchery strains of chinook salmon (*Oncorhynchus tschawytscha*) to infections by *Ceratomyxa shasta*. *Journal of the Fisheries Research Board of Canada* 34, 933–936.

4

Phylum Microspora

I. Dyková

Institute of Parasitology, Czech Academy of Sciences, 370 05 České Budějovice, Branišovská 31, Czech Republic.

INTRODUCTION

Microsporidia are strictly intracellular protozoan parasites which are characterized by the production of spores. The spore has an elaborate extrusion apparatus with a conspicuous extrusible polar tube. They infect animals from protozoa to man. They are usually parasites of invertebrates, but are also widely distributed in teleosts in freshwater, estuarine and marine habitats. Many microsporidian species cause severe diseases in fish in aquaculture facilities, both in marine (e.g. *Glugea stephani* or *Pleistophora priacanthusis*) and freshwater (*G. plecoglossi, Microsporidium takedai* or *P. anguillarum*). In nature, especially in marine habitats, microsporidioses have their greatest impact on young fry and yearling fish. Microsporidioses may constitute one of the limiting factors which affect the growth of fish stocks, and, consequently, reduce productivity in capture fisheries.

MICROSPORIDIA – GENERAL CHARACTERISTICS

Microsporidia (phylum Microspora Sprague, 1977) is a rather isolated group among other protistan phyla. Their rRNA indicates that they are very low on the ladder of eukaryote evolution (Vossbrink *et al.*, 1987). Their phylogenetic origin is not known. Fish-infecting microsporidians are in the class Microspora Delphy, 1963. Their spores have an elaborate extrusion apparatus.

Microsporidian species infecting fishes are widely distributed both geographically and in a variety of hosts. For example, *Glugea stephani* occurs in ten host species (flatfishes in the Atlantic and Pacific Oceans, and in the Mediterranean and North Seas). Some species have even a wider host range e.g. *Pleistophora hyphessobryconis* infects 18 host species of four families and

G. hertwigi has a holarctic distribution. Other species have been known from only one locality and are limited to a host.

Morphology

Most microsporidians are small parasites. The largest developmental stages measure up to 50 µm. They have a very simple structure and lack mitochondria. Their nuclei are either single, isolated, or paired (diplokarya consisting of two closely apposed nuclei).

Spores are infective to fish. They are about 3–10 µm long and are mostly ellipsoidal or egg-shaped (Figs 4.1 SP, and see 4.4A–J below). They have a very complicated structure (Vávra, 1976; Canning and Lom, 1986). The solid, imperforate one-piece spore wall has a thin outer proteinaceous layer (exospore) and a thick inner chitinous layer (endospore). They have an elaborate hatching apparatus. The most conspicuous part is the polar tube (Fig. 4.1 SP–pt). Inserted into the base of the polar cap (or anchoring disc) is a laminar structure at the anterior end of the spore (Fig. 4.3A below). The cap stains with the PAS (periodic acid and Schiff reagent) for polysaccharides, and can be seen as a red granule under light microscopy. The tube extends obliquely from the cap to the posterior half of spore, where it forms a coil beneath the spore wall. The basal part of the tube is embedded in the polaroplast, an organelle consisting of a stack of membranes and of a vesicular posterior part. In spores of species that infect fish the posterior vacuole is very large (Figs 4.1 SP, 4.4A–J below).

The sporoplasm contains either a single nucleus or a diplokaryon and occupies the biconcave space between the polaroplast and the posterior vacuole.

Life cycle

Under an appropriate stimulus in the digestive tract of the host (which can be simulated *in vitro* by hydrogen peroxide), high pressure builds up within the spore. This results in swelling of the polaroplast. The pressure breaks the polar tube and everts it through the thinnest point in the apex of the spore shell. The sporoplasm is propelled through the hollow everting polar tube. The tube is stiff and solid enough to pierce any host cell within its reach and literally injects the sporoplasm into it (Undeen, 1990). The sporoplasm can develop directly in the epithelial cell of the intestine or, more commonly, it finds and enters a macrophage and thus reaches the final specific site in the body.

The ensuing development comprises two sequences – merogony (or schizogony) and sporogony (Fig. 4.1). Merogony serves to produce a great number of parasite stages or meronts. Meronts divide by binary or multiple fission. Thus there are uninucleate cells (Fig. 4.2C) as well as multinucleate plasmodia. Meronts may spread within the infected tissue from one cell to another. This may account for heavy infections with e.g. *Pleistophora* species. These stages were seen in microsporidia that infect insects (Kawarabata and Ishihara, 1984).

Sporogony (Fig. 4.1) produces spores. The intermediate stages are sporonts which are different from meronts. They have an electron dense coat and

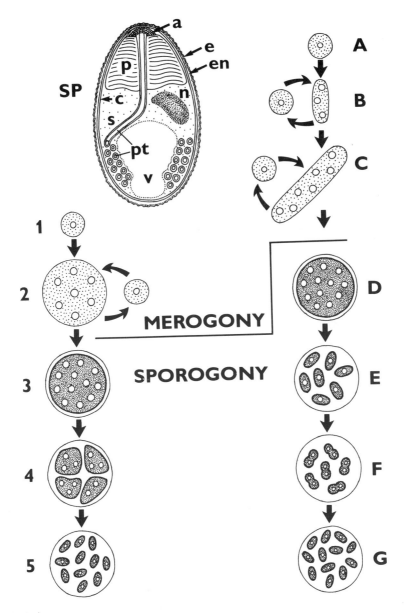

Fig. 4.1. Diagrammatic representation of the spore (SP) and the life cycle of microsporidia. The spore: a = anchoring disc or polar cap; p = polaroplast; s = sporoplasm with the nucleus (n); pt = polar tube; v = posterior vacuole; e = corrugated exospore; en = endospore; c = cell membrane. A to G. Life cycle of *Glugea*, 1 to 5 life cycle of *Pleistophora*; merogony stages are lightly, sporogony stages heavily stipled. 1,A. Released sporoplasms. 2. Merogony by multiple fission of multinucleate meronts. B,C. Merogony by segmentation of two types of cylindrical meronts. 3. Sporogonial plasmodium within a SPV envelope. 4. Its segmentation. 5. Formation of sporoblasts and spores. D. Sporogonial plasmodium within a SPV membrane. E. Cleavage into sporoblast mother cells. F. Their binary division into sporoblasts which mature into spores (G). (Modified after Canning and Lom, 1986.)

produce sporoblasts either by binary fission (disporoblastic sporogony) or they first grow into a multinucleate plasmodial sporont (or sporogonial plasmodium) (Fig. 4.2E). The plasmodial sporont produces sporoblasts by multiple fission or by gradual fragmentation or by producing intermediate stages, the sporoblast mother cells (in the genus *Glugea*). These in turn produce sporoblasts. The sporoblast matures into the spore.

In most genera, a special envelope is produced at the surface of the sporont and during sporogony it detaches from the parasite to demarcate a vesicle-like space around it. This is the sporophorous vesicle (SPV, formerly 'pansporoblast', a term appropriate for myxosporidia). The SPV envelope may be a thick wall or a thin membrane.

In most species, mature spores are of a uniform shape and size. In *Pleistophora*, there may be spore dimorphism (macrospores, microspores and sometimes 'intermediate' spores). They differ chiefly in size. More complicated spore dimorphism among fish-infecting microsporidia is known in *Spraguea lophii*.

Unlike the insect microsporidia (e.g. Becnel *et al.*, 1987), sexual stages have not been proven in fish-infecting species.

Species of some genera stimulate the infected cell to enormous hypertrophy. Such hypertrophic cells with hypertrophic nuclei and surface modifications (microvilli, invaginations or a thick wall) are called xenomas (Weissenberg, 1968; Sprague and Vernick, 1968) (Fig. 4.5A, B below). These cells and their microsporidian parasites integrate morphologically and physiologically and form a separate entity (Weissenberg, 1949) with its own development. Development of microsporidian infection within the fish host depends on many factors. Immunity to microsporidia is largely unexplored. Among environmental factors it is the ambient temperature that is known to influence development. Thus the growth of *Glugea plecoglossi* is extremely retarded at temperatures below 16°C (Takahashi and Egusa, 1977b) and experimental infections with *G. stephani* fail if the host is kept below 15°C (Olson, 1976). In *Microsporidium takedai*, low temperature (below 15°C) does not prevent infection, but the parasite only develops after the ambient temperature is raised above 15°C. Transfer of infected rainbow trout yearlings from the optimal 18°C to 8°C stops parasite development.

Transmission

Fish microsporidia are known to be transmitted perorally and directly between fish. Experimental infections were accomplished by feeding mature spores suspended in water or fed shortly before to planktonic organisms. Transmission by intramuscular or intraperitoneal injection of mature spores was successful in *Pleistophora hyphessobryconis* and *Glugea anomala* Weissenberg, 1968; Lom, 1969). Spores retain their infectivity in water at 40°C for at least one year.

HOST–PARASITE RELATIONSHIPS

The interactions of microsporidia as obligate intracellular parasites with the host have been studied mostly in natural infections (Dyková and Lom, 1980). Since the early studies by Weissenberg (1913, 1921) there have been several recent attempts to elucidate the host–parasite relationship in experimental infections (McVicar, 1975; Olson, 1976; Takahashi and Egusa, 1977a; Berrebi, 1978; Dyková and Lom, 1978; Matthews and Matthews, 1980).

Site selection of microsporidians within the hosts depends on their requirements for a particular cell type within which they can develop. Some species seem to be highly cell specific (e.g. *Spraguea lophii* develops in neurocytes, *Pleistophora mirandellae* in oocytes, other *Pleistophora* species in myocytes). The cell type is difficult to recognize after it has been transformed into a xenoma. This probably explains the ambiguous term 'mesenchymal cells' for migratory and connective tissue cells, which are unrecognizable after microsporidian infection.

Infected cells may become hyperbiotic when xenoma are formed (extensive hypertrophy or tumour-like outgrowth accompanied by fragmentation of the nucleus) or become hypobiotic (regressive changes) when they are infected by non-xenoma forming species. The final changes in both groups are the same, the cells are replaced with mature spores and eventually totally destroyed.

The tissue damage elicited by xenoma forming species is essentially the same as that caused by non-xenoma species. Both changes have regressive characters, i.e. various types of dystrophic changes, atrophy and necrosis. The extent of regressive changes is related to the intensity of infection; after it has reached a critical level the total volume of an organ or tissue is reduced. This seriously impairs their vital functions (Ralphs and Matthews, 1986).

Tissue reactions to microsporidia are essentially directed towards the isolation and elimination of the parasite. Two patterns are recognized according to whether the parasites are localized in xenoma or are diffused in the tissue.

The first type is exemplified by *Glugea* species and comprises three successive stages: a weakly reactive stage, a productive stage with the formation of granulomas, and granuloma involution. The extent of pathological changes depends on the intensity of infection.

The weakly reactive stage is often long-lasting and overlaps with later stages. It is characteristic in infections with young thin-walled xenomas (Fig. 4.2A,B) which contain only schizonts. The host tissue is damaged by pressure atrophy. This elicits proliferation of the connective tissue and results in formation of a thin concentric layer around the xenoma. Large developing xenomas which gradually fill with spores provoke a proliferative inflammation. The host tissue reacts by granulomatous inflammation. The granulation tissue matures and the granuloma undergoes involution. It diminishes in size and the resulting fibrous connective tissue may undergo hyalinization. A gradual repair of the tissue lesions occurs quite often; however, the original function of the organ is not restored.

The second type of host tissue reaction is represented by *Pleistophora* species which infect muscles. Host tissue reaction is surprisingly slight during

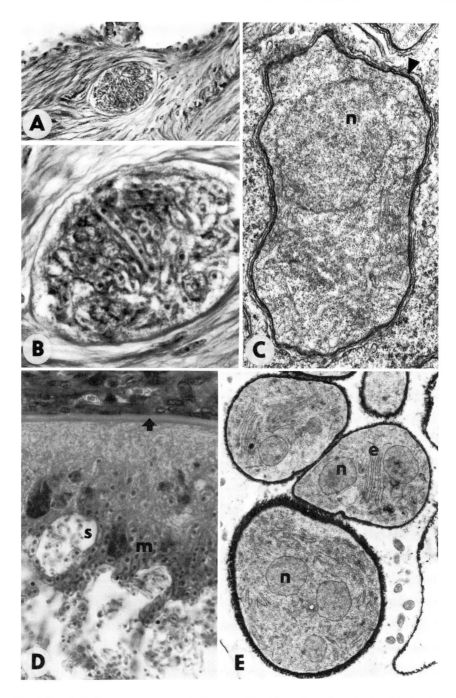

Fig. 4.2. A,B. Young xenoma of *Glugea* with thin cell wall embedded in lamina muscularis of the intestine (A); cylindrical meronts are visible (B) (×330 and ×1100, respectively). C. A *G. anomala* uninucleate meront encased with a cisternae of endoplasmic reticulum (arrowhead); n = nucleus (×26,000). D. Periphery of an advanced xenoma of *G. plecoglossi*; arrow indicates the thick xenoma

merogony and sporogony. There is little evidence that the invaded muscle fibres are isolated by host response. A slight lymphocytic infiltration of myosepta is the first indication of tissue reaction. The productive stage of the tissue reaction occurs when mature spores completely fill the infected muscle fibre. A thick wall of fibroblasts may be formed to demarcate the parasite mass as soon as it undergoes necrotic changes. Phagocytic cells play a crucial role in the host defence mechanism in both types of tissue reactions. Their exact nature and origin are still not clear. In the granulation tissue there are migratory phagocytic cells, macrophages and fibroblasts which ingest and digest spores (up to 30 spores per cell).

Many of the spore-filled macrophages in the centre of the granuloma disintegrate and the resulting spore-containing débris are taken up by other phagocytes. Eventually, the spore content and chitinous spore shells are completely digested (Fig. 4.3B).

The host tissue reaction eventually destroys the spores produced in the course of an infection. The efficiency of the defence response depends upon physiological conditions of the host and environmental conditions. The ambient temperature affects not only humoral response (Corbel, 1975; Roberts, 1975) but also phagocytosis. Low temperatures delay or inhibit macrophage response, fibroplasia and the activity of fixed macrophages of the reticuloendothelial system (Finn and Nielson, 1971; Ferguson, 1976) and also impede the development of the parasite. Thus the histopathological changes are the result of a complicated interaction of the host's response to the parasite and environmental temperature.

DIAGNOSIS OF MICROSPORIDIAN INFECTIONS

Microscopic examination is the only method to detect fish microsporidia. Microsporidia are diagnosed by their spores. Live spores (from lesions such as whitish nodules or streaks within e.g. muscle tissue, or detected by time-consuming search in fresh tissue mounts) can be easily identified by their large posterior vacuole. The posterior vacuole along with their uniform size and shape differentiate them from yeasts, fat droplets and other objects. The polar tube can also be expelled by e.g. 2% hydrogen peroxide. An important diagnostic character is the polar cap at the spore apex which stains as an intense red dot with PAS. In tissue sections spores stain dark blue with Giemsa. The ultimate proof is by using transmission electron microscopy to reveal polar tube coils in the spore and/or the characteristic cell structure in developmental stages. In the case of *Enterocytozoon salmonis*, tiny spores and their preceding stages are in the nuclei of the infected lymphoblasts.

---

wall, m = cylindrical meronts, s = sporoblast mother cell within the SPV membrane (×1100). E. Sporogonial plasmodium in the process of segmentation; n = nucleus, e = endoplasmic reticulum (×4900). A,B,D light micrographs, C,E electron micrographs. (Original.)

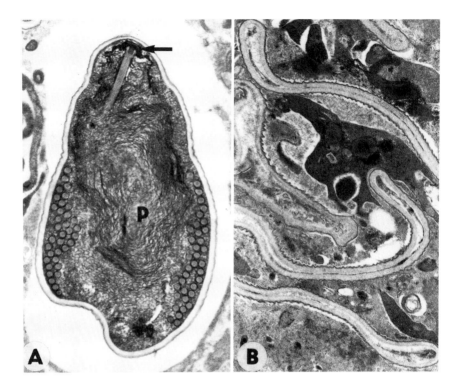

Fig. 4.3. A. Longitudinal section through a mature spore of *Pleistophora mirandellae*; arrow indicates the anchoring disc of the polar tube; p = polaroplast, anteriorly laminar, posteriorly vesicular; due to fixation the polaroplast moved into the posterior vacuole (×14,000). B. Digested spores, with only spore shells persisting of *P. mirandellae* engulfed by host macrophages (×11,000). (Original.)

Species of eleven genera and of the collective group *Microsporidium* are known to infect fish, seven of the genera being specific for fish. Irrespective of genus, all fish-infecting species have spores with an enormously large posterior vacuole, unmatched by microsporidia from other hosts. The classification used in the present review follows that of Canning (1990). Important references to microsporidian taxonomy may be found in Canning and Lom (1986), Larsson (1988) and Sprague (1977, 1982).

Brief definitions of genera of microsporidia parasitic in fishes are as follows:
Enterocytozoon Desportes *et al.*, 1985. Multinucleate merogony and sporogony stages; polar tubes develop in the sporogonial plasmodium. Spores small, hardly above 2 μm in length.
Glugea Thélohan, 1891. Forms large (up to 4 mm in diameter) xenomas with host nuclei fragments at the periphery and with a membrane-like envelope.
Heterosporis Schubert, 1969. Meronts, later together with sporophorous

vesicles are encased within a distinct common wall of parasite origin, the sporophorocyst. Macro- and microspores present.

Ichthyosporidium Caullery and Mesnil, 1905. Xenomas of two kinds: small compartmentalized ones (up to 0.2 mm) and large lobose ones (up to 4 mm), the latter with microvilli on the surface. Spores with diplokaryons.

Loma Morrison and Sprague, 1981. Forms xenomas up to 1 mm in size, with a single central nucleus; xenoma cell membrane is coated with fibrils. Up to eight spores are produced within a thin-walled sporophorous vesicle.

Microfilum Faye, Toguebaye and Bouix, 1991. Forms xenomas up to 0.8 mm in size, with microvilli on the surface. Spores oval with polar tube in form of a short rigid rod (manubrium).

Microgemma Ralphs and Matthews, 1985. Forms xenomas up to 0.5 mm in size with reticulate host cell nuclei at the periphery, their cell membrane extends into microvilli. Sporogony by exogenous budding.

Collective group *Microsporidium* Balbiani, 1881. This group serves as a repository for species that are insufficiently known for an exact generic identification.

Pleistophora Gurley, 1893. Multinuclear meronts and sporonts have a thick amorphous wall, which later becomes the wall of sporophorous vesicle. Macro- and microspores present.

Spraguea Vávra and Sprague, 1976. Forms xenomas (up to 1.5 mm) in infected ganglion cells. Produces two types of spores: stubby ones with a single nucleus and slender ones with a diplokaryon.

Tetramicra Ralphs and Matthews, 1985. Xenoma has a central, single reticulate nucleus and microvillous surface projections. Spores have a conspicuous inclusion in the posterior vacuole.

Representatives of three more genera were found in fish, but their taxonomic assignment or specificity for their fish host may be disputed: *Jírovecia* Weiser, 1977; *Nosemoides* Vinckier, 1975 and *Thelohania* Henneguy, 1892.

MICROSPORIDIAN INFECTIONS IN VARIOUS ORGANS AND TISSUES

Microsporidia differ in their site preferences and the microsporidian infections can be classified accordingly. The full range of organs infected by microsporidian species is presently unknown.

Species that infect multiple organs

Microsporidians that infect multiple organs are members of *Glugea*, *Tetramicra* and *Ichthyosporidium*. Since the host cells infected are connective tissue cells, xenomas can be found in any organ of the body.

Glugea anomala (Moniez, 1887) Gurley, 1893; probable synonyms are *G. weissenbergi* Sprague and Vernick, 1968 and *G. gasterostei* Voronin, 1974. This is the best known species in the genus. It is common in *Gasterosteus aculeatus* and *Pungitius pungitius* in Eurasia and North America, and in some localities its prevalence may exceed 50%.

The spores are elongate and slightly ovoid. The range of their highly variable size is 1.9–2.7 × 3–6 μm. The posterior vacuole extends to the middle of the spore (Fig. 4.4A). The cylindrical meronts divide by constriction into uninucleate daughter cells. Later, sporogonial plasmodium develops and divides into uninucleate sporoblast mother cells, each producing two spores.

The xenomas of *G. anomala* appear as white, usually spherical cysts which develop gradually to a diameter of several mm. Subcutaneously located xenomas may bulge from the organ surface and internal cysts may cause body deformities. At an early stage of infection the host cell starts its hypertrophy (Fig. 4.2A) and later it lays down a thick xenoma wall (Fig. 4.2D). The fragments of hypertrophic host cell nucleus are peripherally located. In young xenomas the meronts are distributed throughout the xenoma. As sporogony begins, the merogonic stages are confined to the periphery and sporogonic stages tend to concentrate in the centre (Fig. 4.5A) where mature spores accumulate. In the aging xenomas, the cytoplasm and fragmented host cell nucleus disappear and the xenoma changes into a 'bag' of spores. This is the stage in which the xenoma provokes host tissue reaction. The xenomas initially cause pressure atrophy to the surrounding tissue, later followed by a granulomatous inflammatory reaction (Fig. 4.5C). In spite of the reparative character of inflammation some changes due to infection (e.g. disappearance of tubular glands) may result in a permanent functional disorder of the infected organ. There is no exact data on morbidity and mortality of infected sticklebacks.

Glugea plecoglossi Takahashi and Egusa, 1977 causes one of the most dangerous diseases of cultured *Plecoglossus altivelis* in Japan. *Oncorhynchus mykiss* can be infected experimentally (Takahashi and Egusa, 1977a). The sites of infection are similar to those of *G. anomala*. The xenomas develop up to a diameter of 2 mm. The development of xenomas is temperature dependent (Takahashi and Egusa, 1977b). The spores are elongate ellipsoidal 2.1 × 5.8 μm and the posterior vacuole occupies about one half of the spore (Fig. 4.4H).

Gross lesions associated with the infection are related to the location of the xenoma. Large xenomas are visible through the skin and may occur as a bulge on the body surface. The sequence of tissue reaction to xenomas developing from early stages (Fig. 4.2A,B) to maturity (Fig. 4.5B) follows the pattern common in *Glugea* infections.

Glugea cepedianae (Putz, Hoffman and Dunbar, 1965) was originally described as a *Pleistophora* species from *Dorosoma cepedianum* in fresh waters in Ohio, USA. In the first year of life, up to 65% of young shad may be infected. Heavy infections result in high mortalities of gizzard shad which is a valuable bait fish in some areas. The spores are slender ovoid, 2.3 × 4.9 μm. Single large xenomas up to 1 cm in diameter were observed to compress the visceral organs against the body wall.

Glugea fennica (Lom and Weiser, 1969) infects subcutaneous tissue and fins of *Lota lota* in fresh waters of Finland and northern Russia. The spores are elongate oval, 2.6–2.7 × 7.1–7.4 μm.

Glugea atherinae Berrebi, 1978 infects the commercially important smelt *Atherina boyeri* in brackish lagoons on the French Mediterranean coast. The prevalence in separate lagoons may range from 1.2 to 21% and the infection is sex dependent. In general, females were found somewhat more frequently infected than males (Berrebi, 1978). The xenomas are basically in the same organs as in other *Glugea* species. The spores are elongate ovoid, 2.9 × 5.7 μm in size and the posterior vacuole occupies one half of the spore length.

The infection is acquired in the first and second year of life. Two types of xenomas are found in yearlings. Large xenomas are up to 14 mm in diameter, occur mainly in the body cavity and cause pressure atrophy of affected and surrounding organs. Numerous small xenomas (up to 0.2 mm in diameter) are found throughout the whole length of the post-oesophageal part of the digestive tract and may cause partial obstruction of the intestine. Experimental infections produced xenomas both in the body cavity and in subcutaneous connective tissue. The xenomas and their effects on organs contribute to mortality of the fish.

Ichthyosporidium giganteum (Thélohan, 1895) Swarczewsky, 1914 was originally described as *Glugea gigantea*. It infects marine fishes *Crenilabrus melops* along the Atlantic coast of France, *C. ocellatus* in the Black Sea, and *Leiostomus xanthurus* along the Atlantic coast of the USA. It has only been recorded once in the first two hosts and its prevalence in the last may reach 25% (Schwartz, 1963). The infected cells which transform into xenomas may be the cells of connective or fat tissue. The diplokaryotic spores are ovoid, 4 × 6 μm in size.

The site of infection is subcutaneous connective tissue, fat tissue and probably also liver parenchyma (Sprague and Hussey, 1980). The infection of *L. xanthurus* is extensive and causes ventral swelling extending from the head to the caudal fin. There are two types of xenomas. The small multilocular cystic formations are up to 0.2 mm in size, and develop simultaneously in enormous numbers while the huge thick-walled lobed xenomas are up to 4 mm in size. Inflammatory reactions develop around both types of xenomas. Proliferative inflammatory response to the large xenomas is characterized by epithelioid cells in a palisade-like arrangement. The early stages of *I. giganteum*, the development of xenoma, the hypothesis of Sprague and Hussey (1980) on the syncytial origin of 'multilocular cysts' as well as the prevalence of this unique parasite, warrant further study.

Tetramicra brevifilum Matthews and Matthews, 1980 was found in feral fish populations of *Scophthalmus maximus* off the Cornwall coast (UK). The prevalence on one-year-old fish reached 10%. In a turbot farm in northwest Spain, *T. brevifilum* was diagnosed in 1990 as an agent responsible for mortalities (11.5%) and low growth rate of survived specimens (Figueras *et al.*, 1992).

In fish with a high intensity of infection, xenomas and aggregates of spores are visible through the body wall and in organs as white nodules; muscle tissue develops a jelly-like consistency.

While in the liver, kidney, spleen and gills, globular xenomas up to 1 mm in size and xenoma-like agglomerations of spores are detected (Fig. 4.6C), elongated or ribbon-like xenomas predominate in the muscle tissue (Fig. 4.6A). These xenomas display a large mass of fragments of the host cell nuclei.

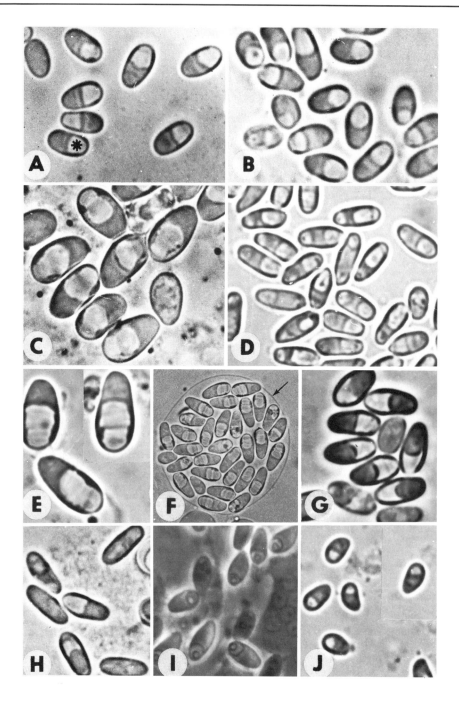

Fig. 4.4. Live and formol-fixed (D only) microsporidian spores. A. *Glugea anomala* from *Gasterosteus aculeatus*; asterisk marks the large posterior vacuole. B. *G. stephani* from *Pseudopleuronectes platessa* (×2700). C. *Pleistophora hyphessobryconis* from *Paracheirodon inessi* (×2800). D. *G. hertwigi* from *Osmerus eperlanus mordax* (×2700). E. Macrospores of *P. mirandellae* from

The ovoid spores are uninucleate, 2 × 4.8 μm in size with a large posterior vacuole (Fig. 4.1I) which is very rare among fish microsporidia because it contains large inclusion. A similar electron-dense body was found in the vacuole of *Nosemoides syaciumi* spore (Faye, 1992).

Except for pressure atrophy of considerable extent due to the large size of xenomas, proliferative inflammation is elicited around large agglomerations of spores in parenchymatous organs. In the muscle tissue giant xenomas displace surrounding muscle fibres in myomeres and provoke regressive changes, atrophy and liquefactive necrosis (Fig. 4.6A,D). The spores released from xenomas are surrounded only by a weak cellular response (Fig. 4.6B). The infection in *S. maximus* shows no organ preference neither in intensitiy nor in prevalence.

Species which infect the intestine

The microsporidians that infect the intestine of marine and freshwater fishes belong to the genus *Glugea*. *Glugea stephani* (Hagenmüller, 1989) Woodcock, 1904 (syn. *Nosema stephani* Hagenmüller, 1899) occurs in flatfish of the genera *Pleuronectes*, *Pseudopleuronectes*, *Parophrys* and *Rhombus* in the northern Atlantic and European seas. The prevalence may slightly exceed 50% and its development is temperature-dependent (see p. 152). Spores from *Pseudopleuronectes platessa* are oval and are about 2.7 × 4.7 μm in size (Fig. 4.4B).

Since xenomas are about 1 mm in diameter, the intestinal lesions can be observed macroscopically. In heavy infections the masses of developing xenomas in subepithelial connective tissue cause thickening of the intestinal wall, atrophy of the epithelium and luminal occlusion. Severe morphological alterations of the intestine along with functional failure may be limiting factors which affect the growth of feral and cultured plaice and flounders.

Glugea hertwigi Weissenberg, 1911 is also highly pathogenic and occurs mainly in the intestine of the euryhaline smelts. It was originally described from *Osmerus eperlanus eperlanus* in the Baltic Sea, and is also common in *O. eperlanus mordax* in Canada and the USA. In north Russian lakes it occurs in *Hypomesus olidus* and in fish of the genus *Coregonus*.

The spores are elongate oval, 2.2 × 4.7 μm in size (Fig. 4.4D). Unlike *G. anomala*, the sporoblast mother cells of *G. hertwigi* do not assume a dumbbell shape.

G. hertwigi forms giant xenomas (up to 4 mm in diameter) mainly in the intestine, though in massive infections almost all organs may be affected. The

Rutilus rutilus (×2700). F. A SPV with medium-sized spores of *P. mirandellae* from *Blicca bjoerkna* (×900). G. *Loma branchialis* from the gills of *Melonogrammus aeglefinus* (×3000). (B,D,G, courtesy of Dr J. Lom.) H. *G. plecoglossi* from *Plecoglossus altivelis* (×1400). I. *Tetramicra brevifilum* from *Scophthalmus maximus* (×2100; courtesy of Dr R.A. Matthews). J. *Microsporidium takedai* from *Oncorhynchus mykiss* (×2400; courtesy of Prof. S. Egusa).

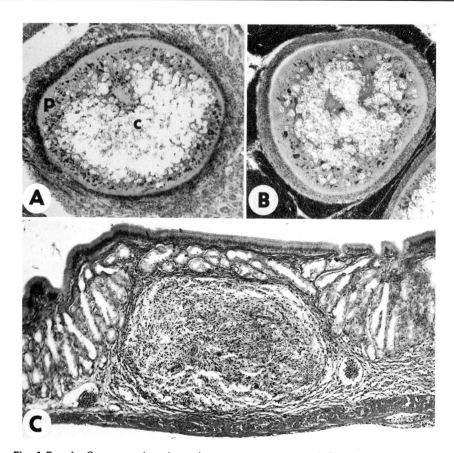

Fig. 4.5. A. Cross section through a mature xenoma of *Glugea anomala* in the subcutaneous connective tissue of the stomach of *Gasterosteus aculeatus*, p = periphery, c = central part of the xenoma replete with spores, (×100). B. *B. anomala* infection in *G. aculeatus*, granuloma in the subepithelial connective tissue of the glandular part of the stomach (×136). C. Xenoma of *G. plecoglossi* in the testis of *Plecoglossus altivelis*. (Original.)

maximum intensity of infection was about 250 xenomas per host (Sinderman, 1963). Severe epizootics with high mortalities were recorded in smelt in Canadian lakes, in New Hampshire (Haley, 1952, 1953, 1954) and in Russia (Anenkova-Khlopina, 1920; Bogdanova 1957; Petrushevski and Shulman, 1958). The effect of *G. hertwigi* on the intestine of the host is basically the same as in *G. stephani* infection. In the anterior part, the most heavily infected portion of the intestine (Delisle, 1972), the intestinal wall may disintegrate completely, and dystrophic changes associated with septicaemia or intoxication may result in death of the host. Infected fish may survive for a period, if organ functions are not severely impaired, and host tissue reactions (i.e. granulomatous inflammatory reaction) take place to eliminate xenomas or to repair the ulcerative lesions. The influence of stress is extremely important. In late spring, mortality is high due to spawning stress and in other seasons it is due to environmental stress (Nepszy and Dechtiar, 1972).

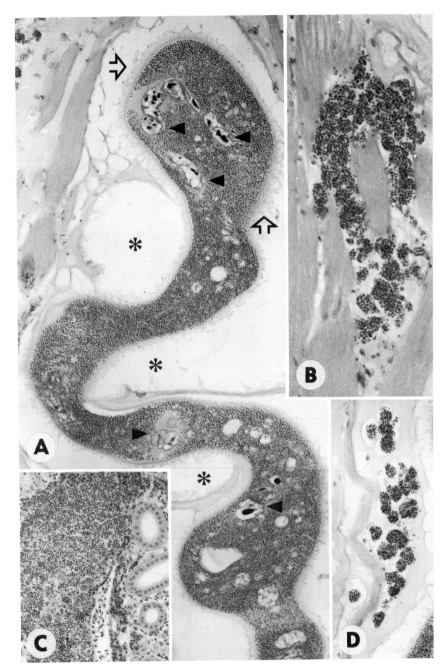

Fig. 4.6. A. *Tetramicra brevifilum* infection. Ribbon-like xenoma in the muscle tissue, not seen in its full length, with hairy surface projections (arrows). Note the multiple fragments of nuclei indicated by arrowheads and the space, probably fluid filled, around the xenoma (asterisk) (×220). B. *T. brevifilum* spores released from xenomas among the muscle fibres (×340). C. Agglomeration of *T. brevifilum* spores in the interstitial tissue of kidney (×220). D. Spores localized in the necrotic muscle tissue surrounding xenoma formation (×242). (Original.)

Glugea luciopercae Dogiel and Bykhovski, 1939, syn. *Glugea dogieli* Gasimagomedov and Issi, 1970, occurs in *Stizostedion lucioperca* in estuarine and freshwater habitats in Europe and Asia. The spores are elongate oval 2.1 × 4.5 μmm with a relatively narrow anterior end.

In the intestine of pike and perch numerous tiny xenomas fuse together to form a confluent layer in the subepithelial connective tissue. According to Shestakovskaya (personal communication) more than 30 xenomas may be fatal to small fry.

Glugea acerinae Jírovec, 1930 is a representative of a group of *Glugea* species which form small xenomas where the host cell nucleus is in a central position. The developmental stages of the parasite are not limited to the periphery of xenoma. The parasite occurs in *Gymnocephalus cernuus* in Central Europe. Spores are ellipsoid 2.7 × 4 μm with a large posterior vacuole.

The xenomas are spherical and do not exceed 0.5 mm in diameter. They are located in the subepithelial connective tissue of the intestine and are sometimes also found in the lamina muscularis. The parasite only causes slight pathological changes in host tissues.

Species which infect the liver

There are few records of microsporidians infecting liver parenchyma. The parasites belong to three genera, *Glugea*, *Microgemma*, and *Microsporidium*. They are known from marine hosts.

Glugea machari (Jírovec, 1934) Sprague, 1977 syn. *Octosporea machari* Jírovec, 1934, has been recorded from the liver of *Dentex dentex* in the Adriatic Sea. The xenomas are up to 400 μm in diameter and are localized near blood vessels.

Microsporidium (Thélohan, 1895) Sprague, 1977 syn. *Glugea ovoidea* Thélohan, 1895 was originally described from *Motella tricirrata*. It also occurs along the French coasts in *Cepola rubescens* and *Mullus barbatus*. Spores are ovoid, 1.5 × 2.5 μm in size. Uninucleate meronts divide by binary fission. Sporogony starts with binucleate cells dividing in two sporonts, each divides to produce two sporoblasts.

Microgemma hepaticus Ralphs and Matthews, 1896 was described from the liver of juvenile *Chelon labrosus* off the coast of Cornwall, UK. The prevalence was about 38%.

Pyriform spores, 2.4 × 4.2 μm size develop within xenomas which are up to 0.5 mm in diameter. Meronts divide by plasmotomy and are enclosed by host membrane. The sporogonic stages are free in the cytoplasm and divide by exogenous budding.

The foci of infection are manifested as white spots in the liver. Xenomas are associated with the walls of blood vessels, bile ducts and sometimes even with capsules of the encysted metacercariae *Bucephalus haimeanus* (see Ralphs and Matthews, 1986). Degenerated xenomas elicit a typical inflammatory reaction leading to granuloma formation. The parasite causes considerable mortalities in O+ age class *C. labrosus*.

Species which infect muscles

The microsporidians that infect muscle tissue are species of *Pleistophora*, *Heterosporis* and *Microsporidium*.

Pleistophora macrozoarcidis Nigrelli, 1946 is common in *Macrozoarces americanus*, in the western part of the northern Atlantic. The prevalence of infection increases with age and size of fish (Sheehy *et al.*, 1974). The microspores are usually ovoid, 3.5 × 5.5 μm and they have a large vacuole at the wider posterior end; the rare sausage-shaped macrospores are up to 8 μm long.

In early infections the parasite foci appear as small white cylinders lying along the axis of the body of the fish host; in advanced stages they appear as tumour-like masses with large areas of body musculature replaced with cell debris and spores. In most cases a thin layer of sarcoplasm remains in each infected muscle fibre. The infected muscle fibres may be encapsulated with concentrically arranged connective tissue (Fig. 4.7D).

Pleistophora ehrenbaumi Reichenow, 1929 is common in *Anarhichas lupus* and *A. minor* in the North Sea. Meyer (1952) reported 10% prevalence in *A. lupus* in the waters west of Iceland.

The spores are ovoid, uninucleate with a large posterior vacuole and distinct coils in the polar tube. Spore size varies from 1.5 × 3 μm (microspores) to 3.5 × 7 μm (macrospores).

In the infected muscle fibres there are mostly small brownish needle-like formations which are filled with spores. Egidius and Soleim (1986) found a giant tumour-like formation (about 8 × 15 × 4 cm) in an emaciated *A. lupus*. In the course of infection muscle fibres are gradually replaced by sporophorous vesicles filled with spores. As long as a thin peripheral layer of sarcoplasm remains no host tissue reaction is apparent. Enhanced phagocytosis of spores occurs when the remnants of muscle fibres disintegrate.

Pleistophora hippoglossoides Bosanquet, 1910, originally described from *Drepanopsetta hippoglossoides*, is also common in *Hippoglossoides platessoides* and *Solea solea*. Muscle tissue of heavily infected hosts is unfit for human consumption, because of consistency and organoleptic qualities. The elongate-ovoid or pyriform spores are monomorphic and are 2.7 × 4.7 μm in size.

The whitish parasitic nodule in the musculature shows through the skin. It consists of a mass of sporophorous vesicles lying in a structureless matrix with the remnants of the muscle fibre. Kabata (1959) observed displacement of the surrounding muscle tissue by these nodules.

Pleistophora hyphessobryconis Schäperclaus, 1941 occurs commonly in fresh water aquarium fish. In addition to its two original hosts *Paracheirodon innesi* and *Hemigrammus erythrozonus*, about 16 other fresh water fishes belonging to four families have been found infected. Some of these are feral fish (Fig. 4.7B). The species was imported in fish from the upper course of the Amazon river. However, it is presently distributed all over the world.

The spores are ovoid, 4 × 6 μm in size and one side is slightly more vaulted. The posterior vacuole occupies more than one half the spore length

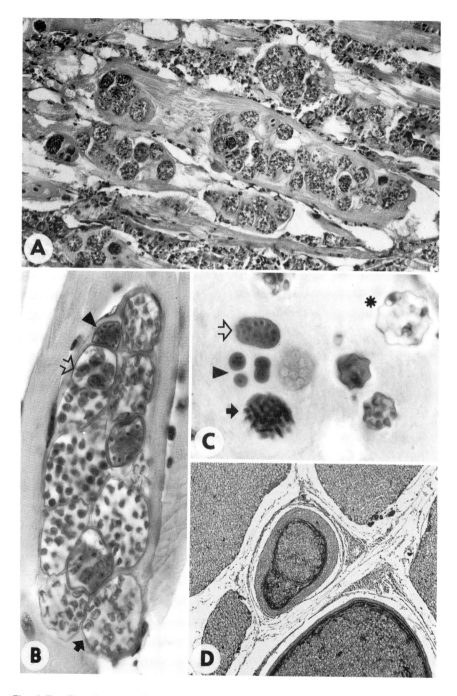

Fig. 4.7. Developmental stages of *Pleistophora hyphessobryconis*, mostly SPVs, pervading the muscle fibres of *Paracheirodon innesi* (×200). B. *P. hyphessobryconis* stages within a myocyte of *Phoxinus phoxinus*; arrowhead marks a sporogonial plasmodium, hollow arrow a segmenting plasmodium, dense arrow a SPV with mature spores (×850). C. Developmental stages of

(Fig. 4.4C). There are up to 34 coils of the polar tube. Sporophorous vesicles with 20 to 130 spores may reach up to 35 µm in diameter.

In the neon and glowlight tetras the infected segments of muscles are clearly visible as greyish or whitish patches under the skin. In heavy infections the parasite may spread to other organs and the body cavity. Developmental stages (Fig. 4.7C). Within the individual muscle fibres may be surrounded by amorphous dystrophic sarcoplasm or by unaffected peripheral sarcoplasm (Fig. 4.7B). As the infection progresses whole muscles are destroyed (Fig. 4.7A). The resulting debris contains masses of sporophorous vesicles and free spores together with host phagocytes with ingested spores.

The genus *Heterosporis* Schubert, 1969 has three species, two of them infect ornamental fishes.

Heterosporis finki Schubert, 1969 is found in the oesophageal wall and skeletal muscles of *Pterophyllum scalare*. The spores are ovoid with a large posterior vacuole. The macrospores are 2.5 × 8 µm, and the microspores 1.5 × 3 µm in size.

The pathogenicity of *H. finki* in juvenile farmed *P. scalare* was studied by Michel *et al.* (1989). The parasite caused significant losses and adult fish appeared to be more resistant. In severe infections fish were emaciated with discoloured or greyish spots on the skin. All the skeletal muscles were milky white and creamy in texture. The striated muscles contained numerous sporophorocysts.

Heterosporis schuberti Lom, Dyková, Körting and Klinger, 1989 infects ornamental fish *Pseudocrenilabrus multicolor* and *Ancistrus cirrhosus*. Massive infection of muscle tissue has been repeatedly found in both fishes and the infection caused up to 95% mortality in *A. cirrhosus*.

The spores from *P. multicolor* vary greatly in shape and size. They may be broadly ovoid microspores (3.1 × 4.2 µm in size), medium size spores (about 6.5 µm in size and 9 per sporophorous vesicle) and macrospores (4.4 = 7 µm). The posterior vacuole occupies about two-thirds of the spore length.

Infected specimens of *P. multicolor* were partly emaciated and showed signs of distress. Large parts of the trunk musculature were pervaded with parasites. The earliest stages identified were early meronts. They were wedged in myocytes and were without a halo of disintegrated sarcoplasm. Small or young sporophorocysts (Fig. 4.8A) were surrounded by muscle fibrils which extended perpendicularly from the cyst surface. In the final stages of infection, the walls of the sporophorous vesicles disappeared within the cyst and a thick wall of connective tissue was built up around the sporophorocysts (Fig. 4.8B). Several sporophorocysts were fused together to form a large cystic structure

P. hyphessobryconis in the dystrophic myocytes of *P. phoxinus*; arrowhead designates the meronts, hollow arrow a sporogonial plasmodium, full arrow a SPV with sporoblasts, and asterisk marks mature spores within a SPV (×860). D. *P. macrozoarcidis* spores replacing all muscle fibres of *Macrozoarces americanus* (×60). (Original.)

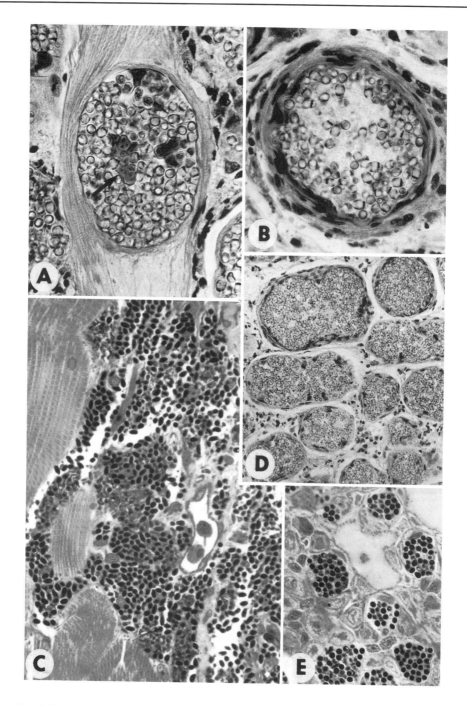

Fig. 4.8.   A,B,D. *Heterosporis schuberti* from *Ancistrus cirrhosus*. A. Sporophorocyst within a muscle fibre; arrow points at merogony stages (×660). B. Sporophorocyst encased with connective tissue cells contains mature spores released from SPVs (×640). D. Sporophorocysts full of spores replacing the

with a common connective tissue wall. Between the sporophorocysts there was moderate cellular infiltration (Fig. 4.8D). Eventually, spores were released into spaces between the myocytes. The muscle was thus converted into a disorganized mass of damaged myocytes, cystic structures, free spores and aggregates of macrophages with ingested spores (Fig. 4.8D).

Heterosporis anguillarum Hoshina, 1951 syn. *Pleistophora anguillarum* Hoshina, 1951 is a serious pathogen of farmed Japanese eel *Anguilla japonica* in Japan and Taiwan. Severe infections localized mainly in skeletal muscles cause growth retardation and/or considerable mortalities. Infected eels may be unsuitable for market. The spores are elongate-ovoid with a huge posterior vacuole occupying almost two-thirds of the spore length. Microspores average 2.4 × 7.8 µm.

The clinical signs include deformities of the body, whitish spots beneath the skin, depressed sites of the trunk musculature and its liquefaction (T'sui and Wang *et al.*, 1988). Infected muscle fibres are gradually replaced with cysts which are full of mature spores. Along with these regressive changes in the muscles the host inflammatory reaction is initiated. In advanced infection the disappearance of muscle fibres or whole myomeres is associated with an extensive fibroblastic response.

Many of the non-xenoma forming *Microsporidium* species infect skeletal muscles. Two of these are important pathogens of cultured commercial fish in Japan.

Microsporidium seriolae Egusa, 1982 infects *Seriola quinqueradiata* in Japanese maricultures. The spores are ovoid or pyriform (2.2 × 3.3 µm in size) and are produced by multiple fission of multinucleate sporogonial plasmodia. Sporophorous vesicles and diplokarya are absent.

Pathological changes are manifested as surface depressions which indicate disintegrated regions of muscles. The amorphous cheese-like tissue debris contains various parasitic stages.

Microsporidium takedai (Awakura, 1974) syn. *Glugea takedai* Awakura, 1974, originally found in *Oncorhynchus mykiss*, infects seven other salmonid species in Japan and Sakhalin. The prevalence may be 100% in *O. mykiss* and *O. gorbuscha*; in *O. masou*, *O. keta* and in *Salvelinus leucomaenis* it may range from 86 to 93%. The infection is temperature dependent and is restricted to the warmer period of year (Urawa, 1989). The parasite grows well at 18°C and at 8°C it stops growing. The spores with subapically attached polar tubes are ovoid, 2 × 3.4 µm in size (Fig. 4.4J).

M. takedai produces in skeletal muscles spindle-shaped cyst-like formations which are up to 6 mm in size. In the heart they are globular, about 2 mm in diameter. Both types of 'cysts' contain proliferating microsporidia in a cytoplasmic mass which lacks host cell nuclei or xenoma wall or any other distinct boundary. The heart muscle 'cysts' prevail in chronic infections and

muscle tissue (×220). C,E. *Microsporidium arthuri* from *Pangasius sutchi*. C. Mass of spores pervading the muscle tissue (×880). E. Isolated groups of spores in macrophages in the epidermis (×820). (Original).

develop at lower temperatures. In the acute disease there is high mortality and enormous numbers of cysts in the trunk musculature (up to 130 'cysts' per gram of tissue). There is a strong negative correlation between the condition of fish and intensity of infection. Mortalities were also observed in wild masu salmon (Urawa, 1989). In experimentally infected yearlings host tissue reaction started on day 11 with inflammatory infiltration. Phagocytic cells appeared on day 17 and their number declined to a minimum on day 30. The phagocytic response was more pronounced in 2-year-old trout.

Microsporidium arthuri Lom, Dyková and Shaharom, 1990 was observed in *Pangasius sutchi* from Thailand. In heavy infections the parasite destroyed parts of the musculature. Round pyriform uninucleate spores averaging 2.1 × 3.1 µm.

Mature spores form large aggregations in the skeletal muscles and in the myosepta of skeletal muscles (Fig. 4.8C). Spore-filled macrophages are scattered among the epithelial and specialized club cells of the epidermis (Fig. 4.8E). Spores ingested in macrophages are transported across the epidermis to the outside milieu.

Species which infect gills

Members of the genus *Loma*, which are morphologically very similar, are restricted to gill filaments or lamellae.

Loma branchialis (Nemeczek, 1911) Morrison and Sprague, 1981 syn. *Nosema branchialis* Nemeczek, 1911 and *Loma morhua* Morrison and Sprague, 1981, are found in *Gadus (Melanogrammus) aeglefinus* and six other species of gadid fish in the North Atlantic and northern seas. The prevalence is about 10 to 60%. Spores are ovoid, 2.3 × 4.8 µm (Fig. 4.4G).

Xenomas are macroscopic. They are white cyst-like formations (about 0.5 mm in size) and are localized in gill filaments or in the lamellae. The reported infections are usually light. Distortions and atrophy or local vascular changes in gill filaments did not seem to affect the host. This tissue reaction which eliminates xenomas and repairs altered gills is similar to the reaction provoked by other xenoma-forming species (Fig. 4.9C). The overall damage to the gills correlates with the intensity of infection.

Loma salmonae (Putz, Hoffman and Dunbar, 1965) Morrison and Sprague, 1981 syn. *Pleistophora salmonae* Putz, Hoffman and Dunbar, 1965 is widespread in feral and hatchery-reared *Oncorhynchus mykiss*, *O. masou* and *O. nerka* throughout North America and Japan. It has been imported into France with *O. kisutch*. Considerable mortalities were recorded in Japan and North America. In one Californian epizootic almost all the fingerlings were lost in a hatchery (Putz, 1964). Spores are pyriform, 2.4 × 7.5 µm in size, two or four are produced in sporophorous vesicles. Uninucleate meronts develop into elongate plasmodia which have at least five nuclei.

The clinical signs of infection are similar to those in *L. branchialis*. In heavy infections gill filaments are distorted, secondary lamellae fuse and there is a marked epithelial hyperplasia in all parts of gills. Wales and Wolf (1955) correlated anaemia with the intensity of *L. salmonae* infection.

Loma fontinalis Morrison and Sprague, 1983, described from *Salvelinus fontinalis* in Canada, is very similar to *L. salmonae*.

Species which infect the gonads

Several species that belong to the genera *Pleistophora* and *Microsporidium* were found to infect gonads of fishes. Most of them are restricted to female hosts. There are also two species of the genus *Thelohania* that infect oocytes (Canning and Lom, 1986).

Pleistophora mirandellae Vaney and Conte, 1901 syn. *Pleistophora longifilis* Schubert, 1910 and *P. oolytica* Weiser, 1949 is common in *Abramis brama*, *Barbus barbus*, *Blicca bjoerkna*, *Leuciscus leuciscus*, *Rutilus rutilus* and other European cyprinids. It is sometimes also found in *Esox lucius* and *Hucho hucho* (see Maurand *et al.*, 1988).

Sporogonial plasmodia arise from oval meronts with one to four nuclei and produce sporophorous vesicles (Figs 4.4F, 4.10A,C). There are macrospores with size range 7.4–12 × 3.9–6.4 µm; (about 30–60 per sporophorous vesicle (SPV) – Fig. 4.4E,F), medium-size spores (about 100 per SPV, size range 4–6.5 × 2–3.5 µm), and microspores (3 × 1.5 µm in size).

The macroscopic white lesions are obvious in the gonads of male and female fish. The infection of oocytes probably occurs before the zona radiata is formed. The parasite develops within the yolk which is eventually replaced by the spores (Fig. 4.10A). Up to 20% of oocytes may be invaded. The oocytes or their remnants are surrounded by a pronounced inflammatory reaction of the proliferative granulomatous type and contain numerous regressively changed spores (Fig. 4.10D). In testes the developmental stages of *P. mirandellae* are in the supporting connective tissue among the seminiferous tubules (Fig. 4.10B). Schubert (1910) reported hypertrophy of infected epithelial cells of the seminiferous tubules followed by their destruction. *P. mirandellae* infection may reduce fecundity in both male and female fish.

Pleistophora ovariae Summerfelt, 1964 is a common parasite of *Notemigonus crysoleucas* and *Pimephales promelas* in the USA with an overall prevalence of about 46% and with a much higher prevalence in bait minnow hatcheries. The prevalence fluctuates with season, being higher in the spawning season (May and June). The sporophorous vesicles, 23 µm in diameter, contain 8 to 20 elongate ovoid spores 8.4 × 4.2 µm in size. The posterior vacuole occupies two-thirds of the spore length.

The development of *P. ovariae* takes place in oocytes. Infected oocytes become prematurely atretic. Spores released from the oocytes are removed by phagocytosis. The course of the inflammatory reaction has not been described. In the post-spawning season a mass of proliferating connective tissue was noted in infected ovaries of fish. Mortalities were not observed but parasitic castration may reduce fecundity of shiners and minnows by more than 40% (Summerfelt, 1964).

Microsporidium sulci (Rašín, 1936) syn. *Cocconema sulci* Rašín, 1936 commonly infects oocytes of *Acipenser ruthenus* and *A. guldenstadti* in the Danube and Volga river basins. Spherical spores, which are 2.5 µm in

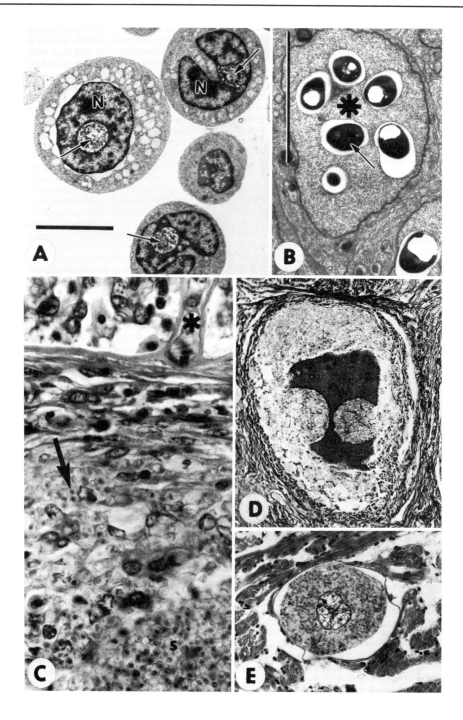

Fig. 4.9. A,B. *Enterocytozoon salmonis* from *Oncorhynchus tshawytscha* (courtesy of Prof. R.P. Hedrick). A. Developmental stages in the nuclei of leucocytes from the kidney. B. Spores within the nuclei of the epithelial cells of a urinary tubule. Bars = 5 μm. C. Periphery of a *Loma branchialis* xenoma

diameter, develop in sporophorous vesicles. Sporogonial stages have diplo-karya, meronts are cylindrical.

Species which infect ganglion neurocytes

Spraguea lophii (Doflein, 1898) Weissenberg, 1976 syn. *Glugea lophii* Doflein, 1898 is a common parasite of *Lophius piscatorius* and of four anglerfish along the coasts of Europe and Iceland, in the North Sea, along the US Atlantic coast and along the coast of Brazil. The prevalences vary from 32% to 100% in *L. piscatorius*, *L. budegassa* and *L. americanus*, and seem to increase with the age of the host (Priebe, 1971).

Spraguea is a dimorphic microsporidian with two developmental sequences, each producing spores of a different type (Loubés *et al.*, 1979). In small xenomas there are uninucleate oval spores of the *Nosemoides*-type. These are 2.5 × 4.5 µm in size. In advanced xenomas there is a peripheral region with different types of developmental stages and slender curved *Nosema*-type spores (1.4 × 3.7 µm in size) with diplokarya.

Pathological lesions associated with *S. lophii* are conspicuous when the cerebrospinal region of the host is examined. Intracellular development of *S. lophii* in the neurocytes of ganglia of the brain and spinal nerves causes hyper-trophy with formation of nodules which resemble a bunch of grapes. There is proliferation of glial cells between the xenomas and this turns the whole spore mass into a granuloma (Fig. 4.9D). The vital functions of the hosts, however, do not seem to be impaired.

Species which infect haematopoietic cells

Enterocytozoon salmonis Chilmonczyk, Cox and Hedrick, 1991 infects blood leucocytes and haematopoietic cells of the lymphocyte lineage in the spleen and kidney of *Oncorhynchus tshawytscha* in the Pacific Northwest of the USA. Similar haematopoietic infections were reported in *O. mykiss* in British Columbia, Canada, and France.

Spores are ovoid, 1.5 × 2 µm in size, and have 8 to 12 turns of the polar tube. All stages develop in direct contact with the host cell nucleo-plasm (Fig. 4.9A,B). There is no diplokaryon at any stage of development (Chilmonczyk *et al.*, 1991).

in the gills of *Melanogrammus aeglefinus* being destroyed by the host tissue reaction. Asterisk marks the lumen of a capillary in the secondary gill lamella; arrowhead points at spores engulfed by a macrophage, s marks the central, still intact mass of spores (× 1100). D. *Spraguea lophii* xenoma being destroyed by the host tissue reaction; the dark central parasitic mass is eroded by macrophages (× 160; from Canning and Lom, 1986). E. Xenoma of *Glugea* sp. from the cardiac muscle of *Ictalurus punctatus*; an example of a group of species forming relatively small xenomas with a central, unfragmented nucleus (× 350; original).

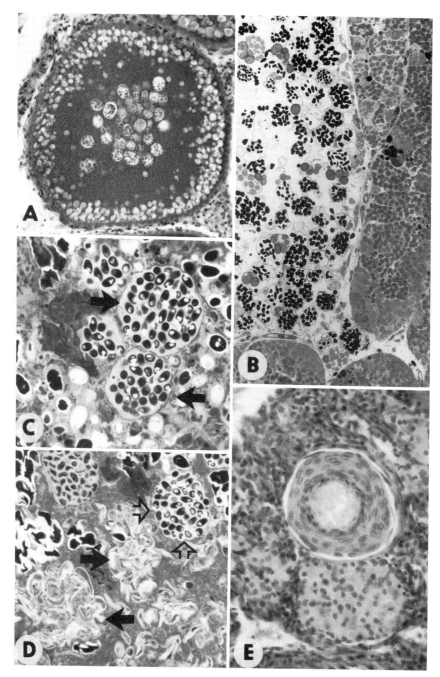

Fig. 4.10. *Pleistophora mirandellae* infection in *Blicca bjoerkna* (A), *Alburnus alburnus* (B), and *Rutilus rutilus* (C,D,E). A. An oocyte with developing SPVs in the centre (×200). B. SPVs of *P. mirandellae* in the testis (×240). C. Full arrow points at the thin membrane of SPVs in the ovary (×1050). D. The full arrow marks masses of spore shells, the hollow arrows indicate SPV (×860). E. Granuloma formation in the ovary infected with *P. mirandellae* (×340). (Original.)

In the chinook salmon *O. tshawytscha* up to 37% of spleen hemoblasts harboured these parasites (Elston *et al.*, 1987). Nonhaematopoietic cells, e.g. epithelial cells of the urinary tubules or mesangial cells of glomeruli were occasionally infected and this was usually associated with a high rate of infection in haematopoietic tissues. The clinical signs are anaemia, lymphoblastosis and leukaemia but these have not been confirmed in experimental infections.

IMMUNODEPRESSION IN FISH INFECTED WITH MICROSPORIDIA

Glugea stephani, a parasite of the intestine of flatfish, induces decrease of the serum IgM level and elicits immunosuppression. The latter is probably caused by soluble factors (prostaglandin and/or leukotrienes) which are released into the serum of the host. Evidence for this is, e.g. the ability to transfer the immunosuppression to healthy fish by injecting them with the serum of exposed fish. Suppression of the immune response defends the parasite's existence in the host, but also impairs the humoral response to other infective agents. *Glugea* interferes with the initiation as well as with other stages of the response. The immune ability recovers under conditions supporting continuous survival of the infected host (Laudan *et al.*, 1989).

PREVENTION AND CONTROL

Prevention is often difficult to control because of direct transmission and longevity of spores. Precautions include proper sanitation of culture vessels, jars and tanks. The construction of special rearing facilities for ornamental fish will have to take into account the prevention of microsporidian infections, entailing systems for strict quarantine of fish (Michel *et al.*, 1989).

Of the many chemicals tested, toltrazuril has thus far been found promising. Schmahl and Mehlhorn (1989) recommended 5 or 20 µg toltrazuril ml^{-1} of water (a symmetric triazinone, Bay Vi, 1942) as a short bath for 4 or 1 hours. This is applied for 6 days at two-day intervals in well aerated water. The chemical only kills the vegetative stages but does not affect mature spores.

Previous attempts at chemotherapy of *Glugea plecoglossi* with fumagillin (Takahashi and Egusa, 1976) and of *Microsporidium takedai* with amprolium (Awakura and Kurahashi, 1967) yielded ambiguous results and hence is not recommended.

CONCLUSIONS AND RECOMMENDATIONS FOR FUTURE STUDIES

Relatively little is known about the biology of fish microsporidia. This is especially true of their relationships with their hosts. Also, their taxonomy needs further studies. This is evidenced by continued discoveries of interesting or economically important new genera and species in this phylum.

Our existing knowledge indicates that microsporidian infections are important in hatcheries and may even control the population growth of feral fishes (e.g., *Glugea hertwigi, G. stephani*) or of cultured fishes (*Enterocytozoon salmonis, Microsporidium takedai*). It is reasonable to expect that with the development of intensive fish culture (especially in regions such as southeast Asia) new microsporidians will emerge as important disease agents.

Future studies should concentrate on devising methods for *in vitro* maintenance of these parasites. This would help to elucidate the immune reactions of fish to microsporidian infection, susceptibility of fish to the parasite and factors affecting the development of microsporidia. Further *in vivo* studies might help to indicate the host cell which transfers the infection from the intestine to target organs and how the merogony stages spread within the host. In contrast to mammalian microsporidia (Canning and Lom, 1986) no one has thus far successfully cultured fish microsporidia outside the host. Establishing an *in vitro* culture in fish cell lines would help to resolve parasite–host cell interactions, the development of xenoma and would facilitate studies on physiology and biochemistry of fish microsporidia.

With the development of *in vivo* and *in vitro* systems to maintain these parasites, a more systematic approach to the development of chemotherapeutic drugs could be adopted.

REFERENCES

Anenkova-Khlopina, N.P. (1920) Contribution to the study of parasitic diseases of *Osmerus eperlanus* (in Russian). Izvestiya Otdela Rybovodstva Nauchno-promyslovykh Issledovanii 1,2.

Awakura, I. and Kurahashi, S. (1967) Studies on the *Plistophora* disease of salmonid fish. II. On prevention and control of the disease. *Scientific Reports of Hokkaido Fish Hatchery* 22, 51–68.

Becnel, J.J., Hazard, E.I., Fukuda, T. and Sprague, V. (1987) Life cycle of *Culicospora magna* (Kudo, 1920) (Microsporidia: Culicosporidae) in *Culex restuans* Theobald with special reference to sexuality. *Journal of Protozoology* 34, 313–322.

Berrebi, P. (1978) Contribution á l'étude biologique des zones saumatres du littoral méditerranéen francais. Biologie d'une Microsporidie: *Glugea atherinae* n.sp. parasite de l'Atérine: *Atherina boyeri* Risso, 1810, (Poisson – Teleostéen) des étangs cotiers. Thesis, Université des Sciences et Techniques du Languedoc, Montpellier, 196 pp.

Bogdanova, E.A. (1957) The microsporidian *Glugea hertwigi* Weissenberg in the stint (*Osmerus eperlanus* m. *spirinchus*) from the lake Ylyua-yarvi (in Russian). In: Petrushevski, G.K. (ed.) *Parasites and Diseases of Fish*. Izvestiya Vsesoyuznogo Nauchno-Issledovatelskogo Instituta Ozernogo i Rechnogo Rybnogo Khozyaistva, Leningrad, p. 328.

Canning, E.U. (1990) Phylum Microspora. In: Margolis, L., Chapman, D., Melkonian, M. and Corliss, J.O. (eds) *The Handbook of Protoctista*. Jones and Bartlett Publishers, Boston, pp. 53–72.

Canning, E.U. and Lom, J. (1986) *The Microsporidia of Vertebrates*. Academic Press, New York, 289 pp.

Chilmonczyk, S., Cox, W.T. and Hedrick, R.P. (1991) *Enterocytozoon salmonis* n.sp.: an intranuclear microsporidian from salmoned fish. *Journal of Protozoology* 38, 264–269.

Corbel, M.J. (1975) The immune response in fish: a review. *Journal of Fish Biology* 7, 539–563.

Delisle, C.E. (1972) Variations mensuelles de *Glugea hertwigi* (Sporozoa: Microsporidia) chez différents tissus et organes de l'éperlan adulte dulcicole et conséquences de cette infection sur une mortalité massive annuelle de ce poisson. *Canadian Journal of Zoology* 50, 1589–1600.

Dyková, I. and Lom, J. (1978) Tissue reaction of the three-spined stickleback *Gasterosteus aculeatus* L. to infection with *Glugea anomala* (Moniez, 1887). *Journal of Fish Diseases* 1, 83–90.

Dyková, I. and Lom, J. (1980) Tissue reactions to microsporidian infections in fish. *Journal of Fish Diseases* 3, 265–283.

Egidius, E. and Soleim, O. (1986) *Pleistophora ehrenbaumi*, a microsporidian parasite in wolffish, *Anarchichas lupus*. *Bulletin of the European Association of Fish Pathologists* 6, 13–14.

Elston, R.A., Kent, M.L. and Harrell, L.H. (1987) An intranuclear microsporidium associated with acute anemia in the chinook salmon, *Oncorhynchus tshawytscha*. *Journal of Protozoology* 34, 274–277.

Faye, N. (1992) Microsporidies des poissons des cotes Senegalaises: faunistique, biologie, ultrastructure. Thesis, University of Montpellier II, 127–140.

Ferguson, H.W. (1976) Studies on the reticulo-endothelial system of fishes. PhD Thesis. University of Stirling, UK.

Figueras, A., Novoa, B., Santarem, M., Martinez, E., Alvarez, J.M., Toranzo, A.E. and Dyková, I. (1992) *Tetramicra brevifilum*, a potential threat to farmed turbot *Scophthalmus maximus*. *Diseases of Aquatic Organisms* 14, 127–135.

Finn, J.P. and Nielson, N.O. (1971) The effect of temperature variation on the inflammatory response of rainbow trout. *Journal of Pathology* 105, 257–268.

Haley, A.J. (1952) Preliminary observations on a severe epidemic of microsporidiosis in the smelt *Osmerus mordax* (Mitchill). *Journal of Parasitology* 38, 183–185.

Haley, A.J. (1953) Observations on a protozoan infection in the freshwater smelt. *Proceedings of the 32nd Annual Session of the New Hampshire Academy of Science* p. 7.

Haley, A.J. (1954) Microsporidian parasite, *Glugea hertwigi*, in American smelt from the Great Bay region, New Hampshire. *Transactions of the American Fisheries Society* 83, 84–90.

Kabata, Z. (1959) On two little-known microsporidia of marine fishes. *Parasitology* 49, 309–315.

Kawarabata, T. and Ishihara, R. (1984) Infection and development of *Nosema bombycis* (Microsporida: Protozoa) in a cell line of *Antherea eucalypti*. *Journal of Protozoology* 44, 52–62.

Larsson, J.I.R. (1988) Identification of microsporidian genera (Protozoa, Microspora) – a guide with comments on the taxonomy. *Archiv für Protistenkunde* 136, 1–37.

Laudan, R., Stolen, J.S. and Cali, A. (1989) The effect of the microsporida *Glugea stephani* on the immunoglobulin levels of juvenile and adult winter flounder (*Pseudopleuronectes americanus*). *Developmental and Comparative Immunology* 13, 35–41.

Lom, J. (1969) Experimental transmission of a microsporidian, *Pleistophora hyphessobryconis*, by intramuscular transplantation. *Journal of Protozoology* 16 (Suppl.), p. 17.

Loubès, C., Maurand, J. and Ormières, R. (1979) Étude ultrastructural de *Spraguea lophii* (Doflein, 1898), microsporidie parasite de la baudroie: essai d'intepretation du dimorphism sporal. *Protistologica* 15, 43–54.

Matthews, R.A. and Matthews, B.F. (1980) Cell and tissue reaction of turbot *Scophthalmus maximus* (L.) to *Tetramicra brevifilum* gen.n., sp.n. (Microspora). *Journal of Fish Diseases* 3, 495–515.

Maurand, J., Loubès, C., Gasc, C., Pelletier, J. and Barral, J. (1988) *Pleistophora mirandellae* Vaney and Conte, 1901, a microsporidian parasite in cyprinid fish of rivers in Hérault: taxonomy and histopathology. *Journal of Fish Diseases* 11, 251–259.

McVicar, A.H. (1975) Infection of plaice *Pleuronectes platessa* L. with *Glugea (Nosema) stephani* (Hagenmüller, 1899) (Protozoa: Microsporidia) in a fish farm and under experimental conditions. *Journal of Fish Biology* 7, 611–619.

Meyer, A. (1952) Veränderung des Fleisches beim Katfish. Fischereiwelt, 57–58.

Michel, C., Maurand, J., Loubés, C., Chilmonczyk, S. and de Kinkelin, P. (1989) *Heterosporis finki*, a microsporidian parasite of the angel fish *Pterophyllum scalare*: pathology and ultrastructure. *Diseases of Aquatic Organisms* 7, 103–109.

Nepszy, S.J. and Dechtiar, A.O. (1972) Occurrence of *Glugea hertwigi* in Lake Erie rainbow smelt (*Osmerus mordax*) and associated mortality of adult smelt. *Journal of the Fisheries Research Board of Canada* 29, 1639–1641.

Olson, R.E. (1976) Laboratory and field studies on *Glugea stephani* (Hagenmüller), a microsporidian parasite of pleuronectid flatfishes. *Journal of Protozoology* 23, 158–164.

Petrushevski, G.K. and Shulman, S.S. (1958) Parasitic diseases in fish in water reservoirs of the USSR (in Russian). In: Petrushevski, G.K. and Polyanski, Yu.I. (eds) *Parasitology of Fishes*. Leningrad University Press, Leningrad, pp. 301–320.

Priebe, K. (1971) Zur Verbreitung des Befalls des Seeteufels (*Lophius piscatorius*) mit *Nosema lophii* auf Fischfangplätzen im östlichen Nordatlantik. *Archiv für Fischereiwissenschaft* 22, 98–102.

Putz, R.E. (1964) Parasites of freshwater fish. II. Protozoa. 1. Microsporidia of fish. Fishery Leaflet 571, 1–4, US Department of the Interior, Washington, DC.

Raabe, H. (1936) Etudes de micro-organismes parasites des poissons de mer. I. *Nosema ovoideum*. Thél. dans le foie des Rougets. *Bulletin de Institut Oceanographique* 695, 1–12.

Ralphs, J.R. and Matthews, R.A. (1986) Hepatic microsporidiosis due to *Microgemma hepaticus* n.g., n.sp. in juvenile grey mullet *Chelon labrosus*. *Journal of Fish Diseases* 9, 225–242.

Roberts, R.J. (1975) The effect of temperature on diseases and their histopathological manifestations in fish. In: Ribelin, W.E. and Migaki, G. (eds) *The Pathology of Fishes*. The University of Wisconsin Press, Madison, Wisconsin, pp. 477–496.

Schmahl, G. and Mehlhorn, H. (1989) Treatment of fish parasites. 6. Effects of sym. triazinone (toltrazuril) on developmental stages of *Glugea anomala*, Moniez 1887 (Microsporidia): a light and electron microscopic study. *European Journal of Protistology* 24, 252–259.

Schwartz, F.J. (1963) A new ichthyosporidium parasite of the spot (*Leiostomus xanthurus*): A possible answer to recent oyster mortalities. *Progressive Fish Culturist* 25, 181–184.

Sheehy, D.J., Sissenwine, M.P. and Saila, S.B. (1974) Ocean pout parasites. *Marine Fisheries Review* 36, 29–33.

Sinderman, C.J. (1963) Disease in marine populations. *Transactions of the North American Wildlife and Natural Resources Conference* 28, 336–356.

Sprague, V. (1977) Classification and phylogeny of the Microsporidia. In: Bulla, L.A., Jr and Cheng, T.C. (eds) *Comparative Pathobiology. Vol. 2. Systematics of the Microsporidia.* Plenum Press, New York, 510 pp.

Sprague, V. (1982) Microspora. In: Parker, S.P. (ed.) *Synopsis and Classification of Living Organisms, Vol. 1.* McGraw-Hill, New York, pp. 589–594.

Sprague, V. and Hussey, K.L. (1980) Observations of *Ichthyosporidium giganteum* (Microsporidia) with particular reference to the host–parasite relations during merogony. *Journal of Protozoology* 27, 169–175.

Sprague, V., Vernick, S.H. (1968) Light and electron microscope study of a new species of *Glugea* (Microsporidia, Nosematidae) in the 4-spined stickleback *Apeltes quadracus. Journal of Protozoology* 15, 547–571.

Summerfelt, R.C. (1964) A new microsporidian parasite from the golden shiner, *Notemigonus crysoleucas. Transactions of the American Fisheries Society* 93, 6–10.

Takahashi, S. and Egusa, S. (1976) Studies of *Glugea* infection of the ayu, *Plecoglossus altivelis.* II. On the prevention and treatment. 1. Fumagilin efficacy as a treatment. *Japanese Journal of Fisheries* 11, 83–88.

Takahashi, S. and Egusa, S. (1977a) Studies on *Glugea* infection of the ayu, *Plecoglossus altivelis* – I. Description of the *Glugea* and a proposal of a new species, *Glugea plecoglossi. Japanese Journal of Fisheries* 11, 175–182.

Takahashi, S. and Egusa, S. (1977b) Studies on *Glugea* infection of the ayu, *Plecoglossus altivelis* – III. Effect of water temperature on the development of xenoma of *Glugea plecoglossi. Japanese Journal of Fisheries* 11, 195–200.

T'sui, W.H. and Wang, C.H. (1988) On the *Pleistophora* infection in eel. I. Histopathology, ultrastructure and development of *Pleistophora anguillarum* in eel, *Anguilla japonica. Bulletin of the Institute of Zoology, Academia Sinica, Taiwan* 27, 159–166.

Undeen, A.H. (1990) A proposed mechanism for the germination of microsporidian (Protozoa: Microspora) spores. *Journal of Theoretical Biology* 142, 223–235.

Urawa, S. (1989) Seasonal occurrence of *Microsporidium takedai* (Microsporida) infection in masu salmon, *Oncorhynchus masou*, from the Chitose river. *Physiology and Ecology Japan.* Spec., Vol. 1, 587–598.

Vávra, J. (1976) Structure of the Microsporidia. In: Bulla, L.A., Jr and Cheng, T.C. (eds) *Comparative Pathobiology, Vol. 1., Biology of Microsporidia.* Plenum Press, New York, pp. 2–85.

Vossbrink, C.R., Maddox, J.V., Friedman, S., Debrunner-Vossbrink, B.A. and Woese, C.R. (1987) Ribosomal RNA sequence suggests microsporidia are extremely ancient eukaryotes. *Nature* 326, 411–412.

Wales, J. and Wolf, H. (1955) The protozoan diseases of trout in California. *California Fish and Game* 41, 183–187.

Weissenberg, R. (1913) Beiträge zur Kenntnis des Zeugungskreises der Mikrosporidien *Glugea anomala* Moniez and *G. hertwigi* Weissenberg. *Archiv für Mikroskopische Anatomie* 82, 81–163.

Weissenberg, R. (1921) Zur Wirtsgewebsableitung des Plasmakorpers des *Glugea anomala* Cysten. *Archiv für Protistenkunde* 42, 400–421.

Weissenberg, R. (1949) Cell growth and cell transformation induced by intracellular parasites. *Anatomical Record* 103, 517–518.

Weissenberg, R. (1968) Intracellular development of the microsporidian *Glugea anomala* Moniez in hypertrophying migratory cells of the fish *Gasterosteus aculeatus* L., an example of the formation of 'xenoma' tumors. *Journal of Protozoology* 15, 44–57.

5

Ichthyophthirius multifiliis and *Cryptocaryon irritans* (Phylum Ciliophora)

H.W. Dickerson and D.L. Dawe

Department of Medical Microbiology, College of Veterinary Medicine, University of Georgia, Athens, Georgia 30602, USA.

PHYLUM CILIOPHORA – GENERAL CONSIDERATIONS

The phylum Ciliophora currently contains 7200 known species and it is likely that there are several thousand more that are as yet undiscovered (Corliss, 1979). Ciliates are found in virtually any body of water ranging from small ponds or streams to open oceans. Most are free-living, but many occur as commensals or symbionts, and a few are parasites of invertebrates and vertebrates. Ciliates range in size from approximately 10 µm to 4500 µm and their body shape varies from nearly spherical to highly ovoid (Corliss, 1979). They have a rigid body form resulting from a distinct fibrous cortex just below the surface plasma membrane. The ciliate cell has a large number of organelles and inclusions.

Some of the distinguishing characteristics of ciliates (Nanney, 1980; Corliss, 1979) are:

1. Nuclear dimorphism. With few exceptions ciliates possess two nuclei; a transcriptionally active macronucleus, and at least one germ-line micronucleus.
2. Cilia or compound ciliary organelles such as cirri. These are often present in large numbers and arranged in distinct patterns. Some species have them only in one stage of the life cycle.
3. Infraciliature located below the surface of the cell. This infrastructure consists of basal bodies and a host of more or less closely associated microtubules and fibrils.
4. A cytostome, or cell mouth, is usually present.

Some ciliates are commensals on the gills or skin of fishes and others (such as *Tetrahymena* spp.) are opportunistic parasites. Ciliates are often found on dead or moribund fishes. Parasitic ciliates which derive benefit from the host to the detriment of the host's survival, are not common. This chapter will describe the two most virulent parasitic ciliates of fishes, *Ichthyophthirius multifiliis* and *Cryptocaryon irritans*.

INTRODUCTION

Ichthyophthirius multifiliis, Fouquet, 1876, and *Cryptocaryon irritans*, Brown, 1951, are pathogenic ciliates that infect fresh- and salt-water fishes, respectively. In this chapter we will review what is known about the pathobiology of both these parasites. In each section we will usually discuss *I. multifiliis* first and then follow with what is known about *C. irritans*. The reason is that there is much more known about *I. multifiliis* than its marine counterpart.

I. *multifiliis*, referred to as Ich, is the most pathogenic of protozoan parasites of fishes. There is no official record on the annual economic loss attributed to ichthyophthiriosis, even though it is considered to be a major problem in aquaculture (Hoffman, 1970). Epizootics were reported in China as early as the tenth century (Dashu and Lien-Siang, cited by Hines and Spira, 1974a). The first major outbreak in North America was described in 1898 by Stiles. Ich was relatively unknown in Russia before 1940, but since then, it has become a serious disease of carp (Bauer, 1970). The significance of the problem will continue to increase with the growth of aquaculture.

Concomitant with the rapid development of mariculture in the last decade, *C. irritans* has also become an increasingly important problem. It was first reported in Japan by Sikama (1938).

GEOGRAPHICAL DISTRIBUTION AND HOST RANGE OF THE PARASITES

Ichthyophthirius multifiliis

Geographical distribution

I. *multifiliis* is a cosmopolitan parasite of fishes. The infection has been reported from virtually all areas where fishes are cultured, including the tropics and subarctic, and in feral fish populations from most continents (Nigrelli *et al.*, 1976; Valtonen and Keränen, 1981). The organism was first described in detail by Fouquet in 1876 in France (Stiles, 1894). According to Stiles, Hildendorf and Paulick published observations on a fish disease caused by a ciliate in Hamburg, Germany in 1869. These were the first detailed descriptions of the parasite and its life cycle. However, the disease was known in fishes in Europe in the Middle Ages (Hoffman, 1970) and there is evidence that it originated in Asia as a parasite of carp (Dashu and Lien-Slang, referenced by Hines and Spira, 1974a).

The parasite appears to have been spread to its present range as a result of transportation of fishes (Nigrelli *et al.*, 1976). Paperna (1972) reported on an outbreak in Uganda and noted that *I. multifiliis* was not found in that country in a previous study. He further indicated that a number of fish species had been imported into Uganda from the United States and Hong Kong. Wurtsbaugh and Tapia (1988) described an epidemic in fishes in Lake Titicaca on the Peru-Bolivian border. They noted that a number of fishes, salmonid and atherineds, had been introduced into Lake Titicaca and its watersheds at various times in the past.

Bragg (1991) found a higher incidence of Ich in salmonid fishes from South African rivers in areas where intensive aquaculture occurred. He suggested that aquaculture contributed to increased infections in both river and hatchery fishes.

Host range

I. multifiliis appears to parasitize all freshwater fishes. There are no records of species with natural resistance (Ventura and Paperna, 1985). However, there are suggestions of variation in the degree of susceptibility between fishes. These variations may depend on such factors as genetic background, physiological status of the fishes, parasite strains, and environmental conditions (Clayton and Price, 1988, 1994).

Epizootics in native fish populations often occur in only one species of fish. Elser (1955) reported on an outbreak in a reservoir in Maryland. The disease affected predominantly yellow perch (*Perca flavescens*). Allison and Kelly (1963) described an epizootic in rivers of northwestern Alabama. The majority of infected fishes were gizzard shad (*Dorosoma cepedianum*) and threadfin shad (*D. petenense*). In an outbreak reported in Kentucky the only fishes infected were blackstripe topminnows (*Fundulus notatus*) (see Kozel, 1976). A recent epizootic in Lake Titicaca on the Peru–Bolivia border primarily affected killifish (*Orestias* spp.) and the majority (93%) of dead were *O. agassii* (see Wurtsbaugh and Tapia, 1988). The occurrence of epizootics in only one or a few species in a mixed fish population may not indicate genetic variation in resistance, but rather different physiologic states which predispose certain individuals or groups to disease. Wurtsbaugh and Tapia (1988) observed that most of the fishes that died in the epizootic were gravid or spent adult *O. agassii*. Pickering and Christie (1980) in a study on parasite infection of brown trout (*Salmo trutta* L.) found Ich more frequently in precocious mature pre-spawning males than in immature fishes. They concluded that sexual maturation was associated with an increase in prevalence and severity of infestation with ectoparasites. Reproductive stress may also be a factor in the apparent variation in susceptibility to infection.

Nigrelli *et al.* (1976) proposed that there are physiological races of *I. multifiliis*. They suggested that the physiological races are related to the temperature tolerance of the host fishes. Thus, there are races of *I. multifiliis* that infect cold (7.2–10.6°C) water fishes such as salmon, and others that infect warm (12.8–16.1°C) water tropical fishes. Fishes with wide ranges of temperature tolerance such as carp and catfish may be susceptible to both cold and warm water parasites. Ich epizootics in arctic fishes suggest that these outbreaks occur when the water temperature reaches a moderate range. Valtonen and Keränen (1981) reported on an epizootic in a salmon hatchery in central Finland in two consecutive years when the water temperature rose to 14°C or higher.

Paperna (1972) stated that *I. multifiliis* in Europe and Asia is highly pathogenic for carp and it has a preference for this species. There was no data to support this speculation. Although ichthyophthiriosis often occurs in carp in Europe and Asia, this could merely be due to the fact that carp are the primary fish raised in these areas, and intensive culture conditions predisposes them to infection.

Seasonal fluctuations in infection

Outbreaks of *I. multifiliis* occur when conditions are favourable for rapid multiplication of the parasite. These include a suitable environment and susceptible fishes. According to McCallum (1982), fish density is not a constraint on the establishment of infection. However, there does appear to be a requirement for some minimum number of fishes before an epizootic occurs. Severe infections occur most commonly in dense populations of fishes.

A critical condition for an outbreak is water temperature. The duration of the developmental cycle of *I. multifiliis* is significantly influenced by temperature (Nigrelli *et al.*, 1976). Generally, as the temperature rises (up to 25–28°C) parasite activity increases and the life cycle is completed in a shorter time than at lower temperatures. In addition, the number of tomites in each cyst varies with the temperature of the water (Nigrelli *et al.*, 1976).

Stress can bring about an outbreak in a fish population. Stress is a complex physiologic reaction that causes the release of steroids from the adrenal glands which in turn decreases the immune function of the host. A wide variety of factors can induce stress in fishes: these include crowding, low dissolved oxygen, chemical pollutants in the water, high temperature, and spawning activities.

Ichthyophthiriosis is most likely to occur when fishes are stressed and the water temperature is relatively warm. Epizootics occur in aquarium-raised fishes when optimum conditions for parasite development and rapid multiplication are present. The parasite reproduces when fishes are under stress from low oxygen and/or crowded conditions. Outbreaks occur in the spring as the water warms and when fishes are spawning (Elser, 1955; Allison and Kelly, 1963). This is very apparent in reports of disease in sub-arctic areas. *I. multifiliis* epizootics in Finland occurred when water temperatures were above 14°C, and stopped as soon as the temperature fell (Valtonen and Keränen, 1988).

The cyclic nature of outbreaks is also influenced by development of immunity in the fish population. It is well documented that fishes infected with *I. multifiliis* develop protective immunity (see Burkart *et al.*, 1990). Epizootics occur when there is a sufficiently large population of susceptible fishes. As the infection progresses, the highly susceptible fishes die while those that are more resistant develop immunity. With time the majority of the surviving fishes will be immune. As these conditions develop, the epizootic wanes and losses stop. The surviving population of fishes breeds and with time the level of disease resistance decreases; this will initiate another epizootic under the right conditions.

How does the parasite survive between outbreaks? The ciliate is an obligate parasite and as such requires susceptible hosts to propagate. There is no evidence of a resistant stage to ensure survival during environmental changes or in the absence of fishes. Hence, it must maintain a low level infection in a population. The dynamics of the host–parasite relationship during the inter-epizootic periods is not understood.

Temperature is an important factor in the persistence of an infection in a fish population. When the water temperature is low, the developmental period of the parasite is increased. The growth period of trophonts on the fish varies from 1 week at 20°C to 20 days at 7°C (Nigrelli *et al.*, 1976). Also, it can occur at temperatures as low as 2–4°C (Bauer, 1953). McCallum (1982) found no density-dependent constraints on the establishment of Ich infections in naive hosts. There was, however, significant variation in the numbers of parasites found on each fish. These differences were attributed to such factors as the amount of mucus produced by individual fishes, nutritional status, and stress. In a subsequent study, McCallum (1985) examined the role of host death on the reproductive potential of *I. multifiliis* in a population of susceptible fishes. He concluded that the low prevalence between epizootics was consistent with the idea that *I. multifiliis* and fish populations are regulated by parasite-induced host mortality. He developed a mathematical model which qualitatively reproduced the observed epizootic behaviour of an *I. multifiliis* infection.

Cryptocaryon irritans

Host range, geographical distribution, prevalence, seasonal fluctuations

Crypcaryon irritans is a parasite of marine fishes. Cryptocaryonosis was first described in Japan in 1938 by Sikama. The disease is now thought to occur worldwide (Sikama, 1961). The parasite is a recurring problem in marine fishes in aquaria and is becoming a concern in commercial mariculture (Colorni, 1985). It was suggested that the parasite spread in the United States through the importation of fishes from Hawaii and the Indo-Pacific area (Nigrelli and Ruggieri, 1966).

The prevalence of the parasite in native fishes in North America is not well documented. The parasite was found on a few fishes in Mission Bay, California (Wilkie and Gordin, 1969). These fishes were examined at the time when there was a concomitant outbreak of cryptocaryonosis in aquaria at the Scripps Institute of Oceanography. It was unclear whether the native fishes were naturally infected, or whether they acquired the infection as a result of contamination with tomonts and theronts released from the Institute.

Cryptocaryon irritans is not fastidious in its host selection (Colorni, 1985). However, the host range does not appear as great as that of *I. multifiliis*. Fishes vary in susceptibility to infection. Elasmobranchs are generally resistant (Wilkie and Gordin, 1969). Also, there are apparent differences in degrees of susceptibility in marine teleosts. Colorni (1985) observed that when outbreaks of *C. irritans* infections occurred in aquaria with mixed fish species, some fishes were more readily infected than others. This confirmed an earlier observation of Wilkie and Gordin in 1969. Nigrelli and Ruggieri (1966) listed 27 species of marine fishes native to the Atlantic or Indo-Pacific areas that had died of *C. irritans* infections. Wilkie and Gordin (1969) reported 93 susceptible and 19 resistant marine fishes. It was suggested that these differences might be related to the ecological niches of the various species (Wilkie and Gordin, 1969).

Cryptocaryonosis appears to be primarily a disease of fishes in tropical environments. It does not occur in water below 19°C (Nigrelli and Ruggieri, 1966; Wilkie and Gordin, 1969) and it is seen most commonly in aquaria kept at 20-25°C (Nigrelli and Ruggieri, 1966). Temperature affects the development of the parasite (Cheung *et al.*, 1979). Many potentially susceptible fishes reside in cold water; these can be infected if the temperature rises above 19°C (Wilkie and Gordin, 1969).

SYSTEMATICS

Ichthyophthirius multifiliis

The first description of the aetiologic agent of white spot disease was by Hilgendorf and Paulicki in 1869 (see Stiles, 1894). They described the life cycle of the protist and placed it in the genus *Pantotricha*. In 1876 Fouquet published a detailed description of the organism and its life cycle. He placed the organism in a new genus and proposed the name *Ichthyophthirius multifiliis*. At present *multifiliis* is the only recognized species of *Ichthyophthirius* (Lee *et al.*, 1985). However, based on differences in gross nuclear morphology between isolates, it was suggested by Nigrelli *et al.* (1976) that there may be more than one. Serotypic strains of *I. multifiliis* exist based on immobilization *in vitro* (Dickerson *et al.*, 1993).

Ichthyophthirius multifiliis is placed in class Oligohymenophora, subclass Hymenostomata, order Hymenostomatida, suborder Ophryoglenina and the family Ichthyophthiridae.

Cryptocaryon irritans

Sikama (1938) described a parasitic holotrichous ciliate as the cause of a disease outbreak in marine fishes and suggested that the disease was similar to ichthyophthiriosis. In 1961 Sikama recognized that this organism was different from *I. multifiliis* and proposed the name *Ichthyophthirius marinus*. However, Brown (1951) had already described a ciliate, *Cryptocaryon irritans*, which caused disease in marine fishes. Brown (1963) published a more detailed description of the cycle of macronuclear development which distinguishes *C. irritans* from *I. multifiliis*.

Although Sikama was the first to describe cryptocaryonosis, the disease may have been observed much earlier by Kerbert (see Stiles, 1894).

Cryptocaryon irritans is placed in class Oligohymenophora, subclass Hymenostomata, order Hymenostomatida, suborder Ophryoglenina, and family Ichthyophthiriidae. However, Cheung *et al.* (1981) in a scanning electron microscopic study of *C. irritans* found that the organism does not have oral accessory membranes (membranelles) similar to those of *I. multifiliis* (see below) and hence questioned the inclusion of *C. irritans* in the order Hymenostomatida. Colorni and Diamant (1993) noted that the similarities between *I. multifiliis* and *C. irritans* are more likely the result of an adaptive convergence of life histories, rather than phylogenetic proximity. These authors suggested a reassessment of the taxonomical status of *C. irritans* as well. Distinct biological and pathological differences have also been found between isolates of *C. irritans* taken from fish in the Red and Mediterranean Seas (Diamant *et al.*, 1991; Colorni and Diamant, 1993).

PARASITE MORPHOLOGY AND LIFE CYCLE

Ichthyophthirius multifiliis

I. multifiliis (Ich) is a holotrichous histophagous ciliate. It possesses a large sausage- or horseshoe-shaped macronucleus and at least one small round micronucleus (Peshkov and Tikhomirova, 1968; Hauser, 1972). The micronuclei of ciliates are transcriptionally inactive and play a role in genetic exchange. While micronuclear exchange (autogamy) occurs in some ciliates, others reproduce indefinitely without it. It is not known whether *I. multifiliis* undergoes autogamy. Our own experience indicates that laboratory isolates maintained by serial passage on fishes gradually lose their infectivity and appear to become senescent. When this occurs, we can usually 'rejuvenate' the infection by introducing a new isolate. We do not know whether the old population is replaced or whether the two isolates undergo a process of genetic mixing.

I. multifiliis cyclically transforms between an obligate fish-associated trophont, and a free-living reproductive stage, the tomont. The theront is the infective stage of the parasite. The nomenclature used in this chapter is from standard references of ciliate taxonomy and protozoan parasitology (Corliss, 1979; Lee *et al.*, 1985; Lom and Dykova, 1992). Refer to Fig. 5.1 for a summary of the various stages involved in the life cycle. The daughter cells (tomites) of the tomont differentiate within a thin-walled cyst (see below) into free-swimming infective theronts. The theront is pyriform to fusiform in shape with a tapered posterior end (see Fig. 5.2). It is about $30 \times 50\,\mu m$ in size, but this varies greatly depending on the initial size of the tomont (MacLennon, 1942; Canella and Rocchi-Canella, 1976; Kozel, 1986; Geisslinger, 1987). The theront is completely covered with cilia each of which measures approximately $5.0\,\mu m$ in length by 0.2 in diameter (Kozel, 1986). A longer cilium, two to three times the length of the others, protrudes from the posterior end of the cell (Canella and Rochi-Canella, 1976; McCartney *et al.*, 1985; Kozel, 1986; Geisslinger, 1987).

Life Cycle of
Ichthyophthirius multifiliis

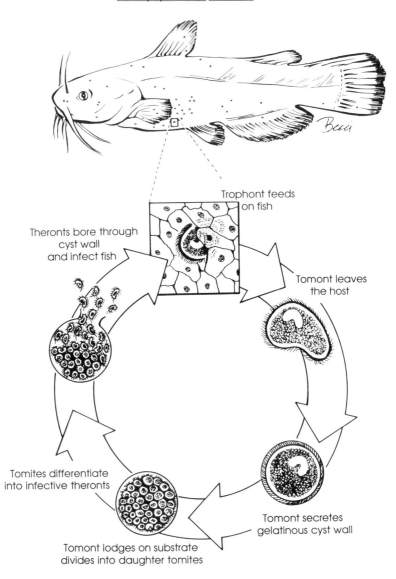

Fig. 5.1. Life cycle of *Ichthyophthirius multifiliis*. The infective theronts attach to susceptible fishes and penetrate through surface mucus and epithelium. Trophonts feed at the basal layers of the skin and gill epithelia. Mature tomonts leave the fishes, secrete gelatinous cysts and divide to form daughter tomites. The tomites differentiate into infective theronts. The rate of development both on and off the host is influenced by temperature. The life cycle of *Cryptocaryon irritans* is similar except that developmental times are longer. See text for details.

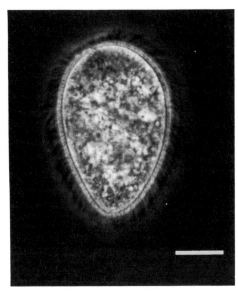

Fig. 5.2. *Ichthyophthirius multifiliis* theront. Phase contrast micrograph of the infective stage of the parasite. Bar = 16 μm.

The theront swims rapidly, frequently changing directions until it contacts a fish (MacLennon, 1935a; Wagner, 1960; Kozel, 1986). The parasite is attracted toward light and is positively chemotactic to unknown substance(s) released from fishes (Lom and Cerkasova, 1974; Wahli and Meier, 1991). Theronts swim for approximately four days, but they may not be infective for the entire period (Parker, 1965; Lom and Cerkasova, 1974).

The theront attaches to the surface epithelium of the skin and gills and penetrates within five minutes to the basal layer (Ewing *et al.*, 1985; Kozel, 1986; Cross and Matthews, 1992; Dickerson, personal observation). Secretory mucocysts concentrated at the apical end of the theront are discharged as it approaches the fish. Ewing and colleagues postulated that the released substance plays a role in attachment (Ewing *et al.*, 1985; Ewing and Kocan, 1992). Geisslinger (1987) found that theronts appeared to stick to epithelia by means of their cilia, and saw no evidence at the scanning EM level that material released by mucocysts was involved. He suggested that theronts attach by means of specialized thigmatic cilia. More work is obviously needed to clarify these observations.

The most important structure associated with penetration is the apical perforatorium (MacLennon, 1935b; Roque *et al.*, 1967; Cannella and Rocchi-Cannella, 1976; Ewing *et al.*, 1985). This organelle was described in detail at the light microscopic level by Cannella and Rocchi-Cannella (1976). It consists of a group of apical ectoplasmic ridges, each arising between ciliary rows, that converge to form a 1.5 to 2 μm protrusion at the tip of the theront. The parasite uses this structure to physically force its way through the epithelial layers (MacLennon, 1935b; Cannella and Rochi-Cannella, 1976).

There is considerable dispute in the literature concerning the rate of development of the cytostome (Cannella and Rochi-Cannella, 1976). The small primitive buccal apparatus of the theront changes in the trophont to a vestibular cavity that is large enough to ingest relatively large cell debris (MacLennon, 1935b; Cannella and Rochi-Cannella, 1976). According to MacLennon (1935b) the theront virtually eats its way into the epithelium using a cell mouth that differentiates immediately upon entering the host. In contrast, Cannella and Rochi-Cannella (1976) contend that the cytostome only develops at 18–32 hours (at a temperature of 26–27°C) after invading the host. In a more recent study, Ewing *et al.* (1985) showed an electron micrograph depicting a young trophont (40 min after infection) in gill epithelium with necrotic cell debris inside a vestibular-like structure. This would tend to support MacLennan's findings that the theront develops a functional buccal apparatus soon after it enters the host. However, since a cytostome is not visible in Ewing's micrograph, it is hard to discern whether the putative buccal apparatus is real or merely a fold in the cell membrane.

The organelle of Lieberkühn, also referred to as the 'watchglass organelle', lies in the oral region of the theront. It is a dense, ovoid, non membrane-bound structure overlaid by an alveolar surface thrown into 5–7 ridges which face the oral cavity (Lynn *et al.*, 1991). The function of this organelle is a mystery although it could possibly be involved in phototaxis. It exists only in the theront stage and disappears soon after the parasite colonizes a fish.

The theront penetrates to the basal layer of the epidermis (Ventura and Paperna, 1985) and undergoes differentiation and growth. Once the vestibular apparatus becomes functional, the trophont increases dramatically in size. The digestive cycle of the parasite has been divided into three main stages based on the ultrastructure of the food vacuoles (Loba-da-Cunha and Azevedo, 1993). The growth rate is positively correlated to increases in the ambient temperature (MacLennon, 1942; Ewing *et al.*, 1986). The parasite rotates and moves in an amoeboid fashion using the perforatorium to scrape cells and cell debris from the edges of the lesion, which it then ingests. The ciliate's surface is continually thrown into folds as the cell rotates and moves in the epithelial tissue. The large horseshoe-shaped macronucleus is readily apparent under a light microscope. The parasite continues to grow by adding ciliary rows as well as membrane and cytoplasmic organelles (including mucocysts, contractile vacuoles, and ribosomes). The single contractile vacuole of the theront multiplies to several hundred in the trophont (Cannella and Rochi-Cannella, 1976; Chapman and Kern, 1983; Ewing and Kocan, 1986).

The trophont creates a tissue space in the epithelial layers. The parasite within this vesicle appears as a white spot about one millimetre in diameter. Large numbers of these are easily visible, hence the common name 'white spot disease' (see Fig. 5.3 and 5.4). Occasionally several parasites occur within the same vesicle. Most researchers have attributed these to theronts penetrating within the same path or comigration within the epithelium (MacLennon, 1935a; Cannella and Rochi-Cannella, 1976). Ewing *et al.* (1988), however, calculated that the number of trophonts within a particular section of epithelium increased during infection, and postulated that parasites multiply within vesicles. Definitive proof of cell multiplication on the fish requires direct demonstration of trophonts in the process of dividing; this has not yet been observed.

Fig. 5.3. Channel catfish fingerling (*Ictalurus punctatus*) infected with *Ichthyophthirius multifiliis*. The fish was exposed to parasites for 7 days at 23°C. Each white spot represents a single trophont.

Fig. 5.4. The same fish as in Fig. 5.2 seen at a higher magnification. Notice how the trophonts within the skin are often raised above the surface of the fish.

The duration of infection is variable and depends on different factors. These include temperature, fish species, physiological state of the host, and body region in which the parasite resides (Cannella and Rochi-Cannella, 1976). If an infected fish dies, the trophont leaves; presumably in response to changes in oxygen tension and the pH of the tissues. Before the parasite can survive however, it must first attain a critical size and stage of differentiation. MacLennon (1942) found that parasites of less than 95 μm were not viable if removed from a fish. More recently, and in agreement with MacLennon, Ewing *et al.* (1986) showed that the predicted minimum diameter necessary for survival ranged between 85 and 104 μm. The time required to reach these sizes is primarily a function of the temperature and the size of the parasite when it initially infected the fish. In MacLennon's studies, trophonts attained these sizes within two days at 27°C (with a calculated growth rate of 5.9% per hour), and three days at 22°C (with a growth rate of 8.3% per hour). Ewing *et al.* (1986) found that close to 100% of the trophonts that leave after two days at 21° and 24°C were able to survive. Ewing *et al.* (1986) and Ewing Kocan and Ewing (1987) also found that parasites associated with the fish for longer periods (four days as opposed to three) were able to exit the epithelium more quickly when the host died. When mature, a trophont can exit the epithelium, secrete a cyst, divide, and produce active, infective theronts. Ewing *et al.* (1986) proposed that before the parasite can survive outside the host, its contractile vacuole must develop so that it can respond to the osmotic changes that occur as the trophont leaves the epithelium.

Dickerson *et al.* (1985) injected theronts into the abdominal cavities of channel catfish (*Ictalurus punctatus*), where they grew for up to three weeks. The trophonts eventually died without secreting a cyst or undergoing divisions, and the dead parasites became enclosed within granulomatous tissue.

The trophont becomes a tomont as soon as it ceases to feed and extricates itself from the epithelium. This is the reproductive stage of the organism. The parasite has usually left the fish at this stage, but sometimes it remains superficially attached to the surface mucus. Secretory mucocysts of the tomont are discharged to produce a gelatinous cyst wall in which the cell undergoes multiple binary divisions (Ewing *et al.*, 1983). The period between leaving the fish and secreting the cyst wall is influenced by temperature. At 21–23°C, the tomont swims for approximately an hour before it secretes the sticky proteinaceous matrix. Tomonts when placed in water at a temperature below 10°C do not secrete a cyst wall or divide (Dickerson, unpublished results). When cooled cells are warmed to 21–23°C, they secrete a cyst and proceed to divide normally.

The tomonts attach on virtually any substrate in the immediate aqueous environment. Occasionally, however, encysted tomonts are found free-floating or in the surface mucus of moribund or dead fishes. The gelatinous capsule anchors the dividing parasite to the substrate. The matrix also appears to prevent entry of bacteria and fungi. The cyst wall has been studied in detail at the light microscopic level by MacLennon (1937) and with the electron microscope by Ewing *et al.* (1983). MacLennon (1937) observed that the wall has inner and outer layers each of which is composed of proteinaceous fibrils.

Ewing *et al.* (1983) confirmed that the cyst consists of two layers, an inner homogeneous layer with the same consistency and electron density of the material found in mucocysts, and a less dense outer layer covered with bacteria and other debris. They found that the wall varied in width with the thickest side at the point of attachment to the substrate. In some instances, secretion of the cyst wall continues after the first or second division, resulting in partitioning into chambers within which the daughter cells divide. MacLennon (1937) found that the cyst wall could be removed following the first division without affecting the viability of the dividing organism.

Shortly after the parasite secretes the cyst wall it begins a series of palintomic divisions. This results in 200–800 tomites. The period between secretion of the cyst and the first cell division is dependent on temperature and tomont size. At 23°C, the first division usually occurs between 30 and 75 min (Canella and Rochi-Cannella, 1979; and Dickerson, personal observation). At the same time the vestibular buccal apparatus and the perforatorium are resorbed. The number and size of tomites varies and depend on the number of cell divisions which is correlated directly with the initial size of the tomont (MacLennon, 1937; Cannella and Rochi-Cannella, 1976).

Divisions are completed in 18–24 hours at 23°C. The tomont usually divides eight to nine times to produce 256 to 516 tomites. Occasionally more divisions will occur producing more daughter cells. These develop into infective theronts. Differentiation of tomites into theronts involves the acquisition of a pyriform cell shape, the development of a rudimentary buccal apparatus, and the formation of a perforatorium. Food vacuoles are not evident in theronts, although small vacuoles with acid phosphatase activity have been observed. (Lobo-da-Cunha and Azevcdo, 1993) The first theronts to differentiate bore their way out of the cyst leaving holes in the wall through which the remaining organisms leave.

Cryptocaryon irritans

C. irritans is a holotrichous ciliate that infects the surface epithelia of marine fishes. It goes through an obligate feeding stage and a fish-free reproductive stage during the course of its life history. The various forms of the parasite reach approximately the same sizes as their *Ichthyophthirius* counterparts. The infective theront (see Fig. 5.5) is pyriform in shape and 25 to 60 μm in length (Brown, 1963; Nigrelli and Ruggieri, 1966; Cheung *et al.*, 1979; Colorni, 1987; Colorni and Diamant, 1993). Feeding trophonts reach diameters of 60 to 450 μm (Cheung *et al.*, 1977; Colorni, 1985; Colorni and Diamant, 1993; Matthews *et al.*, 1993). The mature trophont leaves the fish as a free-swimming tomont (see Fig. 5.6) and secretes a cyst in which it undergoes multiple divisions to produce 200 or more daughter tomites. Tomites differentiate into infective free-swimming theronts that attach and penetrate into the skin and gill epithelia of susceptible fishes.

There have been a number of studies on the morphology of *C. irritans* (see Brown, 1963; Nigrelli and Ruggieri, 1966; Cheung *et al.*, 1981; Colorni

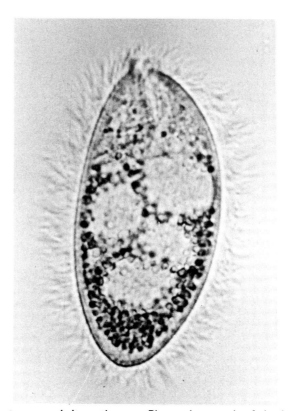

Fig. 5.5. *Cryptocaryon irritans* theront. Photomicrograph of the infective stage of the parasite. The size of the *C. irritans* theront is similar to that of *I. multifiliis* (it is shown here at a greater magnification than the theront in Fig. 5.2). Photomicrograph courtesy of Dr A. Colorni, Israel Oceanographic and Limnological Research Ltd, National Center for Mariculture, PO Box 1212, Elat 881122, Israel.

and Diamant, 1993; Matthews *et al.*, 1993). *C. irritans* has two types of nuclei, a lobed macronucleus and several smaller micronuclei (Brown, 1963; Nigrelli and Ruggieri, 1966; Colorni and Diamant, 1993). The macronucleus of the trophont has four lobes arranged into a crescent (Brown, 1963; Nigrelli and Ruggieri, 1966; Wilkie and Gordon, 1969; Colorni and Diamant, 1993). Each lobe is approximately 10 µm long by 8 µm wide and contains one to two nucleoli (Colorni and Diamant, 1993). The macronuclear membrane has a trilaminar structure with many nuclear pores. The cytoplasm of the *Cryptocaryon* trophont is more opaque than that of *Ichthyophthirius*, making the nucleus of live unstained specimens more difficult to see. The four lobes of the macronucleus fuse into a continuous strand in the encysted tomont before it begins to divide (Brown, 1963; Colorni and Diamant, 1993). The tomont does not multiply by simple palintomic division as in *Ichthyophthirius*, but instead undergoes 'budding'. As first described by Brown (1963), a section of cytoplasm containing part of the macronucleus is pinched off at one end

Fig. 5.6. *Cryptocaryon irritans* tomont. Scanning electron micrograph of the reproductive stage of the parasite after leaving the fish and preceeding secretion of the cyst in which it undergoes division. Bar = 25 µm.

of the cell. This cytoplasmic and nuclear fragment then proceeds to divide into two cells of equal size. The process is repeated with the remaining segment of the tomont until a number of tomites of equal size are produced within the cyst. Nigrelli and Ruggieri (1966) also described the division of the tomont as being asymmetric, resulting in the initial production of a group of daughter cells at one pole followed by division of the rest of the cell.

During cell reproduction the micronuclei proliferate by mitotic division with five to seven eventually being distributed into each of the daughter cells (Brown, 1963; Colorni and Diamant, 1993). Brown indicated that not all of the micronuclei were used, and postulated that some disintegrated and reassociated in the process of autogamy. The time required for tomont division is longer in *Cryptocaryon* than in *Ichthyophthirius* (see below).

The buccal apparatus of the parasite differentiates after nuclear development and before the theront leaves the cyst (Brown, 1963). It consists of a single ring of cirri surrounding the oral opening which is about 20 µm in diameter (Cheung *et al.*, 1981; Colorni and Diamant, 1993). Each cirrus is composed of two adjacent cilia (Keskintepe and Farmer, unpublished results).

Nigrelli and Ruggieri (1966) postulated that this stiff cirrus ring is used to burrow into the epithelium and scrape cells for ingestion. The prominent perforatorium which is important in the infection and feeding processes of *Ichthyophthirius* has not been described in *C. irritans*.

The buccal cavity of the trophont has a protrusile membrane and a smaller one which lies below it (Brown, 1963). It was suggested that the protrusile membrane functions to draw in mucus, blood, and cellular debris as the parasite feeds (Nigrelli and Ruggieri, 1966). In this regard the buccal apparatus of the *C. irritans* trophont differs significantly from that of *Ichthyophthirius*.

Flask-shaped mucocysts are present, but are not differentiated into crystalline and secretory forms as described in *I. multifiliis* by Ewing and Kocan (1992). They do not appear to contribute to cyst formation (Colorni and Diamant, 1993), but may be involved in host invasion (Colorni and Diamant, 1993; Matthews *et al.*, 1993). The encysted parasite is resistant to chemical treatment suggesting that the cyst wall is impermeable to soluble substances in the water (Herwig, 1978). The cyst wall on average is 0.5 to 1 μm thick (Matthews *et al.*, 1993). It has a loose fibrillar outer layer and a multi-layered compact inner wall (Keskintepe and Farmer, unpublished results).

Cheung *et al.* (1981) published electron micrographs depicting tomonts with truncated cilia. Nigrelli and Ruggieri (1966) stated that cilia become absorbed before the parasites secreted cyst walls. Other studies found that the cilia were shed rather than resorbed at this stage (Colorni and Diamant, 1993; and Matthews *et al.*, 1993). Loss of cilia does not occur with *Ichthyophthirius*, and all stages of the parasite are ciliated, including the tomites within the cyst.

Keskintepe and Farmer (unpublished results) found contractile vacuoles to be present, a result that is in agreement with the findings of Sikama (1961) An organelle of Lieberkühn is not present in any stage of the parasite (Colorni and Diamant, 1993; Matthews *et al.*, 1993; Keskintepe and Farmer, unpublished results). The absence of this structure raises doubts as to the classification of *Cryptocaryon* in the same suborder (Ophryoglenina Canella, 1964) as *Ichthyophthirius*.

Development of the parasite

As with *Ichthyophthirius*, the ambient temperature greatly influences growth and development of *Cryptocaryon*. The trophont remains on the fish for 3 to 7 days at optimal growth temperatures (23° to 30°C) (Cheung *et al.*, 1977; Colorni, 1985; Burgess and Matthews, 1994; Yoshinaga and Dickerson, 1994). *C. irritans* differs from its freshwater counterpart in that it generally takes longer to reproduce and differentiate outside the fish. Secretion of the cyst is usually completed within 24 hours at 25°C. Cheung *et al.* (1979) found that only 10% of tomonts held at 7°C secreted cysts. When the organisms were incubated at 25°C the majority (66%) secreted cysts.

At 24 to 27°C, *Cryptocaryon* tomonts take 3 to 28 days to divide and produce infective theronts with most emerging by the eighth day (Nigrelli and Ruggieri, 1979; Cheung *et al.*, 1979; Colorni, 1987; Burgess and Matthews, 1994; Yoshinaga and Dickerson, 1994). In *I. multifiliis* these divisions are completed in 18 to 24 hours. Tomonts divide and develop into theronts at an asynchronous rate (Matthews *et al.*, 1993; Colorni and Diamant, 1993; Burgess and Matthews, 1994; Yoshinaga and Dickerson, 1994). Yoshinaga and Dickerson (1994) discovered that theronts are released from cysts with a circadian periodicity and enter the water between the hours of 0200 to 0900, regardless of the day that they leave the cyst.

Summary of the life cycle of *Cryptocaryon irritans* and comparison to that of *I. multifiliis*

In many respects the life cycle of *Cryptocaryon* is similar to that of *Ichthyophthirius* (see Fig. 5.1). A fish-associated trophont stage and a free-living reproductive tomont stage occur in both parasites. As with *I. multifiliis* the parasitic trophont of *Cryptocaryon* remains on the fish for approximately five to seven days at 25°C. The reproductive stage of *C. irritans*, however, usually takes much longer to develop than in Ich and commonly extends to eight days at 25°C and longer at lower temperatures. Two hundred or more infective theronts emerge from each encysted tomont and each of these penetrates the host by means of specialized ciliary membranelles associated with the buccal apparatus. In contrast, *I. multifiliis* penetrates using a specialized apical structure, the perforatorium.

HOST–PARASITE RELATIONSHIPS OF *ICHTHYOPHTHIRIUS MULTIFILIIS*

Host selection

Ichthyophthirius multifiliis does not appear to have a predilection for any specific group of fishes, although it is believed that the organism originated as a parasite of carp (Hoffman, 1970).

McCallum (1982) indicated that there was a linear relationship between the number of theronts to which a host was exposed and the resultant parasite burden. Wahli and Meier (1991) observed, in agreement with Lom and Cerkasovova (1974), that theronts were positively phototactic. They could not demonstrate that theronts were attracted to fish, a result in disagreement with that of Lom and Cerkasovova who found that theronts were attracted by components of fish blood. Houghton (1987) observed that theronts were attracted to pieces of fish tissue in the water suggesting a possible short-range homing mechanism.

Epizootics of *I. multifiliis* appear to occur uniformly in populations of male and female fishes. However, there are reports where infections occurred predominantly in one sex. Male guppies (*Lebistes reticulatus*) were reported to be more severely infected than females (Paperna, 1972). Similarly, in brown trout (*Salmo trutta*) mature males were more frequently infected than females, with the most severe infections occurring on precociously mature pre-spawning males (Pickering and Christie, 1980). In contrast, in an epizootic in Lake Titicaca, the majority of the dead and infected fish were gravid or spent female killifish (*Orestias agassii*) (see Wurtsbaugh and Tapia, 1988). Infection may not be a function of predilection of the parasite for either sex, but rather that infection occurred on fishes under the greatest stress. Reproductive activity is a significant stress on fishes.

Fishes exposed to *I. multifiliis* develop protective immunity (Hines and Spira, 1974c; Clark *et al.*, 1988), hence survivors of an epizootic are resistant to subsequent infection. Therefore, in native populations, young fishes might be more susceptible to infection than older individuals if the latter were previously exposed to the parasite. In naive feral fish populations, all ages appear to be equally susceptible to infection.

Distribution of parasite on host

Except in very severe infections, *I. multifiliis* is not uniformly distributed on the body of the host fish. The parasite occurs most frequently on the dorsal surface, particularly the head and fins (Hines and Spira, 1973a; Kozel, 1976).

An extremely important site of infection is the gills. The development of trophonts within the gill epithelium is a major factor in the lethal effects of infection. Large numbers of theronts come in contact with the gill epithelium because large quantities of water pass over these surfaces. Since infection is the result of random encounters between the fish and infective theronts, movement of water over the gills increases the probability of theront contact. In addition, gills do not have the same degree of protection against parasite infection as the skin. The surface of the fish is covered with a mucous layer that is part of the defence against parasites. In unstressed fishes there is very little mucus covering the secondary lamella of the gills (Handy and Eddy, 1991). Thus, gills appear to be more susceptible to infection than body surfaces.

Genetic susceptibility

McCallum (1982) noted that in experimental *I. multifiliis* infections only a portion of the infecting theront population developed into trophonts. He concluded that this variability was a result of host factors which might be genetically controlled. One major host resistance factor is the production of surface mucus. An early response to penetration of the fishes' epithelium by infecting theronts is increased mucus production (Hines and Spira, 1974a; Ventura and Paperna, 1985). There may be genetic factors that influence the amount and composition of fish mucus.

Clayton and Price (1992, 1994) demonstrated significant variation in susceptibility to infection among fish species and hybrids and found that by removing effects due to other variables (such as time of infection and temperature) heterosis (hybrid vigour) contributes to resistance to *I. multifiliis*.

Ichthyophthirius multifiliis is believed to have originated as a parasite of carp (Hoffman, 1970). The long host–parasite association may have led to the development of strains of carp that are more resistant to infection. Preliminary genetic studies using scale patterns as genetic markers suggested that strains of carp vary in their resistance to *I. multifiliis* (Clayton and Price, 1988).

Behavioural modifications

In the early stages of disease fishes congregate near water intakes to reduce contact with free-swimming theronts (Kabata, 1985). Fishes also 'flash' or rub their bodies against objects in reaction to skin and gill irritation caused by the theronts (Brown and Gratzek, 1980). Fishes swim more rapidly than normal and often leap out of the water. As the disease progresses they become less active and congregate at the bottom of ponds or aquaria (Hines and Spira, 1973a). Fishes also lie near the edges of ponds moving their gill opercula rapidly in an attempt to obtain more oxygen (Kabata, 1985). This is related to gill damage caused by the parasite. With very heavy infections, fishes become lethargic and stop feeding (Hines and Spira, 1973a).

Gross pathology

In very mild *I. multifiliis* infections the only detectable pathologic changes may be the presence of a few white spots on the surface of the fishes. In more severe cases there are usually large numbers of spots on the skin (see Figs 5.3 and 5.4). Occasionally, however, *I. multifiliis* only infects the gills and hence there are no obvious gross lesions on the body surface (Brown and Gratzek, 1980).

Ulcers develop in the skin of heavily infected fishes and are often the sites of secondary bacterial or fungal infections. The fins become frayed due to loss of tissue between the fin rays (Hines and Spira, 1974a).

Mucus production is increased. Heavily infected carp have a thick lumpy covering of surface mucus (Hines and Spira, 1974a). Increased production of mucus is not a unique response to *I. multifiliis*, however, as it occurs on most fishes that are exposed to irritants or skin parasites.

Infected fishes have enlarged spleens and kidneys and pale mottled livers (Hines and Spira, 1974a). Also, transudate fluids are found in the peritoneal cavities. It is not clear that the changes in the viscera are directly due to the activity of the parasite. These may be secondary effects caused by opportunistic bacterial or fungal infections, and/or anorexia of the fishes in the later stages of the infection. Enlarged gall bladders (a change associated with animals that have not eaten recently) were reported in moribund Ich-infected carp (Hines and Spira, 1974a).

Gross pathology – *Cryptocaryon irritans*

The gross lesions in *C. irritans* infections are similar to those seen in fishes infected with *I. multifiliis*. Petechial haemorrhages occur in the skin (Nigrelli and Ruggieri, 1966; Wilkie and Gordin, 1969). The basic response of the host to the infection is excessive mucus production (Nigrelli and Ruggieri, 1966; Wilkie and Gordin, 1969). With severe infections, skin ulcers occur with secondary *Pseudomonas* spp. bacterial infections (Nigrelli and Ruggieri, 1966).

Clinical signs

The common clinical signs of ichthyophthiriosis are the characteristic white spots. Each spot represents a developing trophont within an epithelial capsule or vesicle (see Fig. 5.4). Visible spots develop several days after initial attachment of theronts (Hines and Spira, 1973a; Ventura and Paperna, 1985). In cases where the infection is restricted to the gills these are not visible (Brown and Gratzek, 1980).

One of the first physiologic responses to infection is an increase in surface mucus production (Hines and Spira, 1973a). Skin penetration by theronts stimulates expanded numbers of mucus-secreting cells in the epidermis. These multiply not only in areas around the parasite, but throughout the epidermis (Hines and Spira, 1974a). In severe infections, mucus may stream off of the posterior edges of the fins and tail.

A careful observer can detect infection before the development of surface lesions by noting changes in fish behaviour (see above). Initially, infected fishes swim more rapidly and rub themselves against objects. As the disease progresses, the fishes surface and gasp for oxygen. They become increasingly lethargic and eventually cease feeding.

In severe infections there is erosion of the epithelia leading to ulcer formation and exposure of the deeper tissues to bacterial and fungal invasion. A common cause of secondary invasions is the fungus, *Saprolegnia* spp., which appears as tufts of 'fuzz' on the skin.

The clinical signs of a *C. irritans* infection are similar to those seen in ichthyophthiriosis. Fishes develop small white cysts on their body surfaces (Wilkie and Gordin, 1969; Yoshinaga and Dickerson, 1994). These lesions are often numerous minute greyish vesicles rather than the larger white spots associated with Ich infections (Nigrelli and Ruggieri, 1966). *C. irritans* infection frequently involves the eyes and leads to corneal clouding and blindness (Nigrelli and Ruggieri, 1966; Wilkie and Gordin, 1969). Marine fishes infected with *C. irritans* produce excessive amounts of surface mucus (Nigrelli and Ruggieri, 1966; Wilkie and Gordin, 1969; Huff and Burns, 1981).

Histopathology

The nature and severity of histopathologic changes seen in *I. multifiliis* infections vary greatly. This variation is influenced by such host factors as stress and nutritional status. However, the parasite load is the major factor contributing to the diverse tissue changes. In general, mild infections elicit minor cellular reactions. This minor host reaction suggests that *I. multifiliis* is, in evolutionary terms, a long-standing parasite of fishes and that, as a result, the host and parasite have a relatively benign relationship. The extensive histo-pathologic changes reported to occur in *I. multifiliis* infections are only seen in severe epizootics or in experimental infections with large numbers of parasites (Ventura and Paperna, 1985).

Histopathology of the skin resulting from mild infections

In primary *I. multifiliis* infections (that is, the first contact of the parasite with the fish) there is little reaction to the penetrating theronts. After 40 hours, most trophonts will have penetrated the epithelial layers and be located next to the basement membrane. The cells between the parasite and the basement membrane become hydropic, vacuolated or necrotic with pyknotic nuclei (Ventura and Paperna, 1985). The growing trophont gradually lifts and displaces the epithelial cell layers until it lies within an epithelial capsule that extends above the skin surface (Ventura and Paperna, 1985). The epithelial layer overlying the trophont expands to cover the parasite during this growth stage. The epithelium retains its architecture of differentiated cell populations (Ventura and Paperna, 1985). Cell damage is observed only in the cellular layers in direct contact with the developing trophont. Host cell debris can be observed in the food vacuoles of the trophont and in the spaces in the epithelial capsule around the parasite. There is evidence of haemorrhage occurring in the skin as a result of parasite invasion (Chapman, 1984; Ventura and Paperna, 1985). Large pale-staining alarm substance cells have been observed in the area of the developing trophont (Chapman, 1984). In mild infections only a few leukocytes are seen in the epithelium.

Histopathology of the skin resulting from heavy infections

In heavy infections a much greater inflammatory reaction occurs in the skin. Penetration by numerous theronts leads to increased epithelial cell hyper-plasia. This cellular proliferation could be a defence mechanism (Ventura and Paperna, 1985). In addition to hyperplasia, there is a generalized increase in the number of mucous cells in the skin (Hines and Spira, 1974a). The epithelium in heavily infected fishes may be up to four times its normal thickness due to proliferation of epithelial and mucous cells (Hines and Spira, 1974a).

Extensive cell necrosis and histolysis occur around trophonts developing in the hyperplastic epithelium. Empty spaces are present along with hydropic, vacuolated and necrotic cells (Ventura and Paperna, 1985). There is conges-tion in the dermal lymphatics leading to oedema and an increased epithelial infiltration of neutrophils, eosinophils and lymphocytes (Hines and Spira, 1974a; Ventura and Paperna, 1985). The outermost epithelial cells degenerate and slough off, eventually exposing the underlying basement membrane (Hines and Spira, 1974a). It was suggested that the extensive cellular reaction observed in the skin of fishes heavily infected with *I. multifiliis* is a hypersen-sitivity reaction (Hines and Spira, 1974a; Ventura and Paperna, 1985).

Histopathology of the gills

Theronts attach to the gills at the middle or base of the gill lamellae. The young trophonts penetrate and displace the interlamellar epithelium until they reach the basement membrane (Ventura and Paperna, 1985). Host cell debris is seen in the cytoplasmic vacuoles of the parasite. Epithelial cells migrate from the apical end of the adjacent lamellae and cover the developing trophont producing a multilayered cap over the parasite. The cell layer over the parasite also includes mucous cells and chloride cells (Hines and Spira, 1974a; Ventura and Paperna, 1985). Epithelial cell proliferation occurs not only adjacent to the parasite but also all along the lamellae (Hines and Spira, 1974a). Mature trophonts may occupy three or four lamellae (Ventura and Paperna, 1985). Mucous cells become prominent in the proliferating epithelium and may make up 50% of the cells in an interlamellar area (Hines and Spira, 1974a). The respiratory epithelium, however, retains its squamous morphology during this proliferation. At this time there is a significant increase in the number of neutrophils in the area. The amount of lymphocyte infiltration is variable (Hines and Spira, 1973b). As the infection progresses the epithelial cells continue to proliferate and eventually fill up the interlamellar space until the lamellae are completely cornified or 'clubbed' (Hines and Spira, 1974a).

In the late stages of heavy infections, complete atrophy of the gill lamella is seen along with areas of necrosis of the gill filaments (Hines and Spira, 1974a).

Histopathology – *Cryptocaryon*

There is limited information on the histopathologic changes that occur with *C. irritans* infections. From the analogies in the infective and feeding stages of *I. multifiliis* and *C. irritans*, it is reasonable to assume that the cellular changes caused by both parasites are similar. The major lesions caused by *C. irritans* are vesicles within the skin. Mucus-producing cells proliferate in the skin and gills. Epithelia of infected gills become hyperplastic and eroded in severe cases.

Clinical pathology

Although the total leukocyte numbers do not change during *I. multifiliis* infections there is a differential shift in the various leukocyte populations. Infected fishes develop a lymphopenia and neutrophilia (Hines and Spira, 1973b). The number of neutrophils in circulation early in the infection may increase five fold with no concurrent rise in total leukocyte numbers. The shift in distribution of leukocyte cell populations is accompanied by an increase in the number of immature blast cells. It was suggested that the leukocyte changes in infected fishes are non-specific stress reactions (see Hines and Spira, 1973b).

Studies on serum levels of Na^+, K^+, Mg^{2+} and blood urea-ammonia indicated significant osmoregulatory disturbances (Hines and Spira, 1974b).

In severe infections there was a marked drop in serum Na^+ and Mg^{2+} levels and a rise in serum K^+ levels. Blood urea-ammonia levels also increased during the course of the infection.

Immune response

It was recognized early on that fishes surviving *I. multifiliis* outbreaks become resistant to reinfection (Bushkiel, 1910). It has been suggested that the duration of immunity depends upon the severity of the initial infection (Bauer, 1953). A number of studies have been done to determine the period of immunity in fishes recovered from natural infections or cured by chemical treatment (Beckert and Allison, 1964; Parker, 1965; Hines and Spira, 1974c; McCallum, 1986). Pyle (1983) found that immunity induced by single exposure of channel catfish to the parasite lasted for at least 3 months. In the natural course of infection, surviving fishes eventually become free of the parasite (Valtonen and Keränen, 1981). Premunition has not been demonstrated although Cross and Matthews (1993a,b) found that a small percentage ($<5\%$) of infecting parasites developed in carp immunized against *I. multifiliis*.

The immune response of any vertebrate host to a parasite is a complex reaction involving the activation of a variety of cell populations and the production of humoral factors such as antibodies. Some host reactions are general, i.e. they react in the same way no matter what parasite is encountered, and some reactions are specific, i.e. they react only with the specific parasite that induced the reaction. The former reactions are called innate, and the later are called adaptive or acquired reactions. In *I. multifiliis* infections both innate and adaptive reactions play significant roles.

Surface mucus is the fishes' first line of defence against infection. In order to infect, the theront must penetrate this mucous layer and burrow into the epithelium. Mucus-secreting cells increase in the skin of infected fishes (Hines and Spira, 1973a; Ventura and Paperna, 1985). Heavily parasitized fishes characteristically have a thick mucous coat and strands of mucus may be seen streaming off their fins and tails. Theronts have limited energy stores which could be depleted during penetration of the thickened mucous layers. Trapped theronts would eventually be swept off as the mucus moves along the fishes' surface. In this regard, the naturally thin mucous layer of gills could be a predisposing factor for infection. One report indicated that the mucous layer does not completely cover the secondary lamellae in unstressed rainbow trout (Handy and Eddy, 1991).

In addition to being a physical barrier, the surface mucus may also contain antiparasite factors. Hines and Spira (1974c) and Wahli and Meier (1985) found that mucus and serum from immune carp and rainbow trout (*Salmo gairdneri*) immobilized trophonts *in vitro*. It was suggested that this immobilizing activity was due to the presence of antibodies and that these prevented the penetration of theronts through the mucus. Xu and Dickerson (unpublished results) using dot blot assays found that Ich-immune channel catfish have mucus antibodies against membrane antigens of *I. multifiliis*. In contrast,

Cross and Matthews (1993b) were unable to demonstrate binding of carp mucus antibodies to thin sections of fixed theronts. The discrepancy in these findings is probably due to the fact that specific antibodies are present in relatively low levels in mucus (as compared to the serum) and may not have been detectable in the experiments of Cross and Matthews.

A basic innate host reaction to *I. multifiliis* is epithelial cell proliferation. In light infections there is little reaction other than the formation of an epithelial cell capsule around the parasite (Ventura and Paperna, 1985). However, in severe or repeated infections extensive epithelial cell proliferation occurs (Hines and Spira, 1974a; Ventura and Paperna, 1985). This epithelial hyperplasia could interfere with penetration of the theront. The skin of infected fishes also becomes infiltrated with neutrophils, basophils, eosinophils, eosinophilic granular cells, macrophages, and lymphocytes (Ventura and Paperna, 1985; Cross and Matthews, 1993a). Increase in eosinophilic granular cells was coupled to the acquired immune response against the parasite (Cross and Matthews, 1993a). Leukocytes (granulocytes) were observed adjacent to parasite surfaces without apparent adherence. How these cells interact with the trophonts in the tissue is not well understood.

Some of the lymphocytes found in the skin of infected fishes may be nonspecific cytotoxic cells (NCC). Graves *et al.* (1984) reported on the occurrence of NCC in the anterior kidney of channel catfish. The NCC were found to be cytotoxic to deciliated *Tetrahymena pyriformis* (a free-living ciliate which often occurs in the same aquatic environment as *I. multifiliis*) and this activity could be blocked by incubating the NCC with formalin-killed *I. multifiliis* (Graves *et al.*, 1985a). The supposition (as yet untested directly) was that NCC also could react with and kill immobilized *I. multifiliis*. It was found that the NCC were mobilized out of anterior kidneys in infected fishes and the circulating NCC had increased target cell affinity and killing capacity (Graves *et al.*, 1985b). There is a decrease in lymphocytes in circulation in Ich infected fishes (Hines and Spira, 1973b). This decrease could be the result of NCC passing out of circulation and into the skin to interact with the immobilized theronts. However, this is only speculation.

As described previously, Hines and Spira (1974c) found that serum and mucus collected from infected fishes immobilized theronts *in vitro*. This was taken as evidence that fishes produced antibodies against the parasite. It was found that various antigens including theronts (Burkart *et al.*, 1990), trophonts (Areerat, 1974) or cilia (Goven *et al.*, 1981b) could be used to induce fishes to produce immobilizing or agglutinating factors in their sera. Immobilization and agglutination were assumed to be the result of antibody activity. However, in early studies the precise character of the factors was not determined. Clark *et al.* (1988) conclusively demonstrated that channel catfish immune to Ich produced antibodies to ciliary antigens of the parasite. They found a strong correlation between antibody levels in the serum as determined by an ELISA test and the ability of the serum to agglutinate live *I. multifiliis* theronts *in vitro*. These results strongly support the conclusion that the serum antiparasite activity is indeed antibody mediated. Subsequent studies by Cross and Matthews (1993b) have also shown binding of fish serum antibodies to surface membranes of *I. multifiliis*.

The relationship between serum antibody and antibody activity in the mucus has not been completely established. It was reported that a secretory form of antibody similar to mammalian secretory antibody occurs in channel catfish (Lobb and Clem, 1981). Lin and Dickerson (unpublished results) protected fish against *I. multifiliis* challenge by intraperitoneal injection of mouse origin immobilizing monoclonal antibody against *I. multifiliis*. These results suggest that the mouse antibody (IgG_1) was transported from the peritoneal cavity to the surface of the fish where it reacted with the parasite. In contrast, transfer of immune fish sera does not protect against *I. multifiliis* infection, suggesting that the teleost serum tetrameric antibody does not pass as readily to the surface as the monoclonal mouse IgG. (Dickerson, unpublished results). Burkart *et al.* (1990) elicited immunity to Ich by the intraperitoneal injection of live theronts. The immunized fishes resisted repeated surface challenges. These results indicate that intraperitoneal vaccination elicits an active surface immune response, the mechanism of which is unknown.

Antibodies appear to be important factors in protecting fishes against *I. multifiliis* infections (Cross and Matthews, 1992). How these function in resistance is just beginning to be understood. It was hypothesized that their basic function is to immobilize theronts in the mucous layer (Hines and Spira, 1974c). Immobilization *in vivo* may be an indirect effect rather than a direct action of the antibodies on theront cilia. Clark *et al.* (1987) found that theronts agglutinated and immobilized by immune serum were surrounded by a mucus-like material and their cilia continued to beat. The authors suggested that antibody binding to theronts caused discharge of parasite mucocysts. The clumping of theronts was the result of their being trapped in this released substance. Cross (1993) demonstrated binding of carp serum antibody to a gelatinous mucus-like material surrounding tomonts. He hypothesized that this material is responsible for immobilization *in vitro* and may form a barrier to further antibody binding *in vivo*. It was proposed by Ewing *et al.* (1985) that discharge of theront mucocysts is also a means of attachment and penetration into the epithelium. Premature discharge of mucocysts could result in theronts that are unable to infect fishes even though they have penetrated the surface mucus and come in contact with the epithelium.

To directly observe the fate of live parasites on fish, Cross and Matthews (1992) placed theronts onto the tails of immune or naive carp and observed them under the microscope. They discovered that theronts penetrated into the skin of both naive and immune fish within minutes. Theronts were unable to remain colonized on immune fish, however, and left within 2 hours. Cross and Matthews argued that immunity was the result of a humoral factor due to the rapidity of the response. There was no evidence of parasites being immobilized or killed in the tissue.

Fishes have subpopulations of lymphocytes similar to those which occur in mammals (Lobb and Clem, 1982; Secombs *et al.*, 1982; Yocum *et al.*, 1985). In mammals the B-lymphocytes develop into antibody producing cells. Analogous cells are elicited in the antibody response against *I. multifiliis*. In mammals the T-lymphocytes are a more complicated cell population being involved in the regulation of the immune system (T-helper and T-suppressor cells) and effector cells (T-cytotoxic cells). The T-cytotoxic cells react with antigens on cell surfaces and kill the antigen-bearing cells. The two model reactions for T-cell activity are delayed type hypersensitivity reactions and graft rejection. Fishes have been shown experimentally to develop delayed hypersensitivity reactions and will reject heterologous scale grafts (Hildeman, 1972; Kikuchi and Egame, 1983; Tataner and Manning, 1983). Whether fishes develop delayed type hypersensitivity or populations of cytotoxic T-cells to *I. multifiliis* is not known. Houghton and Matthews (1986, 1990, 1993) argue for a cell-mediated effector response against *I. multifiliis* based on experiments showing that corticosteroid treatment of Ich-immune carp increases their susceptibility to infection even though serum antibody levels do not drop significantly.

Immune response – *Cryptocaryon irritans*

There is a dearth of information on the immune response of fishes to *C. irritans* infection. It has been suggested (without evidence) that fishes maintain an immunity by premunition (Nigrelli and Ruggieri, 1966). It seems reasonable to postulate that fishes produce antibodies to *C. irritans* both in circulation and possibly in surface mucus. These antibodies could play similar roles to those involved in protective immunity against *I. multifiliis*.

Mechanisms of disease

Fishes infected with small numbers of *I. multifiliis* show few signs of infection other than the development of white spots. Fishes tolerate a few parasites very well. As previously stated, this situation suggests that teleost fishes and *I. multifiliis* have had a long association, and as a result have evolved a balanced host–parasite relationship. Frank disease and/or deaths occur when there is a disturbance in this balanced interaction. This disruption is most frequently the result of infection with large numbers of parasites. When conditions are right (e.g. optimum temperature and sufficient number of hosts) the parasite population increases very rapidly. As a result, fishes are exposed to many parasites in a short period of time.

Infection by *I. multifiliis* triggers hyperplasia of gill epithelial cells. As a result, the gill interlamellar spaces are reduced which significantly limits the surface area available for oxygen interchange. Severe infections thus cause fishes to become oxygen starved. In addition to being the primary site of oxygen uptake, the gills also function in both osmoregulation and excretion of nitrogenous wastes (Smith, 1929). Epithelial hyperplasia interferes with these functions as well. Hines and Spira (1974b) reported that infected carp have decreased levels of serum sodium and magnesium and increased levels of serum potassium. High levels of serum urea-ammonia nitrogen and nitrogenous wastes accumulate in the blood. An additional contributing factor to ionic imbalance is the interplay between nitrogenous waste excretion and ionic exchange. When NH_4^+ ions are excreted through the gills, Na^+ ions are taken up from the water into circulation (Maetz and Garcia-Romea, 1964). Thus, the decrease in NH_4^+ excretion during infection contributes to lowered serum Na^+ concentrations. Increased mucus production in the gills of infected fishes interferes with osmoregulatory functions as fish mucus is relatively impermeable to water and ions (Hughes, 1970).

Proliferative and degenerative changes in the skin also influence osmoregulation. The skin is a major barrier to ion loss and water entry into fishes (Wikgren, 1953). The subepidermal oedema and disruption of the epithelium cause ionic imbalance (Hines and Spira, 1974b).

A final additive factor to osmoregulatory dysfunction is stress. Fishes with severe infections are under stress, and stress reactions have been reported to produce serum ionic imbalances (Stanley and Colby, 1971; Wedemeyer, 1972).

I. multifiliis infections cause severe cellular reactions in the skin. Gradual desquamation, degeneration, necrosis and sloughing of the superficial epithelial cells occurs in the area of the growing parasite (Ventura and Paperna, 1985). This devitalization of the skin reduces the ability of fishes to resist infections by opportunistic pathogens. These secondary infections add stress to an already badly compromised host and increase the mortalities in a population of infected fishes.

Mechanisms of disease – *Cryptocaryon irritans*

The limited body of literature on *C. irritans* does not contain any significant description of disease mechanisms. Both the life cycle of and the pathology associated with *C. irritans* are similar to those of *I. multifiliis*. Thus, it seems reasonable to assume that the mechanisms of disease would be similar for both parasites.

Antigens of *Ichthyophthirius multifiliis*

It was observed as early as the beginning of the twentieth century that fishes which survive mild outbreaks of *Ichthyophthirius* become solidly immune to reinfection (Buschkiel, 1910). It has since been documented in many reports that fishes recovered from infection are protected against subsequent attack by the parasite (Butcher, 1941; Bauer, 1953; Beckert and Allison, 1964; Parker, 1965; Hines and Spira, 1974c; Beckert, 1975). Despite the fact that acquired immunity against Ich is well recognized, the mechanism of the response and the parasite antigens that elicit it remain largely unknown. Nevertheless, the fact that fishes develop such a strong immunity against infection makes it theoretically possible to create protective vaccines for the control and prevention of disease outbreaks.

The use of controlled infections for vaccination
A number of researchers have attempted to vaccinate fishes against *Ichthyophthirius* using controlled infections. Parker (1965) immunized goldfish

(*Carassius auratus*) by exposing them to 0.5 and 1.0 LD_{50} doses of infective theronts. An LD_{50} dose was 108 theronts per fish with a density of 3.7 theronts per ml of water. Immunized fish were resistant to lethal challenges of 8 to 10 LD_{50} doses. Areerat (1974) vaccinated channel catfish (*Ictalurus punctatus*) by exposing them to 20 or 40 encysted tomonts and treating the ensuing infection with 0.1 ppm malachite green. Vaccinated fishes had lower mortality than non-immunized fishes when challenged with 20 tomonts per fish. Hines and Spira (1974c) immunized mirror carp (*Cyprinus carpio*) by exposing them to sub-lethal doses of tomonts. In this study, fishes recovered naturally without treatment and remained free of parasites when challenged with large numbers of tomonts (500 per fish). One hundred percent of naive control fishes died when exposed to tomonts or infected fish in the same system.

Live versus killed vaccines

In general, infection with live parasites produces a more substantial and prolonged protective immune response than vaccination with killed organisms. This is usually attributed to an increase in parasite mass as a result of growth and reproduction on the host. In addition, excretory or secretory products released by the live parasite during infection also stimulate the host's immune response.

Live vaccines usually consist of either a carefully controlled infection with a fully virulent parasite (such as those for *Ichthyophthirius* described above), or an infection with a less virulent or attenuated strain. To date, no attenuated strains of Ich have been described and the use of the virulent organism for vaccination carries the risk of inducing high mortalities in the population one is attempting to immunize. Ich cannot be easily propagated which also makes its use as a live vaccine impractical. The approach has therefore been to immunize fishes against *Ichthyophthirius* using killed parasite preparations. Parker (1965) injected goldfish intraperitoneally with freeze-thawed theronts. When fish immunized in this manner were challenged, they had fewer trophonts on their body surfaces compared to non-immunized control fish. The protection was less than that afforded by vaccination with live organisms. Areerat (1974) injected channel catfish fingerlings intraperitoneally or intramuscularly with lysed tomonts mixed with or without Freund's complete adjuvant. One hundred percent of the immunized fishes survived infection for at least seven days following challenge. Beckert (1975) was able to induce protection after injecting channel catfish intraperitoneally with killed trophonts mixed with Freund's complete adjuvant.

Assessment of protection

It is difficult to compare results of vaccination trials from different laboratories since experiments were often conducted using different antigens, fishes, and challenge procedures. One study directly compared (in the same system) the protection afforded by live vaccines and various killed parasite fractions (Burkart *et al.*, 1990). The vaccination and testing were done on channel catfish under uniform conditions. It confirmed that controlled infection with live organisms afforded higher protection than either injection or immersion with killed parasites. Vaccination by intraperitoneal or surface infection provided complete protection when fishes were challenged 21 days later. These findings agree with other studies where it was found that fishes exposed to non-lethal doses of parasites (Hines and Spira, 1974c; Houghton and Matthews, 1990), or to lethal doses followed by treatment (Beckert and Allison, 1964; Houghton and Matthews, 1986; Clark *et al.*, 1988) were protected against further infection. In contrast, vaccination with killed cell preparations was much less effective; in most cases, all of the vaccinated groups died following challenge (see below). The differences in protection between live and killed parasites suggests that relevant immunogens are either molecules that are rapidly turned over on the live parasite (e.g. membrane proteins or secretory and excretory products), or that they are denatured by the procedures used to treat the parasite in killed vaccines (freezing, deciliation, or formalin fixation).

Theront versus trophont antigens

A strong immunity was elicited in channel catfish following exposure to live theronts (Burkart *et al.*, 1990). However, because theronts change quickly into feeding trophonts once they attach to the fish, it is difficult to determine whether antigens specific to theronts, trophonts, or both were responsible for this protection. To address this question, antigens from both theronts and trophonts were compared in vaccination trials (Burkart *et al.*, 1990). Channel catfish were immunized with: (i) freeze-thawed or formalin-fixed theronts; (ii) isolated theront cilia; and (iii) formalin-fixed trophonts. The group injected with trophont antigen had the highest protection (mean mortality of 51% ± 38). This protection was lower than that reported by Areerat (1974) but suggested that formalin-killed trophonts contain protective antigens. The other groups had 100% mortalities following challenge. These results were in disagreement with previous findings (Goven *et al.*, 1980) where vaccination with theront cilia was found to significantly reduce mortality.

Inoculation with both trophont and theront antigens increased the survival period (days to death) over control fish. It thus appears that killed antigen preparations elicited some protection.

Immobilization antigens

Sera from immune fishes immobilize free-swimming theronts and tomonts *in vitro* (see Fig. 5.7). This was first described by Hines and Spira (1974a) who postulated that it was the result of specific antibodies reacting with the parasite. Immobilization of theronts and trophonts by immune fish sera was subsequently confirmed (Parker, 1965; Areerat, 1975; Beckert, 1975; Wahli and Meier, 1985; Houghton and Matthews, 1986; Clark *et al.*, 1987). Clark *et al.* (1988) using dot blot assays demonstrated that immobilizing antisera from immune channel catfish contained antibodies that reacted specifically with ciliary membrane proteins. This was the first time that a direct correlation was shown between immobilization and the presence of antibodies against parasite antigens.

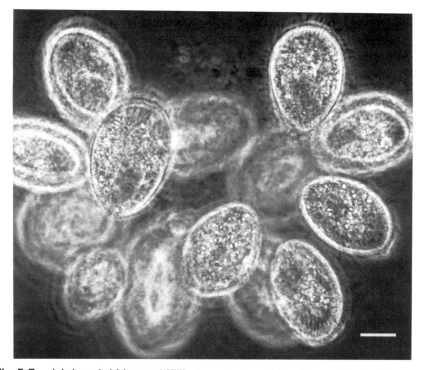

Fig. 5.7. *Ichthyophthirius multifiliis* theronts immobilized by antibody. Theronts were incubated with immobilizing mouse monoclonal antibodies for 30 minutes and photographed under a phase microscope. Bar = 16 μm.

The immobilization antigens of *Ichthyophthirius* have been identified as membrane proteins ranging between 40 and 60 kDa (Dickerson *et al.*, 1989; Lin and Dickerson, 1992; Dickerson *et al.*, 1993). Analysis of these antigens following monoclonal antibody affinity chromatography indicated that they shared common epitopes and that immobilization was dependent on native protein conformation.

It has been postulated that immobilization antigens elicit protective immunity (Hines and Spira, 1974c; Clark *et al.*, 1987). Hines and Spira suggested that immobilization was involved in protection when they observed that naive fishes did not become infected when challenged in the presence of immune fishes. They suggested that this protection resulted from the immobilization of free-swimming parasites by antibodies released from the body surfaces of immune fishes. The hypothesis that antibodies are directly involved in protection has been tested using monoclonal antibodies (Lin and Dickerson, unpublished results). Channel catfish injected intraperitoneally with mouse ascitic fluids containing immobilizing monoclonal antibodies (isotype IgG_1, see Fig. 5.7) were protected from infection when challenged 24 hours later. Control fishes injected with ascitic fluids containing non-immobilizing antibodies (also directed against Ich) succumbed to infection and died. These results clearly indicated that immobilizing murine antibodies provide protection against the parasite. However, immobilizing antisera from Ich-immune fish failed to provide protection in similar adoptive transfer experiments (Dickerson, unpublished). One possible explanation for this discrepancy is that a localized mucosal immune response is responsible for protection and only antibodies produced at the fishes' surface react with the invading parasite.

Concurrent infection and cross immunity

Protective immunity against *Ichthyophthirius* appears to be highly specific. Hines and Spira (1974c), noted that immune carp became infected with other ectoparasites even though they were resistant to reinfection with Ich. This observation was used to argue that immunity against Ich was an acquired specific response. It was of interest, therefore, when Goven *et al.* (1981a,b) suggested that the ciliate *Tetrahymena* shared common antigens with *Ichthyophthirius* and that *Tetrahymena* cilia could be used to vaccinate fish against Ich. *Tetrahymena* is easily grown in culture media, and it appeared that it might be used as a vaccine. Subsequent work by Wolf and Markiw (1982) indicated that trout were also protected after bath immunization in cultures of *Tetrahymena thermophila*. These early successes have not been repeated and the widespread use of *Tetrahymena* for vaccination has not materialized (Dickerson *et al.*, 1984; Houghton, 1987; Burkart *et al.*, 1990). Houghton *et al.* (1992) immunized juvenile carp by intraperitoneal injection with live *Tetrahymena* and found that their sera did not immobilize Ich or protect against infection. Clark *et al.* (1988) found that immobilizing sera from Ich-immune channel catfish reacted with Ich ciliary membrane proteins but not with those of *Tetrahymena*. A possible explanation for the discrepancy is that the protection observed in the earlier studies of Goven and Wolf was the result of non-specific immunity. Wolf found that fish immunized with *Tetrahymena* were resistant to both *Ichthyobodo necator* and Ich. Recently, Ling *et al.* (1993) reported that goldfish (*Carassius auratus*) vaccinated with *Tetrahymena pyriformis* were protected against *I. multifiliis* as well as a number of other protozoan parasites.

Cross protection occurs between different serotypic strains of *I. multifiliis* (Leff *et al.*, 1994). Channel catfish immunized with different immobilization serotypes and challenged with homologous or heterologous strains survive infection.

Antigens of *Cryptocaryon irritans*

Marine fish surviving infection with *Cryptocaryon irritans* become resistant to reinfection (Colorni, 1985; Colorni, 1987). Preliminary results from our laboratory (Dickerson and Yoshinaga, unpublished results) indicate that saltwater-adapted mollies (*Poecilia latipinna*) exposed to controlled *Cryptocaryon* infection and cured by treatment with malachite green and formalin, develop immunity to lethal challenge. The antigens which elicited this response were not identified.

IN VITRO CULTURE AND PROPAGATION OF THE PARASITES

One of the major problems encountered when working with either *Ichthyophthirius* or *Cryptocaryon* is that both organisms are obligate parasites and neither can be grown on artificial media.

We routinely propagate *Ichthyophthirius* by serial passage on channel catfish fingerlings. Briefly, an infected fish is placed in a 20 litre aquarium with five naive fishes. When these fish become heavily infected all but one are removed and placed in 5 litre glass jars containing aerated water. Five naive fish are again added to the aquarium to maintain the infection. The fish in the jars are transferred to fresh jars daily. Trophonts that leave the fish settle to the bottoms of the jars and encyst. The water is gently poured off. The attached tomonts are rinsed in sterile water, allowed to divide overnight, and the free-swimming tomites collected by centrifugation (Dickerson *et al.*, 1989). When trophonts are needed, they are gently dislodged from the surface of heavily infected fishes and placed into ice-cold sterile water in glass centrifuge tubes.

The parasite is cloned by placing single trophonts into one ml of sterile water in individual wells of 24-well plastic tissue culture plates. These develop into theronts overnight at 23–25°C. Theronts from single tomonts are then used to infect susceptible channel catfish fingerlings. These fishes are placed in an aquarium containing naive fishes.

Cryptocaryon culture is maintained in much the same way except that salt water is used and the development times are longer than those of *Ichthyophthirius*. Burgess and Matthews (1994) described a standard method for propagating *C. irritans* on grey mullet (*Chelon labrosus*). Yoshinaga and Dickerson (1994) established a procedure for growing the parasite on saltwater-adapted black mollies (*Poecilia latipinna*).

Cryopreservation

To date, cryopreservation and retrieval of viable, infective organisms has not been possible with either parasite. There have been several attempts with Ich, one of which succeeded in keeping tomonts viable for several months in liquid nitrogen (Beeler, 1980). Although the tomonts divided into theronts following thawing, their infectivity was not tested. We were unable to repeat this work (Dickerson, unpublished).

PARASITE BIOCHEMISTRY AND MOLECULAR BIOLOGY

Very little is known about the nutritional requirements of either *Ichthyophthirius* or *Cryptocaryon* except that they are obligate parasites that grow only by feeding on live fishes. Attempts to culture *Ichthyophthirius* in artificial media have not been successful (Hlond, 1966; Beckert, 1975). Ekless and

Matthews (1993) were able to extend survival of theronts and tomonts in selected monophasic media when compared to controls kept in water.

A number of early studies have been undertaken to examine the biochemical composition of the various developmental stages of *Ichthyophthirius*. Differential staining was used to identify glycogen, nucleic acids, and proteins (MacLennon, 1935b; Uspenskaya, 1963; Ycttehckar and Uspenskaya, 1964). It was found that glycogen, lipids, and proteins accumulated in the cytoplasm of trophonts during the feeding period and that these diminished during the development of the theront. Using SDS-polyacrylamide gel electrophoresis Pyle and Dawe (1985) examined the protein profiles of Ich at various stages of tomont division up to the differentiation of tomites into infective theronts. Differences in protein patterns were found between the trophont and the infective theront.

Differential expression of surface antigen genes in *Ichthyophthirius*

As described above, *Ichthyophthirius* has surface membrane proteins referred to as immobilization antigens (i-ags). The expression of these proteins is differentially regulated during the life cycle of the parasite (Clark *et al.*, 1992). A cDNA library was constructed from trophont mRNA (serotype G1, see Dickerson *et al.*, 1993) and screened with oligonucleotide probes deduced from the N-terminal amino acid sequence of the 48 kD i-ag. A cDNA clone encoding the 48 kD i-ag was isolated and is currently being characterized (Clark *et al.*, 1992; Clark and Dickerson, unpublished results). When RNA from trophonts or theronts was probed in Northern blots using this cDNA, the probe reacted with two mRNA transcripts of 1.9 and 1.6 kbp. These presumably code for the 60 and 48 kD i-ags of serotype G1. The amount of mRNA transcript encoding these proteins is extremely abundant accounting for up to 6% of the total mRNA in the cells. Additionally, the relative amount of the two transcripts varies significantly between the infective and feeding stages. The 1.9 kbp message increases in the theront to 50-fold the amount present in the trophont. At the same time, the 1.6 kbp message increases twofold in the transition from trophont to theront. The fact that i-ag transcript levels increase in parallel with the infectivity of the organism bears on the functional role of these surface proteins (Clark *et al.*, 1992).

DIAGNOSIS OF INFECTION

Fishes infected with either *I. multifiliis* or *C. irritans* usually develop characteristic white spots on their surface (see section on clinical signs above). In mild infections the parasites are not readily seen. The parasites are more difficult to see on fishes with lightly pigmented skin. In early stages of infection, large numbers of theronts can actually kill a fish before the parasite becomes visible and death is caused by massive damage to the gill epithelia.

As mentioned earlier (see section above on host-parasite relationships) infected fishes often have behavioural changes caused by the irritation from parasites in the skin and gills. This behaviour is nonspecific, however, and can also be associated with other bacterial, fungal, and protozoan diseases.

To make a definitive diagnosis of ichthyophthiriosis or cryptocaryonosis it is necessary to microscopically examine tissue from a gill arch, tail fin or body surface. The large (200 to 800 μm) ciliated trophonts are easily seen in unstained wet mounts (10–40 × magnification). The trophont of Ich has a distinctive horseshoe-shaped nucleus which is a pathognomonic sign of *Ichthyophthirius* infection. In early or very heavy infections theronts will also be seen. They are ciliated and pyriform in shape and can easily be confused with *Tetrahymena* spp.

The diagnosis of *Cryptocaryon* is also best made from a wet mount taken from the gills, skin scraping, or fin sections. The presence of a large ciliated protozoan is indicative of cryptocaryonosis. The nucleus of *C. irritans* is crescent-shaped and lobed.

PREVENTION AND CONTROL OF THE PARASITES

Disease control and prevention in an intensive fish production operation depend upon an integrated management programme. The basic elements are: (i) prevention of exposure of fishes to the parasites; (ii) prompt identification of the disease if it occurs; and (iii) treatment of infected fishes and immunization (if possible). Prevention of disease is always more cost effective than treatment.

I. multifiliis and *C. irritans* infections are usually introduced into fish populations by the addition of new infected fishes. In freshwater fish culture no new fishes should be introduced into a facility unless they have been in quarantine for 2 to 3 weeks (Brown and Gratzek, 1980). If possible the quarantined fishes should be held in moderately warm water (24°C). This is the optimal temperature for the development of *I. multifiliis* and in 2 to 3 weeks the parasite would have gone through several infection cycles. Thus, if the fishes were carrying low levels of infection there would be ample time for it to become apparent. The water from the quarantine facility (pond, raceway, aquarium, etc) should not be circulated into any other fish-holding facility.

If an Ich infection develops during the quarantine, the fishes should be treated. Treatments that control the free-swimming theronts break the cycle of infection by preventing reinfection. In addition, it is also possible to treat trophonts that are on the fishes (see chemotherapy section, below).

Physical treatments without the use of chemicals

The objective for the treatment of a disease is to break the cycle of infection. In *I. multifiliis* infection, the most vulnerable stage is the free-swimming

theront. The trophont is located deep in the epidermis of the fish and is protected from most agents that are added to the water.

One simple treatment is dilution or removal of the theronts in the water.
The repeated transfer of fishes to different aquaria is effective. Daily transfer
for 5–7 days is usually sufficient to break the cycle of infection (Brown and
Gratzek, 1980; Houghton and Matthews, 1990). In large fish culture opera-
tions dilution is used if the fishes can be placed in a raceway or shallow pond.
A rapid flow of water is then maintained for a week to dilute out the theronts.

In closed fish culture systems where the water is recycled, Ich infections can be controlled by exposing the water to ultraviolet light. The
recycled water is passed through an ultraviolet light sterilizer and exposed to
91,900 μW s^{-1} cm^{-2} before being cycled back into the aquaria containing
fish (Gratzek *et al.*, 1983).

Theronts are killed at high temperatures (29–30°C). Maintenance of fishes
at these temperatures for a week will eliminate infection (Brown and Gratzek,
1980). The water temperature should be raised gradually to allow the fishes
to acclimate to the change. This treatment is often combined with dilution.
The use of elevated temperatures to control Ich infection is unsuitable for cold
water fishes (Cross, 1972).

Chemotherapy

A variety of chemicals have been used for treating *I. multifiliis*. None of the
treatments is uniformly successful.

Sodium chloride was one of the first to be used. Stiles (1894) reported that
theronts were killed instantly in saturated salt solutions. He proposed creating
a salt concentration gradient for use as treatment. This can be done by placing
salt below a slotted or mesh floor in the aquarium (Cross, 1972). The fishes
tend to stay in the upper portions of the aquarium. When tomonts leave the
fishes, they fall into areas of high salt concentration and are killed. Fishes are
observed to occasionally dive down into the briny water (Stiles, 1894). Various
concentrations of salt (7,000–20,000 ppm) have been suggested for pond
treatments (Kabata, 1985). In general, salt treatment does not rapidly control
the infection and often does not completely remove the parasite from the
environment. The use of salt may have an additional benefit. Fishes infected
with *I. multifiliis* have decreased serum sodium and magnesium (Hines and
Spira, 1974b). The sodium in the water may help fishes maintain osmotic
balance, thus reducing stress. Reduction of stress increases the fishes' defence
reactions.

Formalin is often used to control infections. Fishes in aquaria are treated
with 25 ppm (1 ml formalin to 10 gallon water) of formalin on alternate days
until the infection is cleared. Usually the water is changed on days between
treatments (Brown and Gratzek, 1980). Pond fishes are treated with 15 to
25 ppm (5 to 8 gallons of formalin per acre foot water). Bath treatments are
also used; fishes are treated with 160–250 ppm formalin for 1 hour daily until
mortality stops (Brown and Gratzek, 1980). Care must be taken when formalin

is used in ponds in hot weather. If the formalin kills the algae in the pond, the pond may be depleted of oxygen (Cross, 1972).

Malachite green has also been used either alone or in combination with formalin. The zinc-free oxalate salt of malachite green is the only form of the chemical that is effective against *Ichthyophthirius*. Fish in aquaria are treated with 0.1 ppm at 3- to 4-day intervals until the infection is cleared. Carbon filters in the aquaria should be removed during treatment. Aeration must be maintained (Brown and Gratzek, 1980). Pond fishes are also treated with 0.1 ppm at 3- to 4-day intervals. Bath treatments are done at 2 ppm for 30 minutes. Malachite green is a suspected carcinogen and therefore must be handled with care. It is not approved for use on food fishes in the United States because of the possibility of residual contamination in tissues. When malachite is used in combination with formalin, a stock solution is prepared which consists of 14 g of malachite green dissolved in 1 gallon of formalin. Fishes in aquaria are treated by adding 1 ml of stock solution to 10 gallons of water. The treatment is repeated three times with a 3-day interval between each treatment (Brown and Gratzek, 1980). Again, aeration must be maintained, and carbon filters removed. Raceways are treated (1 ml per 10 gallon) for 6 hours daily and flushed after each treatment. Ponds are treated at 3- to 4-day intervals. The mixture is usually very effective, especially in treating fishes in aquaria. The incorporation of malachite green into fish food has been reported to be effective against trophonts residing on the fish (Schmahl *et al.*, 1992b). The fish must be healthy enough to eat for this treatment to work.

Another chemical that has been successfully used is potassium permanganate. This is recommended for ponds (Brown and Gratzek, 1980). The dose is 2 ppm (or less) for scaleless fishes and 2–5 ppm for scaled fishes. The effectiveness of potassium permanganate is reduced by large amounts of organic matter in the water.

A number of other chemicals have been used; these include copper sulphate (Brown and Gratzek, 1980; Straus, 1993), acriflavine, quinine, mepacrine, mercury compounds and chloramine-T (Cross, 1972). These have not been consistently effective in controlling *Ichthyophthirius* infections.

Attempts have been made to kill trophonts on infected fishes. This approach has been used to eliminate *I. multifiliis* from fishes before they are introduced to a pond, lake, river, aquarium, etc. A major problem is that the mucus and overlying epithelium prevent access of chemicals to the parasite. Three approaches, hyperosmotic shock, vacuum infiltration and surfactant immersion, were tested as a means to deliver drugs to trophonts in the skin. None of the procedures was effective (Post and Vesely, 1983). Toltrazuril was found to reach trophonts imbedded in the skin of fishes (Mehlhorn *et al.*, 1988). Immersion of fishes in water containing 10 µg ml^{-1} of the drug for 4 hours (repeated daily for 3 days) killed fish-associated parasites. This treatment was only effective against trophonts and did not kill theronts. Another triazinone, HOE 092 V, formulated as a water soluble preparation was also effective against trophonts in carp and cardinal tetras (*Paracheirodon axelrodi*) when applied once as a medicated bath at the same concentration and exposure time as toltrazuril (Schmahl *et al.*, 1992a).

Control of *I. multifiliis* outbreaks requires good animal husbandry and management in addition to the use of therapeutic agents. Dead fishes should be removed as soon as they are found because trophonts begin to drop off and encyst almost immediately. Aquaria should be scrubbed and water recycled through a diatomaceous earth filter (Brown and Gratzek, 1980). This will dislodge and remove developing cysts and theronts from the aquaria. In raceways where the water flow can be increased, the bottoms should be swept daily. This will dislodge developing cysts and many will be flushed out in the water flow. In shallow ponds sweeping the bottom will bury some cysts while others attached to the suspended bottom material can be flushed out by increasing the water flow (Brown and Gratzek, 1980).

Aquaria, ponds and raceways should be drained, cleaned and allowed to dry after an outbreak. Drying kills the parasite. The bottom of a dried pond should be treated with lime. If ponds cannot be drained completely, the residual water should be treated with a disinfectant such as calcium hypochlorite (Brown and Gratzek, 1980).

Immunization

The ideal method to prevent infection of fishes with *I. multifiliis* is immunization. Prevention of disease is always preferred to treatment. At the present time there are no practical, commercially available vaccines for use against Ich. A variety of components of *I. multifiliis* have been used to experimentally induce immunity in fishes. However, because *I. multifiliis* is an obligate parasite and can only be collected from live fishes, large scale production of antigenic material for vaccines is extremely difficult. It appeared that this obstacle had been overcome when Goven *et al.* (1980) reported that fishes could be immunized against *I. multifiliis* using the easily cultured ciliate *Tetrahymena pyriformis*. Similar results were reported by Wolf and Markiw (1982), and more recently by Ling *et al.* (1993) However, other work has found *Tetrahymena* ineffective as a protective antigen against Ich (Dickerson *et al.*, 1984; Clark *et al.*, 1987; Burkart *et al.*, 1990; Houghton and Matthews, 1992).

Dickerson and co-workers (Clark *et al.*, 1987, 1988; Dickerson *et al.*, 1989; Burkart *et al.*, 1990) have attempted to find antigens of *I. multifiliis* which induce protective immunity. Immobilization antigens have been identified as putative immunogens, and their genes have been cloned for subsequent expression in suitable vectors (Clark *et al.*, 1992, Clark and Dickerson, unpublished results). Thus, even though there is no vaccine currently available for the immunization of fish against Ich, there is hope that one will be available in the future through the use of biotechnology.

Prevention and control – *Cryptocaryon irritans*

As with *I. multifiliis*, the control of *C. irritans* in marine fishes depends upon breaking the cycle of infection. A variety of treatments have been used to control *C. irritans*.

One basic method is separation of fishes from the encysted tomonts before they develop into infective theronts. In small aquaria this can be done by transferring the fishes to clean aquaria every 3 days. The tanks are cleaned and dried between uses (Colorni, 1987). Another method is to place the fishes in cages in the open sea for at least 10 days. The cages should be several metres above the sea floor. Tomonts fall to the bottom and the released theronts are unable to reach the fishes (Colorni, 1987). In large display tanks with both fishes and invertebrates, chemical treatment is not feasible. The following method was found to be useful. A 1–2 cm layer of clean, fine sand was spread uniformly on the bottom of the tank. After 3 days the sand was removed with suction dredge and replaced with a new layer of sand. The procedure was repeated four times at 3-day intervals (Colorni, 1987).

Chemotherapy

Many of the same chemicals used to treat *I. multifiliis* infections have been tried against *C. irritans*. One chemical that appears to work is copper sulphate. Immersion of fishes in 0.15 to 0.25 ppm copper sulphate for 3 to 10 days is recommended (Herwig, 1978). The treatment may have to be repeated several times. Brown and Gratzek (1980) suggested that continual exposure to copper (0.15 ppm) will control *C. irritans* infections. The solubility of copper is adversely affected by calcium carbonate in the water. It was suggested that citric acid or glacial acetic acid be added to chelate the copper to keep it in solution. This is questionable since the chelated copper may not be available for activity against the parasite (Herwig, 1978).

A mixture of cupric acetate (0.2 ppm) formalin (5.26 ppm) and tris buffer (4.8 ppm) has been an effective treatment (Nigrelli and Ruggieri, 1966). Another variation is to use copper sulphate (0.15 ppm to 0.2 ppm), citric acid and methylene blue stock (1 ml of 1% solution per 2.5 gal water) until the water is a clear blue colour (Nigrelli and Ruggieri, 1966).

The most effective drugs against *C. irritans* infections appear to be quinine derivatives. Quinine hydrochloride, quinine sulphate and quinacrine hydrochloride (atebrine or mepacrine) all are effective against theronts and can be used interchangeably at 8 to 12 mg gallon^{-1}. Quinine hydrochloride is soluble in salt water; quinine sulphate stays effective over a wide range of water pH; and atebrine is the least expensive (Herwig, 1978). During treatment the tanks should be protected from light which inactivates the drugs. Fishes may occasionally show toxic reactions. The clinical signs of drug toxicity include loss of equilibrium, loss of appetite, rapid breathing and lying on the bottom. Treatment of the water with 2–4 drops per gallon of a 0.2% stock solution of potassium permanganate solution generally will counteract the toxicity (Herwig, 1978).

Exposure of the parasite to either hypo- or hypersalinity has been proposed. Colorni (1985) reported that four repeated exposures of 3 hour duration in 10% seawater cleared the infection in 5 to 8 days. Huff and Burns (1981) exposed red snappers to 45% sodium chloride followed by treatment with quinine-HCl. They found that the hypersalinity increased the efficacy of the total treatment regimen. Cheung *et al.* (1979) reported that killifish were cleared of *C. irritans* in seawater diluted $1:1$, $1:2$ and $1:3$ with distilled water. The reduced salinity appeared to affect the trophonts on the fishes as well as the encysted tomonts. The use of hyposalinity may be limited by the ability of the fishes to tolerate the low salt concentrations. Estuary fishes are more likely to tolerate low salt treatments than open sea or reef dwelling fishes.

A suggested treatment is as follows: an initial 5–15 minute bath in freshwater followed by quarantine in a saltwater aquarium containing atebrine at 4–6 mg gallon^{-1}. The freshwater dip is repeated on the second day and the atebrine concentration increased to 8–12 mg gallon^{-1}. This routine is repeated for a total of 5 days. The fishes should be observed for an additional 10 days after treatment (Herwig, 1978).

SUMMARY AND CONCLUSIONS

Ichthyophthirius multifiliis and *Cryptocaryon irritans* are holotrichous ciliates which parasitize freshwater and marine fishes, respectively. Ichthyophthiriosis and cryptocaryonosis are significant problems in fish culturing systems. Outbreaks produce financial losses not only from fish mortalities, but also from the cost of control measures and treatments. There are a variety of treatments or combination of treatments available. An important area of research is the development of effective control methods to prevent outbreaks of these pathogenic ciliates.

Immunization is one of the most cost- and time-effective means of disease prevention. However, there are significant problems related to the development of practical vaccines. One of the major obstacles is that there is no method for cultivation of the organisms. This severely limits the availability of material for use in a vaccine. Future research should address the problem of mass production of protective antigens. One solution would be to develop *in vitro* cultivation systems for *I. multifiliis* and *C. irritans*. A second would be to use genetic engineering and biotechnology for the production of recombinant antigens in bacteria or other easily cultured organisms. The first step in the latter approach is the identification of protective antigens; the genes encoding these can then be cloned and inserted into expression vectors for large-scale production.

Further research is also necessary to understand the immune response against both parasites. Although it has been evident for some time that fishes develop protective immunity against *I. multifiliis*, the effector mechanisms remain obscure. It is necessary to understand this response to take a logical approach toward vaccination.

Another area of research that is needed is the development of techniques to preserve and maintain infective organisms. One of the problems in immunization trials is the lack of standard challenge methods. If large quantities of theronts could be collected and preserved by freezing, then standard isolates could be used for these purposes. Preservation techniques would benefit other areas of research as well. Collection and cryopreservation of isolates from different geographic areas, different fish species and different times would allow comparative studies. This could help to resolve taxonomic questions related to number of species of *Ichthyophthirius* and *Cryptocaryon*. Also, the question of biotypes or strains and host specificity could be investigated in more detail.

Finally, A reexamination of the morphological classification of *Cryptocaryon* might be in order since it lacks the characteristic organelle of Lieberkühn found in *Ichthyophthirius* and other ophryoglenids.

REFERENCES

Allison, R. and Kelly, H.D. (1963) An epizootic of *Ichthyophthirius multifiliis* in a river fish population. *Progressive Fish Culturist* 25, 149–150.

Areerat, S. (1974) The immune response of channel catfish, *Ictalurus punctatus* (Rafinesque), to *Ichthyophthirius multifiliis*. Unpublished Masters Thesis. Auburn University.

Bauer, O.N. (1953) Immunity of fish occurring in infections with *Ichthyophthirius multifiliis* Fouquet, (1876). Akademiîa nauk SSSP. *Doklady Novaîa serviîa* 93, 377–379.

Bauer, O.N. (1970) Parasitic diseases of cultured fishes and methods of their prevention and treatment. In: Dogiel, V.A., Petrushevski, G.K., and Polyanski, Y.I. (eds) *Parasitology of Fishes*. TFH Publications, Hong Kong, pp. 274–275.

Beckert, H. (1975) Observations on the biology of *Ichthyophthirius multifiliis* (Fouquet, 1876). Its susceptibility to ethoxyquin, and some immunological responses of channel catfish to this parasite. Unpublished PhD dissertation, University of Southwestern Louisiana.

Beckert, H. and Allison, R. (1964) Some host responses of white catfish to *Ichthyophthirius multifiliis* (Fouquet). *Proceedings of the Southeastern Assocociation of Game Fish Commissioners* 18, 438–441.

Beeler, C.R. (1981) Controlling cooling rates with a variable vacuum in a Dewar flask. *Cryobiology* 18, 79–81.

Bradshaw, C.M., Richard, A.S., and Sigel, M.M. (1971) IgM antibodies in fish mucus. *Proceedings of the Society of Experimental and Biological Medicine* 136, 1122–1124.

Bragg, R.R., (1991) Health status of salmonids in river systems in Natal. I. Collection of fish and parasitological examination. *The Onderstepoort Journal of Veterinary Research* 58, 59–62.

Brown, E.M. (1951) A new parasitic protozoan, the causal organism of a white spot disease in marine fish *Cryptocaryon irritans* gen and sp.n. *Agenda of Scientific Meetings of the Zoological Society, London, 1950* 11, 1–2.

Brown, E.M., (1963) Studies on *Cryptocaryon irritans* Brown. *Progress in Protozoology, Proceedings of the 1st International Congress on Protozoology*. Academic Press, New York, 284–287.

Brown, E.E. and Gratzek, J.B. (1980) *Fish Farming Handbook. Food, Bait, Tropicals and Goldfish*. AVI Publishing Co., Westport, Connecticut.

Burgess, P.J. and Matthews, R.A. (1994) A standardized method for the *in vitro* maintenance of *Cryptocaryon irritans* (Ciliophora) using the grey mullet *Chelon labrosus* as an experimental host. *Journal of Parasitology* 80, 288–292.

Burkart, M.A., Clark, T.G. and Dickerson, H.W. (1990) Immunization of channel catfish, *Ictalurus punctatus* Rafinesque against *Ichthyophthirius multifiliis* (Fouquet): killed versus live vaccines. *Journal of Fish Diseases* 13, 401–410.

Bushkiel, A.L. (1910) Beitrage zur kenntnis des *Ichthyophthirius multifiliis. Archiv für Protistenkinde* 21, 61–102.

Butcher, A.D. (1941) Outbreaks of white spots of ichthyophthiriasis (*Ichthyophthirius multifiliis* Fouquet, 1876) at the hatcheries of the Ballart fish acclimatization facility with notes on laboratory experiments. *Proceedings of the Royal Society of Victoria* 53, 126–144.

Canella, M.F. and Rocchi-Canella, I. (1976) Biologie des Ophryoglenina (cilies hymeno-stermes, histophages). *Annals of the University of Ferrara* (N.S. Sect. III) 3 (Suppl. 2) 1–510.

Chapman, G.B. (1984) Ultrastructural aspects of the host–parasite relationship in ichthyophthiriasis. *Transactions of the American Microscopical Society* 103, 364–375.

Chapman, G.B. and Kern, R.C. (1983) Ultrastructural aspects of the somatic cortex and contractile vacuole of the ciliate *Ichthyophthirius multifiliis*, Fouquet. *Journal of Protozoology* 30, 481–490.

Cheung, P.J., Nigrelli, R.F. and Ruggieri, G.D. (1979) Studies on cryptocaryoniasis in marine fish: effect of temperature and salinity on reproductive cycle of *Cryptocaryon irritans* Brown, 1951. *Journal of Fish Diseases* 2, 93–97.

Cheung, P.J., Nigrelli, R.F. and Ruggieri, G.D. (1981) Scanning electron microscopy on *Cryptocaryon irritans* Brown 1951, a parasitic ciliate in marine fish. *Journal of Aquaculture* 2, 70–72.

Clark, T.G., Dickerson, H.W., Gratzek, J.B. and Findly, R.C. (1987) *In vitro* response of *Ichthyophthirius multifiliis* to sera from immune channel catfish. *Journal of Fish Biology* 31 (Supplement A), 203–208.

Clark, T.G. , Dickerson, H.W. and Findly, R.C. (1988) Immune response of channel catfish to ciliary antigens of *Ichthyophthirius multifiliis. Developmental and Comparative Immunology* 12, 203–208.

Clark, T.G., McGraw, R.A. and Dickerson, H.W. (1992) Developmental expression of surface antigen genes in the parasitic ciliate *Ichthyophthirius multifiliis. Proceedings of the National Academy of Sciences (USA)* 89, 6363–6367.

Clayton, G.M. and Price, D.J. (1988) Pleiotropic effect on scale pattern genes in common carp: susceptibility to *Ichthyophthirius multifiliis* infection. *Heredity* 60, 312. (Abstract)

Clayton, G.M. and Price, D.J. (1992) Interspecific and intraspecific variation in resistance to ichthyophthiriasis among poeciliid and goodeid fishes. *Journal of Fish Biology* 40, 445–453.

Clayton, G.M. and Price, D.J. (1994) Heterosis in response to *Ichthyophthirius multifiliis* infections in poeciliid fish. *Journal of Fish Biology* 44, 59–66.

Colorni, A. (1985) Aspects of the biology of *Cryptocaron irritans* and hyposalinity as a control measure in cultured gilt-head sea bream, *Sparus aurata. Diseases of Aquatic Organisms* 1, 19–27.

Colorni, A. (1987) Biology of *Cryptocaryon irritans* and strategies for its control. *Aquaculture* 67, 236–237.

Colorni, A. and Diamant, A. (1993) Ultrastructural features of *Cryptocaryon irritans*, a ciliate parasite of marine fish. *European Journal of Protistology* 29, 425–434.

Corliss, J.O. (1979) *The Ciliated Protozoa, Characterization, classification and Guide to the Literature.* Pergamon Press, New York.

Cross, D.G. (1972) A review of methods to control ichthyophthiriasis. *The Progressive Fish Culturist* 34, 165–170.

Cross, M.L. (1993) Antibody binding following exposure of live *Ichthyophthirius multifiliis* (Ciliophora) to serum from immune carp *Cyprinus carpio*. *Diseases of Aquatic Organisms* 17, 159–164.

Cross, M.L. and Matthews, R.A. (1992) Ichthyophthiriasis in carp, *Cyprinus carpio* L.: fate of parasites in immunized fish. *Journal of Fish Diseases* 15, 497–505.

Cross, M.L. and Matthews, R.A. (1993a) Localized leukocyte response to *Ichthyophthirius multifiliis* establishment in immune carp *Cyprinus carpio* L. *Veterinary Immunology and Immunopathology* 38, 341–358.

Cross, M.L. and Matthews, R.A. (1993b) *Ichthyophthirius multifiliis* Fouquet (Ciliophora): the localization of sites immunogenic to the host *Cyprinus carpio* (L.) *Fish and Shellfish Immunology* 3, 13–24.

Diamant A., Issar G., Colorni, A. and Paperna, I. (1991) A pathogenic *Cryptocaryon*-like ciliate from the Mediterranean Sea. *Bulletin of the European Association of Fish Pathologists* 11, 122–124.

Dickerson, H.W., Brown, J., Dawe, D.L. and Gratzek, J.B. (1984) *Tetrahymena pyriformis* as a protective antigen against *Ichthyophthirius multifiliis* infection: comparisons between isolates and ciliary preparations. *Journal of Fish Biology* 24, 523–528.

Dickerson, H.W., Lohr, A.L. and Gratzek, J.B. (1985) Experimental intraperitoneal infection of channel catfish, *Ictalurus punctatus* (Rafinesque), with *Ichthyophthirius multifiliis* (Fouquet). *Journal of Fish Diseases* 8, 139–142.

Dickerson, H.W., Clark, T.G. and Findly, R.C. (1989) *Ichthyophthirius multifiliis* has membrane-associated immobilization antigens. *Journal of Protozoology* 36, 159–164.

Dickerson, H.W., Clark, T.G. and Leff, A.A. (1993) Serotypic variation among isolates of *Ichthyophthirius multifiliis* based on immobilization. *Journal of Eukaryotic Microbiology* 40, 816–820.

Ekless, L.M. and Matthews, R.A. (1993) *Ichthyophthirius multifiliis*, axenic isolation and short-term maintenance in selected monophasic media. *Journal of Fish Diseases* 16, 437–447.

Elser, H.J. (1955) An epizootic of ichthyophthiriasis among fish in a large reservoir. *Progressive Fish Culturist* 17, 132–133.

Ewing, M.S. and Kocan, K.M. (1986) *Ichthyophthirius multifiliis* (Ciliophora) development in gill epithelium. *Journal of Protozoology* 33, 369–374.

Ewing, M.S. and Kocan, K.M. (1987) *Ichthyophthirius multifiliis* (Ciliophora) exit from gill epithelium. *Journal of Protozoology* 34, 309–312.

Ewing, M.S., and Kocan, K.M. (1992) Invasion and development strategies of *Ichthyophthirius multifiliis*, a parasitic ciliate of fish. *Parasitology Today* 8, 204–208.

Ewing, M.S., Kocan, K.M. and Ewing, S.A. (1983) *Ichthyophthirius multifiliis*: morphology of the cyst wall. *Transactions of the American Microscopical Society* 102, 122–128.

Ewing, M.S., Kocan, K.M. and Ewing, S.A. (1985) *Ichthyophthirius multifiliis* (Ciliophora) invasion of gill epithelium. *Journal of Protozoology* 32, 305–310.

Ewing, M.S., Lynn, M.E. and Ewing, S.A. (1986) Critical periods in development of *Ichthyophthirius multifiliis* (Ciliophora) populations. *Journal of Protozoology* 33, 388–391.

Ewing, M.S., Ewing, S.A. and Kocan, K.M. (1988) *Ichthyophthirius* (Ciliophora): Population studies suggest reproduction in host epithelium. *Journal of Protozoology* 35, 549–552.

Farley, D.G. and Heckmann, R. (1980) Attempts to control *Ichthyophthirius multifiliis*

Fouquet (Ciliophora: Ophryoglenidae) by chemotherapy and electrotherapy. *Journal of Fish Diseases* 3, 203–212.

Geisslinger, M. (1987) Observations on the caudal cilium of the tomite of *Ichthyophthirius multifiliis* Fouquet 1876. *Journal of Protozoology* 341, 180–182.

Goven, B.A., Dawe, D.L. and Gratzek, J.B. (1980) Protection of channel catfish, *Ictalurus punctatus* Rafinesque, against *Ichthyophthirius multifiliis* Fouquet by immunization. *Journal of Fish Biology* 17, 311–316.

Goven, B.A., Dawe, D.L. and Gratzek, J.B. (1981a) Protection of channel catfish (*Ictalurus punctatus*) against *Ichthyophthirius multifiliis* (Fouquet) by immunization with varying doses of *Tetrahymena pyriformis* (Lwoff) cilia. *Aquaculture* 23, 269–273.

Goven, B.A., Dawe, D.L. and Gratzek, J.B. (1981b) *In vitro* demonstration of serological cross-reactivity between *Ichthyophthirius multifiliis* Fouquet and *Tetrahymena pyriformis* Lwoff. *Developmental and Comparative Immunology* 5, 283–389.

Gratzek, J.B., Gilbert, J.P., Lohr, A.L., Shotts, E.B. and Brown, J. (1983) Ultraviolet light control of *Ichthyophthirius multifiliis* Fouquet in a closed fish culture recirculation system. *Journal of Fish Diseases* 6, 145–153.

Graves, S.S., Evans, D.L., Cobb, D. and Dawe, D.L. (1984) Nonspecific cytotoxic cells in fish (*Ictalurus punctatus*) I. Optimum requirements for target cell lysis. *Developmental and Comparative Immunology* 8, 293–302.

Graves, S.S., Evans, D.L. and Dawe, D.L. (1985a) Antiprotozoan activity of nonspecific cytotoxic cells (NCC) from channel catfish (*Ictalurus punctatus*). *Journal of Immunology* 134, 78–85.

Graves, S.S., Evans, D.L. and Dawe, D.L. (1985b) Mobilization and activation of nonspecific cytotoxic cells (NCC) in channel catfish (*Ictalurus punctatus*) infected with *Ichthyophthirius multifiliis*. *Comparative Immunology and Microbiology of Infectious Diseases* 8, 43–51.

Handy, P.D. and Eddy, F.B. (1991) The absence of mucus on the secondary lamellae of unstressed rainbow trout, *Oncorhynchus mykiss* (Walbaum). *Journal of Fish Biology* 38, 153–155.

Hauser M. (1972) The intranuclear mitosis of the ciliates *Paracineta limbata* and *Ichthyophthirius multifiliis*. *Chromosoma* 36, 158–175.

Herwig, N. (1978) Notes on the treatment of *Cryptocaryon*. *Drum and Croaker* 18, 6–12.

Hildemann, W.H. (1972) Transplantation reactions of two species of Osteichthyes (Teleostei) from South Pacific coral reefs. *Transplantation* 14, 261–267.

Hines, R.S. and Spira, D.T. (1973a) *Ichthyophthirius multifiliis* (Fouquet) in the mirror carp, *Cyprinus carpio* L. I. Course of infection. *Journal of Fish Biology* 5, 385–392.

Hines, R.S. and Spira, D.T. (1973b) Ichthyophthiriasis in the mirror carp. II. Leukocyte response. *Journal of Fish Biology* 5, 527–534.

Hines, R.S. and Spira, D.T. (1974a) Ichthyophthiriasis in the mirror carp *Cyprinus carpio* L. III. Pathology. *Journal of Fish Biology* 6, 189–196.

Hines, R.S. and Spira, D.T. (1974b) Ichthyophthiriasis in the mirror carp *Cyprinus carpio* (L.) IV. Physiological dysfunction. *Journal of Fish Biology* 6, 365–371.

Hines, R.S. and Spira, D.T. (1974c) Ichthyophthiriasis in the mirror carp *Cyprinus carpio* (L.) V. Acquired immunity. *Journal of Fish Biology* 6, 373–378.

Hlond, S. (1966) Experiments in *in vitro* culture of *Ichthyophthirius*. *FAO World Symposium on Warm-water Pond Fish Culture*. Rome, Italy.

Hoffman, G. (1970) *Parasites of North American Freshwater Fishes*. University of California Press, Berkeley.

Houghton, G. (1987) The immune response in carp, *Cyprinus carpio* L. to *Ichthyophthirius multifiliis*, Fouquet 1876. PhD Dissertation, Department of Biological Sciences, Plymouth Polytechnic, UK.

Houghton, G. and Matthews, R.A. (1986) Immunosuppression of carp to *Ichthyophthirius* using the steroid triamcinolone acetonide. *Veterinary Immunology and Immunopathology* 12, 413–419.

Houghton, G. and Matthews, R.A. (1990) Immunosuppression in juvenile carp, *Cyprinus carpio* L.: the effects of the corticosteroids triamcinolone acetonide and hydrocortisone 21-hemisuccinate (cortisol) on acquired immunity and the humoral antibody response to *Ichthyophthirius multifiliis* Fouquet. *Journal of Fish Diseases* 13, 269–280.

Houghton, G. and Matthews, R.A. (1993) *Ichthyophthirius multifiliis* (Fouquet): survival within immune juvenile carp, *Cyprinus carpio* L. *Fish and Shellfish Immunology* 3, 157–166.

Houghton, G., Healey, L.J. and Matthews, R.A. (1992) The cellular proliferative response, humoral antibody response, and cross reactivity studies of *Tetrahymena pyriformis* with *Ichthyophthirius multifiliis* in juvenile carp (*Cyprinus carpio* L.). *Developmental and Comparative Immunology* 16, 301–312.

Huff, J.A. and Burns, C.D. (1981) Hypersaline and chemical control of *Cryptocaryon irritans* in red snapper, *Lutjanus campechanus*, monoculture. *Aquaculture* 22, 181–184.

Hughes, G.M. (1970) A comparative approach to fish respiration. *Experientia* 26, 113–122.

Kabata, Z. (1985) *Parasites and Diseases of Fish Cultured in the Tropics*. Taylor and Francis, Philadelphia, Pennsylvania.

Kikuchi, S. and Eqame, N. (1983) Effects of α-irradiation on the rejection of transplanted scale melanophores in the teleost, *Oryzias latipes*. *Developmental and Comparative Immunology* 7, 51–58.

Kozel, T.R. (1976) The occurrence of *Ichthyophthirius multifiliis* (Ciliate: Hymenostomatida) in Kentucky waters. *Transactions of the Kentucky Academy of Science* 37, 41–43.

Kozel, T.R. (1986) Scanning electron microscopy of theronts of *Ichthyophthirius multifiliis*: their penetration into host tissue. *Transactions of the American Microscopical Society* 105, 357–364.

Lee, J.J., Hunter, S.H. and Bovee, E.C. (1985) *An Illustrated Guide to the Protozoa*, Society of Protozoologists. Lawrence, Kansas.

Leff, A.A., Yoshinaga, T. and Dickerson, H.W. (1994) Cross immunity in channel catfish, *Ictalurus punctatus* (Rafinesque), against two immobilization serotypes of *Ichthyophthirius multifiliis* (Fouquet). *Journal of Fish Diseases* 17, 429–432.

Lin, T.L. and Dickerson, H.W. (1992) Purification and partial characterization of immobilization antigens from *Ichthyophthirius multifiliis*. *Journal of Protozoology* 39, 457–463.

Ling, K.H., Sin, Y.M. and Lam, T.J. (1993) Protection of goldfish against some common ectoparasitic protozoans using *Ichthyophthirius multifiliis* and *Tetrahymena pyriformis* for vaccination. *Aquaculture* 116, 303–314.

Lobb, C.J. and Clem, L.W. (1981) The metabolic relationship of the immunoglobulins in fish serum, cutaneous mucus, and bile. *Journal of Immunology* 127, 1525–1529.

Lobb, C.J. and Clem, L.W. (1982) Fish lymphocytes differ in the expression of surface immunoglobulin. *Developmental and Comparative Immunology* 6, 473–479.

Lobo-da-Cunha, A. and Azevedo, C. (1993) Processing of food vacuoles in the parasitic ciliate *Ichthyophthirius multifiliis* after exit from the host. *Parasitology Research* 79, 272–278.

Lom, J. and Cerkasova, A. (1974) Host finding in invasive stages of *Ichthyophthirius multifiliis*. *Journal of Protozoology* 21, 457.

Lom, J. and Dykova, I. (1992) *Protozoan Parasites of Fishes*. Elsevier Science Publishers, Amsterdam, The Netherlands.

Lynn, D.H., Frombach, S., Ewing, M.S. and Kocan, K.M. (1991) The organelle of Lieberkühn as a synapomorphy for the Ophryoglenina (Ciliophora: Hymenostomatida). *Transactions of the American Microscopical Society* 110, 1–11.

MacLennon, R.F. (1935a) Observations on the life cycle of *Ichthyophthirius*, a ciliate parasitic on fish. *Northwestern Scientist* 9, 12–14.

MacLennon, R.F. (1935b) Dedifferentiation and redifferentiation in *Ichthyophthirius*. I. *Archives of Protozoology* 86, 191–210.

MacLennon, R.F. (1937) Growth in the ciliate *Ichthyophthirius*. I. Maturity and encystment. *Journal of Experimental Zoology* 76, 423–440.

MacLennon, R.F. (1942) Growth in the ciliate *Ichthyophthirius*. II. Volume. *Journal of Experimental Zoology* 91, 1–13.

Maetz, J. and Garcia-Romea, F. (1964) The mechanism of sodium and chloride uptake by the gills of a fresh-water fish, *Carassius auratus*. II. Evidence of NH_4^+/Na^+ and HCO^-/Cl^- exchanges. *Journal of General Physiology* 47, 1209–1227.

Matthews, B.F., Matthews, R.A. and Burgess, P.J. (1993) *Cryptocaryon irritans* Brown 1951 (Ichthyophthiriidae): the ultrastructure of the somatic cortex throughout the life cycle. *Journal of Fish Diseases* 16, 339–349.

McCallum, H.I. (1982) Infection dynamics of *Ichthyophthirius multifiliis*. *Parasitology* 85, 475–488.

McCallum, H.I. (1985) Population effects of parasite survival of host death: experimental studies of interaction of *Ichthyophthirius multifiliis* and its fish host. *Parasitology* 90, 529–547.

McCallum, H.I. (1986) Acquired resistance of black mollies (*Poecilia latipinna*) to infection by *Ichthyophthirius multifiliis*. *Parasitology* 93, 251–261.

McCartney, J.B., Fortner, G.W. and Hansen, M.F. (1985) Scanning electron microscopic studies of the life cycle of *Ichthyophthirius multifiliis*. *Journal of Parasitology* 71, 218–226.

Mehlhorn, H., Schmahl, G. and Haberkorn, A. (1988) Toltrazuril effective against a broad spectrum of protozoan parasites. *Parasitology Research* 75, 64–66.

Nanney, D.L. (1980) *Experimental Ciliatology, An Introduction to Genetic and Developmental Analysis in Ciliates*. John Wiley and Sons, New York.

Nigrelli, R.F. and Ruggieri, G.D. (1966) Enzootics in the New York Aquarium caused by *Cryptocaron irritans* Brown, 1951 (= *Ichthyophthirius marinus* Sikama, 1961), a histophagous ciliate in the skin, eyes and gills of marine fish. *Zoologica: New York Zoological Society* 51, 97–102.

Nigrelli, R.F., Pokorny, K.S. and Ruggieri, G.D. (1976) Notes on *Ichthyophthirius multifiliis*, a ciliate parasitic on freshwater fishes, with some remarks on possible physiological races and species. *Transactions of the American Microscopical Society* 95, 607–613.

Paperna, I. (1972) Infection by *Ichthyophthirius multifiliis* of fish in Uganda. *Progressive Fish Culturist* 34, 162–164.

Parker, J.C. (1965) Studies on the natural history of *Ichthyophthirius multifiliis* Fouquet 1876 and ectoparasitic ciliate of fish. PhD Dissertation, University of Maryland.

Peshkov, M.A. and Tikhomirova, L.A. (1968) Ultrafine structure of nuclear apparatus in *Ichthyophthirius multifiliis* at the protomite cyst stage. *Doklady Akademiîa Nauk SSSR, Novaîa seriîa* 183, 451–452.

Pickering, A.D. and Christie, P. (1980) Sexual differences in the incidence and severity of ectoparasitic infestation of the brown trout, *Salmo trutta* L. *Journal of Fish Biology* 16, 669–683.

Post, G. and Vesely, K.R. (1983) Administration of drugs by hyperosmotic or vacuum infiltration or surfactant immersion ineffective for control of intradermally encysted *Ichthyophthirius multifiliis*. *Progressive Fish Culturist* 45, 164–166.

Pyle, S.W. (1983) Antigenic and serologic relationships between *Ichthyophthirius multifiliis* Fouquet and *Tetrahymena pyriformis* Lwoff. PhD Dissertation, University of Georgia.

Pyle, S.W. and Dawe, D.L. (1985) Stage-dependent protein composition in the life cycle of synchronous *Ichthyophthirius multifiliis*, a ciliate fish parasite. *Journal of Protozoology* 32, 355–359.

Roque, M., de Puytorac, P. and Lom, J. (1967) L'architecture buccale et la stomatogenese d' *Ichthyophthirius multifiliis* Fouquet, 1876. *Protistologica* 3, 79–189.

Schmahl, G., Raether, W. and Mehlhorn, H. (1992a) HOE 092 V, a new triazine derivative effective against a broad spectrum of fish and crustacean parasites. *Parasitology Research* 78, 702–706.

Schmahl, G., Ruider, S., Mehlhorn, H., Schmidt, H. and Ritter, G. (1992b) 9. Effects of a medicated food containing malachite green on *Ichthyophthirius multifiliis* Fouquet, 1876 (Hymenostomatida, Ciliophora) in ornamental fish. *Parasitology Research* 78, 183–192.

Secombes, C.J., van Groningen, J.J.M. and Egberts, E. (1982) Separation of lymphocyte subpopulation in carp *Cyprinus carpio* L. by monoclonal antibodies: immunohistochemical studies. *Immunology* 48, 165–175.

Sikama, Y. (1938) Uber die Weisspunktchendrankheit bei Seefischen. *Journal of the Shanghai Institute of Science (Section III)* 4, 113–128.

Sikama, Y. (1961) On a new species of *Ichthyophthirius* found in marine fishes. *Scientific Report of the Yokosuka City Museum* 6, 66–70.

Smith, H.W. (1929) The excretion of ammonia and urea by the gills of fish. *Journal of Biological Chemistry* 81, 71–73.

Stanley, J.G. and Colby, P.J. (1971) Effects of temperature on electrolyte balance and osmoregulation in the alewife (*Alosa pseudoharengus*) in fresh and sea water. *Transactions of the American Fisheries Society* 100, 624–638.

Stiles, C.W. (1894) Reports on a parasitic protozoan observed on fish in the aquarium. *Bulletin of the United States Fisheries Commission* 13, 173–190.

Straus, D.L. (1993) Prevention of *Ichthyophthirius multifiliis* infestation in channel catfish fingerlings by copper sulphate treatment. *Journal of Aquatic Animal Health* 5, 152–154.

Tataner, M.F. and Manning, M.J. (1983) The ontogeny of cellular immunity in the rainbow trout, *Salmo gairdneri* Richardson, in relation to the stage of development of the lymphoid organs. *Developmental and Comparative Immunology* 7, 69–75.

Uspenskaya, A.V. (1963) Hyaluronidase in the various stages of the life history of *Ichthyophthirius multifiliis*. *Doklady Akademia Nauk SSSR, Novaîa serviîa* 151, 1031–1033.

Valtonen, E.T. and Keränen, A. (1981) Ichthyophthiriasis of Atlantic salmon, *Salmo salar* L. at the Montta Hatchery in northern Finland in 1978–1979. *Journal of Fish Diseases* 4, 405–411.

Ventura, M.T. and Paperna, I. (1985) Histopathology of *Ichthyophthirius multifiliis* infections in fishes. *Journal of Fish Biology* 27, 185–203.

Wagner, G. (1960) Der entwicklungs zyklus von *Ichthyophthirius multifiliis* Fouquet und der einfluss physikalischer und chemischer aussenfaktoren. *Zeitschrift für Fischerer und der Hilfswissenschaflen* 9, 425–433.

Wahli, T. and Meier, W. (1985) Ichthyophthiriasis in trout: investigation of natural defence mechanisms. In: Ellis, A.E. (ed.) *Fish and Shellfish Pathology*. Academic Press, London, 347–352.

Wahli, T. and Meier, W. (1991) Affinity of *Ichthyophthirius multifiliis* theronts to light and/or fish. *Journal of Applied Ichthyology* 7, 244–249.

Wedemeyer, G. (1972) Some physiological consequences of handling stress in juvenile coho salmon (*Oncorhynchus kisutch*) and steelhead trout (*Salmo gairdneri*). *Journal of the Fisheries Research Board of Canada* 29, 1780–1783.

Wikgren, B.J. (1953) Osmotic regulation in some aquatic animals with special reference to the influence of temperature. *Acta Zoologica Fennica* 71, 1–102.

Wilke, D.W. and Gordin, H. (1969) Outbreak of cryptocaryoniasis in marine aquaria at Scripps Institute of Oceanography. *California Fish and Game* 55, 227–236.

Wolf, K. and Markiw, M.A. (1982) Ichthyophthiriasis: Immersion immunization of rainbow trout (*Salmo gairdneri*) using *Tetrahymena thermophila* as a protective immunogen. *Canadian Journal of Aquatic Science* 39, 1722–1725.

Wurtsbaugh, W.A. and Tapia, R.A. (1988) Mass mortality of fishes in Lake Titicaca (Peru-Bolivia) associated with the protozoan parasite *Ichthyophthirius multifiliis*. *Transactions of the American Fisheries Society* 117, 213–217.

Ycttenckar, A.B. and Uspenskaya, A.V. (1964) Reserve materials, RNA, DNA, and respiratory enzymes at different stages of the life cycle of *Ichthyophthirius multifiliis*. *Acta Protozoologica* 2, 175–194.

Yocum, D., Cuchens, M. and Clem, L.W. (1975) The hapten-carrier effect in teleost fish. *Journal of Immunology* 114, 925–927.

Yoshinaga, T. and Dickerson, H.W. (1994) Laboratory propagation of *Cryptocaryon irritans* Brown, 1951 on saltwater-adapted black mollies (*Poecilia latipinna*). *Journal of Aquatic Animal Health* 6, 197–201.

6

Trichodinidae and Other Ciliates (Phylum Ciliophora)

J. Lom

Institute of Parasitology, Czech Academy of Sciences, 370 05 České Budějovice, Branišovská 31, Czech Republic.

INTRODUCTION

Ciliates are among the most common and widely distributed animal symbionts – parasites, commensals – of fishes. We know quite a bit about those ciliates that cause disease and mortality in fish and some control methods are available. At the same time, many facets of their existence are in urgent need of further study, as evidenced by an even superficial perusal of the ichthyoparasitological or ichthyopathological literature. Although some species have been known since the first half of the last century, most of the *Trichodina* and *Apiosoma*, not to mention other genera, do not have specific identifications. Much of the species diversity is virtually unknown. What is worse, the relations of many ciliates to their hosts are based on speculations. When a ciliate, however inoffensive it may be, is found growing massively on a fish, it is often automatically considered a pathogen and an appropriate '. . . asis' or '. . . osis' is then added to the literature. Also, with the exception of *Tetrahymena*, none of these ciliates has been cultivated *in vivo* and the knowledge on their metabolism and nutritional requirements is nil.

   The aim of this chapter is not only to review the basic information on trichodinids and other ciliates of fishes but also to stimulate interest in their further study. Succinct but thorough information on ciliate morphology and ultrastructure may be found in Lee *et al.* (1985). The ciliates are listed in this review according to their relationship to the host and not according to their systematic position. Under this scheme trichodinids may be both ectocommensals and ectoparasites, and in the case of *Tetrahymena corlissi* we are still not sure if it is an obligatory parasite or a facultative one.

ECTOPARASITIC CILIATES

Ectoparasitic protozoa are among the most important parasites of fish, especially in aquacultures and aquaria. All the ciliates that are discussed in this chapter cause disease in fish when they are stressed. They are transmitted directly and are practically ubiquitous. In feral fishes in good condition the ciliates may occur in low, undetectable numbers. These carrier fishes, when stressed, may relapse with the acute form of the disease. Most of these ciliates can tolerate varied ecological conditions, withstand temperature and salinity fluctuations and have little or no host specificity. Thus they can easily spread to and among most of the fish hosts and proliferate when the resistance of the host is decreased, e.g. fish that find less than optimal conditions in culture. Detailed statistics of direct losses and decreased yield in fisheries are not known, but the economic impact of pathogenic ciliates is well documented by the attention paid to these parasites both in commercial and ornamental fisheries.

Obligatory ectoparasites

These ciliates have no free-living stages and can exceptionally and temporarily be supported by hosts other than fish.

Chilodonella

The genus *Chilodonella* Strand, 1926 comprises many free-living species. Two species infect freshwater fishes. Both species have cosmopolitan distribution, may occur also in estuarine and brackish waters (e.g. in the eastern Baltic Sea) and appear to infect most, if not all, teleost fishes. They cause chilodonellosis, a disease affecting gills and skin, which is well known especially in fish cultures.

These ciliates belong to family Chilodonellidae Deroux, 1970, order Cyrtophorida Fauré-Fremiet in Corliss, 1956, class Kinetophragminophorea de Puytorac *et al.*, 1974.

The body of *Chilodonella* is oval and dorsoventraly flattened. The slightly convex dorsal side is without cilia; there is just a short anterior ciliary row. The flat or slightly concave ventral surface has a ciliature reduced to two longitudinal belts close to the margins of the body. At the ventrally located buccal opening or cytostome, there are three short oral kineties, each comprising two rows of kinetosomes of which only the outer row is ciliated. A conspicuous cytopharyngeal apparatus forms a circlet of a few large rod-like nematodesmata surrounding the cytopharyngeal tube. There is one macronucleus and one micronucleus. An overview of *Chilodonella* ultrastructural morphology may be found in Soltynska (1971).

Chilodonella piscicola (Zacharias, 1894) Jankovski, 1980 (synonym: *Chilodonella cyprini* (Moroff, 1902, see Shulman and Jankovski, 1984; Fig. 6.1A,B) has an asymmetrical oval body, with a notch in the posterior margin, 55 × 43 μm (range 30–80 × 20–62 μm). There are mostly eight to 11 (range seven to 15) kineties in the right arched ciliary band, and 12–13 (range eight to 14) in the straight left band. The cytopharynx, reinforced by 14–16 nematodesmata, is curved at its inner end. It may be slightly extruded for boring into and disrupting epithelial cells of the host. There are two contractile vacuoles, one anteriorly at right, the second posteriorly at left. When removed from the host, the ciliates survive for a period of time (24 hours at 5°C, 1 hour at 20°C). A small number may encyst and the cysts may last for a long time.

C. hexasticha (Kiernik, 1909) Kahl, 1931 differs from the preceding species by the absence of a notch at the posterior body margin, in less numerous and more loosely spaced kineties (five to seven in the right, seven to nine in the left ciliary band). It is smaller (30–65 × 20–50 µm) and its distribution is less widespread than *C. piscicola*. *C. piscicola* tends to infect fingerlings more than adults while *C. hexasticha* prevails on older fish.

Both species may occur simultaneously on the same host. Most of the information on their biology and relations to fish were collected at the time when their specific distinction was not recognized. Prost (1952) was the first to confirm that they are distinct species. The data on their distribution in Eurasia come from this period, while recent records on the occurrence of *Chilodonella* in South Africa, Israel, USA, Australia and Malaysia identified the species *C. hexasticha*.

Chilodonellosis occurs throughout a wide temperature range. Mass infections with *C. piscicola* may be found from slightly above zero, (e.g. at 4°C) to slightly above 20°C, while *C. hexasticha* is known to cause morbidities even at 26 to 31°C (Shariff, 1984). Chilodonellosis accounts for the greatest economic losses due to a single disease in commercial tropical fish stores (Leibovitz, 1980). Since chilodonellosis mostly occurs at low water temperatures when fish are under stress, an assumption prevailed that chilodonellas prefer cold water. There is a single report on massive mortality due to chilodonellosis (*C. hexasticha*) in free waters. This occurred to native fishes in Australia in the winter where the temperature was 8–13°C (Langdon *et al.*, 1985).

Under favourable conditions (e.g. in fish and especially fingerlings that are debilitated in early spring, or by overcrowding), chilodonellas may virtually cover the body surface in a contiguous layer. They disintegrate the body surface using their oral cytoskeletal apparatus and feed on the cell debris. Clinical sights of heavy infections include increase of mucus with an overall dark grey slimy or patchy or mottled grey appearance. Thin films of mucus and cellular debris may become detached from the skin surface. Moribund fish may also show signs of hypoxia, uncoordinated swimming, are emaciated, have opaque eyes and abrased skin.

Pathogenesis in skin lesions has not been studied but has been described in detail in gill infections with *C. hexasticha* (Paperna and Van As, 1983; Shariff, 1984; Langdon *et al.*, 1985). *Chilodonella* initially cause localized hyperplasia of gill epithelium, which later becomes more generalized. Proliferating epithelial cells fill the spaces between secondary lamellae, which may fuse together and coalesce into a single mass. The thin respiratory epithelium is covered by the hyperplastic epithelium and this drastically reduces the respiratory surface of the gills. The epithelium may be infiltrated with lymphocytes and eosinophil granulocytes. Also, there is increased proliferation of mucus and chloride cells.

The hyperplastic epithelium may undergo dystrophic changes with

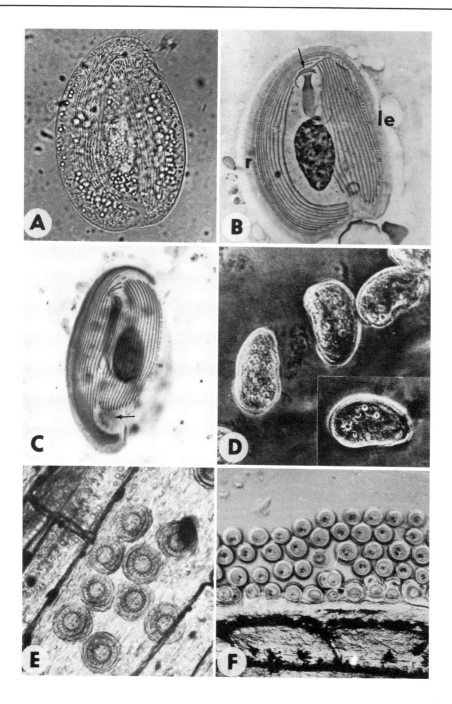

Fig. 6.1. A. *Chilodonella piscicola* in ventral view; subapically notice the circle of nematodesmata (×940; from Lom and Dyková, 1992, with permission of the publisher). B. *C. piscicola*; arrow indicates the three oral kineties above the cytopharynx; r = right and le = left field of somatic kinetics (×980). C. *Brooklynella hostilis*, ventral view; arrow indicates the site of the glandular

dilatation of capillaries, oedema, petechia, and haemorrhages. Complete destruction of the epithelium of primary and secondary lamellae may leave them with only cartilaginous rays. Pathological manifestations may vary and this depends on the intensity of infections. Sometimes large aggregates of melanin can also be seen along some of the primary lamellae undergoing degeneration. Disintegration of the gills and their necrosis renders the gills nonfunctional. Fish lose osmotic balance and suffocate; this is manifested by their increased sensitivity to oxygen deficiency. Heavily infected fish die.

Brooklynella

Brooklynella hostilis Lom and Nigrelli, 1970 (Fig. 6.1C,D) causes brooklynellosis, a severe disease of marine fishes in sea aquaria and in maricultures where it can cause mass mortalities and epizootics were reported repeatedly (e.g. from Kuwait, Singapore). The parasite has cosmopolitan distribution, but is more common in warmer waters. Unlike *Chilodonella*, *Brooklynella* has not been detected in feral fish. In aquaria, it infects most marine teleosts.

The parasite is in the family Hartmanellidae Poche, 1913, and is the only species in the genus. The ciliate is 36–86 × 32–50 µm in size, is kidney-shaped and has a shallow posterior notch. It has a flat ventral and a vaulted dorsal side. The ventral kineties converge anteriorly along a prebuccal suture. There are eight to ten postoral kineties, and a left band of 12–15 kineties which do not reach the posterior margin of the body, stopping short of the glandular organelle. The right band of eight to 11 kineties is long, and arches along the posterior and anterior margin of the body. The middle postoral kinety continues as a series of nonciliated kinetosomes in a spiral course around the posterior glandular organelle which appears as an elevated tubercle. The organelle has a few secretory channels which open into a simple slit and discharges thin secretion. The secretion is associated with attachment of the ciliate to the host. The preoral kinety is quite short while the anterior circumoral kinety is longer than the posterior one. The cytopharyngeal tube, reinforced by six to eight stout nematodesmata, capped by separate beak-like apical ends, is compressed laterally and appears as a broad slit. Descriptions of these and other ultrastructural features may be found in Lom and Corliss (1971). The oval macronucleus is about 18 × 12 µm. There are 13–22 micronuclei and several small contractile vacuoles. Cyst formation has not been recorded.

The infection is limited to the gills. In heavy infections the ciliates destroy the surface tissue by their cytopharyngeal armature, feed on the tissue debris and ingest blood cells. The infected fish suffer from respiratory difficulties, and have haemorrhages and petechiae on the gills. A light infection elicits a

---

organelle (×730; from Lom and Corliss, 1971, with permission of the publisher). D. *B. hostilis*, phase contrast (×310; from Lom and Nigrelli 1970, with permission of the publisher). E. *Trichodina* on the caudal fin of fish fry (×170). F. A mass of trichodinas washed off a fin (×100). A,D,E,F fresh mounts. B,C protargol impregnation. B,E,F originals.

mild inflammatory reaction. The result of the proliferative reaction is fusion of the terminal lamellae. In heavy infections desquamation, haemorrhages, cell proliferation and other tissue reactions are observed. The secondary lamellae may be completely denuded of the epithelial layer and the tissue response is manifested by macrophages and plasma cells. Such severe lesions usually cause fish mortality (Lom and Nigrelli, 1970).

Trichodinids

This is a large assemblage of peritrichous ciliates. They are mostly ectozoic with some endozoic species from several genera. They are the most common ectoparasites of both freshwater and marine fishes and are capable in some cases of inflicting heavy damage to their hosts with resulting mortalities. Trichodinosis is frequent in freshwater and marine fishes that are stressed, e.g. by harsh winter conditions or overcrowding in aquaria.

The parasite is more or less shaped like a flat disc or a hemisphere. The somatic ciliature is reduced to three or four ciliary wreaths around the aboral surface of the body which transforms into an elaborate holdfast organelle, the adhesive disc. This disc has a proteinaceous skeleton which is easily revealed using the Klein's impregnation method (see below). The disc consists of a ring of hollow conical denticles (Figs 6.2B,E, 6.3A,B,C). The centrifugal flat projections of the denticles, mostly semicircular, are the blades and the rod-like centripetal projections are the thorns. The denticles, inserted into each other, are subtended by a corona of fine skeletal rods, called radial pins. The disc is encircled by a movable border membrane, reinforced by fine skeletal rays which appear as fine striations (Fig. 6.3A,C). By means of myofibrillar bundles attached to the skeletal elements, the disc can be vaulted like a sucker and the rim of the border membrane can 'bite' into the host epithelium to firmly attach the ciliate to the surface of gills or skin. Above the border membrane there is a ciliary wreath of a single row of cilia. This is separated by a pellicular fold, called a septum, from a powerful locomotory wreath which is composed of obliquely arranged short rows of six to nine cilia. These cilia are covered by a more or less prominent pellicular fold, the velum, beneath which is inserted a single circle of stiff upwards protruding marginal cilia (Favard et al., 1963; Lom, 1973a; Khan et al., 1974; Hausmann and Hausmann, 1981; Maslin-Leny and Bohatier, 1984).

The vaulted apical surface, which is called the peristomial field, is encircled by two ciliary organelles (Fig. 6.2A). The outer one (homologous with the paroral membrane) is a haplokinety, i.e. a double row of kinetosomes of which only the inner one bears cilia. The second is a polykinety, i.e. a triple row of cilia homologous with a membranelle. These two ciliary organelles are inserted into a rather shallow groove and run closely together counterclockwise, to form an adoral spiral (or adoral ciliary zone) before plunging in a funnel-shaped buccal infundibulum. At its mouth they separate and descend in a spiral course to the cytostomeal opening proper. The three ciliary rows of the polykinety loosen up their close union to form three parallel ciliary rows forming a peniculus. The peniculus is joined by another parallel peniculus and by a very short third peniculus which is perpendicular to the former.

During binary fission the two daughter individuals separate along the longitudinal, apical–antapical axis. This way of cell division seems to be an exception to the transverse cell division in ciliates. The functional apical–antapical polarity, however, is a secondary adaptation to the sessile way of life.

The ciliate undergoes dramatic changes during binary fission. If the macronucleus is horseshoe-shaped it condenses to an ellipsoid shape before division. After division it assumes the original shape again. The adhesive disc separates into two semicircles which then close to form two smaller discs in daughter individuals. The old set of denticles is resorbed and new denticles are formed from thickenings on the radial pins which appear at the ends of the blades of the old denticles. The central conical parts are first formed, followed by the blades and finally the thorns (Fig. 6.2C). Each new radial pin arises between a pair of the old ones. Thus the original size of the disc and its constituents are regained (Kazubski, 1967). In some species, the remains of the thorns of mother individuals form one or two circles in the centre of the disc, i.e. in *Trichodina urinaria* or *T. polycirra*.

Conjugation is rare; there are sexually differentiated micro- and macro-conjugants. In the macroconjugant, a new ring of denticles is formed outside the resorbed denticles (Ahmed, 1977). Only fully mature ciliates with all their constituents and properly formed adhesive disc should be used in taxonomic identification.

There is no cyst formation; transmission takes place when ciliates swim directly from one host to another. They can survive without the fish for hours and perhaps even days.

Trichodinids belong to the family Trichodinidae Raabe, 1959, order Mobilina Kahl, 1933, of subclass Peritricha Stein 1859, class Oligohymenophora de Puytorac *et al.*, 1974. Five genera occur on fishes. The taxonomy is based on the structure of the buccal ciliature, on the appearance of the adhesive disc and numbers and size of its constituents (Fig. 6.3). The adhesive disc can only be studied using the silver impregnation technique of Klein (Lom, 1958; Welborn, 1967). Supplementary data are provided by the con-figuration of nuclei. Since the body shape and size of trichodinids vary enormously (e.g. from flat discoid to hemispheric of dumb-bell shaped) it is impossible to identify a trichodinid without the silver impregnation technique. Buccal ciliature can be studied using the protargol impregnation technique.

Numerous species described without the use of silver impregnation tech-niques are either worthless or nomina nuda if it is impossible to redescribe them using impregnation. Descriptions without the morphology of silver-impregnated adhesive discs are completely misleading (Haider, 1964; Matthes *et al.*, 1988). Some of the important taxonomic papers on this group are by: Welborn (1967); Lom (1970); Lom and Haldar (1976, 1977); Arthur and Lom (1984); Shtein (1984); Van As and Basson (1989); Basson and Van As (1989).

Trichodinids are as a rule commensals. They use their host as a convenient feeding ground to glide over and to which they temporarily attach. They feed on waterborne particles and bacteria as well as detritus from the fish surface. Trichodinids do not occur in large numbers on a healthy fish and hence the

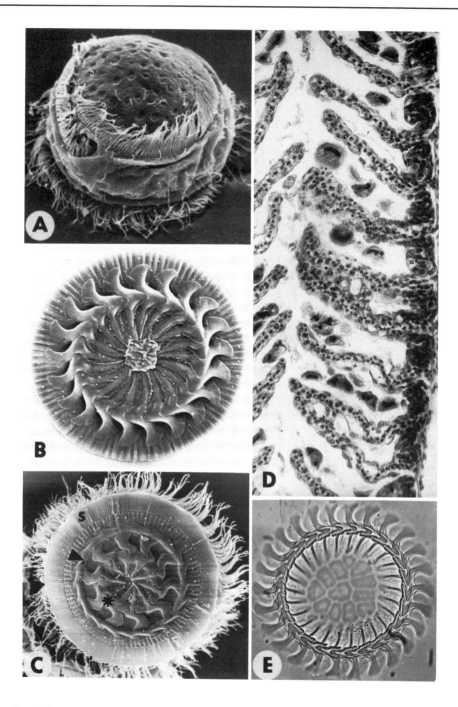

Fig. 6.2. A. An oral view of a *Trichodina* showing the adoral zone spiralling into the infundibulum (×780; from Lom and Dyková, 1992, with permission of the publisher). B. A *Trichodina* denticulate ring with underlying corona of radial pins; the pellicle has been peeled off (×1250). C. An adhesive disc of a *Trichodina* after bipartition: S = striation of the border membrane, arrowhead show the

irritation caused by the attachment of their adhesive discs is quite negligible. When a trichodinid is firmly attached to the host epithelium, the sharp edge of the border membrane is pressed deep into the epithelial cells, and the surface of the cell is vaulted into the hollow of a sucker. In debilitated fish, in fish larvae or in young fry, the natural repellent ability of the skin and gill surface is nonfunctional and the ciliate can massively proliferate. If the surface is covered by a thick layer of ciliates, then their constant attachments and movements would seriously damage the epidermal cells. Under these conditions, the trichodinids behave like true ectoparasites, as they feed on the disrupted cells and can even penetrate into the gill or skin tissue.

Trichodinella and *Tripartiella* are only found on gills. Some species of *Trichodina* occur both on gills and body surface, while many prefer the gills and a minority live on the skin only (Fig. 6.2D). In newly hatched fish larvae, *Trichodinella* and *Tripartiella* are also found on the skin. Trichodinids living on the gills and skin of fish may also be found on the gills and skin of amphibian larvae.

A small number of trichodinids live inside the fish body (Fig. 6.6G).

GENUS *TRICHODINA* EHRENBERG, 1838

Most trichodinids belong to this genus and most of the trichodinid agents which cause mortality in freshwater and marine fishes are *Trichodina*.

The genus is characterized by denticles with massive central conical parts, with flat, often semicircular blades and straight thorns. The adoral spiral is mostly around 360°, while it may be as little as 330° or as much as 540° in some individuals. The diameter of the ciliate is mostly about 50 to 100 µm.

About 190 species were described from fishes, but not all of them were adequately described using the silver impregnation technique. In addition to the morphology of the skeletal elements of the adhesive disc, the following measurements and counts are of primary diagnostic value: diameter of the adhesive disc, diameter of the denticulate ring (Fig. 6.3A), size of denticles (Fig. 6.3B), number of denticles. The diameter of the horseshoe macronucleus and the position of the micronucleus in relation to the macronucleus (Fig. 6.3E) is also of diagnostic value. In the trichodinid species listed below, only these important values are given.

Trichodina species live on aquatic animals from freshwater sponges (Porifera) to amphibian larvae, but most are on fishes. *Trichodina* is not host specific. Species (e.g. *T. cobitis* from *Cobitis taenia*, *T. prowazeki* from *Rutilus rutilus*, or *T. reticulata* from *Carassius carassius*) which are normally on one host can sometimes be found on another fish host. Most *Trichodina* species infect several to very many host species. For example, *T. fultoni, T.*

---

neoformed new denticulate ring, * designates the old denticles in the process of resorption (×800). D. Gill filament of a roach heavily infected with trichodines (H & E; × 100). E. An isolated denticulate ring of *T. reticulata* with the reticulate pattern of the pellicle in the centre (×1050; from Lom and Dyková, 1992, with permission of the publisher). A–C scanning electron micrographs, E fresh mount. B–D originals.

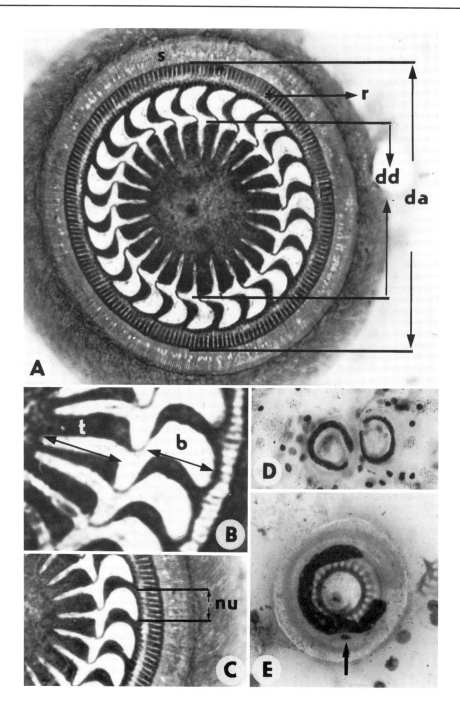

Fig. 6.3. Diagnostically important features in *Trichodina* (and related genera).
A. measurements of the adhesive disc: dd = diameter of the denticulate ring,
da = diameter of the adhesive disc, s = striation of the border membrane,
r = radial pins. B. Measurement of the denticles; t, b = length of thorns and

nigra or *T. acuta* live on various freshwater fishes both in Eurasia and America (Van As and Basson, 1987). An extreme case is that of *T. pediculus*. Its type hosts are freshwater hydrae of the genera *Hydra* and *Pelmatohydra* but it may occur on other aquatic animals (amphibian larvae) and is quite common on freshwater fishes in Eurasia and America. Many species of fish *Trichodina* (e.g. *T. acuta*) also infect freshwater organisms such as *Diaptomus* or tadpoles. In moderate climates this happens in late spring when the ciliates leave the now more resistant fish in search for another susceptible host. These substitute hosts can support a *Trichodina* population for a long time and also elicit morphological changes of the ciliate (Chen, 1963). However, *Trichodina* from copepods in South Africa may be independent host specific species (Basson and Van As, 1991). It was also suggested that this may be the case in other regions.

Low specificity of most species is the reason trichodines are so widely distributed. *T. rectuncinata* is found on the gills of marine gobiid, blenniid, perciform and syngnathiid hosts (Grupcheva *et al.*, 1989). Some species, like *T. jadranica*, occur on both marine and freshwater fishes. In addition, trichodines spread with transcontinental introductions of many of their hosts (common carp, tilapias, herbivorous fishes). Thus common freshwater species such as *T. nigra*, *T. mutabilis*, *T. acuta*, *T. fultoni*, *T. heterodentata* and *Trichodinella epizootica* are found on most families of fishes in Europe, Asia, America and Africa.

Ectozoic trichodinids are on the surface of the body and/or the gills, the opercular cavity and/or nasal pits. On healthy fish, they occur in low numbers. Any decrease in host resistance may result in their rapid proliferation. In temperate regions, there are pronounced seasonal fluctuations in the number of trichodines on a fish. In late winter or spring the parasite populations of parasites may be high on fish that are affected by overwintering stress. The host–parasite balance is often influenced by temperature, e.g. common carp infected with *T. pediculus* will lose the parasite if there is a sudden rise in temperature. However, since various stress factors (e.g. low oxygen in the water) – the nature of which is not always clear – may intervene throughout the year and each species of fish may be affected in a different way, heavy parasitaemias are found throughout the year. Thus, in the same body of water, herbivorous fish are infected heavily in February to April while *Ictiobus bubalus* is heavily infected in the summer (Bauer *et al.*, 1981).

Heavily infected fish may have a greyish-blue coat, which is formed by excessively secreted mucus and peeled epithelia. The fins may be frayed.

Epithelial hyperplasia occurs between the gill lamellae. Although this is a protective reaction the trichodines take advantage of this by feeding on the newly produced cells and cell debris. Cell debris and red blood cells are often

---

blades, respectively, the space between t and b is the width of the central part. C. nu = number of radial pins per denticle. D. Macronuclei in the shape of a horseshoe. E. Position of the micronucleus (arrow). A–C Klein's impregnation method, D,E nuclear stain by Piekarski-Robinow. A–E originals.

found in the food vacuoles. Trichodines may bore into the gill lamellae and even perforate blood vessels. Thus large areas of the gill filament may be denuded with associated bleeding. This may affect the osmotic balance, hampers respiration and may lead to death. Debilitated fish are sluggish, swim near the water surface and cease feeding.

Mortalities associated with heavy trichodinosis have commonly been reported; however, a rigorous assessment of the role of other synergistic or predisposing pathogens has not been done. The chronic impairment of health condition of fish is probably more important in trichodinosis than isolated outbreaks of mortalities. Sanmartin-Durán *et al.*, (1991) and Janovy and Hardin (1987) found that fish infected with *Trichodina* had reduced growth.

Trichodinids commonly occur in association with other ectoparasites, especially monogeneans, on gills and skin (Noble, 1963). Pearse (1972) suggested that the synergistic action of the two parasites cause mortalities. Most, if not all, trichodines are potentially pathogenic to their fish host. Many species are known to be associated with debilitated or diseased condition of marine and freshwater fishes. Some species are known to cause mortalities. Some of these species are listed below.

1. Species with a dark centre in the adhesive disc:
T. nigra Lom, 1961 (Fig. 6.4A) is widely distributed throughout Eurasia and the Philippines. It is mostly on the skin and gills of cyprinid, but also perciform and cichlid hosts. Its morphology is very variable, some of the populations perhaps should be considered separate species. The diameter of the adhesive disc varies from 35 to 65 μm, and that of the denticulate ring from 20 to 39 μm. The number of denticles ranges from 21 to 29.

T. pediculus (O.F. Müller, 1786) Ehrenberg, 1838 (Fig. 6.4B) has a cosmopolitan distribution; it is common on the skin of adult fish and fish fry. It is disc-shaped or hourglass-shaped in lateral view. The parasite has extremely long, needle-pointed thorns. Its adhesive disc ranges from 46 to 68 μm, the denticulate ring ranges from 28 to 38 μm and the number of denticles ranges from 20 to 32.

T. nobilis Chen, 1963 (Fig. 6.4D) is common on the skin and rarely on the gills of *Ctenopharyngodon idella* and other herbivorous fishes in China. It was introduced with the fish to Europe where it is sometimes also found on the common carp. There are deep notches in the proximal anterior border of the blade. The adhesive disc ranges from 61 to 68 μm, the denticulate ring ranges from 37 to 40 μm, and the number of denticles ranges from 24 to 28.

T. mutabilis Kazubski and Migala, 1968 (Fig. 6.4H) is common on the skin and gills of the common carp. It is also found on other cyprinids and fishes from other families. It is distributed throughout Eurasia, and in the Philippines and exhibits great morphological variability, (Kazubski and Migala, 1968). The adhesive disc ranges from 37 to 74 μm, the denticulate ring is from 20 to 46 μm, and the number of denticles ranges from 21 to 32.

T. truttae Mueller, 1937 (Fig. 6.4C) is one of the largest species in the genus. It is a well known pathogen of salmonids (usually on the skin and rarely on the gills) in hatcheries in North America and the former USSR. The

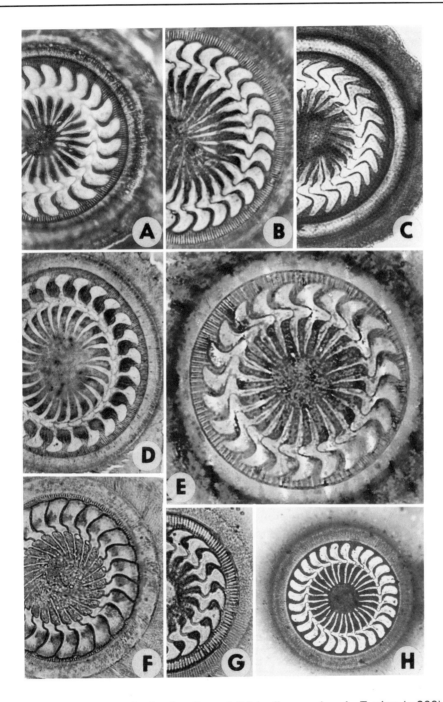

Fig. 6.4. Impregnated adhesive discs of *Trichodina* species. A. *T. nigra* (×900). B. *T. pediculus* (×1000). C. *T. truttae* (×450). D. *T. nobilis* (×770). E. *T. heterodentata* (×1200). F. *T. perforata* (×1100). G. *T. acuta* (×870). H. *T. mutabilis* (×500). A–H originals.

adhesive disc ranges from 76 to 137 µm, the denticulate ring is from 45 to 72 µm, and the number of denticles ranges from 27 to 34. The morphology was studied under scanning electron microscope (Arthur and Margolis, 1984).

T. heterodentata Duncan, 1977 (Fig. 6.4E) was originally described from the cichlids in the Philippines. It has been recorded from other fish families (e.g. Characidae, Morunidae and Cyprinidae) in the Middle East, Taiwan, South Africa and South America. The adhesive disc ranges from 38 to 60 µm, the denticulate ring is from 23 to 51, and the number of denticles ranges from 22 to 30.

T. perforata Lom, Golemansky and Grupcheva (Fig. 6.4F) lives on the gills of *Cyprinus carpio* and *Carassius carassius* in Eurasia. It has deep notches or even holes in the denticle blades. The adhesive disc ranges from 37 to 50 µm, the denticulate ring is from 23 to 29 µm, and the number of denticles ranges from 23 to 26.

Pearse (1972) found an as yet unidentified species which has a dark centre in its adhesive disc to be pathogenic to *Pleuronectes platessa*. It caused higher (up to 50%) mortalities when it occurred with the monogenean *Gyrodactylus unicopula*.

2. Species with a clear centre in the adhesive disc (impregnated with Klein's method):

T. acuta Lom, 1961 (Fig. 6.4G) is widely distributed on the skin and gills of cyprinid, perciform, cichlid and other fishes. It occurs throughout Eurasia, the Philippines and North America. The denticles have strong, curved thorns. The adhesive disc ranges from 35 to 60 µm, the denticulate ring is from 22 to 23 µm, and the number of denticles ranges from 22 to 33.

T. fultoni Davis, 1947 (Fig. 6.5C) is a common parasite on cyprinid, perciform, centrarchid, anguilliform and other fishes in Eurasia and North America. The clear centre in the adhesive disc has argentophilic spots. The adhesive disc ranges from 62 to 82, the denticulate ring from 41 to 55 µm, and the number of denticles range is from 27 to 31. Pathogenicity at 12°C for several North American fishes and cultured eel was established (Hoffman and Lom, 1967; Markiewicz and Migala, 1980).

T. jadranica Raabe, 1958 was detected on the gills of *Mullus barbatus* and other marine fishes. It was later recognized as a common species on the gills of freshwater fishes and as a pathogen in eel cultures.

T. murmanica Polyanski, 1955 (Fig. 6.5B) has a clear centre in its adhesive disc. It is a common species in the North Atlantic, Bering and other northern seas. It occurs on the skin of *Gadus morhua* and coastal gadid and perciform fishes. The adhesive disc and denticulate ring range from 40 to 70 µm and from 41 to 55 µm, respectively; the number of denticles is from 25 to 31.

The pathogenic potential of these two and other marine species has not yet been studied in maricultures.

3. Some species that have a special structure in the centre of the adhesive disc:

T. reticulata Hirschmann and Partsch, 1955 (Fig. 6.5E) is host specific to *Carassius carassius* and *C. auratus* throughout its distribution. The centre of the disc appears compartmentalized; the range of the adhesive disc is 44 to 57 µm, that of the denticulate ring is 25 to 36 µm, and the number of denticles

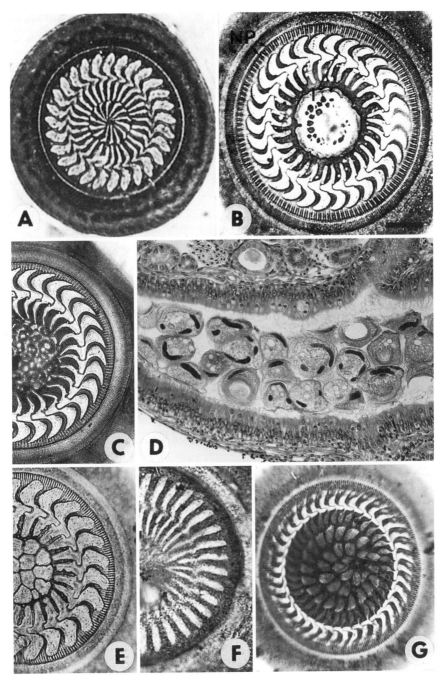

Fig. 6.5. *Trichodina* species. A. *T. centrostrigeata* (×970; from Basson *et al.*, 1983, with permission of the authors and publisher). B. *T. murmanica* (×1000). C. *T. fultoni* (×800). D. *T. polycirra* within a collective urinary duct of the kidney (×170). E. *T. reticulata* (×1000). F. *T. urinaria* (×1060). G. *T. polycirra* (×870). A–C and E–G Klein's impregnation, D histological section, H & E stain. A–F originals.

ranges from 21 to 34. It causes disease and mortalities in *C. carassius, C. auratus* and *Cyprinus carpio* (Ahmed 1976, 1977).

T. centrostrigeata Basson, Van As and Paperna, 1983 (Fig. 6.5A) occurs on carp and tilapia in South Africa. The centre of the disc has the remnants of thorns of the proceeding generation of denticles. The adhesive disc ranges from 32 to 46 μm, and the number of denticles is from 26 to 30.

The number of adequately described marine trichodinids is low compared to those in fresh water. Part of the reason is that it is difficult to obtain good silver impregnation because of the high salt content of sea water.

GENUS *HEMITRICHODINA* BASSON AND VAN AS, 1989

Unlike the *Trichodina*, the blades of denticles in this genus are stunted. *H. robusta* Basson and Van As, 1989 is the only known species and it occurs on the skin of *Marcusenius macrolepidotus* in Transvaal, South Africa.

GENUS *PARATRICHODINA* LOM, 1963

These ciliates sometimes occur in heavy infections. They are small (up to 40 μm in diameter) and shaped like a flat hemisphere. The adoral spiral makes an arch of 150° to 280° but denticles are similar to those in *Trichodina*. Seven species occur on the surface of gills and three species have been described from the urinary tract.

P. incissa (Lom, 1959) Lom, 1963 (Fig. 6.6B) is common on the gills of more than 20 species of freshwater fishes. The hosts are mostly minnows (e.g. *Phoxinus phoxinus*) and other cyprinids. The denticle has a deep notch in the anterior margin of the blade. Its adhesive disc diameter ranges from 16 to 31 μm, and the denticle number is from 19 to 31.

P. corlissi Lom and Haldar, 1977 (Fig. 6.6A) differs from *P. incissa* in having a shallow notch in the blade. The parasite occurs chiefly on the gills of freshwater fishes of the genus *Gobio*. The diameter of the adhesive disc ranges from 19 to 25 μm, and the denticle number is from 18 to 24.

GENUS *TRIPARTIELLA* LOM, 1959

Small ciliates, similar to *Paratrichodina*. The adoral spiral makes an incomplete turn of 180° to 290°. The denticles have a delicate central part and, the backward slanted blades project anteriorly in a thin or knee-like process to interlock with the corresponding notch in the preceding denticle. The delicate thorns may also be slanted back. All 17 described species are from the gills of freshwater fishes.

T. copiosa Lom, 1959 (Fig. 6.6F) is widely distributed in Eurasia. It is known from about 40 species of mainly cyprinid fishes. The rectangular backwards slanted blades have a spike-like anterior projection. The adhesive disc diameter is from 14 to 31 μm, and the number of denticles ranges from 19 to 33.

T. lata Lom, 1963 has been recorded from about ten fish species in Eurasia. The parasite has wide projections on its denticle blades. The adhesive disc ranges from 17 to 25 μm, and the number of denticles is from 20 to 24. The arch of the adoral zone is 270°.

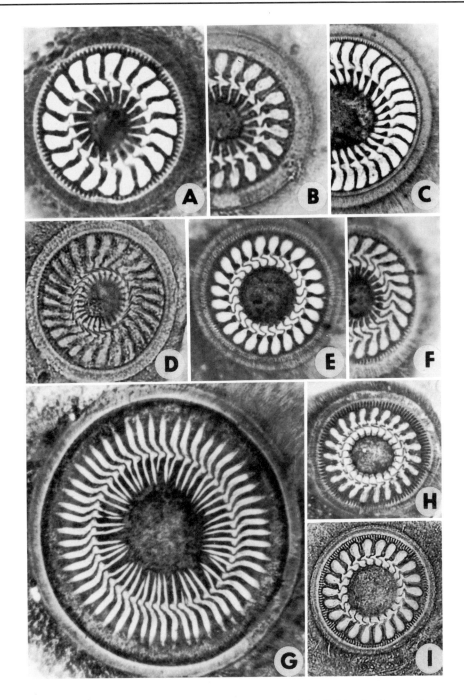

Fig. 6.6. Impregnated adhesive discs of various trichodinids. A. *Paratrichodina corlissi* (×1900). B. *P. incissa* (×1500). C. *P. phoxini* (×1400). D. *Tripartiella bulbosa* (×1500). E. *Trichodinella epizootica* (×1600). F. *Tripartiella copiosa* (×1600). G. *Vauchomia renicola* (×770). H. *Trichodinella subtilis* (×1400). I. *Trichodinella epizootica*, with a result of impregnation different from that of Fig. 6.6E, (×1370). A–I originals.

T. bursiformis (Davis, 1947) is from several species of North American fishes, e.g. *Ambloplites rupestris* and *Acantharchus pomotis*. The parasite has slender triangular blades and closely set denticles. The adoral zone makes a turn of 250°–290°, the adhesive disc ranges from 29 to 36 μm, and the denticle number is from 23 to 29.

T. bulbosa (Davis, 1947) (Fig. 6.6D) is known from a large number of fish hosts especially herbivorous fishes. It is holarctic in distribution and occurs from North America to China. The adhesive disc ranges from 22 to 26 μm, and the denticle number is from 19 to 24.

GENUS *TRICHODINELLA* ŠRÁMEK-HUŠEK, 1953

These are common parasites of freshwater and marine fishes. They sometimes cover the gills in a continuous layer. They are separated from *Tripartiella* by their denticle thorns which are reduced to tiny flat crooks. There are ten known species.

T. epizootica (Raabe, 1950) (Fig. 6.6E,I) is one of the most widely distributed freshwater trichodinids in Eurasia. It is also found in North America and the Pacific region. The type host is *Perca fluviatilis*, but the species was recorded from about 90 fishes from various families. When the host is stressed, the ciliate proliferates massively and becomes highly pathogenic. The adhesive disc ranges from 13 to 47 μm (average 23 μm), and the denticle number is from 16 to 30.

T. subtilis Lom, 1959 (Fig. 6.6H) differs from the preceding species in having two anterior projections at the base of the blade. Typical hosts are freshwater fishes e.g. *Carassius carassius* and *Cyprinus carpio*.

Facultative ectoparasites

This is a large group of essentially free-living ciliates. They are from various taxonomic groups. Some of them are saprozoic, bacteria-feeders while others are histophagous on dead aquatic animals. They infect fishes which are stressed by adverse environmental conditions.

Tetrahymena

The genus *Tetrahymena* Furgason, 1940 includes free-living generally saprozoic ciliates feeding on particulate food including bacteria. Many species are facultative parasites while others are obligate parasites (Elliot, 1973). Such species are aetiological agents of the 'tet' disease and cause mortalities of ornamental fishes.

Tetrahymena are mostly small, pyriform ciliates uniformly covered with meridional kineties. The buccal cavity is equipped with specialized oral ciliary apparatus which consists of the paroral membrane on the right side and of three serially arranged membranelles on the left side of the buccal cavity. The cytostome is at the bottom of the buccal cavity. There are one to three postoral kineties which begin at the posterior margin of the buccal cavity. Comprehensive information on morphology and ultrastructure may be found in Elliot (1973).

The genus is in the family Tetrahymenidae Corliss, 1952, order Hymeno-stomatida Delage and Hérouard, 1896, class Oligohymenophorea de Puytorac *et al.*, 1974. Several species have been encountered on fishes.

Tetrahymena corlissi Thompson, 1955 (Fig. 6.8D) is a parasite of amphibians and freshwater fishes with a (facultatively?) free-living stage. This protozoon infects various fish species. Although it may have a cosmopolitan distribution, records on its occurrence are patchy and they are mostly from the USA (Hoffman *et al.*, 1975).

The body is pyriform (averaging $55 \times 30\,\mu m$ in size), with 25 to 31 kineties. It has one caudal cilium which is more rigid and longer than other somatic cilia, an ellipsoid macronucleus (averages $8 \times 12\,\mu m$), a single micronucleus and a single, posteriorly located contractile vacuole. It forms reproductive cysts, in which the ciliate divides to produce two to eight daughter individuals. In water it moves like a spiralling football.

As a histophagous parasite, *T. corlissi* disintegrates host tissue and feeds on cell debris. In tropical aquarium fishes the infection is manifested by whitish patches due to masses of ciliates in copious amounts of mucus (Johnson, 1978). Fish scales are bristled as the outer skin is lost and the tissue swells. A mass of ciliates may form a rim around the eye orbit ('spectacled eye'). The entry into the body has not been well studied. The ciliates can pervade skin, muscle (Fig. 6.8E) and also organs in the body cavity. Extensive necrotic changes occur in the muscles and subdermal tissue. Inflammatory reaction has only rarely been observed (Dyková, unpublished). The infection results in mortality. The ciliate is important in tropical fish cultures.

Infections with this and other *Tetrahymena* species in fishes may be much more prevalent than reported earlier. Ciliates tentatively identified as *T. pyriformis* (Ehrenberg, 1830), were found to invade fry especially ones with injured integuments. It may not only destroy surface tissues but also internal organs (Shulman and Jankovski, 1984). These ciliates including *T. faurei* (Corliss, 1952) were recorded in the central nervous system of larvae of *Cyprinus carpio*, *Abramis brama*, *Ameiurus* sp. and in spinal canal and musculature of rainbow trout (Corliss, 1960; Stolk, 1960).

T. rostrata (Kahl, 1926) Corliss, 1952 is an edaphic histophagous organism and is a facultative parasite of some invertebrates. It has 27 to 35 kineties and forms both reproductive and resting cysts. It was probably associated with cranial ulcerations in farmed *Salmo salar* in freshwater tanks. Invasion of head tissues was associated with subacute inflammation. The clinical signs were severe fin and tail rot and depigmentation of the head (Ferguson *et al.*, 1987).

Tetrahymena are very easy to culture. Agnotobiotic cultures feed on bacteria and are easy to isolate and maintain in 2.5% proteose peptone. They can also be cultured axenically in chemically defined media (Elliot, 1973).

Uronema marinum

Uronema marinum Dujardin, 1841 is well documented as a parasite of marine fishes in aquaria. It belongs to the order Scuticociliatida Small, 1963. It is equipped with a thigmotactic ciliary field. It is about $30\text{--}50\,\mu m$ in size. It has a long caudal cilium and a buccal area which is half the body length.

Free-living ciliates tentatively identified as this species were found to cause heavy infections of gills, viscera and body muscles in marine fishes (Cheung *et al.*, 1980). Infected fish were predisposed by some environmental stress. The fish initially had white patches which later became lesions or open wounds. Infected fish were nervous, later were listless, had breathing difficulties, and had open, bleeding surface ulcers, which were full of ciliates (Bassleer, 1983). The ciliates had ingested blood cells and cellular debris, and were in internal organs, specifically the kidney, urinary bladder, spinal canal, and blood vessels. There was also extensive damage to the body muscle and fishes died in a relatively short time. Death was attributed to the heavy parasite load on the gills, which interfered with respiration.

Other ciliates

The surface of debilitated and moribund fish may be parasitized by many free-living ciliates. These freshwater ciliates include species of the genus *Coleps* Nitzsch, 1827 and various histophagous ciliates which feed on tissues. These fishes may also be colonized by free-living peritrichous ciliates (e.g. *Vorticella*, *Carchesium* or *Zoothamnium*).

Fish that are seriously stressed but not moribund, are also susceptible to free-living ciliates such as *Chilodonella cucullulus* (O.F. Müller, 1786), *C. uncinata* (Ehrenberg, 1838), *Dexiostoma campylum* (Stokes, 1886) Jankovski, 1967, *Glaucoma scintillans* Ehrenberg, 1830, *Colpidium colpoda* (Ehrenberg, 1831) Stein, 1960, *Frontonia acuminata* Ehrenberg, 1833 and *F. leucas* Ehrenberg, 1838. They may establish transient infections of the skin and gills.

These ciliates live in highly polluted waters, and infect the body surface and gills of pond fish under adverse conditions, e.g. in early spring after a long severe winter. When environmental conditions (oxygen, temperature, food, water quality) improve, they rapidly disappear (Migala and Kazubski, 1972).

Surface wounds on fish, notably on fry, predispose them to invasions by free-living ciliates in the marine environment. *Helicostoma buddenbrooki* Kahl, 1931, *Euplotes* sp. and *Uronema* sp. were recorded on O-group sole and plaice and young *Zoarces viviparus* in mariculture (Purdom and Howard, 1971). They attached to skin abrasions, enlarged the lesions and eventually caused lethal injuries.

ECTOCOMMENSAL AND ENDOCOMMENSAL CILIATES

The ectocommensal ciliates do not feed on fish tissues. They derive their nutrients (organic particles, microorganisms and other protozoa) from the water. Thus they do not inflict serious injury to their hosts and are not the primary agents of disease. However, in debilitated fish they may proliferate and affect normal functions (e.g. respiration) of the fish.

The endocommensal ciliates within the digestive tract feed under normal conditions on particles from the host's diet and bacteria. Under certain conditions they may turn into tissue-invading pathogens.

Suctorians

Suctorians are, in the adult state, sedentary and completely devoid of cilia. They have few to many suctorial tentacles for capture and ingestion of the prey. Each tentacle contains a simple cytopharynx in form of an axial, microtubular tube. At the tip of each tentacle there is a transient cytostome covered by the cell membrane when not in use. The ciliates are attached to the substrate, either directly or by means of a non-contractile secreted stalk. Multiplication is by internal or external budding, producing a migratory ('larval') stage, which bears several locomotory ciliary rows and swims looking for other sites to settle upon.

Suctorians form a subclass Suctoria Claparede and Lachman, 1858 within the class Kinetophragminophorea de Puytorac *et al.*, 1974.

Capriniana piscium (Bütschli, 1889) Jankovski, 1973 (syn. *Trichophrya piscium* Bütschli, 1889) (Fig. 6.7A,B) is a common ectocommensal on the gills of a great variety of freshwater teleosts in Eurasia and North America. It may be abundant in cold as well as warm periods of the year, depending on the condition of the host. Because of morphological differences between populations from different host species, many synonyms (e.g. *T. ictalurus* Davis, 1942, *T. salvelinus* Davis, 1942, *T. micropteri* Davis, 1947 and *T. intermedia* Prost, 1952) were described reflecting the variation range of this ectocommensal. *Capriniana* belongs to the family Trichophryidae Fraipoint, 1878.

The oval or irregularly elongated body is highly variable both in shape and size (about 40–110 × 25–70 µm). It adheres to the secondary gill lamella with a flattened broad attachment surface which secretes a cementing layer about 0.8 µm thick (Lom, 1971). One (sometimes two or three) fascicle of ten to 35 suctorial tentacles, up to 49 µm long, extends from a cone-like elevation of hyaline cytoplasm on the side opposite to the attachment. The capitate ends of tentacles bear numerous haptocysts, i.e. bottle-like organelles 0.4 µm long, containing a toxic substance which, upon contact, impale and attach the prey (ciliates) to be ingested through the tentacle. The tentacles, upon irritation, can be almost completely retracted into the cell. The macronucleus is ellipsoidal, about 13–25 × 15–42 µm in size, with an adjacent spherical micronucleus. There is a single contractile vacuole. The cytoplasm is full of refractile granules and digestive vacuoles. Reproduces by endogenous budding; the discoidal migratory stage bears nine equatorial kineties. *Capriniana* feeds on other ciliates (Lom, 1971), no solid evidence of feeding on host cells has ever been presented. Many authors quote 'trichophryasis' as a disease associated with pathological changes in the gills including necrosis. However, there is no evidence of any serious injuries exerted by *Capriniana*. Massive invasion of gills, possible only in fish predisposed by environmental and other factors, is likely to cause irritation and hampers oxygen exchange.

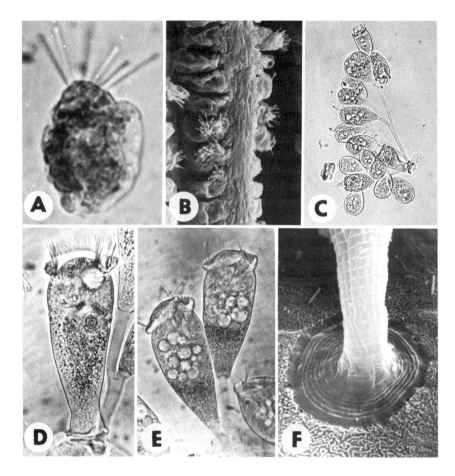

Fig. 6.7. A. A small individual of *Capriniana piscium* detached from gill epithelium (×600). B. A gill filament with attached *C. piscium* (×120). C. A Colony of *Epistylis lwoffi* detached from the fish surface (×140). D. Young *Apiosoma piscicolum* settled on the attachment platelet of the *Epistylis lwoffi* – stalk (×600; from Lom and Dyková, 1992, with permission of the publisher). E. Zooids of *E. lwoffi* (×420). F. The wide scopula of *Apiosoma* attached to the epithelial cell of the host and encircled by a narrow belt devoid of surface patterns (×2000). A,C,D,E living specimens, B,F scanning electron micrograph. A–C,E,F originals.

Sessiline peritrichs

Sessiline peritrichous ciliates, common on the surface of freshwater fishes and less common on marine hosts, sometimes form heavy growths on the skin and gills. Since such a massive growth is usually associated with a diseased condition of the fish, sessilines have often been considered to be pathogens. However, they feed on particles from ambient water and inflict no serious injury to the host.

They have a cylindrical, conical or bell-shaped body, attached permanently to some substrate. Somatic ciliature is reduced to one equatorial ciliary wreath, present usually only in the larval (migratory) stages (telotrochs). The adoral ciliary spiral encircles a vaulted apical surface, the epistomial disc. In the feeding state, the epistomial disc is raised high above a deep groove separating it from an outer peristomial lip. In the groove, cilia of the adoral ciliary zone are inserted. When disturbed, the ciliate retracts the epistomial disc inside the body. Within the infundibulum, the three peniculi run spirally parallel to each other. The non-ciliated surface of the body is raised into numerous fine equatorial pellicular crests or annuli. As yet unsurpassed information on general features of sessiline ultrastructure is to be found in Fauré-Fremiet *et al.* (1962).

Sessiline peritrichs are attached to the substrate by means of a holdfast organelle at the antapical pole, the scopula. The scopula either adheres to the substrates directly, often by means of a thin layer of a sticky substance, or it secretes a stalk. In all sessilines occurring on fish the stalk is non-contractile, without an inner cytoplasmic contractile band. The stalk may bear one ciliate, or if branched, it may bear many ciliates (or zooids) sometimes forming a large colony. In colony-forming genera, the two offspring of a binary fission may build their own branches of the stem stalk, or one of them (always so in the non-colonial forms) may transform into a telotroch. This is disc-shaped, with the epistomial disc retracted and it produces an equatorial locomotory fringe of pectinelles, detaches itself and swims away to another site or host. The sessilines also form protective (resting) cysts to overcome adverse periods.

Sessilines form the suborder Sessilina Kahl, 1935, which includes several families with many genera and is assigned to the order Peritrichida Stein, 1859, class Oligohymenophora de Puytorac *et al.*, 1974.

Sessiline peritrichs are free-living or epizoic on various kinds of animal hosts. Essentially, sessiline peritrichs found on the surface of fish are ectocommensals or symphorionts that use their hosts solely as a moving substrate which facilitates access to a convenient source of food particles – organic debris and waterborne bacteria.

An attached sessiline does no harm to the colonized epidermal or epithelial cells except for the disappearance of their surface ridges at the point of contact (Fig. 6.7E) or an increased occurrence of microfibrils beneath the host cell membrane (Lom, 1973a). Sessilina are not primary disease agents; they never occur in large numbers on a healthy fish. A heavy growth on the surface signifies that the fish has been predisposed by some debilitating factor – environmental and/or infectious – and that the repellent ability of its surface has decreased. If present in a massive number, they may be an additional burden to the host, impairing the respiration and causing surface irritation.

Distinction of families, genera and species in sessiline peritrichs is based on the morphology of the peristomial region and attachment organelles, the scopula and stalk. Guidelines for taxonomy are given in Lom (1966) and Scheubel (1973).

GENUS *APIOSOMA* BLANCHARD, 1885

These are the most common ectocommensals of freshwater fishes (Fig. 6.7D, F). They are solitary ciliates with a circular, sometimes lobed scopula which secretes a thin pad of fibrillar material discernible only in the electron microscope (Lom, 1973a). The macronucleus is mostly conical, rarely ellipsoidal. The genus belongs to the family Epistylidae Kahl, 1935 and comprises almost 70 species, living on the gills and skin.

A. piscicolum Blanchard, 1885 (Fig. 6.7D) is common on the gills and skin of a great number of freshwater fish. The body is plump in young individuas to elongated conical in adult ciliates, up to 110 μm long. Very fine pellicular annular striation covers the body. A similar species, *A. micropteri* (Surber, 1940) was accused of causing mortalities in fingerlings of *Micropterus salmoides* and *M. dolomieu* in North America. Masses of ciliates on the gills allegedly caused suffocation.

GENUS *EPISTYLIS* EHRENBERG, 1830

The genus includes very numerous species either free-living, or ectocommensal on aquatic animals including freshwater fishes. They produce a non-contractile stalk branched to bear a colony of several to many zooids. The macronucleus is horseshoe- to ribbon-shaped. *Epistylis* is the type genus of the family Epistylidae Kahl, 1935.

E. lwoffi Fauré-Fremiet, 1943 (Fig. 6.7C,E) forms colonies with up to eight conical zooids, 40–80 × 20–30 μm in size. It is common on the gills and skin of a variety of freshwater fishes in Eurasia and North America. The stalk is attached to the epithelium by a thin terminal platelet, or it ends in a ring-like loop fastened around the basal body end of an *Apiosoma*.

An *Epistylis* sp. was associated with epizootic outbreaks of 'red sore disease' in the southeastern USA causing heavy mortalities among several species of highly valued fishes of the family Centrarchidae (e.g. *Roccus saxatilis*, *Micropterus salmoides*). These outbreaks occurred especially in water bodies with high levels of organic load, thermal pollution and in combination with the presence of bacterial pathogen, *Aeromonas hydrophila*. There was evidence that the primary causative agent was *A. hydrophila* while *Epistylis* itself was incapable of any lytic or mechanic action resulting in the production of ulceration. Thus *Epistylis* was not the aetiologic agent, but a secondary invader harvesting rich masses of bacteria in the lesions (Esch *et al.*, 1976; Hazen *et al.*, 1978).

E. colisarum (Foissner and Schubert, 1977) was originally described as *Heteropolaria colisarum* by Foissner and Schubert (1977) from *Colisa fasciata* from Eastern Asia. It was also found on North American fish, e.g. *Lepomis gibbosus* or *Ictalurus punctatus*. The stalk is long, up to 1.2 mm and bears up to 16 cylindroid zooids on dichotomous branches. Foissner *et al.* (1985) claim that the terminal platelet of the stalk is embedded into the dermal epithelium. Thus they postulate damage to the host epithelium and lend support to the theory of *Epistylis* as primary invader of the 'red sore disease'. Their findings were thus far not confirmed by other authors.

GENUS *AMBIPHRYA* RAABE, 1952

It comprises solitary ciliates. They are attached to the skin and gills of freshwater and marine fishes directly by their large scopula equipped with short immobile cilia or secreting a layer of sticky material. The body is mostly squat cylindrical and has a permanent motionless equatorial ciliary fringe shielded by a thin pellicular fold. The macronucleus is ribbon-shaped, often sinuous.

A. ameiuri Thompson, Kirkegaard and Jahn, 1947 [syn. *A. miri* Raabe, 1952 and *A. macropodia* (Davis, 1947)] (Fig. 6.8A–C). Although it was described from the gills of North American fishes (*Ameiurus melas melas*, also *Ictalurus punctatus*) it seems to be autochthonous in Europe, where it is common on the fry of cyprinid fish (including *Cyprinus carpio* and the introduced *Ctenopharyngodon idella*). The body is mostly plump conical or cylindrical 50–95 × 40–61 µm, attached to the gills or skin by a scopula in form of a flat disc often exceeding the body width. Telotrochs are disc-shaped with a vaulted oral surface and flat aboral side.

A. ameiuri feeds on water-borne organic particles and inflicts no injury to the colonized epithelial cells except for their slight vaulting due to the pull of the scopula (Fitzgerald *et al.*, 1982). At very high densities, when the surface is covered by a carpet of ciliates, it may irritate the epithelium and impede respiration. Infected fry may then reveal diseased condition and unless treated, large numbers may be lost (Lom and Dyková, 1988).

Balantidium

Ciliates of the genus *Balantidium* Claparède and Lachmann, 1858 infect various homoiotherm and poikilotherm vertebrates; although mostly innocuous, they have a pathogenic potential. The cytostome is located at the base of an anteriorly situated vestibulum. The body is uniformly covered with longitudinal ciliary rows. The genus belongs to the family Balantidiidae Reichenow, 1929, order Trichostomatida Bütschli, 1989, class Kinetophragminophorea de Puytorac *et al.*, 1974.

Balantidium ctenopharyngodoni Chen, 1955 (Fig. 6.8F,G) lives in the intestine of grass carp, *Ctenopharyngodon idella* in China, and also in the grass carp stocks introduced to Europe. The infection is quite common.

The body is ellipsoidal, slightly dorsoventrally flattened, about 40–80 × 25–65 µm. The subapical vestibular groove, almost one third the body long, deepens toward the cytostomeal opening. Its left edge is covered with densely spaced ends of adjacent somatic kineties, bearing longer coalescent cilia. The bean-shaped macronucleus is 15–21 µm long. There is a single adjacent micronucleus. Three contractile vacuoles lie in the middle third of the body. Infection is transmitted by spherical cysts.

Balantidia are essentially harmless endocommensals of the middle and posterior intestine of grass carp. Little known factors, perhaps the lack of starch in the host's food may turn them into histophagous parasites. Then they abrade the epithelium, penetrate into the subepithelial intestinal layer and feed on the tissue cells. Balantidia cause extensive ulceration and elicit granuloma

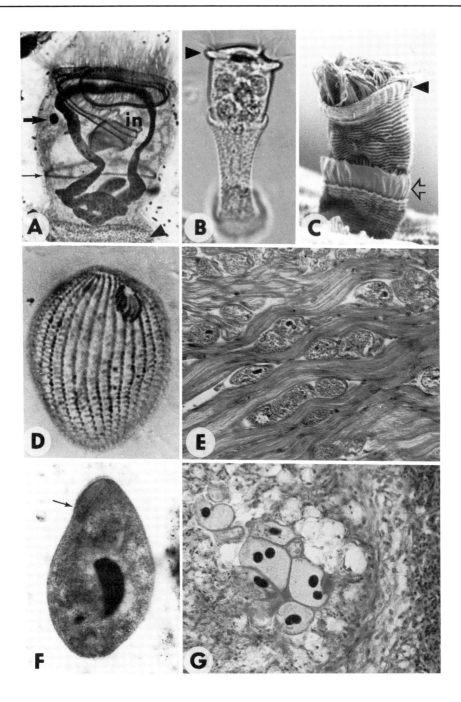

Fig. 6.8. A–C *Ambiphrya ameiuri*. A. A specimen after protargol impregnation, showing the adoral ciliary spiral plunging into the infundibulum (in). The long ribbon-like macronucleus is flanked by a single micronucleus (thick arrow). Thin arrow indicates kinetosomes of the equatorial wreath of cilia, arrowhead indicates the scopula (×750; from Lom and Dyková, 1992, with permission

formation deep in the tissue. *Balantidium*-enteritis is associated with hyper-aemia and inflammatory changes of the intestine. Enteritis may be fatal, and losses in 1–3 year-old grass carp were recorded (Molnár and Reinhardt, 1978).

ENDOPARASITIC TRICHODINIDS

Members of three trichodinid genera have been found in the urinary tract of marine elasmobranchs and freshwater teleosts. The parasites are located from the lumen of the urinary bladder to the collecting renal ducts in the kidney. They always initiate pathologic changes and feed on the cell debris which is produced by irritation, which results in hyperplasia and desquamation of the epithelium.

Their mode of transmission is unknown. It is assumed they are discharged with the urine and somehow find their way to the urinary bladder of another fish.

Endozoic trichodinids as a rule tend to be more host-specific than most of the ectoparasitic species. Also, they have numerous and closely spaced denticles with narrow blades.

GENUS *TRICHODINA* (see pp. 237–244)

T. oviducti Polyanski, 1955 is the largest trichodinid. The adhesive disc ranges from 88 to 271 µm, and the denticle number ranges from 43 to 60. It commonly infects the urogenital system including oviducts, copulatory sacs and seminal grooves of skates, *Raja*, in the Barents Sea and northern Atlantic. It is transmitted venereally and heavy infection is associated with a yellowish mucoid exudate (Khan, 1972; morphology – Khan *et al.*, 1974).

The freshwater species have well developed velum and marginal cilia. Also, the remains of the previous generation of denticles are in the centre of the adhesive disc as in *T. urinaria* Dogel, 1940 (Fig. 6.5F) (syn. *T. algonquinensis* Li and Desser, 1983). It has a holarctic distribution and is in the urinary tract of *Perca fluviatilis*, *P. flavescens*, *Stizostedion lucioperca* and *Notropis cornutus*. The adoral spiral makes one turn and a half and continues another half turn as an impregnable, cilia-less line (Lom, 1960). Its adhesive disc ranges from 40–53 µm and the denticle number is from 29 to 40. Experimental

---

of the publisher). B. A grown ciliate, live, arrow indicates the peristomial lip (×350). C. A stubby form in the scanning electron micrographs. Hollow arrow indicates the pellicular fold shielding the equatorial ciliary wreath. Solid arrow indicates peristomial lip. (×430). D. *Tetrahymena corlissi* as seen through its cell, showing the buccal ciliature, prebuccal suture and somatic cilia (Chatton-Lwoff impregnation method; ×870; from Lom and Dyková, 1992, with permission of the publisher). E. *T. corlissi* pervading skeletal muscles of a guppy (×220). F. *Balantidium ctenopharyngodoni*, arrow indicates the cytostome in the end of the groove (H & E stain; × 870). G. Balantidia within the submucosa of the intestine (×320). B,C,F,G originals.

infection of *Notemigonus crysoleucas* was accomplished by keeping fish in aquaria with infected yellow perch (Li and Desser, 1983).

T. polycirra Lom, 1960 (Fig. 6.5D,G) lives in the urinary tract of *Rutilus rutilus*, *Abramis brama* and sometimes in the common carp. The adhesive disc ranges from 48–58 μm, and the denticle number is from 38 to 55.

GENUS *VAUCHOMIA* MUELLER, 1938

These are large hemispherical or conical ciliates. The adoral ciliary spiral makes 2.5 to 3 turns. Two species have been described from the urinary tract of two species of *Esox* in North America.

V. renicola (Mueller, 1931) lives in *E. niger* in Oneida Lake. It has a flat body with an adhesive disc up to 97 μm, and the denticle number ranges from 44 to 65.

GENUS *PARATRICHODINA* (see p. 244)

The genus has three endoparasitic species. *P. phoxini* Lom, 1963 (Fig. 6.6C) is restricted to *Phoxinus phoxinus* in Europe. It has densely set denticles with massive thorns and rectangular blades without projections. The adoral zone arch varies between 180° and 270°. The adhesive disc ranges from 25–44 μm, and the denticle number varies from 23 to 29.

DIAGNOSIS

Microscopic examination (objective ×10) of wet mounts of excised gills and scrapings from skin, fins and nasal pits is the best available technique to detect ectozoic ciliates. Sessile suctorians are easy to find. Their refractile granules in the cytoplasm make them appear very dark in contrast to surrounding cells. Endocommensal ciliates (*Balantidium* etc.) can be detected by careful examination of intestinal contents and/or mucosal scrapings.

Careful examination of fresh mounts is required to detect light infections. There are no other methods to detect these ciliates.

PREVENTION AND CONTROL

The prevention and treatment strategies are very similar to those taken with most ectozoic protozoan infections. The outbreak of an ectoparasitic protozoan infection always indicates that the defence capabilities of the fish are compromised due to adverse environmental conditions. Hence the host–parasite equilibrium must be shifted back in favour of the fish by changing the environment. This is a prerequisite without which all treatment procedures are only of temporary value.

The guidelines of proper health management for fish culture are summarized as follows:

1. The water should be pathogen-free and it can be either from wells or is sand-filtered. The water must also be free of pollutants, at an appropriate temperature and be physicochemically balanced.

2. Efforts should be made to prevent access of trash fish.

3. Ponds and tanks should be drained at regular intervals and their bottoms sanitized with quicklime (2.5–3 tons ha^{-1}). Raceways and various tanks can be made parasite-free by flushing them with 5% formalin.

4. Ponds should be stocked with parasite-free or previously treated fish.

5. Fish should be fed a dietary balanced food and not overfed.

6. There should be a proper oxygen supply and this should be balanced in relation to temperature.

7. Excessive crowding of fish is to be prevented and there should be minimal handling of the fish.

8. Specific requirements of the fish species should be provided.

9. Fish should be examined regularly for parasites.

Chemical treatment is to be used as an emergency method after the proper management has failed. We recommend sodium chloride, formalin, acriflavine hydrochloride (=trypaflavin), malachite green (p,p-benzylidene bis *N,N*-dimethyl aniline) and/or potassium permanganate (KMnO$_4$) to treat most of the ectozoic ciliates. They can be used in relatively high concentrations in dip baths (**D**) when the fish are immersed in a net for a very short time (seconds to one hour). These chemicals can be used in low concentrations as flush treatment (**F**). Fish are exposed to the chemicals for several hours or days before the water in the tank is replaced with clear water. In a highly diluted state, the chemicals can be left in the tanks for indefinite treatment (**I**). Dip treatment is convenient for small fish, fry and fingerlings, while flush and indefinite treatments are suitable for raceways, tanks, larger water bodies or aquarium systems. The choice always depends on their feasibility in a given situation. The wide range of chemical concentrations recommended by various authors can be explained by the necessity to adjust the concentration to the ambient temperature, oxygen contents and quality of the water, (pH, total hardness, organic load) in relation to the particular fish species and age class. Finally, the resistance of parasite populations to these chemicals should be taken into account. Thus to assess the proper concentration and length of exposure, a trial should precede large-scale treatment. (The concentrations below are given either in percentage or ppm).

For chilodonellas, trichodinids, sessilines, facultative and opportunistic parasites and suctorian and other ectocommensals the following treatments are recommended:

1. Sodium chloride (not for marine or euryhaline ciliates)
   D: 1.5% to 3% for 5 min to 1 hour
   (e.g. sessilines: 3% for 5 to 10 min, or 1.5% for 1 hour; *Chilodonella*: in common carp fry 2% for 15 min at temperatures above 15°C, 1% for 30 min at 20–25°C, in yearling carp 3% for 15–30 min at temperatures above 15°C)
   F: 0.1–0.2% for one to two days

2. Formalin (convenient also for warm water and for marine ciliates)
   D: 150–250 ppm for 30–60 min
   (e.g. in common carp and salmonid yearlings concentration depends on temperature: below 10°C, 250 ppm; 10–15°C, 200 ppm; above 15°C, 150 ppm.

F: 250 ppm for 1 hour
40–50 ppm for 24 hours
I: 15–25 ppm
3. Acriflavine
   F: 50 ppm for one to several days
   I: 10–20 ppm
4. Malachite green
   D: 2.5–5 ppm for 30 seconds
   60 ppm for 10–30 seconds at 10–20°C
   F: 0.15–0.5 ppm for 6 days (e.g. common carp 0.5 ppm, trout fingerlings 0.15–0.2 ppm)
   I: 0.1 ppm
   in combination with formalin:
   0.25 ppm malachite green in 125 ppm formalin for 2–6 hours depending on intensity of oxygenation
5. Potassium permanganate
   D: 0.1% for 30–45 seconds
   50 ppm for 10–15 minutes (e.g. trout fry)
   F: 10 ppm for 60–90 min
   5 ppm for 16 min
   I: 2–3 ppm

For *Brooklynella*, in addition to formalin and acriflavine treatment indefinite immersion in a copper bath is useful: 0.42 ppm cupric acetate in 5.26 ppm formalin in 4.8 ppm Tris buffer.

For *Uronema* and other marine facultative parasitic ciliates, a freshwater dip for 10–15 minutes followed by a 15–20 minute dip in 5 ppm methylene blue or 30 ppm nitrofurazone (repeat twice daily, if necessary)

Raising the water temperature to 25°C for several days may help to restore the defence capability of the fish against *Chilodonella* infections. This may also be useful help against other ciliates, but not against those in the tropics.

CONCLUSIONS AND RECOMMENDATIONS FOR FUTURE RESEARCH

This review is a summary of the basic knowledge of ciliates that live on the surface or inside the fish body. There is an impressive diversity in genera and species, in their form and structure and many species are serious pathogens especially of cultured fish. Chemicals are readily available and are quite effective against the ciliate parasites.

Future studies should centre chiefly on the host–parasite relationships although it is likely that there are many more undescribed species and perhaps genera. To conduct studies on host–parasite relationships we have to have a reliable source of parasites. At the moment there is no *in vitro* culture method nor a dependable *in vivo* technique to maintain these ectozoic ciliates. Once we have a reliable source of parasites we can then start to unravel the host–parasite relationships. The intriguing areas that we should initially pay attention to include the process of infection (e.g. stimuli which attract the ciliate to the fish host), proliferation of the parasite on the fish (e.g. factors that predispose the fish to disease and allow the parasite to proliferate), nutritional requirements of the parasite and the host immune response to the parasite (e.g. protective humoral and cell-mediated responses).

The mechanism of the disease should be further studied. We know little about the pathomorphogenesis in chilodonellosis and trichodinosis and the differences in pathogenicity among various strains and races of *Chilodonella* and *Trichodina*. This basic information is necessary in order to form the basis of an efficient programme on disease prevention.

REFERENCES

Ahmed, A.T.A. (1977) Morphology and life history of *Trichodina reticulata* from goldfish and other carps. *Fish Pathology* 12, 21–31.

Arthur, J.R. and Lom, J. (1984) Some trichodinid ciliates (Protozoa: Peritrichida) from Cuban fishes, with a description of *Trichodina cubanensis* n.sp. from the skin of *Cichlasoma tetracantha*. *Transactions of the American Microscopical Society* 103, 172–184.

Arthur, J.R. and Margolis, L. (1984) *Trichodina truttae* Mueller, 1937 (Ciliophora: Peritrichida), a common pathogenic ectoparasite of cultured juvenile salmonid fishes in British Columbia: redescription and examination by scanning electron microscopy. *Canadian Journal of Zoology* 62, 1842–1848.

Bassleer, G. (1983) *Uronema marinum*, a new and common parasite on tropical saltwater fishes. *Freshwater and Marine Aquarium* 6, 78–79.

Basson, L. and Van As, J.G. (1989) Differential diagnosis of the genera in the family Trichodinidae (Ciliophora: Peritrichida) with the description of a new genus ectoparasitic on freshwater fish from southern Africa. *Systematic Parasitology* 13, 153–160.

Basson, L. and Van As, J.G. (1991) Trichodinids (Ciliophora: Peritrichia) from a calanoid copepod and catfish from South Africa with notes on host specificity. *Systematic Parasitology* 18, 147–158.

Bauer, O.N., Musselius, V.A. and Strelkov, Y.A. (1981) *Diseases of Pond Fishes* (in Russian), 2nd edn. Legkaya i pishchevaya promyshlennost, Moscow.

Chen, C.L. (1963) Studies on ectoparasitic trichodinids from freshwater fish, tadpole and crustacean in China. *Acta Hydrobiologica Sinica* 3, 99–111.

Cheung, P.J., Nigrelli, R.F. and Ruggieri, G.D. (1980) Studies on the morphology of *Uronema marinum* Dujardin (Ciliatea: Uronematidae) with a description of the histopathology of the infection in marine fishes. *Journal of Fish Diseases* 3, 295–303.

Corliss, J.O. (1960) *Tetrahymena chironomi* sp.nov., a ciliate from midge larvae and the current status of facultative parasitism in the genus *Tetrahymena*. *Parasitology* 50, 111–153.

Elliot, A.M. (ed.) (1973) *Biology of Tetrahymena*. Dowden, Hutchinson & Ross, Stroudsburg, Pennsylvania.

Esch, G.W., Hazen, T.C., Dimock, R.V. Jr and Gibbons, J.W. (1976) Thermal effluent and the epizootiology of the ciliate *Epistylis* and the bacterium *Aero-*

monas in association with centrarchid fish. *Transactions of the American Micro-scopical Society* 95, 687–693.

Fauré-Fremiet, E., Favard, P. and Carasso, N. (1962) Etude au microscope électronique des ultrastructures d' *Epistylis anastatica* (Cilié Péritriche). *Journal de Microscopie* 1, 287–312.

Favard, P., Carasso, N. and Fauré-Fremiet, E. (1963) Ultrastructure de l'appareil adhésif des Urcéolaires (Ciliés Péritriches). *Journal de Microscopie* 2, 337–368.

Ferguson, H.W., Hicks, B.D., Lynn, D.H., Ostland, V.E. and Bailey, J. (1987) Cranial ulceration in Atlantic salmon *Salmo salar* associated with *Tetrahymena* sp. *Diseases of Aquatic Organisms* 2, 191–195.

Fitzgerald, M.E.C., Simco, B.A. and Coons, L.B. (1982) Ultrastructure of the peritrich ciliate *Ambiphrya ameiuri* and its attachment to the gills of the catfish *Ictalurus punctatus*. *Journal of Protozoology* 29, 213–217.

Foissner, W. and Schubert, G. (1977) Morphologie der Zooide und Schwärmer von *Heteropolaria colisarum* gen. nov., spec. nov. (Ciliata, Peritricha) einer symphorionten Epistlidae von *Colisa fasciata* (Anabantoidei, Belontiidae). *Acta Protozoologica* 16, 231–247.

Foissner, W., Hoffman, G.L. and Mitchel, A.J. (1985) *Heteropolaria colisarum* Foissner and Schubert, 1977 (Protozoa: Epistylidae) of North American freshwater fishes. *Journal of Fish Diseases* 8, 145–160.

Grupcheva, G., Lom, J. and Dyková, I. (1989) Trichodinids (Ciliata: Urceolariidae) from gills of some marine fishes with the description of *Trichodina zaikai* sp.n. *Folia Parasitologica* 36, 193–208.

Haider, G. (1964) Monographie der Familie Urceolariidae (Ciliata, Peritricha, Mobilina) mit besonderer Berücksichtigung der im süddeutschen Raum vorkommenden Arten. *Parasitologische Schriftenreihe* 17, G. Fischer Jena.

Hausmann, K. and Hausmann, E. (1981) Structural studies on *Trichodina pediculus* (Ciliophora, Peritricha) I. The locomotor fringe and the oral apparatus. – II. The adhesive disc. *Journal of Ultrastructure Research* 74, 131–143 and 144–155.

Hoffman, G.L. and Lom, J. (1967) Observations on *Tripartiella bursiformis, Trichodina nigra* and a pathogenic trichodinid, *Trichodina fultoni*. *Bulletin of the Wildlife Disease Association* 3, 156–159.

Hoffman, G.L., Landolt, M., Camper, J.E., Coast, D.W., Stockey, J.L. and Burek, J.D. (1975) A disease of freshwater fishes caused by *Tetrahymena corlissi* Thompson, 1955, and a key for identification of holotrich ciliates of freshwater fishes. *Journal of Parasitology* 61, 217–223.

Janovy, J. and Hardin, E.L. (1987) Population dynamics of the parasites in *Fundulus zebrinus* in the Paltte river of Nebraska. *Journal of Parasitology* 73, 689–696.

Johnson, S.K. (1978) *Tet Disease of Tropical Fishes and an Evaluation of Correction Techniques*. Texas A & M University Fish Diseases Diagnostic Laboratory F12.

Kazubski, S.L. (1967) Study on the growth of skeletal elements in *Trichodina pediculus* Ehrbg. *Acta Protozoologica* 5, 37–48.

Kazubski, S.L. and Migala, K. (1968) Urceolariidae from breeding carp – *Cyprinus carpio* L. in Zabiniec and remarks on the seasonal variability in trichodinids. *Acta Protozoologica* 6, 137–160.

Khan, R.A. (1972) Taxonomy, prevalence, and experimental transmission of a protozoan parasite, *Trichodina oviducti* Polyanski (Ciliata: Peritricha) of the thorny skate, *Raja radiata* Donovan. *Journal of Parasitology* 58, 680–686.

Khan, R.A., Barber, V.C. and McCann, S. (1974) A scanning electron microscopical study of the surface topography of a trichodinid ciliate. *Transactions of the American Microscopical Society* 93, 131–134.

Langdon, J.S., Gudkovs, N., Humphrey, J.D. and Saxon, E.C. (1985) Death in Australian freshwater fishes associated with *Chilodonella hexasticha* infection. *Australian Veterinary Journal* 62, 409–413.

Lee, J.J., Hutner, S.H. and Bovee, E.C. (eds) (1985) *An Illustrated Guide to the Protozoa.* Society of Protozoologists, Lawrence, Kansas, 629 pp.

Leibovitz, L. (1980) Chilodonelliasis. *Journal of the American Veterinary Medical Association* 177, 222–223.

Li, L.X. and Desser, S.S. (1983) *Trichodina algonquinensis,* a new species of peritrich ciliate from Ontario freshwater fish, and observations on its transmission. *Canadian Journal of Zoology* 61, 1159–1164.

Lom, J. (1958) A contribution to the systematics and morphology of endoparasitic trichodinids from amphibians with a proposal of uniform specific characteristics. *Journal of Protozoology* 5, 251–263.

Lom, J. (1960) On two endozoic trichodinids, *Trichodina urinaria* Dogel 1940, and *Trichodina polycirra* n.sp. *Acta Parasitologica Polonica* 8, 169–180.

Lom, J. (1966) Sessiline peritrichs from the surface of some freshwater fishes. *Folia Parasitologica* 1, 36–56.

Lom, J. (1970) Observations on trichodinid ciliates from freshwater fishes. *Archiv für Protistenkunde* 112, 153–177.

Lom, J. (1971) *Trichophrya piscium*: a pathogen or an ectocommensal? An ultrastructural study. *Folia Parasitologica* 18, 197–205.

Lom, J. (1973) The adhesive disc of *Trichodinella epizootica* ultrastructure and injury to the host tissue. *Folia Parasitologica* 20, 193–202.

Lom, J. and Corliss, J.O. (1971) Morphogenesis and cortical ultrastructure of *Brooklynella hostilis,* a dysteriid ciliate ectoparasitic on marine fishes. *Journal of Protozoology* 18, 261–281.

Lom, J. and Dyková, I. (1992) *Protozoan Parasites of Fishes. Developments in Aquaculture and Fisheries Science,* Vol. 26. Elsevier, Amsterdam.

Lom, J. and Haldar, D.P. (1976) Observations on trichodinids endocommensal in fishes. *Transactions of the American Microscopical Society* 95, 527–541.

Lom, J. and Haldar, D.P. (1977) Ciliates of the genera *Trichodinella, Tripartiella* and *Paratrichodina* (Peritricha, Mobilina) invading fish gills. *Folia Parasitologica* 24, 193–210.

Lom, J. and Nigrelli, R.F. (1970) *Brooklynella hostilis,* n.g., n.sp., a pathogenic cyrtophorine ciliate in marine fishes. *Journal of Protozoology* 17, 224–232.

Markiewicz, F. and Migala, K. (1980) Trichodinid invasion (Peritricha, Urceolariidae) on young eels (*Anguilla anguilla* L.) grown in aquaria. *Acta Hydrobiologica* 22, 229–236.

Maslin-Leny, Y. and Bohatier, J. (1984) Cytologie ultrastructurale de *Trichodina* et *Tripartiella* (Ciliés Péritriches). *Protistologica* 20, 113–132.

Matthes, D., Guhl, W. and Haider, G. (1988) Suctoria und Urceolariidae (Peritricha). In: Mathes, D. (ed.) *Protozoenfauna,* Vol. 7. Gustav Fisher, Stuttgart, New York.

Migala, K. and Kazubski, S.L. (1972) Occurrence of nonspecific ciliates on carps (*Cyprinus carpio*) in winter ponds. *Acta Protozoologica* 9, 329–337.

Molnár, K. and Reihardt, M. (1978) Intestinal lesions in grass-carp *Ctenopharyngodon idella* (Valenciennes) infected with *Balantidium ctenopharyngodonis* Chen. *Journal of Fish Diseases* 1, 151–156.

Noble, E.R. (1963) The relations between *Trichodina* and metazoan parasites on gills of a fish. In: Ludvík, J., Lom J. and Vávra, J. (eds) *Progress in Protozoology.* Proceedings of the First International Congress, Prague, pp. 521–523.

Paperna, I., Van As, J.G. (1983) The pathology of *Chilodonella hexasticha* (Kiernik). Infection in cichlid fishes. *Journal of Fish Biology* 23, 441–450.

Pearse, L. (1972) A note on a marine trichodinid ciliate parasitic on the skin of captive flatfish. *Aquaculture* 1, 261–266.

Prost, M. (1952) Badania nad pierwotniakami pasozytnymi skrzeli ryb. II. *Chilodonella cyprini* Moroff i *Chilodonella hexasticha* Kiernik. *Annales Universitatis M. Curie-Sklodowska*, Lublin – Polonia, 8 (C), 1–3.

Purdom, C.E. and Howard, A.E. (1971) Ciliate infestations: a problem in marine fish farming. *Journal du Conseil International pour Exploration de la Mer* 33, 511–514.

Sanmartin Durán, M.L., Fernandez Casal, J., Tojo, J.L., Santamarina, M.T., Estevez, J. and Urbeira, F. (1991) *Trichodina* sp.: Effect on the growth of farmed turbot (*Scophthalmus maximus*). *Bulletin of the European Association of Fish Pathologists* 11, 89–91.

Scheubel, J. (1973) Die sessilen Ciliaten unserer Süsswasserfische unter besonderer Berücksichtigung der Gattung *Apiosoma* Blanchard. *Zoologische Jahrbücher, Systematik* 100, 1–63.

Shariff, M. (1984) Occurrence of *Chilodonella hexasticha* (Kiernik, 1909) (Protozoa, Ciliata) on big head carp *Aristichthys nobilis* (Richardson) in Malaysia. *Tropical Biomedicine* 1, 69–75.

Shtein, G.A. (1984) Suborder Mobilina Kahl, 1933 (in Russian). In: Shulman S.S. (ed.) *Parasitic Protozoa*, Vol. 1. In: Bauer, O.N. (ed.) *Key to Parasites of Freshwater Fishes of the USSR*, Vol. 140 of Keys to the Fauna of the USSR. Nauka, Leningrad, pp. 321–389.

Shulman, S.S. and Jankovski, A.V. (1984) Phylum Ciliates – Ciliophora Doflein, 1901 (in Russian). In: Shulman, S.S. (ed.) *Parasitic Protozoa*, Vol. 1. In: Bauer, O.N. (ed.) *Key to Parasites of Freshwater Fishes of the USSR*, Vol. 140 of *Keys to the Fauna of the USSR*. Nauka, Leningrad, pp. 252–280.

Soltynska, M.S. (1971) Morphology and fine structure of *Chilodonella cucullanus* (O.F.M.). Cortex and cytopharyngeal apparatus. *Acta Protozoologica* 9, 49–82.

Stolk, A. (1960) *Glaucoma* sp. in the central nervous system of the carp. *Nature* 184, 1737.

Van As, J.G. and Basson, L. (1987) Host specificity of trichodinid ectoparasites of freshwater fish. *Parasitology Today* 3, 88–90.

Van As, J.G. and Basson, L. (1989) A further contribution to the taxonomy of the Trichodinidae (Ciliophora: Peritrichia) and a review of the taxonomic status of some fish ectoparasitic trichodinids. *Systematic Parasitology* 14, 157–181.

Welborn, T.L. (1967) Trichodina (Ciliate: Urceolaridae) of freshwater fishes of the southeastern United States. *Journal of Protozoology* 14, 399–412.

7

Phylum Apicomplexa

K. Molnár

Veterinary Medical Research Institute, Hungarian Academy of Sciences,
H-1581 Budapest, Hungary.

INTRODUCTION

Members of the phylum Apicomplexa Levine, 1970 have a special cell organelle, called the apical complex, which facilitates invasion of the host cell. The majority of species form spores and/or oocysts.

Fish apicomplexans are divided into three groups. Coccidia proper are primarily intestinal parasites and produce resistant oocysts in the host. Adeleid blood parasites (coccidia *sensu lato*) have the merogonic and some gamogonic stages in fish, while spore formation takes place in parasitic annelids or insects. Both groups are characterized by a complete apical complex including the conoid (Conoidasida). The piroplasmorid apicomplexans have an incomplete apical complex and lack the conoid, and there is no spore formation.

Apicomplexans occur in a wide variety of animals (helminths to mammals). Relatively little is known about fish apicomlexans and these include those parasitizing cultured fish species. Coccidioses of the common carp, bighead and silver carp, eel and tilapia are known to cause mortality. Heavy Apicomplexa infections are also common in marine fishes, e.g. swimbladder coccidiosis of the cod caused by *Goussia gadi*, liver coccidiosis of the killifish caused by *Calyptospora funduli*, and testicular infection of the sardine caused by *Goussia sardinae*.

It is likely that the number of described species represents only a small fraction of fish apicomplexans. Lom (1984) recently summarized apicomplexans of marine fishes. In their review on the coccidia proper Dyková and Lom (1983) included 128 species, and this has increased by at least 20 marine and 28 freshwater coccidium species.

The majority of adeleorins and piroplasmorids parasitize marine fishes and only a few species occur in freshwater fishes. The majority of these parasites have not been adequately described, and in most cases only the merogonic and gamogonic stages in blood cells are known. Levine (1988) recorded 60 *Haemogregarina*, two *Cyrilia* and one *Hepatozoon* species in fish, and in Barta's revision (1991), five *Babesiosoma* and two *Dactylosoma* have been added. Fish-parasitic piroplasmorids are represented by five *Haemohormidium* species.

SYSTEMATICS AND TAXONOMIC POSITIONS OF APICOMPLEXANS

A common characteristic of the Apicomplexa is the 'apical complex', a structure demonstrable only by electron microscopy. The organelle appears only at certain stages of development. The apical complex usually consists of a polar ring, a conoid, micronemes, rhoptries and subpellicular tubules. Apicomplexans have a complex life cycle that involves (in most cases) merogony, gamogony and sporogony.

The following classification is based on that of Levine (1988) and is modified by this author:

Phylum: Apicomplexa Levine, 1970
  Class: Conoidasida Levine, 1988
    Order: Eucoccidiorida Leger and Duboscq, 1910
      Suborder: Adeleiorina Leger, 1911
        Family: Haemogregarinidae Neveu-Lemaire, 1901
          Genus: *Hepatozoon* Miller, 1908
          Genus: *Haemogregarina* Danilewsky, 1885
          Genus: *Cyrilia* Lainson, 1981
        Family: Dactylosomatidae Jakowska and Nigrelli, 1955
          Genus: *Dactylosoma* Labbé, 1894
          Genus: *Babesiosoma* Jakowska and Nigrelli, 1956
      Suborder: Eimeriorina Leger, 1911
        Family: Eimeriidae Leger, 1911
          Genus: *Eimeria* Schneider, 1875
          Genus: *Goussia* Labbé, 1896
          Genus: *Crystallospora* Labbé, 1896
          Genus: *Calyptospora* Overstreet, Hawkins and Fournie, 1984
        Family: Cryptosporidiidae Leger, 1911
          Genus: *Cryptosporidium* Tyzzer, 1907
  Class: Aconoidasida Mehlhorn, Peters and Haberkorn, 1980
    Order: Piroplasmorida Wenyon, 1926
        Family: Haemohormidiidae Levine, 1984
          Genus: *Haemohormidium* Henry, 1910

MORPHOLOGY AND DEVELOPMENT

Coccidia proper

Development

The development of fish-parasitic coccidia does not differ essentially from that of coccidia of warm-blooded animals. The development of the Eimeriidae

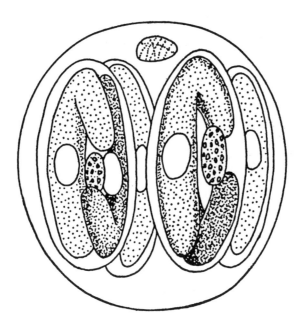

Fig. 7.1. *Goussia acipenseris*. Schematic diagram of the oocyst.

and the Cryptosporidiidae is divided into merogony, gamogony and sporogony.

Merogony begins with invasion of the host cell by the parasite. The intruding sporozoite, enclosed in a parasitophorous vacuole in the cytoplasm or nucleus of the cell becomes rounded, begins to grow and changes into a meront. In the meronts, merozoites are formed which, when released, invade other cells and form second- and third-generation meronts. Gamogony begins when the last generation of merozoites forms macro- and microgamonts. In the microgamonts numerous microgametes develop. After fertilization the macrogamont develops into an oocyst. During sporulation, the zygote in the oocyst divides and forms sporocysts in which sporozoites develop. Sporulation may take place within (endogenous sporulation) or outside (exogenous sporulation) the fish.

Morphological characteristics of the different developmental stages

MEROGONIC STAGES

Merogonic stages in naturally infected fish have been described by numerous workers. However, only Paterson and Desser (1981a) and Hawkins *et al.* (1984a,b) studied meronts and merozoites in experimentally infected fish. The merogony of *Goussia iroquoina* takes place in the intestine, while that of *Calyptospora funduli* is in the liver. According to Hawkins *et al.* (1984b) there were two merozoite generations, while Paterson and Desser (1981a) also found at least two generations. In species with epicellular development, merogony is also epicellular (Molnár and Baska, 1986; Landsberg and Paperna, 1987;

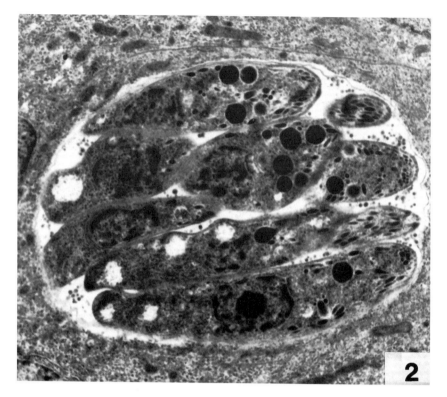

Fig. 7.2. *Goussia balatonica*. Intracellular meront with merozoites (TEM, ×12,000).

Kent *et al.*, 1988; Jastrzebski and Komorowski, 1990). Usually 8–16 merozoites of 8–16 μm in length are formed in the meronts (Fig. 7.2), but in *Goussia
cichlidarum* Landsberg and Paperna (1985) reported meronts containing large
numbers of merozoites. The merozoites of fish coccidia do not differ in structure from those of warm-blooded animals: they have a well-discernible nucleus,
conoid apparatus, and trimembranous pellicle covering the merozoite.

The meronts develop in the cytoplasm, or occasionally in the nucleus,
within a parasitophorous vacuole. In species with epicellular development, the
parasitophorous vacuole is intracellular but extracytoplasmal and it is covered
by a single unit membrane of the host cell. The vacuole of epicellular eimeriids
is complete, which in Molnár and Baska's (1986) view means that its cavity
completely surrounds the parasite. The wall of the vacuole is a single-layered
unit membrane. Good metabolic communication between parasite cytoplasm
and the host cell is ensured by parts of the vacuole protruding into the
cytoplasm of the host cell (Fig. 7.3).

Cryptosporidium have incomplete vacuoles. The parasitophorous vacuole
formed by enterocyte microvilli surrounds the meront only on the side facing
the gut lumen. Between the meront and the host cell cytoplasm there is a
special adhesive zone.

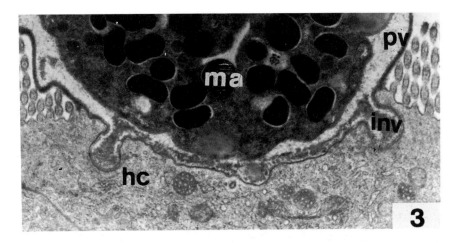

Fig. 7.3. *Eimeria anguillae.* Part of the macrogamont (ma). The parasitophorous vacuole (pv) releases invaginations (inv) into the plasm of the host cell (hc) (TEM, ×18,000).

GAMOGONIC STAGES

Microgamogenesis begins with multiple divisions of the nucleus of the microgamont in the parasitophorous vacuole. The newly formed nuclei gradually migrate to the periphery of the microgamont where they separate a part of the cytoplasm to form microgametes (Fig. 7.4). Microgametes possess two flagella, with the exception of those of cryptosporidia which do not have flagella. The microgamete contains a nucleus, mitochondria, microtubules and a perforatorium at the apical end.

In the majority of species macrogamogenesis takes place in the apical part of the cytoplasm of epithelial cells. However, there are species with intra-nuclear and epicellular development. The macrogamont (Fig. 7.4), which develops from the last merozoite generation, is always surrounded by a parasi-tophorous vacuole. In species with epicellular development the vacuole is extracytoplasmic and this is similar with the meront (Fig. 7.5). Electron micro-scopic studies of the macrogamont cytoplasm (Paterson and Desser, 1981c; Hawkins *et al.*, 1983b) indicate amylopectin, lipid granules and two distinct types of 'wall-forming body' (assumed to be involved in the formation of the thin oocyst wall).

SPOROGONIC STAGES

The oocysts of fish coccidia, except for members of the genus *Cryptospo-ridium*, are rather similar (Figs 7.1, 7.7). The oocysts of the Eimeriidae contain four sporocysts and, occasionally, one to two polar granules. Each sporocyst contains two sporozoites and a residual body. The only important difference between genera is in the structure of the sporocyst.

The oocyst is adapted to the aquatic environment; it has a thin, sensitive

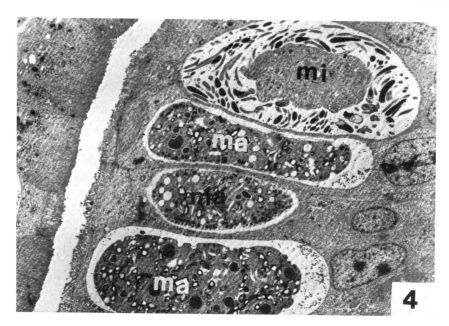

Fig. 7.4. *Goussia balatonica*. Intracellular macrogamonts (ma) and microgamont (mi) in the intestinal epithelium (TEM, ×6000).

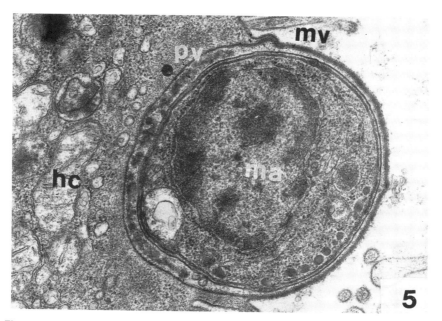

Fig. 7.5. *Eimeria anguillae*. Epicellular macrogamont (ma) in a parasitophorous vacuole (pv) surrounded by microvilli (mv) (TEM, ×30,000).

oocyst wall and does not have a micropyle. The thickness of the one- or three-layered wall varies between 3 and 200 µm.

Each *Cryptosporidium* oocyst contains four naked sporozoites and a residual body. Fish cryptosporidia have been reported by Hoover *et al.* (1981), Pavlasek (1983), Landsberg and Paperna (1986) and Paperna (1987). These authors based their diagnosis on the special adhesive zone of the developmental stages. Sporulated oocysts have only been studied by Paperna (1987) who has found the wall to be thick and double layered.

Sporocysts are round, elliptical, oval or dodecahedral in shape. They differ from each other primarily in the mode of sporozoite release. The wall of the sporocyst is generally thin but is more resistant than that of the oocyst. However, there are species with a relatively thick sporocyst wall, e.g. *Eimeria etrumei* or *Goussia siliculiformis*.

The sporocyst wall is usually composed of two layers. However, some species like *Goussia gadi* and *G. lucida* have a three-layered wall (Odense and Logan, 1976; Daoudi, 1987). The thickness of the wall varies between 30 and 500 µm.

Only a few fish coccidia (e.g. *Eimeria isabellae*) possess a typical *Eimeria* sporocyst (i.e having a Stieda body). In the majority of the species there is only a thickening, plug or cap at one end of the sporocyst. This is where the sporocyst opens and the sporozoites are released in the host or intermediate host.

Goussia-type sporocysts are composed of two equal-sized, round, elliptical or coffin-shaped valves united by a suture. This suture is hardly discernible using light microscopy but can be seen under an electron microscope. Although most fish coccidia have been described as *Eimeria*, the majority of the species undoubtedly belong to the genus *Goussia*. Sometimes the sporocyst is surrounded also by a membraneous veil which is attached to the sporocyst by special membranes. The sporocyst veil has been reported by Lom (1971) for *G. degiustii* and by Baska and Molnár (1989) for *G. sinensis*.

The sporocysts of *Calyptospora* are characterized by a thickening or projection at the caudal end, by sporopodia starting from the spore surface or from the caudal projection, and a sporocyst veil supported by sporopodia and surrounding the sporocyst. The opening of the sporocyst is a longitudinal suture which extends only to the anterior one-third of the sporocyst.

The sporocyst of *Crystallospora crystalloides* is bipyramidal and opens at a suture situated at the foot of the pyramids (Daoudi, 1987).

The sporozoites of fish coccidia are banana-, sausage- or comma-shaped. They usually lie in the sporocyst in a head-to-tail presentation. In the cryptosporidia the sporozoites are side by side and lie in the same direction. The sporozoite has a nucleus easily discernible by light microscopy and is situated in the middle of the body. The conoid apparatus lies at the anterior end, and the posterior end may sometimes be striated.

After the formation of sporocysts, a sporocyst residuum appears within the sporocyst. This residuum is granular in the young oocysts and may be compact in the older ones (Molnár, 1989). Certain species, e.g. *G. siliculiformis* and *G. degiustii* are characterized by a large residuum. In coccidia parasitizing the inner organs, the residuum is used up slowly while it is more rapid in intestinal coccidia.

Infection of fish

It is unclear why the majority of attempts to infect fish with coccidia were unsuccessful. Direct infection with oocysts has been reported by Musselius *et al.* (1965), Zmerzlaya (1966), Steinhagen and Körting (1988) and Steinhagen *et al.* (1989) using *G. carpelli* while Marincek (1973a) was successful with *G. subepithelialis* in the common carp. By feeding oocysts mixed in the feed, Paterson and Desser (1981a,b,c, 1982) produced *G. iroquoina* infection in fathead minnow (*Pimephales promelas*) and common shiner (*Notropis cornutus*) while Landsberg and Paperna (1985) produced *G. cichlidarum* infection in tilapia. In contrast, Molnár (1979) failed to produce experimental *G. carpelli* infection by feeding oocysts mixed in the diet; however, he succeeded if the oocysts were mixed with chicken liver. He consistently achieved massive infection by feeding tubifex collected from natural habitats. Kent and Hedrick (1985) produced *G. carpelli* infection in goldfish only if they had previously fed the oocysts to tubifex or shrimps. Steinhagen and Körting (1990) produced *G. carpelli* infection both by the direct route and through vectors. On the other hand, Solangi and Overstreet (1980) and Fournie and Overstreet (1983) reported that *C. funduli* required a true intermediate host, the grass shrimp (*Palaemonetes pugio*) for its development. The possible role of intermediate hosts in the transmission of fish coccidia was first suggested by Landau *et al.* (1975) who demonstrated the sporozoites of a coccidian parasitizing a muraena (*Gymnothorax moringa*) from the intestinal tissues of a crustacean (Mysidacea). Studies by Fournie and Overstreet (1983) and Steinhagen and Körting (1990) suggest that the invertebrate is a paratenic host in which the sporozoites do not undergo development. The oocysts of fish coccidia undergo rapid sporulation and ageing in the environment. The sporozoites die soon after the residual body of the sporocyst is used up. In tubifex or shrimp vectors, on the other hand, the sporozoites are released in the gut and remain viable for a long time in epithelial cells.

Peculiarities in the development of fish coccidia

Oocysts of fish coccidia are always assumed to be sporulated when they leave the fish. Recent studies, however, have revealed that a number of species shed unsporulated oocysts (Figs 7.6, 7.7). The majority of fish coccidia infect the intestinal epithelium, but a relatively large number of them, especially those with large oocysts, develop at non-intestinal sites like the kidney, spleen, liver, swimbladder and serosae. Most species produce oocysts continuously throughout the year but there are species which follow a one year seasonal cycle and produce oocysts only in the spring.

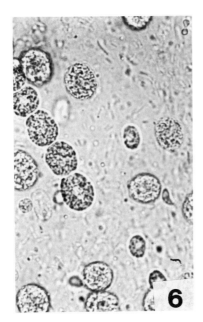
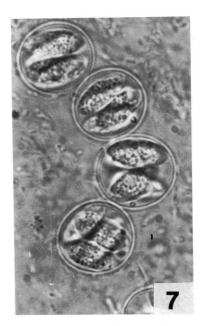

Fig. 7.6. Unsporulated oocysts from the gut of the sterlet. The larger ones are *Goussia vargai* and the smaller ones are *G. acipenseris* (×400).

Fig. 7.7. Extraintestinally sporulated oocysts of *Goussia vargai* (×1000).

Adeleid blood parasites (coccidia *sensu lato*)

Development of adeleid blood parasites in general
Fish-parasitic coccidia *sensu lato* have heteroxenous life cycles which involve two hosts, one being the fish and the other a parasitic leech or insect. The development of coccidia *sensu lato* is divided into merogony, gamogony and sporogony. There is also a special association of the two gamonts prior to encystment (syzygy) and this takes place before sporogony. Although the Haemogregarinidae and the Dactylosomatidae bear striking resemblances (development, morphology and vectors), there are important differences between them in development. Haemogregarinids infect the fish with sporozoites while the infective stage is the merozoite in the dactylosomatids. Also, at sporogony haemogregarinids develop in parasitophorous vacuoles in cells in the definitive host, while dactylosomatids are free in the cell cytoplasm.

Development and morphology of haemogregarinids
The development of fish-parasitic haemogregarinids (Figs 7.8a,b, 7.9) is known primarily from studies of natural infections. Only Khan (1978) described some developmental stages of *Cyrilia uncinata* in experimental infections.

Sporozoites are injected into the blood by leeches. They enter lympho-

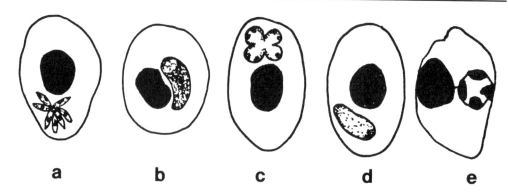

Fig. 7.8. Schematized diagrams of adeleorin and piroplasmorid protozoans in erythrocytes. a, b: meront with seven merozoites (a) and gamont (b) of *Haemogregarina delagei* (after Khan, 1972); c, d: meront of tetrad form (c) and gamont (d) of *Babesiosoma hannesi* (after Paperna, 1981); e: *Haemohormidium beckeri* (after Khan, 1980).

cytes, monocytes, neutrophils or blast cells where they develop into meronts and form merozoites. The first merogony takes place mainly in inner organs, while the second may be in the peripheral leucocytes and erythrocytes. After the merogonic cycles, the vermiform merozoites enter erythrocytes or leucocytes to form macro- and microgamonts. In some species there is some division of gamonts in the erythrocytes, while in others they directly change into micro- and macrogamonts.

Gamonts are taken in during a blood meal. In the intestine of the leech they are released from blood cells, and the micro- and macrogamonts unite in syzygy. During this process they are surrounded by a thin membrane. In *Haemogregarina* and *Cyrilia* the microgamont produces four microgametes, and one of the resulting microgametes fertilizes the macrogamont. Oocyst formation takes place, and depending on the parasite species it may occur either within an enterocyte or on the surface in a parasitophorous vacuole.

Intraleucocytic meronts of *Haemogregarina* spp. (Fig. 7.9a) usually harbour two to eight merozoites and do not significantly enlarge the size of the host cell. The length of the banana- or crescent-shaped merozoites and gamonts varies between 4 and 5 μm; intraleukocytic meronts, however, may reach 26 × 23 μm.

Development and morphology of dactylosomatids
Little is known about the development of dactylosomatids parasitic in fishes. Only the intraerythrocytic merogonic and gamogonic stages (Fig. 7.8c.d) are known. Paperna (1981) suggested that the parasite is transmitted by leeches.

Piroplasmorid blood parasites

Members of the order Piroplasmorida Wenyon, 1926 differ from other apicomplexans in having an apical complex without a conoid. They have no

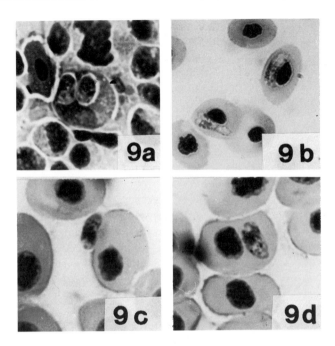

Fig. 7.9. *Haemogregarina* developmental stages in Giemsa-stained blood films. a: intraleucocytic meronts from *Scomber scombrus* (×1500); c: intraerythrocytic gamont of *H. bigemina* (×1500); c: free gamete of *H. acipenseris* in the plasm (×2000); d: intraerythrocytic gamont of *H. acipenseris* (×2000). (Figs a, b: courtesy of Dr A.J. Davies; c, d: courtesy of Dr F. Baska).

spores or sporogonic stages. Piroplasmorids are well-known pathogens of vertebrates. Fish are infected by *Haemohormidium*. Haemohormidia are amoeboid, oval or rounded organisms in the cytoplasm of erythrocytes (Fig. 7.8e). Davies (1980), who studied *H. cotti* using light and electron microscopy, found that the parasites (1.2 to 4.5 µm) in the red blood cells were bounded by a complex membrane like that seen in the pellicle of a typical Apicomplexa. The outer membrane was in direct contact with the host cell cytoplasm. No parasitophorous vacuole was found. Khan (1980), who studied the development of *H. beckeri* in erythrocytes, described its division both by binary fission and merogony. He found no gamogonic stages. The leeches *Platybdella orbiti* and *Johanssonia arctica* were vectors. Dividing stages were found in their gut and proboscis; however, no oocysts were found. Natural transmission occurred when leeches fed on susceptible fish.

HOST SPECIFICITY

Little is known about host specificity of fish coccidia. Fish coccidia seem to be less host specific than mammalian coccidia. A fish coccidium can infect several closely related host species usually of the same genus. However, *G. subepithelialis* of the common carp and *G. sinensis* of silver and bighead carp are host specific. *Eimeria anguillae*, which is found in eel species of both the Atlantic and the Pacific Oceans, and *G. vanasi* and *G. cichlidarum* of tilapia species are good examples of group specificity. Though Shulman (1984) listed more than a dozen hosts for *G. carpelli*, it is likely that *G. carpelli* infects only fish of the genera *Cyprinus* and *Carassius*, and that coccidians in other cyprinids are morphologically similar but less well studied species.

Cryptosporidia are considered to have an extremely broad host range. According to Levine (1984) four species, including the fish-parasitic *Cryptosporidium nasorum*, are valid.

The majority of known adeleids have been described from a single fish species. In contrast, *Haemogregarina bigemina* has been recorded in at least 85 species of fish. This may be explained by the fact that these are not based on experimental studies and that some authors described the meronts or gamonts as new species, while others identified them with the best known species, *H. bigemina*. The majority of investigators (Khan, 1978, 1980; Becker and Overstreet, 1979; Barta and Desser, 1989; Barta, 1991) regarded only leeches as vectors, while Davies and Johnston (1976) as well as Davies (1982) suggested that *H. bigemina* could develop both in the leech *Oceanobdella blennii* and in the isopod *Gnathia maxillaris*.

Barta (1991) suggested that *Babesiosoma* species were less host specific and synonymized some of the described species. He listed 26 fish hosts for the five *Babesiosoma* spp. he considered valid.

AETIOLOGY, HOST–PARASITE RELATIONSHIP

Pathogenicity of apicomplexans

The majority of fish coccidians have relatively low pathogenicity. No mortality was observed even when 85–90% of spleen or liver were infected with *Goussia* spp. or *Calyptospora funduli* (see Molnár, 1976; Solangi and Overstreet, 1980). Lethal infections occur primarily in farm ponds, but severe cases have been reported from natural waters as well. Bauer *et al.* (1981) reported fish mortality caused by *G. carpelli* and *G. sinensis*. Molnár (1976) also recorded deaths due to *G. sinensis*. *Eimeria anguillae* infection caused emaciation and deaths in eels in New Zealand (Hine, 1975). Fiebiger (1913) as well as Odense and Logan (1976) reported mortality in the haddock caused by *G. gadi*. MacKenzie (1978) found a species of *Eimeria* sp. which caused 6–10% reduction in body mass in blue whiting. Pinto (1956) reported parasitic castration as a result of *G. sardinae* infection.

The damage to tissues depends on the intensity of infection. In heavy *E. anguillae* infection Hine (1975) observed partial or total destruction of the intestinal mucosa and submucosa. The epithelial cells became compressed, the submucosal connective tissue vacuolated, and mature oocysts and free sporocysts were passed out with necrotic tissue. Kent and Hedrick (1985) reported high mortality in a lethargic and emaciated 15-day-old goldfish

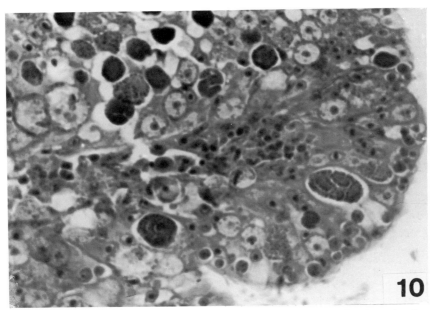

Fig. 7.10. Nodular coccidiosis. Developing macrogamonts (ma) and micro-gamonts (mi) of *Goussia subepithelialis* in the intestinal epithelium of common carp (H. and E., ×500).

population. They found that *G. carpelli* caused chronic enteritis with numerous yellow bodies and inflammatory and necrotic cells in the lamina propria of the gut. The microvillous structure of the intestinal epithelium was also destroyed. Molnár (1976) studied *G. sinensis* in silver carp and also found intensive histological changes in the gut epithelium. Most of the epithelial cells were damaged by meronts or gamonts but neither inflammation nor haemorrhagic lesions developed in the intestinal wall. He concluded that, despite intensive infection, *G. sinensis* was regarded a moderately pathogenic parasite.

More severe changes develop in nodular coccidiosis (Figs 7.10–7.12). Pellérdy and Molnár (1968) reported that agglomerations of *G. subepithelialis* oocysts in common carp were frequently surrounded by inflammatory cells. Marincek (1973b) described second-generation meronts developing basal to the nucleus in the cytoplasm. These exerted a more pathogenic effect on cells than first- and third-generation which were meronts located in the apical region of endothelial cells. The most severe changes were seen during gamogony. The chromatin of infected cell nuclei degenerated, the nuclei became enlarged and infected cells extended above the level of the uninfected cells. The damaged cells changed their cylindrical shape and died. In severe cases the epithelium desquamated and the lamina propria was exposed. The infection was aggravated by inflammation, and lymphocytes and eosinophils appeared in the infected area. The process was characterized by increased mucus production. Large numbers of pinhead-sized yellowish-white nodules were in the enlarged spleen of the gudgeon infected with *G. metchnikovi* (Pellerdy and Molnár, 1968). These nodules contained hundreds of oocysts.

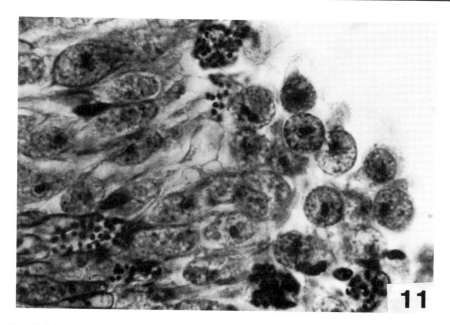

Fig. 7.11. Nodular coccidiosis in tench. Unsporulated oocysts just leaving the infected area (H. and E., ×1000).

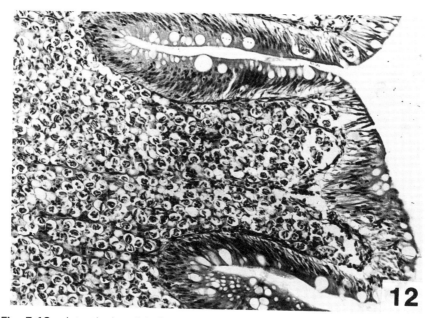

Fig. 7.12. Intestinal nodule in common carp consisting of sporulated *Goussia subepithelialis* oocysts in the subepithelial layer of the gut (H. and E., ×200).

Young stages, mostly macrogametes, were surrounded by inflammatory tissue reaction, while the young oocysts were surrounded by thin connective tissues. In advanced cases the capsule became thick (10 to 15 µm). A similar host reaction was reported in *Calyptospora funduli* infection in killifish (Solangi and Overstreet, 1980). Heavily infected fish had pale white to black lesions densely scattered throughout the liver parenchyma. At the early stage, inflammatory cells (mainly lymphocytes, mononuclear macrophages and eosinophils) accumulated around the gamonts. Later, fibrotic capsules surrounded by inflammatory cells developed around the oocysts. As the process advanced, the inflammatory infiltration disappeared and pigmentation appeared within the fibrotic capsules. Similar pathogenesis and host reaction were observed in the *G. thelohani*-infected liver of a marine fish, *Symphodus tinca* (see Daoudi *et al.*, 1988). Hawkins *et al.* (1981) found few changes in the liver during merogony of *C. funduli*, which they attributed to the formation and release of merozoites. Hepatocyte degeneration, pigment formation and leucocytic infiltration occurred mostly during gamont formation; hepatocytes were replaced by parasites and leucocytes associated with them, including monocytes, heterophils and eosinophils.

The formation of a connective tissue capsule around clumps of oocysts is also common in *Goussia* spp. which infect kidneys. Pellérdy and Molnár (1968) observed a capsule around *G. scardinii* oocysts (Fig. 7.13) in the renal parenchyma, and Morrison and Poynton (1989) reported damage of the renal epithelium and granuloma formation around the affected tubules in *G. spraguei*-infected cod.

The infection of the swimbladder is characterized by severe changes. Both Schäperclaus (1954), who studied *G. gadi* infection in the cod, and Grabda (1983), who investigated *Eimeria jadvigae* infection in *Coryphaenoides holotrachys*, reported thickening of the swimbladder wall. In more severe cases the inner surface of the bladder was covered with a white substance, the wall had a spongy texture and the lumen was filled by a mucous exudate containing large numbers of oocysts.

There is little information on the pathogenic effect of adeleid parasites. In cultured turbot, Ferguson and Roberts (1975) reported a proliferative disease of the haematopoietic tissues; this was later discovered to be caused by a haematozoon (Kirmse and Ferguson, 1976) which Kirmse (1978) described as *Haemogregarina sachai*. Kirmse (1980) found that up to 60% of some populations of *H. sachai*-infected turbots were affected with gross tumours in the musculature and viscera. The lesions consisted of necrotic tissue with a caseous centre. Histologically, there was an accumulation of parasitized reticuloendothelial cells, cell debris and pycnotic nuclei. Parasitaemias of up to 36% of all blood cells were observed; the majority of infected cells were neutrophils and monocytes.

In mackerel caught in certain areas of the Atlantic, 4% of the leucocytes were infected with meronts, and meronts were demonstrated in 100% of impression smears from the spleen. Lesions in the spleen and kidneys contained *Haemogregarina*-like organisms and were often surrounded by a connective tissue capsule (MacLean and Davies, 1990).

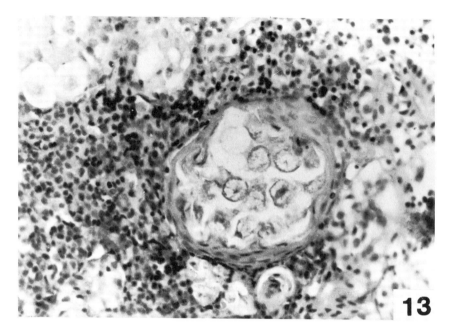

Fig. 7.13. Unsporulated oocysts of *Goussia scardinii* surrounded by a capsule of connective tissue in the kidney of *Chondrostoma nasus* (H. and E., ×1000).

Clinical signs of infection

According to Schäperclaus (1954), Kocylowski and Myaczynski (1963) and Bauer *et al.* (1981), fish infected with *G. carpelli* do not feed properly, group at a certain area of the pond and float in the water with their head downward. In intestinal coccidiosis, faeces is covered with a thick, sometimes reddish mucus layer. In goldfish affected with coccidiosis, Hoffman (1965) observed a long mucus sheath containing masses of oocysts protruding from the rectum. Difficulty in swimming is indicative of swimbladder coccidiosis.

Intensive intestinal infection is always accompanied by abundant mucus production. As a result, the frequently reddened intestinal wall is lined with thick viscous mucus. This is more likely to be derived from the plasm of injured cells than from enhanced secreting activity. In nodular coccidiosis, white nodules, the size of a millet or lentil can be seen through the intestinal wall. Spleen and liver coccidiosis is also indicated by the presence of nodules, and swimbladder coccidiosis by a thickening of the swimbladder wall.

Host–parasite relationship

Fish coccidia and their hosts have a well developed host–parasite relationship. The host usually takes a long time before reacting to the developing parasite. Infected cells are inactive and gradually die. At this time the host response is mainly restricted to replacing the damaged cells (Molnár, 1984). In diffuse coccidiosis, e.g. in *G. carpelli* infection, the inactive epithelial cell which contains the developing macrogamont is gradually surrounded, grown over and pushed to deeper layers by neighbouring epithelial cells. The sporulated oocyst is expelled probably as a result of macrophage activation and necrosis of the epithelium above the oocyst. The oocyst, or two or three oocysts together, are situated in a 'yellow body'. One component of the yellow body is the necrotic host cell, but macrophages engulfing oocysts may also become yellow bodies. According to Dyková and Lom (1981), the yellow body contains mucoid substances and ferrous ions as well as the cytoplasm of the pathologically altered host cell. Also, the yellow body contains lipofuscin and ceroid which may be derived from degenerating cell membranes (Kent and Hedrick, 1985).

Molnár (1984) suggested a new ejection process in *G. subepithelialis* infection. The merogonic and gamogonic stages develop in the epithelial cells and, as Marincek (1973a) pointed out, only a few unsporulated oocysts are excreted from the parasitic nodule (Fig. 7.11). The remaining infected epithelial cells, which contain oocysts, are pushed downwards and grown over by proliferating intact epithelial cells. In this way large numbers of oocysts may be situated and may sporulate close to the submucosa (Fig. 7.12). Another intensive host reaction is mounted against oocysts after the death of host cells. This results in secondary damage of the epithelial cells overlaying the oocysts. Oocysts get into the intestinal lumen as a result of secondary damage to epithelial layer. According to Dyková and Lom (1981), unrejected oocysts in the subepithelium became surrounded by a connective tissue capsule and gradually died. Marincek (1973b, 1978) suggested that a large portion of the infected gut lost its normal function. A similar mechanism was observed in *E. anguillae* infection. Rupture of the host cell membrane releases the majority of the young oocysts into the gut lumen and these sporulate outside the host. However, some of the macrogamonts and young oocysts are surrounded by intact epithelial cells and are driven to deep layers of the epithelium where they sporulate. This must be why Dyková and Lom (1981) and Lacey and Williams (1983) suggested that *E. anguillae* had intracytoplasmic sporogony.

The majority of oocysts of non-intestinal species sporulate within the fish and are released after death of the host. These oocysts, e.g. the oocysts of *G. metchnikovi* in the spleen or those of *G. siliculiformis* on the serous membranes, have a large sporocyst residue and remain viable for a long time after their sporulation. Oocysts scattered in the parenchyma of the spleen are first localized to larger nodules, become surrounded by melanomacrophage cells (Fig. 7.14) and are later destroyed (Molnár, 1984).

Certain coccidian infections persist in fish; this suggests that the mechanism observed in coccidioses of warm-blooded animals occurs also in fish.

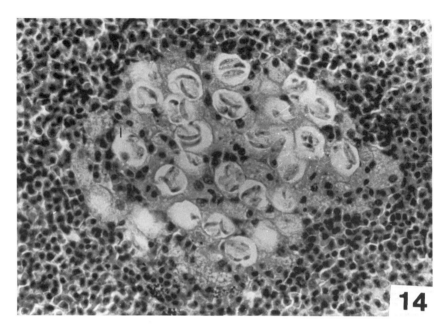

Fig. 7.14. *Goussia metchnikovi* oocysts inside a melanomacrophage centre of the spleen of *Gobio gobio* (×1000).

Nothing is known about humoral immune response of the host. Hawkins *et al.* (1981) and Békési and Molnár (1990) reported a depletion of hepatocytes around *Calyptospora* macrogamonts in the liver and an increase in the number of perisinusoidal cells which aggregated the macrogamonts into islets. Later, the macrogamonts become surrounded by granulocytes, heterophil and other leucocytes. In *G. sardinae* infection, Morrison and Hawkins (1984) observed leucocytes and phagocytes, while in *G. thelohani* infection Daoudi *et al.* (1988) saw lymphocytes and eosinophil cells around the gamonts. According to Molnár (1984), the melanomacrophage cells would gradually destroy *G. metchnikovi* oocysts congregated into islets. In *Calyptospora funduli* infected killifish an infection prevents reinfection (Solangi and Overstreet 1980). On the other hand, Békési and Molnár (1990) observed three generations of *C. tucunarensis* in the liver of tucunare. The developing macrogamonts were surrounded by perisinusoidal cells and oocysts aggregated into nodules. A connective tissue capsule was formed around some oocysts from an earlier infection.

NUTRITION OF FISH APICOMPLEXANS

Apicomplexans with a typical intracytoplasmic parasitophorous vacuole feed through intravacuolar blebs which contain host cell cytoplasm. According to Hawkins *et al.* (1983a,b,c), Morrison and Hawkins (1984) and Jastrzebski (1989), nutrient materials in the cisternae of the host cell endoplasmic reticulum are passed into the space of the parasitophorous vacuole, formed into blebs, and absorbed into the cytoplasm of the parasite. Epiplasmally located coccidia such as *Cryptosporidium* and some *Goussia* and *Eimeria* absorb nutrients only through their attachment zone or through invaginations of the parasitophorous vacuole (Fig. 7.3). In *Cryptosporidium* the parasitophorous vacuole is not complete and there is a special adhesive zone between the cell and the parasite. This is where absorption takes place. Parasite nutrition is ensured by a special feeder organelle which is formed by close contact of the parasitic folds and the plasma membrane at the base of the parasitophorous vacuole. In epicellular *Goussia* and *Eimeria* the parasitophorous vacuole extends invaginations into the host cell cytoplasm (Molnár and Baska, 1986; Daoudi, 1987; Morrison and Poynton, 1989; Jastrzebski and Komorowski, 1990). Paperna and Landsberg (1987), however, reported that these invaginations are substituted by tubular formations in *G. vanasi*.

DIAGNOSIS OF INFECTION

The thin and fragile wall, of the oocyst and sporocyst of fish coccidia do not permit the use of concentration procedures routinely used with coccidia of warm-blooded animals. Fish coccidia are demonstrated exclusively in fresh preparations or by histological methods.

Coccidia in internal organs are detected by examining small pieces of tissue under a coverslip. Low intensity infections are detected if the organs are first digested in 0.25–0.5% trypsin solution.

Intestinal coccidia are demonstrated by microscopic examination of faeces and intestinal scrapings; they are easiest to detect in the mucus from the intestinal epithelium or surface of the faeces (Molnár, 1977). If the fish is fasted for one or two days before the examination, oocysts in the yellow body are easily discernible at $400 \times$ in the gut.

Unsporulated oocysts (small size and thin oocyst wall) (Fig. 7.6) in fresh preparations are often mistaken for granulocytes or algae which are common in the intestinal scrapings. Oocysts are often distinguished only after sporulation. Oocyst sporulation is achieved in several changes of tap-water in a watch glass or in a small Petri dish. Addition of a loopful of penicillin and streptomycin sulphate to the water to control bacterial growth increases the chances of sporulation. The oocyst wall is fragile and this does not permit prolonged storage (Molnár, 1986, 1989). Oocysts in the mucus or faeces are preserved in 1% glutaraldehyde or 4% formalin (Molnár and Hanek, 1974; Molnár, 1977) for a short period. As permanent preparations, fish coccidia are best preserved as histological slides.

Oocysts are relatively easy to detect using histological methods. Mallory's stain is particularly suitable. Mature oocysts stain yellow and stand out in sharp contrast to fish tissues.

PREVENTION AND CONTROL OF APICOMPLEXAN INFECTIONS

Little is known about prevention of losses caused by fish coccidia. Perhaps this is because they only affect fish species (common carp, silver carp, bighead, eel and, possibly, tilapias) cultured in ponds. Only a few of the drugs developed for controlling coccidioses in warm blooded animals have been tried on these fish parasites. Naumova and Kanaev (1962) recommended Osarzol as a therapeutic agent against *G. carpelli* infection of the common carp, but failed to specify the dose. Furazolidone (for every 1000 fingerlings they mixed 300 mg drug to the food) is effective if fed to fish for 3 days (Musselius *et al.*, 1965). Kocylowski *et al.* (1976) reported that amprolium chloride (fed at a dose of $5\,g\,kg^{-1}$ fish at 2-day intervals on three occasions) is quite efficacious against the parasite. To control *E. anguillae*, Hine (1975) successfully treated eels with Furanace (1.0–$1.5\,mg\,kg^{-1}$ of eel) by adding it to the food for 6 days. Reduction in oocyst count in killifish fed monensin (Solangi and Overstreet, 1980) indicated that other anticoccidials might also be efficacious. Bauer *et al.* (1981) suggested that ponds should be disinfected by freezing, drying out, or with addition of lime or chlorinated lime.

CONCLUSIONS AND RECOMMENDATIONS FOR THE FUTURE

Significant progress has been made since the review of Dyková and Lom (1981) and this includes the ultrastructure and intracellular development of fish coccidia. Except for a few pioneering studies, most of the literature is based on natural infections. Hence, a fundamental task in future research into fish coccidia would be to establish conditions for reproducible experimental infections. This would be necessary to better understand the pathogenicity, specificity and immunity to fish coccidia. The possible role of vectors in transmitting these parasites would be another important area of research.

Further studies are needed to elucidate the life cycle and development of Haemogregarinida and Dactylosomatida coccidia. Other areas that need clarification are taxonomy and phylogenetic relationships between groups of related coccidia.

Coccidiosis is presumed to cause substantial losses; however, there is no chemotherapy against fish coccidia in fish farms. Hence research should also be focused on studying the pathogenicity of coccidia and selecting anticoccidials which could be used in fisheries.

REFERENCES

Barta, J.R. (1991) The Dactylosomatidae. *Advances in Parasitology* 30, pp. 1–37.

Barta, J.R. and Desser, S.S. (1989) Development of *Babesiosoma stableri* (Dactylosomatidae; Adeleida; Apicomplexa) in its leech vector (*Bactracobdella picta*) and the relationship of the dactylosomatids to the piroplasms of higher vertebrates. *Journal of Protozoology* 36, 241–253.

Bauer, O.N. Musselius, V.A. and Strelkov, J.A. (1981) *Diseases of Pond Fishes*, (in Russian), 2nd edn. Legkhaya i pischevaya promishlennost, Moscow.

Baska, F. and Molnár, K. (1989) Ultrastructural observations on different developmental stages of *Goussia sinensis* (Chen, 1955), a parasite of the silver carp (*Hypophthalmichthys molitrix* Valenciennes, 1844). *Acta Veterinaria Hungarica* 37, 81–87.

Becker, C.D. and Overstreet, R.M. (1979) Haematozoa of marine fishes from the northern Gulf of Mexico. *Journal of Fish Diseases* 2, 469–479.

Békési, L. and Molnár, K. (1990) *Calyptospora tucunarensis* n.sp. (Apicomplexa: Sporozoea) from the liver of tucunare (*Cichla ocellaris*) in Brazil. *Systematic Parasitology* 18, 127–132.

Daoudi, F. (1987) Coccidies et coccidioses de poissons mediterranens: Systematique, ultrastructure et biologie. Doctoral thesis, Laboratoire d'Ichthyologie et de parasitologie génerale. U.S.T.L. Montpellier.

Daoudi, F., Radujkovic, B., Marques, A. and Bouix, G. (1988) Pathogenicity of the coccidian *Goussia thelohani* (Labbe, 1986), in liver and pancreatic tissues of *Symphodus tinca* (Linne, 1758). *Bulletin of the European Association of Fish Pathologists* 8, 55–58.

Davies, A.J. (1980) Some observations on *Haemohormidium cotti* Henry, 1910, from the marine fish *Cottus bubalis* Euphrasen. *Zeitschrift für Parasitenkunde* 62, 31–38.

Davies, A.J. (1982) Further studies on *Haemogregarina bigemina* Laveran & Mesnil, the marine fish *Blennius pholis* L., and the isopod *Gnathia maxillaris* Montagu. *Journal of Protozoology* 29, 576–583.

Davies, A.J. and Johnston, M.R.L. (1976) The biology of *Haemogregarina bigemina* Laveran & Mesnil, a parasite of the marine fish *Blennius pholis* Linnaeus. *Journal of Protozoology* 23, 315–320.

Dyková, I. and Lom, J. (1981) Fish coccidia: critical notes on life cycles, classification and pathogenicity. *Journal of Fish Diseases* 4, 487–505.

Dyková, I. and Lom, J. (1983) Fish coccidia: an annotated list of described species. *Folia Parasitologica* 30, 193–208.

Ferguson, H.W. and Roberts, R.J. (1975) Myeloid leucosis associated with sporozoan infection in cultured turbot (*Scophthalmus maximus* L.). *Journal of Comparative Pathology* 85, 317–326.

Fiebiger, J. (1913) Studien über die Schwimmblasen-Coccidien der Gadusarten (*Eimeria gadi* n.sp.). *Archiv für Parasitenkunde* 31, 95–137.

Fournie, J.W. and Overstreet, R.M. (1983) True intermediate hosts for *Eimeria funduli* (Apicomplexa) from estuarine fishes. *Journal of Protozoology* 30, 672–675.

Grabda, E. (1983) *Eimeria jadvigae* n.sp. (Apicomplexa: Eucoccidia), a parasite of swimming bladder of *Coryphaenoides holotrachys* (Günther, 1887) off the Falklands. *Acta Ichthyologica et Piscatoria* 13, 131–140.

Hawkins, W.E., Solangi, M.A. and Overstreet, R.M. (1981) Ultrastructural effects of the coccidium, *Eimeria funduli* Duszynski, Solangi et Overstreet (1979) on the liver of killifishes. *Journal of Fish Diseases* 4, 281–295.

Hawkins, W.E., Fournie, J.W. and Overstreet, R.M. (1983a) Organization of sporulated oocysts of *Eimeria funduli* in the gulf killifish, *Fundulus grandis*. *Journal of Parasitology* 69, 496–503.

Hawkins, W.E., Solangi, M.A. and Overstreet, R.M. (1983b) Ultrastructure of the macrogamont of *Eimeria funduli*, a coccidium parasitizing killifishes. *Journal of Fish Diseases* 6, 33–43.

Hawkins, W.E., Solangi, M.A. and Overstreet, R.M. (1983c) Ultrastructure of the microgamont and microgamete of *Eimeria funduli*, a coccidium parasitizing killifishes. *Journal of Fish Diseases* 6, 45–57.

Hawkins, W.E., Fournie, J.W. and Overstreet, R.M. (1984a) Ultrastructure of the interface between stages of *Eimeria funduli* (Apicomplexa) and hepatocytes of the longnose killifish *Fundulus similis*. *Journal of Parasitology* 70, 232–238.

Hawkins, W.E., Fournie, J.W. and Overstreet, R.M. (1984b) Intrahepatic stages of *Eimeria funduli* (Protista: Apicomplexa) in the longnose killifish, *Fundulus similis*. *Transactions of the American Microscopical Society* 103, 185–194.

Hine, P.M. (1975) *Eimeria anguillae* Leger et Hollande, 1922 parasitic in New Zealand eels. *New Zealand Journal of Marine and Freshwater Research* 9, 239–243.

Hoffman, G.L. (1965) *Eimeria aurati* n.sp. (Protozoa: Eimeriidae) from the goldfish (*Carassius auratus*) in North America. *Journal of Protozoology* 12, 273–275.

Hoover, H.M., Hoerr, F.J., Carlton, W.W., Hinsman, E.J. and Ferguson, H.W. (1981) Enteric cryptosporidiosis in a naso tang, *Naso lituratus* Bloch and Schneider. *Journal of Fish Diseases* 4, 425–428.

Jastrzebski, M. (1989) Ultrastructural study on the development of *Goussia aculeati*, a coccidium parasitizing the three-spined stickleback *Gasterosteus aculeatus*. *Diseases of Aquatic Organisms* 6, 45–53.

Jastrzebski, M. and Komorowski, Z. (1990) Light and electron microscopic studies on *Goussia zarnowskii* (Jastrzebski, 1982): an intestinal coccidium parasitizing the three-spined stickleback, *Gasterosteus aculeatus* (L.). *Journal of Fish Diseases* 13, 1–24.

Kent, M.L. and Hedrick, R.P. (1985) The biology and associated pathology of *Goussia carpelli* (Léger and Stankovitch) in goldfish *Carassius auratus* (Linnaeus). *Fish Pathology* 20, 485–494.

Kent, M.L., Fournie, J.W., Snodgrass, R.E. and Elston, R.A. (1988) *Goussia girellae* n.sp. (Apicomplexa: Eimeriorina) in the opaleye, *Girella nigricans*. *Journal of Protozoology* 35, 287–290.

Khan, R.A. (1978) A new haemogregarine from marine fishes. *Journal of Parasitology* 64, 35–44.

Khan, R.A. (1980) The leech as a vector of a fish piroplasm. *Canadian Journal of Zoology* 58, 1631–1637.

Kirmse, P. (1978) *Haemogregarina sachai* n.sp. from cultured turbot *Scophthalmus maximus* L., in Scotland. *Journal of Fish Diseases* 1, 337–342.

Kirmse, P. (1980) Observations on the pathogenicity of *Haemogregarina sachai* Kirmse, 1978, in farmed turbot *Scophthalmus maximus* (L.). *Journal of Fish Diseases* 3, 101–114.

Kirmse, P. and Ferguson, H. (1976) *Toxoplasma*-like organisms as the possible causative agents of a proliferative condition in farmed turbot (*Scophthalmus maximus*). In: Page, L. (ed.), *Wildlife Diseases*. Plenum, New York, pp. 561–564.

Kocylowski, B. and Myaczynski, T. (1963) *Fish Diseases*. Mezôgazdasagi, Budapest.

Kocylowski, B., Zelazny, J., Antychowicz, J. and Panczyk, J. (1976) Incidence of carp coccidiosis and its control. *Bulletin of the Veterinary Institute Pulawy* 20, 12–17.

Lacey, S.M. and Williams, I.C. (1983) *Epieimeria anguillae* (Léger & Hollande, 1922)

Dyková & Lom, 1981 (Apicomplexa: Eucoccidia) in the European eel, *Anguilla anguilla* (L.). *Journal of Fish Biology* 23, 605–609.

Landau, I., Marteau, M., Golvan, Y., Chabaud, A.G. and Boulard, Y. (1975) Hétéroxenie chez les coccidies intestinales de poissons. *Comptes Réndus des Séances de l'Académie des Sciences* 281, 1721–1723.

Landsberg, J.H. and Paperna, I. (1985) *Goussia cichlidarum* n.sp. (Barrouxiidae, Apicomplexa), a coccidian parasite in the swimbladder of cichlid fish. *Zeitschrift für Parasitenkunde* 71, 199–201.

Landsberg, J.H. and Paperna, I. (1986) Ultrastructural study of the coccidian *Cryptosporidium* sp. from stomachs of juvenile cichlid fish. *Diseases of Aquatic Organisms* 2, 13–20.

Landsberg, J.H. and Paperna, I. (1987) Intestinal infections by *Eimeria* (S.L.) *vanasi* n.sp. (Eimeriidae, Apicomplexa, Protozoa) in cichlid fish. *Annales de Parasitologie Humaine et Comparée* 62, 283–293.

Levine, N.D. (1984) Taxonomy and review of the coccidian genus *Cryptosporidium* (Protozoa, Apicomplexa). *Journal of Protozoology* 31, 94–98.

Levine, N.D. (1988) *The Protozoan Phylum Apicomplexa*. 2 vols. CRC Press, Boca Raton, Florida.

Lom, J. (1971) Remarks on the spore envelopes in fish coccidia. *Folia Parasitologica* 18, 289–293.

Lom, J. (1984) Diseases caused by protistans In: Kinne, O. (ed.) *Diseases of Marine Animals*, 4. Biologische Anstalt Helgoland, Hamburg, pp. 114–168.

Mackenzie, K. (1978) The effect of *Eimeria* sp. infection on the condition of the blue whiting *Micromesistius poutassou* (Risso). *Journal of Fish Diseases* 4, 473–486.

MacLean, S.A. and Davies, A.J. (1990) Prevalence and development of intraleucocytic haemogregarines from Northwest and Northeast Atlantic mackerel, *Scomber scombrus* L. *Journal of Fish Diseases* 13, 59–68.

Marincek, M. (1973a) Les changements dans le tube digestif chez *Cyprinus carpio* á la suite de l'infection par *Eimeria subepithelialis* (Sporozoa, Coccidia). *Acta Protozoologica* 20, 217–224.

Marincek, M. (1973b) Dévelopment d' *Eimeria subepithelialis* (Sporozoa, Coccidia) parasite de la carpe. *Acta Protozoologica* 19, 197–215.

Marincek, M. (1978) Uticaj kokcidije *Eimeria subepithelialis* po konstituciju sarana. *Acta Parasitologica Iugoslovenica* 9, 3–12 (in Serbo-Croatian).

Molnár, K. (1976) Histological study of coccidiosis caused in the silver carp and the bighead by *Eimeria sinensis* Chen (1956). *Acta Veterinaria Academiae Scientiarium Hungaricae* 26, 303–312.

Molnár, K. (1977) Comments on the nature and methods of collection of fish coccidia. *Parasitologia Hungarica* 10, 41–45.

Molnár, K. (1979) *Studies on Coccidia of Hungarian Pond Fishes and Further Prospects of their Control.* Proceeding of the International Symposium on Coccidia. Prague pp. 173–183.

Molnár, K. (1984) Some peculiarities of oocyst rejection of fish coccidia. *Symposia Biologica Hungarica* 23, 87–97.

Molnár, K. (1986) Occurrence of two new *Goussia* species in the intestine of the sterlet (*Acipenser ruthenus*). *Acta Veterinaria Hungarica* 34, 169–174.

Molnár, K. (1989) Nodular and epicellular coccidiosis in the intestine of cyprinid fishes. *Diseases of Aquatic Organisms* 7, 1–12.

Molnár, K. and Baska, F. (1986) Light and electron microscopic studies on *Epieimeria anguillae* (Léger et Hollande, 1922), a coccidium parasitizing the European eel, *Anguilla anguilla* L. *Journal of Fish Diseases* 9, 99–110.

Molnár, K. and Hanek, G. (1974) Seven new *Eimeria* spp. (Protozoa, Coccidia) from freshwater fishes of Canada. *The Journal of Protozoology* 21, 489–493.

Morrison, C.M. and Hawkins, W.E. (1984) Coccidians in the liver and testis of the herring *Clupea harengus* L. *Canadian Journal of Zoology* 62, 480–493.

Morrison, C.M. and Poynton, S.L. (1989) A new species of *Goussia* (Apicomplexa, Coccidia) in the kidney tubules of the cod, *Gadus morhua* L. *Journal of Fish Diseases* 12, 533–560.

Musselius, V.A., Laptev, V.I. and Ivanova, N.S. (1965) On the coccidiosis of the common carp II. *Trudi VNIIPRH* 13, 69–78 (in Russian).

Naumova, A.M. and Kanaev, A.I. (1962) An experiment for treating common carp diseased in coccidiosis. *Voprosy Ikhtiologii Akad. Nauk SSSR* 2, 749–751 (in Russian).

Odense, P.H. and Logan, V.H. (1976) Prevalence and morphology of *Eimeria gadi* in the haddock. *Journal of Protozoology* 23, 564–571.

Paperna, I. (1981) *Dactylosoma hannesi* n.sp. (Dactylosomatidae, Piroplasmia) found in the blood of grey mullets (Mugilidae) from South Africa. *Journal of Protozoology* 28, 486–491.

Paperna, I. (1987) Scanning electron microscopy of the coccidian parasite *Cryptosporidium* sp. from cichlid fishes. *Diseases of Aquatic Organisms* 3, 231–232.

Paperna, I. and Landsberg, J.H. (1987) Tubular formations extending from parasitophorous vacuoles in gut epithelial cells of cichlid fish infected by *Eimeria* (s.1.) *vanasi. Diseases of Aquatic Organisms* 2, 239–242.

Paterson, W.B. and Desser, S.S. (1981a) An ultrastructural study of *Eimeria iroquoina* Molnár and Fernando, 1974 in experimentally infected fathead minnows (*Pimephales promelas*, Cyprinidae). 3. Merogony. *Journal of Protozoology* 28, 302–308.

Paterson, W.B. and Desser, S.S. (1981b) An ultrastructural study of microgametogenesis and the microgamete in *Eimeria iroquoina* Molnár and Fernando, 1974, in experimentally infected fathead minnows (*Pimephales promelas*, Cyprinidae). *Journal of Parasitology* 67, 314–324.

Paterson, W.B. and Desser, S.S. (1981c) Ultrastructure of macrogametogenesis, macrogametes and young oocysts of *Eimeria iroquoina* Molnár and Fernando, 1974 in experimentally infected fathead minnows (*Pimephales promelas*, Cyprinidae). *Journal of Parasitology* 67, 496–504.

Paterson, W.B. and Desser, S.S. (1982) The biology of two *Eimeria* species (Protista: Apicomplexa) in their mutual fish hosts in Ontario. *Canadian Journal of Zoology* 60, 164–175.

Pavlasek, I. (1983) *Cryptosporidium* sp. in *Cyprinus carpio* Linné 1758 in Czechoslovakia. *Folia Parasitologica* 30, 248.

Pellérdy, L. and Molnár, K. (1968) Known and unknown eimerian parasites of fishes in Hungary. *Folia Parasitologica* 15, 97–105.

Pinto, J.S. (1956) Parasitic castration in males of *Sardinia pilchardus* (Walb.) due to testicular infestation by the coccidia *Eimeria sardinae* (Thelohan). *Revista de la Faculdade de Ciencias de la Universidade de Lisboa* Serie C 5, 209–214.

Schäperclaus, W. (1954) *Fischkrankheiten*. Akademie Verlag, Berlin.

Shulman, S.S. (1984) Parasitic protozoa. In: Bauer, O.N. (ed.) *Key to Parasites of Freshwater Fish of the USSR* 1. Leningrad (in Russian).

Solangi, M.A. and Overstreet, R.M. (1980) Biology and pathogenesis of the coccidium *Eimeria funduli* infecting killifishes. *Journal of Parasitology* 66, 513–526.

Steinhagen, D. and Körting, W. (1988) Experimental transmission of *Goussia carpelli* (Leger & Stankovitch, 1921; Protista: Apicomplexa) to common carp, *Cyprinus carpio* L. *Bulletin of the European Association of Fish Pathologists* 8, 112–113.

Steinhagen, D., Körting, W. and van Muiswinkel, W.B. (1989) Morphology and biology of *Goussia carpelli* (Protozoa: Apicomplexa) from the intestine of experimentally infected common carp *Cyprinus carpio*. *Diseases of Aquatic Organisms* 6, 93–98.

Steinhagen, D. and Körting, W. (1990) The role of tubificid oligochaetes in the transmission of *Goussia carpelli*. *Journal of Parasitology* 76, 104–107.

Zmerzlaya, E.I. (1966) Temperature effect upon the infestation of carps with coccidia *Eimeria carpelli* Leger et Stankovitsch, 1921. *Zoologicheskhii Zhurnal* 45, 305–308 (in Russian).

8

Monogenea (Phylum Platyhelminthes)

D.K. Cone

Department of Biology, St Mary's University, Halifax, Nova Scotia, Canada B3H 3C3.

INTRODUCTION

Monogeneans are platyhelminths parasitic primarily on fishes, including the Agnatha, the Chondrichthyes, and the Osteichthyes (see Bychowsky, 1957; Malmberg and Fernholm, 1989). Most are ectoparasitic on specific sites on the host such as the head and flank, the fins, crypts of the acoustolateralis system, the surface of the nasal epithelium, or on the branchial arches. At least three lineages of the parasites have adopted an endoparasitic life. *Acolpenteron* and *Kritskyia* occur in the urinary ducts and bladder (Fischthal and Allison, 1942; Kohn, 1990) and *Enterogyrus* attaches to the wall of the foregut (Cone *et al.*, 1987). A species of *Dactylogyrus* is known from the ovipositor of the Amur bitterling (Yukhimenko and Danilov, 1987).

Monogeneans are either sanguiniferous polyopisthocotyleans or tissue grazing monopisthocotyleans. The former do not produce significant tissue damage due to the delicate manner of attachment to secondary gill lamellae and the subtle manner in which blood is drawn from the underlying blood vessels. Members of this group are rarely associated with host mortality (Nigrelli, 1943; Frankland, 1955; Kubota and Takakuwa, 1963; Paperna *et al.*, 1984; Faisal and Imam, 1990). Monopisthocotyleans, on the other hand, promote significant tissue damage through a more disruptive manner of attachment and by grazing on exposed and vulnerable integument. Although the damage is usually minimized because the parasite regularly relocates on the host (Rohde, 1984; Cone and Wiles, 1989), members of the group nevertheless cause significant disease.

Most monopisthocotyleans are oviparous and have a crawling or free-swimming oncomiracidium. Oncomiracidia invade the host and, as poston-comiracidia, migrate over the gills or body to the final site of attachment (Fig. 8.1). Members of one lineage, the gyrodactylids, are viviparous and give birth to individuals containing well-developed embryos. Invasion is by adult parasites transferring directly between adjacent hosts. Both oviparous and

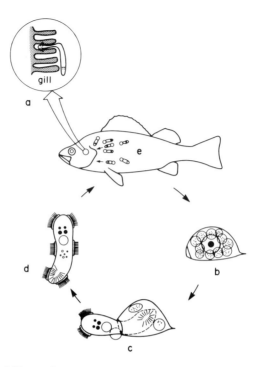

Fig. 8.1. Typical life cycle of an oviparous monogenean, *Urocleidus adspectus* parasitizing the gills of the freshwater perch (*Perca flavescens*). (a) Adult worm on gill lamellae; (b) egg; (c) emerging oncomiracidium; (d) free-swimming oncomiracidium; (e) migrating postoncomiracidium. (From Cone and Burt, 1985.)

viviparous species have high reproductive rates and these promote effective transmission under crowded conditions. Intensities increase rapidly and the resulting infections kill hosts directly (Lester and Adams, 1974; Cusack and Cone, 1986a). Infections have also been suspected of undermining nonspecific defence mechanisms and thus allow invasions by microbial pathogens (Munro, 1982). Monogeneans may be mechanical vectors for viral and bacterial pathogens of fish (Cusack and Cone, 1985, 1986b; Grimes *et al.*, 1985).

It is important to appreciate that pathogenicity is not normal because killing a host does not enhance transmission. In most instances natural selection results in organisms that cause minimal damage to the host. Epizootics are usually the result of a breakdown in the normal host–parasite relationships created by artificial conditions. A minority of monopisthocotyleans facilitate their parasitism through gross disruption of host tissues. Examples include several species of *Dactylogyrus* and related forms that cause epithelial hyperplasia of the gill lamellae (Putz and Hoffman, 1964; Gussev, 1985) as an adaptation that ensures firm attachment to the host and a localized replenishment of food.

Monogeneans are known to cause disease in all forms of fish culture. In spite of this, they are not given serious consideration under fish health

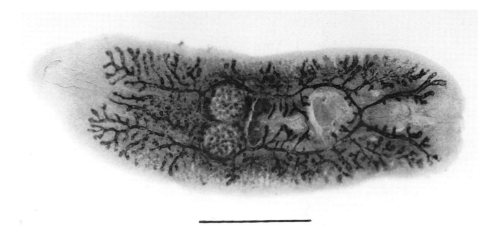

Fig. 8.2. Whole mount of *Dermophthirius carcharhini* from a Galapagos shark held captive in Bermuda. Scale bar = 1 mm. (Courtesy of Dr T. Rand; original.)

guidelines regulating the importation of potential pathogens. There have been numerous instances where introduced parasites have established themselves on feral host populations. The most notable is the establishment of *Gyrodactylus salaris* Malmberg, 1957 from the Baltics to feral and captive populations of *Salmo salar* in Scandinavian rivers. The result has been extended epizootics with significant reduction of feral stocks (Sattaur, 1988). A similar problem has occurred with *Pseudodactylogyrus* Gussev, 1964. This genus is endemic to Pacific anguillid eels and its introduction into Europe has created disease outbreaks involving the particularly susceptible *Anguilla anguilla* (see Buchmann *et al.*, 1987b). There are numerous other examples of how freshwater monogeneans have been distributed around the world by human activities (Hoffman, 1970).

Species frequently associated with disease in fish include members of four distinct lineages: the Microbothriidae, the Capsalidae, the Dactylogyridae, and the Gyrodactylidae. The present review is on the more economically important species within each of these taxa.

FAMILY MICROBOTHRIIDAE

Microbothriids are oviparous ectoparasites of elasmobranchs worldwide; the majority of which are marine species. They are medium to large sized monogeneans with an oval, flattened body (Fig. 8.2). The distinctive feature of the family is the cup-shaped haptor that lacks sclerotized components in the adult. Anteriorly there are two sucker-like organs opening into the oral cavity. There is a well-developed muscular pharynx and two branched, diverticulate intestinal caeca that end blindly. Pigmented light receptors are usually absent. Species of *Dermophthirius* MacCallum, 1926 are important pathogens.

Dermophthirius MacCallum, 1926

Host range

This group occurs on the body surface of elasmobranchs and causes disease when sharks are held under confined conditions (e.g. public aquaria). Problems have been reported at institutions on both the Pacific and Atlantic coasts of North America as well as in Hawaii and Bermuda (Cheung *et al.*, 1982, 1988; Rand *et al.*, 1986). Four species have been described: *D. carcharhini* MacCallum, 1926; *D. maccallumi* Watson and Thorson, 1967; *D. nigrellii* Cheung and Ruggieri, 1983; and *D. melanopteri* Cheung, Nigrelli, Ruggieri and Crow, 1988).

The parasites are common and appear to be host specific. Confirmed reports for *D. carcharhini* include the dusky shark (*Carcharhinus obscurus*) from the New York Bight (Cheung and Ruggieri, 1983) and the Galapagos shark (*Carcharhinus galapagensis*) from Bermuda (Rand *et al.*, 1986). *Dermophthirius nigrellii* is known from the lemon shark (*Negaprion brevirostris*) originating from the Florida Keys (Cheung and Ruggieri, 1983). *Dermophthirius maccallami* is known from *Carcharhinus leucas* in Lake Nicaragua (Watson and Thorson, 1976). *Dermophthirius melanopteri* was described from the Pacific blacktip reef shark (*Carcharhinus melanopterus*) collected at Christmas Island (Cheung *et al.*, 1988).

Systematics and taxonomic position

The original generic diagnosis (MacCallum, 1926) of the genus stands. The amended diagnosis presented by Price (1938) is incorrect with respect to the morphology of the haptor and is not accepted (Cheung and Ruggieri, 1983). Specimens identified by Euzet and Maillard (1967) as *D. carcharhini* from *N. brevirostris* in Senegal were likely to have been *D. nigrellii*.

Parasite morphology and life cycle

Species of *Dermophthirius* are flat, leaf-like parasites (Fig. 8.2). The prohaptor is in the form of two suctorial structures opening into the buccal cavity. The haptor is a small, unarmed cup-shaped organ. The three species encountered in marine aquaria, *D. carcharhini*, *D. nigrellii*, and *D. melanopteri*, are separated by body shape and size and the morphology of the cirrus. *Dermophthirius carcharhini* is fusiform and relatively small (1.8–3.0 mm long by 0.6–1.1 mm wide) (Rand *et al.*, 1986). In contrast, *D. nigrellii* is ovoid and relatively large (2.3–4.0 mm by 2.1–3.5 mm) and has a 'catpaw' cirrus (Cheung and Ruggieri, 1983). *Dermophthirius melanopteri* is small (2.4–2.9 mm by 1.3–2.1 mm) and ovoid but has a spindle-shaped cirrus with one tall and two short tufts of fine spines (Cheung *et al.*, 1988).

Their life cycles have not been studied experimentally. Adults attach to the basal regions of the placoid scales by means of secretory adhesive substances. Suction generated by contraction of the haptoral musculature probably supplements attachment (Rand *et al.*, 1986).

Host–parasite relationships

The disease is characterized by large, localized ulcerated lesions with associated loss of placoid scales. Early in the disease, sharks swim erratically and rub against immersed objects and the bottom of the aquarium. Later, greyish patches occur and open wounds develop. Cheung *et al.* (1988) observed a white lesion on the side of the dorsal fin of a Pacific blacktip reef shark (*C. melanopterus*) held captive for 30 days. Only several specimens of *D. melanopteri* were associated with the wound.

Rand *et al.* (1986) studied the histology of the infection in *C. galapagensis* held in captivity for 3 months. The epidermis is compressed and eroded by the haptor, and goblet cells at the base of the placoid scale on which the parasite is attached are ruptured. It was suspected that parasite activity at the site loosens and dislodges the scales, causing epidermal erosion and thus provides a portal of entry for opportunistic bacteria. Grimes *et al.* (1985) isolated the gram negative bacterium *Vibrio carchariae* from *D. nigrellii*, suggesting that the parasite may serve as a mechanical vector for this pathogen.

Cheung *et al.* (1982) reduced the severity of the disease by lowering the water temperature from 22 to 19°C. Sharks flashed and rubbed less and thus did less physical damage to themselves. It was suggested that the lower water temperature reduces feeding activity of the parasite and hence causes less irritation to the host. Although the parasites were not completely eliminated at the lower temperatures the sharks tolerated the reduced parasite intensities.

The parasites apparently feed on skin tissue within an arc of the attachment site. The parasites are difficult to see without removing the sharks from water for skin scrapings (Cheung *et al.*, 1982). Overstreet (1978) reported that they are much more conspicuous when hosts have been frozen and thawed.

Prevention and control

Cheung *et al.* (1982) assess the effectiveness of various chemicals on *D. nigrellii*. Continuous exposure with 0.5 ppm Dylox (trichlorfon, 2,2,2,-trichloro-l-hydroxyethyl-phosphoric acid dimethyl ester) three times in 10 days at 22–24°C eliminated the parasites with no apparent adverse effects on the fish. Treatment prevented reinfection for 7 months. They noted that Dylox stored for 10 years lost its effectiveness. Treatment with 200 ppm formalin for 15 min at 22–24°C is effective but stresses the sharks. A 15 min dip into fresh water is ineffective.

Skin lesions and associated bacterial infections are treated separately. Cheung *et al.* (1982) obtained effective treatment with topical application of gentamycin in glycerol (500 mg per 10 ml). Success was also obtained with 1% methyl green in 70% ethanol. Wounds healed while the parasites persisted.

FAMILY CAPSALIDAE

These are ectoparasitic on the body surface of predominantly marine teleosts. The body is flat, oval and usually relatively large (Fig. 8.3). Anteriorly there is a pair of disc-like adhesive organs. The haptor is muscular and disc-like in

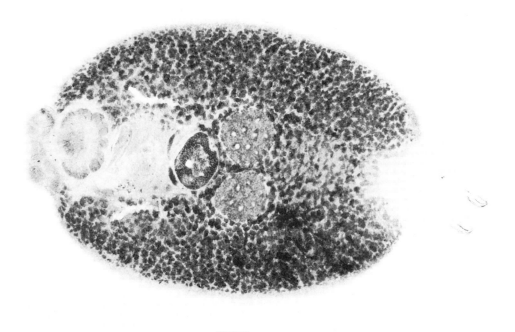

Fig. 8.3. Whole mount of *Neobenedenia melleni* from tilapia *Oreochromis mossambicus* at a facility in Hawaii. Scale bar = 0.25 mm. (Courtesy of Dr J. Kaneko; original.)

shape and sometimes subdivided by distinct septa. There are usually two pairs of hamuli and 14 peripherally located marginal hooks. Some members have secondarily lost the hamuli and marginal hooks. The mouth is ventral and associated with a muscular pharynx. The intestinal caeca are usually diverticulate and end blindly. They have typically two pairs of pigmented light receptors. One species, *Nitzschia sturionis* is known to cause extended destruction of the sturgeon fishery in the Aral Sea (Petruschevski and Schulman, 1961) while *Benedeniella* has caused disease in captive elasmobranch rays (Thoney, 1990). Members of two other genera, *Neobenedenia* Yamaguti, 1963 and *Benedenia* Diesing, 1858, commonly cause disease in marine teleosts.

Neobenedenia melleni (MacCallum, 1927)

Host range

This capsalid is an ectoparasite of numerous marine teleosts (superorder Acanthopterygii). It causes disease in tilapia mariculture (Kaneko *et al.*, 1988) and is a persistent problem with captive fish in public aquaria (Nigrelli and Breder, 1934).

Records of the parasites from feral marine hosts are rare. Nigrelli (1947) reported it from fishes in the West Indies. Baeza and Castro (1975) reported it from a carangid (*Seriola mazatlana*) in Chile. In contrast, it is known from a wide variety of teleosts held captive in public aquaria (Nigrelli, 1943).

Systematics and taxonomic position

MacCallum (1927) first described the parasite as *Epibdella melleni*. Yamaguti (1963) reestablished it as *N. melleni* when the group was revised.

Parasite morphology and life cycle

The adult (Fig. 8.3) measures 2–5 mm by 1.5–3 mm and is usually white with occasional patches of yellow. There are two anterior unarmed adhesive discs; posteriorly the haptor is a smooth ventrally directed disc that contains various sclerites which are of taxonomic importance. The anterior accessory sclerites are 21.6 µm long, while the anterior hamuli are 29.7 µm, the posterior hamuli are 11.8 µm and the marginal hooks are 9.8 µm in length (Jahn and Kuhn, 1932).

Jahn and Kuhn (1932) also studied the early development of *N. melleni*. Adults lay tetrahedral-shaped, tanned operculate eggs that bear a filament 0.8–1.2 mm long. Eggs hatch at room temperature in 5–8 days. The oncomiracidium is fusiform in shape and measures approximately 60 by 225 µm. The prepatent period is not known. Kaneko *et al.* (1988) observed large mats of eggs on the body of infected tilapia, suggesting that retroinfection of this parasite can occur.

Jahn and Kuhn (1932) noted that *N. melleni* did not survive long exposure to the acidic water of New York harbour and that the parasite multiplied rapidly in the neutral water of the aquarium. Apparently, for many years no attempt was made to control the chemical composition of the water at the New York Aquarium, and only after the installation of an efficient water treatment device did the parasites become numerous and a problem. Low pH appears to decrease effective transmission.

Host–parasite relationships

Adults are parasitic on the eyes and epidermis and sometimes within the gill and nasal cavities. Infections produce considerable injury, and if left untreated can kill the host. Young specimens occur all over the body, but adults appear to concentrate near or on the conjunctiva of the eyes. Untreated infections lead to blindness, scale loss, and the development of open wounds that expose underlying muscle and connective tissues. Secondary bacterial infection and death follows. Several thousand parasites have been found on intensely infected fish (Jahn and Kuhn, 1932).

A similar case history involves *Oreochromis mossambicus* cultured in estuarine waters in Hawaii (Kaneko *et al.*, 1988). The parasite was attached to the skin of the anterodorsal portion of the head and to the cornea of the eyes. Heavily infected fish were hyperirritable, produced copious quantities of mucus, and were discoloured (dull grey-green to black). Skin pathology included hyperaemia, multifocal petechial haemorrhages, and sloughing and loss of scales and skin. The eyes were particularly damaged, showing buphthalmos, corneal ulceration, and internal ocular degeneration. Surviving fish were scarred and blind.

Nigrelli (1935, 1937, 1947) and Nigrelli and Breder (1934) examined the host–parasite relationships between *N. melleni* and various marine teleosts held at the New York Aquarium. In several host species, infrapopulation sizes of second infections did not reach those in primary infections. Primary infections were often followed by refractory periods of variable length. Secondly, in specific hosts there was evidence of resistance involving site specific 'skin immunity' in which second infections were limited spatially to sites removed from those occupied during primary infections. Thirdly, infrapopulations developed normally on hosts that had received intraperitoneal injections of dried and freshly killed parasites and passive transfers of serum from previously exposed hosts. Infrapopulations reached different sizes on hosts of different ages.

Nigrelli identified two potential mechanisms; one was the so-called 'Ehrlichs anthreptic immunity', a supposed nutritional limitation imposed on parasites of young hosts, that resulted in reduced parasite growth and development. The other was the toxic potential of host mucus which was suspected to be directly involved somehow with the failure of parasites to establish or grow on the host surface.

Control and prevention

Infections on tilapia raised in saltwater cages (34%) have been controlled by brief freshwater dips (Kaneko *et al.*, 1988). All parasites died after a 2 minute exposure. Host fishes were not visibly stressed by the treatment.

Benedenia monticelli (Parona and Perugia, 1895)

Host range

This parasite occurs on the gills of mugilid fishes in marine waters and is known to cause disease on feral and cultured mullets (Paperna and Overstreet, 1981). Hosts include *Liza carinata*, *Crenimugil crenilabris*, *Mugil auratus*, *Mugil capito*, *Mugil subviridis*, and *Valamugil seheli* from the Mediterranean, the Red Sea, the Gulf of Elat and the Suez.

Systematics and taxonomic position

The parasite was originally described as *Phylline monticelli* but later reclassified as *B. monticelli* by Johnston (1929). *Benedenia monticelli* resembles *N. melleni* in body size and form. However, the two species are easily differentiated by the presence of a vagina in *B. monticelli* and its absence in *N. melleni* (Yamaguti, 1963).

Parasite morphology and life cycle

Benedenia monticelli attaches to the body surface particularly near the head and mouth. Heavily infected fish have eroded snouts and lesions in the mouth. They are often sluggish swimmers and lose the ability to school. Paperna *et al.* (1984) observed a positive correlation between intensity and host length. A similar correlation occurred between intensity and degree of pathology.

Intensely infected fish are emaciated and have lower condition factors than lightly infected fish. During an epizootic in a small Suez lagoon, moribund *L. carinata* (>100 mm in length) had 50 to 300 worms each.

Host–parasite relationships

Paperna and Overstreet (1981) report that capsalids feeding on the epithelium causes erosion and resulting exposure of the underlying dermis. Secondary bacterial infection is associated with inflammation and necrosis of the layer and septicaemia is usually the cause of death. Mortality of mullets infected with *B. monticelli* is often caused by perforation of the gular membrane between the lower jaw. In addition, ocular damage is associated with the infection and ensuing blindness may account for the loss of condition that develops.

Prevention and control

Paperna and Overstreet (1981) reported that infections of *B. monticelli* on *L. carinata* were controlled with a 1 h bath in 10 and 70% sea water. Formalin (15 ppm) is apparently ineffective at controlling monogenean infections on *M. cephalus* (Williams, 1972).

FAMILY DACTYLOGYRIDAE

Members of this family constitute a speciose group of predominantly gill parasites of teleosts in both fresh and salt water. All are oviparous and relatively small. Anteriorly there are a pair of pigmented light receptors and two cephalic lobes which contain secretions of adhesive gland cells. They have a sclerotized penis, the shaped of which is species-specific. They have a muscular pharynx and a tubular, posteriorly confluent intestine. The haptor has two pairs of hamuli and 14 marginal hooks (Fig. 8.4), with both types of sclerites being secondarily reduced or lost in many species.

The most important pathogenic species include members of the genus *Dactylogyrus* Diesing, 1850. They are common gill parasites of cyprinid fishes worldwide. At least seven species are known from carp (*Cyprinus carpio*). Gussev (1985) lists *D. achmerowi* Gussev, 1955; *D. anchoratus* (Dujardin, 1845); *D. crassus* Kulwiec, 1927; *D. extensus* Mueller and Van Cleave, 1932; *D. minutus* Kulwiec, 1927; *D. mrazaki* Ergens and Dulmaa, 1969; *D. vastator* Nybelin, 1924; and *D. yinwenyingae* Gussev, 1962, and provides good taxonomic keys. The two important species are *D. extensus* and *D. vastator* (Musselius, 1987). Other good taxonomic keys include that of Beverley-Burton (1984).

Dactylogyrus extensus **Mueller and Van Cleave, 1932**

Host range

This species occurs on the gills of carp and goldfish. It is considered to be less pathogenic than *D. vastator*. The parasite is endemic in Asia, central

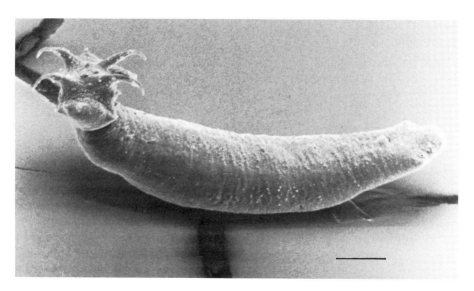

Fig. 8.4. SEM of a dactylogyrid monogenean *Urocleidus adspectus* from the gills of *Perca flavescens*. In this typical ancyrocephaline species the ventral and dorsal pairs of hamuli are well-developed. In other dactylogyrids such as *Dactylogyrus* and *Pseudodactylogyrus* the ventral and dorsal pairs of hamuli are vestigial, respectively. Scale bar = 0.1 mm. (Original SEM micrograph.)

Europe, Israel, Japan, and North America (Mueller and Van Cleave, 1932; Paperna, 1959, 1964a; Prost, 1963; Dechtiar, 1972; Imada *et al.*, 1976; Molnar, 1987). Intensity of infection is highest in the summer months.

Systematics and taxonomic position
The taxonomic status has not changed since the original species description by Mueller and Van Cleave (1932). Ogawa and Egusa (1979) redescribed previously unnoticed features of glands associated with the penis.

Parasite morphology and life cycle
The parasite is a relatively large (990–1584 μm in length and approximately 158 μm wide). It is readily identified by features of the penis and the haptoral sclerites. The hamuli measure 75–88 μm long; the penis measures 54–83 μm and has an accessory sclerite 42–71 μm long (Mizelle and Klucka, 1953).

Prost (1963) studied the life cycle of the parasite under laboratory conditions and on fish in culture ponds in Poland. Eggs laid by adults on the gills leave along with respiratory water currents and they sink. Optimal development in ponds takes 8–9 days at 16–17°C. Details of invasion are not known. At 24–25°C, postoncomiracidia mature and lay eggs in 6–7 days. Parasites on carp in Israel appear to have adapted to local conditions and have an optimal developmental temperature of 24–28°C (Paperna, 1964a).

Host–parasite relationships

This species attaches between adjacent secondary lamellae, principally on those in the middle of primary lamellae located in the ventral region of the arch (Prost, 1963). The parasite feeds on epithelial tissue covering the gill lamellae. Hyaluronidase produced during feeding is thought to contribute to tissue damage (Uspenskaya, 1962).

Young infected carp with only 15 to 35 parasites produce large amounts of mucus (Prost, 1963). Adult parasites cause localized damage to cells at the site of attachment. In instances where the haptoral sclerites penetrate deep into the lamellae there is localized degeneration and formation of necrotic foci. Cellular infiltration is limited to the site of attachment. Atrophy of gill tissue is seen in certain cases. Golovina (1976) reported that intense infections lead to reduced numbers of lymphocytes and increased numbers of monocytes and granulocytes in circulating blood.

Prevention and control

Schmahl and Mehlhorn (1985) assessed the efficiency of praziquantel (Biltricide, Bayer, Leverkusen, Germany) at eliminating infections on carp fingerlings. Infected carp were exposed for 30, 60, 90, 120, and 180 min at 22°C in water containing 0, 1, 5, 10, and 100 µg l^{-1} praziquantel. Exposure for 30 min at 1 µg l^{-1} caused irreversible vacuolization of the tegument in the region of the haptor, leading to dislodgement of the parasite from the host gills. The highest dose of praziquantel (100 µg l^{-1}) caused equilibrium problems in some fingerlings but the fish revived when placed in fresh water.

Dactylogyrus vastator Nybelin, 1924

Host range

This parasite is on the gills of carp and goldfish, and causes disease with young fish raised in ponds. It is endemic to central Asia, but has subsequently been introduced to Europe, Israel, and North America (Paperna, 1963a; Dechtiar, 1972; Molnar, 1984, 1987).

Systematics and taxonomic position

The taxonomic status has not changed since the original species description by Nybelin (1924).

Parasite morphology and life cycle

Dactylogyrus vastator is relatively small (328–388 µm long by approximately 81 µm wide) (Price and Mizelle, 1964). The hamuli are also relatively small and measure only 35–41 µm in total length. Unlike other species in the genus, it has irregularly oval eggs (Paperna, 1963a).

Paperna (1963a) studied the life cycle of *D. vastator*. During May and June the life span at 24–28°C is estimated to be 11–13 days. Development of the egg takes 2–3 days and the postoncomiracidium is sexually mature in 4–5 days. Adult worms live only 5 days. Thus, in the summer (110–130 days) there is the potential for ten parasite generations to be completed in the carp ponds.

The parasite appears to produce two types of eggs (Paperna, 1963a). One type hatches in the normal period. The other, a suspected diapause egg, has delayed development and may represent an overwintering phase which is responsible for infecting young fish in the spring.

Elevated salinity is detrimental to transmission of *D. vastator*. The oncomiracidia are unable to survive at levels greater than 1500–2000 mg Cl 1^{-1}. Adults can tolerate these levels although with reduced life span. Adults die at 6000 mg Cl 1^{-1}. Levels of O_2 and NH_4 within the ponds do not seem to affect parasite transmission and population growth.

Host–parasite relationships

Dactylogyrus vastator attaches to the gill lamellae. Infected gills undergo hyperplasia of the epithelium and mucous cells, with resulting deformation of the apices of the gill lamellae. In young carp this may involve the entire gill arch. Blood vessels and the cartilaginous support rods undergo extensive degeneration leading to respiratory failure and death of the fish. In carp larger than 35 mm, intense infections (300 + parasites) lead to tissue disruption at only the apices of the primary lamellae and thus do not kill the fish (Paperna, 1964b).

Paperna (1964c) showed that *D. vastator* competitively excludes *D. extensus* on carp gills. Carp infected with *D. extensus* and subsequently exposed to *D. vastator* lose infections of *D. extensus* while those of *D. vastator* developed normally. Histopathology in the gills of the carp include epithelial hyperplasia and this is apparently responsible for the loss of *D. extensus* and eventually *D. vastator*. After the gill surface returns to normal, the carp are again susceptible to both species. However, the fish are eventually refractive to further infections of *D. vastator* but not *D. extensus*.

Vladimirov (1971) reported that survival of larvae of *D. vastator* was reduced in serum of intensely infected two-year-old carp compared to survival in serum from naive carp. Bauer (1987) further reported that the most active antiparasitic activity confronting *D. vastator* is lysozyme. Complement and properdin are apparently less active.

The ecology of the disease has been examined in carp ponds in Israel (Paperna, 1963b). Rapidly growing carp appear to tolerate intense infections through rapid regeneration of gill tissue. However, in young fish growing slowly, regeneration of the gill tissue does not maintain itself and respiratory dysfunction is usually lethal. Fish longer than 35 mm can tolerate the infections.

Prevention and control

Sarig *et al.* (1965) prepared 50% solutions of Dipterex (*O,O*-dimethyl-l-hydroxy-2,2,2-trichlorethylphosphate) in xylene and applied it to ponds at a dilution of 0.8 ppm. They found that this dosage effectively controlled infections on carp with no deleterious effects on fingerlings of 12 mm and longer. Prophylactic treatment of two sprays (0.8 ppm) during mid-April to the end of July and 3 weeks thereafter insured against fingerling mortality. The treatments did not affect rotifers, cyclops, and chironomid larvae. Trichlorfon is apparently inefficient at controlling *D. vastator* (Goven and Amend, 1982).

Prost (1963) recommended prophylaxis through interruption of the life cycle of the parasite within carp farms. Farm spawning ponds should be drained annually to destroy potential overwintering eggs in the mud. Water from infected ponds should not be used to supply water to the rearing ponds containing uninfected fry. Yearly production of fry from the spawning ponds should be removed 3 or 4 days after hatching, before they acquire infections.

Pseudodactylogyrus Gussev, 1965

Host range

This group occurs on the gills of anguillid eels in fresh water and has caused significant losses to captive eel stocks (Buchmann *et al.*, 1987b). Infections involve two closely related species, *P. bini* (Kikuchi, 1929) and *P. anguillae* (Yin and Sproston, 1948) (see Buchmann, 1989a).

The parasites presently have widespread geographical distributions on anguillids in Eurasia and Australasia and usually occur as mixed infections. The natural hosts are thought to include only Pacific anguillids, and reports include *Anguilla japonica* in China and Japan and *A. reinhardtii* in Australia (Kikuchi, 1929; Yin and Sproston, 1948; Ogawa and Egusa, 1976; Imada and Muroga, 1977; Chan and Wu, 1984; Horiuchi *et al.*, 1988). Available evidence suggests that both species were introduced initially into the former Soviet Union along with shipments of live *A. japonica* (Buchmann *et al.*, 1987b). Infections appear to have subsequently established on *A. anguilla* and spread among feral and captive stocks of this host in Europe. Single or mixed infections are now known to occur in Hungary, France, Portugal, Denmark, Norway, Sweden, the UK, the former USSR, and probably Italy (Molnar 1983, 1984; Lambert *et al.*, 1984; Golovin, 1987; Mellergaard and Dalsgaard, 1987; Malmberg, 1989; Saraiva and Chubb, 1989). Neither *P. bini* nor *P. anguillae* were found on *Anguilla rostrata* in North America (Hanek and Threlfall, 1970; Hanek and Molnar, 1974; Crane and Eversole, 1980; Field and Eversole, 1984; Dechtiar and Christie, 1988). However, there is a new report of *P. anguilla* in Nova Scotia (Cone and Marcogliese, 1995).

Systematics and taxonomic position

Pseudodactylogyrus microrchis Ogawa and Egusa, 1976 is a junior synonym of *P. anguillae* (see Ogawa *et al.*, 1985).

Parasite morphology and life cycle

Species of *Pseudodactylogyrus* measure 1–2 mm long and 0.1–0.3 mm wide. There are four noticeable pigmented light receptors at the anterior end and frequently a single, tanned egg within the uterus. The genus has several distinct morphological features. The ventral hamuli are prominent (Fig. 8.5) and have an additional sclerite associated with each superficial root. In addition, the dorsal hamuli are small vestiges and essentially of no functional or

Fig. 8.5. SEM of *Pseudodactylogyrus bini* from the gills of *Anguilla anguilla* cultured in fresh water in Denmark. Note the well-developed ventral hamuli. Scale bar = 50 µm. (Courtesy of Dr K. Buchmann; original.)

taxonomic importance. The parasite has 14 marginal hooks in the haptor. Thus, development of the hamuli within the haptor is essentially the opposite of that of *Dactylogyrus*.

Pseudodactylogyrus bini and *P. anguillae* are easily distinguished on the basis of the shape and size of the ventral hamuli. In *P. bini* they are normally 70 µm from the tip of the superficial root to the base of the point, excluding the accessory sclerite. In *P. anguillae* they measure nearly 80 µm.

The life cycle is probably similar to that described for *Dactylogyrus*. Maximum egg production is 9–10 eggs/day at 20°C (Imada and Muroga, 1978). Developmental time is arrestd at low temperatures. Eggs hatch in 3–6 days at 22°C. Chan and Wu (1984) reported that, in the case of *P. bini*, eel mucus is necessary for hatching. The oncomiracidia of both species are shortlived (3–5 h at 20°C). Buchmann (1988a) reported that the rate of infection was higher near the water surface compared to the bottom of the tank. This suggests the oncomiracidia are negatively gravitropic and positively phototropic. The prepatent period for *P. anguillae* is 6–7 days at 28°C (Imada and Muroga, 1978).

Buchmann (1988a) studied aspects of transmission of the parasites under intense culture conditions of *A. anguilla* in Denmark. Settling of oncomiracidia was negatively correlated with abundance of free-living organisms resident in the biofilm on the tank walls. The turbellarian *Stenostomum* sp. was seen to ingest parasite eggs, and may represent a natural predator that could conceivably reduce the numbers of infective oncomiracidia in the tanks.

Host–parasite relationships

Adults attach to the gill lamellae, where the large ventral hamuli impale the secondary lamellae. They are active grazers and feed on mucus and epithelial cells; blood cells leaking from aneurysms in the lamellae may also be ingested (Buchmann *et al.*, 1987a; Buchmann, 1988b). Intensity of infection increased significantly with eel length; 200 parasites occurred on fish 10–12 cm long whereas more than 1000 was found on 40 cm fish (Buchmann, 1989b). When infections are heavy and individual parasites crowd together into groups the hamuli penetrate into tissue as deep as the cartilaginous support tissues. Intense infections lead to epithelial hyperplasia of the buccal and lamellar surfaces, increased mucus production, and physical distortion of the filaments (Chan and Wu, 1984). Although the exact cause of death has not been determined experimentally, the resultant pathology suggests the process involves respiratory failure. *Anguilla anguilla* is particularly susceptible to infections, probably because it is not a natural host of the parasites. Although the two species usually occur as mixed infections, *P. bini* tends to outcompete *P. anguillae* (Buchmann, 1988c).

Buchmann (1988a) examined the possible role of the immune system on establishment of infection by injecting naive hosts with $40\,\mu g\,g^{-1}$ of dexamethasone (fluoromethylprednisolone) and caging the fish in infected tanks. Treated eels were not particularly susceptible to invasion by oncomiracidia.

Prevention and control

Various researchers have examined the efficacy of different chemical baths in controlling the infections (Chan and Wu, 1984; Imada and Muroga, 1979). At a pH of 6–7, 0.05% ammonia reduced infections to 41%. The treatment is stressful to eels as evidenced by increased mucus secretion and respiratory activity. At 0.01% for 5 min, ammonia produced eel mortality and paralysis. However, Horiuchi *et al.* (1988) also examined the use of ammonia in the control of infections; exposure of infected *A. japonica* to 15 ppm ammonia for 18 h at 25°C killed the parasites. Japanese eels tolerated ammonia levels up to 39.5 ppm (pH 6.6 and 7.4).

Treatment with 20 ppm potassium permanganate and with 4% sodium chloride has proved ineffective at controlling infections (Chan and Wu, 1984).

Imada and Muroga (1979) and Chang and Wu (1984) reported that consecutive daily treatment of trichlorfon (Metrifonate) (0.5 ppm) eliminated the parasites. In Denmark, treatment with 1–5 ppm was ineffective against the parasites (Buchmann *et al.*, 1987b). The latter authors suspected that European populations of the parasites may have developed a resistance to the chemical.

Formalin appears to be more effective against infections. In Denmark, elvers used to stock farms are treated for 30 min in 300 ppm formaldehyde (Buchmann *et al.*, 1987b). This treatment reduces the intensity of infection but does not eliminate the parasite. The recommended schedule for an infection of 75–150 parasites per fish is four treatments of 60 ppm at 2–3 day intervals (Buchmann *et al.*, 1987b).

Schmahl and Mehlhorn (1988) report that exposure of infected *A. anguilla* to the anticoccidial drug triazinone (toltrazuril) at 5 μg ml^{-1} for 4–5 h at 20°C killed *P. bini*.

Szekely and Molnar (1987) showed that mebendazole (Vermox) is effective in controlling both *P. bini* and *P. anguillae*. Exposure of fish to 100 mg l^{-1} for 10 min killed the parasites, and was particularly efficient if the eels were first exposed to 5% sodium chloride for 2–5 min. A concentration of 1 mg l^{-1} for 24 hours produced the same results. They noted that there is a delay of six days between treatment and loss of parasites from the gills. At an intensive eel culture system in Denmark, Buchmann and Bjerregaard (1990a) found that mebendazole (1 mg l^{-1}) eradicated infections of both species of parasites within 72 h and concluded that the drug is more efficient under field conditions than in the laboratory. Buchmann and Bjerregaard (1990b) also found that the drug did not influence the physicochemical parameters of water nor the microflora and microfauna in the recirculated tanks. The latter authors concluded that the drug would probably see increased use in the industry but cautioned against indiscriminate use that might lead to tolerance by the parasites. Thiabendazole (Equizole) is ineffective at control of the parasites even at concentrations as high as 100 mg l^{-1} (Buchmann and Bjerregaard, 1990b).

Family Gyrodactylidae

Members of this family include the common viviparous monogeneans parasitic on the gills, fins, or body surfaces of teleost fishes. They have a worldwide distribution. They are relatively small and usually contain developing embryos (Fig. 8.6). Anteriorly there are two cephalic lobes and pigmented light receptors are absent. The oral opening is ventral and associated with a muscular pharynx. The intestinal caeca are simple and end blindly. The penis is bulbous and usually contains one or two rows of small spine-like sclerites. The haptor has 16 peripherally located marginal hooks and a pair of ventrally directed hamuli.

Species of the family and related oogyrodactylids have caused disease problems in all types of fish culture attempted to date (Table 8.1) Most problems involve members of the genus *Gyrodactylus* Nordmann, 1832. Concise keys and supplementary diagnostic features for most species known from Eurasia and North America are available (Gussev, 1985; Beverley-Burton, 1984).

Gyrodactylus anguillae Ergens, 1960

Host range
This is a parasite of eels (*Anguilla rostrata* and *A. anguilla*) and is associated with young hosts held in captivity (Malmberg, 1970). The natural hosts are *A. rostrata* in eastern North America (Crane and Eversole, 1989) and *A. anguilla* in Europe (Ergens, 1960; Malmberg, 1970; Orecchia *et al.*, 1987). The parasite appears to have been introduced into Israel, the former USSR, and Japan with shipments of *A. anguilla* (Paperna and Lahav, 1971; Ogawa and Egusa, 1980; Golovin, 1987).

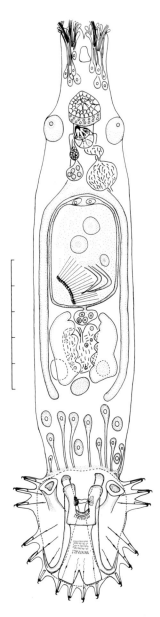

Fig. 8.6. A typical viviparous gyrodactylid *Accessorius* from a freshwater catfish in Peru. Species of this and related genera are identified on the basis of the shape and size of the hamuli and the marginal hook sickles. Scale bar divisions = 10 µm. (From Jara *et al.*, 1991.)

Systematics and taxonomic position

The taxonomy of the parasite has not changed since the original species description.

Parasite morphology and life cycle

Gyrodactylus anguillae is a small parasite. The hamuli are small and stout with well-recurved points. The following measurements are for specimens from hosts in Sweden: hamulus length 36.1 µm; root 11.7 µm; shaft 31.4 µm; point

Table 8.1. Diversity of teleost hosts reported to have disease conditions associated with gyrodactylid monogeneans.

Species	Host group	Reference
Fundulotrema prolongis	Cyprinodontidae	Cone (unpublished)
Gyrodactylus alexanderi	Gasterosteidae	Lester and Adams (1974)
Gyrodactylus anguillae	Anguillidae	Ogawa and Egusa (1980)
Gyrodactylus crysoleucas	Cyprinidae	Lewis and Lewis (1963)
Gyrodactylus derjavini	Salmonidae	Hanzelova and Zitnan (1982)
Gyrodactylus fairporti	Ictaluridae	Hoffman (1979)
Gyrodactylus macrochiri	Centrarchidae	Hoffman and Putz (1964)
Gyrodactylus katharineri	Cyprinidae	Ergens (1983a)
Gyrodactylus pleuronecti	Pleuronectidae	Cone and Wiles (1984)
Gyrodactylus salmonis	Salmonidae	Cone *et al.* (1983)
Gyrodactylus salaris	Salmonidae	Heggberget and Johnsen (1982)
Gyrodactylus stellatus	Pleuronectidae	Kamiso and Olsen (1986)
Gyrodactylus unicopula	Pleuronectidae	MacKenzie (1970)
Gyrodactylus sp.	Cottidae	Tasto (1975)
Gyrodactylus sp.	Clariidae	Amatyakul (1972)
Macrogyrodactylus polypteri	Anabantidae	Khalil (1964)
Oogyrodactylus farlowellae	Loricaridae	Harris (1983)
Paragyrodactyloides superbus	Callichthyidae	Szidat (1973)

16.7 μm. The ventral bar has no anterolateral processes, and the posteriorly directed membrane is triangular in shape and frequently difficult to see microscopically. The marginal hook sickle is 5.6–6.1 μm long, with a thin recurved point (Malmberg, 1970).

Host–parasite relationships

The species is usually on the gills. In intense infections it can be found attached to the walls of the pharynx, nares, and to the fins and skin. It appears to be specific to anguillid fishes. In Sweden the parasite was found only on elvers and it was concluded that it is a euryhaline species. The ecology of transmission is not known.

Malmberg (1970) held elvers captive in 30 l of freshwater at 4–7°C. Within a 4-week period infections had increased dramatically and most fish had died. A fungal infection was associated with the mortalities.

Prevention and control

Golovin (1987) reported treating elvers of *A. anguilla* for ichthyopthiriosis with 0.3% solution of NaCl at 21°C for 11 days. The fish were initially lightly infected with *G. anguilla* but during the salt treatment prevalence and intensity of infection of the parasite increased markedly (up to 72 parasites per fish). Twenty-two days post treatment the prevalence and intensity decreased to near extinction. This seems to indicate that the salt treatment stimulated the euryhaline parasite to multiply faster. If this is confirmed in later studies then infected fish should not be treated with salt solutions.

Gyrodactylus salmonis (Yin and Sproston, 1948)

This is a common parasite of salmonid fishes in North America (Cone *et al.*, 1983). It is likely that this was the the *Gyrodactylus* that caused disease in trout farms in the USA earlier in the century (Atkins, 1901; Pratt, 1919; Moore, 1923; Embody, 1924; Guberlet *et al.*, 1927).

Gyrodactylus salmonis is known from captive *Salmo clarki* and *Oncorhynchus kisutch* in British Columbia; *Oncorhynchus mykiss* in Arkansas, Idaho, Montana, British Columbia, and Nova Scotia; and *Salvelinus fontinalis* and *Salmo salar* in Nova Scotia; *Salmo trutta* in North Carolina (Cone *et al.*, 1983). It is also known from feral *Salmo aquabonita* in California (Heckmann, 1974), and *S. fontinalis* in New Brunswick (Frimeth, 1987) and Nova Scotia (Cone and Cusack, 1988). There is no indication that *G. salmonis* has been transferred to other continents with shipments of *O. mykiss*. The parasite occurs year round on stocks, but is most abundant in spring when water temperature is 8°C (Cone and Cusack, 1988).

Systematics and taxonomic position
Mueller (1936) described gyrodactylids collected from *Salmo clarki* as *Gyrodactylus elegans* 'variety B'. However, it is now known that *Gyrodactylus elegans* Nordmann, 1832 is a parasite of fishes of the genus *Abramis* in Eurasia (Ergens, 1966). Yin and Sproston (1948) listed Mueller's material as *Gyrodactylus elegans salmonis*. Cone *et al.* (1983) elevated Yin and Sproston's subspecies to a specific status and formally established the name *G. salmonis*.

Parasite morphology and life cycle
Gyrodactylus salmonis measures approximately 0.5 mm by 0.15 mm when flattened. The marginal hooks are set away from the perimeter of the haptor (Cone and Odense, 1984; Cone and Cusack, 1988). The hamuli are stout and 53–65 μm long. The ventral bar has small anterolateral projections and a triangular shaped posterior membrane. The marginal hook sickles are also stout, with the following dimensions: length 7–9 μm; proximal width 5–6 μm; distal width 6–7 μm.

Host–parasite relationships
The parasite attaches to the body surface, fins, buccal cavity, and occasionally the surface of the olfactory epithelium. At a farm in Nova Scotia, Cone and Cusack (1988) collected *G. salmonis* from a variety of salmonid hosts. Young-of-the-year *Salvelinus fontinalis* became infected within 8 weeks of transfer to outside rearing ponds. Intensity generally decreased with host age. Nets and buckets are potential vehicles of transmission at fish farms. Water-borne transmission occurs in spring when melting ice frequently causes flooding between holding units. Shipments of fish from other facilities are also potential sources of infection and so is the water supply of the farm.

Juvenile and adult fish do not show clinical signs of disease. Cusack and Cone (1986b) assessed the pathogenicity of *G. salmonis* with fry (2–3 cm) of *S. fontinalis*. Mortalities began 11 days postinfection at 10°C, and continued

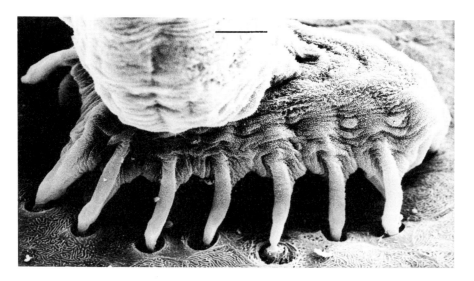

Fig. 8.7. SEM of the haptor of *Gyrodactylus salmonis* attached to the skin of *Oncorhynchus mykiss*, showing the penetration of the marginal hooks into the epithelial surface. Scale bar = 10 μm. (From Cone and Odense, 1984.)

until day 55. The rate of mortality was 1 to 3 fish per day. Throughout the experimental period, intensity gradually rose to a mean of 376 parasites per fish between days 38 to 55 postinfection. By day 60, intensity had suddenly dropped to zero.

Heavily infected fry of *S. fontinalis* are cachexic, lethargic, and overall dark in colour (Cusack and Cone, 1986b). The marginal hooks penetrate into the epithelium of the skin (Fig. 8.7) and result in a reduction in numbers of mucous cells in the skin. Moribund fish have extensive damage to the epithelium lining the renal tubules. Cusack and Cone (1986b) speculated that disruption of the skin surface leads to osmotic stress and ultimately kidney damage.

Cone and Cusack (1988) found no evidence that *G. salmonis* carried infectious pancreatic necrosis virus during an epidemic of IPNV at a trout farm.

Prevention and control

Cone and Cusack (1988) routinely checked salmonid fishes at a trout farm with a sedimentation technique: 60 yearlings were allowed to thrash about for 2 min in 4 l of 1 : 4000 formalin. The fish were removed and the dislodged *Gyrodactylus* were concentrated and preserved in formalin. This is a useful method for screening large numbers of fish.

Early reports (Embody, 1924; Guberlet *et al.*, 1927) about control of gyrodactylosis in trout farms in North America concluded that the disease is very difficult to control and that prevention is more effective. The most effective and safest treatment is to exposure infected fish to a 4.5 to 5% solution of NaCl for 1.5 to 2.5 min and then transfer immediately to fresh running water. A second bath may be necessary after several days. Treatments with acetic acid and NaOH are stressful to infected fish. Prevention includes avoiding overcrowding and maintenance of adequate feeding regimes.

Cone and Cusack (1988) studied the ecology of mixed infections of *G. salmonis* and *G. colemanensis* in a trout farm in Nova Scotia. Yearling and older *S. fontinalis* did not have signs of disease. However, the authors suspected that the parasite would be a problem to young fish in incubating troughs.

Gyrodactylus colemanensis Mizelle and Kritsky, 1967

Host range
This species is also a common parasite of salmonid fishes in North America (Cone *et al.*, 1983). It frequently occurs along with *G. salmonis* as a mixed infection. *Gyrodactylus colemanensis* was originally described from *O. mykiss* at a trout farm in California (Mizelle and Kritsky, 1967). It has since been reported from *O. mykiss* Arkansas, and insular Newfoundland; *S. truttae*, *Salvelinus namaycush* and *S. fontinalis* × *S. namaycush* in Ontario; and *S. fontinalis* in Nova Scotia and Ontario (Cone *et al.*, 1983). Frimeth (1987) reported *G. colemanensis* from feral *S. fontinalis* in New Brunswick.

Systematics and taxonomic position
Gyrodactylus colemanensis is a member of the *G. arcuatus* group, species of which are common on freshwater and coastal marine fishes. The parasite is not related to *G. salmonis* and the other typical gyrodactylid parasites of salmonids in the northern hemisphere. North America is the only region in which this lineage is known to have adapted to salmonid fishes.

Parasite morphology and life cycle
The parasite measures approximately 450 µm long and 110 µm wide. The marginal hooks are set closely to the periphery of the haptor and thus it is not pedunculate (Cone and Cusack, 1988). The haptoral sclerites are relatively small compared to those of *G. salmonis*. The hamuli are 43–51 µm long, with a root of 9–17 µm, a shaft of 32–40 µm, and a point of 19–22 µm. The ventral bar has well-developed anterolateral processes and a gradually tapered posterior membrane. The marginal hook sickles are compact, with the following dimensions; length 5–6 µm, proximal width 4–5 µm, and distal width of 4–5 µm (Cone *et al.*, 1983).

Host–parasite relationships
Cone and Cusack (1989) studied the behaviour of *G. colemanensis* on isolated fry of *S. fontinalis*. The parasite attaches anywhere on the body surface. However, the preferred site is the margin of the fins (Fig. 8.8), particularly the caudal, pectoral and pelvic fins. The parasite regularly relocates around the fin margin and frequently crosses over the body surface to another fin.

Gyrodactylus colemanensis is host specific to salmonid fishes. Cone and Cusack (1988) found that parasites from yearling *O. mykiss* readily reattached

Fig. 8.8. SEM of *Gyrodactylus colemanensis* attached to the periphery of the caudal fin and peduncle of a fry of *Oncorhynchus mykiss*. Scale bar = 1 mm. (From Cone and Wiles, 1989.)

to fry of *S. fontinalis* and developed substantial parasite populations. Similarly, parasites from fry of *S. fontinalis* developed populations on yearling *O. mykiss*.

Intense infections of *G. colemanensis* do not produce obvious clinical signs of disease (Cusack, 1986). The parasite has a delicate mode of attachment. The blade tips of the 16 marginal hooks anchor to an aggregation of microfilaments that remain in necrotic epithelial cells. No inflammatory host response occurs at the attachment site or elsewhere on the fin. Ulceration does not occur (Cone and Wiles, 1989).

Cusack (1986) showed that at 10°C, initial infections of 1–10 worms per *O. mykiss* increased to about 60 on day 22 postinfection and then declined to 0 on all fish by day 85. The parasites were numerically aggregated on individual hosts and there was a negative correlation between intensity and fish size. Infections did not influence growth or survival of the fry and produced no clinical signs of disease.

Infected *S. fontinalis* had a 50% reduction in the number of mucous cells in the epidermis covering the caudal fin in 42 days at 10°C (Wells and Cone, 1990). The changes were first detected 24 days postinfection and became increasingly pronounced during the subsequent 18 day period when parasite numbers declined drastically. It was suggested that parasite activity on the surface indirectly led to reduction in mucous cells through disruption of cell differentiation within the epidermis.

SEM and TEM revealed that bacteria were on the body surface of the parasite, in association with small microvilli projecting from the tegument (Cusack *et al.*, 1988). The bacteria were gram negative rods and resembled *Pseudomonas* and members of the Vibrionaceae. Many of the bacteria were dividing and were particularly abundant at the parasite's site of attachment. However, there was no evidence that the association facilitates bacterial invasion of adjacent host tissues.

Prevention and control
Control methods used to date would include those described above for *G. salmonis*.

Gyrodactylus crysoleucas Mizelle and Kritsky, 1967

Host range
This species parasitizes golden shiner (*Notemigonus crysoleucas*) and causes significant disease in bait fish in the southern United States (Lewis and Lewis, 1970). The parasite occurs on fish throughout the year, with an increase in winter on early spring. Epizootics develop from midwinter until late spring, and may be extended into summer if stocks are transferred to cooler water (Lewis and Lewis, 1970).

Systematics and taxonomic position
Gyrodactylus crysoleucas was originally described from golden shiners in Sacramento, California (Mizelle and Kritsky, 1967).

Many reports of *Gyrodactylus elegans* from *N. crysoleucas* in the United States may be misidentifications of *G. crysoleucas*. The author has examined specimens collected by G. Hoffman during disease outbreaks in shiner ponds throughout the southern USA and all were *G. crysoleucas*.

Parasite morphology and life cycle
Gyrodactylus crysoleucas measures 311–423 µm long by 66–137 µm wide. The hamuli are stout and measure 55–61 µm long. The marginal hook sickles are 6 µm long.

Host–parasite relationships
Eighty to 90% of the parasites occur on the body surface and fins; in intense infections they may also be on the gills (Lewis and Lewis, 1970). Intensity can

exceed 1000 parasites per fish. Parker (1965) concluded that crowding of the hosts and contact with the substrate contributes to high transmission.

Intense infections cause significant damage to the skin, with patches and strings of mucus forming on the body surface. The parasite can kill the fish directly or through secondary fungal infections (Lewis and Lewis, 1970). Infected fish frequently congregate along the edge of ponds. Mortality extends over periods of several weeks, with a few fish dying each day. Fish less than 70 mm are particularly susceptible.

Prevention and control

Formalin and potassium permanganate control the parasite but are often ineffective at complete elimination. Hess (1930) and Allison (1957) reduced parasite intensities with 4 ppm potassium permanganate. Lewis and Lewis (1963) found 25 ppm of formalin to be most effective for treatment of fish in ponds.

A treatment protocol to avoid outbreaks was recommended (Lewis and Lewis, 1963). In the autumn, ponds are treated with 27 pounds of para-formaldehyde per acre-foot of water. This minimizes the development of the parasite during the normal period of population growth. The treatment was effective and eliminated subsequent infection for 2 years. Lewis and Lewis (1970) suggested that a similar prophylactic protocol could be developed for Dylox. Meyer (1968) found that 0.25 ppm concentration of Dylox also controlled infections. A dip for 10 sec in a 350 ppm chlorine solution (sodium hypochlorite) is also effective.

Gyrodactylus ictaluri Rogers, 1967

Host range

This is a parasite of channel catfish (*Ictalurus punctatus*) in the southeastern United States. It is not considered dangerous to adult fish but may kill fry and small fingerlings in fish farms (Hoffman, 1979).

Gyrodactylus ictaluri was originally described from hosts in Florida (Rogers, 1967). It has also been reported from this host species in Mexico (Guzman *et al.*, 1988). In fish farms in Arkansas epizootics were from January through July, with the peak in moralities in April (Meyer, 1970).

Systematics and taxonomic position

The taxonomic position of the parasite has not changed since the original species description (Rogers, 1967).

Parasite morphology and life cycle

Gyrodactylus ictaluri measures 0.4–0.5 mm by 0.1 mm. The haptoral sclerites include stout, short hamuli (40–45 μm long), a ventral bar devoid of antero-lateral processes, and a distinctly rectangular-shaped posterior membrane. The marginal hook sickles have a prominent base and a thin, abruptly terminating blade.

Host–parasite relationships

The parasite occurs on the body surface and is most numerous on the barbels, underside of the head, and fins (Hoffman, 1979). It is normally not on the gills. Intense infections cause frayed fins, reddening of the head ventrally, and amputation of the barbels. Certain fish may become emaciated and die (Hoffman, 1979).

Prevention and control

There have been no specific reports on methods of control of this parasite.

Gyrodactylus katharineri Malmberg, 1964

Host range

This species parasitizes *Cyprinus carpio* throughout Eurasia and has caused disease problems in carp ponds (Ergens, 1983a). Introduction of this fish species throughout Europe, the former USSR, and North America has extended the natural Asian range of the parasite (Kollman, 1968; Malmberg, 1970; Prost, 1980; Hanzelova and Zitnan, 1982; Cone and Dechtiar, 1986; Margaritov, 1986; Miroshnichenko, 1987; Solomatova and Luzin, 1987). It has been reported from a variety of temporary hosts, most of which are cyprinid fishes (Malmberg, 1970; Ergens, 1983a). Hanzelova and Zitnan (1982) reported that *G. katharineri* on carp in Czechoslovakia was most intense during April and May when water temperature was 11°C.

Systematics and taxonomic position

Ergens (1983a) reviewed the taxonomic history of *G. katharineri*. Two species, *Gyrodactylus cyprini* Kollman, 1968 and *Gyrodactylus mizellei* Kritsky and Leiby, 1971 are junior synonyms of *G. katharineri* (see Ergens, 1983a; Cone and Dechtiar, 1986).

Parasite morphology and life cycle

Gyrodactylus katharineri is a relatively large parasite, measuring 0.6–1.1 mm long. The hamuli are also large (70–112 µm long; root 20–35 µm; shaft 51–79 µm; point 27–52 µm). The ventral bar has enormous anterolateral processes (14–34 µm) and a triangular-shaped posterior membrane 18–31 µm long. The marginal hook sickles are stout, and measure 8–10 µm long.

Host–parasite relationships

The parasite is usually found on the body surface and fins, but in cases of intense infections may be found on the gills and on walls of the buccal cavity and pharynx (Ergens, 1983a).

Young carp (fry and yearlings) are frequently and most intensely infected. Solomatova and Luzin (1987) estimated over a million parasites on large adult fish during an outbreak in the former USSR. Heavily infected fish take on a dull blue colour, lose weight and eventually die (Ergens, 1983a). The parasite prefers the scaled form of *C. carpio*; mirror carp appear resistant to heavy infections and do not exhibit clinical signs of disease (Ergens, 1983a).

The pathogenicity appears to arise from mechanical damage exerted by the haptor during normal attachment. Ergens (1983a) reported that infected fish were more susceptible to bacterial and fungal infections. It is speculated that the parasite impedes defence properties of skin mucus. The pathogenicity of *G. katharineri* is reported (Prost, 1980) to be elevated when hosts are concurrently infected with *Dactylogyrus vastator* and *D. extensus*.

Prevention and control

Solomatova and Luzin (1987) treated carp with an ammonia solution (1.5 ml ammonia l^{-1}) and formalin (1 : 2000 for 10 min). The treatments did not eradicate the parasite completely.

Gyrodactylus salaris Malmberg, 1957

Host range

This species parasitizes certain salmonid fishes in fresh water and is thought to be responsible for mortality of *Salmo salar* in Norwegian Rivers (Johnsen, 1978; Heggberget and Johnsen, 1982). There has been considerable research on the geographical distribution of the parasite in Scandinavia (Johnsen and Jensen, 1986, 1988; Halvorsen and Hartvigsen, 1989) and Eurasia in general (Ergens, 1983b). The natural distribution includes freshwaters draining into the Baltic Sea (Malmberg and Malmberg, 1986). It is suggested that the parasite was introduced into Scandinavia from the Baltics (Malmberg, 1989).

Systematics and taxonomic position

Malmberg (1957) described *G. salaris* and placed it originally in the *wagneri* species group. There is considerable confusion between this species and a morphologically similar *G. truttae* Glaser, 1974. Specimens reported to be *G. salaris* by Ergens (1961), Lucky (1963), Rehulka (1973) and Zitnan and Cankovic (1970) were apparently *G. truttae* (see Halvorsen and Hartvigsen, 1989). The report of the parasite from Scotland (Campbell, 1974) may be another misidentification.

Parasite morphology and life cycle

Ergens (1983b) redescribed *G. salaris* using specimens collected from the skin of a nine-month-old *S. salar* caught in Lake Ladoga. The parasite has relatively large and stout hamuli, and a relative small ventral bar, membrane, and anterolateral processes. The following are the dimensions for the haptoral sclerites: hamulus 61–69 µm long (root 20–21 µm, shaft 48–49 µm, point 30–37 µm); marginal hook 34–41 µm long, sickle 8–9 µm long (Ergens, 1983b). Mo and Appleby (1990) obtained haptoral sclerites of *G. salaris* by digesting the worm in artificial gastric juice (0.7 ml concentrated HCl and 0.1 g pepsin in 100 ml distilled water). The hamuli and the marginal hooks can be examined in detail using scanning electron microscopy.

Host–parasite relationships

Gyrodactylus salaris occurs on the body surface and fins of infected salmonids. Attachment by the haptor is apparently relatively superficial compared to *G. salmonis* and tissue damage is principally from feeding activity. The protrusile pharynx leaves microscopic wounds that do not heal normally (Malmberg and Malmberg, 1986). Dying *S. salar* in Norwegian rivers produced excessive amounts of mucus, were discoloured, and had *Saprolegnia* infections.

Bakke *et al.* (1990) compared the susceptibility and resistance of salmon parr from the eastern Atlantic and Baltics to *G. salmonis*. The hatchery-reared Baltic (Neva) stock had both an innate and an acquired resistance to the parasite in contrast to the highly susceptible Atlantic (Alta and Lone) stocks. Genetic differences may account for the differential effect of the parasite. The experimental data support the idea that *G. salaris* is endemic to the Baltic region and recently introduced to Norwegian stocks.

Prevention and control

Mo (1989) concluded that it is almost impossible to get rid of the parasite by chemotherapy and that contaminated holding facilities must be drained. Mo (1989) also concluded that more research is needed to develop ways of increasing survival of parr in infected rivers.

Gyrodactylus turnbulli Harris, 1986

Host range

This species is a common parasite of guppies (*Poecilia reticulata*). Epizootics involving *G. turnbulli* are common in cultured hosts and are of concern to aquarists (Scott, 1985). The parasite has been reported from *P. reticulata* imported from Singapore to England (Harris, 1986) and from New England States to Nova Scotia (Ligou *et al.*, 1991). It is also known from feral *P. reticulata* in Peru (Ligou *et al.*, 1991).

Systematics and taxonomic position

Specimens identified as *Gyrodactylus bullatarudis* Turnbull, 1955 by Scott (1982), Scott and Anderson (1984), Scott and Nokes (1984), Scott and Robinson (1984), and Madhavi and Anderson (1985), were in fact *G. turnbulli* (Harris, 1986).

Parasite morphology and life cycle

Gyrodactylus turnbulli has large stout hamuli, small anterolateral processes, and a marginal hook sickle with a large blade. Ligou *et al.* (1991) reported the following dimensions for the haptoral sclerites: hamulus 57 µm long (root 16 µm, shaft 44 µm, point 25 µm); marginal hook 32 µm long, sickle 8.5 µm long (3.5 µm wide proximally, 4 µm wide distally), handle 25 µm long.

At birth the parasite contains three generations of embryos and, at 25°C, gives birth on days 1, 2, 5, and 3. At least nine consecutive generations are produced at daily intervals without reducing the average birthrate for each generation. The parasite lives an average of 4.2 days. Population growth is temperature dependent, with a peak at 27.5°C. The parasite can survive at 17°C but not at 30°C (Scott, 1985).

Transmission of *G. turnbulli* is by contact. This occurs when fish touch each other or when fish pick at dying, intensely infected fish. Transmission is density-dependent either due to parasites on intensely infected fish actively searching for a new host or because of the increased contact between uninfected hosts and heavily infected ones. Damaged fins caused by infections attract other fish and increased the number of contacts between hosts (Scott, 1985). Successful reattachment of parasites that have fallen off a host is low.

Copulation takes place between older parasites and results in insemination of spermatozoa into a seminal vesicle (Harris, 1989). Newborns and parasites that have only recently given birth are not usually inseminated. The frequency of mating appears to be a density-dependent phenomena as it increases with crowding at peak intensities.

Host–parasite relationships

Gyrodactylus turnbulli parasitizes the skin and fins. Most occur on the caudal peduncle and caudal fin, with fewer on the pectoral, dorsal, pelvic and anal fins (Harris, 1988). However, the distribution of parasites on the host changes during an infection. During the first four days the proportion on the peduncle increases. From day 6 onward, when the numbers of parasites are declining, the proportion on the peduncle decreases. Harris (1988) suggested that this pattern reflects random colonization of the fins followed by migration to the peduncle, and subsequent dispersal during the decline of the infection.

Infected fish are cachexic, with frayed fins and pale coloration. Patches of necrotic epithelium frequently trail in streams and fish often acquire secondary infections of *Ichthyophthirius multifiliis*. Fish near death are usually fed upon by healthier guppies. Infected male fish have a reduced rate of mating display compared to that of less parasitized fish (Kennedy *et al.*, 1987).

Fish infected 1 or 2 weeks with the parasite are less suitable hosts when rechallenged than previously uninfected ones (Scott and Robinson, 1984). On these fish, percentage establishment is lower, the peak burden attained by the challenge infection is lower, and the duration of the challenge is shorter. This period of resistance lasts 2 weeks; it is lost in 4–6 weeks. The functional basis for the refractory period is not known. However, Scott (1985) suspected that it may involve increased cuticular shedding or a specific anti-gyrodactylid immune response mediated through mucus-secreted antibodies. Experiments with hybrid guppies suggest that host response to the parasite is genetically determined, with resistance being dominant over susceptibility (Madhavi and Anderson, 1985).

Prevention and control

Scott and Robinson (1984) successfully treated guppies with a weak (1 : 4000) formalin solution for 1 h. Most aquarists quarantine new stocks for several weeks after treatment with formalin before they are introduced into main holding facilities.

Schmahl and Taraschewski (1987) used a variety of novel drugs (praziquantel, niclosamide, levamisole-HCL; Bayer AG, Leverkusen, Germany for treatment of *Gyrodactylus aculeati* on *Gasterosteus aculeatus*. These may be appropriate for treatment of *G. turnbulli* on the similarly sized *P. reticulata*. Praziquantel (1 µm ml^{-1} for 90 min at 20°C) causes irreversible lesions in the tegument of the parasite. Similarly, niclosamide and levamisole-HCl are effective at 0.075–0.1 µm ml^{-1} for 90 min and 20–50 µm ml^{-1} for 120 min, respectively. However, concentration of niclosamide and levamisole-HCl have to be accurate since the drugs are toxic to fish at higher levels.

GENERAL CONCLUSIONS

Studies on diseases of fish caused by monogeneans traditionally have used descriptive epithets such as capsaliasis, dactylogyriasis, pseudodactylogyriasis, and gyrodactyliasis. These terms unfortunately do not reflect the biological diversity exhibited by individual parasite species and are of limited application. Fundamental differences in site and mode of attachment and feeding behaviour, route of invasion, and seasonal occurrence exist even among members of a single genus. In addition, species specific responses to chemical therapies also exist as do specific host tolerances to the treatments. It is now clear that we must abandon these vague descriptive terms and concentrate more on understanding the diseases caused by individual species of monogeneans.

One aspect that should be given immediate attention is establishing the true relationship between host mortality and the presence of the parasites. It is often the case that a cause-and-effect relationship is only implied and not proven. In these instances there is a need to test, under strict experimental conditions using naive hosts, the degree of pathogenicity of specific species of monogeneans. In addition, there is a need to examine carefully the role that monogeneans play in increasing the susceptibility of fish to bacterial and viral pathogens.

One can view monogeneans as analogues to ticks and lice of domestic animals: their presence on stocks at a farm is an indication of poor husbandry. As the fish culture industry expands it will be important to initiate guidelines to minimize the accidental spread of monogenean parasites with host shipments between farms. The calamities that have surrounded introduction of eel and salmon parasites into Europe and Scandinavia must be prevented.

REFERENCES

Allison, R. (1957) Some new results in the treatment of ponds to control some external parasites of fish. *Progressive Fish-Culturist* 19, 58–63.
Amatyakul, C. (1972) Parasites of pond-raised *Clarias* in Thailand. *FAO Aquaculture Bulletin* 5, 16–17.

An, L., Jara, C. and Cone, D.K. (1991) Five species of *Gyrodactylus* Nordmann, 1832 (Monogenea) from freshwater fishes of Peru. *Canadian Journal of Zoology* 69, 1199–1202.

Atkins, C.G. (1901) The study of fish diseases. *Transactions of the American Fisheries Society* 30, 82–89.

Baeza, H. and Castro, R. (1975) *Benedenia melleni* (MacCallum, 1927 an ectoparasite of Antofagasta fishes (Platyhelminthes, Monogenes). *Apuntes Oceanologicas Chile* 7, 14–22.

Bakke, T.A., Jansen, P.A. and Hansen, L.P. (1990) Differences in the host resistance of Atlantic salmon, *Salmo salar* L., stocks to the monogenean *Gyrodactylus salaris* Malmberg, 1957. *Journal of Fish Biology* 37, 577–587.

Bauer, O.N. (1987) Epizootiological significance of monogeneans. In: Skarlato, O.A. (ed.) *Investigation of Monogeneans in the USSR*. Oxonian, New Delhi, pp. 137–142.

Beverley-Burton, M. (1984) Monogenea and Turbullaria. In: Margolis, L. and Kabata, Z. (eds) *Guide to Parasites of Fishes of Canada. Part 1*. Canadian Special Publication of Fisheries and Aquatic Sciences, No. 74.

Buchmann, K. (1988a) Epidemiology of pseudodactylogyrosis in an intensive eel-culture system. *Diseases of Aquatic Organisms* 5, 81–85.

Buchmann, K. (1988b) Feeding of *Pseudodactylogyrus bini* (Monogenea) from *Anguilla anguilla. Bulletin of the European Association of Fish Pathology* 8, 79–81.

Buchmann, K. (1988c) Interactions between the gill-parasitic monogeneans *Pseudodactylogyrus anguillae* and *P. bini* and the fish host *Anguilla anguilla. Bulletin of the European Association of Fish Pathologists* 8, 98–100.

Buchmann, K. (1989a) Microhabitats of monogenean gill parasites on European eel (*Anguilla anguilla*). *Folia Parasitologica* 36, 321–329.

Buchmann, K. (1989b) Relationship between host size of *Anguilla anguilla* and the infection level of the monogeneans *Pseudodactylogyrus* spp. *Journal of Fish Biology* 35, 599–601.

Buchmann, K. and Bjerregaard, J. (1990a) Mebendazole treatment of pseudodactylogyrosis in an intensive eel-culture system. *Aquaculture* 86, 139–153.

Buchmann, K. and Bjerregaard, J. (1990b) Comparative efficacies of commercially available benzimidazoles against *Pseudodactylogyrus* infestations in eels. *Diseases of Aquatic Organisms* 9, 117–120.

Buchmann, K., Koie, M. and Prento, P. (1987a) The nutrition of the gill parasitic monogenean *Pseudodactylogyrus anguillae. Parasitology Research* 73, 532–537.

Buchmann, K., Mellergaard, S. and Koie, M. (1987b) *Pseudodactylogyrus* infections in eel: a review. *Diseases of Aquatic Organisms* 3, 51–57.

Bychowsky, B.E. (1957) Monogenetic trematodes, their systematics and phylogeny (in Russian). *Akad. Nauk USSR*. 509 pp. English translation by Washington, A.I.B.S., W.J. Hargis, D.C., Jr., (eds) 1987. *Virginia Institute of Marine Science Translation Series*, No. 1, pp. 627.

Campbell, A.D. (1974) The parasites of fish in Loch Leven. *Proceedings of the Royal Society of Edinburgh*, B 74, 347–364.

Chan, B. and Wu, B. (1984) Studies on the pathogenicity, biology and treatment of *Pseudodactylogyrus* for the eels in fish-farms. *Acta Zoologica Sinica* 30, 173–180.

Cheung, P.J. and Ruggieri, G.D. (1983) *Dermophthirius nigrellii* n.sp. (Monogenea: Microbothriidae), an ectoparasite from the skin of the lemon shark, *Negaprion brevirostris. Transactions of the American Microscopical Society* 102, 129–134.

Cheung, P.J., Nigrelli, R.F., Ruggieri, G.D. and Cilia, A. (1982) Treatment of skin lesions in captive lemon sharks, *Negaprion brevirostris* (Poey), caused by monogeneans (*Dermophthirius* sp.). *Journal of Fish Diseases* 5, 167–170.

Cheung, P.J., Nigrelli, R.F., Ruggieri, G.D. and Crow, G.L. (1988) A new microbothriid (Monogenea) causing skin lesions on the Pacific blacktip reef shark, *Carcharhinus melanopterus* (Quoy and Gaimard). *Journal of Aquariculture and Aquatic Sciences* 2, 21–25.

Cone, D.K. and Burt, M.D.B. (1985) Population biology of *Urocleidus adspectus* Mueller (Monogenea) on *Perca flavescens* in New Brunswick. *Canadian Journal of Zoology* 63, 272–277.

Cone, D.K. and Cusack, R. (1988) A study of *Gyrodactylus colemanensis* Mizelle and Kritsky, 1967 and *Gyrodactylus salmonis* (Yin and Sproston, 1948) parasitizing captive salmonids in Nova Scotia. *Canadian Journal of Zoology* 66, 409–415.

Cone, D.K. and Cusack, R. (1989) Infrapopulation dispersal of *Gyrodactylus colemanensis* (Monogenea) on fry of *Salmo gairdneri*. *Journal of Parasitology* 75, 702–706.

Cone, D.K. and Dechtiar, A.O. (1986) On *Gyrodactylus katharineri* Malmberg, 1964, *G. lotae* Gussev, 1953, and *G. lucii* Kulakovskaya, 1952 from host fishes in North America. *Canadian Journal of Zoology* 64, 637–639.

Cone, D.K. and Marcogliese, D. (1995) *Pseudodactylogyrus anguilla* (Yin and Sproston) (Monogenea) on American eel, *Anguilla rostrata*, in Nova Scotia: an endemic or an introduction? *Journal of Fish Biology* (in press).

Cone, D.K. and Odense, P.H. (1984) Pathology of five species of *Gyrodactylus* Nordmann, 1832 (Monogenea). *Canadian Journal of Zoology* 62, 1084–1088.

Cone, D.K. and Wiles, M. (1984) *Ichthyobodo necator* (Henneguy, 1883) from winter flounder, *Pseudopleuronectes americanus* (Walbaum), in the north-west Atlantic Ocean. *Journal of Fish Diseases* 7, 87–89.

Cone, D.K. and Wiles, M. (1989) Ultrastructural study of attachment of *Gyrodactylus colemanensis* (Monogenea) to fins of fry of *Salmo gairdneri*. *Proceedings of the Helminthological Society of Washington* 56, 29–32.

Cone, D.K., Beverley-Burton, M., Wiles, M. and MacDonald, T.E. (1983) The taxonomy of *Gyrodactylus* (Monogenea) parasitizing certain salmonid fishes of North America, with a description of *Gyrodactylus nerkae* n.sp. *Canadian Journal of Zoology* 61, 2587–2597.

Cone, D.K., Gratzek, J.B. and Hoffman, G.L. (1987) A study of *Enterogyrus* sp. (Monogenea) parasitizing the foregut of captive *Pomacanthus paru* (Pomacanthidae) in Georgia. *Canadian Journal of Zoology* 65, 312–316.

Crane, J.S. and Eversole, A.G. (1980) Ectoparasitic fauna of glass eel and elver stages of American eel (*Anguilla rostrata*). *Proceedings of the World Mariculture Society* 11, 275–280.

Crane, J.S. and Eversole, A.G. (1989) Metazoan ectoparasitic fauna of American eels from brackish water. *Proceedings of the Annual Conference of the Southeastern Association of Fish and Wildlife Agencies* 39, 248–254.

Cusack, R. (1986) Development of infections of *Gyrodactylus colemanensis* Mizelle and Kritsky, 1967 (Monogenea) and their effect on fry of *Salmo gairdneri* Richardson. *Journal of Parasitology* 72, 663–668.

Cusack, R. and Cone, D.K. (1985) A report of bacterial microcolonies on the surface of *Gyrodactylus* (Monogenea). *Journal of Fish Diseases* 8, 125–127.

Cusack, R. and Cone, D.K. (1986a) *Gyrodactylus salmonis* (Yin and Sproston, 1948) parasitizing fry of *Salvelinus fontinalis* (Mitchill). *Journal of the Wildlife Disease Association* 22, 209–213.

Cusack, R. and Cone, D.K. (1986b) A review of parasites as vectors of viral and bacterial diseases of fish. *Journal of Fish Diseases* 9, 169–171.

Cusack, R., Rand, T. and Cone, D.K. (1988) A study of bacterial microcolonies associated with the body surface of *Gyrodactylus colemanensis* Mizelle and Kritsky, 1967 (Monogenea) parasitizing *Salmo gairdneri* Richardson. *Journal of Fish Diseases* 11, 271–274.

Dechtiar, A.O. (1972) New parasite records for Lake Erie fish. *Great Lakes Fisheries Commission Technical Report* 17, 20 pp.

Dechtiar, A.O. and Christie, W.J. (1988) Survey of the parasitic fauna of Lake Ontario fishes, 1961-1971. In: Nepszy, S.J. (ed.) *Parasites of Fishes in the Canadia Waters of the Great Lakes.* Great Lakes Fishery Commission Technical Report 51, pp. 66-95.

Embody, G.C. (1924) Notes on the control of *Gyrodactylus* on trout. *Transactions of the American Fisheries Society* 54, 48-53.

Ergens, R. (1960) The helminthofauna of some Albanian fishes. *Ceskoslovenska Parazitologie* 7, 49-90 (in Russian).

Ergens, R. (1961) Zwei weitere Befunde der *Gyrodactylus* art (Monogenoidea) ans der Tschechoslowakei. *Acta Societatis Zoologicae Bohemslov* 25, 25-27.

Ergens, R. (1966) Revision of the helminthofauna of fishes from Czechoslovakia. IV. Group of the species of *Gyrodactylus elegans* Nordmann, 1832 (Monogenoidea). *Folia Parasitologica* 13, 123-126.

Ergens, R. (1983a) A survey of the results of studies on *Gyrodactylus katharineri* Malmberg, 1964 (Gyrodactylidae: Monogenea). *Folia Parasitologica* 30, 319-327.

Ergens, R. (1983b) *Gyrodactylus* from Eurasian freshwater Salmonidae and Thymallidae. *Folia Parasitologica* 30, 15-26.

Euzet, L. and Maillard, C. (1967) Parasites de poissons de mer ouest Africains, recoltes par J. Cadenat. VI. Monogenes de Salaciens. *Bulletin Institut Fondamental Afrique Noire-Serie A Sciences Naturelles* 4, 1435-1493.

Faisal, M. and Imam, E.A. (1990) *Microcotyle chrysophrii* (Monogenea: Polyopisthocotylea), a pathogen for cultured and wild gilthead seabream, *Sparus aurata.* In: Perkins, F.O. and Cheng, T.C. (eds) *Pathology in Marine Science.* Academic Press, New York, pp. 283-290.

Field, D.W. and Eversole, A.G. (1984) Parasites of cultured eels. *World Mariculture Society* 15, 326-332.

Fischthal, J.H. and Allison, L.N. (1942) *Acolpenteron catostomi,* n.sp. (Gyrodactyloidea: Calceostomatidae), a monogenetic trematode from the ureters of suckers, with observations on its life history and that of *A. ureteroecetes. Transactions of the American Microscopical Society* 61, 53-56.

Frankland, H.M.T. (1955) The life history and bionomics of *Diclidophora denticulata* (Trematoda: Monogenea). *Parasitology* 45, 313-351.

Frimeth, J.P. (1987) A survey of the parasites of nonanadromous and anadromous brook char (*Salvelinus fontinalis*) in the Tabusintac River, New Brunswick, Canada. *Canadian Journal of Zoology* 65, 1354-1362.

Golovin, P.P. (1987) Monogeneans of the eel cultured in warm waters. In: Skarlato, O.A. (ed.) *Investigations of Monogeneans in the USSR.* Oxonian, New Delhi, pp. 152-158.

Golovina, N.A. (1976) The changes in the white blood cells of carp during the infection with *Dactylogyrus extensus* (Monogenoidea, Dactylogyridae) in the light of a new classification of blood components (in Russian). *Parazitologiya* 10, 178-182.

Goven, B.A. and Amend, D.F. (1982) Mebendazole/trichlorfon combination: a new antihelminthic for removing monogenetic trematodes from fish. *Journal of Fish Biology* 20, 373-378.

Grimes, D.J., Gruber, S.H. and May, E.B. (1985) Experimental infection of lemon sharks, *Negaprion brevirostris* (Poey), with *Vibrio* species. *Journal of Fish Diseases* 8, 173-180.

Guberlet, J.E., Hansen, G.W. and Kavanagh, J.A. (1927) Studies on the control of *Gyrodactylus. University of Washington Publication in Fisheries* 2, 17-29.

Gussev, A.V. (1985) Multicellular parasites (part one). In: Bauer, O.N. (ed.) *Handbook for Identifying Parasites of Fish of the Fauna of the USSR.* 'Hayka', Leningrad.

Guzman, F.J., Silva, L.G., Salinas, F.S., Fernandez, H.C. and Ebeling, P.W. (1988) *Parasitos y Enfermedades del Bagre (Ictalurus).* Secretaria de Pesca, Mexico, 216 pp.

Halvorsen, O. and Hartvigsen, R. (1989) A review and biogeography and epidemiology of *Gyrodactylus salaris. Norsk Institutt for Naturforskning Utedning* 2, 1–41.

Hanek, G. and Threlfall, W. (1970) Metazoan parasites of the American eel (*Anguilla rostrata* (LeSueur) in Newfoundland and Labrador. *Canadian Journal of Zoology* 48, 597–600.

Hanek, G. and Molnar, K. (1974) Parasites of freshwater and anadromous fishes from Matamek River system, Quebec. *Canadian Journal of Zoology* 31, 1135–1139.

Hanzelova, V. and Zitnan, R. (1982) The seasonal dynamics of the invasion cycle of *Gyrodactylus katharineri* Malmberg, 1964 (Monogenea). *Helminthologia* 19, 257–265.

Harris, P.D. (1983) The morphology and life-cycle of the oviparous *Oogyrodactylus farlowellae* gen. et sp. nov. (Monogenea, Gyrodactylidea). *Parasitology* 87, 405–420.

Harris, P.D. (1986) Species of *Gyrodactylus* von Nordmann, 1832 (Monogenea Gyrodactylidae) from poeciliid fishes, with a description of *G. turnbulli* sp. nov. from the guppy, *Poecilia reticulata* Peters. *Journal of Natural History* 20, 183–191.

Harris, P.D. (1988) Changes in the site specificity of *Gyrodactylus turnbulli* Harris, 1986 (Monogenea) during infections of individual guppies (*Poecilia reticulata* Peters, 1859). *Canadian Journal of Zoology* 66, 2854–2857.

Harris, P.D. (1989) Interactions between population growth and sexual reproduction in the viviparous monogenean *Gyrodactylus turnbulli* Harris, 1986 from the guppy, *Poecilia reticulata* Peters. *Parasitology* 98, 245–251.

Heckmann, R.A. (1974) Parasites of golden trout from California. *Journal of Parasitology* 60, 363.

Heggberget, T. and Johnsen, B.O. (1982) Infestations of *Gyrodactylus* sp. of Atlantic salmon *Salmo salar* L. in Norwegian rivers. *Journal of Fish Biology* 21, 15–26.

Hess, W.N. (1930) Control of external fluke parasites on fish. *Journal of Parasitology* 16, 131–136.

Hoffman, G.L. (1970) Intercontinental and transcontinental dissemination and trans-faunation of fish parasites with emphasis on whirling disease (*Myxosoma cerebralis*) In: Snieszko, S.F. (ed.) *A Symposium on Disease of Fishes and Shell-fishes.* Special publication of the American Fisheries Society No. 5, pp. 69–81.

Hoffman, G.L. (1979) Helminthic parasite. In: Plumb, J.A. (ed.), *Principal Diseases of Farm-raised Catfish. Southern Cooperative Series* No. 225, pp. 40–58.

Hoffman, G.L. and Putz, R.E. (1964) Studies on *Gyrodactylus macrochiri* n.sp. (Trematoda: Monogenea) from *Lepomis macrochirus. Proceedings of the Helminthological Society of Washinhton* 31, 76–82.

Horiuchi, M., Kuwahara, A., Souma, T. and Nakata, M. (1988) Availability of long-hour bathing in ammonia water for the control of pseudodactylogyrosis in cultured eels. *Suisanzoshoku* 35, 259–263.

Imada, R. and Muroga, K. (1977) *Pseudodactylogyrus microrchis* (Monogenea) on the gills of cultured eels – I. Seasonal changes in abundance. *Bulletin of the Japanese Society of Scientific Fisheries* 43, 1397–1401.

Imada, R. and Muroga, K. (1978) *Pseudodactylogyrus microrchis* (Monogenea) on the gills of cultured eels – II. Oviposition, hatching and development on the host. *Bulletin of the Japanese Society of Scientific Fisheries* 44, 571–576.

Imada, R. and Muroga, K. (1979) *Pseudodactylogyrus microrchis* (Monogenea) on the gills of cultured eels – III. Experimental control by trichlorfon. *Bulletin of the Japanese Society of Scientific Fisheries* 45, 25–29.

Imada, R., Muroga, K. and Hirabayashi, S. (1976) *Dactylogyrus extensus* (Monogenoidea) from cultured carp in Japan. *Bulletin of the Japanese Society of Scientific Fisheries* 42, 153–158. (In Japanese).

Jahn, T.L. and Kuhn (1932) The life history of *Epibdella melleni* MacCallum, 1927, a monogenetic trematode parasitic on marine fishes. *Biological Bulletin* 62, 89–111.

Jara, C., An, L. and Cone, D.K. (1991) *Accessorius peruensis* gen. et sp. n. (Monogena: Gyrodactylidea) from *Lebiasina bimaculata* (Characidae) in Peru. *Journal of the Helminthological Society of Washington* 58, 164–166.

Johnsen, B.O. (1978) The effect of an attach by the parasite *Gyrodactylus salaris* on the population of salmon parr in the river Lakselva, Misvaer in northern Norway. *Astarte* 11, 7–9.

Johnsen, B.O. and Jensen, A.J. (1986) Infestation of Atlantic salmon, *Salmo salar*, by *Gyrodactylus salaris* in Norwegian rivers. *Journal of Fish Biology* 29, 233–241.

Johnsen, B.O. and Jensen, A.J. (1988) Introduction and establishment of *Gyrodactylus salaris* Malmberg 1957, on Atlantic salmon (*Salmo salar* L.) fry and parr in River Vetsna, Norway. *Journal of Fish Diseases* 11, 35–45.

Johnston, T.H. (1929) Remarks on the synonymy of certain tristomatid trematode genera. *Transactions of the Royal Society of Southern Australia* 53, 71–78.

Kamiso, H.N. and Olson, R.E. (1986) Host–parasite relationships between *Gyrodactylus stellatus* (Monogenea: Gyrodactylidae) and *Parophrys vestulus* (Pleuronectidae-English sole) from coastal waters of Oregon. *Journal of Parasitology* 72, 125–129.

Kaneko II, J.J., Yamada, R., Brock, J.A. and Nakamura, R.N. (1988) Infection of tilapia, *Oreochromis mossambicus* (Trewavas), by a marine monogenean, *Neobenedenia melleni* (MacCallum, 1927) Yamaguti, 1963 in Kaneohe Bay, Hawaii, USA, and its treatment. *Journal of Fish Diseases* 11, 295–300.

Kennedy, C.E.J., Endler, J.A. Poynton, S.L. and McMinn, H. (1987) Parasite load predicts mate choice in guppies. *Behavioural Ecology and Sociobiology* 21, 291–295.

Khalil, L.F. (1964) On the biology of *Macrogyrodactylus polypteri* Malmberg, 1956, a monogenetic trematode on *Polypterus senegalus* in the Sudan. *Journal of Helminthology* 38, 219–222.

Kikuchi, H. (1929) Two new species of Japanese trematodes belong to the family Gyrodactylidae. *Annotationes Zoologicae Japonensis* 12, 175–186.

Kohn, A. (1990) *Kritskyia moraveci* n.g., n.sp. (Monogenea, Dactylogyridae) from the urinary bladder and ureters of *Rhamdia quelen* (Quoy and Gaimard, 1824) (Pisces, Pimelodidae) in Brazil. *Systematic Parasitology* 17, 81–87.

Kollman, A. (1968) *Gyrodactylus cyprini* n.sp. an *Cyprinus carpio* L. mit einer Bemerkung uber die Mechanik der Randhaken. *Zoologischer Anzeiger* 180, 36–42.

Kubota, S.S. and Takakuwa, M. (1963) Studies on the diseases of marine cultured fishes. I. General description and preliminary discussion of fish in Mie Prefecture. *Journal of the Faculty of the Fisheries, University Mie-Tsu* 6, 107–124 (in Japanese).

Lambert, A., Le Brun, N. and Pariselle, A. (1984) Presence en France de *Pseudodactylogyrus anguillae* (Yin and Sproston, 1948) Gussev, 1965 (Monogenea, Monopisthocotylea, Pseudodactylogyridae n.fam.). *Annales de Parasitologie Humaine et Comparee* 60, 91–92.

Lester, R.J.G. and Adams, J.R. (1974) *Gyrodactylus alexanderi*; reproduction, mortality and effect on its host *Gasterosteus aculeatus. Canadian Journal of Zoology* 52, 827–833.

Lewis, W.M. and Lewis, S.D. (1963) Control of epizootics of *Gyrodactylus elegans* in golden shiner populations. *Transactions of the American Fisheries Society* 92, 60–62.

Lewis, W.M. and Lewis, S.D. (1970) *Gyrodactylus wageneri*-group, its occurrence, importance, and control in the commercial production of the golden shiner. In: Snieszko, S. (ed.) *A Symposium on Diseases of Fishes and Shellfishes*. Special Publication of the American Fisheries Society, Washington, DC, pp. 174–176.

Liguo, A., Jara, C.A. and Cone, D.K. (1991) Five species of *Gyrodactylus* Nordmann, 1832 (Monogenea) from freshwater fishes of Peru. *Canadian Journal of Zoology* 69, 1199–1202.

Lucky, Z. (1963) Fund der Art *Gyrodactylus salaris* Malmberg 1956, auf der Haut der Regenbogenforelle. *Sbornik Vysoke Skolyske Zemedelske Brne* 11, 127–130.

MacCallum, G.A. (1926) Deux nouveaux parasites *Carcharhinus commersoni: Philura orata* et *Dermophthirius carcharhini. Annales de Parasitologie Humaine et Comparee* 4, 162–171.

MacCallum, G.A. (1927) A new ectoparasite trematode, *Epibdella melleni* sp. nov. *Zoopathologica* 1, 291–300.

MacKenzie, K. (1970) *Gyrodactylus unicopula* Glukhova, 1955, from young plaice *Pleuronectes platessa* L. with notes on the ecology. *Journal of Fish Biology* 2, 23–34.

Madhavi, M. and Anderson, R.M. (1985) Variability in the susceptibility of the fish host, *Poecilia reticulata*, to infection with *Gyrodactylus bullatarudis* (Monogenea). *Parasitology* 91, 531–544.

Malmberg, G. (1957) On the occurrence of *Gyrodactylus* on Swedish fishes (In Swedish). *Sodra Sveriges Fiskeriforening, Arsskrift* 1956, 19–76.

Malmberg, G. (1970) The excretory systems and the marginal hooks as a basis for the systematics of *Gyrodactylus* (Trematoda, Monogenea). *Arkiv foer Zoologi* 2, 1–235.

Malmberg, G. (1989) Salmonid transports, culturing and *Gyrodactylus* infections in Scandinavia. In: *Parasites of Freshwater Fishes of North-west Europe*. Proceedings of the International Symposium within the program of the Soviet-Finnish Cooperation, 10–14 January 1988, Petrozavosk, USSR.

Malmberg, G. and Fernholm, B. (1989) *Myxinidocotyle* gen. n. and *Lophocotyle* Braun (Platyhelminthes, Monogenea, Acanthocotylidae) with descriptions of three new species from hagfishes (Chordata, Myxinidae). *Zoologica Scripta* 18, 187–204.

Malmberg, G. and Malmberg, M. (1986) *Gyrodactylus* in salmon and rainbow trout farms. In: Stenmark, A. and Malmberg, G. (eds), *Parasites and Diseases in Natural Waters and Aquaculture in Nordic Countries*. Proceedings of a Zoo-Tax-Symposium, Stockholm, December 2–4, 1986.

Margaritov, N. (1986) Investigation of Gyrodactylidae (Monogenea) in the breeding material of common carp under condition of intensive breeding in Bulgaria. *Godisiahnik na Sofiiskiya Universitet, Biologicheski Fakultet* 77, 64–71 (in Russian).

Meyer, F.P. (1968) Dylox as a control for ectoparasites of fish. Proceedings 22nd Annual Conference Southeast Association of Game and Fish Commissioners, 5 pp.

Meyer, F.P. (1970) Seasonal fluctuation in the incidence of disease on fish farms. In: Snieszko, S.F. (ed.) *A Symposium on Diseases of Fishes and Shellfishes*. Special publication of the American Fisheries Society No. 5, pp. 21–29.

Mellergaard, S. and Dalsgaard, I. (1987) Disease problems in Danish eel farms. *Aquaculture* 67, 139–146.

Miroshnichenko, A.I. (1987) Monogeneans of fresh-water fishes of Crimea. In: Skarlato, O.A. (ed.) *Investigations of Monogeneans in the USSR.* Oxonian Press, New Delhi, pp. 105–114.

Mizelle, J.D. and Klucka, A.R. (1953) Studies on monogenetic trematodes. XIV. Dactylogyridae from Wisconsin fishes. *American Midland Naturalist* 49, 720–733.

Mizelle, J.D. and Kritsky, D.C. (1967) Studies on monogenetic trematodes. XXXIII. New species of *Gyrodactylus* and a key to the North American species. *Transaction of the American Microscopical Society* 86, 390–401.

Mo, T.A. (1989) Parasites of the genus *Gyrodactylus* cause problems in fish culture and fish management. *Norsk Veterinaertidsskrift* 101, 523–527.

Mo, T.A. and Appleby, C. (1990) A special technique for studying haptoral sclerites of monogeneans. *Systematic Parasitology* 17, 103–108.

Molnar, K. (1983) Das Vorkommen von parasitaren Hakensaugwurmern bei der Aslaufzucht in Ungarn. *Zeitschrift für Parasitenkunde* 30, 341–345.

Molnar, K. (1984) Occurrence of new monogeneans of Far east origin on the gills of fishes in Hungary. *Acta Veternaria Hungarica* 32, 153–157.

Molnar, K. (1987) First record of a common carp parasite, *Dactylogyrus molnari* Ergens and Dulma, 1969 (Monogenea) in Hungary. *Parasitologia Hungarica* 20, 41–43.

Moore, E. (1923) Diseases of fish in state hatcheries. *12th Annual Report of the New York State Conservation Commission* (1922), 66–79.

Mueller, J.F. (1936) Studies on the North American Gyrodactyloidea. *Transactions of the American Microscopical Society* 55, 55–72.

Mueller, J.F. and Van Cleave, J. (1932) Parasites of Oneida Lake fishes. Part II. Descriptions of new species and some general taxonomic considerations, especially considering the trematode family Heterophydiae. *Roosevelt Wild Life Annals* 3, 79–137.

Munro, A.L.S. (1982) The pathogenesis of bacterial diseases of fishes. In: Roberts, R.J. (ed.) *Microbial Diseases of Fishes.* Academic Press, London, pp. 115–149.

Musselius, V.A. (1987) Monogeneans of fish farms and their importance in modern methods of pisciculture. In: Skarlato, O.A. (ed.) *Investigations of Monogeneans in the USSR.* Oxonian, New Delhi, pp. 143–151.

Nigrelli, R.F. (1935) On the effect of fish mucous on *Epibdella melleni*, a monogenetic trematode of marine fish. *Journal of Parasitology* 21, 438.

Nigrelli, R.F. (1937) Further studies on the susceptibility and acquired immunity of marine fishes to *Epibdella melleni*, a monogenetic trematode. *Zoologica* 22, 185–192.

Nigrelli, R.F. (1943) Causes of diseases and death of fishes in captivity. *Zoologica* 28, 203–216.

Nigrelli, R.F. (1947) Susceptibility and immunity of marine fishes to *Benedenia* (= *Epibdella*) *melleni* (MacCallum), a monogenetic trematode, III. Natural hosts in the West Indies. *Journal of Parasitology* 33, 25.

Nigrelli, R.F. and Breder, C.M. (1934) The susceptibility and immunity of certain marine fishes to *Epibdella melleni*, a monogenetic trematode. *Journal of Parasitology* 20, 259–269.

Nybelin, O. (1924) *Dactylogyrus vastator* n.sp. *Arkiv foer Zoologi* 16, 1–2.

Ogawa, K. and Egusa, S. (1976) Studies on eel pseudodactylogyrosis – I. Morphology and classification of three eel dactylogyrids with a proposal of a new species, *Pseudodactylogyrus microrchis. Bulletin of the Japanese Society of Scientific Fisheries* 51 381–385.

Ogawa, K. and Egusa, S. (1979) Redescription of *Dactylogyrus extensus* (Monogenea:

Dactylogyridae) with a special reference to its male terminalia. *Japanese Journal of Parasitology* 28, 121–124.

Ogawa, K. and Egusa, S. (1980) *Gyrodactylus* infestations of cultured eels (*Anguilla japonica* and *A. anguilla*) in Japan. *Fish Pathology* 15, 95–99.

Ogawa, K., Chung, H.Y., Kou, G.H. and Imada, R. (1985) On the validity of an eel monogenean *Pseudodactylogyrus microrchis* Ogawa and Egusa, 1976. *Bulletin of the Japanese Society of Scientific Fisheries* 51, 381–385.

Orecchia, P., Bianchini, M., Catalin, N., Cataudella, S. and Paggi, L. (1987) Parasitological study of a population of Tiber River eels (*Anguilla anguilla*). *Parassitologia* 29, 37–47.

Overstreet, R. (1978) *Marine Maladies? Worms, Germs, and Other Symbionts from the Northern Gulf of Mexico*. Blossman Printing, Ocean Springs.

Paperna, I. (1959) Studies on monogenetic trematodes in Israel. I. Three species of monogenetic trematodes of reared carp. *Bamidgeh* 3, 51–67.

Paperna, I. (1963a) Some observations on the biology and ecology of *Dactylogyrus vastator* in Israel. *Bamidgeh* 15, 8–29.

Paperna, I. (1963b) Dynamics of *Dactylogyrus vastator* Nybelin (Monogenea) populations on the gills of carp fry in fish ponds. *Bamidgeh* 15, 31–50.

Paperna, I. (1964a) Adaptation of *Dactylogyrus extensus* (Mueller and Van Cleave, 1932) to ecological conditions of artificial ponds in Israel. *Journal of Parasitology* 50, 90–93.

Paperna, I. (1964b) Host reaction to infection of carp with *Dactylogyrus vastator* Nybelin, 1924 (Monogenea). *Bamidgeh* 16, 129–141.

Paperna, I. (1964c) Competitive exclusion of *Dactylogyrus extensus* by *Dactylogyrus vastator* (Trematoda, Monogenea) on the gills of reared carp. *Journal of Parasitology* 50, 94–98.

Paperna, I. and Lahav, M. (1971) New records and further data on fish parasites in Israel. *Bamidgeh* 23, 43–52.

Paperna, I. and Overstreet, R.M. (1981) Parasites and diseases of mullets (Mugilidae). In Oren, O.H. (ed.) *Aguaculture of Grey Mullets*. International Programme 26, Cambridge University Press, pp. 411–493.

Paperna, I., Diamont, A. and Overstreet, R.M. (1984) Monogenean infestations and mortality in wild and cultivated Red Sea fishes. *Helgolander Wissenschaftliche Meeresuntersuchungen* 37, 445–462.

Parker, J.D. (1965) Seasonal occurrence, transmission, and host specificity of the monogenetic trematode *Gyrodactylus elegans* from the golden shiner (*Notemigonus crysoleucas*). Unpublished PhD Thesis, Southern Illinois University.

Petrushevski, G.K. and Shulman, S.S. (1961) The parasitic diseases of fishes in natural waters of the USSR. In: Dogiel, V.A., Petrushevski, G.K. and Polyanski, Yu,I. (eds) *Parasitology of Fishes*. Oliver and Boyd, pp. 299–319.

Pratt, H.S. (1919) Parasites of fresh-water fishes. Comprising some general considerations. *United States Bureau of Fisheries Economics Circular* 42, 1–10.

Price, C.E. (1938) North American monogenetic trematodes. III. The families Monocotylidae, Microbothriidae, Acanthocotylidae and Udonellidae (Capsaloidea). *Journal of the Washington Academy of Sciences* 28, 183–198.

Price, C.E. and Mizelle, J.D. (1964) Studies on monogenetic trematodes. XXVI. Dactylogyrinae from California with the proposal of a new genus, *Pellucidhaptor*. *Journal of Parasitology* 50, 572–578.

Prost, M. (1963) Investigations on the development and pathogenicity of *Dactylogyrus anchoratus* (Duj., 1845) and *D. extensus* Mueller and v. Cleave, 1952 for breeding carps. *Acta Parasitologica Polonica* 11, 17–47.

Prost, M. (1980) Fish Monogenea of Poland. V. Parasites of the carp, *Cyprinus carpio* L. *Acta Parasitologica Polonica* 27, 125–131.

Putz, R.E. and Hoffman, G.L. (1964) Studies on *Dactylogyrus corporalis* n.sp. (Trematoda: Monogenea) from the fallfish *Semotilus corporalis*. *Proceedings of the Helminthological Society of Washington* 31, 139–143.

Rand, T.G., Wiles, M. and Odense, P. (1986) Attachment of *Dermophthirius carcharhini* (Monogenea: Microbothriidae) to the Galapagos shark *Carcharhinus galapagensis*. *Transactions of the American Microscopical Society* 105, 158–169.

Rehulka, J. (1973) Remarks on the occurrence of *Gyrodactylus salaris* Malmberg 1957 *sensu* Ergens 1961 (Monogenoidea: Gyrodactylidae). *Acta Zoologica Bohemslov* 37, 293–295.

Rogers, W.A. (1967) Six new species of *Gyrodactylus* (Monogenea) from the southeastern US. *Journal of Parasitology* 53, 747–751.

Rohde, K. (1984) Diseases caused by metazoans: helminths. In: Kinne, O. (ed.) *Diseases of Marine Animals. Vol. IV, Part 1. Introduction. Pisces.* Biologische Anstalt Halgoland, Hamburg, pp. 193–320.

Saraiva, A. and Chubb, J.C. (1989) Preliminary observations on the parasites of *Anguilla anguilla* (L.) from Portugal. *Bulletin of the European Association of Fish Pathologists* 9, 88–89.

Sarig, S., Lahav, M., and Shilo, M. (1965) Control of *Dactylogyrus vastator* on carp fingerlings with Dipterex. *Bamidgeh* 17, 47–52.

Sattaur, O. (1988) Parasites prey on wild salmon in Norway. *New Scientist* 120, 21.

Schmahl, G. and Mehlhorn, H. (1985) Treatment of fish parasites. 1. Praziquantel effective against Monogenea (*Dactylogyrus vastator, Dactylogyrus extensus, Diplozoon paradoxum*). *Zeitschrift für Parasitenkunde* 71, 727–737.

Schmahl, G. and Mehlhorn, H. (1988) Treatment of fish parasites. 4. Effects of sym. triazinone (toltrazuril) on Monogenea. *Parasitology Research* 75, 132–143.

Schmahl, G. and Taraschewski, H. (1987) Treatment of fish parasites. 2. Effects of praziquantel, niclosamide, levamisole-HCL, and Metrifonate on Monogenea (*Gyrodactylus aculeati, Diplozoon paradoxum*). *Parasitology Research* 73, 341–351.

Scott, M.E. (1982) Reproductive potential of *Gyrodactylus bullatarudis* (Monogenea) on guppies (*Poecilia reticulata*). *Parasitology* 85, 217–236.

Scott, M.E. (1985) Experimental epidemiology of *Gyrodactylus bullatarudis* (Monogenea) on guppies (*Poecilia reticulata*): short- and long-term studies. In: Rollinson, D. and Anderson, R.M. (eds) *Ecology and Genetics of Host-Parasite Interactions.* Academic Press, London, pp. 21–38.

Scott, M.E. and Anderson, R.M. (1984) The population dynamics of *Gyrodactylus bullatarudis* (Monogenea) within laboratory populations of the fish host *Poecilia reticulata*. *Parasitology* 89, 159–194.

Scott, M.E. and Nokes, D.J. (1984) Temperature-dependent reproduction and survival of *Gyrodactylus bullatarudis* (Monogenea) on guppies (*Poecilia reticulata*). *Parasitology* 89, 221–227.

Scott, M.E. and Robinson, M.A. (1984) Challenge infections of *Gyrodactylus bullatarudis* (Monogenea) on guppies, *Poecilia reticulata* (Peters), following treatment. *Journal of Fish Biology* 24, 581–586.

Solomatova, V.P. and Luzin, A.V. (1987) Gyrodactylosis of carps in fish tanks located on discharged waters of the Kostromsk electric power plant and some problems of the biology of *Gyrodactylus katharineri*. In: Skarlato, O.A. (ed.) *Investigations of Monogeneans in the USSR.* Oxonian, New Delhi, pp. 163–168.

Szekely, C. and Molnar, K. (1987) Mebendazole is an efficacious drug against pseudodactylogyrosis in the European eel (*Anguilla anguilla*). *Journal of Applied Ichthyology* 3, 187–186.

Szidat, L. (1973) Morphology and behaviour of *Paragyrodactylus superbus* gen. et sp. n., cause of fish death in Argentina (in German). *Angewandte Parasitologie* 14, 1–10.

Tasto, R.N. (1975) Aspects of the biology of pacific staghorn sculpin, *Leptocottus armatus* Girard, in Anaheim Bay. In: Lane, D.E. and Hill, C.W. (eds) *The Marine Resources of Anaheim Bay*. State of California Department of Fish and Game Bulletin 165, 123–135.

Thoney, D.A. (1990) The effects of trichlorfon, praziquantel and copper sulphate on various stages of the monogenean *Benedeniella posterocolpa*, a skin parasite of the cownose ray, *Rhinoptera bonasus* (Mitchill). *Journal of Fish Diseases* 13, 385–389.

Uspenskaya, A.V. (1962) Feeding of monogenetic trematodes. *Doklady Akademii Nauk SSSR* 42, 1212–1215.

Vladimirov, V.L. (1971) Immunity of fishes during dactylogyrosis. *Parazitologiya* 5, 51–58.

Watson, D.E. and Thorson, T.B. (1976) Helminths from elasmobranchs in Central American freshwaters. In: Thorson, T. (ed.) *Investigations of the Ichthyofauna of Nicaraguan Lakes*. School of Life Sciences, University of Nebraska, Lincoln, pp. 629–642.

Wells, P.R. and Cone, D.K. (1990) Experimental studies on the effect of *Gyrodactylus colemanensis* and *G. salmonis* on density of mucous cells in the epidermis of fry of *Oncorhynchus mykiss*. *Journal of Fish Biology* 37, 599–603.

Williams, E.H. (1972) Parasitic infections of some marine fishes before and after confinement in feeding cages. *Alabama Marine Resource Bulletin* 8, 25–31.

Yamaguti, S. (1963) *Systema Helminthum. Volume IV. Monogenea and Aspidocotylea*. Interscience Publishers, John Wiley and Sons, New York.

Yin, W.-Y. and Sproston, N.G. (1948) Studies on the monogenetic trematodes of China, Parts 1–5. *Sinensia* 19, 57–85.

Yukhimenko, S.S. and Danilov, V.A. (1987) A new case of transition to endoparasitism in monogeneans. In: Skarlato, O.A. (ed.) *Investigation of Monogeneans in the USSR*. Oxonian, New Delhi, pp. 75–76.

Zitnan, R. and Cankovic, M. (1970) Comparison of the epizootiological importance of the parasites of *Salmo gairdneri irideus* in the two coast areas of Bosnia and Herzegovina. *Helminthologia* 11, 160–166.

9

Digenea (Phylum Platyhelminthes)

I. Paperna

Department of Animal Sciences, Faculty of Agriculture, Hebrew University of Jerusalem, Rehovot 76–100, Israel.

INTRODUCTION

The Digenea (previously termed digenetic trematodes) is one of the three major taxa of parasitic Platyhelminthes, the other two being the Cestoda and the Monogenea. Digeneans are heteroxenous (i.e. they require more then one host to complete their life cycle), and their adult stage is parasitic in vertebrates. All major groups of vertebrates serve as hosts for adult digeneans. Aside from being hosts to adult digeneans, fish may also be infected by the metacercarial larval stage. With one exception, digeneans undergo part or all of their larval development in molluscs; members of the genus *Aporocotyle* complete their larval development in polychaete annelids (Køie, 1982). Very few adult-stage digeneans are known to cause significant harm to the fish host. Notable exceptions are the extraintestinal parasites such as sanguinicoliid blood flukes and possibly also the cyst-forming didymozoids. Metacercarial infection in fish is the main source of disease with subsequent economic loss. Metacercariae may affect growth and survival, or disfigure fish so that they lose their market value as a food or ornamental product (Paperna, 1980, 1991). Some metacercariae in fisheries and aquaculture products (fish and shellfish) are a source for infections in humans and domestic animals (Ito, 1964; Deardorff and Overstreet, 1991). Metacercariae of *Nanophyetus salmonis* transmit rickettsial infection ('salmon poisoning') to dogs (Philip, 1955).

Fish digeneans may be used as biological tags to monitor fish-migration routes and fish-stock composition (Reimer, 1970; MacKenzie, 1983; Lester, 1990), and as indicators of environmental quality. The richness of digenean fauna reflects on the availability of the various host organisms it requires to complete its life cycle. Polluted environments may not offer the necessary host diversity, hence, increased prevalence of digeneans occurs following recovery of the invertebrate community with improved water quality (Overstreet and Howse, 1977).

DISTRIBUTION AND PREVALENCE

Host range and geographical distribution

The most important limiting factor for digenean dispersal is the molluscan host. Molluscs which serve as first intermediate hosts belong to the Gastropoda and Pelecypoda (bivalves), but a few digeneans also occur in Scaphopoda (e.g. an elasmobranch parasite – *Ptychogonimus megastoma*, in *Dentalium* spp. – Palombi, 1941; Wright, 1971). Marine and freshwater prosobranch snails are the most common intermediate hosts of fish digeneans. All known freshwater and marine Bucephalidae develop in bivalve molluscs. Bivalves are also the first intermediate hosts of Fellodistomatidae (*Fellodistomum, Bacciger*), Gorgoderidae (*Phyllodistomum* and *Gorgodera*), and Allocreadiidae (*Allocreadium, Bunodera, Crepidostomum*) (Yamaguti, 1958; Hoffman, 1967; Schell, 1970). Except Cyathocotylidae and *Apatemon* spp., which develop in freshwater prosobranch operculids, all other strigeids, as well as Clinostomatidae and Plagiorchidae, are associated primarily with pulmonate snails (Yamaguti, 1958; Hoffman, 1960, 1967; Finkelman, 1988). However, Sanguinicolidae of marine and freshwater fishes diverge in their choice of hosts, and they include both pulmonates and operculids. The marine members can also be found in bivalves and in polychaete hosts (Smith, 1972). Specificity of individual digenean species to the molluscan host seems to be very restricted, usually to a species, or even sub-species level. However, this has only been demonstrated in non-piscine digeneans, notably in human schistosomes (Paperna 1968a; Wright, 1971). Some digeneans infecting widely distributed fish (such as the grey mullet *Mugil cephalus*), seem to be less specific in their choice of molluscan host (Paperna and Overstreet, 1981). Digeneans with long life spans (Margolis and Boyce, 1969) infecting migratory anadromous or catadromous fishes, are considered useful as 'biological tags' for their hosts (Reimer, 1970; Lester, 1990).

Digeneans in deepwater fish (200–2500 m) are less diverse. When they are in fish in the warmer oceanic regions, they have closer affinities to the fauna of surface-water fishes from colder geographical regions (Campbell *et al.* 1980; Gartner and Zwerner, 1989; Campbell, 1990).

Many digeneans of land vertebrates use fish for their metacercarial stages. Metacercariae are frequently the most common infections in fishes in the marine, estuarine and lacustrine littoral zones. These nutrient-rich waters, are equally attractive to fish, molluscs, and piscivorous birds, hence infections by skin, gill and visceral metacercariae are high. The high natural infections of 0-class plaice with *Cryptocotyle lingua* and *Stephanostomum baccatum* in the northeast Atlantic littoral are well documented (MacKenzie, 1968; MacKenzie and Liversidge, 1975). There are also heavy heterophyiid muscle infections in grey mullets and in juveniles of other fishes in lagoons and inshore marches and estuaries in the eastern Mediterranean sea. These areas are inhabited by high populations of the vector snail, *Pirenella conica* (Fig. 9.1B; Paperna, 1975; Paperna and Overstreet, 1981; Taraschewski and Paperna, 1981; Taraschewski, 1984).

Fig. 9.1. A. *Melanoides tuberculata* (actual size: 20 mm). B. *Pirenella conica* (actual size 15 mm). C. *Bulinus truncatus* (actual size 6–11 mm).

The Syrian-African rift is a major migratory route for birds between Europe and Africa. Water bodies from the Jordan to the East African Great Lakes have common fishes (cichlids, *Clarias* and *Barbus*), snails (*Melanoides tuberculata* (Fig. 9.1A), *Bulinus* (Fig. 9.1C) and *Lymnaea* spp.), and metacercariae whose definitive hosts are herons, cormorants and pelicans. Gill infection with the heterophyiids *Centrocestus* spp. and *Haplorchis* spp. transmitted by *M. tuberculata*, occur in all young-of-the-year cichlids which inhabit shallow waters (Paperna, 1980; Farstey, 1986). *Bulinus* and *Lymnaea* are more fastidious in their habitat demands, and demonstrate greater seasonal and annual variations in abundance (Paperna, 1968b,c), and likewise are the metacercariae they transmit (*Neascus* (black spot), flesh, and brain diplostomatids and clinostomatids – Khalil, 1963, 1969; Paperna, 1964a, 1980; Britz *et al.*, 1985; Yekutiel, 1985; Finkelman, 1988).

In the temperate inland waters of the northern hemisphere, infections by metacercariae of strigeoid digeneans (see Hoffman, 1960) are particularly common. The common infections in North America are 'white grub' (*Posthodiplostomum minimum* encysting in the viscera), 'black grub' (pigmented skin metacercariae, *Uvulifer ambloplitis* and *Crassiphiala bulboglossa*) and muscle metacercaria of *Hysteromorpha triloba* (*Neascus musculicola*) (Hoffman, 1960; Huggins, 1972).

The diploslomatid eye infections are important and common in European and North American freshwater fishes. Dubois (1953) considered the American

Diplostomum flexicaudum to be the European *D. spathaceum*. Both parasites develop to adult stage in gulls (Hoffman, 1960). *D. spathaceum* has been reported from more than 125 fish species (Höglund, 1991). *D. adamsi, D. gasterostei* and *Tylodelphis clavata* also occur in the humour or the retina of the eye (Wooten, 1974; Lester and Huizinga, 1977). Ocular diplostomiasis has been reported to occur in juvenile cichlids and *Barbus* spp. from East and South Africa (Thurston, 1965; Paperna, 1980; Mashego, 1982).

Seasonal distribution

The seasonal occurrences of digeneans in freshwater fishes has been extensively reviewed (Chubb, 1979). However a similar review on marine fish is not available.

Adult stage digeneans

Digenean infections of fish squired by direct cercariae penetration (Sanguinicolidae and Transversotrematidae) are linked to seasonal occurrences of snails and their levels of infection. *Sanguinicola inermis*, which infects the common carp (*Cyprinus carpio*) in Europe, is transmitted during the summer (Lucky, 1964). However, cercariae of *Aporocotyle simplex* have been found in December, and flat-fishes are infected in 6°C water (Køie, 1982). The tube-dwelling polychaete *Lancides vayssieri*, host of larval stage aporocotyles, occurs in Antarctic waters with temperatures not exceeding 1.6°C (Martin, 1952). The ratio of juvenile to adult digeneans enables a determination of time of infection; transmission usually takes place during the summer or early autumn (Chubb, 1979).

In the North Atlantic, *Lecithobothrys bothryophorum* infections in juvenile *Argentina silus* peaks in May (Scott, 1969). The same has been reported for whiting (*Odontagadus merlangus*) infected with *Derogenes varicus, Hemiurus communis, Stephanostomum pristis*, and *Lecithaster gibbosus* (Shotter, 1973). Infection transmitted via the predation of metacercariae-infected intermediate hosts, can be only partially correlated with climatic fluctuations (Scott, 1969; Shotter, 1973). In the North Atlantic, infection by *Derogenes* and *Hemiurus* may be linked to peak abundance of zooplankton in the summer (Shotter, 1973). It becomes more complicated when there are several optional or consecutive intermediate hosts (e.g. *Derogenes varicus*; see Køie, 1979a; 1985). European sea bass (*Dicentrarchus labrax*) become infected with *Bucephalus haimeanus* upon entry into estuarine habitats where both molluscan hosts (*Cardium edule*) and metacercariae-infected gobies occur (Matthews, 1973a). Feeding on plankton has been correlated to *Bunodera luciopercae* infection of *Perca fluviatilis* in fresh water (Chubb, 1979). Seasonal transmission has also been demonstrated for *Allocreadium fasciatus* in tropical waters in India. This was assumed to be related to booms of copepod population following the monsoon season (Madhavi, 1979).

Infections by metacercariae

It is difficult to extrapolate seasonal variations from direct counts of metacercariae on fish, since such infections normally persist for over 12 months
(Dönges, 1969). Differentiation of metacercariae by age is feasible only when
they transform with time (e.g. *Bolbophorus levantinum* Paperna and Lengy,
1963). Seasonal fluctuations in infection can be detected among the young-of-
the-year fish, or in transitory fish, after they enter a new habitat (Lemly and
Esch, 1984a). Active shedding of cercariae and infection of fish commonly
take place during the warmer part of the year in fresh water (Chubb, 1979;
Lemly and Esch, 1984a; Stables and Chappell, 1986), in marine and estuarine
habitats (Køie, 1975; Matthews, 1973a; Cottrell, 1977). In some habitats, infection in snails only reaches peak levels in the autumn (McDaniel and Coggins,
1972). Transmission during the winter is rare since snails do not shed their
cercariae below 10°C (Stables and Chappell, 1986). There is, however, a report
on invasion of *Oncorhynchus mykiss* by *Diplostomum opataceum*. The infection has to be acquired by predation on *Lymnaea* containing precocious
(progenetic) metacercariae (Becker and Brunson, 1966). Another report involves
Oncorhynchus kisutch in Oregon, which hatched in March, and had all become
infected with *Nanophyetus salmincola* by mid-April (Millemann and Knapp,
1970). The favourable ambient temperature in boreal or alpine habitats for
infestation is between 10–17°C (Bauer, 1959; Wooten, 1974). This represents
the low winter temperatures at which activity of pulmonate snails, such as
Bulinus truncatus and *Lymnaea (= Stagnicola) palustris*, is interrupted in
south-east Mediterranean climatic zones (Yekutiel, 1985; Farstey, 1986). The
latter snails live in aquatic habitats fringing Lake Kinneret. *Melanoides tuberculata* which inhabit the lake proper may also be found in deeper waters in the
winter. Some have sporocysts which contain xiphidiocercariae and rediae which
have pleurolophocercous cercariae (Farstey, 1986). The pulmonates inhabiting
the flood pools fringing the lake are more susceptible to transitions in their
habitat. Successive years of drought and flooding resulted in the elimination or
reduction of pulmonate snails, and the disappearance of metacercariae of
Neascus, Bolbophorus levantinum, Clinostomum tilapiae and *Euclinostomum
heterostomum* from the lake-dwelling cichlids for several years (Paperna,
unpublished). However, infections transmitted by lake-inhabiting snails (*M.
tuberculata* transmitting *Centrocestus* and *Haplorchis*, and *L. (= Radix)
auriculata* transmitting *C. complanatum*) are not affected (Paperna, 1964a;
Yekutiel, 1985; Farstey, 1986; Finkelman, 1988; Paperna, unpublished).
Pirenella conica in marine lagoons, fringing the southeastern Mediterranean and
the northern gulfs of the Red Sea continues to shed cercariae (of *Heterophyes*
and others) throughout the winter months when water temperatures of fringing
and land-locked sites may drop to below 10°C (Taraschewski and Paperna,
1981, 1982). Year-round infection by larval digeneans has been reported in
Cerithidea californica from mud-flats in Southern California (Martin, 1955). In
the perennial habitat of the East African lakes, the *M. tuberculata*-transmitted
metacercariae (*Centrocestus* and *Haplorchis*) and a variety of pigmented
skin metacercariae accumulate uninterruptedly till the young cichlids migrate
to deeper waters (Paperna, 1980). A similar year-round recruitment of

Centrocestus (*C. formosanus*, in *Apocheilus panchax*) occurs in India (Madhavi and Rukmini, 1991).

Infections in farmed fishes

Intensive-culture earth ponds, with their heavy organic and nitrogenous load and muddy (eutrophic) bottom are unfavourable habitats for all snails. Omnivorous fish, such as carp and siluroid catfishes, eat thin-shelled snails and snail eggs. Metacercarial infections occur only sporadically in intensive earth-pond systems. These episodes are restricted to a single growing season, and are eliminated when ponds are returned to routine intensive cultivation (Paperna, 1980). Snails can only proliferate in mesotrophic ponds with a firm substrate (earth or gravel) and a low piscine biomass, used for spawning, for holding broodstock, or as a nursery. Extensive systems such as dam reservoirs and similar large water bodies with lower fish biomass, or small ponds or pools with frequent or continuous water exchange fringed by trailing and floating weeds, offer better conditions for snails and are also attractive to piscivorous birds (De Bont and De Bont-Hers, 1952; Lombard, 1968; Paperna, 1968b, 1980). At times, indoor circulation systems, such as raceways and hatcheries, become heavily populated with snails (Stables and Chappell, 1986), but transmission in these systems is often limited to sanguinicolids (Wales, 1958; Hoffman *et al.*, 1985), metacercarial infections are usually prevented when piscivorous birds are excluded from the system with an efficient net system.

Blood flukes infect both cold and warm water fish. *Sanguinicola* spp. have been implicated in massive mortalities of hatchery rainbow, cutthroat (*Salmo clarki*), and brook trout (*Salvelinus fontinalis*) after their snail hosts *Oxytrema* spp., *Flumincola* spp. and *Leptoxis (Mudalia)* spp. became established in the culture system (Davis *et al.*, 1961; Evans, 1974a; Hoffman *et al.*, 1985). *S. inermis* (transmitted by *Lymnaea* spp.) has been reported in pond-reared common carp in eastern Europe (Lucky, 1964). Anderson and Shaharom-Harrison (1986) reported the introduction of *S. armata* with infected bighead carp (*Aristichthys nobilis*) and grass carp (*Ctenopharyngodon idella*) into fish farms in Malaysia. An unidentified blood fluke became established in farmed cichlids in Israel.

In Israel and tropical and southern Africa, massive metacercarial infections have sometimes resulted in mortalities of farmed cichlid fishes. These included gill infections of *Centrocestus*, and subcutaneous *Haplorchis* transmitted by *M. tuberculata* (Sommerville, 1982; Paperna, 1980), skin *Neascus* ('black-spot'), muscle infection with *Bolbophorus levantinum* (Paperna, 1980 and unpublished) and visceral infections of *Clinostomum tilapiae* and *Euclinostomum heterostomum* (Lombard, 1968; Paperna, 1980; Britz *et al.*, 1985) transmitted by *B. truncatus* (Donges, 1974; Finkelman, 1988). Gill infection by *Centrocestus formosanus* resulting in mass mortality has also been reported from farmed eels (*Anguilla japonica*) in Japan (Yanohara and Kagei, 1983). *Lymnaea (Radix)* transmitted *C. complanatum* created heavy infection in farmed loach (*Misgurnus anguillicaudatus*) and ayu (*Plecoglossus altivelis*) in Taiwan, and these caused growth retardation and lowered survival (Liu, 1979; Lo *et al.* 1981).

Mitchell *et al.* (1982) reported heavy mortality of commercially farmed fathead minnows (*Pimephales promelas*), in Missouri (USA) from intensive visceral infection by *Posthodiplostomum minimum minimum*. The snail involved is *Physa*. Trout farms are often troubled by eye infections with *Diplostomum spathaceum* which cause blindness (Shariff *et al.*, 1980; Stables and Chappell, 1986). The snails, *Lymnaea* spp., thrive in both earth ponds and raceways. Ocular diplostomiasis also affects farmed channel catfish (*Ictalurus punctatus*) in the southern USA (Rogers, 1972). Diplostomatid eye infections have also been reported from pond-reared largemouth bass (*Micropterus salmoides*) and rainbow trout (*Oncorhynchus mykiss*) in South Africa (Lombard, 1968).

In the eastern Mediterranean, extensively-used seawater ponds often support dense populations of *Pirenella conica* (Taraschewski and Paperna, 1981) and in the western Mediterranean *Hydrobia* spp. (Maillard *et al.*, 1980). *P. conica* is first intermediate host to a number of heterophyiids including *Heterophyes heterophyes*, which primarily infect grey mullets (Mugilidae), but also cichlids and *Dicentrarchus labrax* (Paperna and Overstreet, 1981). *Sparus auratus* postlarvae, in a culture system in southern France succumbed to massive infection with *Acanthostomum imbutiforme* metacercariae, which were transmitted via *Hydrobia acuta* (Maillard *et al.*, 1980; Euzet and Raibaut, 1985).

EVOLUTION AND TAXONOMY

Evolution

The diversity of forms, structures, developmental strategies and a fairly uniform larval development in the molluscan host have led to the suggestion that the proto-digenean was a molluscan parasite (Pearson, 1972). The phenomenon of progenesis, i.e. larval stages attaining sexual maturity and yielding eggs, while still in the invertebrate host, has been interpreted as a remnant of past life histories (Llewellyn, 1986). In this context an evolutionary relationship has been suggested between Mesozoa and digeneans (see Wright, 1971). The diversity of forms and life histories have been recently subjected to 'phylogenetic analysis' (Brooks *et al.*, 1985), to question the role of adaptive radiation in the digenean evolutionary process.

Taxonomy

Manter (1957) writes of the difficulties in differential diagnosis of Digenea, whereby the great diversity is not only an outcome of evolutionary divergence but also of convergent processes (argued by Brooks *et al.*, 1985). The correct evaluation of a digenean should consider also morphological details of its larval stages and the identity of its intermediate and definitive host (Brooks *et al.*, 1985). The life histories of only a fraction of digeneans are known. Many of the cercariae from molluscan hosts, or metacercariae on fishes cannot be correlated with established species or genera, and many cannot be affiliated with certainity, if at all with a particular parasite family. Alternative methodologies, e.g. biochemical, enzyme detection by electrophorhesis (Bryant and Flockhart, 1986), have not been tested with piscine Digenea, and there is only one communication dealing with molecular taxonomy. Six undescribed species of *Didymozoon* (recognized by morphological criteria) and *Neometadidymozoon helicis* were confirmed to be distinct species using ribosomal DNA (Anderson and Barker, 1993). Biochemical and molecular biology methods are potentially useful in identifying larval forms.

PARASITE MORPHOLOGY AND LIFE CYCLE

Morphology

Adult stages

Adult-stage digeneans (Fig. 9.2) usually have a dorsoventrally flattened oval body with a smooth, spiny or corrugated surface (Fig. 9.17B), a sucker around the anteroventral mouth, and an additional ventral sucker, or acetabulum. Both suckers are used for attachment and locomotion. The digestive system consists of a pharynx connected to the mouth opening, a short oesophagus and two blind intestinal caeca. The excretory system includes flame cells connected to a duct network which is connected to a posteriorly opening bladder. Most trematodes are hermaphroditic (with cross-insemination). The male system consists of testes, vasa efferentia and vas deferens, and a cirrus pouch containing a seminal vesicle, ductus ejaculatorius, and prostate glands. Some also contain a specialized copulatory organ (gonotyl in *Heterophyes* spp.). The female system is comprised of an ovary and paired vitellaria, an oviduct, vitelline ducts, Mëhlis' glands and the receptaculum seminis which open into an ootype. The latter is also connected with a rudimetary vagina – the ductus laureri, and the uterus, through which eggs are evacuated to the genital opening (after Dawes, 1946, Yamaguti, 1958). Eggs are usually oval and operculated (Fig. 9.3).

Considerable structural diversity exists among the digeneans, at the adult (Dawes, 1946; Skrjabin *et al.*, 1964) and larval stages (Dawes, 1946).

In the Gasterostomata (or Bucephalididea) – Bucephalidae, the mouth lies in the middle of the ventral surface. The attachment structure consists of conical muscular sucker ('rhynchus') at the anterior end, simple sac-like gut, and the genital opening not far from the posterior extremity (Fig. 9.4).

Bivesiculidae are small digeneans which do not have both oral and ventral suckers (Fig. 9.5; Pearson, 1968).

Sanguinicolidae (= Aporocotylidae): These are parasites of the blood system. They are slender, spiny, and do not have anterior and ventral suckers and pharynx. The intestinal caeca are short and are either X- or H-shaped

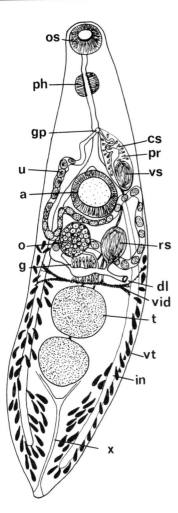

Fig. 9.2. Example of a 'distome' digenean: *Allocreadium isoporum*. (Drawn with reference to Dawes, 1946, actual size 3–5 mm.) Abbreviations: a = acetabulum; cs = cirrus pouch (sac); dl = ductus laureri; g = Mehlis glands; gp = genital opening; in = intestine; o = ovary; od = oviduct; os = oral sucker; ot = ootype; ph = pharynx; pr = prostate; s = sucker; t = testis; vd = vas deferens; vid = vitelline ducts; vs = seminal vesicle; vt = vitellaria; x = excretory duct/bladder.

(Figs 9.6, 9.7). Eggs are thin-shelled and do not have an operculum (Smith, 1972).

Didymozoidae are thread-like tissue parasites, with or without an expanded posterior region. They occur either unencysted (*Atalostrophion, Metanematobothrium*), or live in groups or in pairs in cysts or cyst-like cavities (Fig. 9.8). Many are hermaphroditic (*Didymozoon, Didymocystis*). Equal-sized hermaphroditic individuals of some didymozoids share the same cyst and fuse their hind parts (such as *Colocyntotrema* and *Phacelotrema* spp., and *Neodiplotrema pelamydis* (Dawes, 1946; Yamaguti, 1958; Fig. 9.9). Others show variable degrees of separation into sexes, with intermediate rudimentary hermaphrodites (Gonapodasmiinae and *Nematobothrium* spp.). Their opposite sex systems are reduced, rudimentary or lacking, and only the eggs of the larger 'female' develop (see Yamaguti, 1958).

Strigeoidea has several families which encyst as metacercariae in fish

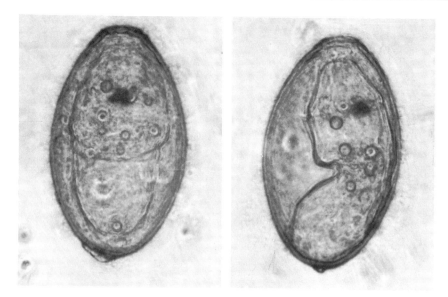

Fig. 9.3. *Clinostomum tilapiae* eggs containing ready-to-hatch miracidium (actual size 125 × 83 µm).

(Strigeidae, Diplostomatidae and Cyathocotylidae). The body is divided into a cup-shaped forebody carrying the suckers, and a cylindrical hindbody which contains the reproductive organs. The ventral sucker's function is taken over by a new holdfast (tribocytic) organ. The copulatory bursa is terminal (Hoffman, 1960; Fig. 9.10).

The known genera of Transversotrematidae – *Transversotrema* and *Prototransversotrema* – are ectoparasitic under the scales of fish. They have leaf-like transversely elongated bodies, with or without an oral sucker, and with pigmented eye spots (Fig. 9.11; Witenberg, 1944; Velasquez, 1958; Angel, 1969). The oral sucker is used for attachment to the host skin (Mills, 1979).

Larval stages

Free-swimming miracidia are pyriform in shape and are covered with cilia-bearing epithelial plates (Køie and Frandsen, 1976). Cilia are absent in miracidia of Hemiuridae, Didymozoidae and Azygiidae (Schell, 1970). The apical papilla at the anterior extremity is devoid of cilia. Connected to this papilla are a median apical gland ('primitive gut') and two lateral clusters of penetration glands. Also present are a pair of pigmented eye spots and posteriorly located clustered and scattered germinal cells (Fig. 9.3).

Sporocysts are tegumental sacs, which enclose germinal cells and developing daughter sporocysts or rediae. The redia has a muscular pharynx connected to a sac-like intestine, a birth-pore located near the pharynx, germinal cells and developing daughter larval stages (Figs 9.12B,C and 9.16). Daughter sporocysts are often oval, but can be elongated, (in strigeids; Fig. 9.12A) or branched (in bucephalids; Dawes, 1946). Rediae are fairly uniformly oval in shape.

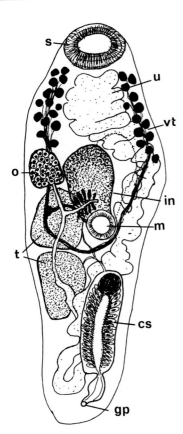

Fig. 9.4. *Bucephalopsis gracilescens* (Bucephalidae). (Drawn with reference to Dawes, 1946 and Bychowskaya-Pavlovskaya *et al.*, 1962; actual size 6 mm; for abbreviations see legend to Fig. 9.2.)

Cercariae (Fig. 9.17A) are formed in either sporocysts or rediae. They have a mouth, pharynx, branched intestine, ventral sucker, excretory vesicles and primordia of the genital organs. Cercariae have a pair of eyes and may also have a tail and fins which are used as locomotive devices. The latter structures are all lost when cercariae transform into metacercariae.

The great structural and functional diversity among cercariae (see Dawes, 1946; Schell, 1970) reflect the numerous strategies employed by cercariae to reach the final hosts (Fig. 9.13). Furcocercariae with forked tails are characteristic of Strigeoidea, Sanguinicolidae and Clinostomatidae; bucephalids have typical gasterostome organization and a bifid tail with a very short and broad tail base (Fig. 9.14) and pleurolophocercariae common to Heterophyiidae have nondivided tails. Some sanguinicolid and clinostomatid cercariae carry a dorsomedian finfold. All these cercariae actively penetrate the integument of their hosts. They have anteriorly opening penetration glands which are absent in cercariae encysting on a substrate, e.g. 'gymnocephalous' group cercariae of Haploporidae (Fares and Maillard, 1974) and those which infect their hosts via ingestion (of Bivesiculidae – Pearson, 1968 and Fellodistomidae – Køie, 1979b). The substrate encysting cercariae are filled with cystogenous glands. Many cercariae employ devices to improve buoyancy. They may have hairs to enlarge the tail surface as in trichocercous

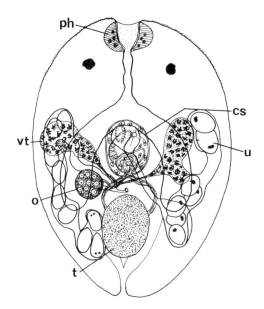

Fig. 9.5. *Paucivitellosus fragilis* (Bivesiculidae). (Drawn with reference to Pearson, 1968; actual size 420 mm; for abbreviations see legend to Fig. 9.2.)

cercariae of Lepocreadidae (*Opechona bacillaris* – Køie, 1975), or a lateral finfold supported by seta-like filaments in cercariae of Megaperidae (Schell, 1970). Tails of *Mesorchis denticulatus* (Echinostomatidae) cercariae become inflated after they emerge from the snail (*Hydrobia ulva* – Køie, 1986). Azygiidae and Bivesiculidae cercariae have thick cylindrical bifurcate tails (Pearson, 1968; Schell, 1970).

The latter, as well as those of Hemiuridae, are cysticercous (cystophorous) cercariae. They are able to retract the body into a large chamber at the basal part of the tail (Pearson, 1968; Køie, 1979a, 1992).

Microcercous cercariae have a variable degree of tail reduction. These cercariae crawl, are capable of limited swimming, and infect their next host when eaten with their molluscan host (*Podocotyle* spp. – Koie, 1981; *Zoogonoides viviparus* – Koie, 1976; *Asymphylodora tincae* – Van den Broak and De Jong, 1979).

In the cotylomicrocercous cercariae of Opecoelidae, the tail apparently functions as an adhesive organ.

Ultrastructural studies

For a general review of the structure and ultrastructure of adult digeneans readers are referred to earlier reviews (Threadgold, 1984; Smyth and Halton, 1983; Erasmus, 1977). Detailed studies on intramolluscan stage fish digeneans are in Køie (1971a,b,c; 1985).

The digenean integument consists of an external layer of cytoplasm joined by cytoplasmic bridges, with nucleated masses of cytoplasm lying internal to

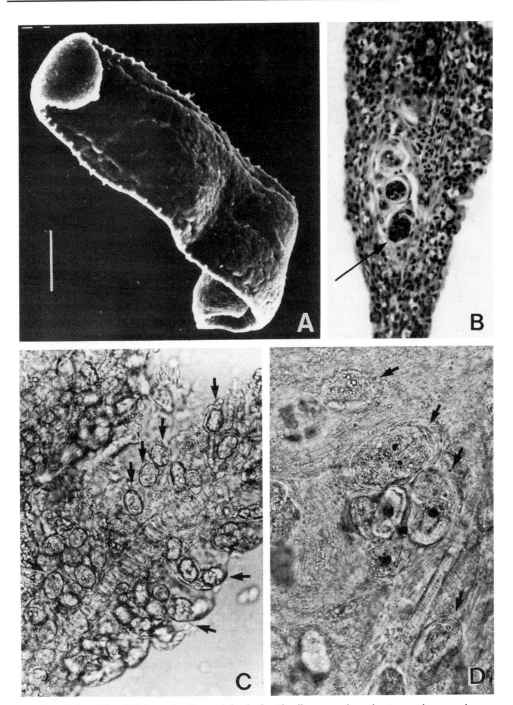

Fig. 9.6. Blood flukes. A. *Sanguinicola fontinalis*, scanning electron micrograph. (From Hoffman *et al.*, 1985, by courtesy of G.L. Hoffman; bar = 100 μm). B. Eggs of *Sanguinicola* sp. in gills of *Oreochromis aurea*, Lake Kinnereth. C. Eggs of a sanguinicolid in gills of juvenile *Liza* sp., Kowie estuary, south-east Cape, South Africa. D. Eggs of *Sanguinicola* sp. accumulated in heart of cultured tilapia (*O. aurea* × *nilotica*) from Israel.

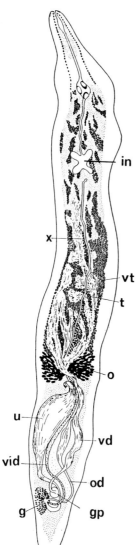

Fig. 9.7. *Sanguinicola dentata*, (Paperna, 1964) Smith 1972, from *Clarias lazera* circulatory system (eggs are released into the kidneys). (Actual size 3.4 mm; for abbreviations see legend to Fig. 9.2.)

the body musculature (Fig. 9.15). Many digeneans (notably Heterophyiidae) have spines protruding through the tegument (Fig. 9.15B). Regional special-izations of the surface are related to functions such as secretion and absorption (for example, the lamellate and highly infolded tegument on the inner face of the adhesive organ of *Apatemon gracilis*; see Erasmus, 1977). Cytochemical uptake, and membrane-transport studies have further clarified the dynamic aspects of the tegument (Lumsden, 1975; Pappas and Read, 1975). Lumsden (1975) discussed the possible role of the external layer, particularly the outer glycocalyx, in permeability control, ionic regulation, protection of the parasites against host enzymes and involvement in antibody/antigen reactions.

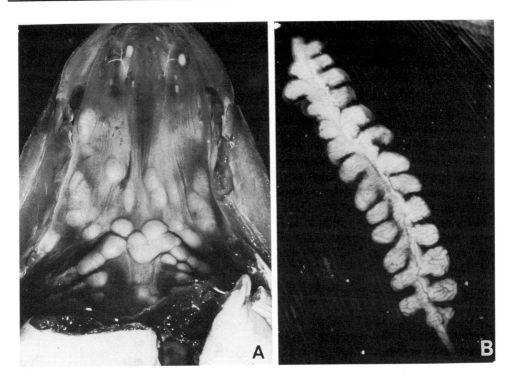

Fig. 9.8. Didymozoidae. A. Palate of *Platycephalus fuscus* infected with *Neometadidymozoon helicis* (in life, bright yellow). (From Lester, 1980, by courtesy of the author.) B. *Lobatozoum multisacculatum* on gills of *Katsumonus pelamis*, New Zealand. (By courtesy of B. Lester.)

The addition of scanning electron microscopic (SEM) studies have provided detailed knowledge of the outer surface (Figs 9.16, 9.17). Tegument surfaces of sporocysts and rediae (Fig. 9.16) show different surface topographies, which probably reflect different nutritive demands (Køie, 1985): *Podocotyle reflexa* sporocysts are covered by long thin microvilli (Køie, 1971a), The *Neophasis lageniformis* redia has numerous folds, whereas that of *Zoogonoides viviparus* has slightly increased surface area (Køie, 1971b, 1985). Rediae of *Aporocotyle simplex* (Sanguinicolidae) have a nearly smooth surface with numerous canal openings on the tegument (Køie, 1982). The surface of the cercariae contains oral or cephalic spines and structures possibly connected with sensory function (Fig. 9.17C) (Køie, 1971c, 1975, 1976, 1977, 1985, 1987). Presumed endocytotic vesicles opening to the surface of *N. lageniformis* cercariae suggest active nutrient absorption by the tegument. These tailless cercariae remain in their rediae until the snail host is eaten by the definitive host (Køie, 1971b, 1973).

All encysting metacercariae are first enclosed in a self-produced, non-cellular granular matrix wall (Fig. 9.15A,C), before becoming enclosed by a cyst wall which is of host origin (Fig. 9.15C) (Halton and Johnston, 1982; Faliex, 1991; Walker and Wittrock, 1992). The parasite wall consists of either

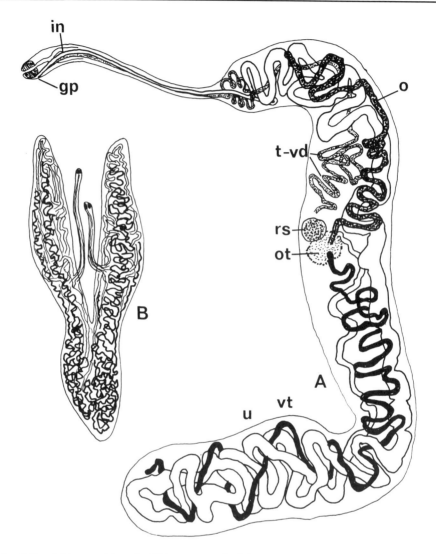

Fig. 9.9. Didymozoidae. A. *Didymozoon faciale* (actual size 16.3 mm). B. *Neodiplotrema pelamydis* – fused pair. (Drawn with reference to Dawes, 1946 (A) and Yamaguti, 1958 (B); for abbreviations see legend to Fig. 9.2.)

acid or neutral mucopolysaccharides and is sometimes two to three layers with different textures (in *Uvulifer ambloplitis* and *Neascus pyriformis* – Wittrock *et al.*, 1991). The tegumental surface of encysting metacercariae changes, and seems to become specialized for absorbtion of nutrients by forming short microvillous-like projections (Køie, 1981) or anastomosing folds (Køie, 1977). Cercaria-type spines (such as those used for penetration) disappear, or are transformed into the spines which are characteristic of the adult stage (Køie, 1985, 1987).

The gut epithelium plasma membrane is complex. In some species,

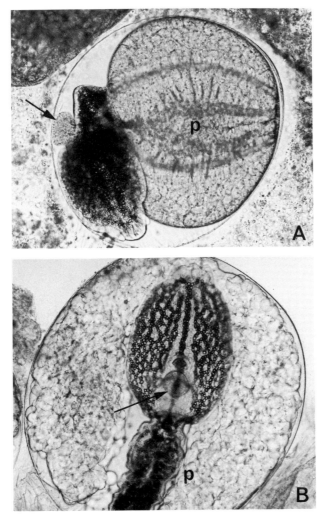

Fig. 9.10. *Bolbophorus levantinum* metacercariae in the cichlid *Oreochromis aurea*, Lake Kinnereth. A. Earlier stage, with expanded posteror half (p). B. Later stage with emptied posterior end, the cyst lumen is filled with the 'reserve' substance (actual size 0.6–0.8 mm). Arrows show holdfast organ.

microvilli with supporting filaments are apparent. The plasma membrane is covered by a granular or filamentous glycocalyx. The ultrastructure of the gastrodermal cytoplasm reflects its metabolic activity: it is rich in granular endoplasmic reticulum (GER), mitochondria, Golgi complexes and various inclusions (Erasmus, 1977).

In the redia, the surface of the oral sucker and the oesophagus are continuations of the external integument and are surrounded by secretory glands. The gut epithelial cell plasma membrane forms large (up to 2 µm) projections, folds or microvilli. Gut cells are linked to each other by long septate desmosomes, and contain granular endoplasmic reticulum, free

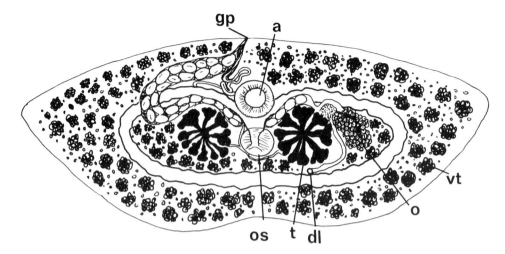

Fig. 9.11. *Transversotrema haasi.* (Drawn with reference to Witenberg, 1944; for abbreviations see legend to Fig. 9.2.)

ribosomes, Golgi complexes and particles of alpha-glycogen. The basal membrane forms tubular invaginations which project into the cell cytoplasm. Beneath the basal membrane is a thin layer of muscle (Køie, 1971; Erasmus, 1977). The cytological features of the cercarial digestive tract resemble that in the redia (Køie, 1971b, 1973).

The energy required for cercarial activity is apparently derived from endogenous sources. Gut cells, however, have the ultrastructural characteristics of a metabolically active epithelium, and in some instances also demonstrate enzyme activity (see Erasmus, 1977). The intestine (as well as the excretory vesicle) of old encysted metacecariae become specialized to store excretory products which are visible as highly refractile bodies (Køie, 1985).

Life cycles

Life span of the adult digenean

Information on the life span of adult digeneans is scanty and has mostly been extrapolated from field data. Some adults apparently live for one or two seasons (spring to autumn – *Rhipidocotyle septpapillata* – Krull, 1934; *Asymphylodora kubanicum* – Evans, 1978), whereas others, mainly in colder habitats, survive for more than a year, and are being replaced annually (Awachie, 1968). Survival is temperature-dependent, overwintering usually prolonges life span. Laboratory studies on *Transversotrema patialense* suggest that, in addition to its dependence on ambient temperatures, its life span is also density-dependent (Mills *et al.*, 1979). The life span of an adult digenean is highly variable, even between members within the same family. Two hemiurids, *Tubulovesicula lindbergi* and *Lecithaster gibbosus*, were studied in captive pacific salmon (*Oncorhynchus* spp.). *T. lindbergi* matured in two to

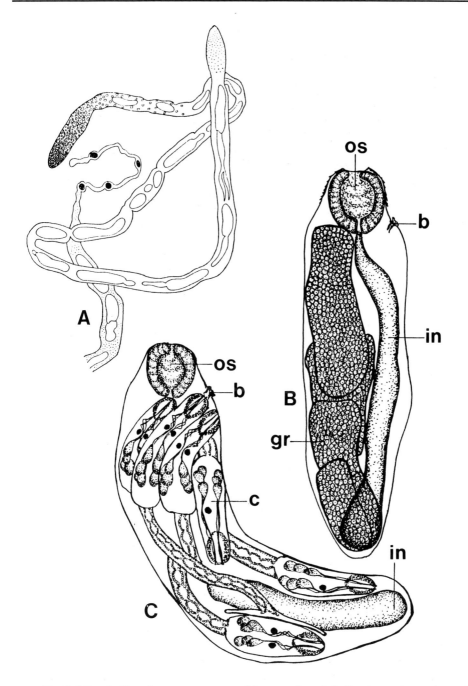

Fig. 9.12. A. Daughter sporocysts with cercariae of *Bolbophorus levantinum*. B. Young and C, mature daughter rediae of *Clinostomum tilapiae*; b = birth pore; c = cercariae; gr = germinating stages; in = intestine; os = oral sucker.

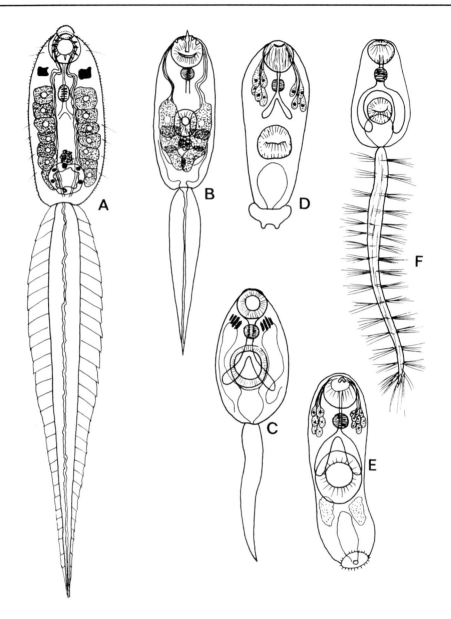

four months and lived at least 31 months, whereas *L. gibbosus* matured in one
to two weeks and lived from two to nine months (Margolis and Boyce, 1969).

Egg release and development into miracidium

Eggs of Clinostomatidae are undeveloped when laid and begin embryonic
development only after evacuation from the host (Dönges, 1974; Finkelman,
1988; Fig. 9.3). However, those of many piscine digeneans contain fully
developed miracidia before being laid. These eggs hatch immediately, or
soon after evacuation from the definitive host (*Asymphylodora tincae* –

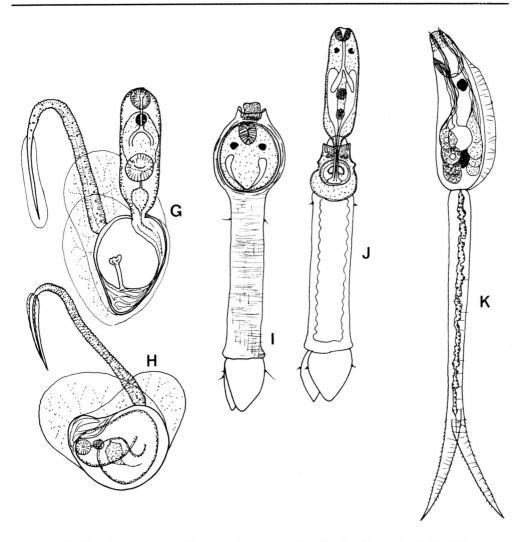

Fig. 9.13. Cercariae. A. Pleurolophocercous (of *Haplorchis* sp.). B. Xiphidiocercous (of Allocreadiidae). C. Gymnocephalous. D. Microcercous. E. Microcercous, of *Zoogonides viviparum*, note terminal attachment-disk crowned with spines. F. Trichocercous. G.H. Extracted and retracted cysticercous, of Hemiuridae, I.J. Retracted and extracted furco-cycticercous cercariae of Bivesiculidae. K. Furcocercous with dorsomedian finfold, of *Clinostomum tilapiae*. (Drawn with reference to Dawes, 1946, Pearse, 1968, Schell, 1970, Fares and Maillard, 1974; Køie, 1992; K, original.)

Van den Broek and de Jong, 1979; *Paucivitellosus fragilis* – Pearson, 1968; and all Haploporidae and Haplosplanchnidae – Manter, 1957; Fares and Maillard, 1974). Hatching in Haploporidae is stimulated by light (Fares and Maillard, 1974).

Eggs produced by digeneans in the kidney or gonads are evacuated from their host with the respective organ's products. Didymozoid eggs seem to remain

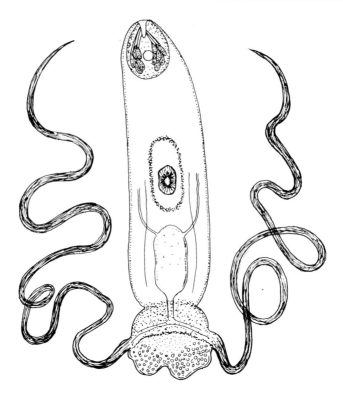

Fig. 9.14. Bucephaliid cercaria. (Drawn by reference to Dawes, 1946 and Matthews, 1973a.)

viable in tissues long after the worms' death (Lester, 1980; Gibson *et al.*, 1981), and are liberated only after death of the host. Eggs of blood flukes (Sanguinicolidae) which contain fully developed miracidia, accumulate in the terminal (distal) blood capillaries. Those that reach blood vessels of the gill filaments release their miracidia, which then actively break through the gill tissue into the water (Davis *et al.*, 1961; Smith, 1972; Fig. 9.6B,C).

Invasion into, and development in the molluscan host

Free-swimming miracidia reach and penetrate the molluscan host, and metamorphose into 'mother' sporocysts. Fully embryonated eggs of Heterophyiidae and Azygiidae do not hatch in the aquatic environment, but infect snail hosts after being ingested (Khalil, 1937; Sillman, 1962; Schell, 1970). Within the mother sporocyst, a new generation of either sporocysts or rediae is produced. These migrate and settle in the molluscan hepatopancreas. Cercariae are formed and they emerge from the snail host. Intramolluscan development of Heterophyiidae, Allocreadiidae, Haploporiidae, Clinostomatidae, Bivesiculidae and Monorchidae (Khalil, 1937; Dawes, 1946; Dönges, 1974; Finkelman, 1988; Pearson, 1968; Fares and Maillard, 1974; Van den Broek and De Jong, 1979) includes both sporocyst and redia stages (Fig. 9.12B,C). Bucephalidae, Sanguinicolidae Plagiorchidae and Strigeoidea (Hoffman, 1960; Smith, 1972; Matthews, 1973a) have their furcocercariae (Fig. 9.13K) or xiphidiocercariae

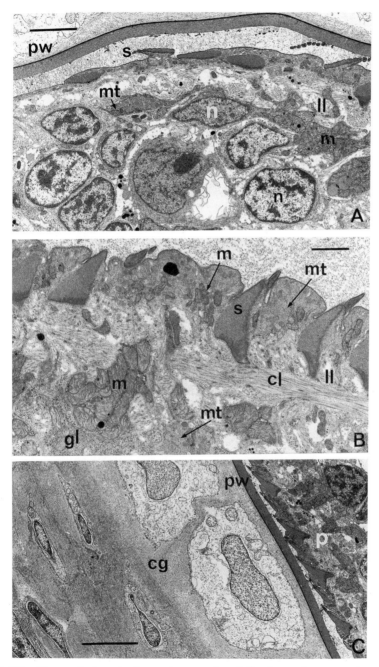

Fig. 9.15. Transmission electron micrographs of *Centrocestus* sp. metacercaria from gills of *Oreochromis aurea*. A. Parasite produced wall and tegument; bar = 3 μm. B. Enlarged view of the tegument; bar = 1 μm. C. The host produced cartilaginous cyst, note onset of degenerative changes in the cartilage-layer cells; bar = 5 μm. Abbreviations: cg. = ground substance of the cartilage; cl = circular muscles; gl = glycogen particles; ll = longitudinal muscles; m = mitochondria with cristae; mt = round tegumental mitochondrial; n = nuclei of the tegument; pw = parasite produced wall; s = spine.

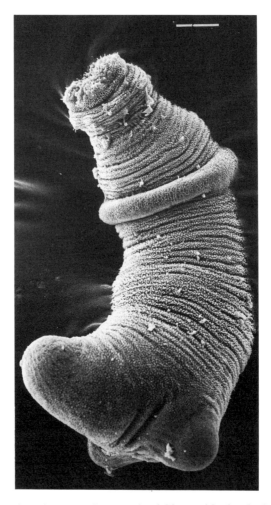

Fig. 9.16. Scanning electron micrograph of *Mesorchis denticulatus* redia; bar = 10 μm. (By courtesy of M. Køie.)

(Fig. 9.13B) formed in the daughter sporocyst stage (Fig. 9.12A). The initial sporocyst dies and daughter sporocysts are released with the rupture of the sporocyst. Daughter stages of rediae escape through a birth pore (Fig. 9.12C); this allows a redia to produce several generations of offspring, and also alternate generations of daughter rediae and cercariae. The intramolluscan infection with rediae-forming digeneans is therefore usually considerably longer than that with sporocyst-forming digeneans. Heterophyiid infections in *Pirenella conica* (Taraschewski and Paperna, unpublished) and in *Melanoides tuberculata* (Farstey, 1986) last over a year, and up to five years in *Cryptocotyle lingua*-infected *Littorina littorea* (Meyerhof and Rothschild, 1940). *C. lingua*-infected *L. littorea* shed 3330 cercariae per day at the early stage of infection, and 830 towards the end of the five year period. Daily, or

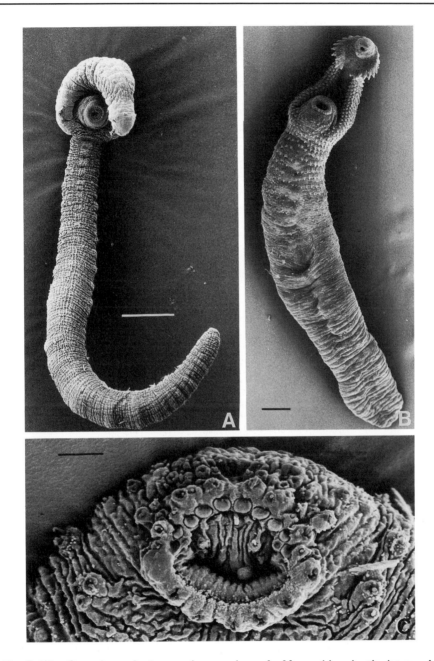

Fig. 9.17. Scanning electron micrographs of *Mesorchis denticulatus*. A. Cercaria, bar = 50 µm; B. Adult worm, bar = 100 µm; C. Anterior end of cercariae, ten flattened spines occur in a semicircle, anterior to the mouth opening, and different type of supposed sensory structures occur around the mouth; bar = 5 µm. (After Køie, 1987, by courtesy of the author.)

periodic, cercarial output in pulmonate snails is often similar (Wright, 1971; Paperna, unpublished), or even higher (Paperna and Lengy, 1963), but the overall production time is shorter for digeneans in pulmonates which have only sporocyst stages (Strigeata, Sanguinicolidae, Plagiorchidae). Cercarial production exhibits daily and periodic fluctuations (Wright, 1971). The periods of low cercarial production correspond to renewals of the germinal cell stocks in the sporocysts or in the rediae (Theron, 1986). Shedding may be stimulated by salinity (tide periods) and temperature, as well as light (day–night cycles), all of which interact with intrinsic periodicity (Wright, 1971). In schistosomes, the timing of the daily emission correlates with periods of definitive host activity (Theron, 1986). An adaptive correlation between cercarial shedding and piscine host behaviour has not been established. *Transversotrema patialense* are shed at night (Rao and Ganapati, 1967), as are cercariae of *Apatemon gracilis* (Wright, 1971) whereas peak emission of *Heterophyes* spp. occurs by middle of the day. In *B. levantinus* the observed daily cycle, however, was not affected by continuous illumination or darkness (Paperna and Lengy, 1963).

Infection of definitive host

Cercariae develop in fish either to adult stages (Sanguinicolidae, Transversotrematidae and *Azygia* spp.) or to an encysted or unencysted metacercaria. Cercariae can penetrate skin or gills (Höglund, 1991; Sommerville and Iqbal, 1991) or are ingested by the fish host (cercariae of Bivesiculidae, Azygiidae and Fellodistomidae – Pearson, 1968; Hoffmann, 1967; Køie, 1979b). These develop to maturity if ingested by a suitable host, or remain a waiting larval stage, which will develop to maturity only if the paratenic host is predated upon by the definitive host (Van Cleave and Mueller, 1934; Ginetsinskaya, 1958; Køie, 1979a,b). Another option is for the cercariae to enter into invertebrate hosts which are subsequently eaten by fish. Cercariae may also attach to plant material or to some other substrate (encysting into metacercariae) to be later browsed by potential host fishes (e.g. Haplosplanchiidae and Haploporidae, Cable, 1954; Fares and Maillard, 1974). Sommerville and Iqbal (1991) found that, although *Sanguinicola inermis* cercariae were able to penetrate common carp over its entire body surface, the majority (68%) penetrated the fins. *Cardicola* (= *Sanguinicola*) *alseae* has been reported to be capable of penetrating only the soft, ventral surface, and *C. klamathensis* can only enter through the less heavily scaled areas in salmonids (Meade and Pratt, 1965; Meade, 1967). *Aporocotyle simplex* easily penetrates through the entire body surface of dabs (*Limanda limanda*), flounders (*Platichthys flesus*) and plaice (Køie, 1982). *S. inermis* remains in fins of carp for over three months following entry while after only five months all worms entered the circulatory system (Sommerville and Iqbal, 1991). *A. simplex* remains in the lymphatic system and among the muscles for over 94 days while their entry into blood vessels is completed only 180 days after exposure to cercariae (Køie, 1982).

Transmission via metacercarial stages

Many digeneans are closely linked with the food preferences of their definitive hosts, and modifications of fish habitat and diet, which occur with age or season, result in changes in the type and level of infection (Scott, 1969; Shotter, 1973).

METACERCARIAL STAGES IN FISH

Most digeneans which encyst as metacercariae in fish (Heterophyiidae, Diplostomatidae, Clinostomatidae) subsequently reach maturity in piscivorous birds. Members of the Diplostomatidae form free (*Diplostomulum*) or encysted (*Neascus*) metacercariae in the eyes (Hoffman, 1960; Donges, 1969). Cercariae of *Diplostomum spathaceum* penetrate mainly through the gills and migration via the blood system to the eyes is completed within 24 hours (Höglund, 1991), although some migrating diplostomulae may also be found in the subcutis and the muscles of the trunk (Ratandart-Brockelman, 1974).

Bucephalidae is the best example of a major group of digeneans with fish as both intermediate and definitive hosts. Definitive hosts are usually predatory fishes (*Esox lucius, Perca fluviatilis, Lophius piscatorius, Conger conger* – Dawes, 1946; *Morone saxatilis* – Matthews, 1973a,b; *Micropterus dolomeiu* and *Lepidosteus* spp. – Hoffman, 1967). However adult bucephalids have been found in grey mullets which are not usually predators (Paperna and Overstreet, 1981). The intermediate piscine hosts are smaller and/or juvenile fishes. Other examples include Macroderoididae, which are parasitic in *Lepidosteus* and *Amia*, with metacercariae in the Poeciliidae (*Gambusia*, etc.), and tadpoles (Hoffman, 1967); and species of *Stephanostomum* (Acanthoclopidae) which infect *Gadus morhua* (cod), *M. saxatilis* (striped bass) and omnivorous fishes (Triglidae, Pleuronectidae and Cottidae), and with metacercariae encysting in gobiid fish and juvenile Pleuronectidae (Ginetsinskaya, 1958; Schell, 1970; Køie, 1978, 1985).

METACERCARIAL STAGES IN MOLLUSCS AND OTHER AQUATIC ORGANISMS

Cercariae may encyst in the tissue of their own molluscan host individuals, or emerge and encyst in another individual of the same species (Ginetsinskaya, 1958; Stunkard, 1959; Van den Broek and de Jong, 1979), or in molluscs of different species and taxa (Hopkins, 1937; Palombi, 1937; DeMartini and Pratt, 1964). When encysting in their own host, cercariae may form metacercariae in the rediae, or in the sporocysts in digeneans which do not have a redia stage. An example is *D. spathaceum* in sporocysts in *Stagnicola emarginata* (Dawes, 1946; Lester and Lee, 1976).

The majority of the digeneans reach their definitive piscine host via planktonic or benthic organisms. In the marine habitats *Fellodistomum felis* second intermediate hosts are brittle stars (Køie, 1980); second intermediate hosts of *Podocotyle atomon* (Opecoelidae) are crustaceans (Amphipoda, Isopoda and Mysidacea – Ginetsinskaya, 1958; Køie, 1985). Metacercariae of marine lepocreadiids, *Opechona bacillaris* and *Lepidapedon rachion*, occur in Coelenterata, Ctenophora, Chaetognatha and Polychaeta (Køie, 1985),

Neopechona cablei occurs in hydrozoan and scyphozoan coelenterates and in ctenophores (Stunkard, 1980), whereas *L. elongatum* occurs in Polychaeta and Mollusca (Køie, 1985). Metacercariae of the zoogoniid, *Zoogonoides viviparus*, have been found in brittle stars, *Ophiura albida*, Polychaeta, lamellibranchs and gastropods (Køie, 1975, 1985). Hemiurid transmission in the marine environment involves planktonic organisms such as *Sagitta* and several medusae, and benthic crustaceans – amphipods, copepods and shrimps (Chabaud and Biguet, 1954).

The greatest versatility is demonstrated in *Derogenes varicus* which has been found in more than 100 fish species, in temperate and sub-polar waters, and in deep waters of the warm seas. The molluscan hosts are *Natica cateria* and *N. pallida*. Copepods (calanoid and harpacticoid) are the common second intermediate hosts, but infections occur in other crustaceans as well, including hermit crabs. *Sagitta* may become infected after consuming copepods. Premature adults in fish may be passed via predation to mature in a predator fish (e.g. cod). Mature *D. varicus* may also re-establish themselves following predation in a new host fish (Køie, 1979a).

Didymozoidae apparently have similar life-histories via planktonic prey-predator chains from copepods via coelenterates, ctenophores, chaetognathes and polychaetes to juvenile fish (Nikolaeva, 1965; Køie and Lester, 1985). Larval forms have also been found in *Lepas* (Cable and Nahhas, 1962).

Common second intermediate hosts of Allocreadiidae (*Crepidostomum* and *Allocreadium* spp.) of freshwater fishes are larvae of aquatic insects: mayflies (Ephemeroptera), caddis-flies (Trichoptera), dragonflies (Odonata) and Chironomidae. Other intermediate hosts include copepods, cladocerans and annelids (Dawes, 1946; Ginetsinskaya, 1958; Hoffman, 1967; Schell, 1970).

Progenetic generations

Sexually mature stages containing fully developed eggs are often found in the first intermediate molluscan host and in other aquatic invertebrates which are hosts of metacercariae. Eggs are released only after disintegration of the host tissue following death or predation. In *Asymphylodora* spp. the progenetic generation in the molluscan host, and also in an annelid second intermediate host (*Nereis diversicolor* (Polychaeta) of the marine species (Reimer, 1973) has been shown to alternate or coexist with the adult stage generation in the definitive host (Stunkard, 1959; Van den Boek and De Jong, 1979). Progenetic stages have also been found in other non-molluscan second intermediate hosts, in amphipods (Hunninen and Cable 1943; De Giusti, 1962) and in crustaceans and chaetognaths (Køie, 1979b). Progenetic metacercariae of *D. varicus* have been found in the parasitic copepod *Lernaeocera lusci* which infects *Gadus luscus* (Dollfus, 1954).

HOST–PARASITE RELATIONSHIPS

Host specificity

Adult stages

According to Manter (1957) piscine digeneans are host specific, since the majority had been from a single host species, or from hosts of the same or related genera. However, his data also showed 35% of the digeneans to have been recovered from more than one host species, and as many as 18% from three or more host species. Data on piscine digeneans from North America (Hoffman, 1967), Africa (Khalil, 1971), and Israel (Paperna, 1964a,b; Fischthal, 1980) indicated that 12 to 39% are 'generalists' (reported from at least two hosts from different fish families). These data should be treated with caution as many single-host records are only single findings, which may be reversed by additional data. Also digeneans are difficult to differentiate in view of the prevailing controversies in species and higher taxa definition (Manter, 1957).

Host specificity at the generic level is even less marked. *Bucephalus, Fellodistomum, Hamacreadium, Phyllodistomum*, and *Stephanostomum* infect a wide variety of fishes from diverse families, and *Halipegus, Phyllodistomum* and *Plagiorchis* also occur in non-piscine hosts. *Deropegus aspina* and *Deropristis inflata* infect both fish and frogs. Most digenean genera contain 'specialists' (i.e. exhibiting host specificity) together with 'generalists'. Only a few genera, and even fewer suprageneric taxa, demonstrate phylogenetic consistency, or predilection to a particular fish taxon (Tubangui, 1931; Paperna, 1964a, Khalil, 1971). All four species of *Proterometra* (Azygiidae) in North America occur in *Lepomis*. Individual species of Didymozoidae are host specific (Yamaguti, 1958, 1970).

Selachians are hosts to 31 species which belong to 11 genera and seven families (Manter, 1957). No digenean family is exclusive to the selachian digeneans, but in no case have digeneans from elesmobranch and teleostean hosts been assigned to the same genus (Manter, 1957; Yamaguti, 1958). No digenean family is also exclusive to any of the pro-teleosteans: accipenserids, Polyodontidae, and holosteans (Bychowskaya-Pavlovskaya *et al.*, 1962; Hoffman, 1967).

Metacercariae

Most digenean metacercariae in fishes have the ability to invade a wide range of hosts. *Diplostomum spathaceum* has been reported from more than 125 fish species (Höglund, 1991). Only a few metacercariae have some degree of predilection to one or a few closely related piscine hosts. Metacercariae of *Clinostomum tilapiae, C. cutaneum* and *Bolbophorus levantinus* occur only in cichlids which mostly belong to *Oreochromis* and *Sarotherodon* (Paperna, 1964b; Ukoli, 1966; Britz *et al.*, 1985; Yekutiel, 1985; Finkelman, 1988). Also, some metacercariae which are in a wide range of hosts nevertheless have preferences for particular fishes. *Heterophyes* spp. in the eastern Mediterranean prefer grey mullet (Mugilidae) over other fish, and among the grey mullets, *Liza ramada* (= *Mugil capito*) is the most preferred species (Paperna and Overstreet, 1981). Development of *D. spathaceum* is completed at 12°C in 28 days in *Phoxinus phoxinus* and *Gobio gobio*, while it takes 35–40 days in *Rutilus rutilus*, 120 days in *Perca fluviatilis* (Sweeting, 1974); and 65–85 days in *Salmo trutta* (Chubb, 1979). The life span of *D. spathaceum* varies in the different hosts and becomes shorter in less compatible ones (Shigin, 1964; Donges, 1969). Sommerville (1981) reported differences in tissue responses among plaice (*Plathichthys platessa*), turbot (*Scophthalmus maximum*), Dover sole (*Solea solea*) and the common dab (*Limanda limanda*) exposed to *Stephanochasmus* (= *Stephanostomus) baccatus* infection. Tissue response was the most intense in turbot which seemed to be the least compatible host and there was mortality of encysted metacercariae.

HABITAT PREFERENCES IN THE HOST

Adult worms in the definitive host

Adult-stage digeneans are primarily parasites of the digestive tract. However, several digenean families have become specialized to extraintestinal habitats. These digeneans have special structural adaptations as well as functional specializations. Sanguinicolidae (e.g. *Aporocotyle, Sanguinicola* and *Cardicola*) are in the blood vessels (Smith, 1972). Transversotrematidae are ectoparasitic, attached under the scales of marine (Witenberg, 1944), brackish-water (Velasquez, 1958; Angel, 1969) and freshwater fish (Mills, 1979). Didymozoidae are in cavities and in tissues. Worms occur in the body cavity in *Mugil cephalus* (Skinner, 1975), and in the orbits of *Labeo* spp. (Khalil, 1969). Worms occur more commonly encysted in the tissue layers near the body cavity, the mouth, the branchial cavity or the fins (see Yamaguti, 1958). *Nematobibothrioides histoldi* lives in the skeletal muscles of ocean sunfish (Thulin, 1980), and *Nematobothrium texomensis* in the ovaries of *Ictobius* (Hoffman, 1967). *Didymocystis palati* and *D. superpalati* are in cysts in the palate of *Neothunnus macropterus*. Their late larval stages and adult stages are in the same host fish (Yamaguti, 1970).

Species of *Phyllodistomum* (Phyllodistominae; Gorgoderiidae) are parasitic in the urinary bladder of fish. Other members of this family (Anaporrhutinae, and Probolitrematinae) occur in the body cavity of elasmobranch fish (see Yamaguti, 1958). *Isoparorchis*, the only genus of Isoparorchiidae, occurs in the air bladder of siluroid fishes (Yamaguti, 1958). *Syncoelium* spp. (Syncoeliidae) occur in the branchial cavity, attached to the apex of a gill raker in sharks, salmonids and tuna (Yamaguti, 1958). Syncoeliidae, of the genus *Paronatrema* are on the skin of the manta ray and in the oviduct of *Squalus* spp. (Manter, 1957).

Other extraintestinal digeneans belong to predominantly gut-inhabiting taxa. The bucephalid *Paurorhynchus hiodontis* occurs in the body cavity of *Hiodon tergisus* (Manter, 1957). The rhynchi, which are characteristic in other bucephalids, are degenerate in this species. One hemiurid, *Gonocerca macroformis* (Manter, 1957), infects cod (*Gadus morhua*) ovaries. *Acetodextra ameiuri* (Cryptogonimidae) infects the swim-bladder, body cavity and ovaries of *Ictalurus* spp. (Hoffman, 1967).

Metacercariae in the piscine intermediate host

Most metacercariae are site or organ-specific. Some encyst only in the skin, beneath the epithelial layer, others in the gill filament cartilage (Farstey, 1986), in the muscles (Paperna 1964a, 1975; Paperna and Lengy, 1963; Paperna and Overstreet, 1981), in the lens and vitreous humour (Larson, 1965; Shariff *et al.*, 1980), and in the retina of the eye (Lester and Huizinga, 1977). *Diplostomulum*, *Ornithodiplostomum*, *Parastictodora*, and *Euhaplorchis* cercariae encyst in the brain (Hoffman, 1967). Of these, *D. baeri eucaliae* (Hoffman and Hundley, 1957), *D. tregenna* (Khalil, 1963), *E. californiensis* (Martin, 1950), and *O. ptychocheilus* (So and Wittrock, 1982) seem to be site specific. Metacercariae of some *Ascocotyle* and *Phagicola* have a predilection for the heart, or truncus arteriosis (Sogandares-Bernal and Lumsden, 1964), but in heavy infections spread to other tissues. Metacercariae of *Haplorchis* are throughout the integumental connective tissues (skin, fins and gills) (Paperna, 1964a; Sommerville, 1982; Yekutiel, 1985). Site specifity may even differ among species of the same genus and closely related genera (Font *et al.*, 1984a,b). *Clinostomum tilapiae* encysts in the connective tissues of the ventral articulation of gill arches, *C. complanatum* in the muscles and *C. cutaneum* beneath the scales (Paperna, 1964a,b; Finkelman, 1988). *C. complanatum* has a preference for red muscles (92%) and a predilection to the branchial and mandibular muscles (62%) (Finkelman, 1988).

Feeding and nutrition

Little is known about nutrition and metabolic function of fish digeneans. The blood-dwelling flukes (Sanguinicolidae), feed on blood, while the encysted didymozoid obtain plasma from the capillaries around the cyst and blood cells are also ingested (Yamaguti, 1970; Lester, 1980). In the ectoparasitic Transversotrematidae, feeding has been suggested to be similar to the skin-inhabiting capsaloid monogeneans (*Benedenia*), i.e. digestion of the host epidermal cells by proteolytic enzymes excreted from the pharynx (Mills, 1979).

The elaborate tegumental surface seen in ultrastructural studies of sporocysts, rediae and encysted metacercariae of piscine digeneans strongly suggests active nutrient absorption (Køie, 1971a,b, 1985). During migration, the metacercariae of *D. spathaceum* have been shown to utilize their glycogen reserves (Ratanart-Brockelman, 1974). In metacercarial cysts of *Bucephaloides gracilescens*, uptake of labelled substances across the cyst wall and subsequent incorporation by the metacercaria indicates that the encysted larva retains communication with the external environment. The degree of encapsulation varies with encystment site and increases with age; substrate incorporation has been found to be lowered in thicker-walled cysts (Halton and Johnston, 1982). The cyst wall of the large clinostomatids is vascularized, which facilitates efficient glucose transport to the encysted worm. It also lacks the acellular, parasite-induced inner wall. Larson *et al.* (1988) showed that the rate of diffusion through the cyst wall exceeded the rate of transport into the worm. Glucose uptake by encysted *Clinostomum marginatum* is not affected by glucose transport inhibitors (phlorizin and phloretin) (Ugelm and Larson, 1987). Apparently, the cyst wall is selectively permeable, permitting rapid diffusion of glucose to the worm surface while restricting the movement of the inhibitors.

Pathology

Adult trematodes

Adult intestinal trematodes are normally considered not to cause disease even when their numbers are high. Extraintestinal parasites are, however potentially pathogenic.

   Blood flukes (sanguinicolids and aporocotyles) cause considerable damage to the gills and impair respiration. Adult worms and eggs can physically obstruct the passage of blood, cause thrombosis and subsequent tissue necrosis (Hoffman *et al.*, 1985; Ogawa *et al.*, 1989). Extensive destruction of the gills by emerging and trapped eggs (Fig. 9.6B,C) appeared to be the direct cause of mortality of infected brook trout (*Salvelinus fontinalis*) in Pennsylvania, USA (Hoffman *et al.*, 1985). In flat-fishes *(Hippoglossoides platessoides, Limanda limanda* and *Pleuronectes platessa*), infected with *Aporocotyle simplex*, fragments of disintegrated worms in the afferent gill filament arteries cause the gill filament to become stunted and necrotic (Køie, 1982). The blood fluke *Paradeontacylix* sp. caused mass mortalities of cage-cultured *Seriola purpurascens*. Infections with up to 1000 worms per single gill filament invoked gill hyperplasia. Eggs in the gills and the ventriculum were encapsulated and papillate proliferation of the endothelium of afferent branchial arteries was induced. No necrosis was found in the gills or elsewhere (Ogawa *et al.*, 1989). Proliferation of the arterial endothelium was also reported in common carp infected with *Sanguinicola inermis* (Prost, M., quoted in Lucky, 1964). Heavy blood loss when miracidia escaped from the gills was apparently the main cause for mortality in trout fingerlings infected with *S. davisi* and *S. klamathensis* (Wales, 1958; Davis *et al.*, 1961). Loss of blood was evidenced by the pale colour of the gills and the decline in packed cell volumes and oxyhaemoglobin levels (Evans, 1974a). Heavily infected cultured carp apparently suffocated during transportation (Lucky, 1964; Smith, 1972). In chronic infections, adult worms disperse and become stranded in the heart, kidneys and caudal vessels. Some eggs become encapsulated, and may also become surrounded with focal granuloma. Nodular foci occur in the heart, head kidney and spleen of *Oreochromis* spp. (Figs 9.6D, 9.18A,B), in the gills, heart and kidneys of carp (Scheuring, 1922), and in the liver of *Paracardicola hawaiensis*-infected *Tetraodon hispidus* (Martin, 1969). Eggs of *S. armata* became widespread in viscera of grass carp (*Ctenopharyngodon idella*) and bighead (*Aristichthys nobilis*) but tissue response was negligible (Anderson

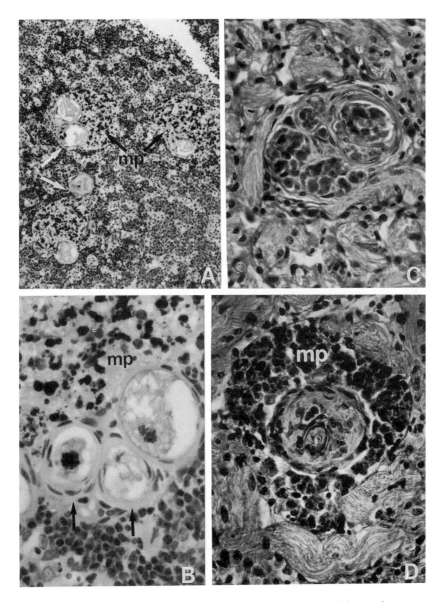

Fig. 9.18. A. Degenerate *Sanguinicola* eggs encased within melanomacrophages (mp) in the head-kidney of *Oreochromis aurea*. B. Encapsulated *Sanguinicola* eggs in spleen of *O. aurea*. C, D. Heart of *Pearsonellum corventum* infected *Plectropomus leopardus*: (C) Several encapsulated egg masses bound as subunits into a single mass and individually infiltrated by macrophages; (D) granuloma encased egg residue, within a melanomacrophage. (From Overstreet and Thulin, 1989, by courtesy of R. Overstreet.)

and Shaharom-Harrison, 1986). Eggs in kidneys of *S. klamathensis*-infected cutthroat trout (*Salmo clarki*) caused hypertrophy or necrosis of the renal epithelium and renal calcifications (Evans, 1974b). In sanguinicolosis of cichlid fishes (Fig. 9.18A,B) and in *Pearsonellum corventum*-infected serranids (*Plectropomus leopardus*), granulomas formed around eggs in the viscera and these were accompanied by melanomacrophage aggregates (Fig. 9.18C,D; Overstreet and Thulin, 1989). Dead *A. simplex* were observed encapsulated in the superficial parts of the liver (Køie, 1982). General effects on growth and condition are not always evident however, and are correlated with the duration and severity of the infection. Smith and Willams (1967) did not find any apparent effect on weight in hake (*Merluccius merluccius*) infected with *Aporocotyle spinosicanalis*.

Encapsulation of didymozoid worms induce no more than limited local reactions (Fig. 9.19; Lester, 1980; Perera, 1992a,b) although at times, there is also haemorrhaging of peripheral capillaries in the buccal dermis of *Platycephalus fuscus* (Figs 9.8A, 9.19C; Lester, 1980). *Nematobothrium spinneri* in the muscles of *Acanthocybium solandri* is enclosed in a thick-walled capsule (Fig. 9.19A,D). Connective tissue and blood capillaries are interdigitated with worm coils which facilitates worm feeding on blood (Lester, 1980). Only natural infections have been recorded, and these are likely to be moderate. It is not known whether these infections are harmful to maricultured fish.

Transversotrema patialense, an ectoparasite on *Brachydanio rerio* integument, leaves pressure and feeding indentations of the body surface. Tissue regeneration occurs soon after the worm changes position (Mills, 1979).

Kidney damage induced by heavy infection of the urethra with *Phyllodistomum umblae* (= *conostomum*) adversely affects survival of char migrating from fresh water to sea water (Berland, 1987). *Asymphylodora kubanica* are pathogenic to *Rutilus rutilus heckeli* in the sea of Azov when they infect the kidneys rather than the intestine (Bauer, 1958).

Metacercariae

Clinical effects of infection are often not obvious. Metacercariae in supposedly sensitive organs such as the brain, cranial nerves or spinal cord, e.g. *Bucephaloides gracilescens* in cod, *Gadus morhua* (Matthews, 1974a), *Diplostomum mashonense* and *D. tregenna*, in *Clarias* spp. (Beverly-Burton, 1963; Khalil, 1963), and *Ornithodiplostomum ptychocheilus* in *Pimephales promelas* (Hoffman, 1958; So and Wittrock, 1982), do not necessarily have debilitating effects on fish; sometimes even when the number of parasites is relatively high, and despite visible structural damage to organs.

Sudden massive outbreak of infection is often fatal. Hoffmann *et al.* (1990) reported a massive *Bucephalus polymorphus* infestation and mortality of cyprinid fishes after the water temperature suddenly increased from 12–14 to 20°C. Exposure to massive numbers of cercariae kill a fry within a few hours (Sommerville, 1982), but such exposures do not normally occur in nature. Cercariae penetrate and encyst deeper in the tissues of small fishes and the relatively larger cysts may interfere with organ function (Fig. 9.20).

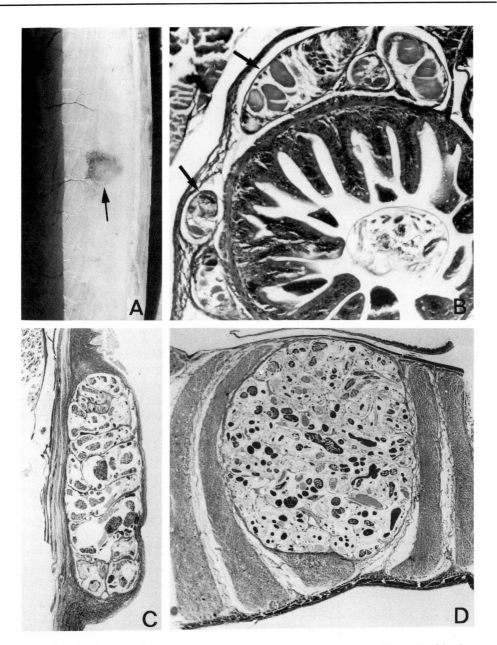

Fig. 9.19. Didymozoids in tissues. A. *Nematobothrium spinneri* in wall of body cavity of *Acanthocybium solandri*, Point-Lookout, Queensland. B. Larval stages encysted on the surface of the intestine of *Favonigobius exquisitus*. C. *N. helicis* capsule in the dermis. D. Section of *N. spinneri* in the body wall muscle. (Photograph courtesy of B. Lester)

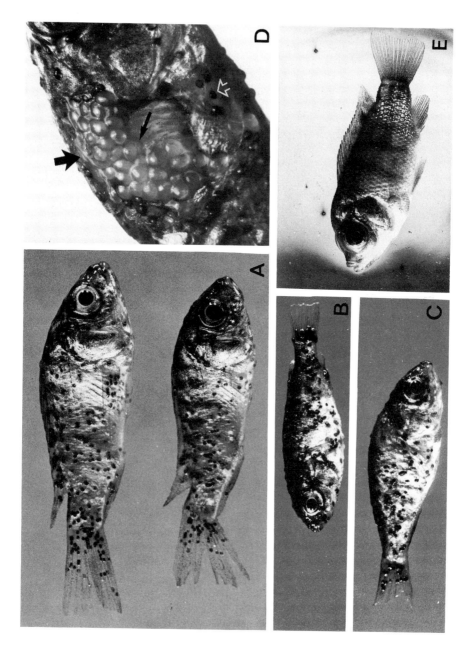

Fig. 9.20. A–D. Juvenile cichlids infected with *Neascus* and *Bolbophorus levantinus*: (A) pond reared *Oreochromis aurea* × *nilotica* (tilapia), 45, 50 mm long; (B,C) *Tristramella simonis* from lake Kinnereth fringe swamps, 35 mm long; (D) skin removed from fish C to demonstrate *B. levantinum* cysts in the muscles (black arrows, white arrow – *Neascus*. E. Deformations induced in young *O. aurea* by exposure to *B. levantinus* infection.

The effects of cercarial infestation was most severe in 0-year class plaice and minimal in 1 + year class (Sommerville, 1981). Massive metacercariae infections in juvenile (0-class) fish have been incriminated as an important cause of natural mortalities, e.g. *Cryptocotyle lingua* infections of plaice *Pleuronectes platessa* (MacKenzie, 1968), *Bolbophorus levantinus* in cichlid fishes (Fig. 9.20; Yekutiel, 1985; Paperna, 1991). Population studies and field observations suggest that fish heavily infected with metacercariae are lost from the host population (Chubb, 1979; Lemly and Esch 1984a,b). Lemly and Esch (1984b) confirmed that heavily *Uvulifer ambloplitis*-infected young-of-the-year bluegill sunfish (>50 per fish) did not survive the winter because of depleted fat reserves. Newly acquired metacercariae significantly stimulated oxygen consumption in experimentally infected fish during the first 60 days after infection. These changes were correlated with the encapsulation of the parasite by the host. Heavy gill infection appears to lower respiratory efficiency: all young cichlids (*Sarotherodon galilaeus*) heavily infected with *Centrocestus* (Fig. 9.21A) (116 ± 48 per fish) succumbed, while all, same size, lightly infected fish (with 15 ± 15) survived (Farstey, 1986) during a three-hour transport.

Cardiac pathology, resulting from thickening due to fibrogranulomatosis of the epicardium, is associated with infection of the pericardium with the strigeid *Apatemon gracilis* (Watson *et al.*, 1992). Tort *et al.* (1987) demonstrated that *in vitro* pumping performance of hearts in infected fish was reduced by as much as 50%. Cardiac infections are also commonly caused by heterophyiids of the genera *Phagicola* and *Ascocotyle* (Fig. 9.21B; Sogandares-Bernal and Lumsden, 1964; Stein and Lumsden, 1971).

Heavy visceral infection of *Posthodiplostomum minimum* ('white grub'; 500–2300 metacercariae in 50–65 mm long fish) in fathead minnow caused abdominal dropsy (Fig. 9.22) with milky ascitic fluid which contained 92% leucocytes and 8% erythrocytes. Granuloma developed around degenerating parasites and parasitic debris (Mitchell *et al.*, 1982).

ENTRY INTO TISSUES AND ENCYSTMENT

Pronounced inflammatory response often accompanies penetration and early migration. This is particularly obvious following heavy exposure. Fish may even die from the penetration wounds caused by *D. spathaceum* (Ferguson and Hayford, 1941), from tissue damage (focal haemorrhages) caused during *H. pumilio* migration in muscles (Sommerville, 1982), and in the brain caused by *D. adamsi* (Lester and Huizings, 1977). Cercariae of *C. lingua* and *Stephanochasmus baccatus* induced temporary epidermal lesions. They may cause also some disruption of connective tissues, inflammatory cell proliferation, sometimes myofibrillar necrosis (associated with bacteria) and reactive swelling of the intermuscular septa (McQueen *et al.*, 1973; Sommerville, 1981). The inflammatory reaction is particularly intense around nonencysting metacercariae (e.g. *Rhipidocotyle johnstonei* and *Prosorhynchus crucibulum* in plaice and turbot) (Matthews 1973b, 1974b), and precede the eventual enclosure of the parasite in a fibrous capsule (Yekutiel, 1985) (Fig. 9.23A). The fibrous capsule produced by the host is superimposed on the acellular wall secreted

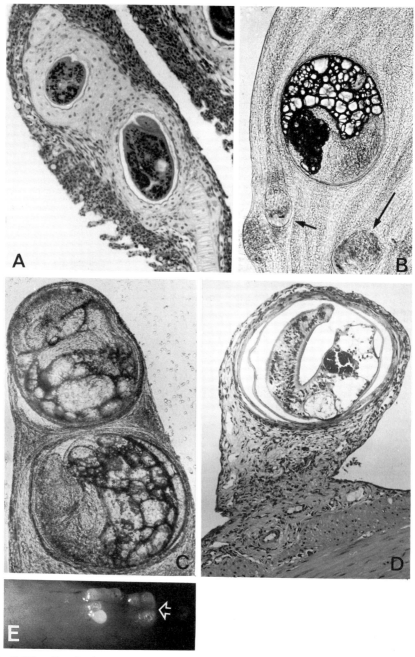

Fig. 9.21. Cartilaginous cyst around *Centrocestus* metacercariae in gills of *Oreochromis aurea*, Lake Kinnereth. B. *Phagicola* sp. and young metacercariae (arrows) in the heart of *Tristramella simonis*, Lake Kinnereth. C–E. *Phagicola nana* metacercarial cysts projecting from the pyloric caeca surface of *Micropterus salmoides*; (C) fresh material under slide pressure; (D) histology, the formed peduncle is infiltrated by vascular and muscle tissue components; (E) view of pyloric caeca surface. (Arrow shows the metacercariae.) (Font *et al.*, 1984, by courtesy of R. Overstreet.)

Fig. 9.22. Abdominal dropsy in *Posthodiplostomum minimum*-infected *Pimephales promelas* (43 mm long). (After Mitchell, *et al.*, 1982, by courtesy of G.L. Hoffman.)

by the encysting cercaria. The peripheral inflammatory cellular infiltrate which precedes fibrous capsule formation consists mainly of macrophages (Sommerville, 1981). Inflammatory cells become stratified into an epitheloid (Fig. 9.23B), while the innermost layer gradually degenerates (Fig. 9.23C). This early capsule, which has the form of a chronic granuloma (and also contains giant cells) is gradually replaced by the fibrous capsule (Sommerville, 1981; Yekutiel, 1985). At 5°C, cyst formation is much slower than at 15°C (McQueen *et al.*, 1973). Metacercariae of *B. levantinum* migrate in tissues before encysting in the muscles (Paperna and Lengy, 1963). The inflammatory response and macrophage proliferation are limited to host tissues damaged in the earlier stages of the encystment (Fig. 9.23; Yekutiel, 1985). Cellular damage around *Heterophyes* spp. infecting grey mullets is minimal even in heavy infections (6000 metacercariae per gram muscle; Paperna, 1975). Under certain circumstances, however, the same metacercariae can induce more extensive cellular changes, particularly during the late stages of infection, when damaged tissues and dead parasites have been accumulated (Font *et al.*, 1984a). In muscles of grey mullets infected with *Stellantchasmus falcatus*, myofibrils degenerating near the parasitic cysts are replaced by large fat cells (Lee and Cheng, 1970). Cellular and deposited ground substances may vary with species, but may also be affected by the host tissues. Collagen may be deposited in the liver (Mitchell, 1974) and in the spleen (Font *et al.*, 1984a), but it is usually absent. When crowded on the serosal surface of the pyloric caeca (Fig. 9.21C–E), metacercariae of *P. nana* evoke intense proliferation of the connective tissues, with a considerable amount of collagen. When the parasite is in the submocosa, the proliferation includes smooth muscle (Font *et al.*, 1984a).

Cysts consolidating around certain dermal metacercariae may incorporate melanophores and exceptionally, other chromophores. These are commonly called 'black spot', and are formed by *Neascus* metacercariae from the genera *Crassophiala, Ornithodiplostomum* and *Uvulifer* (Hoffman, 1960; Wittrock, *et al.*, 1991) (Figs 9.20A–C, 9.24A–C) and heterophyiids of the genera *Cryptocotyle* and *Haplorchis*. Cutaneous melanosis and cloudy eye

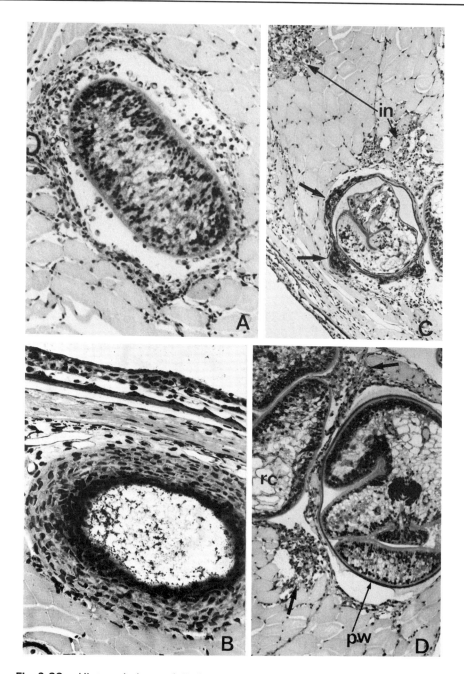

Fig. 9.23. Histopathology of *Bolbophorus levantinus* infection in young *Oreo-chromis aurea*. A. Macrophage infiltration around young unencysted metacer-caria. B. Formation of an epitheloid around the metacercaria. C. Necrotic boundles (arrows) in the fibrous cyst consolidating around the metacercaria and inflam-matory process extending to the muscles (in). D. Mature metacercaria with 'reserve' cells (rc) enclosed in a parasite-derived wall (pw). Note (arrow) focal macrophage proliferations in the muscles.

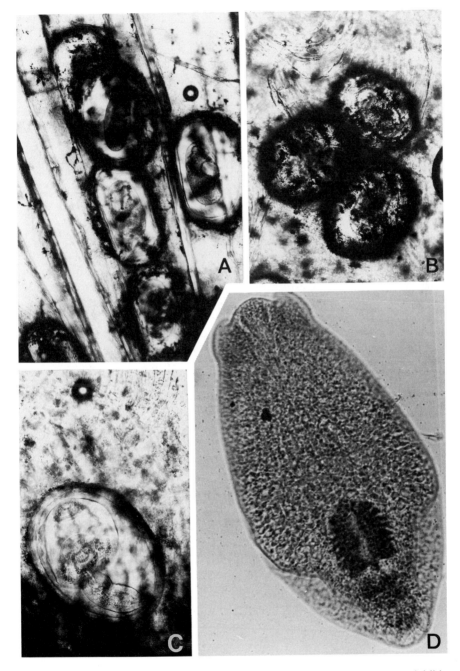

Fig. 9.24. A–C. *Neascus* ('black spot') infection in skin of young cichlids: (C) *Uvulifer*-like sp.; D. *Diplostomum* sp. removed from a visceral cyst in *Clarias gariepinus*, Uganda.

have been reported in flounders, heavily infected with pigmented *C. lingua* (Mawdesley-Thomas and Young, 1967). *Centrocestus* metacercariae on the gills become encysted in cartilaginous capsules (Figs 9.15C, 9.21A). Cercariae penetrate the gill filaments and encyst in the connective tissue adjacent to the cartilage rays of the gill filament. Chondroblasts proliferate from the ray perichondrium and, following deposition of ground substance, the meta-cercaria becomes encapsulated (by day 7 post-infection) within a cartilaginous extension of the filament's ray (Farstey, 1986).

The 'sand-grain' grub of yellow perch is caused by metacercarial cysts of *Apophallus brevis*. It is composed of bone which becomes clustered around the parasite within the peripheral blood vessels. Such cysts have also been identified in ancient fossil perch. The bony structure has two opposite escape canals for the parasites and has lines interpreted as the growth rings of the bone (Sinclair, 1972).

EYE DAMAGE IN METACERCARIAL INFECTIONS

Damage to eyes may either be caused by metacercariae with a predilection for the organ (e.g. *D. spathaceum*), or as a non-specific side effect (e.g. corneal infection by integument-encysting metacercariae). Impairment of vision is aggravated when metacercarial cysts are accompanied by melanophores (black spot). Severe infection leads to exophthalmos, cataracts, and even complete collapse of the eye. Blindness can be uni- or bilateral (Sinderman and Rosenfield, 1954; Mawdesley-Thomas and Young, 1967).

D. spathaceum is a specific pathogen of the eye. The non-encysted metacercariae invade the lens, the vitreous humour and the retina. Detailed pathological and histopathological descriptions of diplostomiasis are in Larson (1965) and Shariff *et al.* (1980). Farmed rainbow trout yearlings with infection limited to a cataract (seen as a small white opacity in the lens with parasites visible beneath the lens capsule), fed normally and remained in good condition. However, in two-year-old trout, the cataract is prominent, with the lens dislocated into the anterior chamber or absent. Fish were dark, apparently blind, emaciated, and fed only from the bottom of the tank. Parasites are most frequently in the cortical region of the lens, within a cavity with fine debris, which result from liquefaction of the cortical fibres. Degeneration of the lens is evident. In the chronic condition, the lens became irregular and wrinkled, formed adherences with the inner components of the eye, and often ruptured. Thickening of the lens capsule lead to exfoliation with proliferation of the epithelial layer. The iris is greatly distorted and it is in contact with the lens. In catfishes (*Ictalurus melas* and *I. nebulosus*) the lens herniates near the point of attachment to the dorsal ligament, with metacercariae spilling into the protrusion and leaving the remainder of the lens clear for visual function (Fig. 9.25; Larson, 1965). Entry of parasites to the subretinal level results in retinal detachment. Cellular response usually becomes evident only in later stage infections. Macrophages aggregate around the debris from ruptured lenses, in the damaged periphery of the iris, in the choroid and around necrotic foci in the optic nerve. The combined effects of cataracts and retinal detach-ment cause blindness (Shariff *et al.*, 1980). Crowden and Broom (1980)

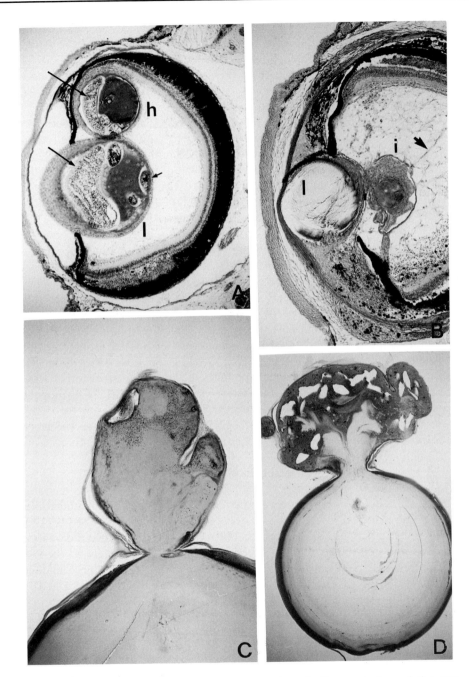

Fig. 9.25. Eye diplostomiasis in *Ictalurus melas*. A. Cross section of the eye showing herniation (h) of the lens (l) and the metacercariae inside (arrow). B. Damaged lens with a proliferated epithelium around a necrotic core (i), note vascularization of the vitreous humour (arrow). C,D, Cross section of herniated lenses, empty spaces contained worms before processing. (After Larson, 1965, by courtesy of the author.)

described loss of efficiency and increase in time devoted to feeding. As the infection progressed, fish spent more time in the surface layer of the water. This would increase their risk of predation by gulls.

When *D. adamsi* cercariae enter (in groups of 20 or more) the retina of *Perca flavescens* and remain unencysted, they form a cavity between the photoreceptor layer and the pigment epithelium. *D. schudderi* becomes established in a similar manner in the retina of *Gasterosteus aculeatus*. In both fish hosts, cytologic damage and alternations occur in both layers of the retina. This suggests loss of vision in the foci of infection in the retina, but an overall minimal visual loss due to the peripheral location of the parasites (Lester and Huizinga, 1977). Eye damage is less obvious when the metacercariae (e.g. *Posthodiplostomum brevicaudatum*) become encysted (by their own wall only) in the vitreous humour and in the retina (Dönges, 1969).

Estimating the extent of mortality in affected fish populations

Estimation of metacercariae-induced host mortality in natural fish populations has been attempted by extrapolating quantitative data on frequency distribution of infection. The decline in variance was considered to result from the loss of heavily parasitized fish. Hence a measure of overdispersion, by comparison to negative binomial distribution, or by calculating the ratio of variance to mean, was advocated as an indirect method of estimating fish mortality (Anderson and Gordon, 1982; Gordon and Rau, 1982; Kennedy, 1984; Lester, 1977, 1984). Seasonal changes in the degree of overdispersion were related to suggested parasite-caused mortality in bluegills during the winter (Lemly and Esch, 1984b). However, extrapolations from field data have met with a variable degree of success, and at times have produced ambiguous results (Kennedy, 1984; Lester, 1984).

Marketing food fish aspects

Fish containing conspicuously visible (large or coloured) free or encysted worms are rejected by consumers (Paperna, 1980; Kabunda and Sommerville, 1984). Fish with badly damaged eyes (trout affected by eye diplostomiasis) are likely to be sold at depreciated prices. The mere suspicion of infection affects demand and depreciates market prices of the entire consignment of the fish species and of fishing catches. Examples are the large yellow clinostomatids, which also tend to excyst and crawl around after the death of the fish (Fig. 9.26), metacercariae of diplostomatids in muscles (Fig. 9.24D) (Paperna, 1980), 'black spot' (Fig. 9.20A–C) and fish containing large didymozoids in skin and gills or in the muscles (Lester, 1980; Figs 9.8, 9.19A).

(For risk to public health see chapter 15 in this book.)

Immune responses

In contrast to the wealth of information that has been accumulated on the immunology of digeneans of higher vertebrates, (Butterworth, 1984), data on fish-infecting digeneans are very preliminary in nature. The main difference

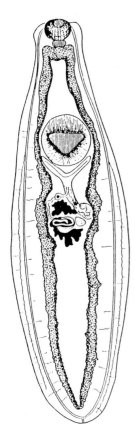

Fig. 9.26. *Clinostomum cutaneum* metacercaria. (Actual size 7 × 2.5 mm; Paperna, 1964b.)

between immune responses of fish and those of mammals and birds is that the ambient temperature affects antibody production in fish (Avtalion *et al.*, 1973). Eosinophilic granulocytes, which play an important role in protective immunity in mammalian digenean (schistosome) infections (Butterworth, 1984) do not seem to have a similar importance in piscine infections (Höglund and Thuvander, 1990). However, it is not certain whether piscine eosinophils are homologous with those of in mammals (Ellis, 1977).

Precipitating antibody against antigen from the intestinal digenean *Telogaster opisthorchis* was demonstrated in the sera and gut mucus of eels, *Anguilla australis schmidtii* and *A. dieffenbachii*, (McArthur, 1978). Wood and Matthews (1987) reported induction of humoral antibodies, sensitized pronephric leucocytes and cytotoxic serum factors in *Chelon labrosus* exposed to *Cryptocotyle lingua* cercariae. Precipitating antibodies were detected, using the Ouchterlony technique, in plaice (*Pleuronectes platessa*) infected with *C. lingua* and *Rhipidocotyle johnstoni*. The immune sera had an immobilizing effect on the cercariae. Antibodies reacted with the tegument and with secretory glands in cercariae of both parasite species. Antibody production was temperature-dependent and did not occur at 5°C (Cottrell, 1977).

The antibodies were macroglobulins, and they resemble mammalian immuno-globulin M (IgM).

Bortz *et al*. (1984), using ELISA (enzyme-linked immunosorbent assay) detected humoral antibodies in trout with natural infections of *Diplostomum spathaceum* and trout injected with sonicated metacercariae. In trout immunized with antigens prepared from cercariae and diplostomulae, anti-cercarial and anti-diplostomulae circulating antibodies were demonstrated using ELISA and immunofluorescence. The two cross-reacted, although the reaction between anti-cercarial sera and diplostomatid antigen was weaker. Fluorescence studies indicated the tail region of the cercariae to be strongly antigenic (Whyte *et al*., 1987). Trout repeatedly exposed to cercariae of *Diplostomum* over a 12 week period developed protective immunity, but specific antibodies could not be detected in the blood (Höglund and Thuvander, 1990). The protection seemed to be due to cell-mediated immunity, or to a nonspecific mechanism of protection against the migrating diplostomulae. In mammals, induction of cellular immunity is required for protection against helminth infections (Butterworth, 1984). Migrating diplo-stomulae induce nonspecific cellular responses which involve infiltration of neutrophils and monocytes (Ratanarat-Brockelmann, 1974). Speed and Pauley (1985) showed that the survival time of immunized, challenged fish increased by four months over that of non-immunized controls.

IN VITRO CULTURE

In vitro studies of piscine digeneans have been concentrated on the induction of sexual maturation of metacercariae, rather than on the long-term main-tenance of larval or adult stages. Metacercariae cannot be used for specific identifications. Stunkard (1930) was the first to attempt cultivation of meta-cercariae to adult stages and, to date, this has been accomplished for a number of species (Kannangara and Smyth, 1974). Metacercariae which already possess genital primordia are more readily cultured to the adult stage which produce eggs capable of hatching (Basch *et al*., 1973; Mitchell *et al*., 1978). Moreover, some metacercariae (*Gynaecotyle adunca*, *Pleurogenoides sitapurii*, *Micro-phallus papillorobustus*) produce eggs after only minimal stimulation, e.g. release from the cyst, exposure to light or changes in temperature (Kannangara and Smyth, 1974). Metacercariae which do not have genital rudiments take longer to develop and are more fastidious in their *in vitro* maintenance needs. *Diplostomum phoxini* was first cultivated in a medium which contained egg yolk (Bell and Smyth, 1958), but subsequent studies demonstrated that the egg yolk had little or no beneficial effect. *D. phoxini* and *D. spathaceum* reached the egg production stage, but eggs of the latter were not viable (Kannangara and Smyth, 1974). In a more recent study, Leno and Holloway (1986) reported cultivation of *D. spathaceum* on chick chorioallantois. Whyte *et al*. (1988) transformed *D. spathaceum* cercariae into metacercariae *in vitro*. These were then further maintained for 72 h. Cercariae were allowed to penetrate via trout skin using a modified version of the Clegg and Smithers (1972) mouse-skin technique for schistosomes. The best results for metacercarial maintenance *in vitro* were obtained with L-15 (Gibco) medium supplemented with added foetal calf serum (Whyte *et al.*, 1988).

PREVENTION AND CONTROL

Transmission control

The best preventative strategy for controlling digenean infections in farmed fish is the elimination of the intermediate host snail. It includes the use of chemical molluscicides, environmental manipulation, and use of molluscophagous fish.

There is extensive literature on the control of snails which are intermediate hosts of schistosomes and *Fasciola* (McCullough and Mott, 1983; Madsen, 1990). Of all the many molluscicides developed to control these snails, only copper sulphate is of practical use in fish ponds and circulation systems. Molluscicidal concentrations of niclosamide (=Bayluscide, Bayer 73) and N-tritylmorpholine (=Frescon, WL 8008, Shell) currently recommended for snail control are toxic to fish (Cowper, 1971). Copper sulphate (5-hydrate) is tolerated by most fish (although some species and the younger fish may be more susceptible). It is inexpensive, it is widely used in fish ponds as an algicide (Sarig, 1971), and it can be safely applied at a dose of 3.5 ppm to seawater and brackish-water ponds, and to neutral and hard freshwater ponds. However, in acid and soft freshwater ponds (pH >6.8, calcium ions >12 ppm) the same or lower concentrations become toxic to fish. Copper salt may be applied continuously at a lower concentration (1 ppm), or as a low-soluble formulation (as copper carbonate or copper oxide) to produce long-term effects (Hoffman, 1970). Treatment of drained ponds or raceways with copper sulphate prior to stocking delayed, but did not prevent, repopulation by snails (Stables and Chappell, 1986).

The environmental limits imposed on snail survival in fish farm systems were discussed earlier and regular weed control, performed manually or with herbicides (Paperna, 1991) can decimate snail populations. Of the many recommended molluscophagous fishes (De Bont and De Bont Hers, 1952; Carothers and Allison, 1968; Paperna, 1980; Taraschewski and Paperna, 1981), only black carp (*Mylopharyngodon piceus*) in fresh waters and sea-breams (*Sparus* spp.) in seawater ponds are compatible with fish-rearing practices and economy.

Parasite control

Praziquantel (Biltricide®, Bayer AG, Germany) is safe and effective against digeneans and cestodes of man and animals (Andrews *et al.*, 1983). Preliminary trials demonstrated its parasiticidal effect on *Diplostomum spathaceum* metacercariae in rainbow trout fed on medicated feed (Bylund and Sumari, 1981). Szekely and Molnar (1991) reported on the elimination of all *D. spathaceum* metacercariae from herbivorous carp which were fed a single dose of drug (300 mg kg^{-1} body mass). Three sequential lower doses of 35–100 mg kg^{-1} yielded 88–100% efficacy, and bath treatments of 1 mg l^{-1} for 9 hours or 10 mg l^{-1} for 1 hour showed 100% and 93–94% efficiency, respectively. Short (30 min–4 h) and long (1–8 days) bath treatments were also effective against *D. pusillum* and *Apatemon gracilis* (Zatkanbayeva and Heckmann, 1990). Mr N. Kraus (personal communication) from Kibbutz Hamaapil fish farm, Israel, used a veterinary formulation of praziquantel (Droncit) to eliminate metacercarial infections of *Centrocestus*, *Haplorchis* and *Bolbophorus levantinus* in juvenile tilapia (>70 mm in length).

Moser and Sakanari (1986) eliminated diplostomatid cercarial infections in *Physa* and *Lymnaea* by 12 hours bath treatment with 5.68 mg l^{-1} praziquantel. In spite of its promising therapeutic qualities, praziquantel is expensive for routine use, except in very special circumstances such as high-priced ornamental fish, breeders or valuable genetic stock. Mr N. Kraus (personal communication) showed that dissolved praziquantel in dip tanks retained its therapeutic efficiency and could be reused for over a month.

CONCLUSIONS

Digeneans which parasitize fish are numerous and diverse in their morphologies and life histories. Studies on adult digeneans and metacercariae require different approaches. The importance of digeneans to the fish culture has long been underestimated, their risk to public health has not received adequate recognition. With the rapid development of warm-water aquaculture as well as mariculture, and the spread of exotic culinary practices to western societies, risks to cultured fish and to consumers of infected fish are likely to become significant (see chapter 15).

Extraintestinal adult digeneans are potentially pathogenic to maricultured fish; however, piscine digeneans are receiving less attention now than in the past. In the last decade the transition in parasitology from an essentially zoological to a biochemical, immunological and molecular science has had only marginal impact on fish digenean research. Future studies should include nutritional physiology, immunology, and taxonomy which make use of DNA analysis or isoenzyme electrophoresis. The prospects of research employing updated methodologies are great. Temperature-dependence of the immune response combined with the peculiarities of the defence mechanisms against helminthic infections offer an attractive and challenging research model. Biochemical and DNA taxonomy are potentially the best options for resolving taxonomic affinities and specific recognition of larval stages.

ACKNOWLEDGEMENTS

I wish to thank my collegues Glenn L. Hoffman, Kearneysville, West Virginia, USA; Johan Höglund, National Veterinary Institute, Uppsala, Sweden; Marianne Køie; Marine Biology Laboratory, Helsingor, Denmark; Omer

R. Larson, University of South Dakota, USA; Bob Lester, University of Queensland, Australia; Paola Oreccia, Universita degli Studi di Roma, Italy; Robin Overstreet, Gulf Coast Research Laboratory, Mississippi, USA; John C. Pearson, University of Queensland, Australia and Darwin Wittrock, the University of Wisconsin Eau Claire, USA, for kindly providing me with published and unpublished data and publications, drawings and photographs and their kind permission to use their illustrations and photographs in this review.

REFERENCES

Anderson, G.R. and Barker, S.C. (1993) Species differentiation in the Didymozoidae (Digenea): restriction fragment length differences in internal transcribed spacer and 5.8S ribosomal DNA. *International Journal for Parasitology* 23, 133–136.

Anderson, I.G. and Shaharom-Harrison, F. (1986) *Sanguinicola armata* infection in bighead carp (*Aristichthys nobilis*) and grass carp (*Ctenopharyngodon idellus*) imported in Malaysia. In: Maclean, J.I., Dizon, I.B. and Hosillos, L.V. (eds) *The First Asian Fisheries Forum*. Asian Fisheries Society, Manila, Philippines, pp. 247–250.

Anderson, R.M. and Gordon, D.M. (1982) Processes influencing the distribution of parasite numbers within host populations with special emphasis on parasite-induced host mortalities. *Parasitology* 85, 373–398.

Andrews, P.H., Thomas, R., Pohlke, R. and Seubert, J. (1983) Praziquantel. *Medical Research Review* 3, 147–200.

Angel, M. (1969) *Prototransversotrema steeri* gen. nov. sp.nov. (Digenea: Transversotrematidae) from a South Australian fish. *Parasitology* 59, 719–724.

Avtalion, R.R., Wojdani, Z., Malik, Z., Shahrabani, R. and Duczyminer, M. (1973) Influence of environmental temperatures on the immune response in fish. *Current Topics in Microbiology and Immunology* 61, 1–35.

Awachie, J.B.E. (1968) On the bionomics of *Crepidostomum metoecus* (Braun, 1900) and *Crepidostomum farionis* (Muller, 1784) (Trematoda: Allocreadiidae) *Parasitology* 58, 307–324.

Basch, P.E., Diconza, J.J. and Johnson, B.E. (1973) Strigeid trematode (*Cotylurus lutzi*) cultured *in vitro*: Production of normal eggs with continuance of life cycle. *Journal of Parasitology* 59, 319–322.

Bauer, O.N. (1958) Relationships between the parasites and their hosts (fishes) In: Dogiel, V.A., Petrushevski, G.K. and Polyanski, Yu.I (eds) *Fundamental Problems of the Parasitology of Fishes*. Izdatelstvo Leningradskovo Universiteta, Leningrad pp. 90–108 (in Russian, English translation: Kabata, Z., 1961, *Parasitology of Fishes*. Oliver and Boyd, Edinburgh, pp. 84–103).

Bauer, O.N. (1959) The influence of environmental factors on reproduction of fish parasites, *Voprosy Ecology* 3, 132–141 (in Russian, English translation – Fisheries Research Board of Canada translation No. 1099).

Becker, C.D. and Brunson, W.D. (1966) Transmission of *Diplostomum flexicaudum* to trout by ingestion of precocious metacercariae in molluscs. *Journal of Parasitology* 52, 829–830.

Bell, E.J. and Smyth, J.D. (1958) Cytological and histochemical criteria for evaluating development of trematodes and pseudophyllidean cestodes *in vivo* and *in vitro*. *Parasitology* 48, 131–148.

Berland, B. (1987) Helminth problems in seawater aquaculture. In: Stenmark, E. and

Malmberg, G. (eds) *Parasites and Diseases in Natural Waters and Aquaculture in Nordic Countries*. Zoo-Tax Naturhistoriska riksmuseet, Stockholm, Sweden, pp. 56–62.

Beverly-Burton, M. (1963) A new strigeid, *Diplostomum (Tylodelphys) mashonense* n.sp. (Trematoda: Diplostomidae) from the grey heron, *Ardea cinerea* L. in southern Rhodesia with an experimental demonstration of part of the life cycle. *Revue de Zoologie et Botanie Africaine* 68, 291–308.

Bortz, B.M., Kenny, G.F., Pauley, G.B., Garcia-Ortigosa, E. and Anderson, D.P. (1984) The immune response in immunized and naturally infected rainbow trout (*Salmo gairdneri*) to *Diplostomum spathaceum* as detected by enzyme-linked immunosorbent assay (ELISA). *Developmental and Comparative Immunology* 8, 813–822.

Britz, J., Van As, J.G. and Saayman, J.E. (1985) Occurrences and distribution of *Clinostomum tilapiae* Ukoli, 1966 and *Euclinostomum heterostomum* (Rudolphi, 1809) metacercarial infections of freshwater fish in Venda and Lebowa, southern Africa. *Journal of Fish Biology* 26, 21–28.

Brooks, D.R., O'Grady, R.T. and Glen, D.R. (1985) Phylogenetic analysis of the Digenea (Platyhelminthes: Cercomeria) with comment on their adaptive radiation. *Canadian Journal of Zoology* 63, 411–443.

Bryant, C. and Flockhart, H.A. (1986) Biochemical strain variation in parasitic helminths. *Advances in Parasitology* 25, 276–319.

Butterworth, A.E. (1984) Cell-mediated damage to helminths. *Advances in Parasitology* 23, 143–207.

Bychovskaya-Pavlovskaya, I.E., Gusiev, A.V., Dubinina, M.N., Izumova, N.A., Smirnova, T.S., Sokolobskaya, I.L., Stein, G.A., Shulman, S.S. and Epstein, V.M. (1962) *Key to the Identification of Parasites of Freshwater Fish of the USSR*. Izdatelstvo Akademie Nauk SSSR, Moscow (in Russian).

Bylund, G. and Sumari, O. (1981) Laboratory tests with Droncit against diplostomiasis in rainbow trout, *Salmo gairdneri* Richardson. *Journal of Fish Diseases* 4, 259–264.

Cable, R.M. (1954) Studies on marine digenetic trematodes of Puerto Rico. The life cycle the family Haplosplanchnidae. *Journal of Parasitology* 40, 71–75.

Cable, R.M. and Nahhas, F.M. (1962) *Lepas* sp., second intermediate host of a didymozoid trematode. *Journal of Parasitology* 48, 34.

Campbell, R.A. (1990) Deep sea parasites. *Annals de Parasitologie Humaine et Comparée* 65 (Supplement 1), 65–68.

Campbell, R.A., Haedrich, R.L. and Munroe, T.A. (1980) Parasitism and ecological relationships among deep-sea benthic fishes. *Marine Biology* 57, 301–313.

Carothers, J.L. and Allison, R. (1968) Control of snails by the redear (snail cracker) sunfish. *FAO Fisheries Reports* 5(44), 399–406.

Chabaud, A.G. and Biguet, J. (1954) Etude d'un trematode Hemiuride. *Annals de Parasitologie Humaine et Comparée* 29, 5–6.

Chubb, J.C. (1979) Seasonal occurrence of helminths in freshwater fishes. Part II. Trematoda. *Advances in Parasitology* 17, 141–313.

Clegg, J.A. and Smithers, S.R. (1972) The effect of the immune rhesus monkey serum on the schistosomula of *Schistosoma mansoni* during cultivation *in vitro*. *International Journal for Parasitology* 2, 79–98.

Cottrell, B. (1977) The immune response of plaice (*Pleuronectes platessa* L.) to the metacercariae of *Cryptocotyle lingua* and *Rhipidocotyle johnstonei*. *Parasitology* 74, 93–107.

Cowper S.G. (1971) *A Synopsis of African Bilharziasis*. H.K. Lewis and Co. London.

Crowden, A.E. and Broom, D.M. (1980) Effect of the eye fluke, *Diplostomum spathaceum*, on the behavior of dace (*Leuciscus leuciscus*). *Animal Behavior* 28, 287–294.

Davis, H.S., Hoffman, G.L. and Surber, E.W. (1961) Notes on *Sanguinicola davisi* (Trematoda: Sanguinicolidae) in the gills of trout. *Journal of Parasitology*, 47, 512–514.

Dawes, B. (1946) *The Trematoda*. The University Press, Cambridge.

Deardorff, T.L. and Overstreet, R.M. (1991) Seafood-transmitted zoonoses in the United States: the fishes, and the worms. In: Ward, D.R. and Hackney, C. (eds) *Microbiology of Marine Food Products, AVI, Van Nostrand Reinold, New York* pp. 211–265.

De Bont, A.F. and De Bont Hers, M.J. (1952) Mollusc control and fish farming in Central Africa. *Nature* 170, 323–324.

De Giusti, D.L. (1962) Ecological and life-history notes on the trematode *Allocreadium lobatum* (Wallin, 1909), and its occurrence as a progenetic form in Amphipods. *Journal of Parasitology* 48, 22.

DeMartini, J.D. and Pratt, I. (1964) The life cycle of *Telolecithus pugetensis* Lloyd and Guberlet, 1932 (Trematoda: Monorchidae). *Journal of Parasitology* 50, 101–105.

Dollfus, R.P. (1954) Miscellanea helminthologica maroccana XIII–XVII. *Archives de l'Institut Pasteur du Maroc* 4, 583–656.

Dönges, J. (1969) Entwicklung – und Lebendauer von Metacercarien. *Zeitschrift für Parasitenkunde* 31, 340–366.

Dönges, J. (1974) The life history of *Euclinostomum heterostomum* (Rudolphi, 1809) (Trematoda: Clinostomatidae). *International Journal for Parasitology* 4, 79–90.

Dubois, G. (1953) Systematique des Strigeida. Completement de la Monographie. *Memoires de la Societe Neuchateloise des Sciences Naturelles* 8, 141 pp.

Ellis, A.E. (1977) The leucocytes of fish: a review. *Journal of Fish Biology* 11, 453–491.

Erasmus, D.A. (1977) The host–parasite interface of trematodes. *Advances in Parasitology* 15, 201–242.

Euzet, L. and Raibaut, A. (1985) Les maladies parasitaires en pisciculture marine. *Symbioses* 17, 51–68.

Evans, N.A. (1978) The occurrence and life history of *Asymphylodora kubanicum* (Platyhelminthes: Digenea: Monorchidae) in the Worcester-Birmingham canal, with special references to the feeding habits of the definitive host, *Rutilus rutilus*. *Journal of Zoology* 184, 143–153.

Evans, W.A. (1974a) Growth, mortality, and hematology of cutthroat trout experimentally infected with the blood fluke *Sanguinicola klamathensis*. *Journal of Wildlife Diseases* 10, 341–346.

Evans, W.A. (1974b) The histopathology of cutthroat trout experimentally infected with the blood fluke *Sanguinicola klamathensis*. *Journal of Wildlife Diseases* 10, 243–248.

Faliex, E. (1991) Ultrastructural study of the host–parasite interface after infection of two species of teleosts by *Labratrema minimus* metacercariae (Trematoda, Bucephalidae). *Diseases of Aquatic Organisms* 10, 93–101.

Fares, A. and Maillard, C. (1974) Recherches sur quelques Haploporidae (Trematoda) parasites des Muges de Mediterranée Occidentale: systematique et cycles evolutifs. *Zeitschrift für Parasitenkunde* 45, 11–43.

Farstey, V. (1986) *Centrocestus* sp. (Heterophyidae) and other trematode infections of the snail *Melanoides tuberculata* (Muller, 1774) and cichlid fish in Lake Kinneret. Unpublished MSc Thesis, Hebrew University of Jerusalem, 1986 (Hebrew text, English summary).

Ferguson, M.S. and Hayford, R.A. (1941) The life history and control of eye flukes. *Progressive Fish Culturist* 54, 1–13.

Finkelman, S. (1988) Infection of Clinostomatidea in the Sea of Galilee fish. Unpublished PhD thesis, Hebrew University of Jerusalem (Hebrew text, English summary).

Fischthal, J.H. (1980) Some digenetic trematodes of marine fishes from Israel's Mediterranean coast and their zoogeography, especially those from Red Sea immigrant fishes. *Zoologica Scripta* 9, 11–23.

Font, W.F., Overstreet, R.W. and Heard, R.W. (1984a) Taxonomy and biology of *Phagicola nana* (Digenea: Heterophyidae). *Transactions of the American Microscopical Society* 103, 408–422.

Font, W.F., Heard, R.W. and Overstreet, R.M. (1984b) Life cycle of *Ascocotyle gemina* n.sp., a sibling species of *A. sexidigita* (Digenea: Heterophyidae). *Transactions of the American Microscopical Society*, 103, 392–407.

Gartner, J.V. and Zwerner, D.E. (1989) The parasite faunas of meso- and bathypelagic fishes of Norfolk Submarine Canyon, Western North Atlantic. *Journal of Fish Biology* 34, 79–95.

Gibson, D.I., MacKenzie, K. and Cottle, J. (1981) *Halvorsenius exilis* gen. nov., sp. nov. a new didymozoid trematode from the Mackerel *Scomber scombrus* L. *Journal of Natural History* 15, 917–929.

Ginetsinskaya, T.A. (1958) Life cycles and biology of the larval stages of parasitic worms of fish. In: Dogiel, V.A., Petrushevski, G.K. and Polyanski, Yu.I. (eds) *Fundamental Problems of the Parasitology of Fishes*. Izdatelstvo Leningradskovo Universiteta, Leningrad, pp. 144–183 (in Russian, English translation: Kabata, Z., 1961, *Parasitology of Fishes*, Oliver and Boyd, Edinburgh, pp. 140–179).

Gordon, D.M. and Rau, M.E. (1982) Possible evidence for mortality induced by the parasite *Apatemon gracilis* in a population of brook sticklebacks (*Culea inconstans*). *Parasitology* 84, 41–47.

Halton, D.W. and Johnston, B.R. (1982) Functional morphology of the metacercarial cyst of *Bucephaloides gracilensis* (Trematoda: Bucephalidae). *Parasitology* 85, 54–52.

Hoffman, G.L. (1958) Studies on the life-cycle of *Ornithodiplotomum ptychocheilus* (Faust) (Trematoda: Strigeoidea) and the 'self cure' of infected fish. *Journal of Parasitology* 44, 416–421.

Hoffman, G.L. (1960) Synopsis of Strigeoidea (Trematoda) of fishes and their life cycles. *Fishery Bulletin of the Fish and Wildlife Service* 60 (*Fishery Bulletin* 175), 436–469.

Hoffman, G.L. (1967) *Parasites of North American Freshwater Fish*. University of California Press, Berkeley.

Hoffman, G.L. (1970) Control methods to snail-born zoonoses. *Journal of Wildlife Diseases* 6, 262–265.

Hoffman, G.L. and Hundley, J.B. (1957) The life-cycle of *Diplostomum baeri eucaliae* (Trematoda: Strigeida). *Journal of Parasitology* 43, 613–627.

Hoffman, G.L., Fried, B. and Harvey, J.E. (1985) *Sanguinicola fontinalis* sp. nov. (Digenea: Sanguinicolidae): a blood parasite of brook trout, *Salnelinus fontinalis* (Mitchill), and longnose dace, *Rhinichthys cataractae* (Valenciennes). *Journal of Fish Diseases* 8, 529–538.

Hoffmann, R.W., Körting, W., Fischer-Scherl, T. and Schaufer, W. (1990) An outbreak of *Bucephalus polymorphus* in fish of the Main river. *Angewandte Parasitologie* 31, 95–99.

Höglund, J. (1991) Ultrastructural observations and radiometric assay on cercarial penetration and migration of the digenean *Diplostomum spathaceum* in the rainbow trout *Oncorhynchus mykiss*. *Parasitology Research* 77, 283–289.

Höglund, J. and Thuvander, A. (1990) Indications of non-specific protective mechanisms in rainbow trout *Oncorhynchus mykiss* with diplostomosis. *Diseases of Aquatic Organisms* 8, 91–97.

Hopkins, S.H. (1937) A new type of allocreadiid cercariae: the cercariae of *Anallocreadium* and *Microcreadium*. *Journal of Parasitology* 23, 94–97.

Huggins, E.J. (1972) Parasites of fishes in South Dakota. *Agricultural Experimental Station, South Dakota State University Brookings and South Dakota Department of Game Fish and Parks, Bulletin* 484, pp. 73.

Hunninen, A. and Cable, R.M. (1943) The life history of *Podocotyle atomon* (Rudolphi) (Trematoda: Opecoelidae). *Transactions of the American Microscopical Society* 62, 57–68.

Ito, J. (1964) *Metagonimus* and other hunman heterophyid trematodes. *Progress of Medical Parasitology in Japan, Meguro Parasitological Museum, Tokyo* 1, 317–392.

Kabunda, M.Y. and Sommerville, C. (1984) Parasitic worms causing the rejection of tilapia (*Oreochromis* species) in Zaire. *British Veterinary Journal* 140, 263–268.

Kannangara, D.W.W. and Smyth, J.D. (1974) *In vitro* cultivation of *Diplostomum phoxini* metacercariae. *International Journal for Parasitology* 4, 667–673.

Kennedy, C.R. (1984) The use of frequency ditribution in an attempt to detect host mortality induced by infections of diplostomatid metacercariae. *Parasitology* 89, 209–220.

Khalil, L.F. (1963) On *Diplostomum tregena*, the diplostomatid stage of *Diplostomum tregena* Nazmi Gohar, 1932, with experimental demonstration of part of the life cycle. *Journal of Helminthology* 37, 199–206.

Khalil, L.F. (1969) Studies on the helminth parasites of freshwater fishes of the Sudan. *Journal of Zoology* 158, 143–170.

Khalil, L.F. (1971) *Check List of the Helminth Parasites of African Freshwater Fishes*. Commonwealth Institute of Helminthology, St Albans.

Khalil, M.B. (1937) The life history of the human trematode parasite *Heterophyes heterophyes*. *Proceeding of the International Congress of Zoology* (Lisbon, 1935) 12, 1989–2002.

Køie, M. (1971a) On the histochemistry and ultrastructure of the daughter sporocyst of *Cercaria buccini* Lebour, 1911. *Ophelia* 9, 145–163.

Køie, M. (1971b) On the histochemistry and ultrastructure of the redia of *Neophasis lageniformis* (Lebour, 1910). *Ophelia* 9, 113–143.

Køie, M. (1971c) On the histochemistry and ultrastructure of the tegument and associated structures of the cercaria of *Zoogonoides viviparus* in the first intermediate host. *Ophelia* 9, 165–206.

Køie, M. (1973) The host-parasite interface of the cercaria and adult *Neophasis lageniformis* (Lebour, 1910). *Ophelia*, 12, 205–219.

Køie, M. (1975) On the morphology and life-history of *Opechona bacillaris* (Molin, 1859) Looss, 1907 (Trematoda, Lepocreadiidae). *Ophelia* 13, 63–86.

Køie, M. (1976) On the morphology and life-history of *Zoogonoides viviparus* (Olsson, 1868) Odhner, 1902 (Trematoda, Zoogonidae). *Ophelia* 15, 1–14.

Køie, M. (1977) Stereoscan studies of cercaiae, metacercariae and adults of *Cryptocotyle lingua* (Creplin 1825) Fischoeder 1903 (Trematoda: Heterophyidae). *Journal of Parasitology* 63, 835–839.

Køie, M. (1978) On the morphology and life history of *Stephanostomum caducum* (Looss, 1901) Manter, 1934 (Trematoda, Acanthocolpidae) *Ophelia* 17, 121–133.

Køie, M. (1979a) On the morphology and life-history of *Derogenes varicus* Muller, 1784) Looss, 1901 (Trematoda, Hemiuridae). *Zeitschrift für Parasitenkunde* 59, 67–78.

Køie, M. (1979b) On the morphology and life-history of *Monascus [=Haplocladus] filiformis* Rudolphi, 1819) Looss, 1907 and *Steringophorus furciger* (Olsson, 1868) Odhner, 1905 (Trematoda, Fellodistomidae). *Ophelia* 18, 113-132.

Køie, M. (1980) On the morphology and life history of *Steringotrema pagelli* (Van Beneden, 1871) Odhner, 1911 and *Fellodistomum felis* (Olsson, 1868) Nicoll, 1909 (syn. *S. ovacutum* (Lebour, 1908) Yamaguti, 1953) (Trematoda, Fellodistomatidae). *Ophelia* 19, 215-236.

Køie, M. (1981) On the morphology and life-history of *Podocotyle reflexa* (Creplin, 1825) Odhner, 1905 (Trematoda, Opecoelidae). *Ophelia* 20, 17-43.

Køie, M. (1982) The redia, cercaria and early stages of *Aporocotyle simplex* Odhner, 1900 (Sanguinicolidae) – A digenetic trematode which has a polychaete annelid as the only intermediate host. *Ophelia* 21, 115-145.

Køie, M. (1985) The surface topography and life-cycles of digenetic trematodes in *Limanda limanda* (L.) and *Gadus morhua* L. PhD Thesis, University of Copenhagen, Marine Biological Laboratory, Helsingor, Denmark, 20 pp.

Køie, M. (1986) The life history of *Mesorchis denticulatus* (Rudolphi, 1802) Dietz, 1909 (Trematoda, Echinostomatidae). *Zeitschrift für Parasitenkunde* 72, 335-343.

Køie, M. (1987) Scanning electron microscopy of rediae, cercariae, metacercariae and adults of *Mesorchis denticulatus* (Rudolphi, 1802) (Trematoda, Echinostomatidae). *Parasitology Research* 73, 50-56.

Køie (1992) Life cycle and structure of the fish digenean *Brachyphallus crenatus* (Hemiuridae). *Journal of Parasitology* 78, 338-343.

Køie, M. and Frandsen, F. (1976) Stereoscan observations of the miracidium and early sporocyst of *Schistosoma mansoni*. *Zeitschrift für Parasitenkunde* 50, 335-344.

Køie, M. and Lester, R.J.G. (1985) Larval didymozoids (Trematoda) in fishes from Moreton Bay, Australia. *Proceedings of the Helminthological Society of Washington* 52, 196-203.

Krull, W.H. (1934) Studies on the life history of the trematode, *Rhipidocotyle septpapillata* n.sp. *Transactions of the American Microscopical Society* 53, 408-415.

Larson, O.R. (1965) *Diplostomulum* (Trematoda: Strigeoidea) associated with herniations of bullhead lenses. *Journal of Parasitology* 51, 224-229.

Larson, O.R., Uglem, G.L. and Kook, J.L. (1988) Fine structure and permeability of the metacercarial wall of *Clinostomum marginatum* (Digenea). *Parasitology Research* 74, 352-355.

Lee, F.O. and Cheng, T.C. (1970) The histochemistry of *Stellantchasmus falcatus* Onji and Nishio, 1915 (Trematoda: Heterophyidae) metacercarial cyst in the mullet *Mugil cephalus* L. and histopathological alternations in the host. *Journal of Fish Biology* 2, 235-243.

Lemly, A.D. and Esch, G.W. (1984a) Population biology and largemouth bass *Micropterus salmoides*. *Journal of Parasitology* 70, 466-474.

Lemly, A.D. and Esch, G.W. (1984b) Effect of the trematode *Uvulifer ambloplites* (Hughes, 1927) on juvenile bluegill sunfish *Lepomis macrochirus*: ecological implications. *Journal of Parasitology* 70, 475-492.

Leno, G.H. and Holloway, H.L. Jr (1986) The culture of *Diplostomum spathaceum* metacercaria on the chick chorioallatois. *Journal of Parasitology* 72, 555-558.

Lester, R.J.G. (1977) An estimate of the mortality in a population of *Perca flavescens* owing to the trematode *Diplostomum adamsi*. *Canadian Journal of Zoology* 55, 288-292.

Lester R.J.G. (1980) Host-parasite relations in some didymozoid trematodes. *Journal of Parasitology* 66, 527-531.

Lester, R.J.G. (1984) A review of methods for estimating mortality due to parasites in wild fish populations. *Helgolander Meeresuntersuchungen* 37, 53–64.

Lester, R.J.G. (1990) Reappraisal of the use of parasites for fish stock identification. *Australian Journal for Marine and Freshwater Research* 41, 885–864.

Lester, R.J.G. and Huizinga, H.W. (1977) *Diplostomum adamsi* sp. n.: description, life cycle, and pathogenesis in the retina of *Perca flavescens. Canadian Journal of Zoology* 55, 64–73.

Lester, R.J.G. and Lee, T.D.G. (1976) Infectivity of the progenetic metacercariae of *Diplostomum spathaceum. Journal of Parasitology* 62, 832–833.

Liu, F.G. (1979) Diseases of cultured loach (*Misgurnus anguillicaudatum*) in Taiwan. *Chinese Aquaculture* 304, 14.

Llewellyn, J. (1986) Phylogenetic inference from platyhelminth life-cycle stages. In: Howell, M.J. (ed.) *Parasitology Quo Vadit?* Proceedings of the Sixth International Congress of Parasitology. Australian Academy of Science, Canberra, pp. 281–289.

Lo, C.F. Huber, F., Kou, G.H. and Lo, C.J. (1981) Studies on *Clinostomum complanatum* (Rudolphi, 1819). *Fish Pathology* 15, 219–227.

Lombard, G.L. (1968) A survey of fish diseases and parasites encountered in Transvaal. *Newsletter of the Limnological Society of South Africa* 11, 23–29.

Lucky, Z. (1964) Contribution to the pathology and pathogenicity of *Sanguinicola inermis* in juvenile carp. In: Ergens, R. and Rysavy, B. (eds) *Parasitic Worms and Aquatic Conditions.* Czechoslovak Academy of Sciences, pp. 153–157.

Lumsden, D.R. (1975) Surface ultrastructure and cytochemistry of parasitic helminths. *Experimental Parasitology,* 37, 267–339.

MacKenzie, K. (1968) Some parasites of O-group plaice, *Pleuronectes platessa* L. under different environmental conditions. *Journal of Marine Research* 3, 1–23.

MacKenzie, K. (1983) Parasites as biological tags in fish population studies. *Advances in Applied Biology* 7, 251–331.

MacKenzie, K. and Liversidge, J.M. (1975) Some aspects of the biology of the cercaria and metacercaria of *Stephanostomum baccatum* (Nicoll, 1907) Manter 1934 (Digenea: Acanthocolpidae). *Journal of Fish Biology* 7, 247–256.

Madhavi, R. (1979) Observations on the occurrence of *Allocreadium fasciatusi* in *Apocheilus melastigma. Journal of Fish Biology* 14, 47–58.

Madhavi, R. and Rukmini, C. (1991) Population biology of the metacercariae of *Centrocestus formosanus* (Trematoda: Heterophyidae) on the gills of *Apocheilus panchax. Journal of Zoology* 223, 509–520.

Madsen, H. (1990) Biological methods for the control of freshwater snails. *Parasitology Today* 6, 227–241.

Maillard, C., Lambert, A. and Raibaut, A. (1980). Nouvelle forme de distomatose larvaire. Etude d'un trematode pathologie pour les alvins de daurade (*Sparus aurata* L. 1758) en encloserie. *Compte Rendu de'Academie des Sciences, Paris* (ser. D) 535–538.

Manter, H.W. (1957) Host specificity and other host relationships among the digenetic trematodes of marine fishes. In: Baer, J.G. (ed.) *Premier Symposium sur la Specifite Parasitaire des Parasites des Vertebres.* International Union of Biological Sciences Serie B, No. 32, pp. 185–197.

Margolis, L. and Boyce, N.P. (1969) Life span, maturation, and growth of two hemiurid trematodes, *Tubulovesicula lindbergi* and *Lecithaster gibbosus,* in pacific salmon (genus *Onchorhynchus) Journal of the Fisheries Research Board of Canada* 26, 839–907.

Martin, W.E. (1950) *Euhaplorchis californiensis* n.g., n.sp. Heterophyidae,

Trematoda, with notes on its life cycle. *Transactions of the American Microscopical Society* 64, 194–209.

Martin, W.E. (1952) Another annelid first intermediate host of a digenetic trematode. *Journal of Parasitology* 38, 356–359.

Martin, W.E. (1955) Seasonal infections of the snail, *Cerithidea californica* Halderman, with larval trematodes. In: *Essays in the Natural Sciences in Honor of Captain Allan Hancock*. University of Southern University Press, Los Angeles, pp. 203–210.

Martin, W.E. (1969) Hawaiian helminths. IV. *Paracardicola hawaiensis* n.gen. n.sp. (Trematoda: Sanguinicolidae) from the balloon fish, *Tetraodon hispidus* L. *Journal of Parasitology* 48, 648–650.

Mashego, S.N. (1982) A seasonal investigation of the helminth parasites of *Barbus* species in water bodies in Lebowa and Venda, South Africa. PhD Thesis, University of the North, South Africa, 1982.

Matthews, R.A. (1973a) The life-cycle of *Bucephalus haimanus* Lacaze-Duthiers, 1845 from *Cardium edule.*, L. *Parasitology* 67, 341–350.

Matthews, R.A. (1973b) The life-cycle of *Prosorhynchus crucibulum* (Rudolphi, 1819) Odner, 1905, and a comparison of its cercariae with that of *Prosorhynchus squamatus* Odner, 1905. *Parasitology* 67, 133–164.

Matthews, R.A. (1974a) The life cycle of *Bucephaloides gracilescens* (Rudolphi 1819) Hopkins, 1954. Digenea: Gasterostomata) *Parasitology* 68, 1–12.

Matthews, R.A. (1974b) Metacercariae and diseases in marine teleosts. *Proceedings of the Third International Congress of Parasitology* 3, 1723.

Mawdesley-Thomas, L. and Young, P.C. (1967) Cutaneous melanosis in a flounder (*Platichthys flesus* L.). *Veterinary Record* October, 7th, 2098.

McArthur, C.P. (1978) Humoral antibody production by New Zealand eels, against the intestinal trematode *Telogaster opisthorchis* Macfarlane, 1945. *Journal of Fish Diseases* 1, 377–387.

McCullough, F.S. and Mott K.E. (1983) The role of molluscicides in schistosomiasis control. World Health Organization Document WHO/VBC/83.879.

McDaniel, J.C. and Coggins, J.R. (1972) Seasonal larval trematode infection dynamics in *Nassarius obsoletus* (Say). *Journal of the Elisha Mitchell Scientific Society* 88, 55–57.

McQueen, A., MacKenzie, K., Roberts, R.J. and Young, H. (1973) Studies on the skin of plaice (*Pleuronectes platessa* L.). III. The effect of temperature on the inflammatory response to the metacercariae of *Cryptocotyle lingua* (Creplin, 1825) (Digenea, Heterophyidae). *Journal of Fish Biology* 5, 241–247.

Meade, T.G. (1967) Life history studies on *Cardicola klamathensis* (Wales, 1958) (Trematoda: Sanguinicolidae). *Proceedings of the Helminthological Society of Washington* 34, 210–212.

Meade, T.G. and Pratt, I. (1965) Description and life-cycle of of *Cardicola alseae* sp.n. (Trematoda: Sanguinicolidae). *Journal of Parasitology*, 51, 575–578.

Meyerhof, E. and Rothschild, M. (1940) A prolific trematode. *Nature* 146, 367–368.

Millemann, R.E. and Knapp, S. (1970) Pathogenicity of the 'salmon poisoning' trematode *Nanophyetus salmincola* to fish. In: Snieszko, S.F. (ed.) *A Symposium on Diseases of Fishes and Shellfishes*. American Fisheries Society, Washington D.C., special publication No. 5, pp. 209–217.

Mills, C.A. (1979) Attachment and feeding of the adult ectoparasitic digenean *Transversotrema patialense* (Sorparkar, 1924) on the zebra fish *Brachydanio rerio* (Hamilton-Buchanan). *Journal of Fish Diseases* 2, 443–447.

Mills, C.A., Anderson, R.M. and Whitfield, P.J. (1979) Density-dependent survival and reproduction within population of the ectoparasitic digenean *Transversotrema patialense* on the fish host. *Journal of Animal Ecology* 48, 383–399.

Mitchell, A.J., Smith, C.E. and Hoffman, G.L. (1982) Pathogenicity and histopathology of an unusually intense infection of white grubs (*Posthodiplostomum m. minimum*) in the fathead minnow (*Pimephalus promelas*). *Journal of Wildlife Diseases* 18, 51–57.

Mitchell, C.W. (1974) Ultrastructure of the metacercarial cyst of *Posthodiplostomum minimum* (MacCallum, 1921). *Journal of Parasitology* 60, 67–74.

Mitchell, J.S., Halton, D.W. and Smyth, J.D. (1978) Observations on the *in vitro* culture of *Cotylurus erraticus* (Trematoda: Strigeidae). *International Journal for Parasitology* 8, 389–397.

Moser, M. and Sakanari, J. (1986) The effect of praziquantel in various larval and adult parasites from freshwater and marine snails and fish. *Journal of Parasitology* 72, 175–176.

Nikolaeva, V.M. (1965) On the developmental cycle of the trematode family Didymozoidae (Monticelli, 1888) Poche, 1907. *Zoologichesky Zhurnal* 44, 1317–1327 (Russian text, English summary).

Ogawa, K., Hattori, K., Hatai, K. and Kubota, S. (1989) Histopathology of cultured marine fish, *Seriola purpurascens* (Carangidae) infected with *Paradeontacylis* spp. (Trematoda: Sanguinicolidae) in the vascular system. *Fish Pathology* 24, 75–81.

Overstreet, R.M. and Howse, H.D. (1977) Some parasites and diseases of estuarine fishes in polluted habitats of Mississippi. *Annals of the New York Academy of Sciences* 298, 427–462.

Overstreet, R.M. and Thulin, J. (1989) Response by *Plectropomus leopardus* and other serranid fishes to *Pearsonellum corventum* (Digenea: Sanguinicolidae), including melanomacrophage centres in the heart. *Australian Journal of Zoology* 37, 129–142.

Palombi, A. (1937) Il ciclo biologico di *Lepocreadium album* Stossich sperimentalamente realizzato. Osservazioni ecologiche e considerazioni sistematiche sulla *Cercaria setifera* (von Jon. Muller) Monticelli. *Revista di Parasitologia* 1, 1–12.

Palombi, A. (1941) *Cercaria dentali* Pelseneer, forma larvale di *Ptychogonium megastoma* (Rud.). Nota preventive. *Rivista Parassitologia* 5, 127–128.

Pappas, P.W. and Read, C.P. (1975) Membrane transport in helminth parasites: a review. *Experimental Parasitology* 37, 469–530.

Paperna, I. (1964a) The metazoan parasite fauna of Israel inland water fishes. *Bamidgeh* 16, 3–66.

Paperna, I. (1964b) Parasitic helminths from inland water fishes in Israel. *Israel Journal of Zoology* 13, 1–20.

Paperna, I. (1968a) Susceptibility of *Bulinus (Physopsis) globosus* and *Bulinus truncatus rohlfsi* from different localities in Ghana to different local strains of *Schistosoma haematobium*. *Annals of Tropical Medicine and Parasitology* 62, 13–26.

Paperna, I. (1968b) Studies on the transmission of schistosomiasis in Ghana. 1. Ecology of *Bulinus (Physopsis) globosus* the snail host of *Schistosoma haematobium* in South East Ghana. *Ghana Journal of Science* 8, 30–51.

Paperna, I. (1968c) Studies on the transmission of schistosomiasis in Ghana. III. Notes on the ecology and distribution of *Bulinus truncatus rohlfsi* and *Biomphalaria pfeifferi* in the lower Volta basin, Ghana. *Ghana Medical Journal* 7, 139–145.

Paperna, I. (1975) Parasites and diseases of the grey mullet (Mugilidae) with special reference to the seas of the Near East. *Aquaculture* 5, 65–80.

Paperna, I. (1980) Parasites, infections and diseases of fish in Africa. Food and Agriculture Organization, UN, CIFA Technical paper No. 7.

Paperna, I. (1991) Diseases caused by parasites in the aquaculture of warm water fish. *Annual Review of Fish Diseases*, 1, 155–194.

Paperna, I. and Lengy, J. (1963) Notes on a new subspecies of *Bolbophorus confusus* (Krause, 1914) Dubois 1935 (Trematoda, Diplostomatidae), A fish-transmitted bird parasite. *Israel Journal of Zoology* 12, 171–182.

Paperna, I. and Overstreet, R.M. (1981) Parasites and diseases of mullets (Mugilidae). In: Oren, O.H. (ed.) *Aquaculture of Grey Mullets*. I.B.P. 26, Cambridge University Press, pp. 411–493.

Pearson, J.C. (1968) Observations on the morphology and life-cycle of *Paucivitellosus fragilis* Coil, Reid and Kuntz, 1965 (Trematoda: Bivesiculidae). *Parasitology* 58, 769–788.

Pearson, J.C. (1972) A phylogeny of life cycle pattern of the Digenea. *Advances in Parasitology* 10, 153–189.

Perera, K.L.M. (1992a) Light microscopic study of the pathology of a species of didymozoan, Nematobothriinae gen. sp., from the gills of slimy mackerel *Scomber australasicus*. *Diseases of Aquatic Organisms* 13, 103–109.

Perera, K.L.M. (1992b) Ultrastructure of the primary gill lamellae of *Scomber australasicus* infected by a didymozoid parasite. *Diseases of Aquatic Organisms* 13, 111–121.

Philip, C.B. (1955) There is always something new under the 'parasitological' sun (the unique story of helminth-borne salmon poisoning disease). *Journal of Parasitology* 41, 125–148.

Rao, K.H. and Ganapat, P.N. (1907) Observation of *Transversotrema patialensis* (Soparkar, 1924) (Trematoda) from Waltair, Andra Pradesh (India) *Parasitology* 57, 661–664.

Ratanart-Brockelman, C. (1974) Migration of *Diplostomum spathaceum* (Trematoda) in the fish intermediate host. *Zeitschrift für Parasitenkunde* 43, 123–134.

Reimer, L.W. (1970) Digene Trematoden und Cestoden der Ostseefische als naturlische Fischmarken. *Parasitologische Schriftenriehe* 20, 40–46.

Reimer, L.W. (1973) Das Auftreten eines Fischtrematoden der Gattung *Asymphylodora* Looss, 1899, bei *Nereis diversicolor* O.F. Muller als Beispiel fur einen Alternativzyklus. *Zoologische Anzeiger* 191, 187–195.

Rogers, W.A. (1972) Southern Cooperative Fish Disease Project, Eighth Annual Report. Department of Fisheries and Allied Aquaculture, Auburn University, Alabama, 101 pp.

Sarig, S. (1971) The prevention and treatment of diseases of warmwater fishes under subtropical conditions, with special emphasis on intensive farming. In: Book 3, Snieszko, S. and Axelrod, H.R. (eds) *Diseases of Fishes*. TFH Publications, Jersey City, New Jersey.

Schell, S.C. (1970) *How to Know the Trematodes*. Wm. C. Brown Company Publishers, Dubuque, Iowa.

Scheuring, L. (1922) Der Lebenscklus von *Sanguinicola inermis* Plehn. *Zoologischer Jahrbucher, Abteilung für Anatomie und Ontogenie der Tiere* 44, 265–310.

Scott, J.S. (1969) Trematode populations in the Atlantic argentine, *Argentina silus*, and their use as biological indicators. *Journal of the Fisheries Research Board of Canada* 26, 879–891.

Shariff, M., Richards, R.H. and Sommerville, C. (1980) The histopathology of acute and chronic infections of rainbow trout *Salmo gairdneri* Richardson with eye flukes, *Diplostomum* spp. *Journal of Fish Diseases* 3, 455–465.

Shotter, R.A. (1973) Changes in the parasite fauna of whiting *Odontogadus merlangus* L. with age and sex of host, season, and from different areas in the vicinity of the Isle of Man. *Journal of Fish Biology* 5, 559–573.

Shigin, A.A. (1964) The life span of *Diplostomum spathaceum* in the intermediate host. *Trudy Gelmintologicheskoi laboratoryi Akademyii Nauk SSSR* 14, 262–272 (in Russian).

Sindermann, C.J. and Resenfield, A. (1954) Diseases of fishes of western North Atlantic III. Mortalities of sea herring (*Clupea harengus*). *Maine Department of Sea Shore Fisheries Bulletin* 21, 1–16.

Sillman, E.I. (1962) The life history of *Azygia longa* (Leidy, 1851), (Trematoda: Digenea), and notes on *A. acuminata* Goldberger 1911. *Transactions of the American Microscopical Society* 81, 43–65.

Sinclair, N.R. (1972) Studies on the heterophyid trematode *Apopallus brevis*, the 'sand grain grub' of yellow perch (*Perca flavescens*). II The metacercaria: position, structure, and composition of the cyst; hosts; geographical distribution and variation. *Canadian Journal of Zoology* 50, 577–584.

Skinner, R. (1975) Parasites of the striped mullet, *Mugil cephalus* from the Biscayne Bay, Florida, with description of a new genus and three new species of trematodes. *Bulletin of Marine Sciences* 25, 318–345.

Skrjabin, K.I. (and his staff) (1964) *Key to the Trematodes of Animals and Man.* English translation, Arai, H.P. (ed.). University of Illinois Press, Urbana.

Smith, J.W. (1972) The blood flukes (Digenea: Sanguinicolidae and Spirorchidae) of cold blooded vertebrates and some comparison with schistosomes. *Helmithological Abstracts, Series A* 41, 161–204.

Smith, J.W. and Williams, H.H. (1967) The occurrence of the blood fluke, *Aporocotyle spinosicanalis* Williams, 1958 in European hake, *Merluccius merluccius* (L.) caught off the British Isles. *Journal of Helminthology* 41, 71–88.

Smyth, J.D. and Halton, D.W. (1983) *The Physiology of Trematodes*, 2nd edn. Cambridge University Press, Cambridge.

So F.W. and Wittrock, D.D. (1982) Ultrastructure of the metacercarial cyst of *Ornithodiplostomum ptychocheilus* (Trematoda: Diplostomatidae) from the brain of fathead minnows. *Transactions of the American Microscopical Society* 101, 181–185.

Sogandares-Bernal, F. and Lumsden, R.D. (1964) The heterophyid trematode *Ascocotyle (A.) leighi* Burton, 1956, from the hearts of certain poecilid and cyprinodont fishes. *Zeitschrift für Parasitenkunde* 24, 3–12.

Sommerville, C. (1981) A comperative study of the tissue response to invasion and encystment by *Stephanochasmus baccatus* (Nicoll, 1907) (Digenea: Acanthocolpidae,) in four species of flatfish. *Journal of Fish Diseases* 4, 53–68.

Sommerville, C. (1982) The pathology of *Haplorchis pumilio* (Looss, 1896) infection in cultured tilapias. *Journal of Fish Diseases* 5, 243–250.

Sommerville, C. and Iqbal, N.A.M. (1991) The process of infection, migration, growth and development of *Sanguinicola inermis* Plehn, 1905 (Digenea: Sanguinicolidae) in carp, *Cyprinus carpio* L. *Journal of Fish Diseases* 14, 211–219.

Speed, P. and Pauly, G.B. (1985) Feasibility of protecting rainbow trout, *Salmo gairdneri* Richardson, by immunizing against the eye fluke *Diplostomum spathaceum*. *Journal of Fish Biology* 26, 739–744.

Stables, J.N. and Chappell, L.H. (1986) The epidemiology of diplostomiasis in farmed rainbow trout from north-east Scotland. *Parasitology* 92, 699–710.

Stein, P.C. and Lumsden, R.D. (1971) An ultrastructural and cytochemical study of metacercarial cyst development in *Ascocotyle pachycystis* Schroeder and Leigh, 1965. *Journal of Parasitology* 57, 1231–1246.

Stunkard, H.W. (1930) The life history of *Cryptocotyle lingua* with notes on the physiology of the metacercaria. *Journal of Morphology* 50, 143.

Stunkard, H.W. (1959) The morphology and life history of the digenetic trematode *Asymphylodora amnicolae* n.sp.; the possible significance of progenesis for the phylogeny of the Digenea. *Biological Bulletin*, 117, 562–581.

Stunkard, H.W. (1980) Successive hosts and developmental stages in the life history of *Neopechona cablei* sp.n. (Trematoda: Lepocreadiidae). *Journal of Parasitology* 66, 636–641.

Sweeting, R.A. (1974) Investigations into natural and experimental infections of freshwater fish by the common eye fluke *Diplostomum spathaceum* Rud. *Parasitology* 69, 291–300.

Szekely, C. and Molnar, K. (1991) Praziquantel (Droncit) is effective against diplostomosis of grasscarp (*Ctenopharyngodon idella*) and silver carp (*Hypophthalmichthys molitrix*). *Diseases of Aquatic Organisms* 11, 147–150.

Taraschewski, H. (1984) Heterophyasis, an intestinal fluke infection of man and vertebrates transmitted by euryhaline gastropods and fish. *Helgolander Meersuntersuchungen* 37, 463–478.

Taraschewski, H. and Paperna, I. (1981) Distribution of the snail *Pirenella conica* in Sinai and Israel and its infection by Heterophyidae and other trematodes. *Marine Ecology Progress Series* 5, 193–205.

Taraschevski, H. and Paperna, I. (1982) Trematode infection in *Pirenella conica* in three sites of a mangrove lagoon in Sinai. *Zeitschrift für Parasitenkunde* 67, 165–173.

Theron, A. (1986) Chronobiology of schistosome development in the snail host. *Parasitology Today* 2, 192–194.

Threadgold, L.T. (1984) Parasitic Platyhelminths. In: Bereiter-Hahn, J., Matoltsy, A.G. and Silvia Richards, K. (eds) *Biology of the Integument 1. Invertebrates.* Springer-Verlag, Berlin, pp. 132–191.

Thulin, J. (1980) Redescription of *Nematobibothriodes histoldi* Noble, 1974 (Digenea: Didymozoidea). *Zeitschrift für Parasitenkunde* 63, 213–219.

Thurston, J.P. (1965) The pathogenicity of fish parasites in Uganda. *Proceedings of the East African Academy* 3, 45–51.

Tort, L., Watson, J.J. and Priede, I.G. (1987) Changes in *in vitro* heart performance in rainbow trout *Salmo gairdneri* Richardson, infected with *Apatemon gracilis* (Digenea). *Journal of Fish Biology* 30, 341–347.

Tubangui, M.A. (1931) Trematode parasites of Philippine vertebrates. III. Flukes from fish and reptiles. *Philippine Journal of Science* 44, 417–423.

Uglem, G.L. and Larson, O.R. (1987) Facilitated diffusion and active transport system for glucose in metacercariae of *Clinostomum marginatum* (Digenea). *International Journal for Parasitology* 17, 847–850.

Ukoli, F.M.A. (1966) On *Clinostomum tilapiae* n.sp. and *C. phalacrocoracis* Dubois, 1931, from Ghana and a discussion of the systematics of the genus *Clinostomum* Leidy, 1856. *Journal of Helminthology* 40, 187–214.

Van Cleave, H.J. and Mueller, J.F. (1934) Parasites' of Oneida lake fishes. Part III. A biological and ecological survey of the worm parasites. *Roosevelt Wildlife Annual* 3, 161–334.

Van den Brock, E. and de Jong, N. (1979) Studies on the life cycle of *Asymphylodora tincae* (Modeer, 1790) (Trematoda, Monorchiidae) in a small lake near Amsterdam. Part 1. The morphology of various stages. *Journal of Helminthology* 53, 79–89.

Velasquez, C.C. (1958) *Transversotrema laruei*, a new trematode of Philippine fish (Digenea: Transversotrematidae) *Journal of Parasitology* 44, 449–451.

Wales, J.H. (1958) Two new blood fluke parasites of trout. *California Fish and Game* 44, 125–136.

Walker, D.J. and Wittrock, D.D. (1992) Histochemistry and ultrastructure of the metacercarial cyst of *Bolbogonotylus corkumi* (Trematoda: Cryptogonimidae). *Journal of Parasitology* 78, 725–730.

Watson, J.J., Pike, A.W. and Priede, I.G. (1992) Cardiac pathology associated with the infection of *Onchorhynchus mykiss* Walbaum with *Apatemon gracilis* Rud. 1819. *Journal of Fish Biology* 41, 163–167.

Whyte, S.K., Allan, J.C., Secombes, C.J. and Chappell, L.H. (1987) Cercariae and diplostomes of *Diplostomum spathaceum* (Digenea) elicit an immune response in rainbow trout, *Salmo gairdneri* Richardson. *Journal of Fish Biology* 31 (supplement), 185–190.

Whyte, S.K., Chappell, L.H. and Secombes, C.J. (1988) *In vitro* transformation of *Diplostomum spathaceum* (Digenea) cercariae and short term maintenance of post-penetration larvae *in vitro*. *Journal of Helminthology* 62, 293–302.

Witenberg, G. (1944) *Transversotrema haasi* a new fish trematode. *Journal of Parasitology* 30, 179–180.

Wittrock, D.D., Bruce, C.S. and Johnson, A.D. (1991) Histochemistry and ultrastructure of the metacercarial cysts of blackspot trematodes *Uvulifer ambloplitis* and *Neascus pyriformis*. *Journal of Parasitology* 77, 454–460.

Wood, B.P. and Matthews, R.A. (1987) The immune response of the thick-lipped mullet *Chelon labrosus* (Risso, 1826), to metacercarial infection of a *Cryptocotyle lingua* (Creplin, 1825). *Journal of Fish Biology* 31 (supplement), 175–183.

Wooten, R. (1974) Observations on strigeid metacercariae in the eyes of fish from Hanningfield Reservoir, Essex, England. *Journal of Helminthology* 48, 73–83.

Wright, C.A. (1971) *Flukes and Snails*. George Allen and Unwin Ltd, London.

Yamaguti, S. (1958) *Systema Helminthum. Vol. I, The Digenetic Trematodes of Vertebrates, Parts I,II*. Interscience Publications, New York.

Yamaguti, S. (1970) *Digenetic Trematodes of Hawaiian Fishes*. Keiga ku Publishing House Co., Tokyo.

Yanohara, Y. and Kagei, N. (1983) Studies on metacercaria of *Centrocestus formosanus* (Nishigori, 1924) – I. Parasitism of metacercariae in gills of young rearing eels, and abnormal deaths of hosts. *Fish Pathology* 17, 237–241 (in Japanese, English summary).

Yekutiel, D. (1985) Metacercarial infections of cichlid fry in Lake Kinnereth. Unpublished M.Sc. thesis, Hebrew University of Jerusalem (in Hebrew, English summary).

Zhatkanbayeva, D. and Heckmann, R.A. (1990) Effectiveness of praziquantel against trematodes of fish. Book of Abstracts *VIIth International Congress of Parasitology*, 20–24 August, 1990, Paris p. 869 (No. 7046).

10

Cestoidea (Phylum Platyhelminthes)

T.A. Dick and A. Choudhury

Department of Zoology, University of Manitoba, Winnipeg, Manitoba, Canada R3T 2N2.

INTRODUCTION

Anyone who has necropsied a fish is immediately aware that tapeworms and fish are an important biological duet. Very heavy infections are commonly reported in the flesh or along the viscera, but remarkably few have been shown to cause mortality of fish. The human fish tapeworm *Diphyllobothrium* has received the most attention, followed by the large larval tapeworms, *Ligula* and *Schistocephalus*, and those of economic importance such as *Triaenophorus*. On the other hand, entire orders, such as the Tetraphyllidea and Tetrarhynchidea, have received relatively little attention. With the exception of the Cyclophyllidea and Aporidea, every order of cestodes has members that mature in fish. While much has been made of the human fish tapeworm, *Diphyllobothrium*, in textbooks and anecdotal reports, it is *Triaenophorus* that has been most extensively studied; not because it is an important human parasite nor due to its pathology in fish (Rosen and Dick, 1983) but rather as a contaminant in the flesh of economically important coregonids.

Historically, most studies on fish cestodes have focused on taxonomy, life cycles, surveys, epizootiology, and population dynamics. Recently, the expansion of the aquaculture industry, as well as basic curiosity about the defence mechanisms of lower vertebrates have made the study of immuno-pathological responses in fish relevant. An increasing worldwide awareness of potential fish-transmitted human pathogens has spurred research into the development of molecular probes to identify fish parasites, particularly in fish inspection laboratories.

This review will focus on the biological uniqueness of fish tapeworms, those cestodes that cause disease in fish of economic importance, their control and suggestions on future research.

HOST RANGE AND GEOGRAPHICAL DISTRIBUTION

Fish tapeworms are ubiquitous and cosmopolitan in distribution and all species of fish have either larval or adult tapeworms. Most of the economically important fish tapeworms are in temperate, north temperate and Arctic regions of the world, with some species (*Diphyllobothrium*, *Triaenophorus*, and *Ligula*) having a circumpolar distribution. This report will focus on selected species (Table 10.1) as there are adequate catalogues of most tapeworms and their hosts in other reviews (Wardle and McLeod, 1952; Bykhovskaya-Pavlovskaya *et al.*, 1962; Hoffman, 1967; Kennedy, 1974; Margolis and Arthur, 1979; Liao and Liang, 1987).

Species diversity of fish tapeworms appears to be greatest in subtropical and temperate regions but this is also the region where most survey work has been done. As one moves North, species diversity declines but the number of parasites are higher, to the point where intestines are distended with masses of adult tapeworms. For example, 1000 encysted larval *Diphyllobothrium* (mean intensity of over 400) and about 700 adult tapeworms of *Proteocephalus* and *Eubothrium* per Arctic char are frequently found (Dick and Belosevic, 1981). Also the prevalences can be up to 97%. One must be careful in drawing conclusions on intensities and prevalence since most fish parasites have seasonal peaks and many potential hosts have not been adequately sampled. Seasonal cycles are more pronounced in temperate, north temperate and Arctic regions and are most evident in the contracted growing season of the Arctic. This seasonality is less pronounced for larval tapeworms than for adult tapeworms. For example, larval *Triaenophorus* tend to accumulate with age of fish (Watson and Dick, 1979), as does *Diphyllobothrium* larvae in whitefish and ciscoe (Dick, unpublished).

ECONOMIC IMPORTANCE

Numerous tapeworms cause disease in fish (Table 10.1) and the problems of tapeworms in aquaculture and fisheries (marine and freshwater) have been reviewed (Sindermann, 1970; Hoffman, 1975). The most important tapeworm in warm-water aquaculture is the Asian pseudophyllidean tapeworm *Bothriocephalus acheilognathi* (see Paperna, 1991). It is an excellent example of translocation of tapeworms, from their original location (in this case, Siberia) with the introduction of their hosts (e.g. cyprinids) into other parts of the world (Bauer and Hoffman, 1976). Similarly, *Proteocephalus ambloplitis* was introduced and established in a warm-water reservoir on the Canadian prairies and had potential for further dissemination through stocking programmes (Szalai and Dick, 1991; Dick and Choudhury, unpublished). *Caryophyllaeus* spp. and *Khawia* spp. are also of concern in Europe and North America, and the effects of *Triaenophorus* on the fisheries is well documented (Lawler, 1970). *T. nodulosus* is considered to be the most

Table 10.1. Pathogenic fish tapeworms, their distribution and hosts.

Order and species	Distribution	Fish host	Location in host
Amphilinidea			
Amphilina foliacea	Eurasia	Sturgeon	Liver, body cavity
A. bipuncta	N. America	White sturgeon	Body cavity
Caryophyllidea			
Caryophyllaeus spp.	Eurasia, N. America	Cypriniforms	Intestine
Khawia sp.	Eurasia	and siluriforms	Intestine
Pseudophyllidea			
Adults			
Bothriocephalus acheilognathi	Eurasia, N. America	Cypriniforms	Intestine
Cyathocephalus truncatus	Eurasia, N. America	Salmonids	Intestine
Eubothrium spp.	Europe, N. America	Arctic char	Intestine
Triaenophorus crassus	Eurasia, N. America	Pike	Intestine
T. nodulosus	Eurasia, N. America	Pike	Intestine
Larvae			
Diphyllobothrium dendriticum	Eurasia, N. America	Salmonids	Viscera
D. latum	Worldwide	Percids	Muscle
D. sebago	N. America	Trout, salmon	Viscera
Ligula intestinalis	Eurasia, N. America	Cypriniforms	Viscera
Schistocephalus sp.	Eurasia, N. America	Sticklebacks	Viscera, muscle
T. crassus	Eurasia, N. America	Whitefish, trout	Muscle
T. nodulosus	Eurasia, N. America	Trout	Viscera, liver
Proteocephalidea			
Adults			
Corallobothrium sp.	N. America	Catfish	Intestine
Proteocephalus exiguus	Europe	Grayling	Intestine
Larvae			
Proteocephalus ambloplitis	N. America	Smallmouth bass	Viscera, ovaries

Pathology caused by larval marine tapeworms has been reviewed by Sinderman (1990) and Williams (1967).

pathogenic of the *Triaenophorus* species in Eurasia (Kuperman, 1973). It causes epizootic outbreaks with heavy mortalities in juvenile rainbow trout held in rearing ponds and under hatchery conditions (Scheuring, 1923; Bauer, 1959; Kuperman, 1973) although natural infections of trout are relatively rare. By contrast, the scourge of freshwater fisheries in North America has always been *T. crassus*, since it causes unsightly infestation in the muscles of whitefish. High *T. crassus* infections can affect the exports of whitefish. Also, the culturing of salmonids and coregonids in fresh water and the increasing use of fish stocking as a management strategy can be affected by *T. crassus*. *T. crassus* was found in juvenile brown trout in 1991 in a hatchery in Manitoba, Canada (authors, unpublished).

SYSTEMATICS AND TAXONOMIC POSITION

The majority of disease-causing cestodes discussed here fall within three orders: Caryophyllidea (*Caryophyllaeus* and *Khawia*), Pseudophyllidea (*Ligula*, *Schistocephalus*, *Diphyllobothrium* and *Triaenophorus*) and Proteocephalidea (*Proteocephalus* spp.). The systematic relationships among various groups of tapeworms have been reviewed by Stunkard (1983). Accordingly, the two lines of descent that derive the Pseudophyllidea (via the Tetrarhynchidea) and the Cyclophyllidea (via the Proteocephalidea) from a common Tetraphyllidean-like ancestor are supported by life-cycle and developmental studies (discussed by Stunkard, 1983). The position of the Caryophyllidea has been the subject of considerable controversy but their historical status as 'primitive' cestodes with affinities to the cestodarians or as monozoic pseudophyllideans has been replaced by a more independent one (Mackiewicz, 1981). Recent phylogenetic analysis of the Platyhelminthes (Brooks, 1989) derives the Amphilinidea (*Amphilina* sp.) and Gyrocotylidea independently from an ancestor common to these two and the remaining tapeworm orders (i.e. the Eucestoda including the Caryophyllidea).

PARASITE MORPHOLOGY AND LIFE CYCLES

The amphilinids and gyrocotylideans have been reviewed by Bandoni and Brooks (1987a,b), Dubinina (1974) and Skrjabina (1974). The morphology and biology of the Caryophyllidea have been reviewed by Mackiewicz (1972). The only common feature between the caryophyllids and cestodarians is their monozoic body plan. Differences include the arrangement of the reproductive apparatus (separation of male and female gonopores and the presence of the uterine pore at the opposite end in the amphilinids) and the hosts they parasitize (amphilinids and gyrocotylids are parasites of more 'ancient' fish such as acipenserids and chimaerids while caryophyllids are typically parasites of cyprinid and catostomid fishes). Plerocercoids of pseudophyllideans often occupy similar sites in the fish host and correct diagnoses require a thorough knowledge of the morphologies of these larvae.

The morphology of plerocercoids of *Schistocephalus* and *Ligula* is well known (Hoffman, 1967; Smyth and McManus, 1989) and they are distinguished by lack of segmentation in *Ligula* (conspicuous in *Schistocephalus*). The plerocercoids of *Triaenophorus* are unique in that they bear tricuspid hooks which vary in size, depending on the species (see review by Kuperman, 1973). The lack of conspicuous morphological differences makes identification of *Diphyllobothrium* plerocercoids difficult. Some researchers use the external tegumental morphology i.e. wrinkled versus smooth tegument and the form/shape of the bothrial grooves and a key has been developed using these characteristics (Andersen *et al.*, 1987; Andersen and Gibson, 1989). The tetrarhynchideans are readily separated from tetraphyllideans by the presence of spiny proboscides in the former. For information on marine fish cestodes causing disease (particularly Tetraphyllidea, Tetrarhynchidea and Diphyllidea), readers are referred to Grabda (1989) and Sindermann (1990).

Ultrastructural studies of plerocercoids and adults of fish cestodes are scarce. Studies show that plerocercoids of *Schistocephalus* and *Ligula* (Charles, 1971) develop thicker outer tegumental layers, surface ridge systems and microtriches characteristic of each species. With age, the number of 'spine precursor' vesicles declines with concomitant increases in specialized vesicles with possible endolysosome-like activity (Charles, 1971). In contrast to the adult tegument, the tegument of *Proteocephalus ambloplitis* plerocercoids possesses short conoid microtriches and lacks vacuolization as well as unicellular tegumental gland cells (Coggins, 1980a). Coggins (1980a) also described the end organ in plerocercoids and demonstrated the presence of proteolytic enzymes (Coggins, 1980b), a structure not present in adult worms and probably used by the plerocercoid in visceral migration. In contrast, the teguments of plerocercoid and adult stages of *Haplobothrium globuliforme* were found to be '... remarkably similar' (although differences in the neck region are noted). This was interpreted as a pre-adaptation of the plerocercoid to the adult existence (MacKinnon and Burt, 1984). Comparative studies on pseudophyllidean plerocercoids have revealed gland cells in the anterior end of the body (head and bothria) in *Eubothrium* and *Triaenophorus* which release their contents via ducts (Kuperman and Davydov, 1982). In contrast, plerocercoids of *Diphyllobothrium latum* have an extensive syncytial gland complex throughout the parenchyma while the Ligulidae possess frontal and lateral glands (Kuperman and Davydov, 1982). Few studies have documented ultrastructural interactions at the host–parasite interface in adult fish tapeworms. McVicar (1972) has shown that unique structures on the tegument of the holdfasts of some adult tetraphyllideans (*Echeneibothrium*, *Phyllobothrium* and *Acanthobothrium*) may be specialized microtriches with osmiophilic ends and presents evidence that the myzorhynchus apical pad of *Echeneibothrium* is involved in direct nutrient uptake from the host submucosa. Additional information on tegumental ultrastructure comes from physiological studies, often *in vitro* (see Smyth and McManus, 1989) and more recently from immunopathological interactions (Hoole and Arme, 1986, 1988; see also section on pathology).

Figure 10.1 illustrates a generalized life cycle of fish tapeworms. Fish

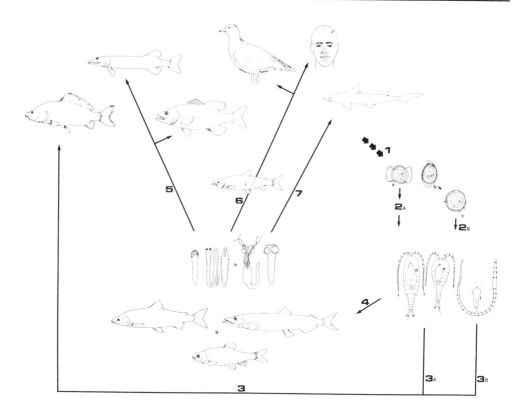

Fig. 10.1 A schematic diagram representing life cycles of important freshwater fish tapeworms. 1. Eggs from definitive host enter water. 2A. Embryonated egg of Proteocephalidea infective to invertebrates. 2B. Hatched coracidium of Pseudophyllidea, Tetraphyllidea, and Tetrarhynchidea infective to invertebrate intermediate hosts. 3. Infection of definitive host via infected invertebrates. 3A. Pseudophyllidea and Proteocephalidea. 3B. Caryophyllidea. 4. Infection of fish intermediate hosts via infected invertebrates (*Diphyllobothrium* spp., *Triaenophorus* spp., *Proteocephalus* spp., Tetraphyllidea, Tetrarhynchidea). 5. Infection of piscivorous fish definitive hosts via infected prey fish (*Proteocephalus* spp., *Triaenophorus* spp.). 6. Infection of homoeotherm definitive host via infected prey fish (*Ligula*, *Schistocephalus*, *Diphyllobothrium* spp. (*D. latum* of man and *D. dendriticum* and *D. ditremum* of gulls also utilize paratenic fish hosts). 7. Infection of definitive cartilaginous fish hosts (sharks, rays etc.) by larvae of Tetrarhynchidea and Tetraphyllidea via infected prey fish. a. Embryonated egg of Proteocephalidea, often with modifications to facilitate floating and dispersal. b. Embryonated operculate eggs of Pseudophyllidea, Tetraphyllidea, Tetrarhynchidea. c. Ciliated motile coracidia released from b. d and e. Procercoids in invertebrate hosts. f. Larva of Caryophyllidea in tubificids. g. Fish intermediate hosts. h. Larval stages (plerocercoids and plerocerci) infective to definitive hosts. i. Paratenic fish host.

tapeworms do not exhibit modifications such as resistant eggs or cysts to withstand abiotic factors and are transferred passively. These tapeworms usually release eggs that complete embryonation in water and commonly hatch into ciliated, motile, and short-lived larvae called coracidia. Non-motile embryos possess egg envelopes modified to maintain position, thereby enhancing the chance of being ingested by invertebrates (also see Jarecka, 1961). In the haemocoel (body cavity) of the invertebrate host the coracidium differentiates and develops into a procercoid stage. Although not much is known about the factors which affect development, we know that space and nutrients are important (Shostak *et al.*, 1984). The migration of the proceroid and its differentiation into a plerocercoid in fish muscle is best illustrated by *Triaenophorus crassus* in coregonids (Rosen and Dick, 1983). The magnitude of size increase is spectacular in larval *Ligula* and *Schistocephalus* where, in the body cavity of fish, the biomass of the parasite may reach 40% of the body weight of the host.

The use of paratenic hosts is well documented in fish parasites (Fig. 10.1). In the genus *Diphyllobothrium*, a piscivorous fish feeds on a smaller infected fish of the same or different species and the plerocercoid stage reencysts in the new host. *D. dendriticum* and *D. ditremum* are good examples of this phenomenon as high numbers of larval stages accumulate in large Arctic char in Arctic lakes. Numerous studies have indicated that certain larval tapeworms (*Ligula* and *Schistocephalus*) influence the behaviour of their fish hosts and enhance transmission and parasitized intermediate hosts may be more susceptible to predation by natural definitive hosts (Mackiewicz, 1988). Transmission to the definitive host (fish, bird or mammal) is through ingestion of the infected intermediate host and can involve consumption of carrion (natural mortality) or offal left by sports and commercial fishermen.

Host specificity of fish parasites is difficult to determine as few studies have adequately evaluated all possible routes of transmission. For example, new hosts are constantly being found for larval tapeworms (Hoffman, 1967; Margolis and Arthur, 1979). Some parasites, e.g Caryophyllidea, are more specific in their invertebrate hosts than others, e.g. Pseudophyllidea. There can be narrow host specificity even in species that have a wide geographical distribution but the distribution is usually closely tied to the distribution of the definitive hosts. For example, *Triaenophorus* has a worldwide distribution that is restricted and closely correlated to its host, *Esox* (pike). *T. nodulosus* is the most widely distributed species with 78 species of intermediate fish hosts. However, *T. crassus* with 39 reported intermediate hosts (Kuperman, 1973, Margolis and Arthur, 1979) has two preferred hosts, whitefish and ciscoe, in North America. Plerocercoids of *Ligula* occur in a wide range of hosts (e.g. catastomids, centrarchids, cyprinids, percids, salmonids) but the definitive host is *Larus*. Host specificity is also evident in *Diphyllobothrium* plerocercoids in fish from Quigly Lake, Canada where *D. dendriticum* occurs in whitefish and ciscoe while *D. latum* occurs in pike and walleye. In summary, there is no clear-cut pattern concerning host specificity of fish tapeworms but in general tapeworms are more specific to their definitive hosts.

As fish are poikilotherms, the effects of abiotic factors are pronounced on their tapeworms where growth and maturation of these parasites are tied to temperature. Most infections of fish occur during the summer and early autumn and adult tapeworms grow and mature during late spring and early summer. In spring and early summer most tapeworms reach sexual maturity and release eggs; this is the time when populations of the invertebrate hosts reach their peak. Differentiation into procercoids is also rapid due to high water temperatures and there is sufficient time for infections of fish to occur prior to winter. Adult tapeworms may be shed during the summer with new recruitment in the late summer and early autumn; *T. crassus* in pike and most proteocephalid species. The synchronization of tapeworm maturation and physiological state of pike is not well understood but we know that loss of *T. crassus* is closely synchronized with the spring spawning of pike. Entire tapeworms are shed at spawning and this may occur in about a week. This has the added advantage that tapeworm eggs are concentrated in shallow, warm water, the same areas that young whitefish and ciscoe frequent during their early stages of development.

    The acquisition of larval stages is similar to that of adults. If there is encystment it usually occurs over a two month period in the first summer and with some associated pathology (see pathology section). Most studies report an accumulation of larval stages of the other tapeworms over time but *D. latum* in the flesh of pike and walleye shows an annual increase in the summer and a decline during the winter months. Studies on *Ligula* suggest an annual recruitment with some intraspecific interaction, apparently tied to space and nutrients in the fish host. However, some of the interactions thought to be expressed through reduced growth of the parasite may actually be new parasites (Szalai *et al.*, 1989).

HOST–PARASITE RELATIONSHIPS

Fish have a well-developed immune system which is often temperature dependent. Most insults to the fish host by a helminth produce a strong cellular response with nonspecific inflammation while the main immunoglobulin (antibody) is of the IgM class. Acute inflammation to heavy metazoan parasite infections commonly produces lesions manifested by cell death and necrosis and distortion of tissues. Generally, migrating parasites (larval forms) produce the most serious reactions; leucocytosis and fibrosis, and more rarely haemorrhage, hyperaemia and necrosis (Arme *et al.*, 1983; Roberts, 1989).

Cellular responses and tissue pathology

The host cellular response to tapeworm infections may result in total (as in some *Diphyllobothrium* spp.) or partial (as in *Ligula*) encapsulation of the parasites. The cellular exudates that initially accompany visceral reactions consist primarily of fibroblasts. Infiltration in tissues or in the body cavity by leucocytes is dominated by macrophages, neutrophils and lymphocytes

(Hoole and Arme, 1983). Sharp *et al.* (1992) found a cellular response to *D. dendriticum* plerocercoids occurred during the first two weeks post infection, followed by an increase in neutrophils and a large influx of macrophages which eventually transformed into epithelioid cells. A blood vascular network eventually developed followed by fibroplasia and the depositon of a collagenous tissue matrix (Sharp *et al.*, 1992). The lymphoid organs of fish such as roach (*Rutilus rutilus*) infected with *Ligula* undergo changes such as increase in melanomacrophage centres in the spleen and increase in 'vacuolated granulocytes' in the pronephros (Taylor and Hoole, 1989). With time, cellular activity decreases as the parasite is encapsulated and isolated. Encapsulation is a general phenomenon and may occur in the viscera (with *Diphyllobothrium* spp., *Triaenophorus nodulosus, Proteocephalus*) or in the muscle (*T. crassus*). Although the parasites are encapsulated, they often live as long as their hosts.

The pathology caused by the plerocercoids of *Ligula* and *Schistocephalus* has been reviewed in detail (see Arme *et al.*, 1983). This includes growth retardation, abdominal distension, as well as displacement and morphological/ physiological alteration of internal organs. Metabolic changes such as inhibition of gametogenesis and reduced liver glycogen, muscle carbohydrates, proteins, blood amino acids, and lipids are frequently associated with infections. Other effects are reduction in oxygen consumption by the host and an increase in serum proteins and alkaline phosphatase. The parasite may increase its own phospholipids at the expense of the host muscle (Arme *et al.*, 1983; Kurovskaya and Kititsyna, 1986; Rzeczkowska and Honowska, 1988). An increase of certain enzymes was also correlated with *T. nodulosus* infection of the liver and lesion production (Scheinert and Hoffmann, 1986). Reduced fecundity in smallmouth bass has been associated with *P. ambloplitis* plerocercoids which cause gonad tissue destruction (McCormick and Stokes, 1982). Outright mortalities can be caused by larval tapeworms such as *Diphyllobothrium* and *Triaenophorus*, especially in aquaculture situations (Rodger, 1991), and mortality may be dependent on the intensity of infections (Rosen and Dick, 1983; Halvorsen and Andersen, 1984). The causes of mortality include destruction of the tissue–water interface, extensive haemorrhaging and secondary microbial infection. The liver is a favourite site of infection and the pathology may include haemorrhaging, necrosis, fibrosis, oedema and discoloration (Arme *et al.*, 1983; Pronina, 1977). Mortality is often attributable to liver dysfunction and blood loss: in salmonids infected with *D. ditremum* (Weiland and Meyers, 1989; Rodger, 1991), in acipenserids infected with *Amphilina* spp. (Popova and Davydov, 1988), in trout infected with *T. nodulosus* (Scheuring, 1923; Bauer, 1959; Kuperman, 1973), in bluegills and basses infected with larvae of *Proteocephalus* (Mitchell *et al.*, 1983; Joy and Madan, 1989).

The early lesions in whitefish muscle infected with *T. crassus* includes haemorrhaging and discoloration. The extent of the pathology due to differentiating and migrating plerocercoids (Fig. 10.2A,B) is usually underestimated. Rosen and Dick (1983) showed by timed and serial sections that damage is quite extensive (Fig. 10.2C).

It has been suggested that the pathogenicity of *D. dendriticum* depends on

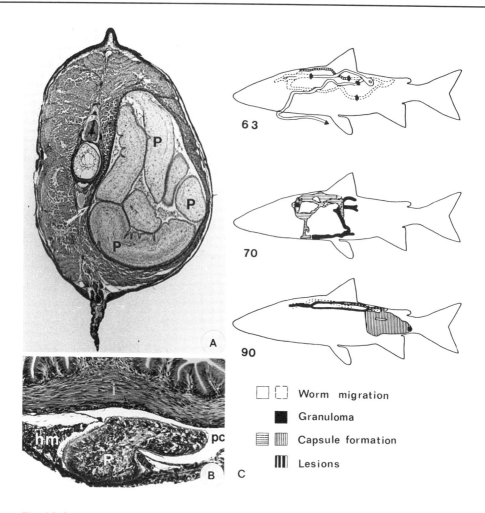

Fig. 10.2. Pathology caused by *T. crassus* in experimentally infected whitefish. A. Cross section through posterior region showing encapsulated plerocercoid (arrow) at 90 days post infection. B. Section through region of small intestine showing the migrating plerocercoid at 5 days post infection. C. Reconstruction of the extensive migration of the plerocercoid and associated tissue changes at 63, 70 and 90 days post infection. i = intestine, hm = hypaxial muscle, P = plerocercoid, pc = peritoneal cavity. (From Rosen and Dick (1983), courtesy of the *Canadian Journal of Zoology*.)

the hosts' defence mechanism. Fish with weak (in salmonids) or no cellular response (in *Coregonus albula*) have more lesions which lead to mortality than fish (such as *C. lavaretus*) which produces a strong cellular encapsulation response (Bylund, 1972). This was also shown with *T. crassus* infections in whitefish (Rosen and Dick, 1983) and in rainbow trout (Rosen and Dick, 1984).

Pathology in the gut

The pathology caused by tapeworms in the gut may cause tissue alteration or destruction, mechanical blockage and nutrient absorption at the expense of the host. With cestodes that penetrate the mucosal layer (such as *Triaenophorus*, and many tetrarhynchideans), the establishment of gut infections usually begins with an initial phase of acute inflammation in the region of contact. The inflammation is characterized by subepithelial oedema, leucocyte infiltration and, in heavy infections, epithelial erosion and necrosis. As infections become more chronic, a relatively low level inflammation is maintained. Also vascularization is decreased to the infected areas and this is accompanied by epithelial hyperplasia (Pronina and Pronin, 1982; Shostak and Dick, 1986). High numbers of parasites (200 worms) did not cause mortality although the diameter of the lumen was reduced by more than 50% and this probably affects movement of food through the intestine (Shostak and Dick, 1986).

The pathology in the gut is often related to the morphology of the holdfast organs of the parasite. Species of Caryophyllidea that possess well-developed/specialized but non-invasive attachment organs elicit little or no pathology, whereas those with developed holdfasts or possessing the terminal introvert on the holdfast are associated with ulcers and nodules (Mackiewicz *et al.*, 1972). Also, the size of the holdfast and hooks determines the extent of pathology as has been shown for *Triaenophorus* spp. (see Kuperman, 1973; Shostak and Dick, 1986).

Heavy gut infections may cause intestinal distension, mechanical obstruction and perforation, as in *Bothriocephalus acheilognathi*. The clinical signs include emaciation and anaemia (Caryophyllidea, *Bothriocephalus*) (Scott and Grizzle, 1979). Nutritional demands made by gut tapeworms may adversely affect growth, condition and fitness; *Eubothrium salvelini* in juvenile coho salmon (Boyce and Yamada, 1977; Boyce, 1979), may cause a change in blood composition, or may increase susceptibility to environmental stress such as zinc. Shostak and Dick (1986) did not find evidence for interspecific nor intraspecific competition for nutrients among helminths in the intestine of pike. This suggests that nutrients were not a limiting factor.

Humoral immunity to cestodes

Although the antibody response in fish to microbial pathogens is well documented, knowledge on specific immunity in helminth infections is sparse (Evans and Gratzek, 1989; Sharp *et al.*, 1992). The demonstration of precipitating antibodies (?) in *Abramis brama* (bream) to *Ligula* (see Molnar and Berczi, 1965) must be viewed with some caution. Precipitation reactions may be confused with C-reactive protein activity which has been demonstrated in turbot with phosphorylcholine-rich extracts from its tapeworm *Bothriocephalus scorpii* (see Fletcher *et al.*, 1980). A complement factor and C-reactive protein have also been implicated in leucocyte adherence to tapeworms (plerocercoid of *Ligula*) (Hoole and Arme, 1986, 1988). Precipitins to phosphorylcholine epitopes which are common to a variety of organisms including microbial pathogens and metazoans, have been observed in lake sturgeon *Acipenser fulvescens* (Choudhury and Dick, 1994). Other studies have only inferred the presence of antibodies to tapeworms (Kennedy and Walker, 1969; McVicar and Fletcher, 1970; Sweeting, 1977). However, recent studies using ELISA have demonstrated unequivocally, the antibody response of rainbow trout to *Diphyllobothrium* spp. (Sharp *et al.*, 1989, 1992).

No detectable cellular response was demonstrated in the gudgeon (*Gobio gobio*) infected with *Ligula* (Arme *et al.*, 1983) and pike and walleye infected with *D. latum* (Dick and Poole, 1985). However, recent information (Taylor and Hoole, 1989) indicated an increase of melanomacrophages in the spleen of infected gudgeon, although cell counts in the pronephros remained unchanged. There was also evidence to suggest that plerocercoids of *Ligula* and *Schistocephalus* adsorbed host proteins which helped them evade host immune responses (Hoole and Arme, 1986).

Reduced pathogenicity is often taken to indicate a long evolutionary period of host–parasite association. *T. nodulosus* causes little pathology in its obligatory hosts, smelt (*Osmerus*) and perch (*Perca*), but severe pathology in trout. Similarly, *Proteocephalus* evokes a weak host response in its natural hosts (*Coregonus lavaretus* and *C. lavaretus migratorius*) but causes persistent inflammation in *Thymallus* (grayling). Experimental infections of rainbow trout with *T. crassus* showed extensive worm migrations. The clinical signs included popeye, extensive haemorrhaging, and death (Dick, unpublished). This may account, in part, for the failure of rainbow trout stocking into pike/whitefish/*Triaenophorus* systems. It is commonly held that host specific helminths are not pathogenic to their hosts. This generalization should be modified as *Amphilina* spp. cause extensive liver pathology in their acipenserid hosts (Popova and Davydov, 1988), for which the genus is specific.

IN VITRO CULTURE

The absence of an intestine in cestodes has made them an interesting challenge for culture studies (Smyth and McManus, 1989; Taylor and Baker, 1987). Culture of cestodes may include the hatching processes (mostly due to physical factors), short-term culture or maintenance (which may or may not include all of the following, i.e. nutrient requirements, physicochemical aspects of the media, and long term culture involving differentiation of the parasite). This discussion will focus on the culture to mature adult stages only. The criteria for development and maturation usually accepted are segmentation, organogeny, gametogenesis, and egg shell formation (Smyth and McManus, 1989). Cestodes also pose more of a challenge for culturists because of their long flat shape and tendency to 'knot-up' in culture. Consequently, in the early days of culture, considerable effort was placed on the physical set-up of the system, i.e. roller tubes vs. continuous flow (Smyth, 1946, 1947, 1982; Berntzen, 1961, 1962, 1965; Berntzen and Voge, 1965).

With the exception of *Schistocephalus* and *Ligula*, little work has been done on fish parasites but these two species are unusual in that the plerocercoid stage is large with extensive energy reserves in the form of glycogen. Since external nutrients were not required for the culture of *Schistocephalus* and *Ligula*, researchers focused on the physicochemical requirements of the culture system (Smyth, 1946, 1947, 1954). Smyth (1982) developed a special culture apparatus with ports to add and replace the culture medium and used cellulose dialysis tubing to compress the worms. Additional culture conditions included a highly buffered media to neutralize acid metabolic byproducts, 100% horse serum or other well buffered culture media, anaerobic or very low pO_2 to prevent premature oxidation of the phenolic precursors in the egg shells, and gentle agitation of the culture medium to ensure diffusion (Smyth and McManus, 1989). Similar conditions were applied successfully to *Ligula* but due to its larger size the culture system was scaled up accordingly. Smyth and McManus (1989) also pointed out that segments of *Ligula* cut from the median region of the plerocercoid will mature *in vitro*. Culturing the less differentiated plerocercoids of *Diphyllobothrium latum* to adults remains a major challenge. Much more work needs to be done on cestode culture in general. Taylor and Baker (1987) summarize the problems and lack of knowledge of *in vitro* culture for cestodes; insufficient details on culture procedures, lack of precise criteria for assessing the quality of the media and culture technique in addition to morphogenesis, lack of defined media, incomplete culture of the entire life cycle, and the absence of cell lines.

NUTRITION AND PHYSIOLOGY

The biochemical composition of cestodes has been compared by Smyth (1976). For most of the fish tapeworms values are as follows; dry weight as a percentage of fresh wet weight varies from 27.0 to 31.8; glycogen as a % of dry weight varied from a low of 13.8 for *Triaenophorus* to a high of 38–50 for *Ligula*; lipid as a percentage of dry weight was about 16%; protein varied from a low of about 35% for *Ligula* to a high of 60% for *D. latum*. These values fall within the range reported for other tapeworms. However, in the case of inorganic compounds the few studies reported on indicate that fish tapeworms have lower values than most other tapeworms (Smyth and McManus, 1989). Smyth and McManus (1989) summarize the general lipid composition of cestodes as being similar to other organisms. As cestodes have lipid compositions similar to their hosts, it is not surprising that fish tapeworms have higher levels (70–80%) of unsaturated fatty acids as their lipids, than most other cestodes. The beta-oxidation enzymes (acyl-CoA synthetase (short and long chains), acyl-CoA dehydrogenase, enoyl-CoA hydratase, 3-hydroxyacyl-CoA dehydrogenase, and acetyl-CoA acyltransferase) are present in *Ligula* and *Schistocephalus*.

The role of carbohydrates as an energy source of cestodes has been extensively studied with the end products of carbohydrate breakdown (aerobic and anaerobic) being acetate and propionate found in *Schistocephalus*, succinate

and lactate for *Diphyllobothrium* and lactate, succinate, acetate, and propionate for *Ligula* (Smyth and McManus, 1989). The glycolytic enzymes
of fish cestodes are similar to those of other cestodes (see Smyth and
McManus for a comparisons). Key glycolytic enzymes such as hexokinase
(in *Bothriocephalus*), phosphofructokinase (in *Schistocephalus*), pyruvate
kinase (in *Ligula* and *Khawia*) and lactate dehydrogenase (in *Schistocephalus*)
have been demonstrated (information on all cestodes is summarized in Smyth
and McManus, 1989). It appears that *Schistocephalus* is capable of utilizing
carbohydrates by a functional TCA cycle (Koerting and Barrett, 1977).
Oxidative phosphorylation may occur in *D. latum* since its mitochondria can
oxidize NADH and succinate and there is some fragmentary evidence for the
pentose-phosphate pathway in *Ligula* and *Schistocephalus*. *Bothriocephalus
scorpii*, in contrast to other cestodes, excretes methyl butyrate as an end
product of respiration (Smyth and McManus, 1989).

The amino acid composition of all cestodes is similar to other organisms.
Collagen is the most common structural protein in tapeworms and has been
characterized from *Bothriocephalus* and *Nybelinia* (Smyth and McManus,
1989). Proteases have been reported from *Schistocephalus* and *Ligula* but their
functions remain obscure as amino acids do not appear to have an important
role in cestode energetics. However, a much more plausible role for proteases
is for tissue penetration by larval cestodes.

There are extensive reviews on the transport of nutrients across the cestode
tegument (Arme and Pappas, 1983; Threadgold, 1984). Although the presence
of a sodium pump has been demonstrated in cestodes, enzymes such as sodium
or potassium ATP-ase have not been detected in most cestodes (Threadgold,
1984; Smyth and McManus, 1989). Pinocytosis as a means of nutrient uptake
has been demonstrated in *Schistocephalus* (Hopkins *et al.*, 1978). Functional
differences along the tegument of *Bothriocephalus gregarius* suggest different
physiological functions (Berrada-Rkhami *et al.*, 1990). Glycoconjugates of the
neck appear to be involved in recognition, adherence, fixation, and protection.
Those glycoconjugates on the strobila are primarily nutritional but may also
protect against digestive enzymes of the host.

IDENTIFICATION OF PARASITES

Adult fish tapeworms are generally identified with the aid of morphological
keys but it is the larval tapeworms (especially the pseudophyllideans) that
create the most problems. Morphological keys are available for cestodes
(Khalil *et al.*, 1994; Schmidt, 1986) and more specifically for *Diphyllobothrium* (Andersen and Gibson, 1989). General references such as Bykhovskaya-
Pavlovskaya *et al.* (1962) and Hoffman (1967) are also useful. Isozymes were
used to distinguish between *D. latum* and *D. dendriticum* (deVos and Dick, 1989)
and between *D. latum* and *D. nihonkaiense* in Japan (Fukumoto *et al.*, 1987).
Populations of *Bothriocephalus* have also been examined using enzyme electrophoresis. *B. barbatus* was found to be homogeneous but *B. gregarius* could
be divided into two groups based on geographical distribution (Renaud *et al.*,

1986). McManus (1985) showed that natural populations of *L. intestinalis* were polymorphic for the enzyme phosphoglucomutase (PGM) and were not correlated to geography or host species, while other enzymes were similar.

Using immunological techniques, antigenic affinities were demonstrated between *D. dendriticum* and *D. ditremum* and between *D. vogeli* and *D. ditremum* whereas *D. latum* showed no cross reactivity with the other three species (Freze *et al.*, 1983).

Molecular techniques utilizing ribosomal DNA probes hybridized against endonuclease restricted genomic DNA on an agarose gel (deVos *et al.*, 1990) were used to identify *Diphyllobothrium* plerocercoids. Recent development of oligonucleotide primers from sequenced intergenic spacer regions of ribosomal DNA from *D. latum* and *D. dendriticum* have been used to identify plerocercoids and adults of *D. latum* and *D. dendriticum* (Dick, unpublished).

PREVENTION AND CONTROL OF TAPEWORMS IN FISH

Numerous studies, including some field applications, have evaluated the role of drugs on fish tapeworms and some were effective, especially those against *Bothriocephalus* and *Proteocephalus* (Table 10.2). For more information on chemotherapy, readers are referred to the excellent account by Schaeperclaus (1992, Vol. 1, pp. 249–252 and Vol. 2; pp. 769–804). Niclosamide is the active ingredient of commercial drugs such as Devermin, Radeverm, Phenasal, Mansonil, Yomesan and medicated feeds such as Zestocarp and Cyprinocestin (Schaeperclaus, 1992). Other synthetic anthelmintics include dibutyl tin oxide used successfully against *Eubothrium crassum* (Schaeperclaus, 1992). Niclosamide preparations have largely replaced herbal drugs such as Kamala (Rottlera), Filixan and tobacco dust (Schaeperclaus, 1992).

The use of vaccines to prevent tapeworm infections in fish is clearly not the approach to take at present due to lack of basic information on fish immune responses to tapeworms.

Aquaculture facilities frequently use nearby natural water bodies as their source of water. These natural systems often have hosts (invertebrate and vertebrate) infected with tapeworms (such as *Diphyllobothrium*, *Proteocephalus*, and *Triaenophorus*). This may lead to the appearance of *Triaenophorus* (larvae) and *Proteocephalus* (adults) in cultured fish (usually trout) with the introduction of infected copepods in the water. Prevention could involve filtration systems (sand, screens) or use of ground water.

Numerous environmental manipulations of natural systems have been attempted and these include extirpation of: (i) fish-eating birds in the vicinity of an aquatic system; (ii) pike in the case of *T. crassus* (Lawler, 1970); and (iii) ciscoe to reduce larval cysts in the commercially valuable whitefish (Miller 1952; Dick, unpublished). An intensive pike removal programme from Hemming Lake, Manitoba succeeded in the extirpation of *T. crassus*.

Table 10.2. Chemotherapy in the control of fish tapeworms.

Parasite	Target host (location)	Chemical	Result	Reference
Bothriocephalus acheilognathi	Carp (China)	Cucurbita, Areca [1] (ground up in feed)	Effective	Nie and Pan, 1985.
	Carp (E. Europe)	Taenifugin carp [2] (minimum 2% of fish wt)	Successful in feed, 3 times over 3 weeks	Zitnan et al., 1981
	Carp (Europe)	Zestocarp [2]	80–100% reduction	Weiroski, 1984
B. gowkongensis	Carp	Mansonil, Yomesan [2]	1 g kg^{-1} fish, 100% effective	Par et al., 1977
	Carp	Taenifugin carp [2]	100% effective	Kral et al., 1980
B. opsariichthydis	Carp	Phenasal	100% effective	Albetova and Michurin, 1984
Bothriocephalus sp.	Carp (S. Africa)	Lintex [3] (50 mg kg^{-1} fish wt)	100% eradication	Brandt et al., 1981
Caryophyllaeus sp.	Carp	'Kamala' [1]	Infection reduced	Bauer, 1959
	Carp	'Rottlera' [1]	Infection reduced	Bauer, 1959
Caryophyllaeus and Khawia	Carp	Zetocarp [2] (Cestocarp)	(>12°C, day 1: 20 g kg^{-1} fish, day 2: 10 g kg^{-1} fish) 93–100% effective	Schaeperclaus, 1992
P. ambloplitis	Bass (N. America)	Mebendazole (100 mg kg^{-1} fish wt per day in feed)	90% reduction	Boonyaratpalin and Rogers, 1984

[1] Herbal extracts.
[2] Active ingredient niclosamide.
[3] Active ingredient 2',5-dichloro-4'-nitrosalicylanilide. See Schaeperclaus (1992) for details.

However, ciscoe were inadvertently extirpated, whitefish were decreased substantially and, while pike size decreased, the total number of pike actually increased. The intensities of *T. crassus* plerocercoids in whitefish were reduced by 42% in whitefish in Quigly Lake through the selective removal of ciscoe (Dick, unpublished). This was conducted prior to spawning in the autumn, but at the same time maintaining a reproductively viable ciscoe population (Dick, unpublished). The removal of invertebrate hosts (copepods) of *T. crassus* was considered but was unacceptable as they are important components of the food chain of many aquatic animals (Miller, 1952).

CONCLUSIONS

Although the emphasis of this review was not on the taxonomy and systematics of fish tapeworms, we would be remiss if we did not point out some of the problem areas that should be resolved. The genus *Proteocephalus* needs a critical and thorough review, particularly the North American species because some of them are not well described, Also, the larval stages of tetraphyllideans and tetrarhynchideans need to be related to their adult stages through life cycle studies and the use of *Scolex pleuronectis* should be discontinued. Also, there is very little known about the intraspecific variability (i.e. morphological, biological, and genetic) in the genus *Diphyllobothrium*.

Morphological identification of larval tapeworms still relies on the local 'expert'. In the future this can probably be done using nucleic acid probes and amplification methods, such as the polymerase chain reaction, especially when intact parasites are unavailable.

The perception of narrow host specificity in fish tapeworms should be viewed with caution. More needs to be known about potential hosts in a given system and the significance of these hosts in transmission. Basic information is lacking for most fish parasite systems especially the significance of biological and genetic variability in a species. Manipulations of portions of the ecosystem (e.g. ciscoe in the case of *Triaenophorus*) to remove parasites have potential, providing the reproductive capacities of the fish species are not compromised; e.g. ciscoe may harbour *T. crassus* but it is also a major food of pike. Parasite numbers can be reduced without drastically altering the fish species composition in a system but an understanding of fish population dynamics as well as parasite transmission is essential. Knowledge of predator–prey interactions, including sports and commercial fishing, must be integrated into the concept of parasite transmission. For example, overfishing of the fast-growing whitefish has contributed to the apparent increase of *Triaenophorus* in whitefish, not because the total parasite levels increased but because it takes more small fish to produce, for example, 100 pounds of fish (the criterion on which parasite levels are based). Knowledge of food chains may help to reduce *D. latum*, *D. dendriticum*, and *D. ditremum* from Arctic and north temperate systems. For example, lakes with lake trout and Arctic char have lower levels of *Diphyllobothrium* than lakes with only char. Although both fish species acquire plerocercoids, lake trout is a less suitable host for *Diphyllobothrium* and hence levels of *Diphyllobothrium* in Arctic char/lake trout lakes are lower. Little is known of the effects on the indigenous fish parasite populations following stocking of an exotic piscivorous fish species. Parasite acquisition by stocked fish is not well understood but if the study by Poole and Dick (1985) is typical, then stocked fish will acquire non-host specific parasites faster than indigenous fish even if they belong to the same species.

As freshwater aquaculture operations expand, particularly cage culture and the use of surface water by hatcheries become more common, we can anticipate additional parasitic tapeworm problems (e.g. mortality, decreased growth and unsaleable products for human consumption). This is due to the difficulty of excluding the copepods from the water supply and increased susceptibility of cultured species (cyprinids, coregonids, percids, salmonids) to indigenous tapeworms. Careful selection of aquaculture sites and monitoring of the water supply will have to be done to minimize some of these problems.

Only a few tapeworm diseases of fish have been closely monitored and carefully documented. In general, cell mediated immunity appears to be the primary response. The pathology is progressive with the formation of large pustules and eventually host encapsulation. It may result in the destruction of the parasite, formation of a granuloma and finally its resorption. The humoral antibody response to tapeworm infections is not well understood. *T. crassus* is a useful model system to study the interaction of cellular and humoral responses since the parasite produces extensive pathology in the intestine of pike and in the muscle of ciscoe and whitefish. The use of chemotherapy and/or vaccines to control tapeworm infections is uncertain. They may have a role in intensive cage culture operations. However, much more will have to be known about the efficacy of the drugs and their persistence in fish tissues. Similarly, considerable basic research is required to understand the fish immune system and its relation to tapeworms. This would include the role of acute phase proteins and the antigenic properties of molecules such as phosphorylcholine moieties.

Finally, there are only a few examples where a migrating and differentiating larval helminth does not seem to elicit a cellular response. In this respect the *D. latum*/pike/walleye system would warrant a closer study.

ACKNOWLEDGEMENTS

The authors thank the Canadian Journal of Zoology for permission to reproduce Fig. 10.2 and Lu Ming Chuan for photography.

REFERENCES

Albetova, L.M. and Michurin, S.M. (1984) Parasites and diseases of carp in the warm waters of the Surgut hydroelectric station. *Sbornik Nauchnykh Trudov Gosudarstvennogo Nauchno-Issledovatel'skogo Instituta Ozernogo i Rechnogo*

Rybnogo Khozyaistva (Bolezni i parazity ryb vodoemov Zapadnoi Sibiri) 226, 3–15.

Andersen, K. (1975) Comparison of the surface topography of three species of *Diphyllobothrium* (Cestoda, Pseudophyllidea) by scanning electron microscopy. *International Journal for Parasitology* 5, 293–300.

Andersen, K. and Gibson D.I. (1989) A key to three species of larval *Diphyllobothrium* Cobbold, 1958 (Cestoda: Pseudophyllidea) occurring in European and North American freshwater species. *Systematic Parasitology* 13, 3–9.

Andersen, K., Ching, H.L. and Vik, R. (1987) A review of the freshwater species of *Diphyllobothrium* with descriptions and the distribution of *D. dendriticum* (Nitzsch, 1824) and *D. ditremum* (Creplin, 1825) from North America. *Canadian Journal of Zoology* 65, 2216–2228.

Arme, C. and Pappas, P.W. (eds) (1983) *Biology of the Eucestoda, Volume 1.* Academic Press, London.

Arme, C., Bridges, J.F. and Hoole, D. (1983) Pathology of cestode infections in the vertebrate host In: Arme, C. and Pappas, P.W. (eds) *Biology of the Eucestoda, Vol. II.* Academic Press, London, pp. 499–538.

Bandoni, S.M. and Brooks, D.R. (1987a) Revision and phylogenetic analysis of the Amphilinidea Poche, 1922 (Platyhelminthes: Cercomeria: Cercomeromorpha). *Canadian Journal of Zoology* 65, 1110–1128.

Bandoni, S.M. and Brooks, D.R. (1987b) Revision and phylogenetic analysis of the Gyrocotylidea Poche, 1926 (Platyhelminthes: Cercomeria: Cercomeromorpha). *Canadian Journal of Zoology* 65, 2369–2389.

Bauer, O.N. (1959) Parasites of freshwater fish and the biological basis for their control. *Bulletin of the State Scientific Research Institute of Lake and River Fisheries* 69, 3–115. (Israel Program for Scientific translations, Jerusalem, 1962.)

Bauer, O.N. and Hoffman, G.L. (1976) Helminth range extension by translocation of fish. In: Page, A. (ed.) *Wildlife Diseases.* Plenum Press, New York, pp. 163–172.

Berntzen, A.K. (1961) The *in vitro* cultivation of tapeworms. I. Growth of *Hymenolepis diminuta* (Cestoda: Cyclophyllidea). *Journal of Parasitology* 47, 351–355.

Berntzen, A.K. (1962) *In vitro* cultivation of tapeworms. II. Growth and maintenance of *Hymenolepis nana* (Cestoda: Cyclophyllidea). *Journal of Parasitology* 48, 785–797.

Berntzen, A.K. (1961) The *in vitro* cultivation of tapeworms. I. Growth of *Hymenolepis diminuta* (Cestoda: Cyclophyllidea). *Journal of Parasitology* 47, 351–355.

Berntzen, A.K. and Voge, M. (1965) *In vitro* hatching of oncospheres of four hymenolepidid cestodes. *Journal of Parasitology* 51, 235–242.

Berrada-Rkhami, O., Leducq, R., Gabrion, J. and Gabrion, C. (1990) Selective distribution of sugars on the tegumental surface of adult *Bothriocephalus gregarius* (Cestoda: Pseudophyllidea). *International Journal for Parasitology* 20, 285–297.

Boonyaratpalin, S. and Rogers, W.A. (1984) Control of the bass tapeworm, *Proteocephalus ambloplitis* (Leidy), with mebendazole. *Journal of Fish Diseases* 7, 449–456.

Boyce, N.P. (1979) Effects of *Eubothrium salvelini* (Cestoda: Pseudophyllidea) on the growth and vitality of sockeye salmon, *Onchorhynchus nerka. Canadian Journal of Zoology* 57, 597–602.

Boyce, N.P. and Yamada, S.B. (1977) Effects of a parasite, *Eubothrium salvelini* (Cestoda: Pseudophyllidea), on the resistance of juvenile sockeye salmon,

Onchorhynchus nerka, to zinc. *Journal of the Fisheries Research Board of Canada* 34, 706–709.

Brandt, F. de W., Van As, J.G., Schoonbee, H.J. and Hamilton-Atwell, V.L. (1981) The occurrence and treatment of bothriocephalosis in the common carp, *Cyprinus carpio* on fish ponds with notes on its presence in the largemouth yellowfish *Barbus kimberleyensis* from the Vaal Dam, Transvaal. *Water SA* 7, 34–42.

Brooks, D.R. (1989) The phylogeny of the Cercomeria (Platyhelminthes: Rhabdocoela) and general evolutionary principles. *Journal of Parasitology* 75, 606–616.

Bykhovskaya-Pavlovskaya, I.E., Gusev, A.V., Dubinina, M.N., Izyumova, N.A., Smirnova, T.S., Sokolovskaya, I.L., Shtein, G.A., Shul'man, S.S. and Epshtein, V.M. (1962) *Key to Parasites of Freshwater Fish of the USSR*. Akademiya Nauk SSSR, Zoologicheskii Institut, Moscow-Leningrad. (Israel Program for Scientific Translations, Jerusalem, 1964.) 919 pp.

Bylund, G. (1972) Pathogenic effects of a diphyllobothriid plerocercoid on its host fishes. *Commentationes Biologicae* 58, 1–10.

Charles, G.H. (1971) The ultrastructure of the developing pseudophyllid tegument (epidermis) with special reference to the larval stages of *Schistocephalus solidus* and *Ligula intestinalis*. *Proceedings of the Second International Congress of Parasitology, Journal of Parasitology, Special Volume* 59(4), 38–39.

Choudhury, A. and Dick, T.A. (1994) Natural anti-phosphorylcholine (PC) antibodies in lake sturgeon, *Acipenser fulvescens* Rafinesque, 1817 (Chondrostei: Acipenseridae). *Fish and Shellfish Immunology* 4, 399–401.

Coggins, J.R. (1980a) Tegument and apical end organ fine structure in the metacestode and adult *Proteocephalus ambloplitis*. *International Journal for Parasitology* 10, 409–418.

Coggins, J.R. (1980b) Apical end organ structure and histochemistry in plerocercoids of *Proteocephalus ambloplitis*. *International Journal for Parasitology* 10, 97–102.

deVos, T. and Dick, T.A. (1989) Differentiation between *Diphyllobothrium dendriticum* and *D. latum* using isozymes, restriction profiles and ribosomal gene probes. *Systematic Parasitology* 13, 161–166.

deVos, T., Szalai, A.J. and Dick, T.A. (1990) Genetic and morphological variability in a population of *Diphyllobothrium dendriticum* (Nitzsch, 1824). *Systematic Parasitology* 16, 99–105.

Dick, T.A. and Belosevic, M. (1981) Parasites of Arctic charr *Salvelinus alpinus* (Linnaeus) and their use in separating sea-run and non-migrating charr. *Journal of Fish Biology* 18, 339–347.

Dick, T.A. and Poole, B.C. (1985) Identification of *Diphyllobothrium dendriticum* and *Diphyllobothrium latum* from some freshwater fishes of central Canada. *Canadian Journal of Zoology* 63, 196–201.

Dubinina, M.N. (1974) The development of *Amphilina foliacea* (Rud.) at all stages of its life cycle and the position of the Amphilinidea in the system of Platyhelminthes. *Parazitologicheskii Sbornik* 26, 9–38.

Evans, D.L. and Gratzek, J.B. (1989) Immune defense mechanisms in fish to protozoan and helminth infections. *American Zoologist* 29, 409–418.

Fletcher, T.C., White, A. and Baldo, B.A. (1980) Isolation of a phosphorylcholine-containing component from the turbot tapeworm, *Bothriocephalus scorpii* (Mueller), and its reaction with C-reactive protein. *Parasite Immunology* 2, 237–248.

Freze, V.I., Sergeeva, E.G. and Kulinich L.I. (1983) Antigenic affinity of cestode species from the genus *Diphyllobothrium* (Cestoidea: Diphyllobothriidae) recorded from the territory of Karelia. *Helminthologia* 20, 121–129.

Fukumoto, S., Yazaki, S., Nagai, D., Takechi, M., Kamo, H. and Yamane, Y. (1987) Comparative studies on soluble protein profiles and isozyme patterns in three related species of the genus *Diphyllobothrium*. *Japanese Journal of Parasitology* 36, 222–230.

Grabda, J. (1989) *Marine Fish Parasitology, An Outline*. VCH Publishers, Weinheim, and Polish Scientific Publishers, Warsaw.

Halvorsen, O. and Andersen, K. (1984) The ecological interaction between arctic charr, *Salvelinus alpinus* (L.), and the plerocercoid stage of *Diphyllobothrium ditremum*. *Journal of Fish Biology* 25, 305–316.

Hoffman, G.L. (1967) *Parasites of North American Freshwater Fishes*. University of California Press, Berkeley.

Hoffman, G.L. (1975) Lesions due to internal helminths of freshwater fishes. In: Ribelin, W.E. and Migaki, G. (eds) *The Pathology of Fishes*. University of Wisconsin Press, Madison, pp. 151–188.

Hoole, D. and Arme, C. (1983) Ultrastructural studies on the cellular response of fish hosts following experimental infection with the plerocercoid of *Ligula intestinalis* (Cestoda: Pseudophyllidea). *Parasitology* 87, 139–149.

Hoole, D. and Arme, C. (1986) The role of leucocyte adherence to the plerocercoid of *Ligula intestinalis* (Cestoda: Pseudophyllidea). *Parasitology* 92, 413–424.

Hoole, D. and Arme, C. (1988) *Ligula intestinalis* (Cestoda: Pseudophyllidea): phosphorylcholine inhibition of fish leucocyte adherence. *Diseases of Aquatic Organisms* 5, 29–33.

Hopkins, C.A., Law, L.M. and Threadgold, L.T. (1978) *Schistocephalus solidus*: pinocytosis by the plerocercoid tegument. *Experimental Parasitology* 44, 161–172.

Jarecka, L. (1961) Morphological adaptations of tapeworm eggs and their importance in the life cycles. *Acta Parasitologica Polonica* 9, 409–426.

Joy, J.E. and Madan, E. (1989) Pathology of black bass hepatic tissue infected with larvae of the tapeworm *Proteocephalus ambloplitis*. *Journal of Fish Biology* 35, 111–118.

Kennedy, C.R. and Walker, P.J. (1969) Evidence for an immune response by dace, *Leuciscus leuciscus*, to infections by the cestode *Caryophyllaeus laticeps*. *Journal of Parasitology* 55, 579–582.

Khalil, L.F., Jones, A. and Bray, R.A. (1994) *Keys to the Cestode Parasites of Vertebrates*. CAB International, Oxon, 768 pp.

Koerting, W. and Barrett, J. (1977) Carbohydrate catabolism in the plerocercoids of *Schistocephalus solidus* (Cestoda: Pseudophyllidea). *International Journal for Parasitology* 7, 411–417.

Kral, J., Sevcik, B., Prouza, A. and Vondrka, K. (1980) [Taenifugin carp, medicated granulate for the treatment of *Bothriocephalus gowkongensis* infections in fish]. *Biologizace e Chemizace Zivocisne Vyroby, Veterinaria* 16, 183–192.

Kuperman, B.I. (1973) *Tapeworms of the Genus Triaenophorus, Parasites of Fishes*. Academy of Sciences of the USSR (Akademiya Nauk, SSSR), Institute of Biology of Inland waters. (Translated from Russian. Amerind Publishing Co., New Delhi, 1981).

Kuperman, B.I. and Davydov, V.G. (1982) The fine structure of glands in oncospheres, procercoids and plerocercoids of Pseudophyllidea (Cestoidea). *International Journal for Parasitology* 12, 135–144.

Kurovskaya, L. Ya. and Kititsyna, L.A. (1986) Physiological–biochemical features of the white amur infected with helminths. *Soviet Journal of Ecology* 17, 168–177.

Lawler, G.H. (1970) Parasites of coregonid fishes. In: Lindsey, C.C. and Woods, C.S. (eds) *Biology of Coregonid Fishes*. University of Manitoba Press, Winnipeg. pp. 279–310.

Liao, X.H. and Liang, Z.X. (1987) Distribution of ligulid tapeworms in China. *Journal of Parasitology* 73, 36–48.

Mackiewicz, J.S. (1972) Caryophyllidea (Cestoidea): A review. Parasitological review. *Experimental Parasitology* 31, 417–512.

Mackiewicz, J.S. (1981) Caryophyllidea (Cestoidea): evolution and classification. *Advances in Parasitology* 19, 139–206.

Mackiewicz, J.S. (1988) Cestode transmission patterns. *Journal of Parasitology* 74, 60–71.

Mackiewicz, J.S., Cosgrove, G.E. and Gude, W.D. (1972) Relationship of pathology to scolex morphology among caryophyllid cestodes. *Zeitschrift für Parasitenkunde* 39, 233–246.

MacKinnon, B.M. and Burt, M.D.B. (1984) The comparative ultrastructure of the plerocercoid and adult primary scolex of *Haplobothrium globuliforme* Cooper, 1914 (Cestoda: Haplobothriodea). *Canadian Journal of Zoology* 63, 1488–1496.

Margolis, L. and Arthur, J.R. (1979) Synopsis of the parasites of fishes of Canada. *Bulletin of the Fisheries Research Board of Canada, 199*. Department of Fisheries and Oceans, Ottawa.

McCormick, H.J. and Stokes, G.N. (1982) Intraovarian invasion of smallmouth bass oocytes by *Proteocephalus ambloplitis* (Cestoda). *Journal of Parasitology* 68, 975–976.

McManus, D.P. (1985) Enzyme analyses of natural populations of *Schistocephalus solidus* and *Ligula intestinalis*. *Journal of Helminthology* 59, 323–332.

McVicar, A.H. (1972) The ultrastructure of the parasite–host interface of three tetraphyllidean tapeworms of the elasmobranch *Raja naevus*. *Parasitology* 65, 77–88.

McVicar, A.H. and Fletcher, T.C. (1970) Serum factors in *Raja radiata* toxic to *Acanthobothrium quadripartitum* (Cestoda: Tetraphyllidea), a parasite specific to *R. naevus*. *Parasitology* 61, 55–63.

Miller, R.B. (1952) A review of the *Triaenophorus* problem in Canadian lakes. *Bulletin of the Fisheries Research Board of Canada*, 95, Department of Fisheries and Oceans, Ottawa.

Mitchell, L.G., Ginal, J., and Bailey, W.C. (1983) Melanotic visceral fibrosis associated with larval infections of *Posthodiplostomum minimum* and *Proteocephalus* sp. in bluegill, *Lepomis macrochirus* Rafinesque, in central Iowa, USA. *Journal of Fish Diseases* 6, 135–144.

Molnar, K. and Berczi, I. (1965) Nachweis von parasitenspezifischen Antikoerpern im Fischblut mittels der Agar-Gel-Praezipitationsprobe. *Zeitschrift für Immunologie Allergie-Forschung* 129, 263–267.

Nei, D.-S. and Pan, J.-P. (1985) Diseases of grass carp (*Ctenopharyngodon idellus* Valenciennes, 1844) in China, a review from 1953–1983. *Fish Pathology* 20, 323–330.

Paperna, I. (1991) Diseases caused by parasites in the aquaculture of warm water fish. *Annual Review of Fish Diseases 1991*, 155–194.

Par, O., Parova, J. and Prouza, A. (1977) [Mansonil – an effective anthelmintic for the treatment of *Bothriocephalus* infections in carp]. *Buletin Vyzkumny Ustav Rybarsky a Hydrobiologicky Vodnany, CSSR* 1, 17–25 (in Russian).

Poole, B.C. and Dick, T.A. (1985) Parasite recruitment by stocked walleye, *Stizostedion vitreum vitreum* (Mitchill), fry in a small boreal lake in Central Canada. *Journal of Wildlife Diseases* 21, 371–376.

Popova, L.B. and Davydov, V.G. (1988) Studies on localization of *Amphilina* spp. (Amphilinidae, Dubinina, 1974) in definitive hosts. *Helminthologia* 25(2), 129–138.

Cestoidea (Phylum Platyhelminthes) 413
Pronina, S.V. (1977) The encapsulation of *Triaenophorus nodulosus* plerocercoids in the liver of *Perca fluviatilis* (histopathology and pathogenesis). *Trudy Buryatskogo Instituta Estestvennykh Nauk* 15, 46–51.

Pronina, S.V. and Pronin, N.M. (1982) The effect of cestode (*Triaenophorus nodulosus*) infestation on the digestive tract of pike (*Esox lucius*). *Journal of Ichthyology* 22, 105–113.

Renaud, F., Gabrion, C. and Pasteur, N. (1986) Geographical divergence in *Bothriocephalus* (Cestoda) of fishes demonstrated by enzyme electrophoresis. *International Journal for Parasitology* 16, 553–558.

Roberts, R.J. (1989) *Fish Pathology*. Ballière Tindall, London.

Rodger, H.D. (1991) *Diphyllobothrium* sp. infections in freshwater reared salmon (*Salmo salar* L.). *Aquaculture* 95, 7–14.

Rosen, R. and Dick, T.A. (1983) Growth and migration of plerocercoids of *Triaenophorus crassus* Forel and pathology in experimentally infected whitefish. *Canadian Journal of Zoology* 62, 203–211.

Rosen, R. and Dick, T.A. (1984) Experimental infections of rainbow trout, *Salmo gairdneri* Richardson, with plerocercoids of *Triaenophorus crassus* Forel. *Journal of Wildlife Diseases* 20, 34–48.

Rzeczkowska, A. and Honowska, M. (1988) Biochemical effect of the plerocercoid of *Ligula intestinalis* (L.) on the bream *Abramis brama* (L). *Wiadomosci Parazitologiczne* 34, 19–27.

Schaeperclaus, W. (1992) *Fish Diseases, Volumes 1 and 2*. A.A. Balkema. Rotterdam, 1398 pp.

Scheinert, P. and Hoffman, R. (1986) Enzymeserologische Untersuchungen an durch *Triaenophorus nodulosus* befallenen Seesaibling (*Salvelinus alpinus* L.) des Koenigsees. *Berliner und Mñchener Tierärztliche Wochenschrift* 99, 383–386.

Scheuring, L. (1923) Studien an Fischparasiten. I. *Triaenophorus nodulosus* (Pall) und die durch ihn im Fischkoerper hervorgerufenen pathologischen Veraenderungen. *Zeitschrift für Fischerei* 22, 93–205.

Schmidt, G.D. (1986) *CRC Handbook of Tapeworm Identification*. CRC Press, Boca Raton, Florida, 675 pp.

Scott, A.L. and Grizzle, J.M. (1979) Pathology of cyprinid fishes caused by *Bothriocephalus gowkongensis* Yea, 1955 (Cestoda: Pseudophyllidea). *Journal of Fish Diseases* 2, 69–73.

Sharp, G.J.E., Pike, A.W. and Secombes, C.J. (1989) The immune response of wild rainbow trout, *Salmo gairdneri* Richardson, to naturally acquired plerocercoid infections of *Diphyllobothrium dendriticum* (Nitzsch, 1824) and *D. ditremum* (Creplin, 1825). *Journal of Fish Biology* 35, 781–794.

Sharp, G.J.E., Pike, A.W. and Secombes, C.J. (1992) Sequential development of the immune response in rainbow trout [*Oncorhynchus mykiss* (Walbalum, 1792)] to experimental plerocercoid infections of *Diphyllobothrium dendriticum* (Nitsch, 1824). *Parasitology* 104, 169–178.

Shostak, A.W. and Dick, T.A. (1986) Intestinal pathology in northern pike, *Esox lucius* l., infected with *Triaenophorus crassus* Forel, 1868 (Cestoda: Pseudophyllidea). *Journal of Fish Diseases* 9, 35–45.

Shostak, A.W., Rosen, R.B. and Dick, T.A. (1984) Orientation of procercoids of *Triaenophorus crassus* Forel in *Cyclops bicuspidatus thomasi* Forbes: effects on growth and development. *Canadian Journal of Zoology* 62, 1373–1377.

Sindermann, C.J. (1990) Diseases caused by helminths and parasitic crustacea. In: Sindermann, C.J. (ed.) *Principal Diseases of Marine Fish and Shellfish*. Academic Press, New York, pp. 52–105.

Skrjabina, E.S. (1974) [*Helminths of Sturgeon (Acipenseridae Bonaparte 1831)*]. 'Nauka', Moscow, 168 pp (in Russian).

Smyth, J.D. (1946) Studies on tapeworm physiology. I. Cultivation of *Schistocephalus solidus in vitro*. *Journal of Experimental Biology* 23, 47-70.

Smyth, J.D. (1947) Studies on tapeworm physiology. II. Cultivation and development of *Ligula intestinalis in vitro*. *Parasitology* 38, 173-181.

Smyth, J.D. (1954) Studies on tapeworm physiology. VII. Fertilization of *Schistocephalus solidus in vitro*. *Experimental Parasitology*. 3, 64-71.

Smyth, J.D. (1976) *An Introduction to Animal Parasitology*. 2nd edn. Hodder and Stoughton, London.

Smyth, J.D. (1982) The insemination-fertilization problem in cestodes cultured *in vitro*. In: Meerovitch, E. (ed.) *Aspects of Parasitology*. McGill University Press, Montreal, pp. 393-406.

Smyth, J.D. and McManus, D.P. (1989) *The Physiology and Biochemistry of Cestodes*. Cambridge University Press, Cambridge.

Stunkard, H.W. (1983) Evolution and Systematics. In: Arme, C. and Pappas, P.W. (eds) *Biology of the Eucestoda*, Vol. 1. Academic Press, London, pp. 1-25.

Sweeting, R.A. (1977) Studies on *Ligula intestinalis*. Some aspects of the pathology in the second intermediate host. *Journal of Fish Biology* 10, 43-50.

Szalai, A.J., Yang, X., and Dick, T.A. (1989) Changes in numbers and growth of *Ligula intestinalis* in the spottail shiner (*Notropis hudsonius*), and their roles in transmission. *Journal of Parasitology* 75, 571-576.

Szalai, A.J. and Dick, T.A. (1990) *Proteocephalus ambloplitis* and *Contracaecum* sp. from largemouth bass (*Micropterus salmoides*) stocked into the Boundary Reservoir, Saskatchewan. *Journal of Parasitology* 76, 598-601.

Taylor, A.E.R. and Baker, J.R. (eds) (1987) *In vitro Methods for Parasite Cultivation*. Academic Press, London, 465 pp.

Taylor, M. and Hoole, D. (1989) *Ligula intestinalis* (L.) (Cestoda: Pseudophyllidea): plerocercoid-induced changes in the spleen and pronephros of the roach, *Rutilus rutilus* (L.), and gudgeon, *Gobio gobio* (L.). *Journal of Fish Biology* 34, 583-596.

Threadgold, L.T. (1984) Parasitic platyhelminthes. In: Bereiter-Hahn, J., Maltoltsy, A.G. and Richards, K.S. (eds) *Biology of the Tegument*. Springer-Verlag, Berlin, pp. 132-191.

Wardle, R.A. and McLeod, J.A. (1952) *The Zoology of Tapeworms*. University of Minnesota Press, Minneapolis.

Watson, R.A. and Dick, T.A. (1979) Metazoan parasites of whitefish, *Coregonus clupeaformis* (Mitchill) and cisco, *C. artedi* LeSueur, from Southern Indian Lake, Manitoba. *Journal of Fish Biology* 15, 579-587.

Weiland, K.A., and Meyers, T.R. (1989) Histopathology of *Diphyllobothrium ditremum* plerocercoids in coho salmon *Oncorhynchus kisutch*. *Diseases of Aquatic Organisms* 6, 175-178.

Weiroski, F. (1984) Occurrence, spread and control of *Bothriocephalus acheilognathi*, in the carp ponds of the German Democratic Republic. In: Olah, J. (ed.) *Fish, Pathogens and the Environment in European Polyculture. Proceedings of International Seminar, 23-27 June, 1981, Szarvas, Hungary. Symposia Biologica Hungary* 24, 149-155.

Williams, H.H. (1967) Helminth diseases of fish. *Helminthological Abstracts* 36, 261-295.

Zitnan, R., Hanzelova, V., Prihoda, J. and Kostan, B. (1981) [Evaluation of the efficacy of Taenifugin carp in the treatment of *Bothriocephalus* infections in carp at low water temperature]. *Biologizace e Chemizace Zivocisne vyroby, Veterinaria* 17(23), 471-477 (in Russian).

11

Phylum Nematoda

T.A. Dick and A. Choudhury

Department of Zoology, University of Manitoba, Winnipeg, Manitoba, Canada R3T 2N2.

INTRODUCTION

Nematodes of fishes are represented by 17 families of which five are unique to fish (Anderson, 1984). They are most frequently encountered as cysts in the flesh, liver, on the surface of the viscera, within the body cavity, in the intestine, and more rarely lying under the tegument. Other sites include the heart, blood vessels, eyes and gonads. Nematodes are more motile than other helminths and therefore one could anticipate more reports on their effects on fish health and on the biotic potential of fish hosts.

Nematodes are usually considered the most economically important helminth parasites of fishes of the world (Hafsteinsson and Rizvi, 1987). The marine ascaridoids (anisakids) have been intensively studied, in part due to the ability of some species to infect humans. Other groups which have received some emphasis are the capillarids and the gnathostomatids which also infect humans. However, most of the knowledge on taxonomy, life cycles, morphology, distribution and transmission dynamics have been from groups which are not of medical importance. Declines in the commercial fish harvest and the increasing production of cultured fish particularly in northern latitudes, has increased interest in nematode parasites of fishes.

This review will focus on the biological aspects of fish nematodes, particularly those of economic importance and causing disease in fish and their control. As there is considerable information on the anisakids infective to humans (Hafsteinsson and Rizvi, 1987, Margolis, 1977) our report will be restricted to their biology in fish.

HOST RANGE AND GEOGRAPHICAL DISTRIBUTION

Nematodes of fish are distributed worldwide with those of greatest economic importance concentrated in the northern hemisphere, largely in the marine environment. With the exception of *Anisakis* and *Pseudoterranova*, which have worldwide distributions, the remaining nematodes of fish infective to humans are in the tropics. This section will focus on selected groups known to cause disease in fish as there are sufficient records of nematodes and their hosts in other publications (Bykhovskaya-Pavlovskaya *et al.*, 1962; Hoffman, 1967; Kennedy, 1974; Margolis and Arthur, 1979; Anderson, 1992).

Capillarids lack host specificity and are the most widely distributed (Hoffman, 1967; Moravec, 1983). Ascaridoids are most abundant in the marine environment but larval anisakids infect fish wherever fish-eating birds and mammals are present. *Anisakis* spp. are common among piscivorous pinnipeds and cetaceans such as the Odontoceti (toothed whales). Among the Mystacoceti (baleen whales), the Balaenopteridae which are 'plankto-ichthyophagi', are common hosts (Delyamure, 1955, Piatt *et al.*, 1989). *P. decipiens* is common in seals while *P. kogiae* and *P. ceticola* are specific to physeterid whales. *Contracaecum* spp. (*C. microcephalum*, *C. spiculigerum*, etc.) are found worldwide in piscivorous birds (cormorants, pelicans etc.) and *C. osculatum* and *Phocascaris* spp. are widely distributed in seals. Larvae of these anisakids show little host specificity and have been reported from numerous species of fish. *Anisakis* spp. are more abundant in pelagic fishes (herring, whiting) while *P. decipiens* is more abundant in demersal and benthic fish (cod, sole, plaice) (Smith and Wootten, 1978; McClelland *et al.*, 1985). These parasites are common in the Atlantic and Pacific waters and rarer in the cold Arctic waters. Larvae of avian *Contracaecum* spp. infect a variety of freshwater fishes (cyprinids, ictalurids, centrarchids, percids) while *Contracaecum/Phocascaris* of seals are present in marine fishes (capelin, cod, whiting etc.). *Raphidascaris acus* has a holarctic freshwater distribution, as adults in piscivorous fishes (typically esocids, but also centrarchids and percids) and as larvae in a variety of fishes (catostomids, coregonids, cyprinids, *Lota*, percids). *Hysterothylacium aduncum* is mainly found in marine piscivorous fish (cod) and larvae occur in a variety of prey fishes including smaller cod and mackerel (Grabda, 1989; Deardorff and Overstreet, 1980a). Larval *Eustrongylides* spp. parasitize holarctic and tropical piscivorous birds (mergansers, cormorants) and a variety of freshwater fishes (Measures, 1988a; Paperna, 1991).

Spirurids parasitize fish mainly as adults and show some specificity but are also widely distributed. Species of *Truttaedacnitis* are typical of 'primitive' groups with diadromous habits (trout, sturgeon, lampreys) and are holarctic in distribution while *Cucullanus* spp. are marine (Anderson, 1992). *Cystidicola* spp. have holarctic distributions but are restricted to salmoniforms. *Spinitectus* spp. are distributed worldwide in marine and freshwater fishes with some species (*S. inermis*, *S. acipenseri*) more host specific than others (*S. carolini*, *S. gracilis*). *Philonema* spp. are typically in anadromous Pacific salmonids and *Philometra* and *Philometroides* are common in freshwater (catostomid) and marine fishes. *Anguillicola* spp. are specific for anguillid eels and the range of *A. crassus* has now been extended to Europe.

ECONOMIC IMPORTANCE

Most adult nematodes are found in the intestine of fish but it is the larval stages in the flesh and viscera which cause disease and economic problems. It is also the larval stages which are infective to humans (Margolis, 1977) and which have the greatest impact on consumer acceptance of fish as a source of protein. While there is ample literature describing the pathology due to nematode infections (see pp. 423–431) there are few documented diseases caused by nematodes in freshwater aquaculture (Molnar, 1986), in warm water aquaculture (Paperna, 1991), and in mariculture (Paperna, 1986). Nematodes such as *Truttaedacnitis*, *Philonema* and *Philometra* cause pathology, mostly in natural fish populations, but the economic impact of these infections are unknown. The increase in number of *Contracaecum* larvae from semi-intensive culture of rainbow trout on the Canadian prairies was due to increasing populations of piscivorous birds such as cormorants and pelicans (Dick *et al.*, 1987). According to Paperna (1991), larval *Contracaecum* and *Eustrongylides*, and adult *Anguillicola* are the most important fish parasites in the tropics. Recently, *Anguillicola crassus* has become a problem in cultured eels in Europe. However, the economic impact of nematodes is most evident in marine fisheries of the northern hemisphere (Moeller, 1989). The greatest effect on the export fishery is in Canada where the estimate of processing costs due to *Pseudoterranova decipiens* infections in marine fish is about $30 million (Malouf, 1986, in Bowen, 1990). However, the direct effect of these marine ascaridoids on fish populations is less well understood.

There is little information on the translocation of nematodes in natural systems but since marine mammals and fish-eating birds are definitive hosts there is a high probability that this has and will continue to occur. The translocation of the eel parasite, *A. crassus* to Europe from its original range in Asia has already generated problems. (Belpaire *et al.*, 1989; Dekker and van Willigen, 1989; Kennedy and Fitch, 1990; Koops and Hartman, 1989; Peters and Hartmann, 1986; Taraschewski *et al.*, 1987).

SYSTEMATICS AND TAXONOMIC POSITION

The evolution and systematics of fish parasitic nematodes have been subjects of considerable controversy and speculation. Anderson (1984) provides an excellent analysis and some of the key features are heteroxeny as a fundamental characteristic of the life cycles, most of the families (12 of a total of 17) are shared with terrestrial vertebrates and nematodes of fish can be plausibly derived from terrestrial rhabditid (for secernenteans) and dorylaimid (for adenophoreans) type ancestors but not from aquatic or marine free living nematodes. This latter point is considered strong evidence for a terrestrial tetrapod origin of these parasites (Anderson, 1984) implying that fish hosts are a more recent acquisition. Alternatively, it has been suggested that parasitic nematodes are derived from sediment dwelling 'thiobiotic' nematodes (free living obligate anaerobes) (Bryant, 1982).

Sprent (1982, 1983), using his extensive knowledge of the ascaridoids, proposed that a heteroxenous aquatic life cycle involving an invertebrate and a vertebrate host was primitive but favoured an original monoxenous life cycle in a crustacean definitive host. Osche (1963) and Sprent (1982) viewed the food chain as important in the evolution of aquatic ascaridoids. In addition to host range expansion, Sprent (1982) proposed 'host succession extension' and suggested that the anisakids of aquatic mammals are more derived than those in fish, a view shared by us. The development and distribution of *Anisakis* and *Pseudoterranova* in crustaceans, fish and mammalian hosts (cetaceans and pinnipeds) suggest close phylogenetic ties and also that fish may have played a pivotal role in the evolution of these anisakids into piscivorous hosts. Gibson (1983), using the morphology of the excretory system, proposed that forms (e.g. acanthocheilids) in 'archaic' hosts such as sharks are basal and other subfamilies such as anisakines and goezines evolved from them. For reviews of the taxonomy of aquatic ascaridoids, the reader is referred to Deardorff and Overstreet (1980a,b) for North American *Goezia* and *Hysterothylacium*, Smith (1983) and Smith and Wootten (1978) for *Anisakis*, Gibson (1983) for anisakids, Smith (1984) for *Raphidascaris*, and the numerous reviews by Sprent (1983 and references therein).

Petter (1974) discussed the evolution of cucullanids with special reference to change in papillar arrangement and considered the genus *Truttaedacnitis* primitive. The genus *Truttaedacnitis* (Petter, 1974) is considered a synonym of *Cucullanus* by Moravec (1979) who also considers *T. stelmioides* of lampreys a synonym of *T. truttae*, a view not shared by Pybus *et al.* (1978a). Anderson (1992) has changed the status of the North American *T. stelmioides* from brook lampreys, described previously by Pybus *et al.* (1978a), to that of 'new species', *T. pybusae*. This change would be more appropriate following a thorough revision of *T. stelmioides* and *T. truttae* from a variety of salmonid and lamprey hosts from North America.

Taxonomic reviews of nematodes in other fish groups include; Moravec and Taraschewski (1988) on *Anguillicola*, Moravec (1983) on Capillaridae, Measures (1988a) on *Eustrongylides*, and Specian *et al.* (1975) on *Proleptus*.

MORPHOLOGY AND LIFE CYCLES

The morphologies of parasitic nematodes in fishes generally conform to the higher taxon (family, superfamily) to which they belong and readers are referred to Chitwood and Chitwood (1950) and Anderson (1992) for a detailed coverage of the development and life histories of these groups.

Adult nematodes are sexually dimorphic. These differences become more pronounced in the advanced larval stages (L3 and L4) and the reproductive systems are well developed in adult males and females after the final moult. The philometrid nematodes are distinctive in that the vulva shows varying degrees of atrophy in mature females. The female reproductive tract of most other nematodes opens externally via a vulva through which eggs/larvae are released. The male copulatory apparatus usually consists of a pair of spicules although *Capillaria* spp. possess only one spicule and spicules are absent in *Anguillicola* spp. Sensory structures (papillae and dierids) are usually more common at the cephalic and posterior extremities of adult nematodes than in larval stages. The gut becomes functional when larvae establish in suitable intermediate hosts. Nematodes whose adults have well developed lips (Ascaridoidea) usually form them between L4 and adult stages. The three well developed lips, typical of adult ascaridoids are greatly reduced in *Goezia* spp. as are the sizes of these worms. The L3 of anisakids in fish are characterized by a prominent anterior boring tooth, and often one or more gut diverticuli. In species such as *P. decipiens*, the boring tooth is reduced in size in larger L3 larvae (Fig. 11.2 below). It is assumed that the boring tooth aids in penetration while the appendix and caecum help in food digestion. In contrast, most spirurids are generally more slender and do not possess gut appendages, although the foregut is often differentiated into two or more parts. Spirurids possess a buccal cavity with a slit-like opening that is reinforced by cephalic plates and collars (*Cucullanus*, *Truttaedacnitis*) or longitudinal rib-like structures (*Camallanus*). These structures are often used in attachment to the host gut and the internalization of host tissue plugs into the buccal cavity. Other spirurids possess reduced lips and sublabia which guard the oral opening (*Cystidicola*, *Spinitectus*). The circular oral opening of *Anguillicola crassus* is bordered by teeth while oesophageal lobes protrude into the oral cavity. The oral opening of dracunculoids is similar. *Eustrongylides* larvae possess a small slit-like oral opening as do adenophoreans such as *Capillaria* spp. The bodies of most fish parasitic nematodes are generally smooth but *Spinitectus* spp. and *Goezia* spp. have spines. Generally, nematodes are whitish-translucent but some may be reddish due to haemoglobin in the pseudocoel (*Pseudoterranova* L3 in cod muscle) which may also be of host origin (*Camallanus*, *Eustrongylides*).

The life cycle strategies of nematodes are diverse as they utilize fish as definitive, intermediate or paratenic hosts. Eggs or larvae produced by adult worms in the gut or tissues of the definitive host (fish or piscivorous homoeotherm) reach the aquatic environment often via host faeces. In most cases, hatched larvae or eggs containing larvae are eaten by an intermediate host or a succession of hosts where development takes place and the infective larvae eventually reach the definitive host where sexual maturity occurs. All nematodes pass through five stages (L1, L2, L3, L4 and an adult stage) with four moults. The timing of the moults in the life cycle often coincides with establishment in a host.

Life cycle patterns of nematodes which cause disease in fish are outlined in Fig. 11.1. *Capillaria* spp. have a direct life cycle; the eggs contain infective larvae. Adult spirurids in the gut (*Camallanus*, *Spinitectus*, *Truttaedacnitis*) and associated structures (swim bladder, *Cystidicola* spp.) release eggs

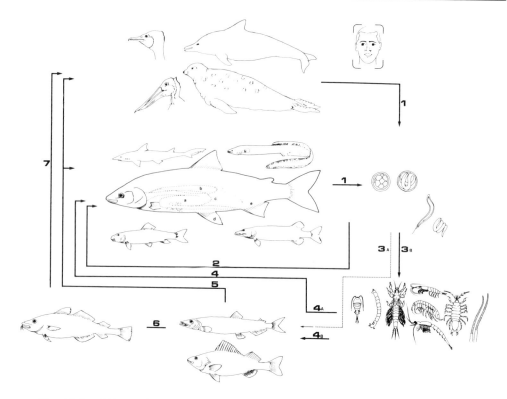

Fig. 11.1. Life cycle patterns of fish parasitic nematodes. 1. Release of eggs and/or larvae from fish or homoeotherm definitive host. 2. Transmission of infective larva within egg to definitive fish host (Capillaridae). 3A. Transmission of larvae directly to fish intermediate host (*Raphidascaris, Contracaecum, Truttaedacnitis*). 3B. Transmission of larvae to invertebrate intermediate host (Ascaridoidea, Spirurida, Dracunculoidea, Dioctophymatoidea). 4A. Transmission of infective larvae to fish definitive host via invertebrates (Spirurida, Dracunculoidea). 4B. Transmission of infective larvae to fish intermediate host via invertebrate hosts (Ascaridoidea, Dracunculoidea, Spirurida). 5. Transmission of infective stages to fish and homoeotherm definitive hosts via fish intermediate/paratenic hosts (*Hysterothylacium, Raphidascaris, Anisakis, Contracaecum/ Phocascaris, Pseudoterranova*). 6. Paratenic fish-fish transmission (*Anisakis, Pseudoterranova*). 7. Transmission to definitive hosts via paratenic fish hosts (*Anisakis, Pseudoterranova*). Location of nematodes in the fish definitive host. a. Gut (Spirurida, Ascaridoidea, Capillaridae). b. Swimbladder (*Cystidicola, Anguillicola*). c. Body cavity and viscera (Ascaridoidea larvae, Capillaridae, Dracunculoidea). d. Subcutaneous tissues particularly fins (Philometridae). e. Subcutaneous abdominal tissue (*Cystoopsis*).

(*Cystidicola, Spinitectus, Truttaedacnitis*) or larvae (*Camallanus*). Development of larvae from unembryonated or partially embryonated eggs takes place in the aquatic environment. Larvae or larvated eggs are ingested by invertebrates where development to L3 occurs in the haemocoel [*Spinitectus* in ephemeropteran naiads; *Cystidicola* in gammarids, *Camallanus* in copepods (see Anderson, 1992)]. The life cycle of *Truttaedacnitis truttae* is more com-

plex. Lampreys are considered important intermediate and occasional definitive hosts (Moravec, 1979). Larvae encysted in the gut wall of European
lampreys (*Lampetra planeri*) developed to adult *T. truttae* when fed to rainbow trout. Hatched larvae of *T. stelmioides* in North America infected
ammocoetes of brook lamprey (*Lampetra lammottenii*), developed in the
intestine to L3 and then migrated to the liver (Pybus *et al.*, 1978b). These larval
nematodes possibly reenter the intestine from the liver and develop to maturity
when ammocoetes transform. Trout fed infected livers of brook lampreys
did not become infected. Interestingly, other species of lampreys (*Lampetra*
spp., *Petromyzon marinus*) harbour larvae of cucullanids encysted in the gut
wall (Pybus *et al.*, 1978b), similar to that observed by Moravec (1979) in
European lampreys. It is not known if *T. truttae* of North American salmonids
develops in lampreys and if larvae from eggs of *Truttaedacnitis* from European lampreys mature in trout.

Dracunculoids have a two host life cycle (Fig. 11.1) and the most interesting aspect is the migration of larvae and adults within the host. *Anguillicola
crassus*, *Philometra*, *Philometroides* and *Philonema* are transmitted to their
definitive fish hosts via copepods which harbour L3. The L3 penetrate the fish
gut, migrate via the body cavity to their final site of infection, often via certain
'stopover sites' such as the peritoneum of the swimbladder. Fertilization of
some species takes place in the peritoneum of the swimbladder. Fertilized
females of some philometrids migrate to reach a final infection site. The
migration of males remains unclear (Molnar *et al.*, 1982). Female philometrids
rupture and release larvae into the body cavity or into the aquatic environment
by emerging from the host (Molnar *et al.*, 1982). In contrast, the larvae of *A.
crassus* penetrate the swimbladder and reach the lumen where adult worms
release eggs. These eggs are voided with host faeces. Some dracunculoids,
particularly those of predatory fish (e.g. *Philometra* of pike, *A. crassus* of
eels) may have small prey fish as paratenic/intermediate hosts (Molnar *et al.*,
1982; Petter *et al.*, 1989; De Charleroy *et al.*, 1990; Haenen and van Banning,
1991).

The life cycles of aquatic ascaridoids are characterized by a definitive host
which is usually the top predator of the food web. They utilize all trophic levels
in their transmission, exhibit low specificity of larvae in intermediate hosts
(invertebrates and vertebrates), plasticity in larval development, and extensive
use of paratenic hosts. The importance of these characteristics varies among
species and according to the nature of the environment. Eggs, released by
gravid females in the gut of definitive hosts, are voided with faeces into the
water where embryonic development occurs and the second or third stage
larvae leave the egg. According to Køie (1993) and Køie and Fagerholm (1993)
(for *Hysterothylacium* and *Contracaecum*) it is the L3 which hatches from the
egg. Invertebrates become infected by ingesting the L2/L3 and subsequently
transmit the larvae to the fish host through the food chain. The infective L3
(to the definitive host) is commonly in the tissues and body cavity of prey fish.
One of the two intermediate hosts (an invertebrate or a fish) may be paratenic.
Larvae of *Anisakis* and *Pseudoterranova* grow and develop both in invertebrate and in fish hosts (Smith, 1983; McClelland *et al.*, 1985; McCelland 1990)
and can also be transmitted from fish to fish (Smith, 1983; Burt *et al.*, 1990).

This results in the accumulation of L3 stages in the larger predatory fish. There is evidence to suggest that euphausiids are capable of transmitting L3 of *Anisakis* to the definitive whale hosts without involving a fish host (Smith, 1983). Infections of *Anisakis* in the piscivorous whale hosts (see p. 416) suggests that this may not be typical. Similarly, L2 of *Hysterothylacium aduncum* infect crustaceans (gammarids, isopods, mysids) where growth and development, apparently to L3, take place (Yoshinaga *et al.*, 1987). Larvae in crustaceans can be transmitted by ingestion to fish such as small cod, smelt, etc. where they reside in the liver and mesenteries. It is unknown if larvae are capable of infecting fish without an intervening invertebrate host or whether the L3 stage in invertebrates is as infective to the definitive fish host as L3 in prey fish. In contrast, L2 of *Raphidascaris acus* do not develop in invertebrates (chironomids and annelids) which only serve as paratenic hosts (Moravec, 1970; Smith, 1984; Torres and Alvarez-Pellitero, 1988) and transmit the parasite to perch and trout, where development occurs. Fish may also acquire the larval stages directly since experimental infections of rainbow trout with larvated eggs or hatched larvae were successful (T. Dick, unpublished). Similarly, *Contracaecum* spp. utilize invertebrates (amphipods, copepods, crab zoea etc.) as paratenic hosts (Huizinga 1966, 1967; Elarifi, 1981; A. Choudhury, unpublished) and growth and development of the larvae to L3 takes place in the fish. As in *Raphidascaris*, L2 of *C. spiculigerum* infect guppies directly (Huizinga, 1966) and L2 of *C. osculatum* develop to the L3 stage when injected into the body cavity of trout (Smith *et al.*, 1990). This indicates that invertebrates, although important for transmission, are not a physiological necessity. Interestingly, some species such as *Hysterothylacium aduncum* and perhaps *Contracaecum spiculigerum* may have both marine and freshwater life cycles (Yoshinaga *et al.*, 1987). The second moult (from L2 to L3) in aquatic ascaridoids is controversial (Smith, 1983) but has been observed in *R. acus* in the gut of intermediate fish hosts (Smith 1984) and in *Anisakis simplex* in euphausiids (Smith, 1983). Køie (1993) provides evidence that larvae of *H. aduncum* hatching from eggs are L3 and not L2. The use of invertebrates and fish as paratenic or intermediate hosts in the life cycle of anisakids cannot be correlated with the definitive host (homoeotherm vs. ectotherm) nor the environment (marine vs. freshwater). However, development to the adult stage occasionally occurs in intermediate hosts, e.g. *R. acus* in the viscera of fish hosts (Smith, 1984) and *Hysterothylacium* in the haemo-coel of shrimp (Margolis and Butler, 1954). This indicates that ascaridoids utilizing fish as the definitive host have greater flexibility regarding stage of development in the intermediate host.

Eustrongylides spp. mature in piscivorous birds and have a life cycle similar to the anisakids (Fig. 11.1). Larvae are ingested by freshwater oligo-chaetes (*Tubifex, Limnodrilus*) where they develope to L3 (Measures, 1988b). Larvae in annelids are ingested by fish where they grow and are lightly encap-sulated on the liver and viscera.

HOST–PARASITE INTERACTIONS

Literature on nematodes which cause diseases in fish is often buried in reviews on the biology of the parasite. The excellent reviews by Margolis (1970) and Schaeperclaus (1992) cover most of the nematodes, including a thorough discussion of the anisakids to that time. Grabda (1989) discusses the less readily accessible European/Russian literature and Paperna (1991) reviews the tropical literature. Table 11.1 lists nematodes which have been reported to cause some degree of disorder to the fish host and Fig. 11.2 describes the general pathology associated with the heart, body cavity, intestine, and liver. The common clinical signs include haemorrhage, inflammation, adhesions (although the effects of these are essentially unknown and may be transient, see Margolis, 1970), oedema, necrosis, encapsulation and/or granuloma formation. The cell types associated with a parasite include lymphocytes, granulocytes such as neutrophils and eosinophils, macrophages, and fibrocytes or fibroblasts. There is clearly a strong cellular reaction to most nematode infections. This is not surprising considering their size and tendency to migrate prior to entering a resting state. It is not known if this cellular response is parasite specific or a non-specific response and whether the extent of pathology is related to a primary infection or numerous reinfections (trickle infections).

Few studies have characterized parasite antigens (Buchmann *et al.*, 1991; Sugane *et al.*, 1992) or the antibody response to nematode infections (Linnik *et al.*, 1989; Buchmann *et al.*, 1991; Priebe *et al.*, 1991). The antigens of larval *A. simplex* include excretory (180 kDa and 40 kDa) and somatic (130 kDa) polypeptides which reacted with IgG in sera of *Anisakis*-infected humans (Sugane *et al.*, 1992). Buchmann *et al.* (1991) showed that *Anguilla anguilla* produced a specific antibody to a 43 kDa parasite antigen of *A. crassus*. Linnik *et al.* (1989) using the IHAT showed the response of carp to *Philometra lusiana* was higher in July–August but was not correlated with the intensity of infection. Priebe *et al.* (1991) found a relationship between antibody titre and *Anisakis* infections and concluded that migration and longevity of *Anisakis* might be determined by a specific immune response in older pollack. Generally, the antibody titre is not correlated with the number of infective larvae nor is it always evident that a challenge infection produces a higher antibody titre.

There are few reported cases of mortality due to nematode infections (Table 11.1). Most of the literature suggests detrimental effects on the host without direct proof of reduced fecundity, growth or behavioural changes. There are examples of a reduction in gonad size and/or sterility (Table 11.1) but often the data on the condition of uninfected fish is sparse or absent. The effect of liver helminthosis in cod is thoroughly reviewed by Margolis (1970) where there is reduced liver size, weight, fat content of liver, total weight of fish, and condition factor of fish. The absence of data on age, length, sex, maturity, spawning condition, and location and time of collection makes interpretation of most field data difficult (Margolis, 1970; Grabda, 1989). A study on the ascaridoid, *R. acus*, which infects livers of yellow perch, showed that the infection level was correlated with a reduction in growth and condition

Table 11.1. Nematodes causing disease in fish.

Species	Host	(Site) and description
TRICHINELLOIDEA		
Capillaria eupomitis	Lepomis gibbosus	(liver) enlarged, anaemia, yellowish[1]
Capillaria sp.	Symphysodon	(intestine) reduced activity, loss of appetite and emaciation, light and dark faecal segments[2]
Capillaria sp.	Scomber japonicus	(liver) atrophy, necrosis[3]
C. petruschewskii	freshwater fishes	(liver) pathogenic[3]
Cystoops acipenseri	Acipenseridae	(abdomen) nodes[4]
	sterlet, sturgeon (Caspian)	(skin, under the bony plates), connective tissue capsules, oedema[5]
DIOCTOPHYMATOIDEA		
Eustrongylides sp.	Bagrus docmac	(skin, muscle, viscera, gonads) inflammatory necrosis, encapsulation, sterility[6]
	Pygocentrus natterei	
	Pseudoplatysoma corruscans	(muscle, spleen, stomach, intestine) encapsulation; lymphocytes, macrophages[7]
	Clarias gariepinus	(mesentery, spleen, gonads, muscle) gonads deformed and degenerate, fat deposits, lesions, muscle digested and infiltrated, emaciation; condition factor reduced; lymphocytes, macrophages in infiltrate[8]
	Haplochromis sp.	
	Salmo gairdneri	
	Paratrygon sp.	(muscle) encapsulation; lymphocytes, macrophages[9]
Eustrongylides sp.	Lepomis gibbosus	(on serosa of intestine and mesentary) granulomatous inflammation; exudates contained erythrocytes and macrophages[10]
Eustrongylides tubifex	Ambloplites rupestris	
	Perca flavescens	
CAMALLANOIDEA		
Camallanus cotti	Poecilia reticulata	(intestine) suppression of sexual display rate and change in feeding behaviour[11]
C. oxycephalus	Stizostedion v. vitreum	(rectum) hyperplasia, inflammation, vascularization, loss of epithelium; eosinophils, lymphocytes, fibroblasts[12]
Procamallanus spiculogubernaculus	Heteropneustes fossilis	(oesophagus, stomach) anaemia[13]

SEURATOIDEA	
Cucullanus (=*Cucullanellus*)	(intestinal wall) estensive pathology [14] (gut wall, mesenteric blood vessels) inflammation, necrosis, dilation of blood vessels, destruction of capillaries, haemorrhage, encapsulation in wall of mesenteric blood vessels; lymphocyte and histiocyte infiltration [3]
Flounders	
Platichthys flesus	
Pleuronectes platessa	
Oncorhynchus mykiss	(pyloric caeca) loss of epithelium, mucosal hyperlasia, haemorrhaging, fibrosis; leucocytosis, granulocytes, fibroblasts [15]
Truttaedacnitis truttae	
Petromyzon marinus	(liver) hepatocytes altered [16]
HABRONEMATOIDEA	
Cystidicola stigmatura	(swimbladder) lesions, ulceration, hyperaemia [17]
Salvelinus namaycush	
S. namaycush	(swimbladder) lesions may be associated with number and size of worms [18]
Coregonus artedii	(swimbladder) inflammation, lesions, enlargement and atrophy of epithelium; histiocytes, granular cells [19]
C. farionis	
Spinitectus sp.	(intestine) inflammation [1]
Lepomis macrochirus	
L. macrochirus	(intestine-submucosa) inflammation, enteritis, haemorrhage; eosinophils, polymorphs [20]
L. macrochirus	(intestine) intestinal penetration, diffuse lesions, aseptic traumatic enteritis, infectious enteritis, desquamative catarrhal inflammation, necrosis, haemorrhage, encapsulation; eosinophils, leucocytes, polymorphonuclear leucocytes, macrophages, fibroblasts [20]
ASCARIDOIDEA	
Acanthocheilus nidifex	(stomach) inflammation, connective tissue formation, vascularization [3]
Galeocerdo cuvieri (=*tigrinus*)	
Anacathocheilus rotundatus	(gut wall, peritoneum, liver) encapsulation in gut wall, liver fibrosis and encapsulation [21]
G. morhua	
Anisakis sp.	(mesenteries, stomach, wall, muscles) inflammation, usually encapsulate ulceration of cod stomach [14]
cod, haddock, redfish, blue whiting, herring	
Clupea harengus pallasi	(pyloric caeca, pancreas, liver, intestine) mechanical compression, granulomatous inflammation, necrosis, trauma of the muscularis externa of the pyloric caeca; macrophages and lymphocytes in exudates [22]

Table 11.1. Nematodes causing disease in fish. (Continued)

Species	Host	(Site) and description
A. simplex	marine fish (cod, redfish)	(liver) destruction of liver tissue, induces small livers[3,4]
Anisakis sp.	Micromesistius potassou	(liver) impaired function, tissue damage[23]
Anisakid larvae	marine bony fish	(liver, gall bladder, intestines, body cavity) emaciation, weakening, death[23]
	Merlangius merlangus	(liver) capsule formation, melanin deposition; neutrophils, macrophages, fibroblasts[24]
Contracaecum sp.	East African fish	(viscera) mortality[6,25]
	Tilapia hybrids	(pericardium)[6]
	cichlids	(pericardium)[6]
	Gadidae	(liver) liver weight, fat content, weight of fish, condition factor decreased with increasing infection[3]
Contracaecum	Merlangius merlangus	(liver) capsule formation, melanin deposits; neutrophils, macrophages, fibroblasts[24]
	Labrus festivus	(liver) played an important role in mortality[3]
	Pagellus erythrinus	
C. bidentatum	sturgeon (Black and Caspian)	(oesphagus, stomach, intestine, swimbladder, body cavity) inflammation[5]
C. microcephalum	Oncorhynchus mykiss[a]	(viscera, body cavity, muscle) inflammation of intestine haemorrhage of muscle[26]
	Fatheads	(atrium of heart, body cavity) enlarged atrium, thrombi[26]
Goezia sinamora	Sticklebacks	(atrium of heart) enlarged atrium, thrombi[26]
	Tilapia aurea	(stomach) nodules, lesions[27]
	Morone saxatilis	(stomach, intestine) nodules, lesions, affects condition factor[27]
G. pelagia	Rachicentron canadum	(stomach) nodules, lesions, haemorrhage, slight autolysis of gastric glands[27]
Hysterothylacium aduncum	Baltic cod	(liver) loss of total weight and liver mass, loss of fat content and oil yield (gut, body cavity) shown to kill[3]
	herring larvae	
	rainbow trout (farmed)	(intestine) mortality[9]

Parasite	Host	Pathology
H. dollfusi	*Polyodon spathula*	(stomach, caeca, intestine) ulcers in stomach, nodules, haemorrhage; infiltrating eosinophilic granulocytes, fibrocytes and macrophages[28]
Pseudanisakis rotundata *Pseudoterranova decipiens*	*Raja radiata* cod	(intestine) inflammation; eosinophilic leucocytes[3]
	Gadus morhua	'.. muscle of badly infected fish becomes loose and spongy ..'[4]
	G. morhua	(muscle) chronic granulomatous inflammation; neutrophils, macrophages, lymphocytes, epitheliod cells, fibroblasts[29]
	Osmerus eperlanus	(muscle) cellular infiltration, encapsulation[30]
Raphidascaris acus	*Anguilla anguilla*	(muscles) reduced swimming performance in circular experimental tanks[31]
	E. lucius	(swimbladder) reduced swimming performance[4]
	Bream	(intestine) inflammation[4] (liver, intestine, gonads) emaciation, lesions in liver and intestine, atrophy of gonads, body cavity with bloody effusion, visceral and lateral peritoneal membranes spongy, pink-red, strongly infiltrated. Liver rusty, pink or yellowish with dark-green hue, congested, impairs growth.[4]
	Bream	emaciation, liver and gall are dirty brown colour, bile exuded into body cavity, gonads degenerate, bloody exudate in body cavity, wall of peritoneum and mesentry become oedematous, loose, reddish in colour, exophthalmia and hydrops of the body cavity indicate acute endotoxicosis[5]
	Lota lota[b] *Stizostedion v. vitreum*[a,b] *Oncorhynchus mykiss*[a,b]	(liver, viscera) lesions, granuloma, necrosis[b] (liver, viscera) lesions, haemorrhage, granuloma, mortality[b]
	Perca flavescens	(liver) lesions, granuloma, necrosis, retardation of sexual maturation, mortality, reduced growth[32,33]
	Coregonus clupeaformis[b] *C. nasus*[b] *S. gairdneri*	(liver, viscera) lesions, nodules[b] mortality[34]

Table 11.1. Nematodes causing disease in fish. (Continued)

Species	Host	(Site) and description
DRACUNCULOIDEA		
Anguillicola crassus	eels	(swimbladder) inflammation, sometimes swimbladder collapses, associated[14]
A. crassus	*Anguilla anguilla*	(swimbladder) Acute: lesions, hyperaemia of wall, epithelial hyperplasia. Chronic: inflammation, oedema, thickening of tunica propria, submucosa and serosa, monocyte infiltration, lymphocytes, histiocytes, plasma cells.[35]
	A. anguilla	(swimbladder) humoral antibody to proteins of MW 60 and 94 kDa[36]
	A. anguilla	(swimbladder) dilation of blood vessels, inflammation, rupture, fibrosis, adhesion to surrounding organs[37]
		(swimbladder) less pathology in experimental than natural infections[38]
		total plasma protein and second and fourth fraction of protein were reduced as infection level increased, lymphocyte/granulocyte ratio changed over the infection period with a drop in lymphocytes and increase in granulocytes[39]
A. globiceps	eels	(swimbladder) haemorrhagic inflammation, reduced growth and ability to withstand transport[6]
Philonema oncorhynchi	*Oncorhynchus nerka, O. keta*	(viscera) mesenteric and viscera adhesions[14]
	O. nerka	(wall of swimbladder) influence orientation of seaward migration of smolt[40]
P. oncorhynchi	*O. nerka*	visceral adhesions[41]
P. agubernaculum	*Salmo salar*	(ovary) encasement of ovary, blockage of genital pore[14,3]
P. rubra	striped bass	(body cavity, viscera) emaciation, sexually impotent[1]
	Morone americana	(viscera) visceral adhesions, oedema, granulomata[14]
P. americana	*Platichthys stellatus*	
	P. stellatus	(fins, subcutaneous tissues)[14]
	Lepidopsetta bilineata	(surface) repulsive-looking nodes[4]

Parasite	Host	Effect (site) and pathology
P. globiceps	Uranoscopus surinamensis	total or partial destruction of gonads [4]
P. translucida	Pseudotolithus senegalensis	(gonads) reduced fecundity by 86–90% [42]
P. saltatrix	P. typus, P. elongatus	(gonads) reproductive potential reduced [43]
	Pomatomus saltatrix	(heart, pericardial cavity) inflammation, blood stasis, necrosis [44]
P. opsalichthydis	Hypomesus olidus	(liver, mesentery, gonads), enlargement of body cavity [34]
Philometra. sp.	Otolithus argentatus	(ovary) destroyed ovaries, atrophy (testes) testes replaced by nematode [14,34,3]
	Chrysophrys auratus	
Philometroides nodulosa	Catostomus commersoni	(cheek galleries) [1]
P. lusiana	carp	(body cavity near swimbladder, gonads, kidneys, gill sacs) immobile, reduced growth, swimbladder function impaired, swollen scale sacs, haemorrhage exophthalmia [5]
P. okeni	sea perch	(eye) exophthalmus [1]
Philometra spp.	blue fill sunfish	(ovary) hypertrophied, fibrotic, atrophy, mortality [14]
P. cephalus	Mugil cephalus	(ovary) granuloma, effects unknown [45]
Philometridae	Carcharhinus limbatus	(ventral and pectoral fins) tumours [34]
Phlyctainophora lamnae	Mustelus mustelus	
PHYSALOPTEROIDEA		
Proleptus obtusus	Scyllium canicola	(intestine, pyloric region) haemorrhage, destruction of epithelium, ulceration, worms may penetrate as deep as the tunica propria and muscularis mucosa; leucocyte infiltration [46,3]

[1]Hoffman (1975), [2]Lo Wing Yat (1988), [3]Margolis (1970) [4]Grabda (1989), [5]Bauer et al. (1977), [6]Paperna (1991), [7]Eiras and Rego (1988), [8]Paperna (1974), [9]Berland (1987), [10]Measures (1988c), [11]McMinn (1990), [12]Authors, unpublished, [13]Sinha and Sinha (1988), [14]Sindermann (1990), [15]Dunn et al. (1983), [16]Eng and Youson (1992), [17]Black (1984), [18]Black (1984), [19]Willers et al. (1991), [20]Jilek and Crites (1982), [21]Kahl (1938b), [22]Hauck and May (1977), [23]Smith and Wootten (1978), [24]Elarifi (1982), [25]Rogers (1976), [26]Dick (1987), [27]Deardorff and Overstreet (1980b), [28]Miyazaki et al. (1988), [29]Ramakrishna and Burt (1991, 1993), [30]Kahl (1938a), [31]Sprengel and Luechtenberg (1991), [32]Poole and Dick (1984), [33]Szalai and Dick (1991), [34]Williams (1967), [35]Molnar et al. (1993), [36]Buchmann et al. (1993), [37]Van Banning and Haenen (1990), [38]Haenen et al. (1989), [39]Boon et al. (1990), [40]Garnick and Margolis (1990), [41]Nagasawa (1989), [42]Ramachandran (1973) in Sindermann (1990), [43]Anyanwu (1983), [44]Cheung et al. (1984) in Sindermann (1990), [45]Rosa-Molinar and Williams (1983), [46]Schurmans-Skekhoven and Botman, 1932, [a]experimental infections [b]Dick (unpublished).

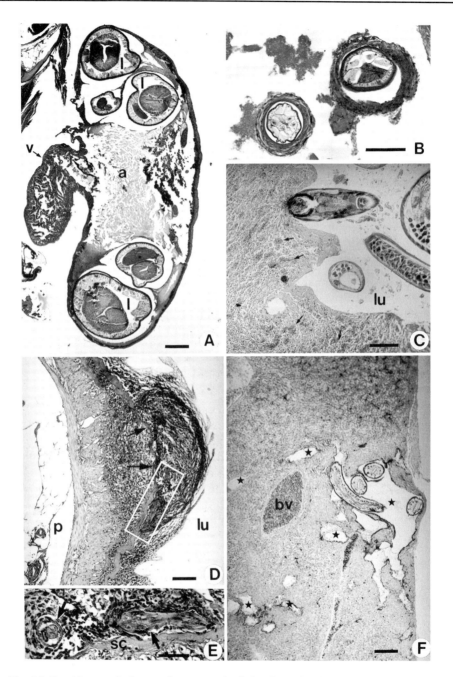

Fig. 11.2. Histopathology of nematode infections in fish. A. Cross-section through the cardiac region of a fathead minnow containing larval *Contracaecum* spp.; atrium is three times larger than normal but there is no evidence of a celluar response to the larva; hematoxylin and eosin (H&E); scale bar = 50 μm. B. Cross-section of a larval *Contracaecum* in the body cavity of a fathead minnow; note the strong celluar reponse in the form of a sleeve of connective tissue around

factor, retardation of sexual maturation, mortality and susceptibility to predation (Szalai and Dick, 1991). Paperna (1991) also reported that parasitism reduces the condition factor of fish. Russell (1979) found that *T. truttae* had little affect on swimming of rainbow trout but reported that decreasing growth rates were correlated with numbers of the parasite.

Nematode infections are known to affect fish behaviour (Garnick and Margolis, 1990; McMinn; 1990; Sprengel and Luechtenberg, 1991). *Philonema oncorhynchi* of *O. nerka* affects the orientation and perhaps survival of smolts and may account for some of the variability in smolt migratory behaviour (Garnick and Margolis, 1990). McMinn (1990) working with *Camallanus cotti* in *Poecilia reticulata* concluded that the sexual (sigmoid) display rate, the most important predictor of female choice, by males, '. . . appears to be the behaviour most sensitive to nematode infection . . .'. While female guppies prefer males with low numbers of parasites, female preference for females (non-sexual behaviour) is not related to parasite intensity. Sprengel and Luechtenberg (1991) concludes that infections of eels with *P. decipiens* and/or *A. crassus* reduce maximum swimming speed and probably increase their susceptiblity to predators. The studies of McMinn (1990) and Sprengel and Luechtenberg (1991) found that even small numbers of worms elicited the same behavioural effects.

IN VITRO CULTURE

The majority of studies on the *in vitro* culture of nematodes has focused on mammalian parasites. Among fish nematodes, only the anisakids have been cultured using defined and semi-defined media (Taylor and Baker, 1968; Hansen and Hansen, 1978; Douvres and Urban, 1987). Notable studies include those on *P. decipiens* in which L3 were cultured to the adult stage (Townsley *et al.*, 1963). There was no mention of the fourth moult nor the release of eggs. *A. simplex* has also been cultured to adults but eggs were not produced (Carvajal *et al.*, 1981). McClelland and Ronald (1974) cultured *Contracaecum*

---

the larva; H&E; scale bar = 10 µm. C. Cross-section of the hind gut of a walleye with *Camallanus oxycephalus*; there is extensive vascularization (arrows) and granulomatous material and villi are absent; Mallory Heidenhain (M-H); scale bar = 50 µm. D. *Raphidascaris acus* from experimental infections of rainbow trout; a single larva has produced an extensive granuloma which extends above and below the stratum compactum (arrows); note absence of villar structure, worms lie within the rectangle (see E); H&E; scale bar = 100 µm. E. enlargement of region lying within the rectangle from D; the stratum compactum appears to act as a barrier to the nematode; H&E; scale bar = 50 µm. F. Section through the liver of trout perch from a natural infection of *R. acus*; there is extensive pathology with worms generating and lying within fluid filled cavities (stars); M-H; scale bar = 100 µm. a = amorphous material, bv = blood vessels, I = larva, lu = lumen side of intestine, p = peritoneal side of the intestine, sc = stratum compactum, and v = ventricle.

osculatum to adults but did not obtain ovigerous females. More recently, *A. simplex* has been cultured from L3 to egg laying adults (Bier and Douvres in Douvres and Urban, 1987), in a roller bottle culture system, using an acidic API-1 medium supplemented with bovine haemin, at 35°C, under a gas phase. A variation of these culture conditions was used to culture *P. decipiens* and *C. osculatum* to adults which produced viable eggs (Likely and Burt 1989, 1992). The main modifications were that the concentration of bovine haem was decreased, the pH lowered, and the medium supplemented with 5 mM L-cysteine for *C. osculatum*. The 5 mM L-cysteine was essential for the third and fourth moults but an increase of L-cysteine to 8 mM was fatal for newly moulted nematodes (Likely and Burt, 1992). The complete ecdysis from the L3 to L4 in anisakids can occur under atmospheric conditions in a semi-defined medium (see Likely and Burt, 1989). However, there is evidence to indicate that less complex culture conditions can be used for some anisakids, since L3 of *Contracaecum/Phocascaris* and *Pseudoterranova* can complete moulting and ecdysis to L4 at 37°C, in 0.9% saline, under atmospheric conditions (Mercer *et al.*, 1986; Dick and Choudhury, unpublished). There are no published reports on the culture of L2 and L3 of anisakids, these stages are commonly found in fish.

NUTRITION AND PHYSIOLOGY

Since nematodes in fish invade tissues similar to those invaded by other parasitic nematodes, i.e. gut, liver, body cavity, gonads, swimbladder, muscles, vascular system, it is unlikely that their basic nutritional requirements differ from nematodes which infect homeotherms. Nematodes require amino acids, simple sugars and lipids (Douvres and Urban, 1987). The two key nutritional ingredients for most nematodes are cholesterol and globular proteins. Nematodes are known to feed on blood (Sinha and Sinha, 1988; Haenen *et al.*, 1989), other tissues (Dunn *et al.*, 1983), and tissue/mucosal plugs (Dunn *et al.*, 1983; Sinha and Sinha, 1988). According to Douvres and Urban (1987) the nutrient requirements for mammalian and fish ascaridoids are quite similar. The fine structure of the intestinal cells of the adult *P. decipiens* from seals are similar to other nematodes but *P. decipiens* has large deposits of glycogen which is similar to that reported for *Ascaris* (Andreassen, 1968). Although little is known about carbohydrate metabolism in fish nematodes, sorbitol dehydrogenase is characterized from *P. decipiens* (Goil and Harpur, 1978). The variation in the colour of *P. decipiens* is correlated with the haemoglobin concentration in the pseudocoelomic fluid but is not related to differences in gene expression at the transcriptional level (Dixon *et al.*, 1993). Using isoelectric focusing and spectrophotometry, Fusco (1978) showed that the haemoglobin from *Spirocamallanus cricotus* and its host, *Micropogonias undulatus*, are different. This indicates that the parasite haemoglobin is not acquired from feeding on blood as is often assumed for camallanid nematodes. Goil and Harpur (1979) found that the non-specific acid phosphomonoesterase from larval *P. decipiens* differed from that of its host, *Gadus morhua*. Volatile ketones and alcohols have been reported from the culture medium of *P. decipiens* but less is known about their origin and function (Ackman, 1976).

The location and distribution of catecholaminergic structures (Goh and Davey, 1976a) and acetylcholinesterases and synapses in the nervous system (Goh and Davey, 1976b) of *P. decipiens* are known and they appear to be similar to other nematodes. Boczon and Bier (1986) investigated the bioenergetic effects of fish muscle mitochondria infected with anisakids and reported that '... Mg^{2+} stimulated adenosine triphosphate activity of muscle mitochondria can be used to estimate the number of nematodes per market fish'. The continuous production of proteolytic enzymes (trypsin-like) from the oesophageal gland of larval *Anisakis simplex* suggest that it may be important for invasion by larvae (Matthews, 1984).

Davey and co-workers used the nematode *P. decipiens* to study moulting. L3, at 37°C in NaCl produced new cuticles but did not ecdyse (Davey and Kan, 1968). However, the addition of insect juvenile hormone produced ecdysis (Davey, 1971). These and further studies led to a working hypothesis that neurosecretory cells associated with the nerve ring is probably associated with ecdysis (Davey and Goh, 1984). Noradrenaline appeared to be associated with ecdysis in L3 (Goh and Davey, 1985). Goh and Davey (1985) suggested that a stimulus in the stomach of seals somehow activates the nervous system to bring about the release of noradrenaline from the nerve ring. There is then a release of a second factor which acts on the excretory cell to cause it to release enzymes, which are part of the moulting fluid. This fluid passes into the space between the two cuticles and ecdysis follows (Goh and Davey, 1985). Mercer *et al.* (1986) reported that temperature stimulated moulting and ecdysis occurred under atmospheric conditions in saline.

NEMATODE IDENTIFICATION

The general reference of Hoffman (1967) for freshwater North American parasites, although dated, is comprehensive and very useful. Also valuable are the works of Mozgovoi (1953), Skrjabin *et al.* (1967), Bykhovskaya-Pavlovskaya *et al.* (1962), Anderson *et al.* (1974–1975) and Sood (1988). Grabda (1989) and Sindermann (1990) provide useful diagrams of nematodes. Readers are referred to the detailed work of Berland (1989) for marine ascaridoids. Larvae (L3) of different genera of ascaridoids occurring in fish are differentiated by the position of the excretory pore, the presence of a mucron, the relative proportions of the oesophagus and ventriculus and when present, of the caecum and appendix (Fig. 11.3).

Identification is usually done with the aid of morphological keys and using light and scanning electron microscopy (SEM). Due to the taxonomic importance of external features such as papillae, alae, striations, the boring tooth, and the oral opening, SEM has been a very useful tool (Fredericksen and Specian, 1980; Valter *et al.*, 1982; Weerasooriya *et al.*, 1986). However, the ascaridoids are not readily identified with SEM as there is often little external morphological difference among larvae. Culturing larval nematodes to adults

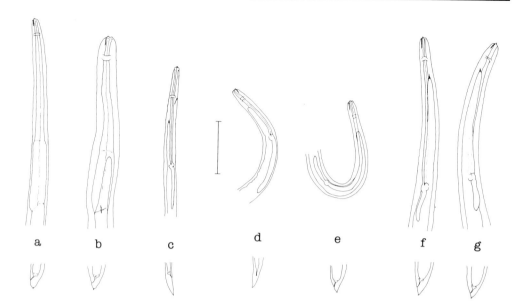

Fig. 11.3. Ascaridoid larvae recovered from fish hosts. a. *Anisakis* sp. L3 from viscera of Pacific herring. b. *Pseudoterranova* sp. L3 from musculature of red snapper (Pacific). c. *Hysterothylacium* sp. L3 from viscera of Pacific herring. d. *Contracaecum/Phocascaris* L3 from stomach of harp seal, Canadian Atlantic. e. *Contracaecum* sp. from heart of fathead minnow, High Rock Lake, Manitoba, Canada. f. *Contracaecum* sp. from body cavity of stickleback, High Rock Lake. and g. *Contracaecum* sp. from body cavity of smallmouth bass, Boundary Reservoir, Saskatchewan, Canada. Scale bar = 1 mm.

in experimental definitive hosts has been used to identify species of *Contracaecum* (Dick *et al.*, 1987; Fagerholm, 1988). *In vitro* culture methods, now available for anisakids (Douvres and Bier, in Douvres and Urban, 1987; Likely and Burt, 1989, 1992), may also aid in diagnosis. However, culturing worms *in vitro* or in experimental hosts is time consuming and depends on obtaining live worms and maintaining experimental hosts. Consequently, electrophoretic methods have been used and they include evaluation of total proteins and enzyme loci using electrophoresis and isoelectric focusing (Nascetti *et al.*, 1986; Orecchia *et al.*, 1986; Appleton and Burt, 1991).

The two major genera where electrophoresis has been applied are *Anisakis* and *Pseudoterranova*. For many years *Anisakis* larvae were identified as type I and type II. Type I is now identified as *A. simplex* and type II larvae have been identified as *A. physeteris* (Orecchia *et al.*, 1986). Matsuura *et al.* (1992) have confirmed this identification using restriction fragment length polymorphisms. The identity of *Anisakis* is further complicated by the discovery of type A and B (referred to as sibling species) from the Mediterranean and North Atlantic, respectively (Mattuicci *et al.*, 1986). In contrast, *P. decipiens* was considered to be one species for many years because there were no morphological differences among adults and larvae from different hosts and geographical regions. Recently, Paggi *et al.* (1991) proposed '. . . three species within' *P. decipiens* based on 16 enzyme loci with the following major distributions; *P. decipiens* A is in grey seals *Halichoerus grypus*, *P. decipiens* B is in harbour seals *Phoca vitulina*, and *P. decipiens* C is in bearded seals *Erignathus barbatus*. *P. decipiens* A was not found in their samples from the North West Atlantic. Genetic variation for *P. decipiens* was also reported by Appleton and Burt (1991) where types I and II L3 were identified in three species of fish from two geographical locations. Clearly, it will be important to know how the variants reported by Appleton and Burt (1991) relate to *P. decipiens* B and C reported from the Canadian Atlantic (Paggi *et al.*, 1991).

CONTROL OF FISH PARASITIC NEMATODES

Nematodes that utilize intermediate hosts in their life cycles may be controlled, particularly in aquaculture conditions, by preventing access of intermediate hosts or definitive hosts to the water or by eliminating these hosts. For parasites such as *Contracaecum* and *Eustrongylides*, where the definitive hosts are piscivorous birds, elimination of these hosts may not be desirable or possible. Since most fish nematodes require intermediate invertebrate hosts, screening or treating the water supply in an aquaculture facility may be a solution, especially if the source has infected hosts. Also, all wild fish should be examined for internal and external parasites before stocking (Bauer and Hoffman, 1976; Schaeperclaus, 1992). A few anthelmintics have been applied, primarily with *A. crassus* infections in eels. Taraschewski *et al.* (1988) tested five drugs but only levamisole–HCl brought about complete recovery of infected eels and killed *A. crassus*. Hartmann (1989) found that levamisole (at 2 and 5 mg l^{-1}) administered in a water bath caused the paralysis, death and expulsion of preadult and adult *A. crassus* in the lumen of the swimbladder. Earlier (developing) stages in the wall of the swimbladder were less sensitive to the treatment while eggs and larvae seemed unaffected. Hartmann (1989) suggested that treatment should be carried out several times over an extended period of time until all adults, maturing at different times from larvae in the wall of the swimbladder, are killed. Furthermore, precautions must be taken since eggs and larvae that are released during the treatment period are infective to copepods. Djajadiredja *et al.* (1983) recommended a 1 hour bath in formalin (0.25–0.33 ppt) for fish infected with *Camallanus* although the efficacy of the treatment was not discussed.

Treatment of the discus (*Symphysodon*, Cichlidae) infected with *Capillaria* has been reviewed by Lo Wing Yat (1988, see references therein). Anthelmintics such as garlic oil, piperazine, trichlorphon and levamisole, administered in the diet, were either ineffective or the effect varied greatly among fish. Trichlorphon in a bath method was effective when used in concentrations of 4 ppm and 7 ppm but produced variable results as well as side-effects (loss of appetite and colour, and sensitivity to bright light) (Lo Wing Yat, 1988). However, Pena *et al.* (1988) found that fresh minced garlic (200 mg l^{-1} water) and its hexane extract was 100% effective in treating *Capillaria* in carp while ammonium-potassium tartrate ($1.5\,\mathrm{mg\,l^{-1}}$, twice daily) was 86% effective.

CONCLUSIONS

There are numerous taxonomic problems to be resolved in fish nematodes, with major problems in the spirurids and the ascarids. The species of the genus *Truttaedacnitis* in lamprey and fish is still controversial and some of the current species need to be reevaluated. The family Cystidicolidae needs further study and a complete review of the group using SEM is needed. Unquestionably the group requiring the most taxonomic work is the Ascaridoidea, both the freshwater and marine forms. The fish–bird transmitted *Contracaecum* spp. require taxonomic and life cycle studies, both in North and South America and in the Africa/Eurasia axis. The increasing populations of fish-eating birds have had an impact on both commercial and sport fisheries, and will continue to do so, but we are still unable to identify the larvae of most ascaridoid species in fish. With the application of molecular biological techniques, the marine ascaridoids will undoubtedly be separated into more distinct genetic groups but how this information will be used to reduce their effect on commercial fisheries remains unclear. Even if it is shown that variants or siblings species of ascaridoids and other nematodes are transmitted through a select group of intermediate/definitive hosts in a relatively well-defined region, the only plausible economic solution resulting from these studies will probably be a reduction in population numbers of the definitive hosts, either through removal or reduced reproduction.

Transmission dynamics of most fish nematodes will be important in the future for the wild caught fishery but with increasing emphasis on extensive and intensive aquaculture. A good example of progress is the eel parasite, *A. crassus*, as research involves basic knowledge and control measures. Areas where more information is needed include: the importance of invertebrates in the transmission of *Truttaedacnitis* and other cucullanids; the significance of fish predation and paratenic hosts on the transmission of the ascaridoids and larval survival time in the fish hosts; the significance of low fish host-specificity in the ascaridoids and the effect of a loss or a decline in numbers of one species of fish on the parasite population levels. Perhaps *Raphidascaris acus* could be used as a model system to address many of these questions since it is readily accessible, present in simple and complex freshwater systems, easily maintained in experimental hosts and exhibits most of the characteristics of the ascaridoids.

Considering the extensive knowledge of the species of hosts infected and pathology induced by fish nematodes, and the effect of unsightly parasites to the consumer, few studies document the detrimental effects on the fish host. There is remarkably little information on the effects of marine ascaridoids on fish production, with the exception of liver quality. More studies are needed, on wild and cultured fish, to document sublethal effects of parasites on swimming, orientation and on growth and sexual maturation. With a decline in the wild caught fisheries and an increase in fish culture, parasitologists will eventually have to address this cost to production prior to the implementation of control measures. Documented effects on production will probably stimulate research in more comprehensive methods of control, i.e. the role of predation/parateny/reservoir fish hosts within fish species transmission complexes of wild populations. This approach should include the manipulation of some or all species of fish to decrease parasite numbers while maintaining the reproductive potential of the commercially important species of fish.

Other approaches to control will require an understanding of antigenic properties of parasites, presence of toxins produced by them, their susceptibility to drugs and other chemicals, and the feasibility of vaccine development and application (probably of use only in intensive aquaculture for high value fish). However, since fish are the most diverse of all vertebrate groups, but with the least known about their immune systems, progress in this area will be slow and probably limited to a few important cultured species.

ACKNOWLEDGEMENTS

The authors thank the *Journal of Wildlife Diseases* for permission to reproduce Fig. 2A and B. We also acknowledge the contribution of histological sections from students Joseph Carney (Fig. 2D and E) and Patrick Nelson (Fig. 2F).

REFERENCES

Ackman, R.G. (1976) Volatile ketones and alcohols of codworms (*Terranova decipiens*) from axenic culture and fresh fish muscle. *Journal of the Fisheries Research Board of Canada* 33, 2819–2821.

Anderson, R.C. (1984) The origins of zooparasitic nematodes. *Canadian Journal of Zoology* 62, 317–328.

Anderson, R.C. (1992) *Nematode Parasites of Vertebrates: Their Development and Transmission*. CAB International, 578 pp.

Anderson, R.C., Chabaud, A.G. and Willmott, S. (eds) (1974, 1975) *CIH Keys to the Nematode Parasites of Vertebrates*. Nos. 1, 2 and 3. Commonwealth Agricultural Bureaux, Farnham Royal, UK.

Andreassen, J. (1968) Fine structure of the intestine of the nematode, *Ancylostoma caninum* and *Phocanema decipiens*. *Zeitschrift für Parasitenkunde* 30, 318–336.

Anyanwu, A.O. (1983) Parasitic infestations of *Pseudotolithus* spp. off the coast of Lagos, Nigeria. *Journal of Fish Biology* 22, 29–33.

Appleton, T.E. and Burt, M.D.B. (1991) Biochemical characterization of third stage larval sealworm, *Pseudoterranova decipiens* (Nematoda: Anisakidae), in Canadian Atlantic waters using isoelectric focusing of soluble proteins. *Canadian Journal of Fisheries and Aquatic Sciences* 48, 1800–1803.

Bauer, O.N. and Hoffman, G.L. (1976) Helminth range extension by translocation of fish. In: Page, L.A. (ed.) *Wildlife Diseases*. Plenum Press, New York, pp. 163–172.

Bauer, O.N., Musselius, V.A., Nikolayeva, V.M. and Strelkov, Yu.A. (1977) *Ichthyopathology*. Pischchevaya Promyshlennost' Press, 432 pp. English translation by Teaneck, N.J. Ansel Translations, 1980.

Belpaire, C., De Charleroy, D., Thomas, K., van Damme, P. and Ollevier, F. (1989) Effects of eel restocking on the distribution of the swimbladder nematode *Anguillicola crassus*. *Journal of Applied Ichthyology* 5, 151–153.

Berland, B. (1987) Helminth problems in sea-water aquaculture. In: Stenmark, E. and Malmberg, G. (eds) *Parasites and Diseases in Natural Waters and Aquaculture in Nordic Countries*. Zoo-tax, Naturahistoriska riksmuseet, Stockholm, Sweden, p. 56–62.

Berland, B. (1989) Identification of larval nematodes from fish. In: Moeller, H. (ed.) *Nematode Problems in North Atlantic Fish*. Report from a workshop in Kiel, 3–4 April, 1989, pp. 16–22.

Black, G.A. (1984) Swimbladder lesions in lake trout (*Salvelinus namaycush*) associated with mature *Cystidicola stigmatura* (Nematoda). *Journal of Parasitology* 70, 441–443.

Boczon, K. and Bier, J.W. (1986) *Anisakis simplex*: Uncoupling of oxidative phosphorylation in the muscle mitochondria of infected fish. *Experimental Parasitology* 61, 270–279.

Boon, J.H., Cannaerts, V.M.H., Augustijn, H., Machiels, M.A.M., De Charleroy, D. and Ollevier, F. (1990) The effect of different infection levels with infective larvae of *Anguillicola crassus* on haematological parameters of European eel (*Anguilla anguilla*). *Aquaculture* 87, 243–253.

Bowen, W.D. (ed.) (1990) Population biology of sealworm (*Pseudoterranova decipiens*) in relation to its intermediate and seal hosts. *Canadian Bulletin of Fisheries and Aquatic Sciences* 222, 306 pp.

Bryant, C. (1982) The biochemical origins of helminth parasitism. In: Symons, L.E.A., Donald, A.D. and Dineen, J.K. (eds) *Biology and Control of Endoparasites*. Academic Press, Sydney, pp. 29–52.

Buchmann, K., Pedersen, L.O. and Glamann, J. (1991) Humoral immune response of European eel *Anguilla anguilla* to a major antigen in *Anguillicola crassus* (Nematoda). *Diseases of Aquatic Organisms* 12, 55–57.

Burt, M.D.B., Campbell, J.D., Likely, C.J. and Smith, J.W. (1990) Serial passage of larval *Pseudoterranova decipiens* (Nematoda: Ascaridoidea) in fish. *Canadian Journal of Fisheries and Aquatic Sciences* 47, 693–695.

Bykhovskaya-Pavlovskaya, I.E., Gusev, A.V., Dubinina, M.N., Izyumova, N.A., Smirnova, T.S., Sokolovskaya, I.L., Shtein, G.A., Shul'man, S.S. and Epshtein, V.M. (1962) *Key to the Parasites of Freshwater Fish of the USSR*. Izdatelstvo, Akademii Nauk SSSR. Moskva-Leningrad. [Israel Program for Scientific Translations, Jerusalem, 1964, 919 pp.].

Carvajal, J., Barros, C., Santander, G. and Alcade, C. (1981) *In vitro* culture of larval anisakid parasites of the Chilean Hake *Merluccius gayi*. *Journal of Parasitology* 67, 958–959.

Chitwood, B.G., and Chitwood, M.B. (1950) *An Introduction to Nematology. Section 1. Anatomy*. Chitwood Publisher. Place.

Davey, K.G. (1971) Moulting in a parasitic nematode, *Phocanema decipiens*. VI. The mode of action of insect juvenile hormone and farnesyl methyl ether. *International Journal for Parasitology* 9, 121–125.

Davey, K.G. and S.L. Goh. (1984) Ecdysis in a parasitic nematode: direct evidence for an ecdysial factor from the head. *Canadian Journal of Zoology* 62, 2293–2296.

Davey, K.G. and Kan, S.P. (1968) Molting in a parasitic nematode, *Phocanema decipiens*. IV. Ecdysis and its control. *Canadian Journal of Zoology* 46, 893–898.

De Charleroy, D., Grisez, L., Thomas, C., Belpaire, F. and Ollevier, F. (1990) The life cycle of *Anguillicola crassus*. *Diseases of Aquatic Organisms* 8, 77–84.

Deardorff, T.L. and Overstreet, R.M. (1980a) Taxonomy and biology of North American species of *Goezia* (Nematoda: Anisakidae) from fishes, including three new species. *Proceedings of the Helminthological Society of Washington* 47, 192–217.

Deardorff, T.L. and Overstreet, R.M. (1980b) Review of *Hysterothylacium* and *Iheringascaris* (both previously = *Thynnascaris*) (Nematoda: Anisakidae) from the Northern Gulf of Mexico. *Proceedings of the Biological Society of Washington* 93, 1035–1079.

Dekker, W. and van Willigen, J. (1989) Short note on the distribution and abundance of *Anguillicola* in The Netherlands. *Journal of Applied Ichthyology* 1, 46–47.

Delyamure, S.L. (1955) *Helminthofauna of Marine Mammals (Ecology and Phylogeny)*. Skrjabin, K.I. (ed.) Izdateltsvo Akademii Nauk SSSR. 1968. Translation, Israel Programme for Scientific Tranlations, Jerusalem, IPST Cat. No. 1886.

Dick, T.A. (1987) The atrium of the fish heart as a site for *Contracaceum* spp. larvae. *Journal of Wildlife Diseases* 23, 328–330.

Dick, T.A., Papst, M.H. and Paul, H.C. (1987) Rainbow trout (*Salmo gairdneri*) stocking and *Contracaecum* spp. *Journal of Wildlife Diseases* 23, 242–247.

Dixon, B., Kimmins, W. and Pohajdak, B. (1993) Variation in colour of *Pseudoterranova decipiens* (Nematoda: Anisakidae) larvae correlates with haemoglobin concentration in the pseudocoelomic fluid. *Canadian Journal of Fisheries and Aquatic Sciences* 50, 767–771.

Djajadiredja, R., Panjaitan, T.H., Rukyani, A., Sarono, A., Satyani, D. and Supriyadi, H. (1983) Country reports – Indonesia. In: Oswald, E. and Hulse, J.H. (ed.) *Fish Quarantine and Fish Diseases in Southeast Asia*. IDRC, Canada, pp. 19–30.

Douvres, F.W. and Urban, J.F. (1987) Nematoda except parasites of insects. In: Taylor, A.E.R. and Baker, J.R. (eds) *In Vitro Methods for Parasite Cultivation*. Academic Press, London, pp. 318–378.

Dunn, I.J., Russell, L.R. and Adams, J.R. (1983) Caecal histopathology caused by *Truttaedacnitis truttae* (Nematoda: Cucullanidae) in rainbow trout, *Salmo gairdneri*. *International Journal for Parasitology* 13, 441–445.

Eiras, J.C. and Rego, A.A. (1988) Histopatologia da parasitose de peixes do Rio Cuiaba (Mato Grosso) por larvas de *Eustrongylides* sp. (Nematoda, Dioctophymidae). *Revista Brasileira de Biologia* 48, 273–280.

Elarifi, A.E.A.O. (1981) Aspects of the biology of larval *Contracaecum osculatum* (Rudolphi, 1802), from *Merlangius merlangus* (L.) in Scottish waters. PhD thesis, University of Aberdeen, Aberdeen, UK, 124 pp.

Elarifi, A.E. (1982) The histopathology of larval anisakid nematode infections in the liver of whiting, *Merlangius merlangus* (L.), with some observations on the blood leucocytes of the fish. *Journal of Fish Diseases* 5, 411–419.

Eng, F. and Youson, J.H. (1992) Morphology of the lives of the brook lamprey, *Lampebra lamoltenii* before and during infection with the nematode, *Truttaedacnitis stelmiodes*, hepatocytes, sinusoids and persinusoidal cells. *Tissue and Cell* 24, 575–592.

Fagerholm, H.-P. (1988) Incubation in rats of a nematodal larva from cod to establish its specific identity: *Contracaecum osculatum*, (Rudolphi). *Parasitology Research* 75, 57–63.

Fredericksen, D.W. and Specian, R.D. (1980) The value of cuticular fine structure in identification of juvenile anisakine nematodes. *Journal of Parasitology* 67, 647–655.

Fusco, A.C. (1978) *Spirocamallanus cricotus* (Nematoda): Isoelectric focusing and spectrophotometric characterization of its hemoglobin and that of its piscine host, *Micropogonias undulatus*. *Experimental Parasitology* 44, 155–160.

Garnick, E. and Margolis, L. (1990) Influence of four species of helminth parasites on orientation of seaward migrating sockeye salmon (*Oncorhynchus nerka*) smolts. *Canadian Journal of Fisheries and Aquatic Sciences* 47, 2380–2389.

Gibson, D.I. (1983) The systematics of ascaridoid nematodes – a current assessment. In: Stone, A.R., Platt, H.M. and Khalil, L.F. (eds) *Concepts in Nematode Systematics*. Academic Press, London, pp. 321–338.

Goh, S.L. and Davey, K.G. (1976a) Localization and distribution of catecholaminergic structures in the nervous system of *Phocanema decipiens* (Nematoda). *International Journal for Parasitology* 6, 403–411.

Goh, S.L. and Davey, K.G. (1976b) Acetylcholinesterase and synapses in the nervous system of *Phocanema decipiens* (Nematoda): a histochemical and ultrastructural study. *Canadian Journal of Zoology* 54, 752–771.

Goh, S.L. and Davey, K.G. (1985) Occurrence of noradrenaline in the central nervous system of *Phocanema decipiens* and its possible role in the control of ecdysis. *Canadian Journal of Zoology* 63, 475–479.

Goil, M.M. and Harpur, R.P. (1978) Studies on sorbitol dehydrogenase from the parasitic nematode larvae of *Phocanema decipiens*. *Zeitschrift für Parasitenkunde* 57, 117–120.

Goil, M.M. and Harpur, R.P. (1979) A comparison of the non-specific acid phosphomonoesterase activity in the larva of *Phocanema decipiens* (Nematoda) with that of its host the codfish (*Gadus morhua*). *Zeitschrift für Parasitenkunde* 60, 177–183.

Grabda, J. (1991) *Marine Fish Parasitology. An Outline*. Verlagsgesellschaft mbH, Weinheim, 306 pp.

Haenen, O.L.M. and van Banning, P. (1991) Experimental transmission of *Anguillicola crassus* (Nematoda, Dracunculoidea) larvae from infected prey fish to the eel *Anguilla anguilla*. *Aquaculture* 92, 115–119.

Haenen, O.L.M., Grisez, L., De Charleroy, D., Belpaire, C. and Ollevier, F. (1989) Experimentally induced infections of European eel *Anguilla anguilla* with *Anguillicola crassus* (Nematoda, Dracunculoidea) and subsequent migration of larvae. *Diseases of Aquatic Organisms* 7, 97–101.

Hafsteinsson, H. and Rizvi, S.S. (1987) A review of the sealworm problem: Biology, implications and solutions. *Journal of Food Protection* 50, 70–84.

Hansen, E.L. and Hansen, J.W. (1978) Nematoda parasitic in animals and plants. In: Taylor, A.E.R. and Baker, J.W. (eds) *Methods of Cultivating Parasites In Vitro*. Academic Press, London, pp. 227–277.

Hartmann, F. (1989) Investigations on the effectiveness of Levamisol as a medication against the eel parasite *Anguillicola crassus* (Nematoda). *Diseases of Aquatic Organisms* 7, 185–190.

Hauck, A.K. and May, E.B. (1977) Histopathologic alterations associated with *Anisakis* larvae in Pacific herring from Oregon. *Journal of Wildlife Diseases* 13, 290–293.

Hoffman, G.L. (1967) *Parasites of North American Freshwater Fishes*. University of California Press, Berkeley, 486 pp.

Hoffman, G.L. (1975) Lesions due to internal helminths of freshwater fishes. In: Ribelin, W.E. and Migaki, G. (eds) *The Pathology of Fishes*. University of Wisconsin Press, Madison, pp. 151–187.

Huizinga, H.W. (1966) Studies on the life cycle and development of *Contracaecum spiculigerum* (Rudolphi, 1809) (Ascaridoidea: Heterocheilidae) from marine piscivorous birds. *Journal of the Elisha Mitchell Scientific Society* 52, 181–195.

Huizinga, H.W. (1967) The life cycle of *Contracaecum multipapillatum* (Von Drasche, 1882) Lucker, 1941 (Nematoda: Heterocheilidae). *Journal of Parasitology* 53, 368–375.

Jilek, R. and Crites, J.L. (1982) Intestinal histopathology of the common bluegill, *Lepomis macrochirus* Rafinesque, infected with *Spinitectus carolini* Holl, 1928 (Spirurida: Nematoda). *Journal of Fish Diseases* 5, 75–77.

Kahl, W. (1938a) Nematoden in Seefischen. I. Erhebungen ueber die durch Larven von *Porrocaecum decipiens* Krabbe in Fischwirten hervorgerufenen geweblichen Veraenderungen und Kapselbildungen. *Zeitschrift für Parasitenkunde* 10, 415–431.

Kahl, W. (1938b) Nematoden in Seefischen. II. Erhebungen ueber den befall von Seefischen mit larven von *Anacanthocheilus rotundatus* (Rudolphi) und die durch diese Larven hervorgerufenen reaktionen des Wirtsgewebes. *Zeitschrift für Parasitenkunde* 10, 513–534.

Kennedy, C.R. (1974) A checklist of British and Irish freshwater fish parasites with notes of their distribution. *Journal of Fish Biology* 13, 255–263.

Kennedy, C.R. and Fitch, D.J. (1990) Colonization, larval survival and epidemiology of the nematode, *Anguillicola crassus*, parasitic in the eel, *Anguilla anguilla*, in Britain. *Journal of Fish Biology* 36, 117–131.

Køie, M. (1993) Aspects of the life cycle and morphology of *Hysterothylacium aduncum* (Rudolphi, 1802) (Nematoda, Ascaridoidea, Anisakidae). *Canadian Journal of Zoology* 71, 1289–1296.

Køie, M. and Fagerholm, H.-P. (1993) Third-stage larvae emerge from eggs of *Contracaecum osculatum* (Nematoda, Anisakidae). *Journal of Parasitology* 79, 777–780.

Koops, H. and Hartmann, F. (1989) *Anguillicola*-infestations in Germany and in German eel imports. *Journal of Applied Ichthyology* 1, 41–45.

Likely, C.G. and Burt, M.D.B. (1989) Cultivation of *Pseudoterranova decipiens* (sealworm) from third-stage larvae to egg-laying adults *in vitro*. *Canadian Journal of Fisheries and Aquatic Sciences* 46, 1095–1096.

Likely, C.G. and Burt, M.D.B. (1992) *In vitro* cultivation of *Contracaecum osculatum* (Nematoda: Anisakidae) from third stage larvae to egg-laying adults. *Canadian Journal of Fisheries and Aquatic Sciences* 49, 347–348.

Linnik, V. Ya, Beznos, T.V. and Shimko, V.V. (1989) [*Philometra lusiana* infection in carp; possibilities of using immunological diagnostic methods.] *Veterinarnaya Nauka - Proizvodstvu* 27, 110–112.

Lo Wing Yat, S. (1988) The control of *Capillaria* infections in *Symphysodon*. *Freshwater and Marine Aquarium* 11, 68–83.

Margolis, L. (1970) Nematode diseases of marine fishes. In: Snieszko, S.F. (ed.) *A Symposium on Diseases of Fishes and Shellfishes*. Special Publication No. 5. American Fisheries Society. pp. 190–208.

Margolis, L. (1977) Public health aspects of 'codworm' infection: A review. *Journal of the Fisheries Research Board of Canada* 34, 887–898.

Margolis, L. and Arthur, J.R. (1979) Synopsis of the parasites of fishes of Canada. *Bulletin of the Fisheries Research Board of Canada* 199, 1–269.

Margolis, L. and Butler, T.H. (1954) An unusual and heavy infection of a prawn, *Pandalus borealis* Kroyer, by a nematode, *Contracaecum* sp. *Journal of Parasitology* 40, 649–655.

Matiucci, S., Nascetti, G., Bullini, L., Orecchia, P. and Paggi. L. (1986) Genetic structure of *Anisakis physeteris*, and its differentiation from the *Anisakis simplex* complex (Ascaridida: Anisakidae). *Parasitology* 93, 383–387.

Matsuura, T., Sun, S. and Sugane, K. (1992) The identity of *Ansiakis* type II larvae with *Anisakis physeteris* confirmed by restriction fragment length polymorphism analysis of genomic DNA. *Journal of Helminthology* 66, 33–37.

Matthews, B.E. (1984) The source, release and specificity of proteolytic enzyme activity produced by *Anisakis simplex* larvae (Nematoda: Ascaridida) *in vitro*. *Journal of Helminthology* 58, 175–178.

McClelland, G. (1990) Larval sealworm (*Pseudoterranova decipiens*) infections in benthic macrofauna. In: Bowen, W.D. (ed.) *Population Biology of Sealworm* (Pseudoterranova decipiens) *in Relation to its Intermediate and Seal Hosts. Canadian Bulletin of Fisheries and Aquatic Sciences* 222, pp. 47–65.

McClelland, G. and Ronald, K. (1974) *In vitro* development of the nematode *Contracaecum osculatum* Rudolphi 1802 (Nematoda: Anisakinae). *Canadian Journal of Zoology* 52, 847–855.

McClelland, G., Misra, R.K. and Martell, D.J. (1985) Variations in abundance of larval anisakines, sealworm (*Pseudoterranova decipiens*) and related species in cod and flatfish. *Canadian Technical Report of Fisheries and Aquatic Sciences* 1392, 57 pp.

McMinn, H. (1990) Effects of the nematode parasite *Camallanus cotti* on sexual and non-sexual behaviour in the guppy (*Poecilia reticulata*). *American Zoologist* 30, 245–249.

Measures, L. (1988a) Revision of the genus *Eustrongylides* Jaegerskoeld, 1909 (Nematoda: Dioctophymatoidea) of piscivorous birds. *Canadian Journal of Zoology* 66, 885–895.

Measures, L. (1988b) The development of *Eustrongylides tubifex* (Nematoda: Dioctophymatoidea) in oligochaetes. *Journal of Parasitology* 74, 294–304.

Measures, L. (1988c) Epizootiology, pathology, and description of *Eustrongylides tubifex* (Nematoda: Dioctophymatoidea) in fish. *Canadian Journal of Zoology* 66, 2212–2222.

Mercer, J.G., Munn, A.E., Smith, J.W. and Rees, H.H. (1986) Cuticle production and ecdysis in larval marine ascaridoid nematodes *in vitro*. *Parasitology* 92, 711–720.

Miyazaki, T., Rogers, W.A. and Semmens, K.J. (1988) Gastro-intestinal histopathology of paddlefish *Polyodon spathula* (Walbaum), infected with larval *Hysterothylacium dollfusi* Schmidt, Leiby and Kritsky, 1974. *Journal of Fish Diseases* 11, 245–250.

Moeller, H. (1989) (ed.) *Nematode Problems in North Atlantic Fish. Report from a Workshop in Kiel, 3–4 April, 1989.* International Council for the Exploration of the Sea, Copenhagen. 58 pp.

Molnar, K. (1986) Solving parasite-related problems in cultured freshwater fish. In: Howell, M.J. (ed.) *Parasitology – Quo Vadit? Proceedings of the Sixth International Congress of Parasitology*, Australian Academy of Science, Canberra, pp. 319–326.

Molnar, K., Baska, F., Csaba, Gy., Glavits, R. and Szekely, C.S. (1993) Pathological and histopathological studies of the swimbladder of eels *Anguilla anguilla* infected by *Anguillicola crassus* (Nematoda: Dracunculoidea). *Diseases of Aquatic Organisms* 15, 41–50.

Molnar, K., Chan, G.L. and Fernando, C.H. (1982) Some remarks on the occurrence and development of philometrid nematodes infecting the white sucker, *Catostomus commersoni* Lacepede (Pisces: Catostomidae), in Ontario. *Canadian Journal of Zoology* 60, 443–451.

Moravec, F. (1970) Studies on the development of *Raphidascaris acus* (Bloch, 1779) (Nematoda: Heterocheilidae). *Vestnik Ceskoslovenské Společnosti Zoologické* 34, 33–49.

Moravec, F. (1979) Observations on the developemnt of *Cucullanus (Truttaedacnitis) truttae* (Fabricius, 1794) (Nematoda: Cucullanidae). *Folia Parasitologica* 26, 295–307.

Moravec, F. (1983) Taxonomic problems in capillarid nematodes parasitic in cold-blooded vertebrates. In: Stone, A.R., Platt, H.M. and Khalil, L.F. (eds) *Concepts in Nematode Systematics*. The Systematics Association Special Volume No. 22. Academic Press, London, pp. 361–373.

Moravec, F. and Taraschewski, H. (1988) Revision of the genus *Anguillicola* Yamaguti, 1935 (Nematoda: Anguillicolidae) of the swimbladder of eels, including descriptions of two new species, *A. novazealandiae* sp.n. and *A. papernai* sp.n. *Folia Parasitologica* 35, 125–146.

Mozgovoi, A.A. (1953) [Ascaridata of animals and man and the diseases caused by them] In: Skrjabin, K.I. (ed.) *Principles of Nematology Volume II, Parts 1 and 2*. Izdateltstvo Akademii Nauk SSSR.

Nagasawa, K. (1989) Notes on parasites of aquatic organisms – 4. Visceral adhesions of high-seas sockeye salmon caused by the nematode *Philonema oncorhynchi*. *Aquabiology* 11, 320–321.

Nascetti, G., Paggi, L., Orecchia, P., Smith, J.W., Matiucci, S. and Bullini, L. (1986) Electrophoretic studies on the *Anisakis simplex* complex (Ascaridida: Anisakidae) from the Mediterranean and North-east Atlantic. *International Journal for Parasitology* 16, 633–640.

Orecchia, P., Paggi, L., Mattiucci, S., Smith, J.W., Nascetti, G. and Bullini, L. (1986) Electrophoretic identification of larvae and adults of *Anisakis* (Ascaridida: Anisakidae). *Journal of Helminthology* 60, 331–339.

Osche, G. (1963) Morphological, biological and ecological considerations in the phylogeny of parasitic nematodes. In: Dougherty, E.C., Brown, Z.N., Hanson, E.D. and Hertman, W.D. (eds) *The Lower Metazoa: Comparative Biology and Phylogeny*. University of California Press, Berkeley, pp. 283–302.

Paggi, L., Nascetti, G., Cianchi, R., Orecchia, P., Mattiucci, S., D'Amello, S., Berland, B., Brattey, J., Smith, J.W. and Bullini, L. (1991) Genetic evidence for three species within *Pseudoterranova decipiens* (Nematoda, Ascaridida, Ascaridoidea) in the North Atlantic and Norwegian and Barents seas. *International Journal for Parasitology* 21, 195–212.

Paperna, I. (1991) Hosts, distribution and pathology of infections with larvae of *Eustrongylides* (Dioctophymidae, Nematoda) in fishes from East African lakes. *Journal of Fish Biology* 6, 67–76.

Paperna, I. (1986) Solving parasite-related problems in cultured marine fish. In: Howell, M.J. (ed.) *Parasitology – Quo Vadit? Proceedings of the Sixth International Congress of Parasitology*. Australian Academy of Sciences, Canberra. pp. 327–336.

Paperna, I. (1991) Diseases caused by parasites in the aquaculture of warm water fish. *Annual Review of Fish Diseases* 1991, 155–194.

Pena, N., Auro, A. and Sumando, H. (1988) A comparative trial of garlic, its extract and ammonium-potassium tartrate as anthelmintics in carp. *Journal of Ethnopharmacology* 24, 199–203.

Peters, G. and Hartmann, F. (1986) *Anguillicola*, a parasitic nematode of the swim bladder spreading among eel populations in Europe. *Diseases of Aquatic Organisms* 1, 229–230.

Petter, A.J. (1974) Essai de classification de la familie des Cucullanidae. *Bulletin du Museum National d'Histoire Naturelle, Paris, 3e Serie, no. 255, Zoologie* 177, 1469-1490.

Petter, A.J., Fontaine, Y.A. and Le Belle, N. (1989) Etude du developpement larvaire de *Anguillicola crassus* (Dracunculoidea, Nematoda) chez un Cyclopidae de la region Parisienne. *Annales de Parasitologie Humaine et Comparée* 64, 345-355.

Piatt, J.F., Methven, D.A., Burger, A.E., McLagan, R.L., Mercer, V. and Creelman, E. (1989) Baleen whales and their prey in a coastal environment. *Canadian Journal of Zoology* 67, 1523-1530.

Poole, B.C. and Dick, T.A. (1984) Liver pathology of yellow perch, *Perca flavescens* (Mitchill), infected with larvae of the nematode *Raphidascaris acus* (Bloch, 1779). *Journal of Wildlife Diseases* 20, 303-307.

Priebe, K., Huber, C., Maertlbauer, E. and Terplan, G. (1991) Nachweis von Antikoerpern gegen Larven von *Anisakis simplex* beim Seelachs *Polachius virens* mittels ELISA. *Journal of Veterinary Medicine Series B* 38, 209-214.

Pybus, M.J., Anderson, R.C. and Uhazy, L.S. (1978a) Redescription of *Truttaedacnitis stelmioides* (Vessichelli, 1910) (Nematoda: Cucullanidae) from *Lampetra lamottenii* (Lesueur, 1827). *Proceedings of the Helminthological Society of Washington* 45, 238-245.

Pybus, M.J., Uhazy, L.S. and Anderson, R.C. (1978b) Life cycle of *Truttaedacnitis stelmioides* (Vesichelli, 1910) (Nematoda: Cucullanidae) in American brook lamprey (*Lampetra lamottenii*). *Canadian Journal of Zoology* 56, 1420-1429.

Ramakrishna, N.R. and Burt, M.D.B. (1991) Tissue response of fish to invasion by larval *Pseudoterranova decipiens* (Nematoda; Ascaridoidea). *Canadian Journal of Fisheries and Aquatic Sciences* 48, 1623-1628.

Ramakrishna, N.R. and Burt, M.D.B. (1993) Cell-mediated immune response of rainbow trout (*Oncorhynchus mykiss*) to larval *Pseudoterranova decipiens* (Nematoda; Ascaridoidea) following sensitization to live sealworm, sealworm extract, and nonhomologous extracts. *Canadian Journal of Fisheries and Aquatic Sciences* 50, 60-65.

Rogers, W.A. (1976) Helminthic diseases of North American freshwater fishes. In: Page, L.A. (ed.) *Wildlife Diseases*. Publisher, Place, pp. 153-162.

Rosa-Molinar, E. and Williams, C.S. (1983) Larval nematodes (Philometridae) in granulomas in ovaries of black-tip sharks, *Carcharhinus limbatus* (Valenciennes). *Journal of Wildlife Diseases* 19, 275-277.

Russell, L.R. (1979) Effects of *Truttaedacnitis truttae* (Nematoda: Cucullanidae) on growth and swimming of rainbow trout, *Salmo gairdneri*. *Canadian Journal of Zoology* 58, 1220-1226.

Schaeperclaus, W. (1992) *Fish Diseases, Volumes 1 and 2*. A.A. Balkema. Rotterdam, 1398 pp.

Schuurmans-Stekhoven, J.H. Jr. and Botman, T.P.J. (1932) Zur Ernaehrungsbiologie von *Proleptus obtusus* Duj. und die von diesem Parasiten hervorgerufenen reaktiven Aenderungen des Wirtsgewebes. *Zeitschrift für Parasitenkunde* 4, 220-239.

Sindermann, C.J. (1990) *Principal Diseases of Marine Fish and Shellfish. Volume 1. 2nd Edition*. Academic Press, London, 521 pp.

Sinha, A.K. and Sinha, C. (1988) Macrocytic hypochromic anaemia in *Heteropneustes fossilis* (Bl.) infected by the blood sucker nematode *Procamallanus spiculoguber-naculus* (Agarwal). *Indian Journal of Parasitology* 12, 93-94.

Skrjabin, K.I., Sobolev, A.A. and Ivashkin, V.M. (1967) *Osnovi Nematodologii. Volume XVI. Spirurata of Animals and Man and the Diseases Caused by them.*

Part 4. Thelazioidea. Izdatel'stvo 'Nauka', Moscow. [Israel Program for Scientific Translations, Jerusalem, 1971, 624 pp.]

Smith, J.D. (1984) Development of *Raphidascaris acus* (Nematoda, Anisakidae) in paratenic, intermediate and definitive hosts. *Canadian Journal of Zoology* 62, 1378–1386.

Smith, J.W. (1983) *Anisakis simplex* (Rudolphi, 1809, det. Krabbe, 1878) (Nematoda: Ascaridoidea): Morphology and morphometry of larvae from euphausiids and fish, and a review of the life history and ecology. *Journal of Helminthology* 57, 205–224.

Smith, J.W. and Wootten, R. (1978) *Anisakis* and anisakiasis. *Advances in Parasitology* 16, 93–163.

Smith, J.W., Elarifi, A.E., Wootten, R., Pike, A.W. and Burt, M.D.B. (1990) Experimental infection of rainbow trout, *Oncorhynchus mykiss*, with *Contracaecum osculatum* (Rudolphi, 1802) and *Pseudoterranova decipiens* (Krabbe, 1878) (Nematoda: Ascaridoidea). *Canadian Journal of Fisheries and Aquatic Sciences* 47, 2293–2296.

Sood, M.L. (1988) *Fish Nematodes for South Asia.* Kalyani Publishers, New Delhi.

Specian, R.D., Ubelaker, J.E. and Dailey, M.D. (1975) *Neoleptus* gen. n. and a revision of the genus *Proleptus* Dujardin, 1845. *Proceedings of the Helminthological Society of Washington* 42, 14–21.

Sprengel, G. and Luechtenberg, H. (1991) Infection by endoparasites reduces maximum swimming speed of European smelt *Osmerus eperlanus* and European eel *Anguilla anguilla. Diseases of Aquatic Organisms* 11, 31–35.

Sprent, J.F.A. (1982) Host-parasite relationships of ascaridoid nematodes and their vertebrate hosts in time and space. *Memoires du Museum d'Histoire Naturelle, Serie A, Zoologie* 128, 255–263.

Sprent, J.F.A. (1983) Observations on the systematics of ascaridoid nematodes. In: Stone, A.R., Platt, H.M. and Khalil, L.F. (eds) *Concepts in Nematode Systematics.* Academic Press, London, pp. 303–320.

Sugane, K., Quing, L. and Matsuura, T. (1989) Restriction fragment length polymorphism of Anisakinae larvae. *Journal of Helminthology* 63, 269–274.

Szalai, A.J. and Dick, T.A. (1991) Role of predation and parasitism in growth and mortality of yellow perch in Dauphin Lake, Manitoba. *Transactions of the Fisheries Society* 120, 739–751.

Taraschewski, H., Moravec, F., Lamah, T. and Anders, K. (1987) Distribution and morphology of two helminths recently introduced into European eel populations: *Anguillicola crassus* (Nematoda, Dracunculoidea) and *Paratenuisentis ambiguus* (Acanthocephala, Tenuisentidae). *Diseases of Aquatic Organisms* 3, 167–176.

Taraschewski, H., Renner, C. and Melhorn, H. (1988) Treatment of fish parasites. 3. Effects of levamisole-HCl, metrifonate, febendazole, mebendazole, and ivermectin on *Anguillicola crassus* (nematodes) pathogenic in the air bladder of eels. *Parasitology Research* 74, 281–289.

Taylor, A.E.R. and Baker, J.R. (1968) *Methods of Cultivating Parasites in vitro.* Academic Press, London.

Torres, P. and Alvarez-Pellitero, M-P. (1988) Observaciones sobre el ciclo vital de *Raphidascaris acus* (Nematoda, Ansiakidae): desarrollo *in vitro* de los estadios iniciales e infestaciones experimentales de macroinvertebrados y *Salmo gairdneri. Revista Iberica de Parasitologia* 48, 41–50.

Townsley, P.M., Wight, H.G., Scott, M.A. and Hughes, M.L. (1963) The *in vitro* maturation of the parasite nematode, *Terranova decipiens*, from cod muscle. *Journal of the Fisheries Research Board of Canada* 20, 743–747.

Weerasooriya, M.V., Fujino, T., Ishii, Y. and Kagei, N. (1986) The value of external morphology in the identification of larval anisakid nematodes: a scanning electron microscope study. *Zeitschrift für Parasitenkunde* 72, 765–778.

Wilers, W.B., Dubielzig, R.R. and Miller, L. (1991) Histopathology of the swim bladder of the cisco due to the presence of the nematode *Cystidicola farionis* Fischer. *Journal of Aquatic Animal Health* 3, 130–133.

Williams, H.H. (1967) Helminth diseases of fish. *Helminthological Abstracts* 36, 261–295.

Valter, E.D., Popova, T.I. and M.A. Valovaya (1982) Scanning electron microscope study of four species of anisakid larvae (Nematoda: Anisakidae). *Helminthologia* 19, 195–209.

Van Banning, P. and Haenen, O.L.M. (1990) Effects of the swimbladder nematode *Anguillicola crassus* in wild and farmed eel, *Anguilla anguilla*. In: *Pathology in Marine Science*. Academic Press, London, pp. 317–330.

Yoshinaga, T., Ogawa, K. and Wakabayashi, H. (1987) Experimental life cycle of *Hysterothylacium aduncum* (Nematoda: Anisakidae) in fresh water. *Fish Pathology* 22, 243–251.

12

Phylum Acanthocephala

B.B. Nickol

School of Biological Sciences, University of Nebraska-Lincoln, Lincoln, Nebraska 68588–0118, USA.

INTRODUCTION

Acanthocephalans are endoparasitic worms (approximately 1000 species), and about one-third of the species are found as adults in the intestine of fishes. Juvenile worms of many other species occur in the viscera, especially the mesentery and liver, of fishes that act as paratenic hosts. Epizootics are known from hatcheries (Bullock, 1963) and have been linked to local extinction of natural populations (Schmidt et al., 1974). There is little knowledge of the more subtle effects the parasite has on host populations.

HOST AND GEOGRAPHICAL DISTRIBUTION

Freshwater and marine fishes from the Arctic (Van Cleave, 1920) to the Antarctic (Zdzitowiecki, 1986) and from surface waters to the depths of the ocean (Noble, 1973; Paggi et al., 1975; Gartner and Zwerner, 1989) harbour acanthocephalans. Fishes of most systematic groups are parasitized. Although these parasites are rare in agnathans and elasmobranchs, a few species are known to occur in lampreys (Hoffman, 1967), rays (McVicar, 1977; Buckner et al., 1978), and sharks (Yamaguti, 1961). Sturgeons, primitive ray-fin fishes (Chondrostei), are parasitized by six species (Yamaguti, 1961; Hoffman, 1967; Golvan and de Buron, 1988) and bowfins and gars, intermediate ray-fin fishes, harbour another few species (Hoffman, 1967; Golvan and de Buron, 1988). Most of the acanthocephalans of piscine hosts are found in the bony fishes (Teleostei). More species of Acanthocephala are found in the cypriniform families Cyprinidae and Catostomidae than in any other piscine family.

SEASONAL FLUCTUATIONS

A review of seasonal occurrence of acanthocephalan species in piscine hosts (Chubb, 1982) reveals many differences among species and geographical localities. Kennedy (1972) found that feeding response was an important factor in controlling the intensity of *Pomphorhynchus laevis* in dace (*Leuciscus leuciscus*) and that water temperature influenced this response. Unless negated by another factor, such as temperature-dependent rejection (Kennedy, 1972), prevalence and intensity should be greatest during times when intermediate hosts with infective larvae constitute an appropriate portion of the diet of definitive hosts. When this occurs throughout the year, there may be no seasonal periodicity in the occurrence of acanthocephalans, For example, *Echinorhynchus salmonis* shows no seasonal periodicity in prevalence, intensity, or development and maturation in rainbow smelt (*Osmerus mordax*) from Lake Michigan where the amphipod intermediate host is available to fish throughout the year (Amin, 1981).

Absence of seasonal periodicity does not imply that rates of recruitment and mortality are constant throughout the year. High water temperatures (under laboratory conditions) reduce the success with which *Pomphorhynchus laevis* establishes in goldfish (*Carassius auratus*) hence Kennedy (1972) suggested that increased feeding by fish in the summer balanced a lower rate of parasite establishment. Thus, seasonal differences in rates of recruitment and mortality (from failure to establish in fish) could occur even if parasite occurence shows no periodicity.

There are instances in which appropriate invertebrate intermediate hosts are consumed by definitive hosts throughout the year, but there are definite seasonal cycles in infections. Awachie (1965) demonstrated that, even though infective larvae of *Echinorhynchus truttae* occur in *Gammarus pulex* and that the amphipods are an important part of the food of brown trout (*Salmo trutta*) all year, seasonal differences in densities of infective larvae lead to fluctuating occurrence in fish. Variations in habits of hosts during the year, for instance changes in diet, also cause seasonal differences in prevalence and intensity of infections (Halvorsen, 1972).

Kennedy (1970) suggested that, when maturation of helminths from freshwater fishes shows seasonal cycles, peak egg production almost always occurs late in spring or early in summer. This generalization applies to many, but not all, species of Acanthocephala. Annual cycles in fishes frequently begin with recruitment in spring, and most transmission between fish and invertebrate hosts occurs during summer and early autumn. Development in intermediate hosts infected late in the season is retarded by low temperatures (DeGiusti, 1949; Awachie, 1966) until spring. The ability to resume or accelerate development is not always related to temperature alone, and development may be faster after longer periods below a temperature threshold. This prevents development to the infective larval stage during sporadic intervals of warm weather in early autumn (Tokeson and Holmes, 1982).

Some acanthocephalans show adaptations directly related to seasonal events in the life histories of fishes. Peak egg production by by *Fessisentis friedi* apparently occurs as pickerel (*Esox americanus*) move into shallow water to spawn (Muzzall, 1978). Maximum intensity and egg production of *Pomphorhynchus bulbocolli* in white suckers (*Catostomus commersoni*) have also been related to migration and spawning (Muzzall, 1980). In Lake Michigan, *Echinorhynchus salmonis* reaches peak sexual maturity in chinook salmon (*Oncorhynchus tshawytscha*) during spawning (Amin, 1978). These seasonal patterns result in peak egg production by the acanthocephalans when fishes are in shallow waters where appropriate invertebrate intermediate hosts are most abundant.

A common type of cycle consists of spring recruitment by fishes after the environment has warmed and fish and invertebrate hosts have become active, summer transmission between fishes and invertebrates, slow development in invertebrates during the winter, and completion of larval development with rising temperatures in the spring.

Deviations from the spring 'recruitment cycle' often are related to deviations in fish–invertebrate relationships and environmental disturbances. Prevalence of *Echinorhynchus salmonis* in yellow perch (*Perca flavescens*) in Lake Ontario rises in the autumn after which it declines until the parasite is absent from fish during the summer (Tedla and Fernando, 1970). Watson and Dick (1979) also found that *E. salmonis* was most common in whitefish (*Coregonus clupeaformis*) from Southern Indian Lake, Manitoba, in late autumn when the amphipod intermediate host was a prominent whitefish food item.

In instances where invetebrates are rare or absent during winter, the cycle may be altered. This results in recruitment to the definitive host population in the autumn. Maturation then is slow with egg production occurring in the spring. Eure (1976) found this for *Neoechinorhynchus cylindratus* in a heated reservoir in South Carolina. Recruitment into largemouth bass (*Micropterus salmoides*) occurs in the autumn while ostracod numbers decrease. By November, ostracods are scarce and most acanthocephalans are in bass. Egg production is delayed until spring when ostracods again are abundant.

Environmental disturbances might not affect host relationships of all parasites similarly. Contrary to Eure's (1976) findings, Boxrucker (1979) detected no seasonal difference in prevalence or intensity of *Pomphorhynchus bulbocolli* in a thermally heated area in Laka Monona, Wisconsin, but instead he found that, in an unheated area, this species displayed the familiar pattern of increasing prevalence and intensity during the spring followed by declining numbers in the autumn. Elevated winter temperatures may have resulted in more rapid parasite development in intermediate hosts and fish also feed more extensively on them than during winter in unheated environments.

SYSTEMATICS AND ORIGIN

Some zoologists regard the invaginable proboscis, general body form, and the pseudocoelom as evidence of affinity to the nematodes, priapulids and kinorhynchs (Meyer, 1933; Morris and Crompton, 1982). Van Cleave (1941) emphasized that the life cycle with no free-living stage points to an ancestral group that was already parasitic. He considered embryology, the types of intermediate hosts, and anatomical features to indicate a close relationship with cestodes.

The phylum comprises four classes totaling nine orders (Table 12.1). Members of the class Archiacanthocephala do not occur in fish. The entire class Polyacanthocephala contains only four species, primarily parasites of South American caimans. However, at least two of the species occur as adults in fishes (Machado, 1947) and another as larvae in the liver of fishes that apparently act as paratenic hosts (Schmidt and Canaris, 1967). Nearly all of the species found in fishes belong to the classes Eoacanthocephala and Palaeacanthocephala. With the exception of eight species found in North American turtles, all eoacanthocephalans are parasites of fishes. Among the palaeacanthocephalans, all of order Echinorhynchida are parasites of fishes except for a few species that occur as adults in amphibians. Acanthocephalans in the other palaeacanthocephalan order, Polymorphida, occur as adults in birds and mammals, but larvae of many species are found in viscera of fishes that act as paratenic hosts.

MORPHOLOGY

Acanthocephalans are bilaterally symmetrical, dioecious, pseudocoelomate worms that lack an alimentary canal. They are characterized by a spined proboscis that is invaginable and retractable into a saccular proboscis receptacle (Fig. 12.1). Body length of adults varies greatly among species, ranging from less than 2 mm to greater than 700 mm. Acanthocephalans of most species are about 10 mm long.

Attachment to the host intestine is by means of a proboscis (Fig. 12.2). Hooks and spines on the proboscis are arranged in a variety of patterns that serve as important taxonomic characters. The neck (Fig. 12.1) is an unarmed region immediately behind the proboscis, bounded anteriorly by the proximalmost proboscis hooks and posteriorly by an invagination of body wall (Van Cleave, 1947). The neck usually is short and inconspicuous, but in some genera it is extremely elongated and may be inflated to form a bulb. The remainder of the body is a trunk that, in some species, bears tegumental spines. As all ontogenetic stages lack alimentary canals, a genital pore is the only opening visible by light microscopy. The genital pore occurs posteriorly on the trunk, and in males it is surrounded by an invaginable bursa (Fig. 12.1).

Aside from the anterior lemnisci and proboscis receptacle, the only conspicuous internal organs are those of the reproductive system. In males there are two testes followed posteriorly by 1–8 cement glands (Fig. 12.1). In members of the class Eoacanthocephala there is a separate cement reservoir, whereas in those of other classes ducts from cement glands may be dilated to store secretions of the cement glands. Efferent ducts from the testes and cement glands, or cement reservoir, join posteriorly and discharge through a small copulatory organ.

Acanthocephalan ovaries fragment early in development to form ovarian balls that float free in the pseudocoelom. As development proceeds, the pseudocoelom fills with eggs that have four membranes, which surround the developing larva (acanthor). Eggs are passed through a funnel-shaped uterine bell to a selector apparatus that allows only fully formed eggs into the uterus (Fig. 12.1). Other eggs are directed back to the pseudocoelom through a series of canals and pores. The female system terminates in a vagina, regulated by a vaginal sphincter, that discharges through a genital pore.

The body is covered by a thin glycocalyx (Beermann *et al.*, 1974) beneath which is a bilayered plasma membrane. Numerous foldings of the plasma membrane form canals (about 15 nm in diameter) in the outer layer of the tegument (Stranack *et al.*, 1966). The fibrous syncytial tegument, sometimes called hypodermis, comprises three layers: an outer striped layer pierced by numerous crypts formed by the infolded plasma membrane (Fig. 12.3), a middle felt layer that contains large concentrations of glycogen (von Brand, 1939), and an inner radial layer that contains the tegumental nuclei and a network of unlined channels that form the lacunar system. In the neck, the radial layer extends into the pseudocoelom as a pair of projections called lemnisci (Fig. 12.1).

Beneath the tegument, the body wall consists of an outer layer of circular and an inner layer of longitudinal muscles, a series of tubular canals that may have a circulatory function, and the rete system, which is associated with the body wall musculature (Miller and Dunagan, 1976).

The nervous system consists of a cerebral ganglion located in the proboscis receptacle, nerves branching from the cerebral ganglion, and, in males, a pair of genital ganglia and a bursal ganglion. Sensory receptors are simple and inconspicuous. An apical sense organ occurs at the tip of the proboscis. The apical sense organ varies in form, from a papilla to a depression surrounded by a small elevated rim (Miller and Dunagan, 1985). Paired lateral sense organs appear as depressions on the neck, but not all species possess apical or lateral sense organs. Males possess sense organs on the bursa.

Most acanthocephalans, including all of those found in fishes, do not possess an excretory system. When an excretory system is present, the protonephridial organs discharge through ducts that enter the vas deferens or uterus.

LIFE CYCLE AND TRANSMISSION

Acanthocephalans require vertebrate animals as definitive hosts and arthropods as intermediate hosts. Those in terrestrial definitive hosts usually have insect intermediate hosts whereas those of aquatic definitive hosts usually require microcrustaceans (amphipods, copepods, isopods, or ostracods). All known acanthocephalan intermediate hosts were listed by Schmidt (1985).

Intermediate hosts are infected by ingestion of eggs voided with the faeces of definitive hosts. Eggs hatch in the alimentary tract of the intermediate host. The acanthor, armed with an aclid organ, penetrates the gut and continues to develop through several ontogenetic stages (known as acanthellae) in the

Table 12.1. Characterization of the classes and orders of the phylum Acanthocephala.

Taxon	Key recognition features [a]	Principal definitive hosts	Principal intermediate hosts [b]
Class Polyacanthocephala	Main lacunar canals dorsal and ventral; cement glands 8, elongate with multiple large nuclei; proboscis receptacle closed sac with single muscle layer; tegumental nuclei numerous, amitotic fragments	Crocodilians	Unknown
Order Polyacanthorhynchida	Only order of the class		
Class Archiacanthocephala	Main lacunar canals dorsal and ventral; cement glands 8, spherical with single nucleus; proboscis receptacle single walled, often with ventral cleft; tegumental nuclei few, elongate or branched, if fragmented, fragments close together	Birds and mammals	Terrestrial insects
Order Apororhynchida	Proboscis globular, not retractable, spineless or with rootless spines usually not reaching surface	Birds	Unknown
Order Gigantorhynchida	Proboscis conical, retractable, with distinct regions of large anterior hooks and small posterior spines	Birds and mammals	Orthoptera, Coleoptera
Order Moniliformida	Proboscis cylindrical, retractable, with strongly recurved hooks diminishing in size anterior to posterior and arranged in longitudinal rows	Mammals	Cockroaches
Order Oligacanthorhynchida	Proboscis spherical, retractable, with few large hooks arranged in spirals	Birds and mammals	Orthoptera, Coleoptera

Taxon	Characteristics	Definitive hosts	Intermediate hosts
Class Palaeacanthocephala	Main lacunar canals lateral; cement glands 2–8, with fragmented nuclei; proboscis receptacle closed sac with 2 muscle layers; tegumental nuclei numerous amitotic fragments or few highly branched fragments	Fishes, amphibians, birds and mammals	Crustacea
Order Echinorhynchida	Palaeacanthocephala of lower vertebrates	Fishes and amphibians	Amphipods, isopods
Order Polymorphida	Palaeacanthocephala of higher vertebrates	Birds and mammals	Amphipods, isopods, decapods
Class Eoacanthocephala	Main lacunar canals dorsal and ventral; cement gland single syncytium with large nuclei; proboscis receptacle closed sac with single muscle layer; tegumental nuclei few, large	Fishes	Crustacea
Order Gyracanthocephala	Eoacanthocephalans with trunk spines	Fishes	Copepods
Order Neoechinorhynchida	Eoacanthocephala without trunk spines	Fishes	Ostracods

[a] For a complete key to classes, orders, families, and subfamilies, see Amin (1987).
[b] For a complete list of known intermediate hosts, see Schmidt (1985).

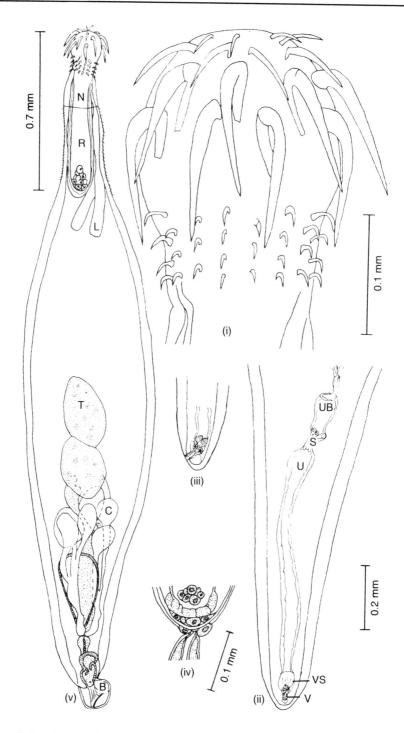

Fig. 12.1. Camera lucida drawings of *Arhythmacanthus paraplagusiarum* from the tonguefish *Paraplagusia guttata*. (i). Proboscis. (ii). Ventral view of posterior extremity of female. (iii). Lateral view of posterior extremity of female. (iv). Posterior tip of proboscis receptacle. (v). Male, entire. B = bursa; C = cement

serosa. During these stages, primordia of adult organs are laid down. Late in acanthella development the proboscis invaginates, and all structures of adult worms are present. This juvenile, or cystacanth, is the infective stage for a definitive host. Details of larval development were described by Ward (1940) for *Neoechinorhynchus cylindratus*, a common acanthocephalan of basses, and by DeGiusti (1949) for *Leptorhynchoides thecatus*, a species reported from at least 122 piscine species.

Ingestion of a cystacanth-containing intermediate host results in infection of a definitive host. Paratenic hosts occur in the life cycle of some species. In paratenic hosts, cystacanths from an intermediate host penetrate the intestinal wall and localize in mesenteries or visceral organs, frequently the liver, but maturity does not occur until the paratenic host is consumed by an appropriate definitive host.

Acanthocephalans of many species are known to alter behaviour of intermediate hosts in a manner that could increase likelihood of transmission to definitive hosts. Isopods (*Asellus intermedius, Lirceus lineatus*) infected with *Acanthocephalus dirus*, parasitic in a variety of fishes, spend significantly less time under leaves and more time 'wandering' than do uninfected conspecifics (Muzzall and Rabalais, 1975b; Camp and Huizinga, 1979), and ostracods (*Cypridopsis vidua*) infected with *Octospiniferoides chandleri*, parasitic in mosquito fish (*Gambusia affinis*) form aggregations milling about at the water surface whereas uninfected ones concentrate at the bottom (DeMont and Corkum, 1982). Amphipods (*Gammarus pulex*) infected with *Pomphorhynchus laevis*, parasitic in several piscine species, display a higher degree of positive phototrophism than do uninfected conspecifics, spend less time resting on the substrate, move more often toward the surface, and rest more on surface vegetation (Kennedy *et al.*, 1978).

Although the degree to which behavioural modifications actually promote transmission is largely untested experimentally, there are examples where this has been demonstrated, e.g. infected amphipods (*Gammarus pulex* infected with *Pomphorhynchus laevis*) are eaten by fish in significantly greater proportions than are uninfected individuals, even when infected amphipods constitute a very small fraction of the amphipod population (Kennedy *et al.*, 1978).

HOST–PARASITE RELATIONSHIPS

Site selection

After ingestion by the definitive host, the cystacanth everts its proboscis and attaches to the small intestine. Many species apparently remain at the activation site and develop to maturity. Others migrate and localize in specific

gland; L = lemniscus; N = neck; R = proboscis receptacle; S = selector apparatus; T = anterior testis; U = uterus; UB = uterine bell; V = vagina; VS = vaginal sphincter. (From Nickol, 1972.)

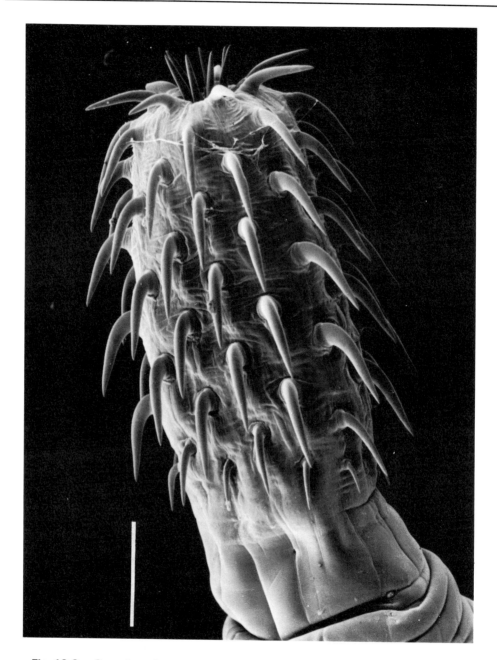

Fig. 12.2. Scanning electron micrograph of the proboscis and neck of *Acanthocephalus lucii*, a parasite of European fishes of several species. Scale bar = 0.2 mm. (Unpublished micrograph courtesy of D.W.T. Crompton and O.L. Lassière.)

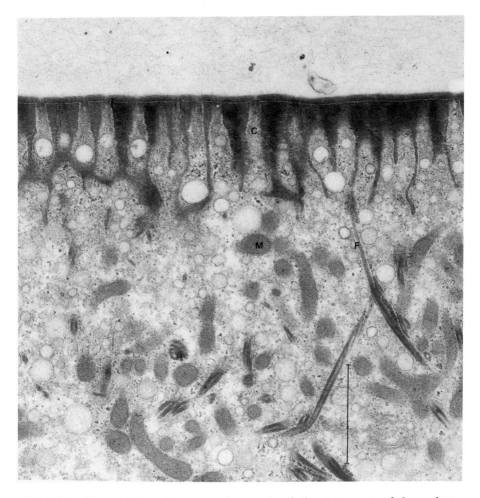

Fig. 12.3. Transmission electron micrograph of the tegument of *Leptorhyn-choides thecatus* from a green sunfish (*Lepomis cyanellus*) showing crypts (C), fibres (F) of the felt layer, and mitochondria (M). Scale bar = 5 μm. (Unpublished micrograph courtesy of J.A. Ewald.)

regions of the alimentary tract. In most cases it is unknown whether differences in sites occupied result from differences in sites of activation, differential mortality, or emigration by the parasite. It was suggested that localization of *Leptorhynchoides thecatus* in the caeca of green sunfish (*Lepomis cyanellus*) results from differential mortality (Uznanski and Nickol, 1982), whereas *Acanthocephalus dirus* (=*Acanthocephalus jacksoni*) moves posteriorly with maturation in several species of trout (Bullock, 1963). Crompton (1975) summarized sites of Acanthocephala in the alimentary tract of vertebrates, and Kennedy (1985) studied site segregation of the Acanthocephala from all but one species of the British freshwater fishes.

Course of infection

Copulation in the definitive host may occur within 24 hours of infection (Muzzall and Rabalais, 1975a). For most species, egg production starts between four and eight weeks after infection, and it continues for approximately two months. The number of eggs produced is related to size of the worm. At peak production, a female *Macracanthorhynchus hirudinaceus* (large worms found in swine) may produce about 260,000 eggs per day (Kates, 1944) and a *Polymorphus minutus* (small worms parasitic in waterfowl) produces about 1700 eggs per day (Crompton and Whitfield, 1968).

Male worms have shorter life spans than do females; death of males and subsequent loss from the host may begin shortly after copulation. Although females of some species, especially those parasitic in mammals, may live longer than a year, most in poikilothermic hosts live about one season, but there may be more than one generation each year.

Clinical signs and gross pathology

Acanthocephalans in moribund or dead animals are frequently assumed to indicate deleterious effects; worms have been observed protruding from the rectum of infected fish (Schmidt *et al.*, 1974) (Fig. 12.4). There are, however, numerous instances of exceedingly heavy infections in animals that do not show any obvious disease.

Bullock (1963) described signs of malnutrition, and these included underdeveloped musculature, heads that appeared disproportionately large, and a slightly concave dorsolateral surface, in several species of trout infected with *Acanthocephalus dirus*.

The most obvious signs of infection are fibrotic nodules produced by some species on the surface of the intestine (Fig. 12.5). At times the viscera may be discolored or the infected intestine enlarged and inflamed. Pyloric caeca of the green sunfish (*Lepomis cyanellus*) infected with *Leptorhynchoides thecatus* are nearly twice the diameter of those in uninfected fish of the same size (de Buron and Nickol, 1994) (Fig. 12.6). Acanthocephalans occasionally perforate the intestinal wall and protrude into the coelom or attach their proboscis to another organ.

Histopathology

Acanthocephalans embed their spiny proboscis into the mucosal epithelium and frequently between villi. At the site of attachment cells are destroyed, and fibroblasts, lymphocytes, and macrophages are mobilized below the lamina propria (Dezfuli *et al.*, 1990) where chronic fibrinous inflammation, resulting in an increased amount of connective tissue, causes a thickening (Bullock, 1963). In some species, fibroplasia extends to layers of the muscularis mucosa (de Buron and Nickol, 1994).

Fig. 12.4. *Acanthocephalus dirus* attached to the prolapsed rectum of a mottled sculpin (*Cottus bairdi*). Photograph = actual size. (From Schmidt *et al.*, 1974.)

Worms of the genus *Pomphorhynchus* possess very long necks and bulbous proboscides that anchor them deep within the gut wall of their piscine hosts. Dezfuli (1991) described the cellular structure of a fibrotic tunnel that forms around the neck and proboscis bulb of *Pomphorhynchus laevis*. The tunnel terminates in a capsule, covered by serosa and mesentery, that protrudes several millimetres into the coelomic cavity (Fig. 12.7). Sometimes these capsules persist as conspicuous fibrinous nodules on the external surface of the alimentary canal, or the proboscis perforates the capsule to emerge free in the coelom or to penetrate the liver. Similar findings were reported for *Pomphorhynchus bulbocolli* in rainbow darters (*Etheostoma caeruleum*) (McDonough and Gleason, 1981) and two species of catostomids (Chaicharn and Bullock, 1967). *Acanthocephalus anguillae* is also known to perforate the intestine of its host and attach to the liver. In goldfish (*Carassius auratus*) large portions of this organ are replaced by proliferative tissue, often with patches of pancreas, surrounding the embedded proboscis (Taraschewski, 1989a).

Most studies of acanthocephalan-induced lesions reveal areas along the trunk of the worm where mucosal epithelium is compressed or eroded

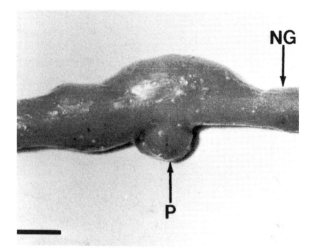

Fig. 12.5. External view of intestine from a quillback (*Carpiodes cyprinus*) showing parasite-induced nodule (P) and expansion of the intestine compared to the normal gut (NG) from a fish infected with *Neoechinorhynchus carpiodi*. Scale bar = 5 mm. (From Szalai and Dick, 1987.)

(Fig. 12.6). In these cases, mucosal folds and tips of villi may be absent and the paramucosal lumen contains large amounts of mucoid material. Because the mucoid material gave the same histochemical reactions as goblet cells, Bullock (1963) concluded that goblet cells are the major source of the mucoid material. In green sunfish (*Lepomis cyanellus*) infected with *Leptorhynchoides thecatus*, there is a significantly greater number of goblet cells in parasitized pyloric caeca than in unparasitized caeca in the same fish (de Buron and Nickol, 1994). This suggests a parasite-induced response that might lessen damage from the erosive nature of the worms.

Pathophysiology

Destruction of intestinal villi and the necrotic and degenerative changes in mucosal epithelium adversely affect the absorptive efficiency of the fish intestine. This may affect the general health and growth of the host. According to Bristol *et al.* (1984), there is a negative correlation between numbers of acanthocephalans and body lipid in trout and Buchmann (1986) demonstrated negative correlation between the number of *Echinorhynchus gadi* present and energy stores in Baltic cod (*Gadus morhua*). Trout infected with *Pomphorhynchus laevis* have lower muscle protein than uninfected fish (Wanstall *et al.*, 1982). Seasonal factors, such as host spawning or reduced winter feeding, exacerbate the protein depletion. Wanstall *et al.* (1982) suggested that *P. laevis* induces mobilization of endogenous protein for energy in preference to endogenous carbohydrates or lipids. This phenomenon is not uncommon in

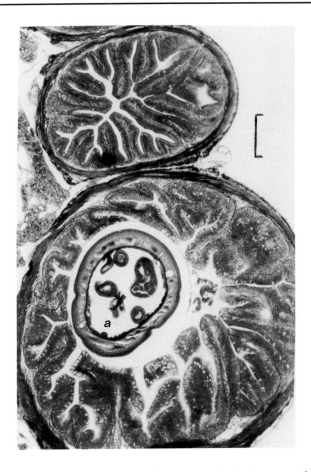

Fig. 12.6. Transverse section through the region of pyloric caeca of a green sunfish (*Lepomis cyanellus*) infected with *Leptorhynchoides thecatus* showing unparasitized caecum (top) and enlarged parasitized caecum (bottom). Note the compressed and eroded villi along the trunk of the parasite (a) compared to normal villi in the unparasitized caecum. Scale bar = 0.1 mm. (Unpublished micrograph courtesy of I. de Buron.)

fishes under stress. Further, the acanthocephalan *Plagiorhynchus cylindraceus*, considered of little pathological consequence, has a significant detrimental effect on the flow of food energy through the host and alters its basal metabolism (Connors and Nickol, 1991).

Attempts to relate acanthocephalan infections to the condition factor or suvivorship of fishes have not been convincing; however, even minor effects on nutritional status may influence fish populations. Theoretically, few hosts, if any, need to die as the result of parasitism for parasites to influence abundance or density (Anderson and May, 1978; May, 1985).

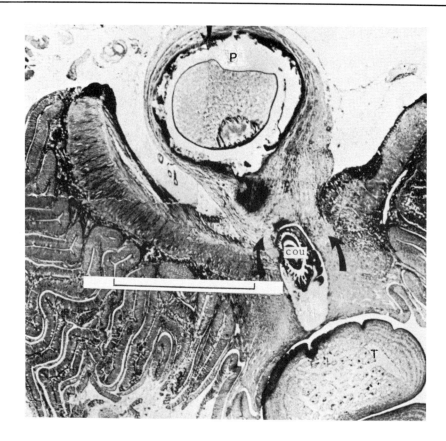

Fig. 12.7. Section of chub (*Leuciscus cephalus*) intestine showing penetration of *Pomphorhynchus laevis*. Note hyperplasia of the lamina propria (arrows) around the neck (cou) and bulbous proboscis (P) of the worm, forming a nodule in the coelom. An anterior portion of the trunk (T) of the worm shows in the intestinal lumen. Scale bar = 1 mm. (Unpublished micrograph courtesy of I. de Buron.)

Mechanism of disease

It has been suggested that acanthocephalans secrete toxic substances that paralyse or kill their hosts or that promote other patent pathological changes such as prolapse of the rectum (Fig. 12.4). No such substance has been isolated, however, and Schmidt *et al.* (1974) attributed such damage to localized tox-aemia and chronic fibrinous inflammation, which result from laceration of cells at the site of attachment.

It has been assumed until recently that intestinal penetration and destruction of cells were entirely mechanical due to the action of the spined proboscis. Miller and Dunagan (1971) described a pore-like opening and groove on hooks of certain acanthocephalans and postulated delivery of a secretion via the hooks. Taraschewski (1989b) demonstrated proteolytic enzymes that are incorporated in osmiophilic material associated with the proboscis of *Paratenui-sentis ambiguus* in eels (*Anguilla anguilla*) and postulated they are discharged from the worms through pores in the proboscis spines. Such enzymes with trypsin-like activity secreted by *Pomphorhynchus laevis* in chub (*Leuciscus cephalus*) degrade collagen (Polzer and Taraschewski, 1994) and thus are capable of degrading one or more major components of the host's gastro-intestinal tissue.

HOST IMMUNE RESPONSE

Cellular

The histopathological effects of acanthocephalans are characterized by increased numbers of eosinophils, neutrophils, and monocytes at the attach-ment site. Mobilization of leucocytes occurs regardless of whether the fish species is suitable for development of the parasite, and in most cases the role of cellular immunity is unknown.

Hamers *et al.* (1992) found interspecific differences in the response of leucocytes in fishes parasitized by *Paratenuisentis ambiguus*. In the eel (*Anguilla anguilla*), a suitable definitive host, the response was much less intense than in the carp (*Cyprinus carpio*) or the rainbow trout (*Oncorhynchus mykiss*), both unsuitable hosts that expel the acanthocephalans within a few days. The leucocytes damage acanthocephalan tegument extensively in carp, hence they speculated that cellular defence is a factor in determining host specificity of *P. ambiguus*.

Humoral

Precipitating antibodies to acanthocephalan antigens have been reported from sera (Harris, 1970; Szalai *et al.*, 1988) and intestinal mucus (Harris, 1972) of infected fishes. Sera from chub (*Leuciscus cephalus*) held at 10°C and with no history of exposure to *Pomphorhynchus laevis* do not have anti-*P. laevis* precipitins, but precipitins were detected within 160 days after infection. Parenteral injection of *P. laevis* antigen also induces a similar response by 150 days after injection (Harris, 1972). Anti-*P. laevis* precipitins are not found in four other piscine species in which *P. laevis* occurs, but unlike in *L. cephalus*, does not reach maturity. This lead Harris (1972) to speculate that the antigenic substance is produced only by mature worms.

Despite a humoral response, chub (*L. cephalus*) have large numbers of *P. laevis* in a variety of stages of development. Harris (1972) found no evidence of expulsion of worms, even from fish with mature worms.

Cross immunity

Concurrent infections with more than one species of acanthocephalan and/or with helminths of other phyla are frequent. *Pomphorhynchus bulbocolli* and *Acanthocephalus dirus* occur concurrently in rainbow darters (*Etheostoma caeruleum*) where they occupy sites in close proximity to one another. There appears to be no synergistic effect as both species cause damage as if they were in single species infections (McDonough and Gleason, 1981). In contrast, numbers of the cestode *Proteocephalus exiguus* and *Neoechinorhynchus* sp. show an inverse relation in ciscoes (*Coregonus artedii*), although the two helminth species occupy different intestinal sites. Cross (1934) believed that nonspecific immunity limits either *P. exiguus* or *Neoechinorhynchus* sp.

Very little is known about the nature of precipitins in serum of acantho-cephalan-infected fishes, and seldom has specific immunoglobulin been demonstrated. Szalai *et al.* (1988), however, confirmed that anti-*Neoechinorhynchus carpiodi* precipitins in serum from infected quillback (*Carpiodes cyprinus*) were not complement-reactive protein or the alpha migrating factor. Partial characterization of chub (*Leuciscus cephalus*) antibody to *Pomphorhynchus laevis* indicated it is an IgM type (Harris, 1972). Consequently, precipitating antibody occurs in at least one acanthocephalan infection but its role in limiting infection and the degree of species specificity are not known.

IN VITRO CULTIVATION

Studies on nutritional requirements, development, and response to chemo-therapeutic agents would be facilitated greatly by successful *in vitro* cultivation. Hamers *et al.* (1991) obtained axenic growth of *Paratenuisentis ambiguus* taken from intermediate hosts, but it was substantially slower than *in vivo* growth and maximum survival was 63 days. In no case have immature worms been cultured to egg-producing maturity.

There is only one report of culturing acanthocephalans *in ovo*, a procedure highly successful for digenetic trematodes (Fried and Stableford, 1991). Young and Lewis (1977) obtained growth of *Corynosoma constrictum* cystacanths, normally infective to waterfowl, on the chicken chorioallantoic membrane; however, they survived for only three days. Cystacanths of two other species, also parasitic in birds, survived up to nine days, but they did not grow.

Crompton and Lassière (1987) reviewed reported maintenance techniques and listed medium constituents and protocols for those where there was some indication of success.

PARASITE NUTRITION, PHYSIOLOGY AND BIOCHEMISTRY

Feeding mechanisms

Feeding, nutrition, and metabolism in Acanthocephala were reviewed by Starling (1985). Most of the information is based on two species from mammals and one from waterfowl, but there is no reason to believe that acanthocephalans from fishes differ significantly.

Uptake of nutrients occurs through the tegument. Intimate contact between the acanthocephalan proboscis and neck and the flattening or erosion of host mucosal cells along the trunk of the worms lead to a frequent assumption that nutrients are obtained as a result of leakage from host tissue. Bullock (1963) suggested that, in addition to uptake of secretions from the host mucosa, worms absorb nutrients from the copious mucus frequently produced in response to infection and which contains end products of host digestion. It is known that at least some species obtain nutrients from dietary contents in the intestinal lumen of the host (Edmonds, 1965) and that hydrolytic enzyme activity occurs at the tegumental surface of the worm (Uglem *et al.*, 1973). Although cytological localization of surface hydrolases is unknown, phosphatase (Byram and Fisher, 1974) and aminopeptidase (Uglem *et al.*, 1973) activities are associated with the tegumental crypts.

The tegumental crypts, formed by infolded plasma membrane, provide an extensive amplification of the surface area for nutritent uptake, but Starling (1985) concluded that the crypts are unstirred regions comparable to intervillar spaces of vertebrates and, as such, might make a relatively minor contribution to uptake of solutes such as glucose and free amino acids. They should, however, greatly increase the facility with which products of hydrolytic enzymes are absorbed.

Nutritional requirements

In addition to dietary protein (Crompton *et al.*, 1983), hydrolysis of secretory digestive proteins, degradation of moribund mucosal cells, and probably exchange between the mucosal epithelium and intestinal lumen are sources of free amino acids in the intestinal lumen of a vertebrate animal (Starling, 1985). Amino acids from host tissues contribute significantly to the pool available to acanthocephalans, which assimilate certain of these both by diffusion and by active transport (Uglem and Read, 1973). This exchange between host tissue, diet, and worms *in vivo* and the inability to culture acanthocephalans satisfactorily *in vitro* have impeded attempts to define nutritional requirements for amino acids. Largely for the same reasons, lipid requirements are likewise not known.

Energy metabolism

Glycogen and trehalose appear to be the principal endogenous carbohydrates in acanthocephalans. Their tissues are rich in glycogen and depleted stores are replenished rapidly (Read and Rothman, 1958). Glycogen synthase occurs in both dephosphorylated and glucose 6-phosphate-dependent phosphorylated forms that apparently can be interconverted (Starling, 1985). Trehalose is present in pseudocoelomic fluid, reproductive tissues, and the body wall (McAlister and Fisher, 1972). Presumably it contributes to osmolality of body tissues and extracellular fluids, but the role of trehalose is largely undetermined in acanthocephalans. It is unlikely to serve as a primary storage of carbohydrate for energy (Laurie, 1959). Starling and Fisher (1978) believed trehalose may trap glucose within acanthocephalan tegument after its absorption and carry glucose moieties to non-tegumental tissues.

Most tissues, especially those of the lemnisci and neck, of adult acanthocephalans possess lipid deposits, some of which contain dissolved carotenoid pigments that give individuals of certain species an obvious coloration, e.g. the orange, red, brown, or yellow described by Ravindranathan and Nadakal (1971) for *Pallisentis nagpurensis* from the freshwater fish *Ophiocephalus striatus*. There are a few isolated studies on lipid metabolism in acanthocephalans (see Starling, 1985), but almost nothing is known about lipid biosynthesis, turnover, or biological significance.

Carbohydrate is the main source of energy, and it is reasonably clear that glucose and glycogen are metabolized via the conventional Embden-Meyerhof pathway to phosphoenol pyruvate. Further metabolism probably involves anaerobic reactions common to other helminths, but the steps are still uncertain (Starling, 1985). Lactate and succinate are the main end products although Crompton and Ward (1967) found extensive ethanol excretion by *Moniliformis moniliformis*. Starling (1985) reviewed the glycolytic enzymes and metabolic end products of acanthocephalans.

Feeding-induced pathogenesis

Deep penetration by the proboscis results in fibrinous nodules that it may project several millimetres into the coelom (Fig. 12.5). Szalai and Dick (1987) reported extensive vascularization of such nodules in quillback (*Carpiodes cyprinus*) induced by *Neoechinorhynchus carpiodi* and demonstrated increased leakage of proteins from the blood in the region of the nodules. They suggested that the lesion might ensure a limited, but steady, supply of nutrients for the parasites.

Pomphorhynchus laevis is another species that induces nodule formation (Fig. 12.7). Polzer and Taraschewski (1994) found that the worm releases an aminopeptidase and a trypsin-like collagenase into the culture medium. They concluded that acanthocephalan aminopeptidases have a nutritional role but that the collagenase facilitates rapid and deep penetration by the worm into the intestinal tissues. This suggests that proteolytic enzymes degrade peptides at the body surface. In some species, the histolytic action permits rapid and deep penetration, which leads to formation of nutrient-supplying nodules.

PREVENTION AND CONTROL OF INFECTION

Diagnosis

At least some species of fish respond to acanthocephalan infections (natural and experimental) by production of specific antibodies (Harris, 1972). However, there is insufficient information for immunodiagnosis. Acanthocephalans can be diagnosed from eggs passed out in the faeces of the host, but this is not convenient and there is little necessity for the technique. Most infections are detected during postmortem examination.

Chemotherapy

The antidiarrhoeic drug loperamid was efficacious in treating rainbow trout (*Salmo gairdneri = Oncorhynchus mykiss*) infected with *Echinorhynchus truttae*; 50 mg kg^{-1} doses were administered orally mixed with food pellets on three consecutive days (Taraschewski *et al.*, 1990). This treatment might be useful in commercial cultivation of fishes.

Prophylaxis

Prevention of exposure is the most effective method of limiting acanthocephalan infections. These parasites are good colonizers and expand their ranges by movements of infected animals, but transfaunations due to anthropochore fish movements, including those that result from stocking and management of freshwater fishes, are of growing concern (Kennedy, 1993). Introductions such as that seen in *Neoechinorhynchus rutili* in rainbow trout (*Oncorhynchus mykiss*) of the Plitvice Lakes National Park (in the former Yugoslavia) (Bristol *et al.*, 1984) serve as warnings for aquaculture endeavours. Acanthocephalans can be introduced with either definitive or intermediate hosts. If colonization of the habitat is successful, the resulting parasitism can be costly (Taraschewski *et al.*, 1990).

SUMMARY AND CONCLUSIONS

Acanthocephalans occur globally in most fishes, but epizootics and significant economic losses as a result of infection are unusual. Acanthocephalans require vertebrate animals for definitive hosts and arthropods for intermediate hosts. Isopods, amphipods, and ostracods are the usual intermediate hosts for aquatic species. Infection occurs when a definitive host consumes the infective cystacanth stage contained in an arthropod or in a paratenic host. Worms are typically recruited into fish populations during the spring, with maturation, egg production, and transmission to intermediate hosts in the summer and early autumn. Adult worms usually live about one season.

Acanthocephalans attach by a spiny proboscis in the intestine of definitive hosts. Mucosal tissue is damaged at the attachment site, resulting in fibroplasia that may extend through the submucosa and into the muscularis. Occasionally perforation of the gut wall occurs. The mucosal epithelium frequently is compressed or eroded along the trunk of the worm, and the tips of the villi may be absent. Destruction of intestinal villi and necrotic and degenerative changes in mucosal epithelium almost certainly will reduce the absorptive efficiency of the fish intestine.

Some acanthocephalans are known to induce antibody formation in piscine species. Haemorrhage at the attachment site and copious amounts of mucus in parasitized fishes provide opportunity for antibody–worm interaction, but it is not known if protective immunity develops.

Acanthocephalans lack an alimentary tract and hence uptake of nutrients, derived both from leakage of host tissues and from dietary contents in the intestinal lumen of the host, is through the tegument. Hydrolytic enzyme activity at the tegumental surface probably assists in obtaining nutrients and in rapid penetration by the worm. Some acanthocephalans penetrate deeply and induce formation of a nodule that extends into the coelom of the host. Such nodules are extensively vascularized. Increased leakage of proteins from the blood into the nodules ensures a steady supply of nutrients for the parasites.

Carbohydrate, in the form of glycogen, is the main substrate for energy and glycolysis occurs in the presence or absence of oxygen. Lactate and succinate are the main end products, but large amounts of ethanol are excreted by at least one species.

All media in which significant growth has been achieved contain undefined components. Further research for a defined medium is necessary before appreciable progress can be made in determining nutritional requirements, especially for amino acids and lipids.

The consequences of acanthocephalan-induced reductions in energy efficiency and altered basal metabolism of hosts are likely to be focused more sharply with increasing emphasis on aquaculture. Suitable methods for *in vitro* cultivation are necessary to elucidate responses of worms to chemotherapeutic agents. Known detrimental effects of species such as *Acanthocephalus anguillae*, *Echinorhynchus truttae*, and *Neoechinorhynchus rutili* in commercially cultured fishes are likely to stimulate research directed at testing for chemotherapeutic compounds. When efficacious drugs are found, their toxicity to fishes, aquatic invertebrates, and humans must be assessed.

REFERENCES

Amin, O.M. (1978) Effect of host spawning on *Echinorhynchus salmonis* Muller, 1784 (Acanthocephala: Echinorhynchidae) maturation and localization. *Journal of Fish Diseases* 1, 195–197.

Amin, O.A. (1981) The seasonal distribution of *Echinorhynchus salmonis* (Acanthocephala: Echinorhynchidae) among rainbow smelt, *Osmerus mordax* Mitchell, in Lake Michigan. *Journal of Fish Biology* 19, 467–474.

Amin, O.M. (1987) Key to the families and subfamilies of Acanthocephala, with the erection of a new class (Polyacanthocephala) and a new order (Polyacanthorhynchida). *Journal of Parasitology* 73, 1216–1219.

Anderson, R.M. and May, R.M. (1978) Regulation and stability of host-parasite population interactions. I. Regulatory processes. *Journal of Animal Ecology* 47, 219–247.

Awachie, J.B.E. (1965) The ecology of *Echinorhynchus truttae* Schrank, 1788 (Acanthocephala) in a trout stream in North Wales. *Parasitology* 55, 747–762.

Awachie, J.B.E. (1966) The development and life-history of *Echinorhynchus truttae* Schrank 1788 (Acanthocephala). *Journal of Helminthology* 40, 11–32.

Beermann, I., Arai, H.P. and Costerton, J.W. (1974) The ultrastructure of the lemnisci and body wall of *Octospinifer macilentus* (Acanthocephala). *Canadian Journal of Zoology* 52, 553–555.

Boxrucker, J.C. (1979) Effects of a thermal effluent on the incidence and abundance of the gill and intestinal metazoan parasites of the black bullhead. *Parasitology* 78, 195–206.

Bristol, J.R., Mayberry, L.F., Huber, D. and Ehrlich, I. (1984) Endoparasite fauna of trout in the Plitvice Lakes National Park. *Veterinarski Arhiv* 54, 5–11.

Buchmann, K. (1986) On the infection of Baltic cod (*Gadus morhua* L.) by the acanthocephalan *Echinorhynchus gadi* (Zoega) Müller. *Nordisk Veterinaermedicin* 38, 308–314.

Buckner, R.L., Overstreet, R.M. and Heard, R.W. (1978) Intermediate hosts for *Tegorhynchus furcatus* and *Dollfusentis chandleri*. *Proceedings of the Helminthological Society of Washington* 45, 195–201.

Bullock, W.L. (1963) Intestinal histology of some salmonid fishes with particular reference to the histopathology of acanthocephalan infections. *Journal of Morphology* 112, 23–44.

Byram, J.E. and Fisher, F.M. (1974) The absorptive surface of *Moniliformis dubius* (Acanthocephala) II. Functional aspects. *Tissue and Cell* 6, 21–42.

Camp, J.W. and Huizinga, H.W. (1979) Altered color, behavior and predation susceptibility of the isopod, *Asellus intermedius*, infected with *Acanthocephalus dirus*. *Journal of Parasitology* 65, 667–669.

Chaicharn, A. and Bullock, W.L. (1967) The histopathology of acanthocephalan infections in suckers with observations on the intestinal histology of two species of catostomid fishes. *Acta Zoologica* 48, 19–42.

Chubb, J.C. (1982) Seasonal occurrence of helminths in freshwater fishes Part IV. Adult Cestoda, Nematoda and Acanthocephala. *Advances in Parasitology* 20, 1–292.

Connors, V.A. and Nickol, B.B. (1991) Effects of *Plagiorhynchus cylindraceus* (Acanthocephala) on the energy metabolism of adult starlings, *Sturnus vulgaris*. *Parasitology* 103, 395–402.

Crompton, D.W.T. (1975) Relationships between Acanthocephala and their hosts. In:

Jennings, D.H. and Lee, D.L. (eds) *Symposia of the Society for Experimental Biology, 29.* Cambridge University Press, Cambridge, pp. 467–504.

Crompton, D.W.T. and Lassière, O.L. (1987) Acanthocephala. In: Taylor, A.E.R. and Baker, J.R. (eds) In Vitro *Methods for Parasite Cultivation.* Academic Press, London, pp. 394–406.

Crompton, D.W.T. and Ward, P.F.V (1967) Production of ethanol and succinate by *Moniliformis dubius* (Acanthocephala). *Nature* 215, 964–965.

Crompton, D.W.T. and Whitfield, P.J. (1968) The course of infection and egg production of *Polymorphus minutus* (Acanthocephala) in domestic ducks. *Parasitology* 58, 231–246.

Crompton, D.W.T., Keymer, A., Singhvi, A. and Nesheim, M.C. (1983) Rat dietary fructose and the intestinal distribution and growth of *Moniliformis* (Acanthocephala). *Parasitology* 86, 57–81.

Cross, S.X. (1934) A probable case of non-specific immunity between two parasites of ciscoes of the Trout Lake region of northern Wisconsin. *Journal of Parasitology* 20, 244–245.

de Buron, I. and Nickol, B.B. (1994) Histopathological effects of the acanthocephalan *Leptorhynchoides thecatus* in the ceca of the green sunfish, *Lepomis cyanellus. Transactions of the American Microscopical Society* 113, 161–168.

DeGiusti, D.L. (1949) The life cycle of *Leptorhynchoides thecatus* (Linton), an acanthocephalan of fish. *Journal of Parasitology* 35, 437–460.

DeMont, D.J. and Corkum, K.C. (1982) The life cycle of *Octospiniferoides chandleri* Bullock, 1957 (Acanthocephala: Neoechinorhynchidae) with some observations on parasite-induced, photophilic behavior in ostracods. *Journal of Parasitology* 68, 125–130.

Dezfuli, B.S. (1991) Histopathology in *Leuciscus cephalus* (Pisces: Cyprinidae) resulting from infection with *Pomphorhynchus laevis* (Acanthocephala). *Parassitologia* 33, 137–145.

Dezfuli, B.S., Grandi, G., Franzoi, P. and Rossi, R. (1990) Histopathology in *Atherina boyeri* (Pisces: Atherinidae) resulting from infection by *Telosentis exiguus* (Acanthocephala). *Parassitologia* 32, 283–291.

Edmonds, S.J. (1965) Some experiments on the nutrition of *Moniliformis dubius* Meyer (Acanthocephala). *Parasitology* 55, 337–344.

Eure, H. (1976) Seasonal abundance of *Neoechinorhynchus cylindratus* taken from largemouth bass (*Micropterus salmoides*) in a heated reservoir. *Parasitology* 73, 355–370.

Fried, B. and Stableford, L.T. (1991) Cultivation of helminths in chick embryos. *Advances in Parasitology* 30, 107–165.

Gartner, J.V., Jr. and Zwerner, D.E. (1989) The parasite faunas of meso- and bathypelagic fishes of Norfolk submarine canyon, western North Atlantic. *Journal of Fish Biology* 34, 79–95.

Golvan, Y.J. and de Buron, I. (1988) Les hôtes des acanthocéphales. II-Les hôtes définitifs. 1. Poissons. *Annales de Parasitologie* 63, 394–375.

Halvorsen, O. (1972) Studies on the helminth fauna of Norway XX. Seasonal cycles of fish parasites in the River Glomma. *Norwegian Journal of Zoology* 20, 9–18.

Hamers, R., Taraschewski, H., Lehmann, J. and Mock, D. (1991) *In vitro* study of the impact of fish sera on the survival and fine structure of the eel-pathogenic acanthocephalan *Paratenuisentis ambiguus. Parasitology Research* 77, 703–708.

Hamers, R., Lehmann, J., Sturenberg, F. and Taraschewski, H. (1992) *In vitro* study of the migratory and adherent responses of fish leucocytes to the eel-pathogenic acanthocephalan *Paratenuisentis ambiguus* (Van Cleave, 1921) Bullock et Samuel, 1975 (Eoacanthocephala: Tenuisentidae). *Fish and Shellfish Immunology* 2, 43–51.

Harris, J.E. (1970) Precipitin production by chub (*Leuciscus cephalus*) to an intestinal helminth. *Journal of Parasitology* 56, 1035.

Harris, J.E. (1972) The immune response of a cyprinid fish to infections of the acanthocephalan *Pomphorhynchus laevis*. *International Journal for Parasitology* 2, 459–469.

Hoffman, G.L. (1967) *Parasites of North American Freshwater Fishes*. University of California Press, Berkeley.

Kates, K.C. (1944) Some observations on experimental infections of pigs with the thorn-headed worm *Macracanthorhynchus hirudinaceus*. *American Journal of Veterinary Research* 5, 166–172.

Kennedy, C.R. (1970) The population biology of helminths of British freshwater fish. *Symposia of the British Society for Parasitology* 9, 145–159.

Kennedy, C.R. (1972) The effects of temperature and other factors upon the establishment and survival of *Pomphorhynchus laevis* (Acanthocephala) in goldfish, *Carassius auratus*. *Parasitology* 65, 283–294.

Kennedy, C.R. (1985) Site segregation by species of Acanthocephala in fish, with special reference to eels, *Anguilla anguilla*. *Parasitology* 90, 375–390.

Kennedy, C.R. (1993) Introductions, spread and colonization of new localities by fish helminth and crustacean parasites in the British Isles: a perspective and appraisal. *Journal of Fish Biology* 43, 287–301.

Kennedy, C.R., Boughton, P.F. and Hine, P.M. (1978) The status of brown and rainbow trout, *Salmo trutta* and *S. gairdneri* as hosts of the acanthocephalan, *Pomphorhynchus laevis*. *Journal of Fish Biology* 13, 265–275.

Laurie, J.S. (1959) Aerobic metabolism of *Moniliformis dubius* (Acanthocephala). *Experimental Parasitology* 8, 188–197.

Machado, D.A.F. (1947) Revisão do gênero '*Polyacanthorhynchus*' Travassos, 1920 (Acanthocephala, Rhadinorhynchidae). *Revista Brasileira de Biologia* 7, 195–201.

May, R.M. (1985) Host–parasite associations: their population biology and population genetics. In: Rollinson, D. and Anderson, R.M. (eds) *Ecology and Genetics of Host–Parasite Interactions*. Academic Press, London, pp. 243–262.

McAlister, R.O. and Fisher, F.M. (1972) The biosynthesis of trehalose in *Moniliformis dubius* (Acanthocephala). *Journal of Parasitology* 58, 51–62.

McDonough, J.M. and Gleason, L.N. (1981) Histopathology in the rainbow darter, *Etheostoma caeruleum*, resulting from infections with the acanthocephalans, *Pomphorhynchus bulbocolli* and *Acanthocephalus dirus*. *Journal of Parasitology* 67, 403–409.

McVicar, A.H. (1977) Intestinal helminth parasites of the ray *Raja naevus* in British waters. *Journal of Helminthology* 51, 11–21.

Meyer, A. (1933) Acanthocepala. In: *Dr. Bronn's Klassen und Ordnungen des Tierreichs, Vol. 4*. Akademische Verlagsgesellschaft, Leipzig, pp. 333–582.

Miller, D.M. and Dunagan, T.T. (1971) Studies on the rostellar hooks of *Macracanthorhynchus hirudinaceus* (Acanthocephala) from swine. *Transactions of the American Microscopical Society* 90, 329–335.

Miller, D.M. and Dunagan, T.T. (1976) Body wall organization of the acanthocephalan *Macracanthorhynchus hirudinaceus*: A reexamination of the lacunar system. *Proceedings of the Helminthological Society of Washington* 43, 99–106.

Miller, D.M. and Dunagan, T.T. (1985) Functional morphology. In: Crompton, D.W.T. and Nickol, B.B. (eds) *Biology of the Acanthocephala*. Cambridge University Press, Cambridge, pp. 73–123.

Morris, S.C. and Crompton, D.W.T. (1982) The origins and evolution of the Acanthocephala. *Biological Reviews of the Cambridge Philosophical Society* 57, 85–115.

Muzzall, P.M. (1978) The host-parasite relationships and seasonal occurrence of
Fessisentis friedi (Acanthocephala: Fessisentidae) in the isopod (*Caecidotea communis*). *Proceedings of the Helminthological Society of Washington* 45, 77–82.

Muzzall, P.M. (1980) Ecology and seasonal abundance of three acanthocephalan
species infecting white suckers in SE New Hampshire. *Journal of Parasitology* 66,
127–133.

Muzzall, P.M. and Rabalais, F.C. (1975a) Studies on *Acanthocephalus jacksoni*
Bullock, 1962 (Acanthocephala: Echinorhynchidae). I. Seasonal periodicity and
new host records. *Proceedings of the Helminthological Society of Washington* 42,
31–34.

Muzzall, P.M. and Rabalais, F.C. (1975b) Studies on *Acanthocephalus jacksoni*
Bullock, 1962 (Acanthocephala: Echinorhynchidae). III. The altered behavior of
Lirceus lineatus (Say). *Proceedings of the Helminthological Society of Washington*
42, 116–118.

Nickol, B.B. (1972) Two species of Acanthocephala from Australian fishes with
description of *Arhythmacanthus paraplagusiarum* sp.n. *Journal of Parasitology*
58, 778–780.

Noble, E.R. (1973) Parasites and fishes in a deep-sea environment. *Advances in Marine
Biology* 11, 121–195.

Paggi, L., Orecchia, P. and Della Seta, G. (1975) Su di un nuovo acanthocefalo
Breizacanthus ligur sp.n. (Palaeacanthocephala: Arhythmacanthidae Yamaguti,
1935) parassita di Alcune specie di pesci bentoici del Mar Ligure. *Parassitologia*
17, 83–94.

Polzer, M. and Taraschewski, H. (1994) Proteolytic enzymes of *Pomphorhynchus
laevis* and in three other acanthocephalan species. *Journal of Parasitology* 80,
45–49.

Ravindranathan, R. and Nadakal, A.M. (1971) Carotenoids in an acanthocephalid
worm *Pallisentis nagpurensis*. *Japanese Journal of Parasitology* 20, 1–5.

Read, C.P. and Rothman, A.H. (1958) The carbohydrate requirement of *Moniliformis*
(Acanthocephala). *Experimental Parasitology* 7, 191–197.

Schmidt, G.D. (1985) Development and life cycles. In: Crompton, D.W.T. and Nickol,
B.B. (eds) *Biology of the Acanthocephala*. Cambridge University Press, Cambridge, pp. 273–305.

Schmidt, G.D. and Canaris, A.G. (1967) Acanthocephala from Kenya with descriptions of two new species. *Journal of Parasitology* 53, 634–637.

Schmidt, G.D., Walley, H.D. and Wijek, D.S. (1974) Unusual pathology in a fish
due to the acanthocephalan *Acanthocephalus jacksoni* Bullock, 1962. *Journal of
Parasitology* 60, 730–731.

Starling, J.A. (1985) Feeding, nutrition and metabolism. In: Crompton, D.W.T. and
Nickol, B.B. (eds) *Biology of the Acanthocephala*. Cambridge University Press,
Cambridge, pp. 125–212.

Starling, J.A. and Fisher, F.M. (1978) Carbohydrate transport in *Moniliformis dubius*
(Acanthocephala). II. Post-absorptive phosphorylation of glucose and the role of
trehalose in the accumulation of endogenous glucose reserves. *Journal of Comparative Physiology* 126, 223–231.

Stranack, F.R., Woodhouse, M.A. and Griffin, R.L. (1966) Preliminary observations
on the ultrastructure of the body wall of *Pomphorhynchus laevis* (Acanthocephala). *Journal of Helminthology* 40, 395–402.

Szalai, A.J. and Dick, T.A. (1987) Intestinal pathology and the site specificity of the acanthocephalan *Neoechinorhynchus carpiodi* Dechtiar, 1968, in quillback, *Carpiodes cyprinus* (Lesueur). *Journal of Parasitology* 73, 467–475.

Szalai, A.J., Danell, G.V. and Dick, T.A. (1988) Intestinal leakage and precipitating antibodies in the serum of quillback, *Carpiodes cyprinus* (Lesueur), infected with *Neoechinorhynchus carpiodi* Dechtiar, 1968 (Acanthocephala: Neoechinorhynchidae). *Journal of Parasitology* 74, 415–420.

Taraschewski, H. (1989a) *Acanthocephalus anguillae* in intra- and extraintestinal positions in experimentally infected juveniles of goldfish and carp and in sticklebacks. *Journal of Parasitology* 75, 108–118.

Taraschewski, H. (1989b) Host–parasite interface of *Paratenuisentis ambiguus* (Eoacanthocephala) in naturally infected eel and in laboratory-infected stickle-backs and juvenile carp and rainbow trout. *Journal of Parasitology* 75, 911–919.

Taraschewski, H., Mehlhorn, H. and Raether, W. (1990) Loperamid, an efficacious drug against fish-pathogenic acanthocephalans. *Parasitology Research* 76, 619–623.

Tedla, S. and Fernando, C.H. (1970) Some remarks on the ecology of *Echinorhynchus salmonis* (Muller, 1784). *Canadian Journal of Zoology* 48, 317–321.

Tokeson, J.P.E. and Holmes, J.C. (1982) The effects of temperature and oxygen on the development of *Polymorphus marilis* (Acanthocephala) in *Gammarus lacustris* (Amphipoda). *Journal of Parasitology* 68, 112–119.

Uglem, G.L. and Read, C.P. (1973) *Moniliformis dubius*: Uptake of leucine and alanine by adults. *Experimental Parasitology* 34, 148–153.

Uglem, G.L., Pappas, P.W. and Read, C.P. (1973) Surface aminopeptidase in *Moniliformis dubius* and its relations to amino acid uptake. *Parasitology* 67, 185–195.

Uznanski, R.L. and Nickol, B.B. (1982) Site selection, growth, and survival of *Leptorhynchoides thecatus* (Acanthocephala) during the prepatent period in *Lepomis cyanellus*. *Journal of Parasitology* 68, 686–690.

Van Cleave, H.J. (1920) Acanthocephala. *Report of the Canadian Arctic Expedition 1913–18* 9, 1E–11E.

Van Cleave, H.J. (1941) Analysis of distinctions between the acanthocephalan genera *Filicollis* and *Polymorphus*, with description of a new species of *Polymorphus*. *Transactions of the American Microscopical Society* 66, 302–313.

Van Cleave, H.J. (1947) Relationships of the Acanthocephala. *The American Naturalist* 75, 31–47.

von Brand, T. (1939) The glycogen distribution in the body of Acanthocephala. *Journal of Parasitology* 25, 22S.

Wanstall, P.W., Robotham, P.W.J. and Thomas, J.S. (1982) Changes in the energy reserves of two species of freshwater fish during infection by *Pomphorhynchus laevis*. *Parasitology* 85, xxvii.

Ward, H.L. (1940) Studies on the life histiory of *Neoechinorhynchus cylindratus* (Van Cleave, 1913) (Acanthocephala). *Transactions of the American Microscopical Society* 59, 327–347.

Watson, R.A. and Dick, T.A. (1979) Metazoan parasites of whitefish, *Coregonus clupeaformis* (Mitchill) and cisco *C. artedii* Lesueur from Southern Indian Lake, Manitoba. *Journal of Fish Biology* 15, 579–587.

Yamaguti, S. (1961) *Systema Helminthum, Vol. 5. Acanthocephala.* Interscience Publishers, New York.

Young, B.W. and Lewis, P.D., Jr (1977) Growth of an acanthocephalan on the chick chorioallantois. *Proceedings of the Montana Academy of Sciences* 37, 88.

Zdzitowiecki, K. (1986) Acanthocephala of the Antarctic. *Polish Polar Research* 7, 79–112.

13

Phylum Arthropoda

R.J.G. Lester and F.R. Roubal

Department of Parasitology, The University of Queensland, Brisbane, Queensland 4072, Australia.

INTRODUCTION

Arthropod parasites of fish have been recognized by man since the time of Aristotle, who recorded a pennellid on tunny and swordfish in the Mediterranean. About 2000 species have since been described, the majority of which belong to the class Copepoda. Some of the arthropods that affect wild fish are of commercial significance as they affect host survival or cause unsightly changes in the flesh. Others cause ongoing problems in aquaculture such as the anchor worm, *Lernaea cyprinacea*. The biology of most arthropods parasitic on fish is poorly known. However, the relatively recent development of raising salmon in sea cages has resulted in an explosion of papers relating to sea-lice, *Lepeophtheirus salmonis*.

In this chapter, we discuss those arthropods for which pathological changes in fish have been described.

COPEPODA: ERGASILIDAE

Introduction

Ergasilid copepods damage the gills and cause commercially significant epizootics in cultured and wild populations of fish. *Ergasilus sieboldi* affects tench, pike, whitefish, eel and flounder in Europe (Dogiel *et al.*, 1961; Grabda, 1991). *Ergasilus lizae* is important in mullet culture in the Mediterranean and Middle East (Paperna, 1975, 1991) and *Pseudergasilus zacconis* infects cultured ayu, *Plecoglossus altivelis*, in Japan (Nakajima *et al.*, 1974). Kelly and Allison (1962) implicated *E. lizae* in the death of largemouth bass, *Micropterus salmoides*, in a brackish lake. Recently, *Ergasilus labracis* killed large numbers of Atlantic salmon, *Salmo salar*, parr (85–125 mm long) over a 4 day period in New Brunswick, Canada (Hogans, 1989).

Host range, distribution and seasonality

Most species in the family Ergasilidae belong to the genus *Ergasilus* of which 65 species are parasitic on freshwater fish and 33 species on marine teleosts (Kabata, 1979). *Ergasilus sieboldi* infects over sixty species of fish throughout much of the former USSR (except the Pacific province). The main hosts are *Tinca tinca, Esox lucius, Acerina cernua, Lucioperca lucioperca,* and *Silurus glanis* (Dogiel *et al.*, 1961; Zmerzlaya, 1972). Several thousand parasite individuals may occur on the gills of one fish. Cyprinids and salmonids, other than *Coregonus* spp., are less severely infected (Abrosov and Bauer, 1959; Dogiel *et al.*, 1961). The parasite has spread to many parts of Europe with the introduction of infected fish. In Germany 39 of 79 species of fish are host to *E. sieboldi,* the most important being *Esox lucius* and *Abramis brama* (see Dogiel *et al.*, 1961). *E. sieboldi* is also common in British freshwater fishes (Abdelhalim *et al.*, 1991).

In North America, *Ergasilus luciopercarum* infects chinook salmon, yellow perch, rainbow trout and brown trout, but *E. nerkae* infects only coho salmon (Buttner and Hamilton, 1976). *Ergasilus labracis* infects a wide range of marine species, including salmonids (Hogans, 1989). Ergasilids on Amazonian fishes are very host-specific according to Thatcher and Boeger (1983).

Larvae of *E. sieboldi* occurred in Lake Arakul, USSR, from the end of May to November with two peaks in summer (August and September) during which more than 98% of the larvae died (Kashkovsky and Kashkovskaya-Solomatova, 1986). Zmerzlaya (1972) found that new infections by *E. sieboldi* did not occur during winter. Females that overwintered produced egg sacs in April (water temperature 5.7–7°C), and the first generation of females settled on the host in late June. A second generation appeared from mid-August to mid-September. There was no recruitment after October when egg sac formation ceased and no egg sacs were present on female copepods over winter. According to Reichenbach-Klinke and Landolt (1973) there was no development in summer if water temperature dropped below 14°C. Kuperman and Shulman (1977) found that egg sac formation by *E. sieboldi* was stimulated by temperature in spring and early summer, but in late summer and autumn photoperiod rather than temperature limited development.

Ergasilus labracis infected more than 90% of striped bass, *Morone saxatilis,* from the lower Chesapeake Bay in salinities from 0.1 to 32 ppt. The parasite was present and reproductively active year-round with only a temporary slow down in egg production during November and December. Laboratory-held eggs hatched in temperatures as low as 5°C (Paperna and Zwerner, 1976a).

Tuuha *et al.* (1992) compared the prevalence and intensity of infection by *Ergasilus briani* (16.9%, 0.5), *Neoergasilus japonicus* (15.6%, 0.4) and *Paraergasilus longidigitus* (2.1%, 0.002) on 1255 roach, *Rutilus rutilus,* and *E. sieboldi* (9.9%, 0.1) and *P. longidigitus* (4.9%, 0.05) on 866 perch, *Perca fluviatilis,* from four lakes in central Finland. The prevalence of *E. briani* depended largely on temperature with greater prevalence in summer (24.2%)

than in winter (11.3%). The occurrence of *E. sieboldi* was determined mostly by water quality, the parasite was more abundant in oligotrophic than in eutrophic lakes. *Neoergasilus japonicus* was influenced by year, season and lake type, apparently due to the longer duration of the free-living stage.

Ergasilus lizae is a cosmopolitan species which is restricted largely to the Mugilidae (Paperna and Overstreet, 1981; Kabata, 1992), though it has been found on eels, tilapia, carp as well as mullet in aquaculture ponds in Israel where it occurred in salinities from 0.2–21 ppt (Paperna, 1975, 1977, 1991). Low salinity and high water temperatures were associated with high levels of infection by *E. lizae* on mullets (Raibaut *et al.*, 1975; Ben Hassine *et al.*, 1983). *E. lizae*, removed from silver mullet, *M. curema*, survived for four hours in salinities ranging from 0 to 37 ppt (Conroy and Conroy, 1986). The intensity of infection of *E. lizae* and *E. versicolor* on 0-class *Mugil cephalus* was high in summer but declined in autumn (Rawson, 1977). Roubal (1990) found that the prevalence and intensity of *Dermoergasilus acanthopagri* and *Paraergasilus acanthopagri* fluctuated in synchrony irrespective of season (temperature, salinity) unlike the seasonal cycles of other ectoparasites on the marine bream, *Acanthopagrus australis*.

Systematics

Keys to identify ergasilid copepods are in Roberts (1970) for North American species, Kabata (1979) for European species, and Do (1982) for Japanese species. New species have been described recently from Australia (Kabata, 1992) and India (Ho *et al.*, 1992).

Morphology and life cycle

Ergasilid copepods are less modified than most fish-parasitic copepods and resemble free-living copepods in segmentation. Only fertilized females are parasitic; males and developmental stages live in the water column.

The mature female of *Ergasilus sieboldi* (Fig. 13.1) is 1.5 mm long, has a cephalothorax formed from the cephalosome and the first leg-bearing segment. Four leg-bearing segments follow, the last intimately associated with the swollen genital segment which gives rise to paired multiseriate egg sacs, each with more than 100 eggs (Kabata, 1979; Grabda, 1991). The large second antennae (a2) are used for attachment; they either pierce or encircle a gill filament. The a2 cuticle is swollen in genera such as *Dermoergasilus*. In *Ergasilus* and *Sinergasilus* spp. the a2 end in a large curved spine; in *Diergasilus* and *Thersitina* spp. the a2 end in two spines; in *Paraergasilus* three spines.

Mouthparts are on the mid-ventral surface of the cephalothorax, and consist of a two-segmented mandible with a falcate terminal segment, a short first maxilla with two distal setae and a falcate second maxilla with dentiform setae (Kabata, 1979). The naupliar feeding apparatus is typical for

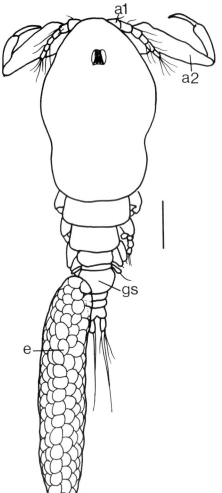

Fig. 13.1. *Ergasilus sieboldi*, adult female with one egg sac removed. a1, a2 = first and second antennae, e = egg sac, gs = genital segment. Scale bar = 200 µm.

planktotrophic copepod nauplii and the adult feeding apparatus (mandibles, maxillae 1 and 2) appears at the first copepodid stage (Abdelhalim *et al.*, 1991). The stomach extends anterior and posterior to the mouth parts, merging posteriorly with the straight intestine that leads to the anus at the end of the abdomen between the paired uropods. During sexual maturation, uterine processes that contain oocytes fill a large part of the cephalothorax. The oviducts lead to the genital segment to join with paired cement glands.

The life cycle of *Ergasilus sieboldi* (see Abdelhalim *et al.*, 1991) and *Neoergasilus japonicus* (see Urawa *et al.*, 1980a,b) consist of six nauplii and five copepodid stages and one adult stage. Sexual differentiation is evident in the fourth copepodid stage (Zmerzlaya, 1972; Abdelhalim *et al.*, 1991). It takes 22 days for the eggs of *E. sieboldi* to develop into free-living males and females at 17.7–20.1°C (Zmerzlaya, 1972). After fertilization, the female attaches to a fish. Zmerzlaya (1972) observed that overwintered females of *E. sieboldi* produced eggs in the spring which meant that sperm was stored for several months. At least two generations per year, including one that over-winters, were suggested by Tuuha *et al.* (1992) for *E. sieboldi*, *E. briani* and *Neoergasilus japonicus* in Finnish lakes. Non-ovigerous females of *E. labracis* transferred from a winter temperature of 7.2°C to the laboratory and the temperature increased slowly to 20°C produced egg sacs (Paperna and Zwerner, 1976a).

Host–parasite relationships

Site and host selection, course of infection

Ergasilus sieboldi attaches near the base of the gill filament and prefers (in decreasing order) gill arches 3,2,4 and 1 (Abrosov and Bauer, 1959; Zmerzlaya, 1972). It generally infects lake-dwelling fish over two years old (Dogiel *et al.*, 1961). Non-ovigerous females leave the host and attach to another host, but ovigerous females apparently sink more readily making this transfer less successful (Zmerzlaya, 1972). Abrosov and Bauer (1959) kept whitefish in nets in different parts of a lake and found that those on the bottom away from the shore became the most heavily infected.

Ergasilus lizae attach near the base of gill filaments of *Mugil cephalus* (see Ben Hassine and Raibaut, 1981) and prefer gill arches 2 and 3 (Rawson, 1977). *E. labracis* also attach near the base of the gill filament. Young females occur on gill arches 2 and 3 but most mature females are on gills 1 and 4 (Paperna and Zwerner, 1981).

The numbers of *Ergasilus colomesus* increase with fish size (Thatcher and Boeger, 1983). Those of *Ergasilus centrarchidarum*, *E. lizae*, and *E. auritus* decrease in large fish (Noble *et al.*, 1963; Joy, 1976; Cloutman and Becker, 1977); factors involved in this age immunity have still to be elucidated.

Clinical signs and histopathology

Fish such as *Tinca tinca* that are heavily infected with *Ergasilus sieboldi* die, especially in summer when water temperatures are highest. There is extensive gill damage and severe haemorrhage with inflammation and exsanguination associated with the attachment and feeding of the parasite. Blood vessels in the gill filaments are blocked and this leads to atrophy of gill tips (Dogiel *et al.*, 1961). Similar histopathology is associated with *Pseudergasilus zacconis* (Fig. 13.2) (Nakajima *et al.*, 1974). Infection of *Abramis brama* with *E. sieboldi* results in the disturbance of blood cell maturation with lymphopaenia and reactive granulocytosis in heterophil and basophil leuco-cytes (Einszporn-Orecka, 1973a,b). Rakova (see Dogiel *et al.*, 1961) reported a 20% drop in lymphocytes and an increase in monocytes, polymorphonuclear agranulocytes and neutrophils in *Tinca tinca* and *Leuciscus idus* infected with *E. sieboldi*. *Mugil cephalus* infected with several hundred *E. lizae* were emaciated (Paperna, 1975), and Atlantic salmon parr infected with *E. labracis* exhibited slow, weak swimming, gulping of air at the surface and darkening

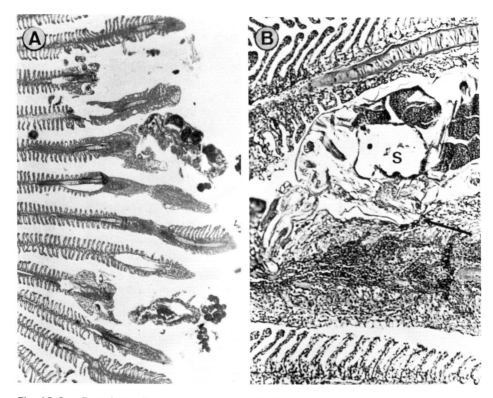

Fig. 13.2. *Pseudergasilus zacconis* A. On gill filaments of *Plecoglossus altivelis* showing extensive hyperplasia and necrosis of the gill. B. Section through infested filament showing deep penetration of second antenna, and mouth parts (arrow) closely apposed to gill epithelium. s = copepod.

of the skin (Hogans, 1989). Paperna and Zwerner (1981) reported an increase in number of mucous cells, epithelial hyperplasia and infiltration of macrophages, lymphocytes and eosinophils in gill filaments of *Morone saxatilis* infected by *E. labracis*.

Ergasilus australiensis causes epithelial hyperplasia in the basal half of the gill filaments and pseudobranchs of *Acanthopagrus australis* (see Roubal, 1986b, 1989a). *Dermoergasilus acanthopagri*, which attaches to the tip of the gill filament by second antennae that encircle the filament, causes localized swelling and occlusion of the filament arterial vessels (Roubal, 1987, 1989a). The much smaller *Paraergasilus acanthopagri*, which attaches by small spines to the nasal and opercular cavities, causes a localized epidermal proliferation. Increase in number of mucous cells, fusion of lamellae and filaments due to epithelial proliferation, lymphocyte infiltration and exsanguination was reported by Rogers (1969) for *E. cyprinaceus*. A similar effect was described by Thatcher and Boeger (1983) for *E. colomesus*.

Mechanism of disease

Zmerzlaya (1972) considered that the histopathology associated with *Ergasilus sieboldi* was caused by attachment and feeding and the severity of damage was proportional to the intensity of infection. Paperna and Zwerner (1981) found that tissue proliferation associated with *E. labracis* was initiated adjacent to the mouth parts and was not caused by attachment. A similar view was given by Oldewage and Van As (1987) for *E. mirabilis*.

Gill damage resulted in loss of gill surface area for respiration and led to suffocation, particularly at high water temperatures (Grabda, 1991). Hogans (1989) suggested that osmoregulatory failure might be important. Löpmann (1940) found that no more than 50% of gill surface area could be affected before the fish died.

Effects on host

Heavy infections cause death. Sublethal infections result in decreased condition (Grabda, 1991; Schäperclaus *et al.*, 1991). Infection of *Tinca tinca* and *Coregonus peled* with *Ergasilus sieboldi* was associated with reduced weight, total length and condition factor, reduced total fat content, and increased water content (Dogiel *et al.*, 1961; Abrosov and Bauer, 1959; Kabata, 1970).

Concurrent infection

Wounds in pond-cultured *Tinca tinca* caused by *Ergasilus sieboldi* were secondarily infected with *Saprolegnia* sp. (see Reichenbach-Klinke and Landolt, 1973). No secondary bacterial infection was found in the gills of cyprinid fishes in Alabama infected with *Ergasilus cyprinaceus* (see Rogers, 1969). Donoghue (1986) found that *Pungitius pungitius* already infected by *Thersitina gasterostei* were more likely to be infected by a second generation of the parasite.

Parasite nutrition and physiology

Halisch (1940) suggested that attached ergasilid copepods fed by external digestion and showed that female copepods confined in small volumes of water released factors that dissolved gelatin. The nature of these enzymes has still to be determined. The diet of *Ergasilus sieboldi* was shown by Einszporn (1964, 1965a,b) to consist of mucus, epithelium and blood. She proposed that blood was a major component of the diet, and cast some doubt on the validity of Halisch's data, suggesting that bacterial contamination resulted in dissolved gelatin. However, extrabuccal digestion is now considered a major part of feeding by ergasilids (Kabata, 1984). Donoghue (1989) found mucus to be the principal diet of *Thersitinia gasterostei*. The food and feeding behaviour of free-living stages are unknown (Abdelhalim *et al.*, 1991).

Diagnosis

The parasites are readily seen with the naked eye or with the dissecting microscope.

Prevention and control

Hogans (1989) reported the successful treatment of *Ergasilus labracis* on Atlantic salmon parr with Neguvon (0.25 mg l^{-1}). *Ergasilus ceylonensis* on the gills of the Asian cichlid *Etroplus suratensis* in Sri Lanka was treated successfully with 3–5 ppm potassium permanganate or 3–5 ppm 3,4-dichloropropioanilide but infected fish were unable to tolerate 50–500 ppm formalin, 5 ppm gammexane or 0.1–5.0 ppm fenitrothione (Wijeyaratne and Gunawardene, 1988). Fresh water and hyposaline sea water were ineffective against *E. lizae* (see Conroy and Conroy, 1986).

Kabata (1985) suggested several treatments that include: (i) copper sulphate and ferric sulphate in a 5 : 2 ratio; (ii) Dipterex (Dylox, Chlorophor, Foschlor, Neguvon, Masoten) at 0.15 ppm for 6 hours; (iii) Bromex (dimethyl 1,2-dibromo-2,2-dichloroethyl phosphate) 0.125–0.15 ppm in water of greater than 400 ppm chlorinity (fish in fresh water are more sensitive thus reducing significantly the safety margin); (iv) Malathion (*O,O*-dimethyl *S*-(1,2-dicarbethoxyethyl) phosphorothioate) at 0.2 ppm for 6 hours; and (v) the dyes Brilliant Green and Violet K at 0.1–1 mg l^{-1} which killed nauplii in 3–24 hours.

Prophylactic measures for confined areas suggested by Schäperclaus *et al.* (1991) include: (i) strict quarantine to prevent the import of infected fish and water; (ii) netting of lakes to reduce levels of carrier fish such as *Abramis brama*; (iii) growth of aquatic vegetation to restrict the preference by *Ergasilus sieboldi* for open water; (iv) replace tench by eel or common carp; and (v) free tench from infection by dip bath (10 minutes) in trichlorphon (5 mg l^{-1}).

COPEPODA: *SARCOTACES*

Introduction

These bizarre copepods cause cysts several centimetres long in the muscle under the skin. The cysts are exposed when fish are filleted, and cut cysts release copious black fluid over the fillets.

Sarcotaces verrucosus Olsson 1872, the type species and subject of this section, has a worldwide distribution, from abyssal depths to tropical reefs (Heegaard, 1947a; Kuitunen-Ekbaum, 1949; Moser *et al.*, 1985). Hosts include members of the families Moridae, Macrouridae, Scorpaenidae, Antennariidae, Triglidae, Acanthuridae and Serranidae. Two other species, *S. japonicus* and *S. shiinoi*, were described from two Japanese fish by Izawa (1974). The genus is usually included in the family Phylichthyidae, many of which parasitize the lateral line and other sensory canals of teleosts.

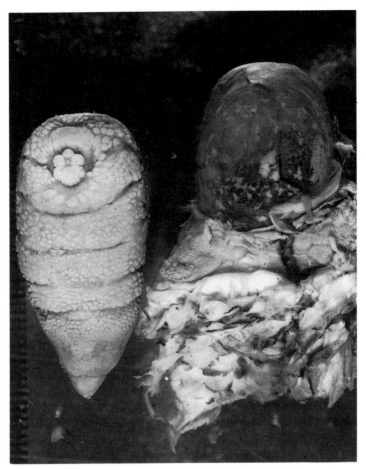

Fig. 13.2. *Pseudergasilus zacconis* A. On gill filaments of *Plecoglossus altivelis* showing extensive hyperplasia and necrosis of the gill. B. Section through infested filament showing deep penetration of second antenna, and mouth parts (arrow) closely apposed to gill epithelium. s = copepod.

Morphology and life cycle

Female up to 45 mm long in cold temperate waters, smaller in tropics, body broadly oval, surface verrucose in bands corresponding to segments, double rosette around minute mouthparts, tail pointed with spine (Fig. 13.3). Most of the bulk of the body is formed from the gut which typically contains a thick black fluid. Male up to 3 mm (Fig. 13.4), 0 to 4 per female (Aitken, 1942).

Eggs embryonate around the female in the pouch formed by the host tissue. They hatch as nauplii, pass out of the sac into the water, and develop through five naupliar stages within 45 hours at 20–22°C. The first copepodid is infective, but can survive at least a week without a host (Izawa, 1973). Developmental stages in fish have not been described.

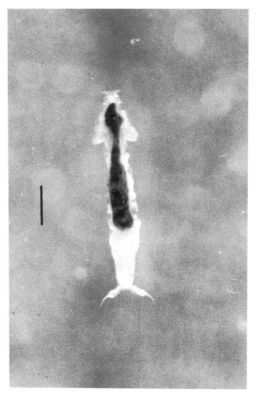

Fig. 13.3. *Sarcotaces verrucosus*, female, ventral view, alongside host capsule (from *Mora moro*, Tasmania). Scale : units 1 mm.

Host–parasite relationships

In Moridae and Serranidae parasites tend to occur beneath the lateral line. In *Lepidion eques* and *Peristedion amissus* the parasites occur on the snout, in the moray *Gymnothorax kidako* they are in the buccal cavity, in *Sebastes* spp. and *Promylantor nezumi* are usually close to the anus, and in *Antennarius* sp. they are scattered over the head and body.

Moser *et al.* (1985) found that the parasites were more prevalent (11/49) on small *Sebastes ciliatus* (less than 345 mm; up to 12 years old) than on large fish (13/116; 350–500 mm; 12–27+ years).

The parasites lie in the dermis and project into the muscle or body cavity. Infections can be difficult to detect externally except when females are over a bony surface in which case they cause a prominent swelling in the skin. The female lies within a pyriform sac of host tissue, usually with a small opening to the outside and the opening is maintained by the tail. The male lies in the same sac, attached to the inner wall and is mobile.

In *Sebastes* spp. the wall of the sac around the anterior end of the female

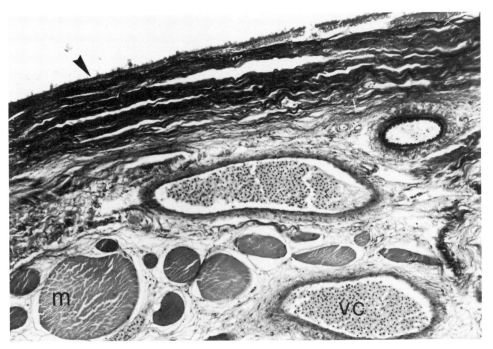

Fig. 13.4. *Sarcotaces verrucosus,* male (from *Epinephelus undulostriatus,* Queensland). Scale bar = 0.5 mm.

is thin and almost transparent. It becomes thicker and opaque towards the posterior end (Kuitunen-Ekbaum, 1949). In serranids the sac wall is uniformly thick and opaque. Histologically, the wall is composed of collagen which is well vascularized on the outer surface (Fig. 13.5). Large degenerate cysts contain dark masses claylike in texture; small degenerate cysts contain pale fibrous tissue only (Moser *et al.*, 1985).

Parasite nutrition

Nauplii and first copepodids are lecithotrophic and lack functional mouth-parts (Izawa, 1974). The source of nutrient for the parasitic stages is not clear. The gut of females contains hemosiderin, a breakdown product of haemoglobin, and erythrocytes of fish have been found in young parasites though not in old (Kuitunen-Ekbaum, 1949). However, there is no evidence of haemorrhage in the wall of the sac and no inflammation that can be related to the mouth parts. Hjort (1895) speculated that parasites initially grew rapidly and expanded filling with blood and when the cyst wall was fully formed they ceased feeding and survived on stored gut contents.

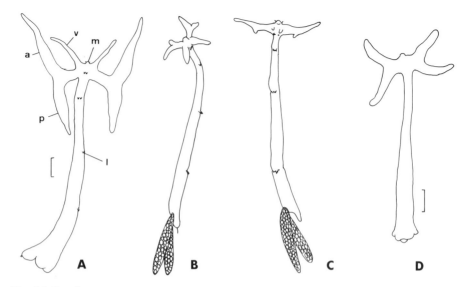

Fig. 13.6. *Lernaea* spp. morphology of female. A and B. *L. cyprinacea.* C. *L. polymorpha.* D. *L. cruciata.* a = anterior process of dorsal horn, l = third pair of legs, m = mouth, p = posterior process of dorsal horn, v = ventral horn. Scale bars 1 mm. (A redrawn from Kabata, 1979; B and C redrawn from Shariff and Sommerville, 1986a; C redrawn from Kabata, 1988.)

COPEPODA: *LERNAEA*

Introduction

Lernaeids or anchor worms are common pests in freshwater aquaculture of cyprinids and to a lesser extent of salmonids and other fish. Epizootics in cultured fish are often associated with high mortality. The parasites also cause problems in man-made lakes (Anon, 1980; Berry *et al.*, 1991) and in commercial aquaria (Shariff *et al.*, 1986; Dempster *et al.*, 1988). They are particularly pathogenic to small fish because of their relatively large size.

Host range, distribution, seasonality

Lernaea cyprinacea (Fig. 13.6A,B) has a worldwide distribution, typically on *Carassius carassius* though it has been reported from 45 other species of cyprinids. It is also found on other Cypriniformes (Catostomidae, Cobitidae), and on fish belonging to the following orders: Acipenseriformes (Acipenseridae), Anguilliformes (Anguillidae), Channiformes (Channidae), Cyprinodontiformes (Cyprinodontidae, Poecillidae), Gadiformes (Gadidae), Gasterosteiformes (Gasterosteidae), Perciformes (Anabantidae, Apogonidae, Centrarchidae, Cichlidae, Gobiidae, Mastacembalidae, Osphronemidae, Percidae), Salmoniformes (Esocidae, Plecoglossidae, Salmonidae, Umbridae) and Siluriformes (Bagridae, Ictaluridae, Siluridae) (Kabata, 1979). In a tropical display aquarium in Malaysia, 23 out of 58 species of fish became infected (Shariff *et al.*, 1986). In North America, the parasite has also been reported on tadpoles (Tidd and Shields, 1963).

Distribution of the parasite is partly dictated by water temperature. Its optimum temperature is 26 to 28°C (Shields and Tidd, 1968). Thus, in temperate climates the parasite is commonest in late summer. It is more prevalent in still or slowly flowing water than in fast flowing streams (Hoffman, 1976; Bulow *et al.*, 1979).

Lernaea polymorpha (Fig. 13.6C) is an economically important parasite on bighead carp, *Aristichthys nobilis* and silver carp, *Hypophthalmichthys molitrix* in southeast Asia (Shariff and Sommerville, 1986a).

Systematics

Kabata (1983) gives a key to genera of Lernaeidae. In the genus *Lernaea* there are over 40 species of which about half are from Africa. The taxonomy is based on the shape of the anchors, a difficult character not only because of their inherent variability (compare Fig. 13.6A and B) but also because their shape is modified by bone or other resistance the anchors encounter during their development in the fish (Fryer, 1961; Thurston, 1969).

Shariff and Sommerville (1990) recommended that identification should be based on a range of specimens. They found, in experimental infections, that *L. polymorpha* (Fig. 13.6C) showed less variability than *L. cyprinacea* and was distinguished by the 'T' bar shape of its anchor, and by the relatively short ventral horns which grew to face each other. Bauer (1991) suggested that *L. cyprinacea* is specific to *Carassius* species, and that it is *L. elegans*, a morphologically similar parasite, that is found on a range of fishes.

Lernaea cruciata, a parasite of *Lepomis*, *Ambloplites*, *Micropterus* and *Morone* spp. in North America, lacks ventral horns (Kabata, 1988) (Fig. 13.6D).

Morphology and life cycle

The adult female *Lernaea cyprinacea* has a small semispherical cephalothorax which contains the mouth. Behind it is a well-developed holdfast normally consisting of a bifurcate dorsal process and a simple ventral process (Fig. 13.6A). The elongate neck and trunk carry the four pairs of legs of the premetamorphosed female. The abdomen is short. *In situ* the holdfast and part of the trunk are buried in the host while most of the trunk and the abdomen project into the water.

Lernaea cyprinacea requires only one host to complete its life cycle. It has three free-living naupliar stages and five copepodid stages which are usually on the gills and are relatively immobile although they are not permanently attached. The male is free-moving and the post-metamorphosed female attached. At 27°C the larval stages take 12 to 17 days to develop into adult males or premetamorphosed females (Shields, 1978; Shariff and Sommerville, 1986b). Adult males die within 24 hours. Females are fertilized and either attack the same host or swim to another host. They chew and bore their way into the host tissues as they metamorphose into adults. Within 1 day and before they have fully metamorphosed they produce their first batch of eggs. These hatch 24 to 36 hours later and the egg sacs are then shed. A new pair of egg sacs are produced within 1 to 3 days (Shields and Goode, 1978; Shariff and Sommerville, 1986b). The largest egg sacs are produced 5 to 10 days post metamorphosis and the parasites die within 30 days at 28 to 32°C.

Development is greatly reduced at lower temperatures; at 20°C nauplii take 7 days to moult into copepodids and cease development altogether below this temperature. Shields and Tidd (1968) found that some metamorphosed females survived at 10°C for over 3 months and grew slowly, though many were lost from the fish. Others were lost when the temperature was raised to 25°C, possibly because they had not burrowed deeply enough during the cold period. Of 71, only six survived to produce eggs. Nevertheless, in cold temperate climates, the parasite probably overwinters as metamorphosed females.

Lernaea polymorpha also requires only one host, *Aristichthys nobilis*, to complete its life cycle (Shariff and Roberts, 1989). Some lernaeids apparently use two. Copepodids of *L. variabilis* were found on short-nosed gars *Lepisosteus platostomus*, whilst adult females occurred on bluegills *Lepomis pallidus* (see Wilson, 1917). In Africa, catfish *Bagrus* spp. carry copepodids of both *Lernaea cyprinacea* and *L. barnimiana* prior to the development of metamorphosed females on *Tilapia* sp. (Fryer, 1966; Thurston, 1969).

Host–parasite relationships

Site and host selection, course of infection

Copepodids of *Lernaea cyprinacea* are generally found attached to gills of fishes, especially cyprinids (Haley and Winn, 1959; Khalifa and Post, 1976). They are also found on the gills of other fish, such as catfish *Bagrus docmac* and even in the branchial chamber of tadpoles *Rana* spp. (Fryer, 1966; Shields and Tidd, 1974). Wilson (1917) collected copepodids of *L. variabilis* from the gills of *Lepisosteus platostomus*.

Copepodids of *L. cyprinacea* will attach and develop elsewhere on the body (Shields, 1978). In an experimental infection on goldfish, Shields and Tidd (1974) found that only 1% were on the gills whilst 77% were on the fins. They suggested that water current was the main factor in determining where the copepodids settled. Shariff and Roberts (1989) observed that the fins of *Aristichthys nobilis* were continually moving once nauplii of *L. polymorpha* were added to the tank suggesting that the parasites were attaching to and irritating them.

Premetamorphosed females usually attack the general body surface and lodge in the superficial layers of the body musculature. Those of *L. cyprinacea*

Fig. 13.7. *Lernaea cyprinacea.* Skin lesions caused by parasites in rainbow trout *Oncorhynchus mykiss* (arrows). (From Berry *et al.*, 1991.)

attach all over the body of cyprinids in still water, whereas in moving water they are principally near the bases of fins. On rainbow trout, a third of them may attach on gills or buccal cavities (McNeil, 1961). On *Tilapia* sp. and *Anguilla anguilla*, the parasites are found principally in the buccal cavity (Fryer, 1966; Ghittino, 1987).

Clinical signs and histopathology

Copepodids of *Lernaea cyprinacea* on small cyprinids cause disruption and necrosis of gill epithelium. Khalifa and Post (1976) reported that large numbers of larvae on the gills caused the death of fish.

Penetration of metamorphosing females is generally associated with punctate haemorrhages which increase to 5 mm across as the parasite grows (Khalifa and Post, 1976). Such lesions both with and without parasites are common in epizootics (Haley and Winn, 1959; Uehara *et al.*, 1984; Berry *et al.*, 1991; Fig. 13.7).

Muscle necrosis is evident at the anterior end of penetrating parasites. Epithelial tongues extend into the muscle alongside the canal, and beneath these and around the head are infiltrating leucocytes. Extravasated blood is frequently present between the parasite cuticle and the epithelial layer. This may ooze into the water together with cellular debris. Eventually a thick connective tissue sheath envelopes the part of the parasite within the host (Fig. 13.8; Khalifa and Post, 1976; Shields and Goode, 1978; Berry *et al.*, 1991). Inflammation in goldfish was not evident at 10°C and resumed when the temperature was raised to 25°C (Shields and Tidd, 1968). The point of entry is eventually surrounded by a collar of hyperplastic epithelium and

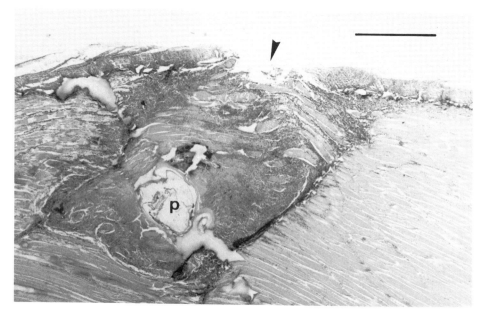

Fig. 13.8. *Lernaea cyprinacea* in rainbow trout *Oncorhynchus mykiss*. Histological section of infected skin showing lesion around the holdfast (p) and ulcer at surface (arrow). Scale bar = 0.5 mm. (From Berry *et al.*, 1991.)

thickened dermis. In *Catostomus commersoni*, the epidermal collar contained large numbers of an eosinophilic blood granulocyte, and the dermal collar contained many neutrophils (Lester and Daniels, 1976).

In tadpoles, the parasite caused more tissue destruction but stimulated less response. Epithelial tongues were less evident, and the connective tissue sheath was less dense (Shields and Goode, 1978).

Lernaea cruciata in white bass *Morone chrysops* caused an initial mild necrosis which initiated a mild oedema and infiltration of neutrophils around the anchor. This was followed by macrophage infiltration and phagocytosis of dead neutrophils and other cell debris. Fibroblasts then proliferated and eventually matured into fibrocytes with collagen deposition. They formed part of a chronic granuloma which adhered tightly to the anchor processes and cephalothorax. Growing anchors were able to penetrate scales (Joy and Jones, 1973). Burrowing females in largemouth bass *Micropterus salmoides* were completely concealed beneath the scales, and caused yellow to red local discoloration. Host reaction was primarily a mononuclear infiltrate containing eosinophilic granular cells. This was mostly confined to the scale pocket containing the parasite (Noga, 1986).

Punctate haemorrhages appeared when female *L. polymorpha* began to penetrate bighead carp *Aristichthys nobilis*. The transparent body of the parasite became milky and finally an ivory colour. Heavily infected fish became sluggish and died (Shariff and Roberts, 1989).

In naive fish, penetrating females caused disruption of the epidermis and dermis, necrosis and haemorrhage. This was followed by acute inflammation and finally by a highly vascular chronic granulomatous fibrosis, whereby collagen fibres encapsulated the horns of the parasite. The penetration site was first sealed around the parasite by epidermal migration, then by inflammatory exudate and finally by a collar of fibrous tissue. In resistant fish, the breach in the epidermis was relatively small but there was extensive haemorrhage in the dermis. The thickened epidermis and oedematous dermis contained far more eosinophilic granule cells and lymphocytes than naive fish, apparently the result of immune hypersensitivity (Shariff and Roberts, 1989).

Lernaea cyprinacea in the eyes of trout caused blindness (Uzmann and Rayner, 1958). The holdfast in the anterior chamber of the eye of bighead carp caused extensive mechanical damage and a severe inflammatory response. The antlers were surrounded by bands of fibrous tissue, the cornea was opaque, the lens capsule ruptured, and an inflammatory exudate of mononuclear cells and neutrophils filled the anterior chamber (Shariff, 1981).

Host immune response

Shields (1978) suspected that goldfish acquired immunity to *L. cyprinacea* because he was unable to reinfect fish that had once had a heavy infestation. Shields and Goode (1978) observed that half of the parasites on experimentally infected goldfish were rejected by the host 1 week after maturation of the first egg sacs. Wounds left by rejected females healed rapidly.

Shariff *et al.* (1986) found that after an epizootic of *L. cyprinacea* in a display aquarium, 18 out of 23 fish species had apparently acquired resistance by the time of a second outbreak 6 months later. Woo and Shariff (1990) demonstrated experimentally that few new infections of *L. cyprinacea* developed in *Helostoma temmincki* which had recovered from an earlier infection. Fish that became infected lost their infections rapidly. Parasites on resistant fish produced fewer eggs and the resultant larvae were less infective than those from females on naive fish.

Shariff (1981) proposed that bighead carp developed immunity to *Lernaea* sp. because initial infections occurred on the skin, and later infections in the eye. In experimentally infected fish the prevalence of *L. polymorpha* first increased then after 4 months declined to zero (Shariff and Sommerville, 1986c). Degenerating females were found in haemorrhagic ulcers of resistant fish (Shariff and Roberts, 1989).

Noga (1986) suggested that haemorrhagic ulcers in largemouth bass *Micropterus salmoides* which contained degenerating *Lernaea cruciata* were the result of immune hypersensitivity.

Mechanism of disease

The mechanical destruction of epidermis and dermis caused by grazing copepodids stimulates inflammation (Shields and Tidd, 1974). When the parasites are abundant on the gills, this presumably precipitates respiratory distress through oedema and slowed blood circulation in the lamellae.

Host mortality caused by adult females is generally the result of the physical destruction of tissues. They also induce pressure necrosis, and may secrete a histolytic and/or digestive enzyme (Shields and Goode, 1978; Shariff and Roberts, 1989). Small fish (4–6 cm) and tadpoles suffer more than large fish because the head of the parasite often enters the body cavity and penetrates the liver, or penetrates the brain, spinal cord or other vital organ. Tadpoles apparently died from blood loss through haemorrhages caused by the penetration of metamorphosing females (Tidd and Shields, 1963).

Effects on host

Bighead carp *A. nobilis* in ponds infested with *L. polymorpha* were up to 35% lighter than carp in uninfested ponds (Shariff and Sommerville, 1986c). Six or more females per fingerling slowed the growth of cultured *Cyprinus carpio* and *Ctenopharyngodon idella* (Faisal *et al.*, 1988). *Lernaea cyprinacea* hindered the feeding and growth of eels (*Anguilla anguilla*) (Ghittino, 1987).

During an outbreak of *L. cyprinacea* in a display aquarium, an overall mortality of 7–9% was associated with the parasite (Shariff *et al.*, 1986). High mortality in small cyprinids, salmonids and tadpoles has frequently been associated with infection by *L. cyprinacea* (Tidd and Shields, 1963; Khalifa and Post, 1976; Anon, 1980).

Concurrent infection and cross immunity

Khalifa and Post (1976) observed unspecified bacterial and fungal infections in the wound caused by the penetration of *L. cyprinacea* into cyprinids. Noga (1986) investigated early 'red-sore' lesions on the skin of largemouth bass *Micropterus salmoides* and found that at least 35% of them contained the remains of *Lernaea cruciata*. The parasites were almost all metamorphosing females. Bacteria, particularly *Aeromonas hydrophila*, were present in some of the early lesions and bacteria of several species predominated in more advanced lesions. He suggested that the wounds caused by the copepod allowed in bacteria and fungi, and thus initiated red-sore disease.

Wilson (1917) observed that the gills of *Lepisosteus platostomus* carrying copepodids of *Lernaea variabilis* frequently did not have glochidia, and vice versa.

In vitro culture

Egg sacs, either separate or still attached to excised females, develop normally in dishes in non-aerated water. Nauplii hatch and moult three times to the infective first copepodid in the dish (Grabda, 1963; Shields, 1978).

Parasite nutrition and physiology

Naupliar stages do not feed. Copepodids apparently graze on epidermis and dermis (Shields and Tidd, 1974). Metamorphosed females ingest tissue debris and erythrocytes (Khalifa and Post, 1976) released by mechanical damage from their mouthparts and possibly from the secretion of digestive enzymes (Shariff and Roberts, 1989).

The osmolarity of the haemolymph of attached metamorphosed females is similar to that of the host, though the parasites died within 24 hours in 90% sea water. Attached females in 15% or 30% sea water survived for at least 6 days but failed to produce viable eggs. They resumed normal reproduction within 48 hours of return to fresh water. In 10% sea water the parasite successfully completed its life cycle. Excised parasites showed little ability to osmoregulate (Shields and Sperber, 1974).

Diagnosis of infection

Adult females can be seen macroscopically; copepodids require the use of a dissecting microscope.

Prevention and control

Elimination of the parasite usually requires treatment over several weeks to break the life cycle at the larval stage because embedded females are difficult to kill.

Organophosphate insecticides, particularly trichlorphon (Dipterex, Neguvon, Masoten) are commonly employed. As copepodids take 8–9 days to develop (at 27°C), treatment is carried out every 7 days to prevent reinfection until all the adult females have died (within 30 days at 27°C). Trichlorphon at 0.25 ppm will kill the copepodid stage but not the nauplii or adults (Sarig, 1971; Kabata, 1985). It is less effective at high temperatures. Bromex (dimethyl 1,2-bromo-2,2-dichlorethyl phosphate) at 0.12–0.15 ppm kills both nauplii and copepodids (Sarig, 1971).

Ghittino (1987) treated eels with trichlorphon (0.25–0.3 ppm) in 3 to 6 weekly treatments. Treatment was discontinued at least 1 month before eels were harvested for human consumption.

Shariff *et al.* (1986) successfully treated infested fishes with 0.16 ppm Unden (2-isopropoxy-phenyl-*N*-methylcarbamate) at weekly intervals for 4 weeks then a final treatment of 0.16 ppm trichlorphon (Dipterex) in week 5. Nauplii were not affected by either drug. Unden initially immobilized copepodids within 2 hours and killed them, but by the 4th treatment half the copepodids were still alive. Both drugs caused fish to cease feeding.

A safer non-residual chemical that has been used effectively is sodium chlorite (Dempster *et al.*, 1988). When the chlorite ion ClO_2^- was maintained at a concentration of 20–40 mg l^{-1} and the pH above 6, the chlorite eradicated *L. cyprinacea* from a commercial aquarium and was non-toxic to fish. Initially bacteria in the biological filters ceased to function and water had to be exchanged during the first 2 weeks to keep the ammonia and nitrite levels down. However, chlorite resistant strains of bacteria developed later and the filters again became biologically active.

Hoffman and Lester (1987) reported that Dimilin (UniRoyal Chemical Co., USA), a growth regulator, at 0.03 ppm eradicated *L. cyprinacea* from golden shiner (*Notemigonus crysoleucas*). Other chemicals that have been used in the past are in Hoffman and Meyer (1974) and Hoffman (1976). Kabata (1985) gives details of herbal remedies such as teaseed cake that have been used in China.

For small numbers of fish, adult female parasites can be removed by hand, larval stages killed chemically, and the fish returned to clean water. Scott and Fogle (1983) bathed adult infested carp (*Cyprinus carpio*) for 5 min in 1% trichlorphon (Dipterex; 10 g active ingredient per litre), anaesthetized the fish, removed visible copepods, then put the fish in a clean tank. Thereafter the water was treated with 0.2–0.25 ppm trichlorphon monthly. Faisal *et al.* (1988) eradicated *L. cyprinacea* from carp broodstock by mechanically removing metamorphosed females and placing the fish in potassium permanganate ($25\,\mathrm{g\,m^{-3}}$) for 30 min to kill other stages.

To avoid environmental and other side effects of chemical treatments, several biological methods have been proposed. Kabata (1985) reported that the copepod *Mesocyclops* preyed on free-swimming larval *Lernaea* and suggested that planktonic predators could be used in biological control.

Shariff and Sommerville (1986b) proposed that all fish should be removed from the pond for a minimum of 7 days (at 25–29°C) as the copepodids die on the 6th day if no host is available. Woo and Shariff (1990) suggested that no naive fish be introduced for a period as the resistance that recovered fish acquire may be sufficient to break the life cycle. Shields (1978) observed that goldfish removed maturing parasites from each other. Increasing the rate of water exchange will help to reduce the numbers of larvae.

COPEPODA: CALIGIDAE

Introduction

The sea lice, *Lepeophtheirus salmonis* (Fig. 13.9) and *Caligus elongatus* (Fig. 13.10) are the main parasites of sea-caged Atlantic salmon, *Salmo salar*, in the northern hemisphere (Wootten *et al.*, 1982; Pike, 1989). They cause skin erosion, most often on or near the head. Heavily infected fish die. Smolts newly introduced to sea water are the most susceptible (Wootten *et al.*, 1982). Outbreaks of *L. salmonis* in farmed salmon first occurred in the mid 1960s in Norway and in the mid 1970s in Scotland (Wootten *et al.*, 1982). Schram and Haug (1988) consider *L. hippoglossi* as a potential threat to the culture of halibut, *Hippoglossus hippoglossus* (L.). Other species of *Lepeophtheirus* are found on a wide range of host species (Kabata, 1979) but no economic impact has been attributed to them.

Other species of *Caligus* have caused mortalities in cultured and wild populations of fish and are potential threats to marine aquaculture. *Caligus patulus* killed cultured milkfish, *Chanos chanos*, in the Philippines

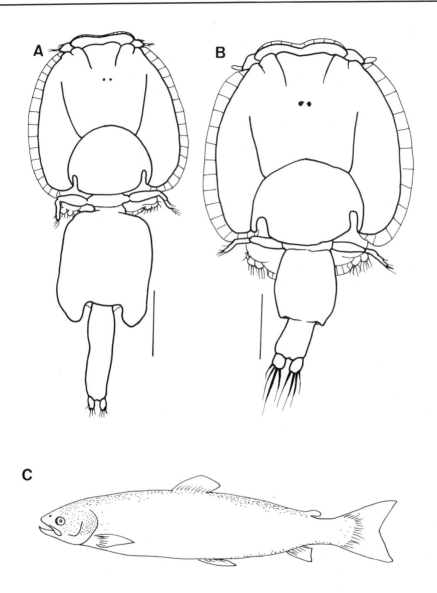

Fig. 13.9. *Lepeophtheirus salmonis*. A, female. B, male. (Based on Kabata, 1979.) Scale bars = 1 mm. C, distribution on Atlantic salmon (after White, 1940.)

(Lavina, 1977; Jones, 1980; Lin, 1989), and *Pseudocaligus apodus* killed cultured mullet in Israel (Paperna and Lahav, 1974). *Caligus epidemicus* (Fig. 13.11) killed wild fish in southern Australia (Hewitt, 1971) and tilapia, *Oreochromis mossambicus*, in Taiwan (Lin and Ho, 1993). Death of rainbow trout, *Oncorhynchus mykiss*, was caused by *C. orientalis* (see Urawa and Kato, 1991).

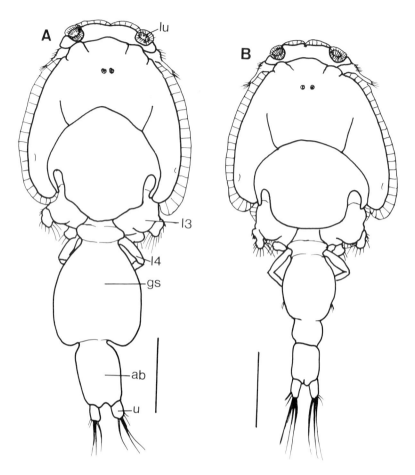

Fig. 13.10. *Caligus elongatus*. A. Female. B. Male. Note the large genital segment (gs) in the female. ab = abdomen, l3 and l4 = third and fourth pairs of legs, lu = lunules, u = uropod. Scale bar = 1 mm. (Based on Kabata, 1979.)

Host range, geographical distribution, seasonality

Lepeophtheirus salmonis has a circumpolar distribution in the northern hemisphere where it is largely restricted to the Salmonidae, especially the genera *Salmo, Salvelinus* and *Oncorhynchus* (see Kabata, 1979; Egidius, 1985). The prevalence and intensity of infection on wild fish are usually low, although White (1940) found grilse of *Salmo salar* entering the estuary of the Moser River, Nova Scotia, to be heavily infected and exhibited severe lesions. Berland (1993) found the prevalence and intensity of *L. salmonis* and *C. elongatus* on wild salmon in west Norway to be similar in 1973 and 1988, but significantly increased in 1992. In northwest and northeast Ireland, all production of *L. salmonis* larvae is from wild stocks of *S. trutta*, but in the west 94% of *L. salmonis* larvae originated from farmed salmon (Tully

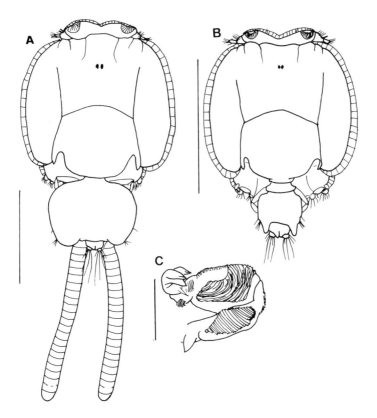

Fig. 13.11. *Caligus epidemicus.* A. Female. B. Male. Scale bars = 1 mm. Note the sexual dimorphism. C. Second antenna of male. Scale bar = 0.1 mm.

and Whelan, 1992). These authors estimated that 1–38 million larvae were produced per day from single salmon farms. In addition, higher sea water temperatures between 1989 and 1992 decreased generation times and increased production of copepod larvae (Tully *et al.*, 1993). *Lepeophtheirus salmonis* is absent from sites with lowered salinity. Farms in Norway are in sites with a consistently higher salinity than in Scotland and have more severe *L. salmonis* infections (Pike, 1989). Salmon farms in the Bay of Fundy, Canada, are more adversely affected by *C. elongatus* than by *L. salmonis* (Hogans and Trudeau, 1989a,b). During a survey of *L. salmonis* on salmon in the northern Pacific, Nagasawa *et al.* (1993) found 78% and 15% of all individuals in this species on pink salmon (*Oncorhynchus gorbuscha*) and chinook salmon (*O. tschawytscha*), respectively, with fewer on steelhead trout (*O. mykiss*), coho (*O. kisutch*), chum (*O. keta*) and sockeye (*O. nerka*) salmon.

Lepeophtheirus pectoralis infects pleuronectids such as plaice, *Pleuronectes platessa*, flounder, *Platichthys flesus*, and dab, *Limanda limanda*. It is found in the north-east Atlantic Ocean, western Baltic Sea and the White Sea (Boxshall, 1974c). Zeddam *et al.* (1988) have shown *L. thompsoni* to infect only turbot, *Psetta maxima*, whereas *L. europaensis* infected brill,

Scophthalmus rhombus, and flounder, *Platichthys flesus*. Meeüs *et al.* (1990) infected turbot with both *L. thompsoni* and *L. europaensis* and hybridized the two species of parasite experimentally. They attributed natural reproductive isolation to host specificity of infective stages and the low probability of heterospecific mating. The haematophagous *L. thompsoni* has larger clutches and its eggs and larvae tolerate greater salinity change than the mucophagous *L. europaensis*. Within the Mediterranean Sea there is an adaptive heterogeneity in *L. europaensis*, which occur at higher densities and with greater clutch sizes on flounder than on brill; eggs and free-living stages tolerate a greater salinity range if obtained from flounder (Meeüs *et al.*, 1993a,b).

Caligus elongatus is a cosmopolitan species and has been found on over 80 species of fish in 17 orders and 43 families (including salmonids, pleuronectids, scombrids, clupeids, gadids and elasmobranchs). It is the most common species of parasitic copepod in British waters (Parker, 1969; Boxshall, 1974c; Kabata, 1979), is rare on wild salmon but is more common on wild sea trout, *Salmo trutta* (Wootten *et al.*, 1982), and has been recorded on cultured brook trout, *Salvelinus fontinalis*, and rainbow trout, *O. mykiss*, in eastern Canada (Hogans and Trudeau, 1989b). Eggs and adults of *C. elongatus* were found to be significant dietary items of 0+ year-class cod, *Gadus morhua*, and haddock, *Melanogrammus aeglefinus* (see Neilson *et al.*, 1987).

Seasonality

Gravid females and other stages of *Lepeophtheirus salmonis* occur year-round on salmon farms in northern Europe, with a succession of generations during the year (Wootten *et al.*, 1982). The greatest numbers of *L. salmonis* occurred in late summer and autumn due to an accumulation of the parasite from successive generations, but Tully (1989) found that the total intensity of infection by *L. salmonis* did not rise cumulatively because mature lice disappeared from fish before maturation of the next generation was complete. Although reproduction appears to be temperature-dependent in cold water sites (Wootten *et al.*, 1982), there was little reproduction during the summer in Ireland until the onset of winter because high summer water temperatures adversely affected a parasite adapted to the colder water of higher latitudes (Tully, 1989). Adult females from winter generations are significantly larger, produce larger egg sacs and more but smaller eggs than in summer (Ritchie *et al.*, 1993).

Lepeophtheirus pectoralis has its greatest prevalence (>95%) and intensity (5 per fish) of infection on flat fish in summer (August–September) (Boxshall, 1974a). There is a bimodal egg production in May and August, the first peak resulting from gravid females that have overwintered (parasite longevity of 10 months), but die in spring after producing 2–3 pairs of egg strings. The resulting larvae produce a summer generation which develop rapidly (but are shorter lived than the winter generation) with a peak of egg production in August (Boxshall, 1974a). The abundance of *L. hospitalis* on the starry flounder, *Platichthys stellatus*, in Yaquina Bay, Oregon, increased in summer, decreased with the winter rain period, and increased again in spring with the migration of other infected starry flounder through the bay to spawn upstream (Voth, 1972).

In Scotland, *Caligus elongatus* infects sea-caged salmon year-round with the greatest number in autumn, but with an apparent lack of well marked successive generations (Wootten *et al.*, 1982). Although large invasions by larvae do occur, relatively few copepods seem to mature; the parasite may die or adult parasites enter the plankton (Wootten *et al.*, 1982). Possible reservoirs of infection in wild fish include saithe, *Pollachius virens*, and herring, *Clupea harengus* L. (Wootten *et al.*, 1982; MacKenzie and Morrison, 1989; Bruno and Stone, 1990).

The abundance of *C. elongatus* on cultured Atlantic salmon in the lower Bay of Fundy, Canada was highest in late summer and autumn (average 17 per fish) and lowest in winter (1–3 per fish) (Hogans and Trudeau, 1989b). Market-size Atlantic salmon and smolt have similar numbers of parasites because the annual cycle of infection prevents accumulation in older fish (Hogans and Trudeau, 1989b). The generation time in laboratory culture of *C. elongatus* is 5 weeks at 10°C. This allows 4–8 generations annually in the water temperature range of 5–14°C in the Bay of Fundy (Hogans and Trudeau, 1989a). The parasite leaves the host at temperatures lower than 6°C (Stuart, 1990). *Caligus curtus* was found rarely on these salmon, and was found usually on gadoid fish such as cod, haddock and pollack (Hogans and Trudeau, 1989b). *Caligus clemensi* is the only species in marine waters of British Columbia where it infects a wide range of clupeiform, perciform, gasterosteiform and gadiform fish in surface waters (Parker and Margolis, 1964).

Caligus epidemicus infects four species of marine bream, *Acanthopagrus australis*, *A. berda*, *A. butcheri* and *A. latus*, from around Australia (Byrnes and Rohde, 1992) as well as other species of fish. The parasite was abundant on *A. australis* in an estuary in northern New South Wales during one year (2.8 per fish) but not the next (0.5 per fish) although salinity and temperature were similar in both years; apparently the low level of host specificity permits other hosts to be utilized (Roubal, 1990). An epizootic of *C. epidemicus* within the lower Mitchell river, Victoria, Australia caused the death of bream, *A. butcheri*, and mullet, *Mugil cephalus*, *Aldrichetta forsteri*, *Liza argentea* and *Myxus elongatus*; the epizootic was associated with high salinity up to 28 ppt and temperature up to 21.8° during a severe drought (Hewitt, 1971). An epizootic of *C. pageti* on a fish farm in Egypt was also associated with a high salinity up to 45 ppt (Hewitt, 1971). *Caligus orientalis* is a euryhaline parasite and infects a wide range of hosts in Japan, including *Hucho perryi*, *Tribolodon hakonesis*, *Hypomesus transpacificus* and *Mugil cephalus*; these fish may act as reservoir hosts for infection of cultured rainbow trout (Urawa and Kato, 1991).

Caligus shistonyx, *C. epidemicus* and other *Caligus* spp. have been found in plankton trawls (Cowles, 1930; Hewitt, 1971). *Caligus epidemicus* has been recorded also on the telson and legs of the penaeid prawn, *Penaeus monodon* (see Ruangpan and Kabata, 1984).

Systematics

The family Caligidae contains 23 genera. The general body shape and relative size of body tagma are the primary features used in identification, although setation of appendages is also helpful. Keys are provided by Yamaguti (1963), Cressey (1967), Kabata (1979) and Margolis *et al.* (1975).

Morphology and life cycle

Adult caligids usually show sexual dimorphism; the female is usually larger than the male, and male appendages, particularly the first maxillae and second antennae, are modified to aid attachment during copulation. Adults of *Lepeophtheirus salmonis* have a large, rounded, flat cephalothorax characteristic of most caligids. The female, 10–18 mm long, has a more prominent genital segment than the male (5–7 mm long) (Kabata, 1979) (Fig. 13.9). Paired egg strings up to 2 cm long and bearing a total of 700 eggs (107–1220, see Wootten *et al.*, 1982; Costello, 1993) are produced by the female from the posterior end of the genital segment. The limbs are similar in both sexes but the male has striated ventral surfaces on the second antennae to enhance attachment to the posterodorsal surface of the female during mating. The male mates predominantly with the second of the two pre-adult female stages, but also mates with the first pre-adult and adult females (Wootten *et al.*, 1982); a similar observation was reported for *L. polyprioni* (Hewitt, 1964). However, experimental studies by Anstensrud (1990d) found that whereas mating of adult males with adult females of *L. pectoralis* resulted in successful spermatophore transfer such was not the case between adult male and pre-adult female matings.

Caligus spp. have frontal lunules that are absent in *Lepeophtheirus*. Frontal lunules are evident in the late chalimus stages of *Caligus*. Male and female *Lepeophtheirus* can be distinguished at the first pre-adult stage. However, young stages of the parasites are more difficult to identify, especially in mixed infections. The chalimus stages of *C. elongatus* have a longer frontal filament and paired eyespots whereas *L. salmonis* has a single eyespot (Wootten *et al.*, 1982).

Adult females of *C. elongatus* (Fig. 13.10A) are 6–8 mm long; males (Fig. 13.10B) have a slimmer genital segment and are about 5 mm long. The parasite is golden-brown or yellow in colour (Hogans and Trudeau, 1989b). Two egg sacs each contain about 30 eggs (Mackinnon, 1992). The appendages of *C. epidemicus* are shown in Figure 13.12.

The internal anatomy of the Caliginae, especially the reproductive system, has been described by Wilson (1905a) and Gnanamuthu (1950). The ultrastructure of sensory structures was described by Bron *et al.* (1993a) and Gresty *et al.* (1993) and the frontal filament by Pike *et al.* (1993). Unlike other copepods, the midgut is not divided into different zones (Nylund *et al.*, 1992). Lewis (1969) discussed the homology of the maxillae, and subsequent studies on the musculature of this and other appendages in *L. pectoralis* were given

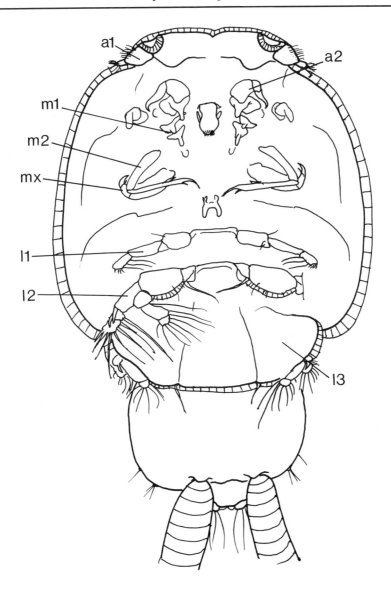

Fig. 13.12. *Caligus epidemicus*, ventral view showing appendages. a1, a2 = first and second antennae, l1, l2, l3 first, second and third leg, m1 m2 = first and second maxillae, mx = maxilliped.

by Boxshall (1990). Locomotion of adult caligids was described by Kabata and Hewitt (1971).

Kabata (1972b) proposed that the life cycle of caligid copepods consists of five phases and ten stages. The life cycle of *L. salmonis* has the typical caligid complement of two free-living naupliar stages (N1 and N2), an infective copepodid stage (C), four attached chalimus stages (Ch1–Ch4), two free-living, pre-adult stages (PA1, PA2) and one adult stage (Johnson and

Albright, 1991a,b). Descriptions of the whole or part of the life cycle of *Lepeophtheirus* spp. can be found in White (1942), Lewis (1963), Voth (1972), Boxshall (1974a), Johannessen (1978) and Schram (1993). Studies of various completeness on the life cycles of *Caligus* spp. include Wilson (1905b) Heegaard (1947b), Hwa (1965), Izawa (1969) Hewitt (1971) and Jones (1980). However, nine stages (2N, 1C, 4Ch, 1Pa, 1A) have been recorded for most *Caligus* spp., with 11 stages (6Ch) for *C. epidemicus* (see Lin and Ho, 1993) and eight (4Ch, 0PA) for *C. punctatus* (see Kim, 1993).

Temperature and duration of development stages

The generation time for *Lepeophtheirus salmonis* is 8–9 weeks at 6°C, 6 weeks at 9–12°C and 4 weeks at 18°C (Wootten *et al.*, 1982; Stuart, 1990). Up to four generations can occur between May and October in Scotland with a summer water temperature of 9–14°C (Wootten *et al.*, 1977, 1982). Tully (1989) found a generation time (ovigerous female to ovigerous female) of 56 days at 13.6°C (males took 52 days) for *L. salmonis* in an experimental cage in Ireland; Johnson and Albright (1991a) reported a generation time of 7.5–8 weeks (at 10°C) in the laboratory. Duration of egg development varies from 17.5 days at 5°C to 5.5 days at 15°C, with durations at these respective temperatures of 52 hours and 9.2 hours for the N1 stage and 170 hours and 35.6 hours for the N2 stage. Durations of the copepodid (up to 10 days), Ch1 (5 days), Ch2 (5 days), Ch3 (9 days), Ch4 (6 days), PA1 (10 days and PA2 (12 days) stages have been observed (Johnson and Albright, 1991b).

Acclimatization of ovigerous females of *L. salmonis* to 11.5°C promotes successful hatching at a higher temperature (22°C) than does acclimatization at a lower temperature (Johannessen, 1978). Tully (1989) found that lower water temperature promoted larger body size, longer developmental time but greater fecundity. The period from fertilized egg to first egg laying by *L. europaensis* is 44 days (15°C), with a maximum life span of 135 days (Meeüs *et al.*, 1993a).

The copepodid stage of *Caligus elongatus* lasts 50 hours at 13°C, the NI and N2 stages last 15–30 hours and 35 hours, respectively, at 10°C, with no moults below 3°C (Hogans and Trudeau, 1989b). Cabral (1983) reported that *C. minimus* develops from copepodid to ovigerous female in 18 days at 20°C, 25 days at 18°C and 35 days at 15°C. at 35 ppt salinity. The developmental cycle of *C. epidemicus* from hatching to ovigerous female is 17 days at 24.5°C and 10–11 days for *C. pageti* (see Lin and Ho, 1993). Durations of developmental stages of *C. epidemicus* are N1 (6 hours), N2 (14.5 hours), C (up to 3–4 days), Ch2–6 (each approx. 1 day) (Lin and Ho, 1993).

Host–parasite relationships

Site and host selection, course of infection

Lepeophtheirus salmonis nauplii and copepodids are positively phototactic, swim upwards in response to pressure, but not to chemical cues. Host contact is by a burst swimming response to water flow or mechanical vibration

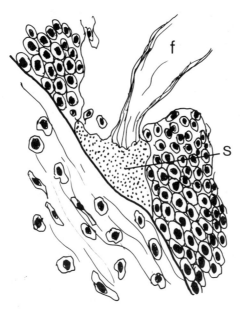

Fig. 13.13. *Lepeophtheirus salmonis.* Section through the attachment site of a copepodid recently attached to *Salmo salar*. The frontal filament (f) adheres to the basal membrane by an adhesive secretion (s). (Redrawn from Bron *et al.*, 1991.)

generated by a host (Bron *et al.*, 1993a). The copepodid grips the surface of the host with its clawed antennae. The surface is examined by the antennules that bear high threshold contact chemoreceptors; non-salmonid hosts are rejected and the copepodids reenter the water column (Bron *et al.*, 1993a). On suitable salmonid hosts, the antennae penetrate the epidermis and dermis and the anterior end of the cephalothoracic shield is pushed into the epidermis causing it to separate from the basement membrane. The frontal filament then attaches to the basement membrane by an adhesive secretion (Fig. 13.13), and the larvae moult into the attached chalimus stages (Bron *et al.*, 1991; Jones *et al.*, 1990). Voth (1972) noted that the copepodid of *L. hospitalis* makes rapid jabs to the skin, releases a droplet of adhesive and the copepodid then backs away to draw out the frontal filament which hardens almost immediately; the process takes 5 minutes to complete. The frontal filament of *C. elongatus* attaches directly to the fish scale without an adhesive secretion (Pike *et al.*, 1993).

Chalimus stages of *L. salmonis* are permanently attached by a frontal filament to dorsal and pectoral fins and around the anus of wild and caged fish (Wootten *et al.*, 1982), but under confined experimental conditions, the fins as well as the buccal cavity and gills were infected (Bron *et al.*, 1991). Following experimental infection, most copepodid and chalimus larvae were found on the gills of *Salmo salar* (see Johnson and Albright, 1991b), a site apparently not examined in previous studies. Large numbers of adults were found in clusters on the skin near the anal fin of *Oncorhynchus gorbuscha* and *O. nerka* (see Nagasawa, 1987).

Pre-adult and adult stages of *L. salmonis* move freely over the surface of the host and are capable of leaving the host to infect other fish; they are abundant on the head and dorsal surface and to a lesser extent on the posterior ventral surface (White, 1940; Wootten *et al.*, 1982). They attach by suction generated with the cephalothorax and sealed by its marginal membrane and third pair of legs (Kabata and Hewitt, 1971). During ecdysis, pre-adults of *L. pectoralis* attach by a fine frontal filament arising from the 'median sucker' of Wilson (1905a) (Anstensrud, 1990d).

Nauplii of *L. pectoralis* are positively phototactic; copepodids are positively phototactic initially but later sink to the bottom whereupon they display positive rheotaxis (Boxshall, 1976). Copepodids attach where they contact a suitable host but do not attach to unsuitable hosts, implying a chemical recognition. *Lepeophtheirus pectoralis* larvae from flounder preferentially settled on flounder whereas *L. pectoralis* larvae from plaice preferred plaice. Infection occurs on or near the base of the body fins. Mobile post-chalimus stages occur on the general body surface where copulation occurs also. Males, non-ovigerous females, and juveniles of *L. pectoralis* attach to the body, fins and walls of the branchial chamber but ovigerous females prefer the undersurface of pectoral and pelvic fins (Boxshall, 1976, 1977). *Lepeophtheirus pectoralis* had an overdispersed frequency distribution on plaice, which suggested to Boxshall (1974b) that fish acquired several copepodids at once due to the aggregated distribution of free-swimming infective larvae.

Caligus elongatus adults and larvae attach to most areas of the body of salmon, but adults prefer the dorsal and lateral surfaces of the head, the anterior portion of the abdomen between the opercula, and the base of the caudal fin (Hogans and Trudeau, 1989b). The ventrolateral surfaces near the pelvic fin, around the anus and the base of the dorsal fin are also preferred sites (Hogans and Trudeau, 1989a). The copepodids and chalimus stages of *C. epidemicus* are found on the body surface and fins of bream, *Acanthopagrus australis*, and the adults move about on the body surface; attached stages of *C. epinepheli* occur on the gill arch and the adult parasites within the buccal cavity (Roubal, 1981). This suggests an active site selection by the copepodids. We have observed adult *C. epidemicus* that had settled on the bottom of a container to respond with rapid upward swimming when a shadow passed over the container.

Fraile (1986) reported that larvae of *C. minimus* were attracted by the mucus of its host, the sea bass *Dicentrarchus labrax*; the parasite displayed positive photaxis and positive rheotaxis at fast water currents but not in low water currents. Larvae of *C. elongatus* (syn. *rapax*) infected young whiting in the shade of floating logs and medusae, primarily *Cyanea* spp. and *Rhizostoma* spp., in the sea off Plymouth; this suggests a preference of the parasite larvae for these habitats (Russell, 1933; Shotter, 1973).

Clinical signs and histopathology

Initially, whitish spots across the neck and along the base of the dorsal fins indicate sites of feeding by *Lepeophtheirus salmonis*. At higher levels of

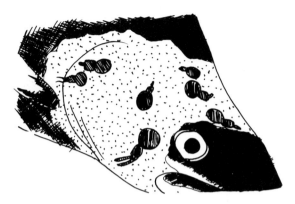

Fig. 13.14. *Lepeophtheirus salmonis* on Atlantic salmon *Salmo salar* showing severe erosion of the epidermal and deeper tissues of the head and operculum. (Drawn from Pike, 1989.)

infection these become skin lesions and then large open wounds (Fig. 13.14); there may be subepidermal haemorrhaging and erosion of the skin to expose the cranial bones (Wootten *et al.*, 1982). The large open wounds may be associated with secondary bacterial infections (Egidius, 1985). Secondary fungal infections may ensue if fish with exposed wounds are returned to fresh water (Hastein and Bergsjo, 1976). Oedema, hyperplasia, sloughing of epidermal cells and inflammation are caused by attachment and feeding of pre-adult and adult *L. salmonis* (Jonsdottir *et al.*, 1992). Chinook and coho salmon are more resistant to *L. salmonis* than Atlantic salmon and respond to infection by extensive epithelial hyperplasia and inflammation (Johnson and Albright, 1992a). Coho salmon implanted with hydrocortisone ($0.5 \, \text{mg g}^{-1}$ body weight) produced a diminished epithelial hyperplasia and inflammation, and were more susceptible to infection by *L. salmonis* than were control coho salmon. This suggested to Johnson and Albright (1992b) that nonspecific host defence mechanisms are important in the resistance to *L. salmonis*.

Jones *et al.* (1990) described the histopathology associated with early developmental stages of *L. salmonis*. Initial mechanical damage caused by copepodid and attachment and feeding by Ch1 and 2 causes a mild epidermal hyperplasia. Later stages cause damage by feeding with a focus of irritation around the periphery of the lesion associated with the frontal filament. The greatest damage is associated with the remnant of the frontal filament following detachment by the Ch4 stage. Lesions, 0.5 cm in diameter, have an outer ring of heavily pigmented tissue and a depressed core of white skin. The basement membrane is reorganized over the remains of the filament and there is dermal fibrosis with inflammatory infiltration.

Damage caused by *L. pectoralis* is usually epidermal hyperplasia restricted to one side of the fin (Boxshall, 1977). Sessile stages such as adult females and, more significantly, Ch3 and Ch4 cause the greatest damage that frequently extends into the dermis. Dermal insult leads to fibroplasia, cellular infiltration and eventually dense granulation tissue (Boxshall, 1977). Heavy infections by *L. pectoralis* result in lacerated and bleeding skin with partial destruction of the pectoral fin (Mann, 1970).

Caligus elongatus leaves a 'grazing trail' over the body surface from the original point of attachment to its 'preferred site' (Hogans and Trudeau, 1989a). When several adults occur at one site there is extensive damage to epidermal and dermal tissue.

Caligus orientalis causes skin ulcers on *Liza akame* (Urawa *et al.*, 1979), and *C. macarovi* leaves small round scars in the epidermis of *Cololabis saira* (see Hotta, 1962). *Caligus uruguayensis* in the mouth cavity of *Trichurus savala* is partly encapsulated by the epithelial and connective tissue hyperplasia that it stimulates (Radhakrishnan and Nair, 1981a). Adipose fins of *Oncorhynchus gorbuscha* were eroded by the attached chalimus stages of *C. clemensi* (see Parker and Margolis, 1964). In contrast, adults of *C. epidemicus* on *Acanthopagrus australis* cause little damage. Its copepodids erode the epidermis during initial settlement, and the chalimus stages are associated with only minor epidermal proliferation and inflammation in the dermis where the frontal filament is attached. However, after the final chalimus moult, the cast off frontal filament provokes a localized chronic inflammatory response characterized by epidermal and fibroblast proliferation, lymphocyte and macrophage infiltration and giant cell formation (Roubal, 1994).

Host immune response

No serum antibody response was found in naturally infected rainbow trout, but a serum response to an antigen (>200 kDa) was associated with the apices of gut epithelium folds in *Lepeophtheirus salmonis* from naturally infected Atlantic salmon (Grayson *et al.*, 1991). These authors suggested that the lack of response by rainbow trout may be the result of lower levels and shorter duration of infection compared to Atlantic salmon. They also found different antigens in adult and chalimus stages of *L. salmonis*. Atlantic salmon immunized with crude extracts of either adult *Caligus elongatus* or *L. salmonis* produced humoral antibodies that reacted with antigens in these extracts as well as fewer antigens from crude extracts of chalimus stages and eggs (Reilly and Mulcahy, 1993). A nonspecific cellular response was suggested by Johnson and Albright (1992a,b).

Mechanism of disease

The primary cause of pathology associated with adult caligid copepods results from parasite feeding; the extent of damage depends on the number of parasites. Five adult *Lepeophtheirus pectoralis* cause significant pathology on salmon smelts, but up to 2000 parasites have been recorded on a fish (Brandal and Egidius, 1977). Twelve to 15 *Caligus elongatus* will kill salmon smolt, and 40–50 of these parasites will kill an adult salmon (Stuart, 1990). Roubal (unpublished data) found that the sparid *Acanthopagrus australis* confined in 1 m^3 experimental sea cages acquired infections in excess of 6000 individuals of *C. epidemicus* compared to <10 in most wild fish. The heavily infected fish had damaged fins and the body surface was covered with mucus;

the fish were in poor condition and the cornea of some fish was infected with several chalimus and adult copepods.

The damage caused by the chalimus stages results from attachment by the frontal filament and consequently the limited feeding radius (Boxshall, 1977). There is no evidence that enzymes are released on to the surface of the host as has been reported for the external digestion by *Ergasilus* copepods (Halisch, 1940; Kabata, 1970).

Osmoregulatory failure through extensive skin damage appears to be the main cause of death though secondary bacterial infection has been suggested also (Wootten *et al.*, 1982).

Effects on host physiology

Subclinical effects are unknown although it has been suggested that *Lepeophtheirus salmonis* irritate the host and cause it to jump from the water or rub against solid objects resulting in skin abrasion (Wootten *et al.*, 1982). Neilson *et al.* (1987) found no evidence for loss of condition factor in infected juvenile cod or haddock.

Concurrent infection and cross-immunity

Caligus elongatus occurs frequently with *Lepeophtheirus salmonis* in farmed Atlantic salmon (Wootten *et al.*, 1982), though there is no evidence that infection by one species has any effect on infection by the other. Unidentified bacteria found in the gut of *L. salmonis* have been proposed as a source of disease in caged salmonids (Nylund *et al.*, 1991, 1992). *L. salmonis* may act as a vector for *Aeromonas salmonicida* and infectious salmon anaemia (Nylund *et al.*, 1993).

In vitro culture

Egg sacs are either removed from the mature female copepods or are left attached to the female and allowed to hatch in clean, filtered sea water that is changed regularly. A suitable host for the copepodid is necessary for any further development.

Parasite nutrition

Caligids feed by scraping the skin with a divided dentigerous bar, the strigil, located on the posterior lip of the mouth tube. The mandibles aid the passage of food into the mouth tube (Kabata, 1974). Naupliar stages lack a gut and anus whereas the copepodid has a mouth cone but lacks the strigil (Johnson and Albright, 1991b). The first chalimus stage of *Lepeophtheirus salmonis* has well developed mouth parts and a functional alimentary canal and appears to be the first feeding stage in the life cycle (Jones *et al.*, 1990; Bron *et al.*, 1991).

Mucus and epidermis appear to be the main diet (Boxshall, 1977; Wootten *et al.*, 1982). However, adult females of *Lepeophtheirus salmonis* and adults of other species have been found with blood in the gut in situations where either blood vessels or haemorrhaging tissue occur near the surface (Brandal *et al.*, 1976). *Caligus elongatus* had higher levels and a greater diversity of proteases in the gut than did *L. salmonis* (see Ellis *et al.*, 1990); the difference was attributed to the wider host range utilized by *C. elongatus*. Lipase was found in the gut of *L. salmonis* by Grayson *et al.* (1991).

Diagnosis

The large female parasites though semitransparent and of often cryptic coloration are usually visible to the discerning eye on the gills, fins or body of fish or in the buccal and opercular cavities.

Prevention and control

McLean *et al.* (1990) reported that most *Lepeophtheirus salmonis* are lost in 2 days from Atlantic salmon immersed in fresh water but some remained attached for up to 60 days. According to Wootten *et al.* (1982) the parasite may survive for up to 25 days in fresh water, but it will not reproduce at this low salinity. Eggs would not hatch at a salinity of 10 ppt (12°C), and most aborted at 11.5 ppt (5–12°C); the lower limit for successful development is 16 ppt (Berger, 1970; Johannessen, 1978; Wootten *et al.*, 1982) with the preferred salinity >25 ppt (Stuart, 1990). Johnson and Albright (1991b) found maximum survival of *L. salmonis* copepodids at 25 ppt salinity and 10°C in experiments at 5, 10, 15°C and salinities of 15, 20, 25, 30 ppt; at each of the other salinities survival was greatest at 15°C. Female *L. salmonis* can maintain their hyperosmotic state down to a salinity of 12.4 ppt (see Hahnenkamp and Fyhn, 1985). Adult male *L. salmonis* had the highest tolerance to lowered salinity (3.5–5.3 ppt). Higher water velocity 'flushed out' more first pre-adult stage of *L. salmonis* than later stages, and more males than females (Jaworski and Holm, 1992).

The ability of *L. salmonis* to survive for several days in fresh water precludes fresh water as a treatment. However, *Caligus elongatus* on pond-reared red drum, *Sciaenops ocellatus*, were controlled by a 20–30 minute dip in fresh water, but 'Copper Control' (Argent; 8.5% copper chelated with mono- and triethanolamine) at 6 mg l^{-1} (18 hours) and 150 mg l^{-1} (30 minutes), formalin at 10 mg l^{-1} (18 hours) or 250 mg l^{-1} (30 minutes), and trichlorphon at 0.25 mg l^{-1} (18 hours) were ineffective (Landsberg *et al.*, 1991). *Caligus elongatus* is difficult to control in sea-cages because the adult parasite readily leaves the caged salmonids and lives in the water column or infects other fish in the vicinity (Wootten *et al.*, 1982).

Chemotherapeutics used against sea lice include formaldehyde, organophosphate insecticides (malathion, trichlorphon, dichlorvos, azamethiphos), ivermectin, pyrethrum, carbaryl, diflubenzuron, hydrogen peroxide and aquatic extracts of onions and garlic (Roth *et al.*, 1993; Costello, 1993).

Formaldehyde (400 mg l^{-1}) is marginally effective but the low safety margin and large amounts needed preclude large scale usage. Onions and garlic were largely ineffective.

Organophosphates (OP) inhibit acetylcholinesterase (AChE) activity in cholinergic nervous systems. Experimental trials with malathion (*S*-1,2-bis (ethoxycarbonyl)ethyl *O*,*O*-dimethyl phosphorodithioate) 5 mg l^{-1} (1 hour) was 100% efficient in removing lice, but the narrow safety margin prevented commercial application (Roth *et al.*, 1993).

Baths in trichlorphon (dimethyl 2,2,2-trichloro-1-hydroxyethylphosphonate) (Neguvon, Dipterex, Tugon, Dylox, Masoten) were effective against preadult and adult lice, but not chalimus stages (Lavina, 1977; Wootten *et al.*, 1982; Stuart, 1990). The compound decomposed into the more toxic dichlorvos, dependent on temperature and pH, which resulted in fish death. Although recommended concentrations (15 mg l^{-1} at 14–18°C to 300 mg l^{-1} at <6°C) have been determined, this compound has been superseded by preparations based on dichlorvos (Roth *et al.*, 1993).

Dichlorvos (*O*,*O*-dimethyl-2,2-dichlorovinyl phosphate (DDVP)) (Nuvan (50/50 formulation of DDVP and dibutyl phthalate), Aquaguard, Vapona, Apavap) at 0.2–2.0 mg l^{-1} DDVP, dependent on temperature, is effective against post-chalimus stages (see Roth *et al.*, 1993). The second preadult stage is the most susceptible to delousing with Nuvan and adult males are more susceptible than females (Jaworski and Holm, 1992). Males are more active and leave the fish more readily than females. If applied orally, the variable feeding response within the caged population of fish means that some may receive an excess dose (Brandal and Egidius, 1979). Repeated exposure at 3–6 day intervals resulted in cumulative AChE depression in fish (Salte *et al.*, 1987; Roth *et al.*, 1993). Low dissolved oxygen levels (DO$_2$) in water promotes increased ventilation volume and blood flow which causes greater absorption of the chemical. Hoy *et al.* (1992b) found that inhibition of trout brain AChE activity was increased 13% at DO$_2$ of 3 mg l^{-1} compared to 15 mg l^{-1}.

Dobson and Tack (1991) found a rapid dispersal of Nuvan from cages such that there was no detectable level (2 μg^{-1}) outside a 25 m range of the cages during and after treatment. The half-life of dichlorvos in well oxygenated water is 3.9 days at 13.5°C, pH 8.0 while the half life of trichlorphon is 29 hours (Samuelsen, 1987; Pike, 1989). Nuvan persists to some extent (Hoy and Horsberg, 1990), and is toxic to some invertebrates at low levels (Egidius and Moster, 1987; Mattson *et al.*, 1988). Certain aspects of the environmental management and legislation associated with the use of dichlorvos are discussed by McHenerey *et al.* (1992) and Buchanan (1992). Copepods repeatedly exposed to OP can develop resistance (Jones *et al.*, 1992), as has occurred with insects. This means that stronger solutions have to be used to treat the parasite with a concomitant reduction in the safety margin (Roth *et al.*, 1993). However, no resistance by *C. elongatus* was detected by Bron *et al.* (1993b).

Recent trials with azamethiphos (*S*-6-chloro-2,3-dihydro-2-oxo-1,3-oxazolo(4,5-*b*)pyridin-3-ylmethyl *O*,*O*-dimethyl phosphorothioate) (Alfacron) have found it to kill 100% of post-chalimus stage lice at 0.01 mg l^{-1} (1 hour), but larval stages appear to survive. Azamethiphos does not cause cumulative AChE depression from repeated exposure, is better tolerated by fish, though thorough studies of its potential ecological impact have to be done (Roth and Richards, 1992; Roth *et al.*, 1993).

Pyrethrum kills susceptible organisms by interfering with the closure of sodium channels in nervous membranes. Oil containing 4% Ph-Sal (=1% pyrethrum) and 4% piperonyl butoxide was dispersed onto the surface of sea cages to expose copepods on fish that leapt from the water (see Roth *et al.*, 1993). Results were variable (0–70% of lice killed), although a 3 second spray in air killed 90% of lice. An apparatus to deliver Py-Sal has been advertised (Anon, 1992). Pyrethrum and synthetic pyrethroids are relatively toxic to fish and invertebrates; their persistence in the marine environment has yet to be fully assessed (Roth *et al.*, 1993).

Ivermectin (a neuroactive macrocyclic lactone) was tested in experiments by Palmer *et al.* (1987) as a 0.2 mg kg^{-1} body weight oral dose after coating with 5% gelatin on to food pellets. The treatment reduced the number of lice and prevented damage to the host though there was a narrow safety margin. Hoy *et al.* (1992a) concluded that ivermectin was not well suited to oral treatment because high concentrations reached the central nervous system and the drug was excreted slowly. There is at present only limited information about its toxicity to target and non-target marine organisms (Roth *et al.*, 1993).

Carbaryl (1-naphthyl *N*-methylcarbamate) (Sevin) inhibits AChE, does not accumulate in fish and shellfish tissues, but was unsatisfactory because of its persistence in the environment and its toxic degeneration product (1-naphthol) (Bruno *et al.*, 1990; Roth *et al.*, 1993).

Diflubenzuron (1-(4-chlorophenyl)-3(2,6-difluorobenzyl)-urea (DFB)) (Dimilin) inhibits chitin synthesis in insects, is not readily water soluble, is stable in light, and has a half-life of 29 days at 10°C, pH 7.7. DFB was effective at an oral dose of 75 mg kg^{-1} over 14 days, is relatively non-toxic to fish (rainbow trout 96h–LD$_{50}$ = 140 mg l^{-1}), but is extremely toxic to marine crustacea (0.5 µg l^{-1}). This toxicity as well as its stability make it an unlikely candidate for routine use in the marine environment (Roth *et al.*, 1993).

Exposure to 1.5 g l^{-1} hydrogen peroxide in a bath for 20 minutes removed 85–100% of lice, and was effective against chalimus and mobile stages of sea lice (Thomassen, 1993). Toxicity to salmon increases with temperature and treatments at temperatures greater than 14°C were not recommended. Exposure times of 60 minutes or at 10 g l^{-1} for 30 minutes caused damage to the gill epithelium (Thomassen, 1993).

Non-chemical control methods for sea lice are receiving attention. Vaccination has been mentioned already (p. 506). Fallowing of salmon farms for periods greater than 30 days (7–8°C) prevents carryover of infectious agents and allows the sea bed to recover (Grant and Treasurer, 1993).

Wilson (1905b) suggested that *Caligus*, *Lepeophtheirus* and allied genera could be controlled by '. . . a plentiful introduction of small fish . . . which will eat up the larvae of parasitic copepods'. Cleaner wrasse, particularly

Fig. 13.15. *Lepeophtheirus salmonis* being removed by a goldsinny wrasse *Ctenolabrus rupestris* from the head of *Salmo salar*. (Photo A. Bjorndal, reproduced by permission.)

rockcook, *Centrolabrus exoletus*, and goldsinny, *Ctenolabrus rupestris*, are used now to control sea lice infestation in salmon farms (Fig. 13.15). In 1989, Norwegian farmers used 50,000 wrasse (65% goldsinny, 15% rockcook) to treat 2.3 million smolts in 115 cages. At ratios of goldsinny:salmon of 1 : 52 and rockcook:salmon of 1 : 8 most lice were removed; a ratio of 1 : 158 resulted in more lice remaining but they caused no harmful effects; at 1 : 260 the wrasse were not effective (Bjordal, 1988, 1990; Costello and Bjordal, 1990; Bjordal, personal communication). The number of mobile lice, but not chalimi were lower on salmon in cages with wrasse (2–11 lice per fish) than in cages without wrasse (<50 lice per fish) (Treasurer, 1993). Cages in which wrasse have been introduced require little or no chemical treatment to control sea lice. No *L. salmonis* were found on wrasse from Scottish fish farms and *Caligus centrodonti* from some wrasse were not found on salmon, which suggested to Bron and Treasurer (1992) that transfer between wrasse and salmon did not pose a problem. Wrasse were killed by a typical strain of *Aeromonas salmonicida* (see Treasurer, 1993), but an atypical *A. salmonicida* found in some wrasse was nonpathogenic to salmon (Frerichs *et al.*, 1992).

COPEPODA: PENNELLIDAE

Introduction

Pennellids are widespread and highly visible parasites of marine fish. Aristotle recorded them on tunny and swordfish in 330 BC (McGregor, 1963). Most cause localized changes in adjacent tissues and some result in loss of condition or reduced gonad development.

Lernaeocera branchialis has been estimated to cause reductions of more than 1000 tonnes per year in the gadoid catch around Scotland through loss of condition (Kabata, 1970). In mariculture, infected cod, *Gadus morhua*, farmed in sea cages in Newfoundland show poor weight gain and increased mortality (Khan *et al.*, 1990). *Lernaeocera lusci* has caused mortality in cultured sole, *Solea solea* in Britain (Slinn, 1970; Kirmse, 1987).

Dark muscle lesions caused by *Pennella hawaiiensis* delayed the marketing of 169,000 tons of boarfish *Pentaceros richardsoni* until the source of the lesion was identified (Kurochkin, 1985).

Host range, distribution and seasonality

Lernaeocera spp. are restricted to the North Atlantic.

Adult females of *L. branchialis* are found primarily on gadoids, particularly *Gadus morhua, G. ogac, Boreogadus saida, Pollachius pollachius, Merlangius merlangus*, and *Eleginus navaga*, but also occur on fishes in other families, including Merlucciidae (*Merluccius merluccius*), Serranidae (*Dicentrarchus labrax, Serranus cabrilla*) and Callionymidae (*Callionymus lyra*), in the Arctoboreal region (Kabata, 1979). They are often associated with coastal fish, possibly because here there are hosts carrying juvenile and male parasites. These stages are found primarily on lumpfish *Cyclopterus lumpus* in Newfoundland, lemon sole *Microstomus kitt* in the northern North Sea, and flounder *Platichthys flesus* in the southern North Sea (Kabata, 1979; Khan *et al.*, 1990).

Adult female *L. lusci* occur primarily on the gadoid *Trisopterus luscus* in northwest Europe, and on *T. minutus* and *Merluccius merluccius* in the Mediterranean (Bastide-Guillaume *et al.*, 1987). Slinn (1970) found juvenile stages on *Solea solea*. Adult females of *L. minuta* occur on the goby *Pomatoschistus minutus*; however, its first host is not known.

Lernaeocera branchialis lives for 1 to 1.5 years. In the North Sea, juveniles and males reach peak abundance on flounder in April/May followed by a peak in abundance of females on cod and haddock in mid to late summer (Stekhoven, 1936; Kabata, 1958). Off Newfoundland, juveniles and males are commonest on lumpfish in July/August (Templeman *et al.*, 1976). In the English Channel there is little seasonal change (Sproston and Hartley, 1941a). Kabata (1958) observed high annual variability in prevalence of adult females on haddock near Aberdeen, UK. In consecutive years, prevalence dropped from 23% to 5%.

Fig. 13.16. *Haemobaphes diceraus,* second antenna. (After Kabata, 1988.)

There are indications that other pennellids survive for a year or less. Hughes (1973) observed that 17% of Pacific saury *Cololabis saira* were infected with *Pennella* sp. along the coast of Oregon and Washington in August/September 1970. In 1971 in the same months less than 1% of the same age cohorts were infected, though many bore scars caused by previous infection. Nagasawa *et al.* (1988) observed in the sea off Japan that 60% of large *C. saira* were infected with *Pennella* sp. in August, and this had dropped to 5% by late October.

Systematics

A key to genera of pennellids based on adult females is in Kabata (1972a). Juvenile females of *Lernaeocera, Lernaeenicus, Trifur, Peniculus,* and possibly *Pennella,* can be identified to genus based on the ornamentation on the posterior wall of the mouth cone (Romero and Kuroki, 1986, 1989). El Gharbi *et al.* (1985) concluded that the shape of the holdfast and forebody of *Lernaeenicus sprattae* depended on its location in the host.

Males and copepodids are insufficiently known. However, these stages can be recognized as members of the family Pennellidae from the characteristic second antennae which is typically subchelate with a strong opposable claw (Fig. 13.16). The chalimus stage of pennellids typically has a very short frontal filament, sometimes completely embedded in the tissues of the fish (Kabata, 1988). Schram (1979) gives features which distinguish juveniles of *Lernaeenicus sprattae* from those of *Lernaeocera branchialis.*

There are three species of *Lernaeocera* (Kabata, 1979): *L. branchialis, L. lusci* and *L. minuta.* Adult females are identified from the shape of the holdfast and by their size. The holdfast of *L. branchialis* (Fig. 13.17) has one dorsal and two lateral thoracic lobes. Holdfasts of *L. lusci* and *L. minuta*

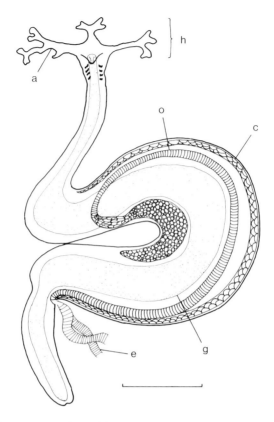

Fig. 13.17. *Lernaeocera branchialis*. Adult female (diagrammatic). (Based on Sproston and Hartley, 1941b, and Capart, 1948.) a = antler, c = cement gland, e = egg string, g = gut, h = holdfast, o = oviduct. Bar = 5 mm.

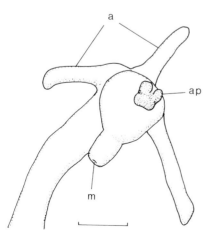

Fig. 13.18. *Lernaeocera lusci*, young specimen from *Trisopterus minutus* showing the antennary processes (ap). a = antlers, m = mouth. Bar = 1 mm. (Drawn from Bastide-Guillaume *et al.*, 1987.)

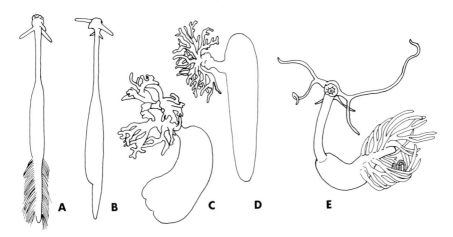

Fig. 13.19. Pennellidae, morphology of adult females. A, *Pennella*; B, *Lernaee-nicus*; C, *Phrixocephalus*; D, *Peroderma*; E, *Lernaeolophus*; (A–D redrawn from Kabata, 1979; E from Grabda, 1991.)

incorporate two additional lobes, the antennary processes, that arise on the dorsal side between the mouth and the holdfast proper (Fig. 13.18). *Lernaeocera lusci* is larger than *L. minuta*; its total length, measured in a straight line from anterior to posterior end, is 6–15 mm versus 4–7.5 mm for *L. minuta*.

Two forms of *Lernaeocera branchialis* are recognized (Kabata, 1979), *L. branchialis* f. *obtusa* where the body is curved in the shape of a 'U' and *L. branchialis* f. *branchialis* in which the body has more bends and forms an 'omega'. The morphological differences may be environmentally determined as *L. b. obtusa* is frequently found on large haddock where more of the body is used to reach the ventral aorta than in *L. b. branchialis* on immature cod or whiting.

Morphology and life cycle

Adult females develop elaborate holdfasts and show little sign of segmentation (Fig. 13.19). Their life cycles require one or two species of hosts.

Lernaeenicus sprattae, a parasite of sprat and pilchards, requires one host species. Eggs hatch into the first naupliar stage which is free-swimming. After a day it moults into a second nauplius and a day later into a copepodid (Schram and Anstensrud, 1985). The copepodid is attracted to the surface at night and fastens to a sprat or young pilchard by its second antennae. It probably moults to first chalimus before producing the frontal filament, moults through three more chalimus stages and in its final moult changes into a mobile male or premetamorphosis female. Fertilization occurs on the fish, then the female may swim to a second fish or remain on the first fish to complete her metamorphosis (Schram, 1979). El Gharbi *et al.* (1985) found that in the Golf du Lions, Mediterranean Sea, copepodids attached to larval pilchards on the breeding grounds during the winter. They developed on them as they moved inshore into lagoons during the spring. By the summer the fertilized premetamorphosed females were ready to infest the same or another host and completed their development on that host as it moved back out to the breeding grounds in the autumn. Thus the cycle takes about a year.

The egg string is in the form of a tube. Embryos develop to nauplii, then the intact egg leaves through a slit in the lateral wall of the egg string and, soon after, hatches. An excised adult female spontaneously shed its empty egg strings and immediately produced new ones which it filled with eggs over the next 24 hours (Schram, 1979).

Haemobaphes intermedius, a parasite of cottids, probably completes its life cycle on one host also (Roth, 1988).

Some pennellids are unusual among the copepods in that they require two hosts. *Lernaeocera branchialis* uses a non-gadoid fish as its first host and then the female tranfers to a gadoid to complete her development. The adult female is a dark red sigmoid worm often with yellow egg strings (Fig. 13.17). Eggs hatch in 12 days at 10°C to release a nauplius which moults to a metanauplius and then to an infectious copepodid within 2 days (Capart, 1948; Whitfield *et al.*, 1988). Copepodids search for a host for up to a week. They attach to gills of flounder *Platichthys flesus* or lumpfish *Cyclopterus lumpus*, moult through four chalimus stages to adult males and subadult females. This takes about 25 days at 10°C (Whitfield *et al.*, 1988). The males, which develop faster than females, seek out a female chalimus larva and attach to adjacent gill tissue. As soon as the female completes her final moult and while still attached (now by her second antennae) the male grips her with his second antennae and moves back until he is attached at the genital segment. He secretes two spermatophores, curls his abdomen under the female and attaches them over the openings of her two oviducts. This takes 2–3 minutes. Over the next few hours the spermatophores empty into receptaculum seminis located just inside the genital opening (Capart, 1948) and then the spermatophore envelope is shed. He may fertilize the female a second time, or she may be fertilized by other males several more times before she leaves the flounder. Males remain on flounder for over 5 weeks after maturity and thus accumulate on their gills (Anstensrud, 1989, 1990a,b,c).

The metamorphosing female becomes free-swimming for several days until it locates the second host and her abdomen elongates by straightening folds in the cuticle (Smith and Whitfield, 1988). Four to 5 months after attachment to the second host she produces a batch of eggs and dies after 1.5 years (Kabata, 1958). The eggs, held in coiled strings by a mesentery-like membrane (Heegard, 1947b) appear to be fertilized as they are extruded (Wilson, 1917). Though second batches of eggs have been produced *in vitro* by excised females (Heegaard, 1947b; Whitfield *et al.*, 1988), suggesting several batches of eggs could be produced in the field, Khan (1988) found that egg strings were not lost by females *in situ* but increased in size from December to June then shrank and eventually dropped off when the parasite died. Kabata (1958) suggested that eggs were laid more or less continuously, were liberated from the end of the egg string and sank to the bottom where they hatched. Liberation of the egg does not appear to be a prerequisite for hatching as nauplii hatch directly from the egg string *in vitro* (Sproston, 1942; Heegard, 1947b; Whitfield *et al.*, 1988).

Cardiodectes medusaeus, a parasite of myctophids, uses an invertebrate first host and has three fewer instars than *Lernaeenicus sprattae* or *Lernaeocera branchialis*. The egg hatches directly into a copepodid which attaches to the mantle or gill of a pelagic gastropod (heteropod, pteropod, or *Janthina*). After three chalimus stages it moults to an adult male or a premetamorphosis female. Fertilized females leave the snail and seek out and attach to fish (Perkins, 1983).

Four chalimus stages and males of *Pennella varians* occur on the gills of *Sepia* and *Loligo*. The adult female is unknown (Rose and Hamon, 1953).

Host–parasite relationships

Site and host selection, course of infection

Pennellid juveniles and males attach to the surface of the skin or gills. Adult females penetrate into tissues and are usually attached near a blood vessel. Some, such as *Peniculisa wilsoni* (Fig. 13.20), penetrate only a short distance, others, such as *Lernaeenicus hemirhamphi*, burrow deeply into all organs seeking areas with a rich blood supply (Natarajan and Nair, 1973). In all cases at least the tip of the body and egg strings protrude from the fish.

Juveniles of *L. sprattae* occur primarily on the fins, particularly the pectoral fins, of pilchards and sprat. Copepodids initially attach anywhere on the surface, then over the next 24 hours they move to the fins before moulting into a chalimus (Anstensrud and Schram, 1988). Adult females on pilchards tend to burrow into muscle, particularly near the base of the dorsal fin (El Gharbi *et al.*, 1985). They occasionally pierce the heart and may enter the wall on the other side (Rousset and Raibaut, 1989). Females on sprat primarily attack the eye, particularly the upper rear quadrant (Anstensrud and Schram, 1988).

Juveniles and males of *Lernaeocera branchialis* attach and grow near to the tips of gill filaments of the first host. Metamorphosing females swim into the gill cavity of the second host, attach to gill filaments and work their way along the filament and down the arch (Kabata, 1958). They usually penetrate near the ventral end of the third or fourth arch, and develop their anchors and mouthparts in the wall of either the bulbus arteriosus (in North Sea gadoids; Stekhoven, 1936; Kabata, 1979), or the ventral aorta (in Newfoundland cod; Khan, 1988).

Maturing females of *Phrixocephalus cincinnatus* (Fig. 13.19C) develop within the eye of flounder *Atheresthes stomias*. The juvenile female breaks through the cornea causing little damage and develops internally unlike most other pennellids (possibly because the eye is relatively large and the tiny female cannot reach the retina to feed without complete entry). During development the female penetrates through the retina into the choroid and produces

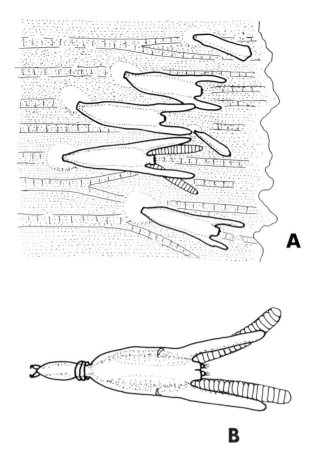

Fig. 13.20. *Peniculisa wilsoni*. A. Adult females attached to fin of *Diodon histrix*. B. Entire parasite. (After Radhakrishnan, 1977.)

cephalic and thoracic lateral outgrowths. The holdfast is well formed before the genital area begins to grow and pushes out of the eye through the cornea, possibly by pressure necrosis or by lytic secretions from its anus. Thoracic outgrowths of mature females spread throughout the vitreous humour (Kabata, 1969b).

Lernaeolophus sultanus (Fig. 13.19E), a pennellid with a worldwide distribution parasitic on scombrids and other fishes, typically attaches to the palate. Its cephalic processes grow up into the eye socket while its genital segment and abdomen lie free in the buccal cavity (Grabda, 1972). *Lernaeolophus aceratus* on wrasse attaches in the gill cavity near the dorsal end of gill arches, and embeds its head between the liver and vertebral column (Ho and Honma, 1983).

Two pennellids insert their heads directly into blood vessels. *Haemobaphes diceraus*, a relatively large pennellid 60 mm or more long, attaches in the gill cavity of *Theragra chalcogramma* (Fig. 13.21). Its 'S'-shaped genital segment and abdomen lie on the posterior side of a gill arch between the two rows

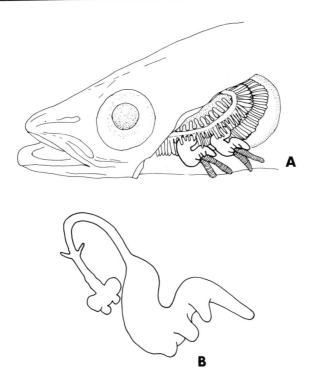

Fig. 13.21. *Haemobaphes diceraus.* A. Adult females attached to gill of *Theragra chalcogramma* (after Grabda, 1975, 1991). B. Entire parasite (after Kabata, 1979.)

of gill filaments. The neck enters the lumen of the branchial artery and passes against the blood flow into the dorsal aorta and eventually into the conus arteriosus so that its head is adjacent to the valves of the ventricle. The length of the neck varies according to where the parasite settles in the gills (Grabda, 1975).

Adult female *Cardiodectes medusaeus* are attached ventral to the heart of myctophids and penetrate more or less directly into the bulbus arteriosus (Fig. 13.22). Metamorphosing females have been found encapsulated in the wall of the bulbus. Their mode of entry into the fish is not known (Perkins, 1983).

Clinical signs and histopathology

The head and anterior thorax (cephalothorax) of most pennellids is buried deeply in the host tissues. The head is usually enveloped in a large chronic granuloma composed of fibrous and necrotic tissue, with some vascularization near the mouth of the parasite. The fibrous tissue is often dense and may form a hard cartilage-like capsule as in *Cardiodectes longicervicus* (see Shiino, 1958). A sleeve of hyperplastic fibrous connective tissue surrounds the thorax from the granuloma to the surface of the fish.

The posterior thorax or genital complex generally lies outside the host

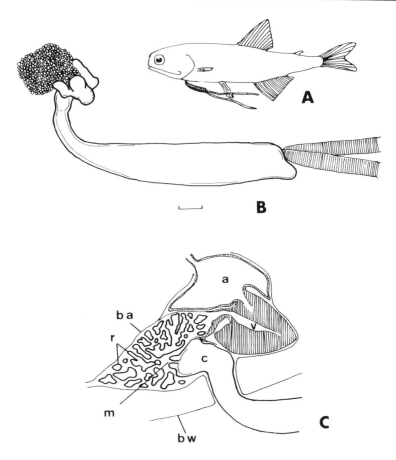

Fig. 13.22. *Cardiodectes medusaeus.* A, Adult female attached to myctophid (after Perkins, 1983). B, Entire parasite, scale bar = 1 mm (after Shiino, 1958). C, Section through heart of myctophid with parasite embedded in the bulbus arteriosus. a = atrium, ba = bulbus arteriosus, bw = body wall, c = copepod, m = mouth, r = rhizoid holdfast processes, v = ventricle (after Kabata, 1981.)

and is usually free of any host tissue response. However, in *Lernaeolophus aceratus* in the wrasse *Halichoeres tenuispinis*, the sheath extends outwards into the gill cavity so that it wraps around most of the trunk of the parasite and leaves only the abdomen with its branching processes free (Ho and Honma, 1983). *Peroderma pacifica*, which is attached to the mid body of the codlet *Bregmaceros japonicus*, penetrates through to the vertebral column, and is completely buried in host tissue except for the tip of the posterior end and its egg sacs (Izawa, 1977).

Juvenile *Lernaeocera branchialis* cause the ends of gill filaments to thicken and lamellae to fuse as a result of tissue proliferation (Kabata, 1958). Infections have been associated with a decrease in the condition factor of estuarine flounder (Moller and Anders, 1986). Cod with adult female parasites show hyperactivity, erratic swimming and a tendency to remain at the water surface.

The head of the parasite is encapsulated by a mass of vascularized fibrous tissue which distends the wall of the blood vessel. The antlers as they grow sometimes break into the lumen but the mouthparts always appear to be lodged in the granuloma in the wall. Local tissue necrosis occurs around the head; adjacent muscles may be discoloured, fragile and necrotic. In multiple infections, parasites are sometimes shed leaving open haemorrhagic lesions which develop necrotic margins and caseous exudates (Stekhoven, 1936; Khan, 1988).

Peniculisa wilsoni, a parasite which tends to occur in groups on the fin of *Diodon hystrix* (Fig. 13.20), grips the fin ray by its second antennae. Dense fibrous tissue, partly calcified, develops around the distal segment of each antenna fusing them to the bone. The ray element may break and become displaced within the tissue, and soft tissue around the site of attachment proliferates and grows into a conspicuous tumour which envelopes the head and part of the thorax (Radhakrishnan, 1977; Radhakrishnan and Nair, 1981b,c).

Species that invade the eye stimulate granuloma formation but otherwise do surprisingly little damage while they are alive. In *Phrixocephalus cincinnatus*, a haematoma forms around the holdfast and this displaces the retina towards lens. When the parasite dies, decomposition of the holdfast destroys the posterior chamber of the eye and sometimes also the lens (Kabata, 1969b).

The skull of fish with *Lernaeolophus sultanus* is eroded near the body of the parasite, the olfactory epithelium in the nasal cavity is disrupted by the holdfast and small haemorrhages occur in the tumour around the oral opening of the parasite (Grabda, 1972).

Gill filaments adjacent to *Haemobaphes diceraus* are usually shortened or absent (Fig. 13.21). Parasite and tissue response partly block the lumen of the branchial artery and cause damage to the inner layer of the conus arteriosus (Grabda, 1975). Parts of dead *Haemobaphes cyclopterina* were found in the heart and branchial artery of the flounder *Lepidorhombus whiffiagonis* by Bristow and Berland (1988).

The lumen of the conus arteriosus of myctophids is greatly expanded by the two sets of anterior processes and the two pairs of lateral lobes of *Cardiodectes medusaeus* (Fig. 13.22). The wall of the conus adjacent to the copepod is devoid of muscle and a fibrous collar surrounds the neck of the parasite at its point of exit on the fish (Perkins, 1983; Moser and Taylor, 1978).

Host immune response

Previous infections of *Lernaeocera branchialis* do not protect against reinfection (Stekhoven and Punt, 1937; Khan, 1988) though recently attached females are more likely to be shed from the fish if a parasite is already present. In experimental exposures, small cod were more frequently infected than large cod (Khan, 1988).

No antibody to *Lernaeenicus radiatus* was found in an infected sciaenid, *Leiostomus xanthurus* by Thoney and Burreson (1988) using an ELISA test.

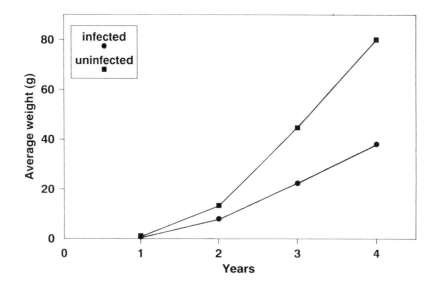

Fig. 13.23. Effect of *Haemobaphes diceraus* on the weight of *Limanda herzens-teini*. (From data in Nagasawa and Maruyama, 1987.)

Faisal *et al.* (1990) reported that infestation by *Phrixocephalus cincinnatus* suppressed parts of the cellular immune response in flounder.

Mechanism of disease

The disease caused by *Lernaeocera branchialis* is primarily the result of anorexia, stress and blood loss. Sudden mortality occurs when part of the holdfast enters the vessel lumen resulting in thrombi and blockage of major blood vessels (Khan, 1988). *Cardiodectes medusaeus*, a parasite which lies within a blood vessel, does not cause such thrombi, perhaps because it injects an anticoagulant into the bloodstream (Kabata, 1981, 1984).

Effects on host

Parasites attached to fins or in muscle seem to have little effect on the general health of the fish. For example, Nagasawa *et al.* (1985) and Watanabe *et al.* (1985) could detect no change in the condition or sexual maturity of saury, *Cololabis saira*, infected with *Pennella* sp. in the western Pacific. However, Hughes (1973) observed a 17% loss of weight in the same fish infected by *Pennella* sp. in the eastern Pacific.

Pennellids in more critical sites appear to have a greater effect on their host. Sole, *Limanda herzensteini*, infected with *Haemobaphes dicereus* in their branchial arteries were shorter than uninfected fish of the same age and lighter in weight (Nagasawa and Maruyama, 1987; Fig. 13.23).

Hemirhamphus xanthopterus parasitized by *Lernaeenicus hemirhamphi* in the kidney and elsewhere, had smaller and darker livers than normal. Their fat contents dropped from 15% wet weight to 8% in fish with two to five parasites. In contrast, the liver fat in fish with immature parasites increased

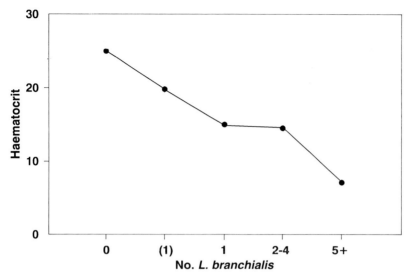

Fig. 13.24. Effect of *Lernaeocera branchialis* on the haematocrit of *Merluccius merluccius*. (From data in Guillaume *et al.*, 1985.)

to 20% (Natarajan and Nair, 1976), a response similar to that in gadoids with *Lernaeocera branchialis* described below.

Cardiodectes medusaeus in the conus arteriosus of the myctophid, *Stenobrachius leucopsarus*, caused sterilization and an increased growth rate (Moser and Taylor, 1978). *Peroderma cylindricum* retards gonad development in *Sardina pilchardus*, the European sardine (Kabata, 1970).

Lernaeocera branchialis is clearly detrimental to its host, especially if the host is young (Khan *et al.*, 1990). However, in the first month of infection before emaciation sets in, fish show increased growth, increased haemoglobin and increased fat content compared to unparasitized fish (Kabata, 1958; Khan, 1988; Khan and Lee, 1989).

Fish with established infections become anaemic. Kabata (1958) recognized infected haddock *Melanogrammus aeglefinus* by their pallor, particularly in the skin around the eyes and showed that their blood had decreased haemoglobin content. Mann (1952) found that the blood of infected whiting *Merlangius merlangius* contained only 20–22% haemoglobin compared to 38–40% in controls. There was a marginal decrease in erythrocyte count, and their oxygen uptake decreased by 30%. Infected cod *Gadus morhua* had less haemoglobin and lower serum protein than uninfected controls (Khan *et al.*, 1990), and in hake *Merluccius merluccius* the erythrocyte count, haematocrit and haemoglobin content decreased in relation to the number of parasites (Fig. 13.24; Guillaume *et al.*, 1985).

Infected fish lose condition. In the North Sea the average loss of weight of whiting infected with *L. branchialis* was 5–10% and in cod and haddock 20–30% (Mann, 1952; Hislop and Shanks, 1981). However, Guillaume *et al.* (1985) and Sherman and Wise (1961) detected no change in the condition of infected hake or cod respectively.

Kabata (1958) found that haddock with one immature parasite had a higher condition factor than controls, and this relationship was reversed in fish with one or more adult parasites. Using experimentally infected cod, Khan and Lee (1989) showed that adults with young parasites ate more food and gained more weight than controls for the first few weeks of the infection, though over a 16 month period conversion efficiency and condition factor were lower than controls. Young infected fish did not show the initial growth spurt; they ate less and gained less weight then controls over the whole period.

The total fat content of infected gadoids is less than that of uninfected fish (Kabata, 1958; Moller and Anders, 1986), particularly in whiting (Mann, 1970). The parasite also affects the development of the gonads. Hislop and Shanks (1981) estimated that *L. branchialis* was responsible for a 21% reduction in egg production in haddock. Khan (1988) found that infected cod had a gonadosomatic index about half that of controls and spawning was retarded.

Naturally infected cod have lower survival rates after capture (Moller, 1984; Khan *et al.*, 1990). In experimental infections, about one third of infected cod died within 8 months of infection, half of these within the first 2 months (Khan, 1988; Khan *et al.*, 1990). Additional stress caused either by exposure to fractions of crude oil or by infection with the trypanosome *Trypanosoma murmanensis* increased mortality (Khan, 1984, 1988).

Concurrent infection

Bib, *Trisopterus luscus* with *L. lusci* frequently had *Cryptocotyle lingua*, another coastal parasite (Evans *et al.*, 1983). Khan (1984) suggested that *Lernaeocera branchialis* was absent from offshore cod (Newfoundland/ Labrador) because of mortality associated with concurrent infection with *Trypanosoma murmanensis*.

Parasite nutrition and physiology

Nauplii, metanauplii and early copepodid stages of *Lernaeocera branchialis* and *Lernaeenicus sprattae* do not feed and can be readily raised in sea water (Heegard, 1947b; Schram, 1979; Schram and Anstensrud, 1985). The larvae of *L. sprattae* do not survive low salinities and copepodids move to surface waters at night (Schram and Anstensrud, 1985).

Pennellid chalimus larvae presumably feed on epidermal cells like the chalimus of caligoids. Mature males of *Lernaeocera branchialis* feed on epidermis (Anstensrud, 1989).

Adult female pennellids generally feed on the blood and lymph of fish from haemorrhage and inflammation within the granuloma. Blood cells have been found within the gut of *Lernaeenicus sprattae*, *Phrixocephalus cincinnatus*, *Cardiodectes medusaeus* and *Lernaeolophus aceratus* by Rousset and Raibaut (1989), Kabata (1969b, 1981) and Ho and Honma (1983) respectively.

Adult females of *Lernaeocera branchialis* feed intermittently (Sproston and Hartley, 1941b). As they age they become darker, turning from red to black, possibly because of accumulated biproducts from blood digestion. The anus appears to be non-functional in mature specimens (Sproston and Hartley, 1941b). Heegaard (1947b) kept excised females alive and unfed for at least 2 months in sea water at 6–8°C, during which time they laid eggs which hatched. The haemolymph is normally hypertonic to fish blood and comes close to that of sea water in parasites that have not recently fed (Panikkar and Sproston, 1941).

Digestion of blood in the midgut of *L. branchialis* and *Lernaeolophus aceratus* is associated with the release of large vesicles from the epithelium. These vesicles, which sometimes form chains, move into the lumen and burst presumably to release digestive enzymes (Stekhoven and Punt, 1937; Capart, 1948; Ho and Honma, 1983).

Capart (1948) suggested that digestion in *Lernaeocera branchialis* took place over about 8 days during which the intestine, capacity 0.1 ml, changed colour from red to clear (Sproston and Hartley, 194la).

The cuticle on the ventral side of the head of *Pennella antarctica* and *P. elegans* contains pores which may be involved in the absorption of nutrients (Wilson, 1917; Kannupandi, 1976). Perkins (1985) found that the anterior processes of *Cardiodectes medusaeus* contained masses of mitochondria, suggesting that small molecules pass through the cuticle from the host blood. The processes also contained ferritin crystals which she suggested were a result of iron detoxification. It seems likely that in other pennellids anterior branching processes are associated with nutrition.

Prevention and control

Slinn (1970) anaesthetized broodstock infected with *Lernaeocera lusci* and removed or damaged the developing female parasites with forceps.

COPEPODA: LERNAEOPODIDAE

Introduction

Lernaeopodids are highly modified copepods parasitic chiefly on marine and freshwater teleosts. Adult females are attached permanently by a unique structure, the bulla, which is implanted in the host tissue. Dwarf, short-lived adult males attach to the female.

Members of the genus *Salmincola*, especially *S. californiensis* (Fig. 13.25), are a potential threat to salmonids farmed in fresh water. The salmon gill-maggot, *S. salmonea* (Fig. 13.26), has been known from Atlantic salmon, *Salmo salar*, since 1766 and was known to Linnaeus (Friend, 1940).

Alella macrotrachelus (Fig. 13.27) is one of the most harmful gill parasites of the sparid, *Acanthopagrus schlegeli*, cultivated in Japan (Ueki and Sugiyama, 1979; Kawatow *et al.*, 1980). *Achtheres percarum* infects the gills

Fig. 13.25. *Salmincola californiensis* female with small male attached (after Kabata and Cousens, 1973). Scale bar = 0.5 mm.

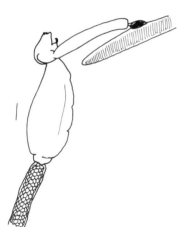

Fig. 13.26. *Salmincola salmonea* attached to gill filament of *Salmo salar* (after Friend, 1940). Scale bar = 1 mm.

of freshwater fish and cause mortality in severe infections (Reichenbach-Klinke and Landolt, 1973). *Tracheliastes maculatus* infects mostly *Abramis brama* in Poland and heavy infections are reported to kill the host (Grabda, 1991).

Host range, distribution, seasonality

The family Lernaeopodidae contains over 260 species most of which are marine. About one-tenth of the species infect freshwater fish. Those that infect elasmobranchs belong to the small *Charopinus* branch of the family (Kabata, 1984).

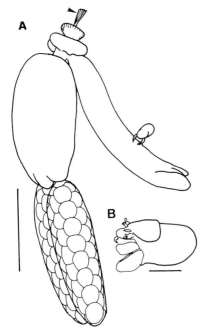

Fig. 13.27. *Alella macrotrachelus*. A, Adult female with small male attached to kneck. Thorn-like bulla (arrow) arises from the second maxilla. B, Male. Scale bar = 0.2 mm.

Kabata (1966) proposed that evolution and zoogeography of the Lernaeo-podidae proceeded along marine and fresh water lines. The evolution of the fresh water branch was associated with salmonid fish from which the para-site spread to ecologically related fishes. The main branch is *Salmincola* although the genus *Achtheres* represents an earlier more primitive branch. He suggested that the centre of origin was the Palearctic Region, with the fresh-water branch spreading through the land masses of the northern hemisphere, but not below the equator. Thus, freshwater species are not in the southern continents.

Salmincola spp. are restricted largely to salmonids and coregonids (Hoffman, 1977). *Salmincola californiensis* (Fig. 13.25) is a native of North American streams that empty into the northern Pacific Ocean (Kabata, 1969a). It infects all species of Pacific salmon and trout and has been introduced into the eastern United States with the transport of live fish and eggs (Hoffman, 1977, 1984). It was imported into Iowa with farmed trout from Missouri (Sutherland and Wittrock, 1985). *Salmincola edwardsii* has a holarctic distribution and is restricted to char, *Salvelinus* spp., especially brook trout, *Salvelinus fontinalis*, in North America (Black *et al.*, 1983). *Salmincola salmonea* occurs only on *Salmo salar* (Friend, 1940). Species that infect non-salmonid fish are *S. lotae* on gadids and *S. cottidarum* on several cottids (Lasee *et al.*, 1988).

Although *Salmincola* spp. are restricted largely to fresh water, Black *et al.* (1983) reported that *S. edwardsii*, *S. salmonea* and *S. californiensis* can survive on salmonids in the sea. Friend (1940) found that *S. salmonea* could survive for several months, and possibly up to two years, on Atlantic salmon at sea. The parasite was able to grow but not reproduce. New infection occurred only when the salmon returned to fresh water to spawn. The parasites that survived at sea were able to produce viable eggs and infect other salmon. Most *S. californiensis* died within two months following the transfer of its host, *Oncorhynchus nerka*, from fresh to sea water, but some survived for up to one year (Bailey *et al.*, 1989).

Little is known about seasonal variations in infection by *Salmincola*. High levels of infection by *S. edwardsii* were found on brook trout, *Salvelinus fontinalis*, leaving the Riviere a la Truite, Canada in spring during their seaward migration, but low levels were found on fish in the estuary (Black *et al.*, 1983). Friend (1940) found heaviest infections of *S. salmonea* on Atlantic salmon following their spawning in fresh water. Frimeth (1987) found foci of *S. edwardsii* in freshwater nonanadromous *Salvelinus fontinalis* in the Tabusintac River, New Brunswick.

Black (1982) found that the prevalence and intensity of infection by *S. edwardsii* on the gills of char, *Salvelinus fontinalis*, increased with host size. The size of the host has an important bearing on the microhabitat of attached larval and adult parasites.

Alella macrotrachelus infects the gills of *Diplodus sargus*, *D. annularis*, *D. vulgaris* and *Charax puntazzo* in the Mediterranean Sea, and *Acanthopagrus schlegeli* and *Sparus longispinis* in Japan (Cabral, 1983). A species that resembles closely *A. macrotrachelus* infects the gills of *Acanthopagrus australis*, *A. butcheri*, *A. latus* and *A. berda* in Australia. *Alella pagelli*, also from the gills of sparid fish, occurs in the North Sea, the Mediterranean and South Africa (Kabata, 1979). *Alella ditrematis* infects Japanese embiotocid fish and *A. pterobrachiata* parasitizes *Epinephelus merra* in Australia (Ho, 1983).

The intensity of *A. macrotrachelus* increased on *Acanthopagrus schlegeli* farmed in Japan with a decrease in water temperature from September to February, declined in March and April, increased again in May and June with rising water temperature, and declined again from June until the fish were harvested in September (Muroga *et al.*, 1981). There was a peak of infection by all stages of *A. macrotrachelus* on wild populations of *Acanthopagrus australis* in northern New South Wales, Australia, in summer although the parasite was present at low levels year-round (Roubal, 1990). Small fish were infected more heavily than large fish (Roubal, unpublished data). Cabral (1983) found *A. macrotrachelus* on *Diplodus sargus* mainly during the summer in the north-west Mediterranean. Most chalimus stages occurred in spring and this resulted in large numbers of ovigerous females in summer. There were no chalimus or copepodid stages in June, suggesting that ovigerous females carry the infection over winter.

The genus *Clavella* contains approximately 19 species (Kabata, 1979) which parasitize four orders of fishes in the Pacific and Atlantic Oceans. *Clavella adunca* (syn. *uncinata*) (Fig. 13.28) infects 51.9% of inshore whiting, *Merlangius merlangius*, near the Isle of Man, but only 28.7% in the open sea; the prevalence and intensity of infection decreases with age of fish and the movement of fish away from inshore areas (Shotter, 1971).

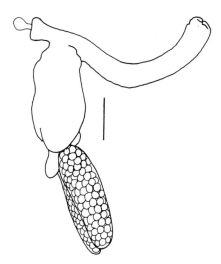

Fig. 13.28. *Clavella adunca* adult female. Scale bar = 1 mm. (From Kabata, 1979.)

Systematics

The Lernaeopodidae has five branches: *Salmincola, Lernaeopoda, Charopinus, Brachiella*, and *Clavella*, (see Kabata, 1979, 1980, 1981). The *Salmincola* branch, which is associated with freshwater fishes, is the most primitive and *Clavella*, the most advanced (Kabata, 1984).

Kabata (1979) re-defined *Alella* (within the *Clavella* branch) with *A. pagelli* as the type species and admitted *A. ditrematis* and *A. pterobranchiata*. He considered *A. acanthari* and *A. macrotrachelus* as conspecific with the type species. Kawatow *et al.* (1980) and Ho (1983) resurrected *A. macrotrachelus* and this was supported by Cabral (1983). There are now four species (*A. ditrematis, A. macrotrachelus, A. pagelli* and *A. pterobrachiata*). *Alella macrotrachelus* from Australian *Acanthopagrus australis* may be a distinct species based on the spination of the maxilliped (Cabral, 1983).

Parasite morphology and life cycle

Body proportions, size and shape of appendages and size and shape of the bulla varies widely between genera within the family.

The short cephalothorax of the adult female *Salmincola* is flexed at an angle to the trunk (Figs 13.25, 13.26). Second maxillae are longer than the cephalothorax and arise where the latter meets the trunk and meet distally where they join the button-shaped bulla. Paired, multiseriate egg sacs arise near the rounded posterior or distal end of the trunk. The female of *Allella* (Fig. 13.27) is similar except that the second maxillae are much shorter than the neck, and the bulla is thorn-shaped. The small salmincolid adult male consists of a cephalothorax in line with the slightly longer trunk. The posteroventral part of the male cephalothorax is enlarged and forms a base for the second maxillae and maxillipeds are used for attachment (Wilson, 1911; Kabata and Cousens, 1973). The adult male of *Alella* is an inverted U-shaped (Fig. 13.27), and is attached to either the neck or near the genital openings of the female (Roubal, 1981).

The bulla, produced by the maturing female in the frontal region of its head, consists of a stalk-like manubrium that expands into an anchor and is inserted permanently in the host tissue. Kabata and Cousens (1972) and Kabata (1981) recognized three types of bulla based on the internal arrangements of ducts. These types correspond to bullae from (i) freshwater teleosts, (ii) marine teleosts and (iii) elasmobranchs.

The mouth parts, located at the end of the neck, consist of large second antennae, small first antennae and maxillae, and large maxillipeds. The mandibles are located within the mouth cone as is typical of the siphonostomes. The alimentary tract consists of a short oesophagus, a long intestine and an anus. Glands associated with the bulla are located at the base of the second maxillae. Paired ovaries are located in the trunk and developing oocytes fill most of this region. The wall of the oviduct is the cement gland. A spermatophore is attached by the male to each of the paired vaginal openings, and paired uteri lead to the seminal receptacle between each of the oviducts. The adult male lernaeopodid copepodid has a blind ending alimentary tract which suggests that it does not feed once it attaches to the female (Wilson, 1915; Rigby and Tunnell, 1971).

Kabata (1981) recognized two types of life cycle within the lernaeopodid family. The most common type is illustrated by *Salmincola*, and *Alella*, whereas the other, abbreviated type is exhibited by *Clavella*.

The life cycle of *S. californiensis* (Dana 1852) consists of three phases and six stages: one copepodid, four chalimus (Ch1–Ch4) and one adult (Kabata and Cousens, 1973).

Nauplii of *S. californiensis* develop in the egg sacs and moult simultaneously to the copepodid upon hatching (Kabata and Cousens, 1973). Larvae of *S. salmonea* are released from the egg sac within a membrane and sink to the bottom whereupon they hatch (Friend, 1940). Copepodids attach to the host by a frontal filament and moult to the Ch1 stage (Kabata and Cousens, 1973). They undergo three more moults. During the Ch4 stage the female develops a bulla in the anterior space on the cephalothorax previously occupied by the frontal filament and uses this to attach permanently to the host following the final moult. The pre-adult then metamorphoses, without moulting, to an adult. The young male does not re-attach after moulting from the Ch4 stage but seeks out a female (which may still be a Ch4) and attaches to the genital region. Spermatophores are introduced into the orifices of the oviducts and once inseminated, a cement substance seals the vaginal openings to prevent further copulation. The male dies after copulation (Kabata and Cousens, 1973). After 1–2 weeks the adult female produces a pair of egg sacs (60–300 eggs) and the copepodids hatch after 2–3 weeks. A second pair of egg sacs is produced 2–3 weeks after the first eggs hatch and the parasite dies after this second lot of eggs hatch. The life cycle may take 1–6 months. This depends on temperature and species. The parasite may overwinter as copepodids (Hoffman, 1977).

The approximate times of developmental stages of *S. californiensis* at 11–12°C are: copepodid 24 hours (may live for up to 2 days without a host), Ch1 12 hours, Ch2 12 hours, Ch3 female 48 hours, Ch3 male 24 hours, Ch4 female 2 hours up to 2 weeks, Ch4 male 40 hours (Kabata and Cousens, 1973). Friend (1940) found that the copepodid of *S. salmonea* could live for up to 6 days. There was a developmental period of about 6 months between settlement and the production of larvae by adult females in the winter–spring host spawning season and a period of 5 months in the summer–autumn season. There was a period of 15 days between production of egg sacs and hatching at 12–13°C. Cool water temperature reduced body size and the number of eggs produced by *S. salmoneus* (see Johnston and Dykeman, 1987).

The life cycle of *A. macrotrachelus* was described by Kawatow *et al.* (1980) and Caillet (see Raibaut, 1985). It consists of one nauplius, one copepodid, four chalimus and an adult stage. Once the Ch4 stage moults into a juvenile female (attached by the bulla) the parasite grows and metamorphoses, without moulting, into an adult. Sexual dimorphism is evident at the Ch1 stage. The short-lived male is attached to immature and mature females. According to Caillet there is a pupal stage after the copepodid attaches to the gill filaments and preceding the Ch1 stage, but Kawatow *et al.* present it as a late Ch1. The life cycle of *Tracheliastes maculatus* also has one nauplius, one copepodid, four Ch stages and the developing adult stage. Sexual dimorphism is evident at the Ch3 stage (Piasecki, 1989).

Studies in our laboratory on the development of *A. macrotrachelus* showed that the embryo hatched as a nauplius and after swimming actively for less than 5 minutes the nauplius moulted rapidly (a few seconds) to the copepodid (S. Manion, personal communication). The copepodid could survive for up to four days at 25°C if no host was available. Juvenile females appeared 7 days after settlement, and the eggs took 10 days to develop within the egg sac. Kawatow *et al.* (1980) found a period of 10 days between copepodid settlement and the female stage (well-developed neck). Cabral (1983) reported a period of 10 days between settlement and a pre-adult female, and a life span of less than 2–4 months. A similar life span was suggested by the field data of Roubal (1990).

Clavella adunca has a life cycle devoid of the chalimus stage. A single nauplius moults to the copepodid stage which attaches and becomes a pupa. The pupa emerges and grows without moulting to an adult (Heegaard, 1947b; Shotter, 1971). The life span of the parasite is no more than 6 months (Shotter, 1971).

Host–parasite relationships

Site and host selection, course of infection

The early development of *Salmincola edwardsii* is more rapid at 12°C compared to 9.5°C irrespective of either photoperiod or fluctuation in temperature (Poulin *et al.*, 1990a). Once hatched, copepodids spend some time swimming in short bursts (Kabata and Cousens, 1977) but most of their time is spent on the bottom where shadows and shock waves trigger increased duration and speed of swimming (Poulin *et al.*, 1990b).

Copepodids move over the surface of the host using their second antennae and maxillae. The maxillipeds make a cavity in the host's surface and the large adhesive terminal plug of the frontal filament is inserted into the cavity (frequently onto underlying skeletal structures). The copepodid 'walks' backwards and uncoils the frontal filament; this evagination may take 3–5 hours (Kabata and Cousens 1973).

Following the Ch4 stage, the female inserts the distal end of the bulla into a cavity in the host's surface and the second maxillae are inserted into the basal end and manubrium of the bulla. The bulla is expanded by secretory substances from the maxillary glands passing through the maxillae and openings in the maxillary plugs and into the manubrial ducts (Kabata and Cousens, 1973).

The distribution of the parasite on the host is determined by the stage of the parasite as well as the size of the host. Larval stages of *S. californiensis* attach preferentially to the fins (37.3%), fin base (31.5%) and skin (24.9%), and only rarely (6.2%) in the branchial chamber of fry (3.2.–5.8 mm) of sockeye salmon, *Oncorhynchus nerka*. Fin bases are preferred (62.5%) by adult parasites (Kabata and Cousens, 1977). Larval parasites are on the gill filaments of juvenile (10.2–27 cm) sockeye salmon, but after the Ch4 moult they move to the walls of the branchial chamber. Large numbers of adult parasites occur on the gill filaments of adult sockeye salmon (Kabata and Cousens, 1977).

The copepodids of *S. salmonea* attach near the tip of the gill filament. Following the moult to the juvenile adult, the parasite moves further up the filament where the bulla is butted onto the lateral surface of the gill filament (Friend, 1940) and only the outer surface of the hemibranch is infected.

Salmincola edwardsii most often occurs on the gills but is found also on the opercula, fins and body surface (Black, 1982). At low levels of infection, larvae of *S. edwardsii* attach to the adipose fin on juvenile brook trout, *Salvelinus fontinalis*, but at high densities they prefer the dorsal fin (Black *et al.*, 1983). Larvae of *S. californiensis* attach to the base of the gill filaments on large rainbow trout, *O. mykiss*, but adult copepods attach to the tips of the filaments. Adult *S. californiensis* occur on the body surface and opercular and oral cavities at high levels of infection (Sutherland and Wittrock, 1985). *Salmincola yamame* occurs on the inner opercular surface of *O. masou* (see Hoshina and Suenaga, 1954), and *S. lotae* on the roof of the mouth of *Lota lota* (see Lasee *et al.*, 1988).

Several copepodids of *A. macrotrachelus* may attach to the tip of a primary gill filament of *Acanthopagrus australis*, but only one adult female parasite is found on a filament (Roubal, 1981). The parasite prefers the first gill arch and filaments at the ends of a gill arch. Caillet and Raibaut (1979) found that copepodids of *A. macrotrachelus* were morphologically alike but sexually pre-determined. Male copepodids settle in the vicinity of adult females already on the gills.

Clavella adunca enters the mouth of whiting *Merlangius merlangius* and

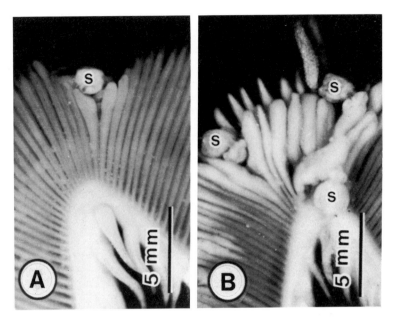

Fig. 13.29. *Salmincola californiensis* (S) attached to gill filaments of *Oncorhynchus mykiss* showing crypting (A) and clubbing (B). (From Sutherland and Wittrock, 1985, with permission.)

settles on the gill arch. Kabata (1960) found 61.7% of 285 *C. adunca* on the first gill arch. Both juvenile and adult parasites of adult fish were found in this position, but in juvenile whiting most adult *C. adunca* were on the posterior opercular rim. In cod, *Gadus morhua*, the parasite prefers the first gill arch but also occurs in the buccal cavity. On the other hand, *C. dubia* has no preference for any gill arch (Kabata, 1960). Shotter (1976) reported that two-thirds of adult *C. adunca* were attached to the posterior opercular rim of 3–6 month old whiting and to the gill rakers, especially of the first gill arch, in older whiting. Pelagic (larval) whiting are not infected but they become infected once they settle to the bottom (Kabata, 1960). Shotter (1971) reported that the nauplius of *C. adunca* was positively phototactic for 2–3 hours post-hatching but then settled on the bottom of the container.

Clinical signs and histopathology

The pathology associated with lernaeopodid copepods depends on the tissue infected, the species of parasite, its size and the type of bulla. Damage by *Salmincola* to the gills can be extensive. Kabata and Cousens (1977) found that less damage was caused by the implanted terminal plug of the frontal filament than by the bulla of young *S. californiensis*. Pressure on the tips of the gill filaments due to large adult parasites on the inner opercular wall caused a reduction in length of several adjacent primary gill filaments. This 'crypting' was also caused by adult *S. californiensis* attached to tips of gill filaments of *Oncorhynchus mykiss* (Fig. 13.29A), but Sutherland and Wittrock (1985)

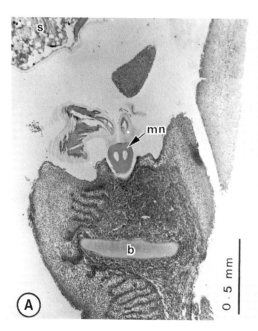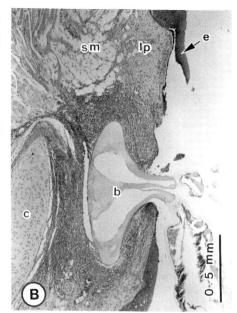

Fig. 13.30. *Salmincola californiensis.* Sections of *Oncorhynchus mykiss* through sites of attachment at the gill tip (A) and pharyngeal tissue (B). b = bulla, c = cartilaginous supporting tissue, e = epithelium, lp = lamina propria, mn = manubrium, s = *Salmincola*, sm = submucosa. (From Sutherland and Wittrock, 1985, with permission.)

attributed it to retarded filament growth. The extent of crypting on the gills of Atlantic salmon caused by *S. salmonea* may be a measure of the period since infection; older infections are associated with deeper crypts than more recent infections (Friend, 1940).

The gill epithelium of *O. nerku* affected by *S. californiensis* shows hyperplasia and hypertrophy leading to fusion of filaments; aneurysms also appear (Kabata and Cousens, 1977). Attachment by the larval frontal filament causes only a limited proliferative response. Insertion of the bulla into subepithelial sites causes only a mild reaction of the connective tissue. In muscular attachment sites compaction and disintegration of myotomes occurs especially near the subanchoral surface of the bulla (Kabata and Cousens, 1977). When the parasite is attached to the operculum there may be osteogenic activity in the vicinity of the bulla leading to a build up of spongy bone but continued insult results in the degeneration of the bone (Kabata and Cousens, 1977).

Sutherland and Wittrock (1985) found hyperplasia of the gill epithelium of *O. mykiss* (Fig. 13.30A) with fusion of up to 40 adjacent gill filaments caused by *S. californiensis* (Fig. 13.29B). Extravasation and infiltration of eosinophilic granule cells was also noted. There was loss of epidermis where the bulla of *S. californiensis* penetrated into the pharyngeal tissue of *O. mykiss*

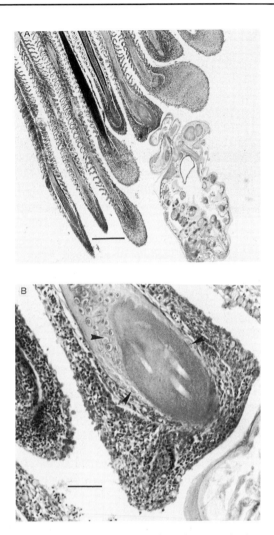

Fig. 13.31. *Alella macrotrachelus*, adult female attached to gills of *Acanthopagrus australis* showing (A) proliferated epithelium on adjacent filaments and reduced length of filament to which the parasite is attached, and (B) bulla surrounded by proliferated chondrocytes (arrows) from cartilaginous bar. Scale bars, A = 200 μm, B = 40 μm. (From Roubal, 1989b.)

and the lamina propria was replaced by infiltrating leucocytes (Fig. 13.30B). Distortion of gill filaments and pronounced epithelial hyperplasia was caused by *S. yamame* (see Hoshina and Suenaga, 1954).

Alella macrotrachelus causes swelling and/or crypting of several gill filaments of *Acanthopagrus australis* (Fig. 13.31A), but *Neobrachiella lata*, which attach to the mucosa of the bulla cavity, and *Clavellopsis parasargi*, which attach to the epithelium on the gill arch of this host, are associated with only small nodules that surround the bulla. Recently-attached *Alella macrotrachelus* are associated with the swelling of the tips of one or two filaments but as the parasite grows, the longer neck permits feeding over a wider area. Filaments on adjacent hemibranchs and, in the case of small fish, adjacent holobranchs are affected (Roubal, 1986a, 1987).

Penetration of the frontal filament during initial attachment by *A. macrotrachelus* is associated with epithelial hyperplasia and cellular necrosis in the vicinity of the filament as a result of mechanical damage. Neutrophils, indicative of early inflammation, are evident in the vicinity of the filament (Roubal, 1989b). Penetration of the bulla also causes localized damage and epithelial hyperplasia, but the infiltrating cells are dominated by lymphocytes and macrophage-like cells. Giant cells and foci of necrotic epithelial tissue are seen occasionally. When the bulla abuts onto the filament cartilaginous supporting bar there is a proliferation of chondrocytes that engulfs the distal end of the bulla and ensures permanent attachment (Fig. 13.31B) (Roubal, 1989b). Muroga *et al.* (1981) reported aneurysms in the secondary lamellae of *Acanthopagrus schlegeli* in the vicinity of attached *A. macrotrachelus*.

Only a small 'tumour of attachment' is caused when *Clavella adunca* attaches to the gill filaments of *Gadus morhua* but the same parasite causes a large tumour when attached to the gill filaments of *Melanogrammus aeglefinus* (see Kabata, 1984). Shotter (1971) described the tissue of whiting, *Merlangius merlangus*, around the bulla of *C. adunca* to be hypertrophied and devoid of pigment but no such sign when attached to gill rakers or dermal denticles of the mouth. Heavy infections resulted in some occlusion of the branchial and buccal cavities.

Host immune response

An immune response directed against these parasites has still to be determined. Friend (1940) attributed the low infections of *Salmincola salmonea* on kelts (recently spawned for first time and not returned to sea) of Atlantic salmon in spring to result from an immune response although kelts in late autumn and winter had high levels of infection.

Mechanism of disease

Attachment causes greater damage to the host than does feeding (Kabata, 1984; Kabata and Cousens, 1977; Sutherland and Wittrock, 1985). Attachment to the gill filaments as well as pressure from parasites within the buccal cavity damages the gills leading to a loss of respiratory surface area. Both attachment and feeding determine the extent of damage by *Alella* to the gills of *Acanthopagrus australis* since the number and length of filaments affected are related to the length of the neck (age of parasite) (Roubal, 1986a, 1987, 1989b).

Attachment sites may also provide portals of entry for secondary invaders. Friend (1940) pointed out that fungal infection of gill filaments infected by *Salmincola salmonea* follows spawning and poor condition of Atlantic salmon.

Effects on host physiology

Brook trout, *Salvelinus fontinalis*, with gill *Salmincola edwardsii* had a lower resistance to high water temperature than trout without the parasite, but there was no effect on fish fecundity or condition factor (Vaughan and Coble, 1975). Allison and Latta (1969) found that *S. edwardsii* had no effect on condition factor of brook trout, but the fecundity of rainbow trout in a California hatchery was lowered by *S. edwardsii* infection (Gall *et al.*, 1972).

Sutherland and Wittrock (1985) reported that 'runt' rainbow trout were heavily infected with *S. californiensis* but a causal relationship could not be established. Up to 25% of gill surface area was lost through crypting associated with heavy infections of *S. californiensis* in the buccal cavity of juvenile sockeye salmon (Kabata and Cousens, 1977). Roubal (1986a, 1987) found that adult female *Alella macrotrachelus* affected more gill filament length and surface area in younger *Acanthopagrus australis* (Fig. 13.32).

The biological implications of such losses on the growth performance and survival of captive and wild fish have not been ascertained. Janusz (1980) found that young cod, *Gadus morhua*, infected by *Clavella adunca* had a higher average weight and condition factor than uninfected cod. He attributed this to the earlier recruitment of larger juvenile fish into the adult population and their consequent infection by *Clavella*; thinner juvenile fish are not recruited as soon and thus remain uninfected for longer.

In vitro culture

Detached egg sacs are allowed to develop and hatch. Copepodids are provided with a suitable host to permit further development.

Nutrition and feeding

Lernaeopodids feed on mucus and epithelial tissue. The mandibles rake across the skin of the host and passage of food toward the mouth opening is then assisted by contraction of the mouth tube (Chandran and Nair, 1988). Adults and larvae of *Salmincola californiensis* apparently feed by browsing on the host epithelium and, where blood vessels are close to the surface, blood is also ingested (Kabata and Cousens, 1977). Bare tips on gill filaments adjacent to where *S. salmonea* attached was attributed to feeding (Friend, 1940).

The role of the bulla, apart from attachment, is unclear. Cousens showed that radioactive-labelled amino acids passed through the bulla into the parasite, but whether substances pass from the bulla into the host is unresolved (see Kabata, 1984). Substances secreted by the parasite may facilitate the passage of the bulla through the host tissue during attachment.

Diagnosis of infection

The large female parasites are usually visible with the naked eye on the gills, fins, body or in the buccal and opercular cavities of fish.

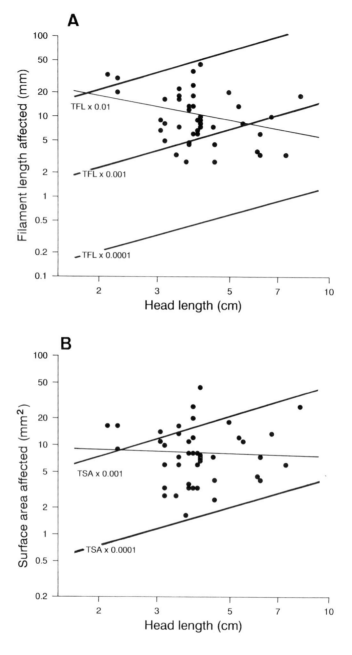

Fig. 13.32. Effect of *Alella macrotrachelus*, adult females, on filament length (A) and gill surface area (B) of *Acanthopagrus australis*. TFL, total filament length; TSA, total surface area. (Modified from Roubal, 1987.)

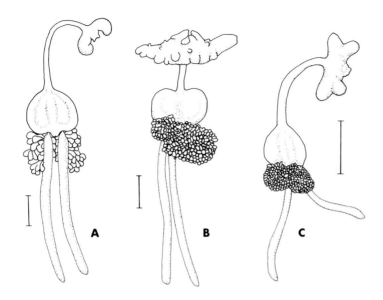

Fig. 13.33. *Sphyrion* species, adult females. A, *S. lumpi.* B, *S. laevigatum* from *Genypterus capensis.* C, *S. quadricornis* from *Coelorhynchus braueri.* Scale bars = 10 mm. (A after Grabda, 1991, B and C after Gayevskaya and Kovaleva, 1984.)

Prevention and control

Hoffman (1977) reported that chemicals such as 0.85% calcium chloride, 0.2% copper sulphate, 1.7% magnesium sulphate, 0.2% potassium chlorate and 1.2% sodium chloride were ineffective against adult *Salmincola* but were useful against copepodids at 3- or 4-day intervals as long as deemed necessary.

Bailey *et al.* (1989) suggested the removal of *S. californiensis* by sea water, but the ability of the parasite to survive for long periods in this environment makes such an approach unfeasible. McGladdery and Johnston (1988) suggested the physical removal of *S. salmona* and holding Atlantic salmon at 2–3°C to prevent re-infection.

COPEPODA: *SPHYRION*

Introduction

Sphyrion lumpi (Fig. 13.33A) commonly infects 5–10% of redfish *Sebastes mentella* in the North Atlantic and its prevalence appears to be increasing (Reimer, 1989). In Germany, regulations prohibit the sale for human consumption of fillets with more than 5% affected by the parasite (Jungnitz, 1989). Heavily infected fish go for fishmeal or pet food. The parasite is estimated to increase the cost of processing redfish fillets by 80% (Hargis, 1958).

Host range, distribution and seasonality

Sphyrion lumpi is found on deepwater pelagic fish, especially *Sebastes mentella*, to a lesser extent *Sebastes marinus*, in the temperate north and south Atlantic Ocean. It also parasitizes other scorpaenids, and gadids, morids, macrourids, and pleuronectids (*Reinhardtius hippoglossoides*) in the same habitat (Ho, 1989). It has a patchy distribution; centres of infection in *Sebastes* spp. in the North Atlantic are off Maine (Nigrelli and Firth, 1939), off south Labrador, near Bear Island and in the Irminger Sea (Bakay, 1989). Ho and Kim (1989) report finding the parasite in the Sea of Japan.

In the Irminger Sea, prevalence peaked in summer. Gravid females were particularly abundant in July/August when the redfish aggregated. At this time newly attached parasites were also in evidence. Most infections occurred in fish over 33 cm long (Bakay, 1989).

There are two other species in the genus. *Sphyrion laevigatum* (Fig. 13.33B) is found on deepwater Ophidiidae and Merlucciidae in southern temperate waters. On the ophid *Genypterus capensis* it occurred only on fish over 60 cm, though smaller fish were found on the same grounds (Payne, 1986). *Sphyrion quadricornis* (Fig. 13.33C) parasitizes macrourids in temperate waters worldwide (Ho, 1989).

Systematics

The three species can be separated by their holdfasts (Fig. 13.33). The holdfast of *Sphyrion laevigatum* is 2–3 times as wide as long, and that of *S. quadricornis*, a smaller species, has a distinct bifurcation at the lateral ends of the otherwise smooth holdfast (Gayevskaya and Kovaleva, 1984). The Sphyriidae probably evolved on macrourids (Ho, 1989).

Morphology and life cycle

Adult females of *S. lumpi* have the anterior cephalothorax expanded into a sphyra or 'hammer' which is embedded in the fish muscle, a narrow 'neck' which projects from the fish, and a broad genital complex which is subspherical when the dorsoventral muscles are relaxed and flattened when they are contracted. Attached to the complex are branching posterior processes. These increase in complexity with increasing age. Tiny males, 2 mm long, similar in morphology to males of the Clavella-branch of the Lernaeopodidae (Kabata, 1988), are attached near genital pore of about 5% of females (Squires, 1966) and produce spermatophores. In the female a small seminal vesicle is present in the median line on the ventral side of the intestine just anterior to the vulvae (Wilson, 1919). At least two and possibly more batches of eggs may be produced as Squires (1966) observed that eggs in egg sacs developed simultaneously, and females with egg sacs normally contained developing ova in the ovary. Two naupliar stages are recognizable within the egg before it hatches to release a copepodid (Wilson, 1919; Jones and Matthews, 1968). The life cycle is unknown.

The life span of an adult female may be at least a year (Squires, 1966). It is less than that of the host because scars are often present. The number of scars increases with fish size and large fish have more scars than parasites. Scars are left by *S. laevigatum* indicating that it too has a life span shorter than its host (Payne, 1986).

Host–parasite relationships

Site and host selection, course of infection

In *Sebastes mentella* 56% of *Sphyrion lumpi* were located in the dorsal muscles, 21% at the anus, 19% in the ventral part, 3% in the caudal peduncle and 1% in the head (Bakay, 1989). Parasites are often attached in groups in fish with multiple infestations (Williams, 1963). Juveniles eat through tissue and produce a cavity 1 cm or so across.

Clinical signs and histopathology

Juvenile females cause a dark lump in the skin with a tiny opening. As the parasite grows, the opening becomes larger until in some cases the skin and muscle are turned out to form large flaps. The parasite causes intense inflammation, as evidenced by oedema, leucocyte infiltration and a fibrin network. Adjacent blood vessels become extremely dilated. The area is eventually isolated by a wall of connective tissue which, in older infections, is thick, fibrous and encapsulates the cephalothorax of the parasite. Melanophores and erythrophores may be present in the surface of the abscess (Nigrelli and Firth, 1939).

When the parasite dies, the part external to the fish is quickly lost. The holdfast and neck which remain in the muscle degenerate in a viscid exudate which discharges through the skin. In the exudate Nigrelli and Firth (1939) found flagellates and chains of alga-like organisms $18 \times 12 \, \mu m$ in size. After an unknown period, the abscess subsides and much of the remaining necrotic tissue is removed by macrophages. In the final stages, melanophores may move from the capsule wall into surrounding tissue and produce a diffuse brownish black mark often visible on the internal surface of the fillet. The mark apparently exists for a shorter period of time than the abscess and eventually disappears (Priebe 1986, 1989).

Parasite nutrition and physiology

Developing females take blood, muscle and other tissues into their capacious gut. Most of their growth takes place before the abscess has fully developed. The parasites darken with age, presumably a result of accumulation of haematin and other waste products. The posterior processes may have a respiratory function (Kabata, 1988), and their development may be linked to the encapsulation of the cephalothorax.

Diagnosis of infection

By visual inspection using the unaided eye for adults; juveniles require a dissecting microscope. In fillets, abscesses are detected by candling or ultra-violet light (Jungnitz, 1989).

BRANCHIURA

Introduction

Argulids have been recognized as pests of cultured trout in Europe and carp in China since the 17th century (Wilson, 1902; Kabata, 1985). They cause mortalities of fish in aquaria, in lakes and estuaries, and occasionally cause problems in sea-caged salmonids (Menezes *et al.*, 1990; Stuart, 1990). Secondary infections by fungi and bacteria reduce the commercial value of parasitized carp and goldfish (Shimura, 1983a). Argulids can transmit spring viraemia of carp (Ahne, 1985).

The three best known species are *Argulus foliaceus*, *A. japonicus* and *A. coregoni.*

Host range, distribution and seasonality

Argulus foliaceus is common on carp in Europe and Asia. It occurs in cold temperate climates and infests a wide range of hosts including Cyprinidae, Salmonidae, Gobiidae, Gasterosteidae, Acipenseridae, frogs and toads (Yamaguti, 1963). It is in eutrophic rather than oligotrophic lakes (Valtonen *et al.*, 1987).

Argulus japonicus has a worldwide distribution. It has been introduced with aquarium fish from the orient and now is common wherever goldfish are found (Cressey, 1978). In North America it infests primarily goldfish but has been found on *Cyprinus* and *Ictalurus* (see Amin, 1981). In Europe it infests *Carassius, Cyprinus* and other genera including *Esox, Perca, Tinca* and *Sardinus*. Its distribution overlaps with *A. foliaceous*, but is generally in warmer waters (Stammer, 1959).

Argulus coregoni occurs primarily on salmonids and also infests cyprinids and other hosts. It occurs in rivers and large lakes, generally in cooler water than *A. foliaceus*, in Europe, China, Japan and North America (Gurney, 1948; Yamaguti, 1963). It has not been reported from Canada (Kabata, 1988).

These three species reach peak abundance during the summer and autumn. In severe climates they overwinter as eggs (Razmashin and Shirshov, 1981; Shimura, 1983a). In more moderate climates, *A. foliaceus* survives the winter as adults (Kimura, 1970) and may breed all year (Bower-Shore, 1940). In South Africa, gravid females of *A. japonicus* have been found during the winter (Shafir and Van As, 1986).

Argulus alosae, an estuarine species, was found recently on cultured salmon, *Salmo salar*, and trout, *Oncorhynchus mykiss*, in three marine fish farms in eastern Canada. Fish mortality associated with the infestation occurred when the water temperature rose above 16°C. Trout were more severely affected than the salmon (Stuart, 1990; McGladdery, personal communication).

Chonopeltis spp. occur only in Africa; *Dolops* spp. are in Africa and South America (Fryer, 1968a). *Dolops ranarum* infests a wide variety of hosts. Its females overwinter on *Oreochromis* and *Clarias* species (Avenant and Van As, 1986).

Systematics

Branchiura differ from Copepoda in that they have compound eyes, continue to moult after maturity, lay eggs singly, and develop without a true naupliar stage. Spermatophores are not normally produced and, when they are, they contain products from both testes. Argulid DNA shows similarities to that of pentastomes (Abele *et al.*, 1989).

Of the 150 or so species of Branchiura, about 100 belong to the genus *Argulus* (Kabata, 1985). Most branchiurans are freshwater parasites, the remainder are found in estuarine or marine environments (Kabata, 1984).

For keys to Branchiura and guide to early taxonomic literature see Yamaguti (1963). More recent specialist keys include Cressey (1978) for species from the northeastern United States, Kabata (1988) for species in Canada, and Fryer (1982) for those in Great Britain.

Parasite morphology and life cycle

The most obvious feature of argulids is the sucker-like first maxillae (Fig. 13.34) (Overstreet *et al.*, 1992). These are mobile structures on a slender stalk. The parasite can detach them from the fish alternately and so 'walk' about the fish surface.

On either side of the carapace are two respiratory areas; their positions are taxonomically useful. The cuticle there is thinner and more permeable than rest of cuticle. It is devoid of spines, and is adjacent to a blood sinus (Martin, 1932; Sutherland and Wittrock, 1986).

The proboscis is formed from the bases of the mandibles and the interlocked labrum and labium. At the entrance to the mouth are two conical projections, the 'labial spines' or 'siphons', which have a pore at the tip. They probably secrete enzymes that aid digestion. Within the proboscis lie two large serrate mandibles. These can be everted through the mouth and presumably lacerate host tissue. Three delicate lamellae guard the opening of the oesophagus. They are finely serrated and act as a fine filter which appears to be why blood cells are not found in the gut (Martin, 1932). Under the scanning electron microscope, the surface of the labrum appears smooth, and the surface of the labium is covered in scattered rows of blunt teeth in

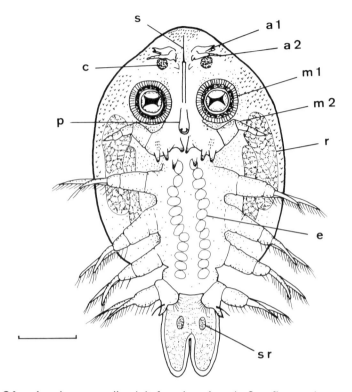

Fig. 13.34. *Argulus monodi*, adult female. a1 and a2 = first and second antennae, c = compound eye, e = egg, m1 and m2 = first and second maxillae,
p = proboscis, r = respiratory area, s = stylet, sr = seminal receptacle. Scale
bar = 1 mm. (After Fryer, 1959.)

those species that have been examined (*A. japonicus*, *A. coregoni* and
A. appendiculosus; Shimura, 1983b; Sutherland and Wittrock, 1986).

The proboscis of *Dolops ranarum* differs from those in *Argulus* in that
it is relatively short and lacks the two labial spines. In place of the lamellae
at the mouth of the oesophagus are densely packed setae (Avenant-Oldewage
and Van As, 1990).

The stylet in argulids lies anterior to the mouth tube and is separate from
the digestive tract (Fig. 13.34, s). It is a delicate structure consisting of a long
tapering hollow spine and a broad sheath into which the spine can be retracted.
Near the tip of the stylet are two openings, one of which appears to be
secretory and the other sensory (Shimura, 1983b).

Females grow larger than males. They can be recognized usually well
before they are mature by the prominent seminal receptacles on the abdomen
(Fig. 13.34, sr).

The basal segments of the second, third and fourth swimming legs of
the male are modified for clasping the female. During copulation the male
A. foliaceus attaches to the dorsal side of the female and clasps her last pair
of legs between his last two pairs, interlocking his legs using a socket on the

basal segment of the last leg and a peg on the basal segment of the third leg. The abdomen of the male is twisted under that of the female first to one side then to the other. In this way the male genital opening is brought into direct contact with the spermathecae of the female (Martin, 1932). Males of *D. ranarum* produce spermatophores (Fryer, 1960).

Adult argulids can survive for several days away from fish. They lay their eggs on an inert object one at a time side by side in single, double or triple columns. Eggs are fertilized at the time of deposition (Wilson, 1902), the ovum shell being pierced by a spine in *D. ranarum* (Fryer, 1960). Female *A. japonicus* reattach to fish and feed in between laying batches of eggs (Shafir and Van As, 1986).

Eggs of *A. japonicus* hatch after 10 days at 35°C and after 61 days at 15°C. Hatching is asynchronous and unrelated to the position of the egg in the egg mass (Shafir and Van As, 1986). In *D. ranarum*, eggs at the periphery of the egg mass hatch sooner than those in the centre (Fryer, 1964).

Eggs of *A. coregoni* laid in August hatched in September, those laid September, October and November hatched in May, June and July the following year. Thus there were two generations per year (cf. three generations in *A. foliaceus* and *A. japonicus*) (Shimura, 1983a).

Eggs of *A. americanus* hatched in 35–37 days at 15–16°C; hatched females were gravid by day 49 (Shimura and Asai, 1984). Eggs of *A. varians*, a marine species from *Sphaeroides testudineus*, hatched in 25 days at 25°C (Bouchet, 1985).

The stage that hatches from the egg of argulids is immediately parasitic and, other than in *Chonopeltis* (see below), usually attaches to the host species on which it will mature. The hatchlings of *A. foliaceus, A. japonicus, A. coregoni, A. americanus, A. catostomi, A. lepidostei* and *A. maculosus* are copepodid-like with long second antennae and mandibular palps (Fig. 13.35). After several days they moult to the second stage which resembles an adult without suckers. After a further five to six moults (eight in *A. coregoni*) at intervals of 2 to 6 days, they mature in about 4 weeks depending on the temperature (Wilson, 1902; Hindle, 1949; Schluter, 1979; Shimura, 1981).

Eggs of *A. funduli, A. megalops, A. puthenveliensis, A. stizostethii* and *D. ranarum* hatch into juvenile adults (Wilson, 1902; Fryer, 1968a; Shimura, 1981).

The hatchling of *Chonopeltis* species bears some resemblance to the copepodid-like larva of argulids. However, its three pairs of segmented appendages, second antennae and first and second maxillae lack setae and are modified for grasping. The first maxillae are the largest, and are tipped by two barbed spines, one lying in a groove of the other, which deeply penetrate into the tissues of the host. Larvae of *C. australis* and *C. brevis* use a small catfish or cyprinid as an intermediate host and attach under the operculum. Adults are parasitic on body or fins of other cyprinids. Their mode of transfer from host to host is not clear as neither larvae nor adults are able to swim (Fryer, 1956, 1968a; Nierkerk and Kok, 1989).

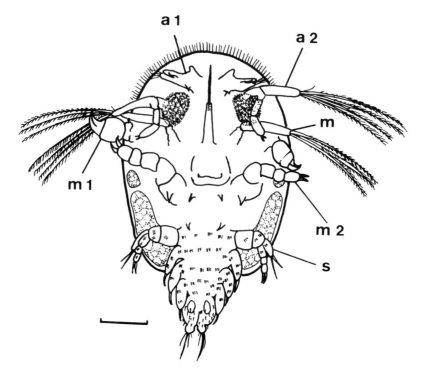

Fig. 13.35. *Argulus coregoni* first stage larva. a1 and a2 = first and second antennae, m = mandibular palp, m1 and m2 = first and second maxillae, s = first swimming leg. Scale bar = 0.1 mm. (After Shimura, 1981.)

Host–parasite relationships

Site and host selection, course of infection

Argulus foliaceus and *A. japonicus* attach primarily to the caudal peduncle of carp in culture ponds (Bazal *et al.*, 1969). Site preference is less marked for *A. foliaceus* on *Xiphophorus helleri*. At 28°C most parasites were on the flank, caudal fin and pectoral fins; at lower temperatures many of the large parasites (over 2.8 mm) were on the surface of the operculum (Schluter, 1978). Small individuals of *A. coregoni* attached all over *O. masou*. Large individuals preferred the skin behind the bases of the pectoral and pelvic fins, and to a lesser extent the adipose fin. The parasites spread all over dead fish, suggesting that site specificity is related to water flow. Few occurred in the buccal or gill cavities except for recently hatched parasites (Shimura, 1983a). The branchial cavity is the primary site for attachment of many branchiurans, including *A. catostomi*, *A. amazonicus*, *A. juparanaensis*, *Dolops ranarum*, *D. geayi* and some *Chonopeltis* species (Wilson, 1902; Fryer, 1968a; Malta, 1982; Malta and Silva 1986).

Poulin and FitzGerald (1989c) found that *A. canadensis* attached more frequently to fish that already had *Argulus* than to uninfected fish.

Fig. 13.36. *Argulus* sp. lesion on the sciaenid *Cynoscion regalis*. e = epidermal hyperplasia around the border of the lesion. (From Noga *et al.*, 1991.)

Clinical signs and histopathology

Infested fish are lethargic, stay in the corners of tanks, cease feeding and lose condition (Hindle, 1949; Das *et al.*, 1980). Carp with *A. foliaceous* initially try to remove the parasite by rubbing against the substrate. In chronic heavy infections, the skin becomes opaque, the fins frayed, and the fish is listless (Stammer, 1959).

Allum and Hugghins (1959) reported an epizootic of *A. biramosus* (= *A. appendiculosus*) on catostomids, cyprinids, ictalurids and other fishes in two lakes in the USA. The 300–400 argulids per fish were principally attached near the bases of dorsal, anal and caudal fins. The skin was inflamed, scales loosened, and in severe cases, the fins were frayed and almost gone.

The first feeding sites of argulids are often marked by haemorrhagic spots. Under low magnification they appear as craters formed by hyperplasia of epidermis at the margins of the wound (Fig. 13.36). Histologically, the craters may be restricted to the epidermis, especially on large fish with a thick epidermis, or they may penetrate through to the stratum spongiosum of the dermis and even to the stratum compactum beneath. The dermis becomes oedematous. Mucous and club cells are absent from any epidermis remaining in the crater but are abundant in tissue at the margin of the crater. In terminal cases, the epithelium over the whole fish becomes thin and may be missing from parts of the body and fins (Stammer, 1959).

Bower-Shore (1940) noted that six months after an *A. foliaceus* had fed on a small carp, the site was still plainly visible as a pale area.

Mechanism of disease

Though there are reports of fish particularly sensitive to *Argulus* infection, and either thrash wildly (Kroger and Guthrie, 1972a) or are stunned by a single puncture, fish can generally carry many *Argulus* with little sign of disease (Stammer, 1959). Mortalities are usually associated with hundreds of *Argulus* per fish and these counts may be underestimates because argulids rapidly leave a host when the host is caught (Shimura, 1983a; Stuart, 1990). Such mortalities are probably due to the breakdown of the epithelial integrity and the resultant loss of ionic and osmotic homiostasis. Furthermore, open lesions in the dermis allow fungal and bacterial infections to establish. These, together with anorexia, contribute to the mortality.

Effects on host

Shimura *et al.* (1983a) showed an anaemia in young *Oncorhynchus masou* infested with 250 *Argulus coregoni* per fish. Leucocyte counts, plasma total protein, cholesterol and calcium were also lowered significantly. Statistically significant changes were not observed in numbers of immature erythrocytes or thrombocytes, or in the plasma concentrations of glucose and phosphate.

Three species of sticklebacks (Gasterosteidae) held in aquaria had a significantly higher mortality in tanks with *A. canadensis* than in tanks without the parasite (Poulin and FitzGerald, 1987). Sticklebacks in infested tanks tended to keep away from the bottom (Poulin, and FitzGerald, 1989a). They also formed larger shoals than in uninfested tanks and this behavioural change decreased the average number of attacks received by individual fish (Poulin and FitzGerald, 1989b).

Antigenicity

Inflammation at the feeding sites is not a major component of the histological changes. This suggests that the secretions of the parasite have low antigenicity. Ferguson (1989) indicated that a break in the dermis normally caused water-logging of muscle with minimal inflammation.

Concurrent infection

Argulus foliaceus has been shown to be a mechanical vector for the virus that causes spring viraemia of carp (SVCV; Ahne, 1985). Argulids can serve as intermediate hosts for nematodes belonging to the Anguillicolidae and Skrjabillanidae (Moravec, 1978; Tikhomirova, 1983). They also carry epiphytes (Van As and Viljoen, 1984; Viljoen and Van As, 1985; Sutherland and Wittrock, 1986) though these are not apparently infective to the fish.

Infections are often concurrent with *Saprolegnia* (Bower-Shore, 1940; Stammer 1959; Allum and Hugghins, 1959).

Oncorhynchus masou exposed to water contaminated with *Aeromonas salmonicida* had a higher mortality rate in tanks which also contained small numbers of *Argulus coregoni* (10 per fish) than in tanks without the parasite.

However, there was little correlation between the location of furunculosis lesions and the sites of attachment of the crustacean (Shimura *et al.*, 1983b).

Parasite nutrition and physiology

Argulids use the stylet to probe the tissue before feeding. They feed by sucking up the products of extracellular digestion that apparently results from the secretions of the stylet and the labial spines, and macerations by the mandibles.

Shimura and Inoue (1984) injected an extract of the mouth parts and stylet of *A. coregoni* into the muscle of rainbow trout and observed a strong haemorrhagic response. *In vitro* no haemolysis was evident with trout red blood cells and no cytotoxic effects were seen in RTG-2 cultured cells. They suggested the haemorrhagic response facilitated the ingestion of blood by the parasite. Bower-Shore (1940) used histological stains to show that there were traces of haemoglobin in the gut of *A. foliaceus*, though intact blood cells were not found (Minchin, 1909; Stammer, 1959).

Chonopeltis australis and *C. brevis* feed on mucus. They have brush-like structures on the labium, and *C. australis* has a radula-like disc with spines just inside the mouth, all presumably to aid in mucus collection (Nierkerk and Van As, 1986; Nierkerk and Kok, 1989).

Dolops ranarum feeds on blood. It produces a deep wound, apparently using the first maxillae (which are not sucking discs in this genus) as the mandibles are relatively small (Fryer, 1968a; Avenant-Oldewage and Van As, 1990).

Respiration in argulids probably takes place not only through the 'respiratory areas' but also through the abdominal lobes. The lobes have a brisk blood circulation and are very large in *Chonopeltis schoutedeni*, a parasite of benthic fish in muddy areas. *Dolops ranarum* has haemoglobin as a respiratory pigment and can survive in waters of low oxygen (Fryer, 1968a).

Diagnosis of infection

Argulus spp. can be seen scuttling over the surface of the fish with the naked eye.

Prevention and control

Insecticides are commonly used to treat fish with argulosis. Gemmexane (hexachlorocyclohexane, benzene hexachloride, BHC or 666) is particularly effective (Hindle, 1949; Singhal *et al.*, 1986) though it is highly toxic to man and some fish (Siluridae) and not easily degraded. Ponds are treated with 0.1–0.2 ppm in two or three applications at weekly intervals. The parasite develops resistance to it fairly quickly (Kabata, 1985).

Other chemical treatments include: Pyrethrum (20–100 ppm for 10–20 minutes); Malathion (0.25 ppm for 6 hours); Dipterex (100 ppm for 1 hour, longer at lower dose); trichlorphon (Tr-technical grade or Tr-crystalline (Neubert, 1984) at 0.25–5 ppm for several hours); Neguvon (up to 50,000 ppm for short baths, 0.25 ppm for long treatments); and the antimalarial drugs quinine hydrochloride (13.5 ppm) or Atebrine (10 ppm) for several days (Kabata, 1985). DTHP or 0.0-dimethyl 2,2,2-trichloro-1-hydroxyethylphosphonate was used by Puffer and Beal (1981) at 2.5 mg l^{-1} for 1 hour. Details of other treatments that sometimes work, such as salt or formalin, are given by Kabata (1985).

To prevent the introduction of argulids, incoming water can be filtered or taken from a source free of fish. Incoming fish can be quarantined and if necessary treated, either by chemicals or by physically removing the *Argulus* with forceps.

For biological control in mud ponds, remove all hard objects from the pond and introduce substrate such as wooden slats to which eggs can be attached. The slats can be removed weekly and eggs laid on them destroyed (Kabata, 1985).

To avoid problems with *A. foliaceus* in southwestern Siberia, Razmashin and Shirshov (1981) suggest that lakes should not be stocked with coregonid fry until the water has been at 14°C or more for 4–5 days. By then, overwintering eggs will have hatched and the larvae will have perished.

Argulids are eaten by *Gambusia*, angel fish (*Pteropyllum scalare*) and sticklebacks (Gasterosteidae) (Kabata, 1970, 1985; Poulin and FitzGerald, 1987). Predation on *A. catostomi* by small dace and roach was proposed by Wilson (1902) to control its numbers on fish in a dam in Massachussetts.

ISOPODA: CYMOTHOIDAE

Introduction

Most isopods seen on fish are cymothoids. Of the 300 cymothoids described, over 80% are from tropical and subtropical seas, many from the Indo-Malayan archipelago (Avdeev, 1985). A few occur in fresh water though none has been found in fresh water in Europe or North America.

The skin parasite *Nerocila orbignyi* and the gill parasite *Mothocya parvostis* have been associated with poor fish growth and increased fish mortality in cultured *Dicentrarchus labrax* in the Mediterranean, and cultured *Girella punctata* in Japan, respectively (Bragoni *et al.*, 1983, 1984; Hatai and Yasumoto, 1982a). Salmon farms in Chile and Australia are plagued by the buccal parasites *Ceratothoa gaudichaudii* and *Ceratothoa* cf. *imbricata* (Schafer *et al.*, 1989; Bruce, personal communication).

Host range

Nerocila orbignyi occurs on at least ten families of fish including Mugilidae, Sparidae, Carangidae, Molidae and even Holocephalidae. It is found in the Mediterranean, the tropical and southern Atlantic and in Australasia (Trilles, 1975; Bruce, 1987). *Mothocya parvostis* is currently only known from Japan (Bruce, 1986a). *Ceratothoa gaudichaudii* occurs along the South Atlantic and Pacific coasts of the Americas. Of other common species, *Olencira praegustator* parasitizes menhaden along the east coast of the USA from New Jersey to Florida (Kroger and Guthrie, 1972b), and *N. acuminata* is found in up to 40 species of fish along the west coast of the USA (Brusca, 1981).

Host specificity tends to be low in species of *Nerocila* and high in *Mothocya* and *Renocila* (Bruce, 1990). In temperate waters *Anilocra* spp. appear to have low specificity; *A. physodes*, from the north-east Atlantic and Mediterranean, is reported from 25 genera of fish in 13 families (Trilles, 1975). In the tropics, not only do *Anilocra* spp. show high host specificity (Williams and Williams, 1981) but the species of host preferred may vary with locality (Williams *et al.*, 1982). Adult *A. pomacentri* on the Great Barrier Reef parasitize the pomacentrid *Chromis nitida* at Heron Island (Adlard, 1990), two *Pomacentrus* spp. at Lizard Island, and midway between these two sites at Palm Island they are found only on three other pomacentrid species, though the first three pomacentrids occur there (Bruce, 1986b).

Cymothoids are chiefly parasites of teleosts. Many of the records from elasmobranchs probably represent trawl transfers (Brusca, 1981). However, the large isopod *Elthusa splendida* from the coast of Brazil (Sadowsky and Moreira, 1981) has been found only in the mouth of the dogfish, *Squalus cubensis*.

Systematics

Identification of cymothoids can be difficult because of unclear early literature and the inherent variability in the animals. Bruce (1986a), for example, lists 22 published records of '*Irona melanosticta*' all of which were incorrect. Keys to selected genera and species are given by: Schultz (1969), Kussakin (1979), Brusca (1981), Brusca and Iverson (1985), Castro and Silva (1985), Bruce (1986a, 1987), Bruce and Bowman (1989), Kensley and Schotte (1989), and Williams and Williams (1992). Bruce (1990) gives a review of marine species that have been assigned to the genus *Lironeca*.

Parasite morphology and life cycle

The chief morphological features used in identification are illustrated in Fig. 13.37.

Females are normally broader than males. The width/length ratio × 100 (the Montalenti index; Montalenti, 1948) has been used to sex cymothoids

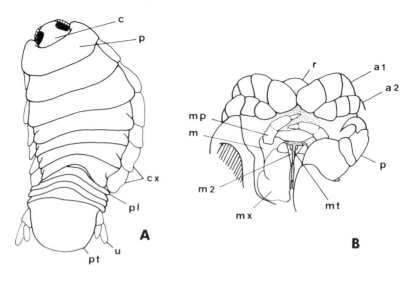

Fig. 13.37. Cymothoid morphology. A, *Mothocya rosea* dorsal view. c = cephalon, cx = coxae, p = first pereonite, p1 = second pleonite, pt = pleotelson, u = uropod rami (after Bruce, 1986a). B, *Ceratothoa guttata* mouthparts. a1 a2 = first and second antennae, m = mandible, m2 = second maxilla, mp = mandibular palp, mt = mouth, mx = maxilliped, p = first periopod (leg), r = rostrum (after Bruce and Bowman, 1989.)

though some species such as *Ceratothoa gaudichaudii* do not show such sexual dimorphism (Szidat, 1966). Females have two gonopores on the ventral surface near the bases of the sixth pair of legs and develop plate-like oostegites that grow from the bases of the legs and which interleave to cover the marsupium. Males have two penes near the midline between the seventh (last) pair of legs. Usually the endopods of their second pleopods are drawn out on the inner side to form an 'appendix masculina'.

To copulate, the male of *Anilocra physodes* moves beneath the female and turns over (Legrand, 1952). The male of *Elthusa vulgaris* moves from the gills to the buccal cavity to reach the female (Brusca, 1978b). Copulation occurs before the oostegites cover the gonopores (Szidat, 1966; Brusca, 1978a).

All cymothoids closely examined, which include members of the genera *Cymothoa*, *Nerocila*, *Anilocra* and *Lironeca*, have been shown to be protandrous hermaphrodites (Bullar, 1876; Mayer, 1879; Legrand, 1951, 1952; Fryer, 1968b). Sex reversal occurs in a single moult in *N. acuminata* at one of several moults (Brusca, 1978a).

Mothocya bohlkeorum and *A. prionuri* probably have multiple broods as large mature females are found with and without oostegites (Williams and Williams, 1982, 1986). During its life span of less than a year, *A. pomacentri* has three broods (Adlard, 1990).

Fertilized ova pass into the marsupium where they embryonate, hatch and moult (Hatai and Yasumoto, 1980). The resultant 'manca' larva has six rather than seven pairs of legs, two large compound eyes, and setose pleopods, uropods and telson. They leave the marsupium in groups and swim rapidly, frequently against the current and towards the light at least for the first day (Sandifer and Kerby, 1983; Williams and Williams, 1985b; Segal, 1987; Adlard, 1990). They attach to a variety of fish (Lindsay and Moran, 1976; Waugh *et al.*, 1989) and attempt to feed over the next 2 to 3 days. Mancae of *Elthusa vulgaris* appear to be attracted by fish mucus (Moser and Sakanari, 1985). Mancae of *A. pomacentri* would only feed on *Chromis nitida* less than 30 mm long as the manca was unable to pull out the scales of larger fish to reach the muscle beneath. In those isopods which fed, muscle fibres were visible in the gut within 2 days and blood cells within 3 days (Adlard, 1990). Mancae of *N. acuminata* fed on fish then dropped off and moulted on the bottom of the tank (Segal, 1987). Like other isopods they moult in two halves, the posterior half first.

The stages following mancae have not been confirmed experimentally. However, in many species, such as many *Anilocra* spp., the males retain their swimming capability and micropredatory habit, and are rarely found on host fish. Males of *O. praegustator* and *L. redmanii* (as *L. ovalis*) were rare on fish in the Delaware estuary, though they were common in plankton tows, 47 male *O. praegustator* being taken in one 5 minute tow (Kroger and Guthrie, 1972b; Lindsay and Moran, 1976). These males only lose their ability to swim when they start to change into females. In many other species mature males are found permanently attached to fish alongside females. Adult females are permanently attached non-swimmers.

The length of females is usually strongly correlated with host length (Montalenti, 1948; Menzies *et al.*, 1955; Weinstein and Heck, 1977; Maxwell, 1982) suggesting early infection and long life. Some males show a similar correlation. Trilles (1964a) used the strength of the male correlation in six species as an indication of whether males moved between hosts and hence of their swimming ability.

Estimates of the life span of adult females vary from up to the life of the fish, e.g. 9 years for *C. imbricatus* (see Maxwell, 1982), to 1 year or less for *L. redmanii*, *N. orbignyi* and *A. pomacentri* (Sadzikowski and Wallace, 1974; Bragoni *et al.*, 1984, Adlard, 1990).

Host–parasite relationships

Site and host selection

Adult females are either attached to the skin, gills and buccal cavity or burrow into the fish and develop in a pouch. In all cases, they almost invariably show strong site specifity. Many species on the skin are above the eye, below the eye, or on the caudal peduncle. Those in the gills are frequently asymmetrically shaped to fit the curve of the gill chamber.

Adult females of buccal species are usually attached to the floor of the mouth though some, such as the giant isopods *Elthusa splendida* and *E. neocyttus*, are attached to the roof (Sadowsky and Moreira, 1981; Stephenson, 1987). *Olencira praegustator* is twisted, the anterior part attaching to the roof and the posterior to the side (Turner and Roe, 1967). *Ceratothoa oestroides* sometimes attaches to the tongue and but more often to the roof (Vu-Tan-Tue, 1963; Trilles, 1964b).

Most buccal isopods face towards the mouth opening. A few face posteriorly, such as *Enispa convexa* and *Asotana magnifica* (Menzies *et al.*, 1955; Thatcher, 1988). *Elthusa splendida* faces sometimes anteriorly and sometimes posteriorly (Sadowsky and Moreira, 1981), and in *O. praegustator* immature females face posteriorly while adults face anteriorly (Kroger and Guthrie, 1972b).

Pouch dwelling forms normally penetrate somewhere on the ventral surface of the fish and project into the body cavity leaving their pleopods in contact with the outside.

Juvenile stages frequently show little host specificity and will feed on a range of fishes (Lindsay and Moran, 1976; Segal, 1987; Waugh *et al.*, 1989; Adlard and Lester, 1994). Juveniles of *E. vulgaris* on their final host attached first to the body then over the next few hours to days crawled into the gill cavity (Brusca, 1978b).

The prevalence of the parasites may be underestimated by trawl samples as Robinson (1981) showed that small males of *E. vulgaris* rapidly abandoned fish in trawls.

Clinical signs and pathology

BEHAVIOUR

Fish frequently react violently to mancae and juveniles that are attempting to attach. Sandifer and Kirby (1983) found that mancae from *Lironeca redmanii* caused striped bass *Morone saxatilis* 60–70 mm long to leap and thrash about the tank. Segal (1987) described how seven species of fish when attacked broke surface, swam rapidly, thrashed body, rubbed against objects and dived into the sand to dislodge the parasites. He also observed that the reactions were more violent if the fish had already experienced a manca attack. Adlard and Lester (1994) reported that *Chromis nitida* attacked by mancae of *Anilocra pomacentri* rubbed so violently against objects that skin haemorrhages appeared. Longer term behavioural changes have been reported by Guthrie and Kroger (1974) who found that menhaden parasitized by *Olencira praegustator* tended to stay in nursery areas and had difficulty avoiding surface trawls. Adult isopods generally cause little irritation except when they move, as for copulation (Legrand, 1952).

GROSS PATHOLOGY

There is rarely an open wound associated with permanently attached parasites, apparently because most parasites feed intermittently and the feeding lesions heal over between meals.

Healing lesions have been reported from under the mouthparts of several skin-dwelling isopods. Morton (1974) found that *Nerocila phaeopleura* on the caudal peduncle of *Sardinella* was associated with a shield-shaped wound. Williams and Williams (1981) observed that *A. myripristis*, attached near the eye of a soldier fish *Myripristis jacobus*, caused erosion of scales and loss of pigment, presumably a result of dermis removal and repair. Beneath *A. haemuli* the skin of *Haemulon* spp. had turned from brown to orange and some bone deformation was observed. *Nerocila acuminata* (as *N. californica*) on anchovy caused tissue damage several square centimetres across exposing muscle and occasionally an entire fin was eroded away (Brusca, 1978b). The same parasite on the caudal peduncle of a serranid was associated with skin lesions up to 0.5 mm across which Rand (1986) concluded were probably old feeding wounds. He also observed loss of pigment in skin around the pereopods.

Isopods in the gill cavity are frequently associated with loss of gill filaments presumably as a result of the feeding actions of the parasites (Menzies *et al.*, 1955; Turner and Roe, 1967; Kroger and Guthrie, 1972b; Guthrie and Kroger, 1974; Sadzikowski and Wallace, 1974; Williams and Williams, 1982, 1985a). Filaments take longer to regenerate than dermis and damage to filaments has been used to indicate former infestation (Stephenson, 1976). In *Gymnothorax eurostus* almost all the gill filaments were missing from gill chambers occupied by *Ichthyoxenus puhi*; this extreme damage was presumably why only one gill cavity per fish was found parasitized (Bowman, 1960).

Ceratothoa imbricatus caused a callus-like thickening (dystrophic calcification) over the inner edges of the gill rakers and branchial arch in *Trachurus declivis* (Stephenson, 1976). *Cuna insularis* in the gill chamber of *Abudefduf* in the West Indies eroded the wall to such an extent it exposed the heart (Williams and Williams, 1985a).

The damage done by isopods in the buccal cavity is usually less noticable but more chronic. Weinstein and Heck (1977) reported a little scar tissue on the tongue under *Cymothoa excisa*. Sadowsky and Moreira (1981) observed a darkened area beneath the mouth of *Elthusa splendida* but no break in the skin. Stephenson (1987) noticed slight skin abrasion together with a patch of clotted blood under the mouthparts of *E. neocyttus*. Thatcher (1988) noted only a small indentation in the floor of the mouths of piranhas with *Asotana magnifica*.

However, some species cause the tongue to degenerate. *Cymothoa excisa* and *C. exigua* caused partial and total degeneration respectively in lutjanids (Weinstein and Heck, 1977; Brusca and Gilligan, 1983). Romestand and Trilles (1977a) described the regression of the tongue of the sparid *Boops boops* infected with *Ceratothoa oestroides*. The tongue first became narrow and then shorter until 50% of the tongue was lost.

Boops boops with either *C. oestroides* or *C. parallela* attached to the roof of the mouth developed a ventrally curved rather than flat vomer bone and a large patch of vomerine teeth which are not found in unparasitized fish (Vu-Tan-Tue, 1963; Romestand and Trilles 1977a). Similar changes occurred in *Spicara* spp. with the same parasites (Trilles, 1964b), and in *Maena smaris* with *Emetha audouini* (Romestand, 1979).

The pouch of pouch-dwelling genera is formed from an invagination of the body wall. It develops over several months as the growing isopod pushes its way into the skin. The pouch is kept open to the outside by the hind-most legs and the continual movement of pleopods (Huizinga, 1972). When *Artystone minima* penetrates near a pectoral fin, the fin is eventually lost (Thatcher and Carvalho, 1988).

Manca larvae and juveniles are aggressive feeders. Within 48 hours of attachment, manca larvae of *Anilocra pomacentri* produced an ulcer up to 2.5 mm across which penetrated through the epidermis, dermis and into the muscle (Adlard, 1990). A larva of *Artystone trysibia* entered the orbital cavity of an experimentally exposed catfish causing severe haemorrhage which destroyed the eye (Huizinga, 1972).

HISTOPATHOLOGY

Inflammation is typically localized and adjacent to the mouthparts. Adult female *Renocila heterozota* on anemone fish caused localized damage to scales, epidermis, dermis and muscle (Bowman and Marischal, 1968). Adult females of *Anilocra physodes* on *M. maena* were associated with scale absence, loss of dermal layers, and influx of erythrocytes, eosinophils and lymphocytes into the epidermis, dermis and superficial layers of muscle (Romestand *et al.*, 1977).

Romestand and Trilles (1977a) describe the histological changes associated with the regression of the tongue of *Boops boops* parasitized by *Ceratothoa oestroides*, including osteoclasts and granuloma formation.

In those species that form pouches within the fish such as *Ourozeuktes, Ichthyoxenus* and *Artystone*, the wall of the pouch is thin and membranous, apart from near the mouthparts where it is roughened and thickened. It is formed from skin components and contains epidermis, fibrous connective tissue, muscle and in some cases pigment cells and scale primordia (Hale, 1929; Akhmerov, 1939; Huizinga, 1972; Avdeev, 1983).

Host immune response

Acquired precipitating antibodies to the haemolymph of *Anilocra, Nerocila, Mothocya*, and especially the buccal isopods *Emetha* and *Ceratothoa* spp. were demonstrated in the blood of their hosts by Romestand and Trilles (1975). These had low specificity as the serum of some infected hosts reacted to the haemolymph of cymothoids of other genera. The authors suggest the antibodies were partly protective as fishes that produced them usually carried only one or two isopods whereas *Mugil cephalus*, which produced no demonstrable antibody, carried up to 12 *N. orbignyi* per fish.

Effect on host

ANAEMIA

Romestand and Trilles (1977b) and Romestand (1979) found that *Anilocra physodes* and *Ceratothoa* spp. caused haematocrit and erythrocyte counts to decrease, and erythroblast count and spleen size to increase in three species of fish. Infected fish also had reduced lipid in the serum and liver (Romestand, 1979). *Nerocila orbignyi* on caged sea bass caused hypochromic macrocytic anaemia, a decrease in blood protein, blood lipids and triglycerides, and an increase in urea. The conditions reversed when the parasites were lost (Bragoni *et al.*, 1983).

CONDITION

Though some cymothoids apparently have little effect on host weight, such as *Cymothoa excisa* in the buccal cavity of lutjanids (Weinstein and Heck, 1977) and *Ceratothoa imbricatus* in jack mackerel (Maxwell, 1982), many species occur on hosts that weigh less than uninfested fish of the same length. Examples are: *Leuciscus waleckii* with *Ichthyoxenus amurensis* (see Petrushevsky and Shulman, 1961), *Hemirhamphus sajori* with *Irona melanosticta* (probably *Mothocya sajori*, see Bruce, 1986a; Hattori and Seki, 1956), *Morone americana* with *Lironeca redmanii* (see Sadzikowski and Wallace, 1974), *Girella tricuspidata* with more than two *C. imbricatus* (see Lanzing and O'Connor, 1975), *Pagellus erythrinus* with *Anilocra physodes* (see Romestand and Trilles, 1979), *Boops boops* with *C. oestroides* or *C. parallela* (see Romestand and Trilles, 1979), and *Dicentrarchus labrax* with *Nerocila orbignyi* (see Bragoni *et al.*, 1983).

Parasitized fish grow more slowly than those uninfected (Sadzikowsky and Wallace, 1974; Romestand, 1979; Romestand and Trilles, 1979, Hatai and Yasumoto 1981, 1982a; Adlard, 1990) (Fig. 13.38).

MORTALITY

Krykhtin estimated than 13% of the coregonids in the Amur River died before reaching marketable size as a result of *Ichthyoxenos amurensis* (see Petrushevsky and Shulman, 1961). Bragoni *et al.* (1984) found that 7 and 18% of mortalities in two cages of sea bass were attributable to *Nerocila orbignyi* that had apparently been carried into the vicinity by migrating mullet. Other adult isopods seem to have little direct effect on mortality, but the larvae of many species are highly pathogenic to small fish (Lindsay and Moran, 1976; Sandifer and Kirby, 1983; Segal, 1987; Adlard, 1990).

STERILIZATION

Hattori and Seki (1956) found that the drop in condition factor of adult female *H. sajori* parasitized by ?*Mothocya sajori* was caused by a decrease in the size of the gonad. Adlard and Lester (1994) observed that few oocytes were produced by *Chromis nitida* carrying *Anilocra pomacentri*. Their computer simulation of the parasitized population showed that the reduction in fecundity had a greater impact on host numbers than the increased mortality caused by feeding manca stages. Presumably the relatively large amounts of blood taken by the parasites produce a nutritional drain on the host and this inhibits full development of the ovaries.

Concurrent infection

The rapid healing of skin feeding sites is probably the reason few secondary infections occur, though Rand (1986) noticed a mixed population of gram negative bacteria near the edge of one lesion. Lawler *et al.* (1974) in Mississippi found lymphocystis in the gills of 21 perch, 20 of which had *Lironeca*

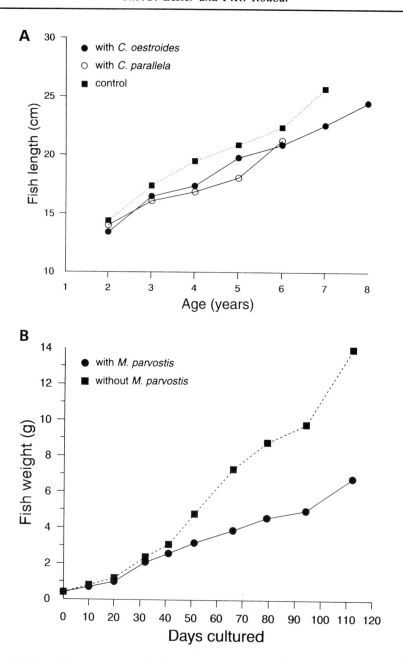

Fig. 13.38. Effect of cymothoids on fish growth. A, *Ceratothoa oestroides* and *C. parallela* on *Boops boops* (after Romestand and Trilles, 1979). B, *Mothocya parvostis* on *Girella punctata* (after Hatai and Yasumoto, 1982a). C, *Anilocra pomacentri* on *Chromis nitida* (from Adlard and Lester, 1994). D, *Anilocra* sp. on *Apogon multilinearis* (from Warburton, unpublished).

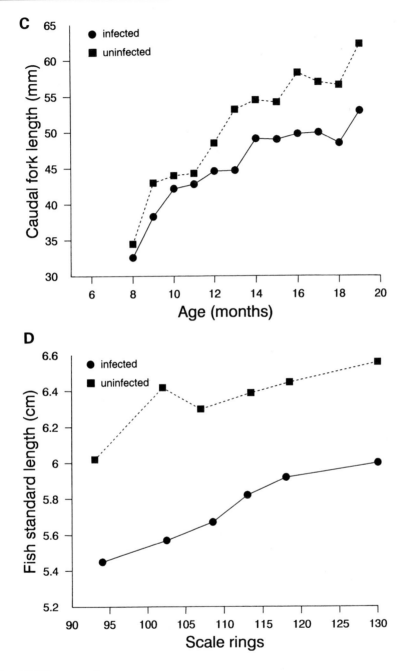

Fig. 13.38. *contd.*

redmanii. They concluded that the isopod was probably not a vector but that it irritated the skin and thus allowed the virus to enter. Rodgers and Burke (1988) suggested that isopods might provide the initial lesion for 'red-sore' to develop.

Fish with isopods are less able to withstand other forms of stress. Keys (1928) observed that killifishes, *Fundulus parvipinnis*, with *Elthusa californica* died more quickly under low oxygen conditions or in low salinity than uninfested fish. Lewis and Hettler (1968) found that menhaden with *Olencira praegustator* and related gill damage did not survive at 34°C as long as non-parasitized fish.

In vitro culture

Eggs removed from brood pouches of *Mothocya parvostis* were raised in seawater and hatched to produce larvae infective to fish (Hatai and Yasumoto, 1980).

Parasite nutrition and physiology

Cymothoids feed on blood and macerated tissue. They use their sturdy maxillipeds to tear into the host to release blood which they suck up with their muscular oesophagus. An anticoagulant is produced by the lateral oesophageal glands (Romestand and Trilles, 1976; Romestand, 1979). In *Anilocra physodes* haematin, a by-product of blood digestion, was found in the intestinal caeca (hepatopancreas) though not in the intestine or posterior intestinal diverticula (Romestand, 1979).

Cymothoids feed intermittently (Romestand and Trilles, 1977b, 1979), sometimes only after a long interval in adult females. Female *Ceratothoa oestroides* fed after the larvae had left the brood pouch and before its ovary enlarged for the next brood. No significant feeding was thought to occur at other times. The intestinal caeca, swollen and dark brown at the start of the reproductive cycle, gradually became clearer and smaller as the eggs developed (Romestand *et al.*, 1982). In *A. pomacentri*, 3 days after larvae were released the parent moulted and shortly thereafter host blood was observed within the three pairs of intestinal caeca. A new brood of eggs appeared in the brood pouch 18 days later without further ingestion. The entire reproductive cycle took 2 months (Adlard, 1990).

Prevention and control

Formalin has been used to treat cultured pompano infested with cymothoids (Brusca, 1981). Hatai and Yasumoto (1980) used trichlorphon at 100 ppm for 5 minutes to remove *Mothocya parvostis* from *Seriola quinqueradiata*. Another pesticide, methyl isoxathion, was used by Hatai and Yasumoto

(1982b) at a dose of 1 ppm for 2 to 5 minutes to remove the same parasite from *Girella punctata*.

Bragoni *et al.* (1984) recommended using fine mesh nets in the vicinity of aquaculture cages to keep out mullet believed to be the source of *Nerocila orbignyi* infection in cultured bass *Dicentrarchus labrax*.

Salmon cultured in sea cages in Tasmania are given freshwater baths in the summer to control amoebic gill disease. This also controls the isopods (*Ceratothoa* cf. *imbricata*) which tend to attach in late summer but leave the fish in freshwater baths. Alternatively one treatment with Dichlorvos in late summer (1.5 ppm at 15.5°C or 2.0 ppm at 11°C for 1 hour) has proved effective (C. Foster, personal communication).

Williams and Williams (1979) observed that the cleaner shrimp *Periclimenes pedersoni* removed juvenile *Anilocra* sp. from French grunt *Haemulon flavolineatum* in the laboratory.

OTHER FLABELLIFERA

Within the isopod suborder Flabellifera, four families other than the Cymothoidae are sometimes found on or in fish: the Aegidae, Corallanidae, Tridentellidae and Cirolanidae. They differ from the cymothoids in that the last four pairs of legs are not prehensile. In the Aegidae more so than the others, the first three pairs are strongly prehensile. Aegids do not appear to be protandrous hermaphrodites (Wagele, 1990). For keys to families and genera see Kensley and Schotte (1989), Delaney (1989) and Delaney and Brusca (1985).

The non-cymothoid Flabellifera are mostly free-living. A few, particularly in the Aegidae, appear to be true parasites. On the Great Barrier Reef two species, an aegid *Aega lethrina* and a corallanid *Argathona macronema*, are common in the nasal passages of serranids and lutjanids (Bruce, 1983). Some Flabellifera are micropredators that temporarily attach to fish to feed. The aegid, *Rocinela belliceps*, attacks, takes blood, and kills young pink salmon, *Oncorhynchus gorbuscha*, on the Pacific coast of the USA (Novotny and Mahnken, 1971). Another aegid, *Alitropus typus*, is widespread in the tropical IndoPacific in brackish water and attacks the gills, fins and body surface of many fishes including *Chanos chanos*, *Channa striatus*, and species of *Tilapia, Mugil, Gerres* and *Lebistes* (Nair and Nair, 1983; Kabata, 1985; Ho and Tonguthai, 1992). In the laboratory, young isopods attached to *C. striatus* for 1–2 days during which they fed. They then left the fish for 1–2 days, reattached, fed again and left the fish a second time. The fish developed macrocytic hypochromic anaemia during the feeding periods and began to recover afterwards. Small fish in the field die from the attacks (Nair and Nair, 1983). Fenitrothion, Dichlorvos, Malathion, Ediphenphos and Phosphomidon killed 50% of the parasites in 2 days at dosages of 0.001, 0.009, 0.05, 0.28 and 5.19 ppm respectively (Nair and Nair, 1982).

No cirolanids are true parasites, though *Cirolana* spp. are widespread scavengers and are sometimes encountered deep in the tissues of dead and dying teleosts and elasmobranchs (Bird, 1981; Berland, 1983).

ISOPODA: GNATHIIDAE

Introduction

Gnathids are blood-feeding marine isopods that have been reported to cause mortality in captured eels *Anguilla anguilla* and sea-caged mullet *Crenimugil crenilabris* (Mugridge and Stallybrass, 1983; Paperna and Overstreet, 1981). Damage to fish skin reduces the marketable value of infested fish (Paperna and Por, 1977).

Gnathids infest many species of marine and estuarine teleosts and elasmobranchs. In the laboratory, larvae of *Paragnathia formica* infected fish of the families: Ammodytidae, Anguillidae, Gasterosteidae, Labridae, Cyprinidae, Triglidae, Cottidae, Gobiidae, Pleuronectidae, Trachinidae and Callionymidae, and even a freshwater amphibian (*Rana temporaria*). It did not attack several invertebrates or man (Monod, 1926). Some species of *Gnathia* infest only elasmobranchs (Paperna and Por, 1977; Honma *et al.*, 1991).

Species are identified using characters of the free-living adult males. The parasitic larval or 'praniza' stages are difficult to identify except that praniza of *Paragnathia* spp. have a rounded telson with a smooth margin, and those of *Gnathia* spp. have a triangular telson with a toothed margin (Mugridge and Stallybrass, 1983). It is sometimes possible to keep praniza *in vitro* in seawater until they moult into adult males which can then be identified (Hesse, 1855; Stoll, 1962; Paperna and Por, 1977; Wagele, 1988, 1990).

Morphology and life cycle

Eggs of *Paragnathia formica* and *Gnathia calva* hatch into the first of three praniza stages, all of which are parasitic (Stoll, 1962; Wagele, 1988; Fig. 13.39). The eggs of *G. piscivora* hatch into a stage that does not feed but moults after 7 days into the first praniza larva which is parasitic (Paperna and Por, 1977). Praniza swim very rapidly using pleopods. The first praniza of *P. formica* are negatively phototropic and survive for at least 3 weeks without a host (Monod, 1926). The last praniza stage leaves the fish and moults into an adult female or male (premale then male in *G. calva*; Wagele, 1988). Adults are free-living on the sea bed in small colonies of several females and one male. The duration of the life cycle varies widely between species. It takes about a year in *P. formica*; the first two praniza last 6 to 7 weeks each, the third up to 4 months and adults live 12 to 20 months (Stoll, 1962; Amanieu, 1963). In *G. calva*, an Antarctic species, the life cycle takes several years. The eggs are incubated for about a year and the larvae take 4 to 5 year to reach maturity. Males may survive a further 2 years as the guardian of a harem (Wagele, 1988). The duration of the third praniza stage of *Gnathia piscivora* at 24°C is 7–10 days (Paperna and Por, 1977).

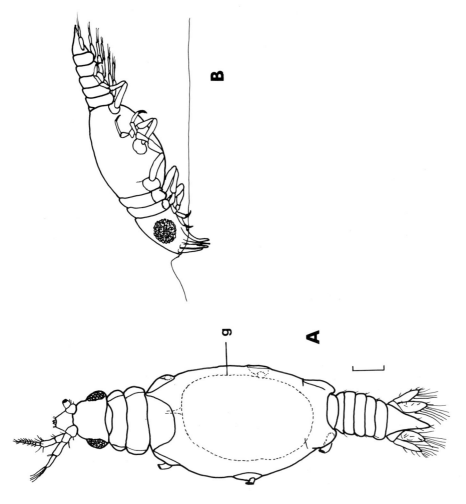

Fig. 13.39. Praniza of gnathids. A, *Gnathia calva*, g = gut distended with blood, scale bar 250 μm (after Wagele, 1987). B, *Paragnathia formica* attached to host (after Monod, 1926).

Fig. 13.40. *Gnathia* sp., praniza larvae on *Mugil subviridis*. (Photograph by courtesy of I. Paperna)

Host–parasite relationships

Site and host selection, course of infection

Praniza larvae that attack elasmobranchs attach to gills and gill septa (Paperna and Por, 1977; Lester and Sewell, 1989; Honma and Chiba, 1991). Those that attack teleosts will attach almost anywhere if scales are absent or are poorly developed, otherwise they attach to the fins, and buccal or gill cavities (Monod, 1926; Mugridge and Stallybrass, 1983).

Gnathia piscivora only attack fish at night and usually in water less than 2 m deep. Infective larvae were recovered from inshore plankton samples. Those attached to the skin left the host within 2–4 hours, those attached to gills and the walls of the gill cavity remained for 1 or more days. Fish unaccustomed to their cages were more likely to be attacked than less stressed fish (Paperna and Por, 1977; Paperna and Overstreet, 1981; Fig. 13.40).

Adult gnathids have generally been found in burrows, crevices or dead barnacles. Adults of *G. calva* occur in sponges (Wagele, 1988).

Clinical signs and histopathology

Fish behave erratically to dislodge newly attached parasites though this reaction only lasts a few minutes. The mouthparts of the parasites penetrate the epidermis and cause haemorrhage from vessels in the dermis. Some species feed for only a few hours. In Israel, fish in cages or entangled in gill nets developed dermal haemorrhages after attack by praniza. They were stressed and became anaemic (Paperna and Por, 1977). The skin of *Anguilla anguilla* with 100 *Paragnathia* praniza developed loose folds and the eel looked dehydrated (Mugridge and Stallybrass, 1983). The penetration site on a young eel parasitized 4 days previously was raised 0.5 mm diameter as a result of epithelial hyperplasia and an accumulation of lymphocytes and eosinophils (Monod, 1926).

Gnathids on rays remain in position for a longer time and inflammation develops around their feeding site. Stingrays *Dasyatis akajei* and *D. matsubarai* with praniza in Japan showed reddish marks at the site. The adjacent dermis was vacuolated and filled with masses of intact and lysing erythrocytes; macrophages, neutrophils, lymphocytes and thrombocytes were abundant (Honma *et al.*, 1991; Honma and Chiba, 1991).

Parasite nutrition

The praniza is attached by its hooked rostrum, by the hooked gnathopods (= pylopod of Wagele) and by the hooked and serrated but immobile mandibles. Skin penetration is performed by the maxillules which make bursts of rapid tearing movements. A strong muscular oesophagus draws blood into the proventriculus and anterior hind gut which rapidly inflates. *Gnathia maxillaris* on blennies took 2–24 hours to feed (Monod, 1926; Davies, 1981). *Gnathia calva* has salivary glands which produce anticoagulants, two sac-like digestive glands, and symbiotic bacteria up to 10 μm long in the rectal vesicle (Juilfs and Wagele, 1987).

Prevention and control

Paragnathia sp. on *Anguilla anguilla* was controlled by changing the incoming water from sea to fresh (Mugridge and Stallybrass, 1983).

CIRRIPEDIA: ANELASMA

Introduction

Anelasma squalicola is a lepadid barnacle highly modified for parasitism. It occurs on the small shark *Etmopterus spinax*. Five per cent of *E. spinax* to the west of Ireland were found by Hickling (1963) to be infected. It was generally on small individuals (20–40 cm) in waters over 150 fathoms.

Parasite morphology and life cycle

The body of the parasite, up to 35 mm long, is in two parts. The 'capitulum' contains a prosciform mouth, six pairs of biramous appendages, penis and egg sacs. These are all partly covered by a tough leathery mantle which is devoid of plates. The 'peduncle' is soft and delicate, has scattered 'roots' over

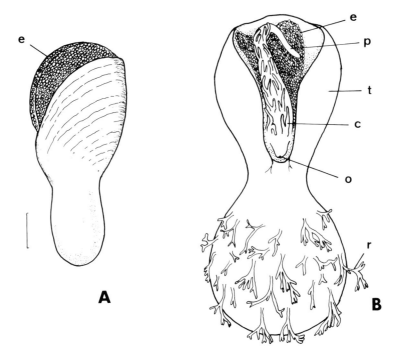

Fig. 13.41. *Anelasma squalicola.* A, Lateral view. e = egg sac (after Darwin, 1851). B, Anterior view. c = biramous appendage (cirrus), o = mouth, p = penis, r = roots, t = mantle. Scale bar = 5 mm. (After Johnstone and Frost, 1927.)

its surface and contains the ovaries (Fig. 13.41). Egg sacs are white; the rest of the capitulum is purplish brown (Johnstone and Frost, 1927).

Eggs are found throughout the year. They hatch to release naupliar larvae which contain yolk, no eyespot, no functional mouth and are poor swimmers (Frost, 1928; Hickling, 1963). The remainder of the life cycle is unknown.

Host–parasite relationships

Most parasites are attached to the ring of tissue around the base of the first dorsal spine. Parasites usually occur in pairs (of 86 infected fish 70 had two parasites), are usually of similar size, and their size is positively related to the size of the fish (Hickling, 1963).

The parasite lies at an angle in the skin. The granuloma that develops around the peduncle causes an obvious swelling. Adjacent muscle is replaced by fibrous connective tissue. The epithelium around the parasite is intact and epithelial tongues grow down into the cavity. The parasite's 'roots' may extend 10 mm into the muscle; adjacent to them are degenerating muscle fibres, an increased number of blood vessels and fibrous tissue typical of chronic inflammation (Johnstone and Frost, 1927).

Livers of infested fish 41–44 cm long were 10% lighter than those of uninfested fish of the same length. In mature female fish, 98% of those parasitized had inactive ovaries compared to 42% in non-parasitized fish. In mature males, 86% had 'reduced' testes compared to 0.05% in unparasitized fish. The low prevalence in large fish plus the absence of evidence of parasite loss suggests that parasitized fish may have a higher mortality rate than those not parasitized (Hickling, 1963).

Parasite nutrition and physiology

The thin cuticle over the peduncle, the branching root system, and the changes in infected hosts, strongly suggest that the parasite absorbs nutrient from the host. Furthermore, the lack of spines or setae on the legs suggest these do not carry out the normal cirripedian function of filtering water for food particles. Nevertheless the parasite has a prominent mouth and a functional gut leading Darwin (1851) to conclude that it fed on organisms crawling by on the surface of the shark.

SUMMARY, CONCLUSIONS AND RECOMMENDATIONS

Details of the biology of many economically important arthropods parasitic on fish remain unknown. Methods for their control are frequently ineffective.

Biology

Lernaea cyprinacea has been a problem in aquaculture for centuries and is likely to be around for many more years. Details of its biology that remain clouded include the host specificity of the copepodid stage. Is it weak, as suggested by Grabda (1963) or strong as suggested by the data of Haley and Winn (1959)? Wilson (1917) stated that females were fertilized at the last copepodid stage but Grabda (1963) found no copepodids with spermatophores, only premetamorphosed females. Fryer (1966) suggests that in *L. bagri* they are fertilized not in the gills but on the final host and he recovered lernaeid males from a plankton sample. It is not known whether males feed.

The complex head structures in many pennellids evidently serve for much more than attachment. A good laboratory model is needed to explore their function. Similarly, the function of the abdominal processes of *Pennella* and *Sphyrion* species are unknown. In *Sphyrion* do contractions of the prominent dorsoventral muscles cause haemolymph to flow into the branched appendages?

Some copepods cause general changes in the host, the cause of which is unknown. *Lepeophtheirus* infection results in morbidity and both lernaeids and pennellids decrease the growth rate. The mechanism is particularly interesting because new infections of *L. branchialis* are associated with an increased growth rate.

Sphyrion lumpi has a low host specificity and thus may lend itself to ready establishment of a laboratory population. This will quickly reveal its life cycle, the biology of the female, particularly how it penetrates and whether it only feeds in the initial phases of infection. A short laboratory study plus analysis of field data will reveal how long females survive and how long the scars persist. Analysis of growth rings on scales or otoliths of infected and uninfected fish may indicate whether the parasite has any effect on the growth rate of the fish.

Sarcotaces verrucosus is another copepod with low host specificity. Laboratory work would establish its life cycle and possibly mode of nutrition and effect on the fish.

The reason for the strong host specificity in *Lernaeocera* species is not clear, particularly in view of Slinn's (1970) observation that *L. lusci* matured on its first host under experimental conditions.

Though the parasitic barnacle *Anelasma squalicola* has been difficult to obtain, the next observer may be able to shed light on the function of its root system, particularly if its early growth stages are found to be completely enclosed within the host, like the juveniles of some whale barnacles (Samaras and Durham, 1985).

Small numbers of argulids cause intense irritation to certain fish, and this can lead to their death. Other fish die only with very heavy infestations. We need to elucidate the physiological changes in infested fish, especially those that are particularly sensitive to small numbers, and those that die with very heavy burdens. It would also be interesting to determine the function of the argulid stylet. *Chonopeltis inermis* lacks a stylet and apparently feeds primarily on mucus (Fryer, 1956).

The taxonomy of parasitic isopods, primarily a problem in aquaculture in the tropics and subtropics, needs revision, particularly the genera *Ceratothoa*, *Cymothoa* and related forms. No cymothoid life cycle has been completed in the laboratory. The biology of the juvenile and male stages of cymothoids, their mobility, feeding preferences, and when and where copulation occurs is unknown for almost all species.

All cymothoids appear to be protandrous hermaphrodites, full development of which is environmentally determined. This aspect has not yet been assessed in gnathid isopods and it is not clear whether maturation of males is inhibited in the presence of a mature male. Gnathid isopods are likely to become a problem in some areas with the expansion of marine cage culture.

Control

In wild fish stocks, the main tool we have for controlling infection is fishing pressure. An increase of this will decrease both the population density and the average age of the fish and this may be sufficient to eliminate the parasite (Forrester, 1956). In many cases increasing fishing pressure is not an option.

In other cases, it needs to be applied selectively, perhaps to a reservoir host or only to areas where the parasite is abundant.

In aquaculture, there are more opportunities for disease control. Husbandry methods, such as quarantine procedures and adjustment of water flow, are widely employed and could be extended with more information on the ecology of the parasites. For example, the factors that control the numbers of argulids in natural fish populations are unknown. Multispecies systems, such as wrasse in salmon farms to control sea lice, may have applicability in the control of parasites such as argulids and lernaeids.

Chemicals are widely used to control infections. The rapidity with which lernaeids, argulids, and to a less extent caligoids, develop resistance to insecticides suggests we need to look for alternate ways of controlling infections, especially as some naupliar stages seem to be totally resistant. For lernaeids the demonstration that fish can acquire immunity, though in what form is not yet known, opens up what may be the best long term solution: the enhancement of resistance in fish. This may be through controlled exposures to the parasite, through vaccination, or through genetic transformation as discussed below.

The copepodid stages of species of lernaeids and pennellids develop on gills. It is not known whether immunity to this stage develops in fish. If it can be stimulated, either naturally or through vaccination, it could help to control the parasite. The high rejection rate of metamorphosing females of *L. branchialis* in multiple infections suggests a hypersensitive reaction. If one is present, it may lead to ways to immunologically protect the fish.

In the long term, the best solution appears to be the introduction of engineered resistance in the fish. Manufactured vaccines based on antigens in the parasite not normally exposed to the fish, such as part of the gut, may cause the parasite to leave the fish after it has taken in fish tissue containing antibody. Ideally this would happen soon after the parasite first contacted the fish.

Fish antibodies themselves may not be very effective in removing the parasite, even if active against a hidden antigen, because the parasites have evolved ways to deal with teleost antibody. We could be more successful introducing a gene from a vertebrate that manufactures a different type of antibody. Genetic transformation opens up other possibilities. The epidermal cells, the first point of contact for a parasitic copepod, could be modified to produce a secretion distasteful to the copepod or to block receptors on the copepod used in host recognition. The transformation could be applied topically by dipping, perhaps using an epithelial virus as a vector, or by producing transgenic offspring in which all the epithelial cells would be modified from birth. The large number of eggs produced by many fish makes them an ideal candidate for electroporation and biolistic methods of introducing DNA. These methods have already been used to introduce other genes, such as the gene for growth hormone. In the next decade, these and similar techniques are likely to be applied to develop resistance in fish through genetic transformation and should lead to effective control of at least *Lepeophtheirus salmonis* in *Salmo salar*.

ACKNOWLEDGEMENTS

We thank P.T.K. Woo for editorial help, and N.L. Bruce and V.E. Thatcher for information on isopods. Papers were kindly translated by J. Tang (Japanese), E. Weinstein (German) and F. Wood (Russian). The work was supported in part by a fellowship to FRR from the Australian Research Council.

REFERENCES

Abdelhalim, A.I., Lewis, J.W. and Boxshall, G.A. (1991) The life-cycle of *Ergasilus sieboldi* Nordmann (Copepoda: Poecilostomatoida), parasitic on British freshwater fish. *Journal of Natural History* 25, 559–582.

Abele, L.G., Kim, W. and Felgenhauer, B.E. (1989) Molecular evidence for inclusion of the phylum Pentastomida in the Crustacea. *Molecular Biology and Evolution* 6, 685–692.

Abrosov, V.N. and Bauer, O.N. (1959) Ergasilosis of the 'peled' whitefish (*Coregonus peled*) in the Pskov region. *Bulletin of the State Scientific Research Institute of Lake and River Fisheries* 49, 222–226.

Adlard, R.D. (1990) The effects of the parasitic isopod *Anilocra pomacentri* Bruce (Cymothoidae) on the population dynamics of the reef fish *Chromis nitida* Whitley (Pomacentridae). PhD Thesis, University of Queensland, 118 pp.

Adlard, R.D. and Lester, R.J.G. (1994) Dynamics of the interaction between the parasitic isopod, *Anilocra pomacentri*, and the coral reef fish *Chromis nitida*. *Parasitology* 109, 311–324.

Ahne, W. (1985) *Argulus foliaceus* L. and *Piscicola geometra* L. as mechanical vectors of spring viraemia of carp virus (SVCV). *Journal of Fish Diseases* 8, 241–242.

Aitken, A. (1942) An undescribed stage of *Sarcotaces*. *Nature* 150(3797), 180–181.

Akhmerov, A.K. (1939) On the ecology of *Lironeca amurensis*. *Annals of Leningrad University* 43, 11.

Allison, L.N. and Latta, W.C. (1969) Effects of gill lice (*Salmincola edwardsii*) on brook trout (*Salvelinus fontinalis*) in lakes. *Research and Development Report, Michigan Department of Natural Resources* 189.

Allum, M.O. and Hugghins, E.J. (1959) Epizootics of fish lice, *Argulus biramosus*, in two lakes of eastern South Dakota. *Journal of Parasitology* 45(2), 33–34.

Amanieu, M. (1963) Evolution des populations de *Paragnathia formica* (Hesse) au cours d'un cycle annuel. *Bulletin of the Institute of Oceanography, Monaco* 60(261), 1–12.

Amin, O.M. (1981) On the crustacean ectoparasites of fishes from southeast Wisconsin. *Transactions of the American Microscopical Society* 100, 142–150.

Anon. (1980) ACT lake fish killed by parasite. *Australian Fisheries* 39(6), 13.

Anon. (1992) Lice chute from Innovac. *Fish Farming International* August, 38.

Anstensrud, M. (1989) Experimental studies of the reproductive behaviour of the parasitic copepod *Lernaeocera branchialis* (Pennellidae). *Journal of the Marine Biological Association, UK* 69, 465–476.

Anstensrud, M. (1990a) Effects of mating on female behaviour and allometric growth in the two parasitic copepods *Lernaeocera branchialis* (L., 1767) (Pennellidae) and *Lepeophtheirus pectoralis* (Muller, 1776) (Caligidae). *Crustaceana* 59, 245–258.

Anstensrud, M. (1990b) Male reproductive characteristics of two parasitic copepods,

Lernaeocera branchialis (L.) (Pennellidae) and *Lepeophtheirus pectoralis* (Mueller) (Caligidae). *Journal of Crustacean Biology* 10, 627–638.

Anstensrud, M. (1990c) Mating strategies of two parasitic copepods (*Lernaeocera branchialis* (L.) (Pennellidae) and *Lepeophtheirus pectoralis* (Mueller) (Caligidae)) on flounder: polygamy, sex-specific age at maturity and sex ratio. *Journal of Experimental Marine Biology and Ecology* 136, 141–158.

Anstensrud, M. (1990d) Moulting and mating in *Lepeophtheirus pectoralis* (Copepoda: Caligidae). *Journal of the Marine Biological Association, UK* 70, 269–281.

Anstensrud, M. and Schram, T.A. (1988) Host and site selection by larval stages and adults of the parasitic copepod *Lernaeenicus sprattae* (Sowerby) (Copepoda, Pennellidae) in the Oslofjord. *Hydrobiologia* 167, 587–594.

Avdeev, V.V. (1983) On the formation of zoocecidium in fishes under the influence of parasitic isopods of the family Cymothoidae. *Parazitologiya, Leningrad* 17, 420–422.

Avdeev, V.V. (1985) Specific features of the distribution of marine isopod crustaceans of the family Cymothoidae (Isopods, Flabellifera). In: Hargis, W.J. (ed.), *Parasitology and Pathology of Marine Organisms of the World Ocean*, NOAA Technical Report NMFS 25, pp. 89–92.

Avenant, A. and Van As, J.G. (1986) Observations on the seasonal occurrence of the fish ectoparasite *Dolops ranarum* (Stuhlmann, 1891) (Crustacea: Branchiura) in the Transvaal. *South African Journal of Wildlife Research* 16, 62–64.

Avenant-Oldewage, A. and Van As, J.G. (1990) The digestive system of the fish ectoparasite *Dolops ranarum* (Crustacea: Branchiura). *Journal of Morphology* 204, 103–112.

Bailey, R.E., Margolis, L. and Workman, G.D. (1989) Survival of certain naturally acquired freshwater parasites of juvenile sockeye salmon, *Oncorhynchus nerka* (Walbaum) in hosts held in fresh and sea water, and implications for their use as population tags. *Canadian Journal of Zoology* 67, 1757–1766.

Bakay, Y.I. (1989) On infestation of marine redfishes (*Sebastes* genus) of the North Atlantic by the copepod *Sphyrion lumpi* (Kroyer, 1845). In: *Proceedings of the Workshop on Sphyrion lumpi*, Padogogisclie Hochschule, Gustrow, pp. 29–36.

Bastide-Guillaume, C., Douellou, L., Romestand, B. and Trilles, J.P. (1987) Comparative study of two *Lernaeocera* parasites of *Merluccius merluccius* and *Trisopterus minutus capelanus* from the Lion Gulf (Sete, France). Morpho-biometry, antigenic communities, enzymatic polymorphism. *Revue des Traveaux de l'Institut des Peches Maritimes, Nantes* 49, 143–154.

Bauer, O.N. (1991) Spread of parasites and diseases of aquatic organisms by acclimatization: a short review. *Journal of Fish Biology* 39, 679–686.

Bazal, K., Lucky, Z. and Dyk, V. (1969) Localisation of fish-lice and leeches on carps during the autumn fishing. *Acta Veterinaria (Brno)* 38, 533–544.

Ben Hassine, O.K. and Raibaut, A. (1981) Realisation experimentale du cycle evolutif de *Ergasilus lizae* Krøyer, 1863, copepode parasite de poissons mugilides. Premiers resultats de l'infestation. *Archives de l'Institut Pasteur de Tunis* 58, 423–430.

Ben Hassine, O.K., Braun, M. and Raibaut, A. (1983) Etude comparative de l'infestation de *Mugil cephalus cephalus* Linne, 1758 par le copepode *Ergasilus lizae* Kroyer, 1863 dans deux alagunes du littoral Mediterraneen francais. *Rapports et Proces-Verbeaux des Reunions Commission Internationals pour l'Exploration Scientifique de la Mer Mediterranee, Monaco* 28, 379–384.

Berger, V.Y. (1970) Effect of sea water of different salinities on *Lepeophtheirus salmonis* (Kroyer) – an ectoparasite of salmon. *Parazitologiya* 4, 136–138.

Berland, B. (1983) The presence of the isopod *Cirolana borealis* and the amphipod

Tmetonyx cicada in the roes of cod (*Gadus morhua*) and saithe (*Pollachius virens*). *Fiskers Gang* 6/7, 175–179 (Abstract).

Berland, B. (1993) Salmon lice on wild salmon (*Salmo salar* L.) in western Norway. In: Boxshall, G.A. and Defaye, D. (eds) *Pathogens of Wild and Farmed Fish: Sea Lice*. Ellis Horwood, New York, pp. 179–187.

Berry, C.R., Babey, G.J. and Shrader, T. (1991) Effect of *Lernaea cyprinacea* (Crustacea: Copepoda) on stocked rainbow trout (*Oncorhynchus mykiss*). *Journal of Wildlife Diseases* 27, 206–213.

Bird, P.M. (1981) The occurrence of *Cirolana borealis* (Isopoda) in the hearts of sharks from Atlantic coastal waters of Florida. *Fishery Bulletin* 79, 376–382.

Bjordal, A. (1988) Cleaning symbiosis between wrasses (Labridae) and lice infested salmon (*Salmo salar*) in mariculture. *International Council for the Exploration of the Sea C.M.* F:17, 1–8.

Bjordal, A. (1990) Wrasse as cleaner-fish for farmed salmon. *Progress in Underwater Science* 16, 17–28.

Black, G.A. (1982) Gills as an attachment site for *Salmincola edwardsii* (Copepoda, Lernaeopodidae). *Journal of Parasitology* 68, 1172–1173.

Black, G.A., Montgomery, W.L. and Whoriskey, F.G. (1983) Abundance and distribution of *Salmincola edwardsii* (Copepoda) on anadromous brook trout, *Salvelinus fontinalis*, (Mitchill) in the Moisie River system, Quebec. *Journal of Fish Biology* 22, 567–575.

Bouchet, G.C. (1985) Redescription of *Argulus varians* Bere, 1936 (Branchiura, Argulidae) including a description of its early development and first larval stage. *Crustaceana* 49, 30–35.

Bower-Shore, C. (1940) An investigation of the common fish louse, *Argulus foliaceus* (Linn.). *Parasitology* 32, 361–371.

Bowman, T.E. (1960) Description and notes on the biology of *Lironeca puhi*, n.sp. (Isopoda: Cymothoidae), parasite of the Hawaiian moray eel, *Gymnothorax eurostus* (Abbott). *Crustaceana* 1, 83–91.

Bowman, T.E. and Mariscal, R.N. (1968) *Renocila heterozota*, a new cymothoid isopod, with notes on its host, the anemone fish, *Amphiprion akallopisos*, in the Seychelles. *Crustaceana* 14, 97–104.

Boxshall, G.A. (1974a) The population dynamics of *Lepeophtheirus pectoralis* (Müller): seasonal variation in abundance and age structure. *Parasitology* 69, 361–371.

Boxshall, G.A. (1974b) The population dynamics of *Lepeophtheirus pectoralis* (Muller): dispersion pattern. *Parasitology* 69, 373–390.

Boxshall, G.A. (1974c) Infections with parasitic copepods in North Sea marine fishes. *Journal of the Marine Biological Association, UK* 54, 355–372.

Boxshall, G.A. (1976) The host specificity of *Lepeophtheirus pectoralis* (Muller, 1776) (Copepoda: Caligidae). *Journal of Fish Biology* 8, 255–264.

Boxshall, G.A. (1977) The histopathology of infection by *Lepeophtheirus pectoralis* (Muller) (Copepoda: Caligidae). *Journal of Fish Biology* 10, 411–415.

Boxshall, G.A. (1990) The skeletomusculature of siphonostomatoid copepods, with an analysis of adaptive radiation in structure of the oral cone. *Philosophical Transactions of the Royal Society of London* B 328, 167–212.

Bragoni, G., Romestand, B. and Trilles, J.-P. (1983) Parasitism by cymothoids among sea-bass (*Dicentrarchus labrax* Linnaeus) in rearing. II. Parasitic ecophysiology in Diana Pond, Corsica. *Annales de Parasitologie Humaine et Comparée* 58, 593–609.

Bragoni, G., Romestand, B. and Trilles, J.-P. (1984) Parasitoses a cymothoadien chez le loup, *Dicentrarchus labrax* (Linnaeus, 1758) en elevage. I. Ecologie parasitaire dans le cas de l'Etang de Diana (Haute Corse) (Isopoda, Cymothoidae). *Crustaceana* 47, 44–51.

Brandal, P.O. and Egidius, E. (1977) Preliminary report on oral treatment against sea lice, *Lepeophtheirus salmonis* with Neguvon. *Aquaculture* 10, 177–178.

Brandal, P.O. and Egidius, E. (1979) Treatment of salmon lice (*Lepeophtheirusalmonis* Kroyer, 1838) with Neguvon – description of methods and equipment. *Aquaculture* 18, 183–188.

Brandal, P.O., Egidius, E. and Romslo, I. (1976) Host blood: A major food component for the parasitic copepod *Lepeophtheirus salmonis* Kroyeri, 1838 (Crustacea: Caligidae). *Norwegian Journal of Zoology* 24, 341–343.

Bristow, G.A. and Berland, B. (1988) *Haemobaphes cyclopterina* (Fabricius, 1780) (Copepoda: Lernaeoceridae): a new host record, *Glyptocephalus cynoglossus* (L.) with notes on the ecology of host and parasite. *Sarsia* 73, 287–290.

Bron, J.E. and Treasurer, J.W. (1992) Sea lice (Caligidae) on wrasse (Labridae) from selected British wild and salmon-farm sources. *Journal of the Marine Biological Association, UK* 72, 645–650.

Bron, J.E., Sommerville, C., Jones, M. and Rae, G.H. (1991) The settlement and attachment of early stages of the salmon louse, *Lepeophtheirus salmonis* (Copepoda: Caligidae) on the salmon host, *Salmo salar*. *Journal of Zoology* 224, 201–212.

Bron, J.E., Sommerville, C. and Rae, G.H. (1993a) Aspects of the behaviour of copepodid larvae of the salmon louse *Lepeophtheirus salmonis* (Kroyer, 1837). In: Boxshall, G.A. and Defaye, D. (eds) *Pathogens of Wild and Farmed Fish: Sea Lice*. Ellis Horwood, New York, pp. 125–142.

Bron, J.E., Sommerville, C., Wootten, R. and Rae, G.H. (1993b) Influence of treatment with dichlorvos on the epidemiology of *Lepeophtheirus salmonis* (Kroyer, 1837) and *Caligus elongatus* Nordmann, 1832 on Scottish salmon farms. In: Boxshall, G.A. and Defaye D. (eds) *Pathogens of Wild and Farmed Fish: Sea Lice*. Ellis Horwood, New York, pp. 263–274.

Bruce, N.L. (1983) Aegidae (Isopods: Crustacea) from Australia with descriptions of three new species. *Journal of Natural History* 17, 757–788.

Bruce, N.L. (1986a) Revision of the isopod crustacean genus *Mothocya* Costa, in Hope, 1851 (Cymothoidae: Flabellifera), parasitic on marine fishes. *Journal of Natural History* 20, 1089–1192.

Bruce, N.L. (1986b) Australian *Pleopodias* Richardson, 1910, and *Anilocra* Leach, 1818 (Isopods: Cymothoidae), crustacean parasites of marine fishes. *Records of the Australian Museum* 39, 85–130.

Bruce, N.L. (1987) Australian species of *Nerocila* Leach, 1818, and *Creniola* n.gen. (Isopods: Cymothoidae), crustacean parasites of marine fishes. *Records of the Australian Museum* 39, 355–412.

Bruce, N.L. (1990) The genera *Catoessa, Elthusa, Enispa, Ichthyoxenus, Idusa, Lironeca* and *Norileca* n.gen. (Isopods, Cymothoidae), crustacean parasites of marine fishes, with descriptions of eastern Australian species. *Records of the Australian Museum* 42, 247–300.

Bruce, N.L. and Bowman, T.E. (1989) Species of the parasitic isopod genera *Ceratothoa* and *Glossobius* (Crustacea: Cymothoidae) from the mouths of flying fishes and halfbeaks (Beloniformes). *Smithsonian Contributions to Zoology* 489, 1–28.

Bruno, D.W. and Stone, J. (1990) The role of saithe *Pollachius virens* as a host for *Lepeophtheirus salmonis* and *Caligus elongatus*. *Aquaculture* 89, 201–208.

Bruno, D.W., Munro, A.L.S. and McHenery, J.G. (1990) The potential of carbaryl as a treatment for sea lice infestations of farmed salmon, *Salmo salar* L. *Journal of Applied Ichthyology* 6, 124–127.

Brusca, R.C. (1978a) Studies on the cymothoid fish symbionts of the eastern Pacific (Crustacea: Isopoda: Cymothoidae). II. Biology and systematics of *Lironeca vulgaris*. *Occasional Papers of the Allan Hancock Foundation* 2, 1–19.

Brusca, R.C. (1978b) Studies on the cymothoid fish symbionts of the Eastern Pacific (Isopoda, Cymothoidae) I. Biology of *Nerocila californica*. *Crustaceana* 34, 141–154.

Brusca, R.C. (1981) A monograph on the Isopoda Cymothoidae (Crustacea) of the eastern Pacific. *Zoological Journal of the Linnean Society* 73, 117–199.

Brusca, R.C. and Gilligan, M.R. (1983) Tongue replacement in a marine fish (*Lutjanus guttatus*) by a parasitic isopod (Crustacea: Isopoda). *Copeia* 1983, 813–815.

Brusca, R.C. and Iverson, E.W. (1985) A guide to the marine isopod Crustacea of Pacific Costa Rica. *Revista de Biologia Tropical* 33 Supplement 1, 1–77.

Buchanan, J.S. (1992) The management of environmental risk: a case study based on the use of dichlorvos to control sea lice infestations on Atlantic salmon. In: Michel, C. and Alderman, D.J. (eds) *Chemotherapy in Aquaculture: from Theory to Reality*. Office International des Epizooties, Paris, pp. 187–194.

Bullar, J.F. (1876) The generative organs of the parasitic Isopoda. *Journal of Anatomy and Physiology* 11, 118–123.

Bullock, A.M., Phillips, S.E., Gordon, J.D.M. and Roberts, R.J. (1986) *Sarcotaces* sp., a parasitic copepod infection in two deep-sea fishes, *Lepidion eques* and *Coelorhynchus occa*. *Journal of the Marine Biological Association, UK* 66, 835–843.

Bulow, F.J., Winningham, J.R. and Hooper, R.C. (1979) Occurrence of the copepod parasite *Lernaea cyprinacea* in a stream fish population. *Transactions of the American Fisheries Society* 108, 100–102.

Buttner, J.K. and Hamilton, W. (1976) *Ergasilus* (Copepoda: Cyclopoida) infestation of coho and chinook salmon in Lake Michigan. *Transactions of the American Fisheries Society* 105, 491–493.

Byrnes, T. and Rohde, K. (1992) Geographical distribution and host specificity of ectoparasites of Australian bream, *Acanthopagrus* spp. (Sparidae). *Folia Parasitologica* 39, 249–264.

Cabral, P. (1983) Morphologie, Biologie et Ecologie des Copepodes Parasites du Loup *Dicentrarchus labrax* (Linne, 1758) et du Sar raye *Diplodus sargus* (Linne, 1758) de la region Languedocienne, PhD Thesis, Université des sciences et Techniques du Languedoc, Montpellier.

Caillet, C. and Raibaut, A. (1979) Observations experimentales sur la sexualite du Copepode Caligoide *Clavellodes macrotrachelus* (Brian, 1906), parasite branchial du Sar *Diplodus sargus* (Linne, 1758). *Compte Rendu de l'Academie des Sciences, Paris Ser. D* 288, 223–226.

Capart, A. (1948) Le *Lernaeocera branchialis*. *Cellule* 52, 159–212.

Castro, A.L. and Silva, J.L. (1985) Isopoda. In: Schaden, R. (ed.), *Manual de Identificacao de Invertebrados Limnicos do Brasil*. CNPq, Brasilia, pp. 1–10.

Chandran, A. and Nair, N.B. (1988) Functional morphology of the mouth tube of a lernaeopodid *Pseudocharopinus narcinae* (Pillai, 1962) (Copepoda: Siphonostomatoida). *Hydrobiologia* 167/168, 629–634.

Cloutman, D.G. and Becker, D.A. (1977) Some ecological aspects of *Ergasilus centrarchidarum* Wright (Crustacea: Copepoda) on largemouth and spotted bass in Lake Fort Smith, Arkansas. *Journal of Parasitology* 63, 372–376.

Conroy, G. and Conroy, D.A. (1986) The salinity tolerance of *Ergasilus lizae* from silver mullet (*Mugil curema* Val., 1836). *Bulletin of the European Association of Fish Pathologists* 6, 108–109.

Costello, M.J. (1993) Review of methods to control sea lice (Caligidae: Crustacea) infestation on salmon (*Salmo salar*) farms. In: Boxshall, G.A. and Defaye, D. (eds) *Pathogens of Wild and Farmed Fish: Sea Lice*. Ellis Horwood, New York, pp. 219–252.

Costello, M. and Bjordal, A. (1990) Wrasse. How good is this natural control on sealice? *Fish Farmer* May/June, 44–46.

Cowles, R.P. (1930) A biological study of the offshore waters of Chesapeake Bay. *Bulletin of the Bureau of Fisheries* 46, 346–349.

Cressey, R.F. (1967) *Caritus*, a new genus of caligoid copepod, with a key to the genera of Caliginae. *Proceedings of the US National Museum* 123, 1–8.

Cressey, R.F. (1978) Marine flora and fauna of the northeastern United States. Crustacea: Branchiura. *NOAA Technical Report NMFS Circular* 414, 1–10.

Darwin, C. (1851) *A Monograph on the Subclass Cirripedia, vol. I.* Ray Society, London.

Das, P., Kumar, D., Ghosh, A.K., Chakraborty, D.P. and Bhaumik, U. (1980) High yield of Indian major carps against encountered hazards in a demonstration pond. *Journal of the Inland Fisheries Society of India* 12, 70–78.

Davies, A.J. (1981) A scanning electron microscope study of the praniza larva of *Gnathia maxillaris* Montagu (Crustacea, Isopoda, Gnathiidae), with special reference to the mouthparts. *Journal of Natural History* 15, 545–554.

Delaney, P.M. (1989) Phylogeny and biogeography of the marine isopod family Corallanidae (Crustacea, Isopoda, Flabellifera). *Contributions in Science, Natural History Museum of Los Angeles County* 409, 1.

Delaney, P.M. and Brusca, R.C. (1985) Two new species of *Tridentella* Richardson, 1905 (Isopoda: Flabellifera: Tridentellidae) from California, with a rediagnosis and comments on the family, and a key to the genera of Tridentellidae and Corallanidae. *Journal of Crustacean Biology* 5, 728–742.

Dempster, R.P., Morales, P. and Glennon, F.X. (1988) Use of sodium chlorite to combat anchor worm infestation of fish. *Progressive Fish-Culturalist* 50, 51–55.

Do, T.T. (1982) *Paraergasilus longidigitus* Yin, 1954 (Copepoda, Poecilostomatoida) parasitic on Japanese freshwater fishes, with a key to Japanese Ergasilidae. *Fish Pathology* 17, 139–145.

Dobson, D.P. and Tack, T.J. (1991) Evaluation of the dispersion of treatment solutions of dichlorvos from marine salmon pens. *Aquaculture* 95, 15–32.

Dogiel, V.A., Petrushevski, G.K. and Polyanski, Y.I. (eds) (1961) *Parasitology of Fishes*. Oliver and Boyd, London.

Donoghue, S. (1986) *Thersitina gasterostei* (Pagenstecher, 1861) (Copepoda: Ergasilidae) infecting the stickleback *Pungitius pungitius* L. at Chalk Marshes, Gravesend, Kent. *Annals de Parasitologie Humane et Comparée* 61, 673–682.

Donoghue, S. (1989) Histopathology of ten-spined stickleback, *Pungitius pungitius* L., infected by the parasitic copepod *Thersitina gasterostei* (Pagenstecher) with observations on the contents of the parasite gut. *Bulletin of the European Association of Fish Pathologists* 9, 19–21.

Egidius, E. (1985) Salmon lice, *Lepeophtheirus salmonis*. *Journal of Animal Morphology and Physiology* leaflet 26, 1–4.

Egidius, E. and Moster, B. (1987) Effect of Neguvon and Nuvan treatment on crabs (*Cancer pagurus*, *C. maenas*), lobster (*Homarus gammarus*) and blue mussel (*Mytilus edulis*). *Aquaculture* 60, 165–168.

Einszporn, T. (1964) Observations on the kind of food taken up by *Ergasilus sieboldi*
    Nordmann. *Wiadomosci Parazytologiczne* 10, 527–529.

Einszporn, T. (1965a) Nutrition of *Ergasilus sieboldi* Nordmann. I. Histological struc-
    ture of the alimentary canal. *Acta Parasitologica Polonica* 13, 71–80.

Einszporn, T. (1965b) Nutrition of *Ergasilus sieboldi* Nordmann, II. The uptake of
    food and the food material. *Acta Parasitologica Polonica* 13, 373–380.

Einszporn-Orecka, T. (1973a) Changes in the picture of peripheral blood of tench
    Tinca tinca (L.) under the influence of *Ergasilus sieboldi* Nordm. I. Composi-
    tion of morphotic blood constituents of uninfected fish in annual cycle. *Acta
    Parasitologica Polonica* 21, 29.

Einszporn-Orecka, T. (1973b) Changes in the picture of peripheral blood of tench
    Tinca tinca (L.) under the influence of *Ergasilus sieboldi* Nordm. II. Changes in
    the leucocyte system. *Acta Parasitologica Polonica* 21, 38.

El Gharbi, S., Rousset, V. and Raibaut, A. (1985) Biology of *Lernaeenicus sprattae*
    (Sowerby, 1806) and its pathogenic effects on pilchard populations from the
    coasts of Languedoc-Roussillon (French Mediterranean). *Revue des Traveaux de
    l'Institut Peches Maritimes, Nantes* 47, 191–201.

Ellis, A.E., Masson, N. and Munro, A.L.S. (1990) A comparison of proteases
    extracted from *Caligus elongatus* (Nordmann, 1832) and *Lepeophtheirus salmonis*
    (Kroyer, 1838). *Journal of Fish Diseases* 13, 163–165.

Evans, N.A., Whitfield, P.J., Bamber, R.N. and Espin, P.M. (1983) *Lernaeocera lusci*
    (Copepoda: Pennellidae) on bib (*Trisopterus luscus*) from Southampton Water.
    Parasitology 86, 161–173.

Faisal, M., Easa, M.S., Shalaby, S.I. and Ibrahim, M.M. (1988) Epizootics of *Lernaea
    cyprinacea* (Copepoda: Lernaeidae) in imported cyprinids to Egypt. *Tropenland-
    wirt* 89, 131–141.

Faisal, M., Perkins, P.S. and Cooper, E.L. (1990) Infestation by the pennellid copepod
    Phrixocephalus cincinnatus modulates cell mediated immune responses in the
    Pacific arrowtooth flounder, *Atheresthes stomias*. In: Perkins, F.O. and Cheng,
    T.C. (eds) *Pathology in Marine Science*. Academic Press, London, pp. 471–478.

Ferguson, H.W. (1989) *Systemic Pathology of Fish*. Iowa State University Press,
    Ames.

Forrester, C.R. (1956) The relation of stock density to 'milkiness' of lemon sole in
    Union Bay, B.C. *Journal of the Fisheries Research Board of Canada* 105, 11.

Fraile, L. (1986) Experimental demonstration of an attraction of the copepodus
    and adult stages of the parasite *Caligus minimus* (Copepoda: Caligidae) for the
    bass *Dicentrarchus labrax* (Teleostei, Serranidae). *European Aquaculture Society,
    Special Publication* 9, 185. (Abstract)

Frerichs, G.N., Millar, S.D. and McManus, C. (1992) Atypical *Aeromonas salmonicida*
    isolated from healthy wrasse (*Ctenolabrus rupestris*). *Bulletin of the European
    Association of Fish Pathologists* 12, 48–49.

Friend, G.F. (1940) The life-history and ecology of the salmon gill-maggott *Salmincola
    salmonea* (L.) (copepod crustacean). *Transactions of the Royal Society of
    Edinburgh* 60, 503–541.

Frimeth, J.P. (1987) Potential use of certain parasites of brook charr (*Salvelinus
    fontinalis*) as biological indicators in the Tabusintac River, New Brunswick,
    Canada. *Canadian Journal of Zoology* 65, 1989–1995.

Frost, W.E. (1928) The nauplius larva of *Anelasma squalicola* (Loven). *Journal of the
    Marine Biological Association, UK* 15, 125–128.

Fryer, G. (1956) A report on the parasitic Copepoda and Branchiura of the fishes of
    Lake Nyasa. *Proceedings of the Zoological Society of London* 127, 293–344.

Fryer, G. (1959) A report on the parasitic Copepoda and Branchiura of the fishes of Lake Bangweulu (Northern Rhodesia). *Proceedings of the Zoological Society of London* 132, 517–550.

Fryer, G. (1960) The spermatophores of *Dolops ranarum* (Crustacea, Branchiura): their structure, formation and transfer. *Quarterly Journal of Microscopical Science* 101(4), 407–432.

Fryer, G. (1961) Variation and systematic problems in a group of lernaeid copepods. *Crustaceana* 2, 275–285.

Fryer, G. (1964) Further studies on the parasitic Crustacea of African freshwater fishes. *Proceedings of the Zoological Society of London* 143, 79–102.

Fryer, G. (1966) Habitat selection and gregarious behaviour in parasitic crustaceans. *Crustaceana* 10, 199–209.

Fryer, G. (1968a) The parasitic Crustacea of African freshwater fishes; their biology and distribution. *Journal of Zoology* 156, 45–95.

Fryer, G. (1968b) A new parasitic isopod of the family Cymothoidae from clupeid fishes of Lake Tanganyika – a further Lake Tanganyika enigma. *Journal of Zoology* 156, 35–43.

Fryer, G. (1982) *The Parasitic Copepoda and Branchiura of British Freshwater Fishes*, Freshwater Biological Association, Ambleside.

Gall, G.A.E., McClendon, E.L. and Schafer, W.E. (1972) Evidence on the influence of the copepod (*Salmincola californiensis*) on the reproductive performance of a domesticated strain of rainbow trout (*Salmo gairdneri*). *Transactions of the American Fisheries Society* 101, 345–346.

Gayevskaya, A.V. and Kovaleva, A.A. (1984) Crustacea of the genus *Sphyrion* (Copepoda, Sphyriidae) in Atlantic fishes. *Hydrobiological Journal, Kiev* 20, 41–46.

Ghittino, C. (1987) Positive control of buccal lernaeosis in eel farming. *Revista Italiana di Piscicoltura e Ittiopatologia* 22, 26–29.

Gnanamuthu, C.P. (1950) Sex differences in the chalimus and adult forms of *Caligus polycanthi*, sp. nov. (Crustacea, Copepoda) parasitic on *Balistes maculatus* from Madras. *Zoological Survey Records of India* 47, 159–170.

Grabda, J. (1963) Life cycle and morphogenesis of *Lernaea cyprinacea* L. *Acta Parasitologica Polonica* 11, 169–199.

Grabda, J. (1972) Observations on penetration of *Lernaeolophus sultanus* (Milne Edwards, 1840) (Lernaeoceridae) in organs of *Pneumatophorus colias* (Gmelin, 1788). *Acta Ichthyologica et Piscatoria* II, 115–125.

Grabda, J. (1975) Observations on the localization and pathogenicity of *Haemobaphes diveraus* Wilson, 1917, (Copepoda: Lernaeoceridae) in the gills of *Theragra chalcogramma* (Pallas). *Acta Ichthyologica et Piscatoria* 5, 13–23.

Grabda, J. (1991) *Marine Fish Parasitology*. VCH, New York.

Grant, A.N. and Treasurer, J.W. (1993) The effects of fallowing on caligid infestation on farmed salmon (*Salmo salar* L.) in Scotland. In: Boxshall, G.A. and Defaye, D. (eds) *Pathogens of Wild and Farmed Fish: Sea Lice*. Ellis Horwood, New York, pp. 255–260.

Grayson, T.H., Jenkins, P.G., Wrathmell, A.B. and Harris, J.E. (1991) Serum responses to the salmon louse, *Lepeophtheirus salmonis* (Kroyer, 1838), in naturally infected salmonids and immunised rainbow trout, *Oncorhynchus mykiss* (Walbaum), and rabbits. *Fish and Shellfish Immunology* 1, 141–155.

Gresty, K.A., Boxshall, G.A. and Nagasawa, K. (1993) Antennulary sensors of the infective copepodid larva of the salmon louse, *Lepeophtheirus salmonis* (Copepoda: Caligidae). In: Boxshall, G.A. and Defaye, D. (eds) *Pathogens of Wild and Farmed Fish: Sea Lice*. Ellis Horwood, New York, pp. 83–98.

Guillaume, C., Doueellou, L., Romestand, B. and Trilles, J.P. (1985) Influence of a haematophagenous parasite: *Lernaeocera branchialis* (L., 1767) (Crustacea, Copepoda, Pennellidae), on some erythrocytic constants of the host fish: *Merluccius merluccius. Revue des Traveaux de l'Institut des Peches Maritimes, Nantes* 47, 55–61.

Gurney, R. (1948) The British species of fish-louse of the genus *Argulus. Proceedings of the Zoological Society of London* 118(3), 553–558.

Guthrie, J.F. and Kroger, R.L. (1974) Schooling habits of injured and parasitized menhaden. *Ecology* 55, 208–210.

Hahnenkamp, L. and Fyhn, H.J. (1985) The osmotic response of salmon louse, *Lepeophtheirus salmonis* (Copepoda: Caligidae), during the transition from sea water to fresh water. *Journal of Comparative Physiology B* 155, 357–365.

Hale, H.M. (1929) *The Crustaceans of South Australia*, Govt. SA, Adelaide.

Haley, A.J. and Winn, H.E. (1959) Observations on a lernaean parasite of freshwater fishes. *Transactions of the American Fisheries Society* 88, 128–129.

Halisch, W. (1940) Anatomie und bilogie von *Ergasilus minor. Zeitschrift für Parasitenkunde* 11, 284–330.

Hargis, W.J. (1958) Parasites and fishery problems. *Proceedings of the Gulf and Caribbean Fishery Institute* 1958, 50–75.

Hastein, T. and Bergsjo, T. (1976) The salmon lice *Lepeophtheirus salmonis* as the cause of disease in farmed salmonids. *Revista Italiana Piscicoltura e Ittiopatologia* 11, 3–5.

Hatai, K. and Yasumoto, S. (1980) A parasitic isopod, *Irona melanosticta* isolated from the gill chamber of fingerlings of cultured yellowtail, *Seriola quinqueradiata. Bulletin of Nagasaki Prefectural Institute of Fisheries* 6, 87–96.

Hatai, K. and Yasumoto, S. (1981) Some notes on the ironasis of cultured young yellowtail, *Seriola quinqueradiata. Bulletin of Nagasaki Prefectural Institute of Fisheries* 7, 77–81.

Hatai, K. and Yasumoto, S. (1982a) Effects of *Irona melanosticta* on the growth of young rudderfish *Girella punctata. Bulletin of Nagasaki Prefectural Institute of Fisheries* 8, 75–79.

Hatai, K. and Yasumoto, S. (1982b) The effects of methyl isoxathion in eliminating the parasitic isopod *Irona melanosticta. Aquaculture, Japan* 30, 147–150.

Hattori, J. and Seki, M. (1956) An isopod, *Irona melanosticta* parasitic on *Hemirhamphus sajori* (T.& S.) and its influence on the hosts. *Dobutsugaku Zasshi* 65, 422–425.

Heegaard, P. (1947a) Discussion of the genus *Sarcotaces* (Copepoda) with a description of the first known male of the genus. *Kungliga Fysiografiska Sallskapets i Lund Forhandlingar* 17, 122–129.

Heegaard, P. (1947b) Contributions to the phylogeny of the arthropods: Copepoda. *Spolia Zoologica Musei Hauniensis* 8, 1–236.

Hesse, E. (1855) Sur les noms d'Ancee et de Pranize donnes a des Crustaces consideres comme des especes distinctes, et n'etant reelement que des individus dune meme espece a differents ages. *Comptes Rendus de l'Academie des Sciences, Paris* 41.

Hewitt, G.C. (1964) The postchalimus development of *Lepeophtheirus polyprioni* Hewitt, 1963 (Copepoda : Caligidae). *Transactions of the Royal Society of New Zealand, Zoology* 4, 157–159.

Hewitt, G.C. (1971) Two species of *Caligus* (Copepoda, Caligidae) from Australian waters, with a description of some developmental stages. *Pacific Science* 25, 145–164.

Hickling, C.F. (1963) On the small deep-sea shark *Etmopterus spinax* L., and its cirripede parasite *Anelasma squalicola* (Loven). *Journal of the Linnean Society* 45, 17–24.

Hindle, E. (1949) Notes on the treatment of fish infected with *Argulus*. *Proceedings of the Zoological Society of London* 119(1), 79–81.

Hislop, J.R.G. and Shanks, A.M. (1981) Recent investigations on the reproductive biology of the haddock, *Melanogrammus aeglefinus*, of the northern North Sea and the effects on fecundity of infection with the copepod parasite *Lernaeocera branchialis*. *Journal du Conseil* 39, 244–251.

Hjort, J. (1895) Zur Anatomie und Entwicklungsgeschichte einer im Fleisch von Fischen schmarotzenden Crustacée (*Sarcotaces arcticus* Collett). *Videnskabsselskabets Skrifter, I. Mathematisk-naturv* 2, 1–14.

Ho, J.-S. (1983) Copepod parasites of Japanese surfperches: their inference on the phylogeny and biogeography of Embiotocidae in the Far East. *Annual Report of the Sado Marine Biological Station, Niigata University* 13, 31–62.

Ho, J.-S. (1989) Historical biogeography of *Sphyrion* (Copepoda: Sphyriidae). In: *Proceedings of the Workshop on* Sphyrion lumpi, Padagogische Hochschule, Gustrow, pp. 4–22.

Ho, J.-S. and Honma, Y. (1983) *Lernaeolophus aceratus*, a new species of copepod parasitic on rainbowfish from the Sea of Japan, with notes on food and feeding. *Journal of Crustacean Biology* 3, 321–328.

Ho, J.-S. and Kim, I.H. (1989) *Lophoura* (Copepoda: Sphyriidae) parasitic on the rattails (Pisces: Macrouridae) in the Pacific, with note on *Sphyrion lumpi* from the Sea of Japan. *Publications of the Seto Marine Biological Laboratory* 34, 37–54.

Ho, J.-S. and Tonguthai, K. (1992) Flabelliferan isopods (Crustacea) parasitic on freshwater fishes of Thailand. *Systematic Parasitology* 21, 203–210.

Ho, J.-S., Jayarajan, P. and Radhakrishnan, S. (1992) Copepods of the family Ergasilidae (Poecilostomatoida) parasitic on coastal fishes of Kerala, India. *Journal of Natural History* 26, 1227–1241.

Hoffman, G.L. (1976) Parasites of freshwater fish. IV. The anchor parasite (*Lernaea elegans*) and related species. *US Fish and Wildlife Service Fish Disease Leaflet* No. 46, 8 pp.

Hoffman, G.L. (1977) Copepod parasites of freshwater fish: *Ergasilus, Achtheres*, and *Salmincola. US Fish and Wildlife Service, Fish Disease Leaflet* No. 48, 10 pp.

Hoffman, G.L. (1984) *Salmincola californiensis* continues the march eastward. *American Fisheries Society FHS Newsletter* 12, 4.

Hoffman, G.L. and Lester, R.J.G. (1987) Workshop 4F: Crustacean parasites of fish. *International Journal for Parasitology* 17, 1030–1031.

Hoffman, G.L. and Meyer, F.P. (1974) *Parasites of Freshwater Fishes*. TFH, Neptune City.

Hogans, W.E. (1989) Mortality of cultured Atlantic salmon, *Salmo salar* L., parr caused by an infection of *Ergasilus labracis* (Copepoda: Poecilostomatoida) in the lower Saint John River, New Brunswick, Canada. *Journal of Fish Diseases* 12, 529–531.

Hogans, W.E. and Trudeau, D.J. (1989a) *Caligus elongatus* (Copepoda: Caligoida) from Atlantic salmon (*Salmo salar*) cultured in marine waters of the lower Bay of Fundy. *Canadian Journal of Zoology* 67, 1080–1082.

Hogans, W.E. and Trudeau, D.J. (1989b) Preliminary studies on the biology of sea lice, *Caligus elongatus, Caligus curtus* and *Lepeophtheirus salmonis* (Copepoda: Caligoida) parasitic on cage-cultured salmonids in the lower Bay of Fundy. *Canadian Technical Report of Fisheries and Aquatic Sciences* 1715, 1–14.

Honma, Y. and Chiba, A. (1991) Pathological changes in the branchial chamber wall of stingrays, *Dasyatis* spp., associated with the presence of juvenile gnathiids (Isopoda, Crustacea). *Fish Pathology* 26, 9–16.

Honma, Y., Tsunaki, S., Chiba, A. and Ho, J.-S. (1991) Histological studies on the juvenile gnathiid (Isopoda, Crustacea) parasitic on the branchial chamber wall of the stingray, *Dasyatis akejei*, in the Sea of Japan. *Report of the Sado Marine Biological Station, Niigata University* 21, 37–47.

Hoshina, T. and Suenaga, G. (1954) On a new species of parasitic copepods from Yamame (salmoid fish) of Japan. *Journal of the Tokyo University of Fisheries* 41, 75–79.

Hotta, H. (1962) The parasitism of saury (*Cololabis saira*) infected with parasitic Copepoda, *Caligus macarovi* Gussev, during fishing season in 1961. *Bulletin of the Tohoku Regional Fisheries Research Laboratory* 21, 50–56.

Hoy, T. and Horsberg, T.E. (1990) Residues of dichlorvos in Atlantic salmon (*Salmo salar*) after delousing. *Journal of Agricultural and Food Chemistry* 38, 1403–1406.

Hoy, T., Horsberg, T.E. and Nafstad, I. (1992a) The disposition of ivermectin in Atlantic salmon (*Salmo salar*). In: Michel, C. and Alderman, D.J. (eds) *Chemotherapy in Aquaculture: from Theory to Reality*. Office International des Epizooties, Paris, pp. 461–467.

Hoy, T., Horsberg, T.E. and Wichstroem, R. (1992b) Inhibition of acetylcholinesterase in rainbow trout following dichlorvos treatment at different dissolved oxygen levels. In: Michel, C. and Alderman, D.J. (eds) *Chemotherapy in Aquaculture: from Theory to Reality*. Office International des Epizooties, Paris, pp. 206–211.

Hughes, S.E. (1973) Some metazoan parasites of the Eastern Pacific saury, *Cololabis saira*. *Fishery Bulletin* 71, 943–952.

Huizinga, H.W. (1972) Pathobiology of *Artystone trysibia* Schioedte (Isopoda: Cymothoidae), an endoparasitic isopod of South American freshwater fishes. *Journal of Wildlife Diseases* 8, 225–232.

Hwa, T.K. (1965) Studies on the life cycle of a fish louse (*Caligus orientalis* Gussev). *Acta Zoologica Sinica* 17, 48–63.

Izawa, K. (1969) Life history of *Caligus spinosus* Yamaguti, 1939 obtained from cultured yellow tail, *Seriola quinqueradiata* T. & S. (Crustacea: Caligoida). *Report of Faculty of Fisheries, Prefectural University of Mie* 6, 127–157.

Izawa, K. (1973) On the development of parasitic Copepoda I. *Sarcotaces pacificus* Komai (Cyclopoida: Phylichthyidae). *Publications of the Seto Marine Biological Laboratory* 21, 77–86.

Izawa, K. (1974) *Sarcotaces*, a genus of parasitic copepods (Cyclopoida: Phylichthyidae), found on Japanese fishes. *Publications of the Seto Marine Biological Laboratory* 21, 179–191.

Izawa, K. (1977) A new species of *Peroderma* Heller (Caligoida: Lernaeoceridae), parasitic on the fish *Bregmaceros japonicus* Tanaka. *Pacific Science* 31, 253–258.

Janusz, J. (1980) An influence of the parasite *Clavella adunca* (Strom 1762) (Copepoda parasitica: Lernaeopodidae) on the cod (*Gadus morhua* L.) from north-west Atlantic waters. *Acta Ichthyologica et Piscatoria* 10, 103–118.

Jaworski, A. and Holm, J.C. (1992) Distribution and structure of the population of sea lice, *Lepeophtheirus salmonis* Kroyer, on Atlantic salmon, *Salmo salar* L., under typical rearing conditions. *Aquaculture and Fisheries Management* 23, 577–589.

Johannessen, A. (1978) Early stages of *Lepeophtheirus salmonis* (Copepoda, Caligidae). *Sarsia* 63, 169–176.

Johnson, S.C. and Albright, L.J. (1991a) The developmental stages of *Lepeophtheirus salmonis* (Kroyer, 1837) (Copepoda: Caligidae). *Canadian Journal of Zoology* 69, 929–950.

Johnson, S.C. and Albright, L.J. (1991b) Development, growth, and survival of *Lepeophtheirus salmonis* (Copepoda: Caligidae) under laboratory conditions. *Journal of the Marine Biological Association, UK* 71, 425–436.

Johnson, S.C. and Albright, L.J. (1992a) Comparative susceptibility and histopathology of the response of naive Atlantic chinook and coho salmon to experimental infection with *Lepeophtheirus salmonis* (Copepoda: Caligidae). *Diseases of Aquatic Organisms* 14, 179–193.

Johnson, S.C. and Albright, L.J. (1992b) Effect of cortisol implants on the susceptibility and histopathology of the response of naive coho salmon *Oncorhynchus kisutch* to experimental infection with *Lepeophtheirus salmonis* (Copepoda: Caligidae). *Diseases of Aquatic Organisms* 14, 195–205.

Johnston, C.E. and Dykeman, D. (1987) Observations on body proportions and egg production in the female parasitic copepod (*Salmincola salmoneus*) from the gills of Atlantic salmon (*Salmo salar*) kelts exposed to different temperatures and photoperiods. *Canadian Journal of Zoology* 65, 415–419.

Johnstone, J. and Frost, W.E. (1927) The cirripede fish parasite *Anelasma squalicola* (Loven): its general morphology. *Report of the Lancashire Sea-Fisheries Laboratory* 35, 29–91.

Jones, D.H. and Matthews, J.B.L. (1968) On the development of *Sphyrion lumpi* (Kroyer). *Crustaceana* Suppl. 1, 177–185.

Jones, J.B. (1980) A redescription of *Caligus patulus* Wilson, 1937 (Copepoda, Caligidae) from a fish farm in the Philippines. *Systematic Parasitology* 2, 103–116.

Jones, M.W., Sommerville, C. and Bron, J. (1990) The histopathology associated with the juvenile stages of *Lepeophtheirus salmonis* on the Atlantic salmon, *Salmo salar* L. *Journal of Fish Diseases* 13, 303–310.

Jones, M.W., Sommerville, C. and Wootten, R. (1992) Reduced sensitivity of the salmon louse, *Lepeophtheirus salmonis*, to the organophosphate dichlorvos. *Journal of Fish Diseases* 15, 197–202.

Jonsdottir, H., Bron, J.E., Wootten, R. and Turnbull, J.E. (1992) The histopathology associated with the pre-adult and adult stages of *Lepeophtheirus salmonis* on the Atlantic salmon, *Salmo salar* L. *Journal of Fish Diseases* 15, 521–527.

Joy, J.E. (1976) Gill parasites of the spot *Leiostomus xanthurus* from Clear Lake, Texas. *Transactions of the American Microscopical Society* 95, 63–68.

Joy, J.E. and Jones, L.P. (1973) Observations on the inflammatory response within the dermis of a white bass, *Morone chrysops* (Rafinesque), infected with *Lernaea cruciata* (Copepoda: Caligidae). *Journal of Fish Biology* 5, 21–23.

Juilfs, H.B. and Wagele, J.W. (1987) Symbiotic bacteria in the gut of the bloodsucking Antarctic fish parasite *Gnathia calva* (Crustacea: Isopoda). *Marine Biology* 95, 493–499.

Jungnitz, H.-A. (1989) Some food hygienical aspects of the invasion by *Sphyrion lumpi* from the point of view of the veterinary control organisation of the GDR. In: *Proceedings of the Workshop on* Sphyrion lumpi, Padagogische Hochschule, Gustrow, pp. 65–68.

Kabata, Z. (1958) *Lernaeocera obtusa* n.sp. Its biology and its effects on the haddock. *Scottish Home Department Marine Research* 3, 1–26.

Kabata, Z. (1960) Observations on *Clavella* (Copepoda) parasitic on some British gadoids. *Crustaceana* 1, 342–352.

Kabata, Z. (1966) Comments on the phylogeny and zoogeography of Lernaeopodidae (Crustacea: Copepoda). *International Congress of Parasitology* 1 E12, 1082–1083.

Kabata, Z. (1969a) Revision of the genus *Salmincola* Wilson, 1915 (Copepoda: Lernaeopodidae). *Journal of the Fisheries Research Board of Canada* 26, 2987–3041.

Kabata, Z. (1969b) *Phrixocephalus cincinnatus* Wilson, 1908 (Copepoda: Lernaeoceridae): morphology, metamorphosis, and host-parasite relationship. *Journal of the Fisheries Research Board of Canada* 26, 921–934.

Kabata, Z. (1970) Crustacea as enemies of fishes. In: Snieszko, S.F. and Axelrod, H.R. (eds) *Diseases of Fishes, Book 1.* TFH Publ., New Jersey.

Kabata, Z. (1972a) Copepoda parasitic on Australian fishes, XI. *Impexus hamondi* new genus, new species with a key to the genera of Lernaeoceridae. *Proceedings of the Biological Society of Washington* 85, 317–322.

Kabata, Z. (1972b) Developmental stages of *Caligus clemensi* (Copepoda: Caligidae). *Journal of the Fisheries Research Board of Canada* 29, 1571–1593.

Kabata, Z. (1974) Mouth and mode of feeding of Caligidae (Copepoda), parasites of fishes, as determined by light and scanning electron microscopy. *Journal of the Fisheries Research Board of Canada* 31, 1583–1588.

Kabata, Z. (1979) *Parasitic Copepoda of British Fishes.* Ray Society, London.

Kabata, Z. (1980) Evolution and systematics of parasitic Copepoda. *Bulletin of the Canadian Society of Zoologists* 11, 26–31.

Kabata, Z. (1981) Copepoda (Crustacea) parasitic on fishes: problems and perspectives. *Advances in Parasitology* 19, 1–71.

Kabata, Z. (1983) Two new genera of the family Lernaeidae (Copepoda: Cyclopoida) parasitic on freshwater fishes of India. In: *Selected Papers on Crustacea.* Rabindranath, Krishna Pillai Farewell Committee, Trivandrum, pp. 69–76.

Kabata, Z. (1984) Diseases caused by metazoans – crustaceans. In: Kinne, O. (ed.) *Diseases of Marine Animals Vol. IV Pt 1.* Biologische Anstalt Helgoland, Hamburg, pp. 321–399.

Kabata, Z. (1985) *Parasites and Diseases of Fish Cultured in the Tropics.* Taylor & Francis, London.

Kabata, Z. (1988) Copepoda and Branchiura. In: Margolis, L. and Kabata, Z. (eds) *Guide to the Parasites of Fishes of Canada, Part II – Crustacea.* Department of Fisheries and Oceans, Ottawa, pp. 3–127.

Kabata, Z. (1992) Copepods parasitic on Australian fishes, XV. Family Ergasilidae (Poecilostomatoida). *Journal of Natural History* 26, 47–66.

Kabata, Z. and Cousens, B. (1972) The structure of the attachment organ of Lernaeopodidae (Crustacea: Copepoda). *Journal of the Fisheries Research Board of Canada* 29, 1015–1023.

Kabata, Z. and Cousens, B. (1973) Life cycle of *Salmincola californiensis* (Dana 1852) (Copepoda: Lernaeopodidae). *Journal of the Fisheries Research Board of Canada* 30, 881–903.

Kabata, Z. and Cousens, B. (1977) Host–parasite relationships between sockeye salmon, *Oncorhynchus nerka*, and *Salmincola californiensis* (Copepoda: Lernaeopodidae). *Journal of the Fisheries Research Board of Canada* 34, 191–202.

Kabata, Z. and Hewitt, G.C. (1971) Locomotory mechanisms in Caligidae (Crustacea: Copepoda). *Journal of the Fisheries Research Board of Canada* 28, 1143–1151.

Kannupandi, T. (1976) Cuticular adaptations in two parasitic copepods in relation to their modes of life. *Journal of Experimental Marine Biology and Ecology* 22, 235–248.

Kashkovsky, V.V. and Kashkovskaya-Solomatova, V.P. (1986) Ecology of larvae of *Ergasilus sieboldi* (Copepoda parasitica) in the Lake Arakul. *Parazitologiya* 20, 32–38.

Kawatow, K., Muroga, K., Izawa, K. and Kasahara, S. (1980) Life cycle of *Alella macrotrachelus* (Copepoda) parasitic on cultured black sea-bream. *Journal of the Faculty of Applied Biological Sciences* 19, 199–214.

Kelly, H.D. and Allison, R. (1962) Observations on the infestation of a fresh water fish population by a marine copepod (*Ergasilus lizae* Kroyer 1863). *Proceedings of the Sixteenth Annual Conference of the Southeastern Association of Game and Fish* 16, 236–239.

Kensley, B. and Schotte, M. (1989) *Guide to the Marine Isopod Crustaceans of the Caribbean*. Smithsonian Institution, Washington.

Keys, A.B. (1928) Ectoparasites and vitality. *American Naturalist* 62, 279–282.

Khalifa, K.A. and Post, G. (1976) Histopathological effect of *Lernaea cyprinacea* (a copepod parasite) on fish. *The Progressive Fish-Culturist* 38, 110–113.

Khan, R.A. (1984) Concurrent infections of two parasites, *Lernaeocera branchialis* (Copepoda) and *Trypanosoma murmanensis* (Protozoa) in Atlantic Cod, *Gadus morhua*. *4th European Multicolloquium of Parasitology*.

Khan, R.A. (1988) Experimental transmission, development, and effects of a parasitic copepod, *Lernaeocera branchialis*, on Atlantic cod, *Gadus morhua*. *Journal of Parasitology* 74, 586–599.

Khan, R.A. and Lee, E.M. (1989) Influence of *Lernaeocera branchialis* (Crustacea: Copepoda) on growth rate of Atlantic cod, *Gadus morhua*. *Journal of Parasitology* 75, 449–454.

Khan, R.A., Lee, E.M. and Barker, D. (1990) *Lernaeocera branchialis*: potential pathogen to cod ranching. *Journal of Parasitology* 76, 913–917.

Kim, I.-H. (1993) Developmental stages of *Caligus punctatus* Shiino, 1955 (Copepoda: Caligidae). In: Boxshall, G.A. and Defaye, D. (eds) *Pathogens of Wild and Farmed Fish: Sea Lice*. Ellis Horwood, New York, pp. 16–29.

Kimura, S. (1970) Notes on the reproduction of water lice (*Argulus joponica* Thiele). *Bulletin of Freshwater Fisheries Research Laboratory* 20, 109–126.

Kirmse, P. (1987) Important parasites of Dover sole (*Solea solea* L.) kept under mariculture conditions. *Parasitology Research* 73, 466–471.

Kroger, R.L. and Guthrie, J.F. (1972a) Occurrence of the parasitic branchiuran, *Argulus alosae*, on dying Atlantic menhaden, *Brevoortia tyrannus*, in the Connecticut River. *Transactions of the American Fisheries Society* 101, 559–560.

Kroger, R.L. and Guthrie, J.F. (1972b) Incidence of the parasitic isopod, *Olencira praegustator*, in juvenile Atlantic menhaden. *Copeia* 1972, 370–374.

Kuitunen-Ekbaum, E. (1949) The occurrence of *Sarcotaces* in Canada. *Journal of the Fisheries Research Board of Canada* 7, 505–512.

Kuperman, B.I. and Shulman, R.E. (1977) On the influence of some abiotic factors on the development of *Ergasilus sieboldi* (Crustacea, Copepoda). *Parazitologiya* 11, 117–121.

Kurochkin, Y.V. (1985) Applied and scientific aspects of marine parasitology. In: Hargis, W.J.J. (ed.), *Parasitology and Pathology of Marine Organisms of the World Ocean*. NOAA Tech.Rept. NMFS 25, pp. 15–18.

Kussakin, O.G. (1979) *Marine and Brackishwater Isopod Crustacea, Suborder Flabellifera*. Academy of Sciences USSR, Leningrad.

Landsberg, J.H., Vermeer, G.K., Richards, S.A. and Perry, N. (1991) Control of the parasitic copepod *Caligus elongatus* on pond-reared red drum. *Journal of Aquatic Animal Health* 3, 206–209.

Lanzing, W.J.R. and O'Connor, P.F. (1975) Infestation of luderick (*Girella tricuspidata*) populations with parasitic isopods. *Australian Journal of Marine and Freshwater Research* 26, 355–361.

Lasee, B.A., Sutherland, D.R. and Moubry, M.E. (1988) Host-parasite relationships between burbot (*Lota lota*) and adult *Salmincola lotae* (Copepoda). *Canadian Journal of Zoology* 66, 2459–2463.

Lavina, E.M. (1977) The biology and control of *Caligus* sp., an ectoparasite of the adult milkfish *Chanos chanos* Forskal. *SEAFDEC Quarterly Research Report, Aquaculture Department* 1977, 12–13.

Lawler, A.R., Howse, H.D. and Cook, D.W. (1974) Silver perch, *Bairdiella chrysura*: new host for lymphocystis. *Copeia* 1974, 266–269.

Legrand, J.-J. (1951) Etude statistique et experimentale de la sexualite d'*Anilocra physodes* L. (Crustace Isopode Cymothoide). *Bulletin de la Société d'Histoire Naturelle de Toulouse* 86, 176–183.

Legrand, J.-J. (1952) Contribution a l'etude experimentale et statistique de la biologie d'*Anilocra physodes* L. *Archives de Zoologie Experimentale et Generale* 89, 1–56.

Lester, R.J.G. and Daniels, B.A. (1976) The eosinophilic cell of the white sucker, *Catostomus commersoni*. *Journal of the Fisheries Research Board of Canada* 33, 139–144.

Lester, R.J.G. and Sewell, K.B. (1989) Checklist of parasites from Heron Island, Great Barrier Reef. *Australian Journal of Marine and Freshwater Research* 37, 101–128.

Lewis, A.G. (1963) Life history of the caligid copepod *Lepeophtheirus dissimulatus* Wilson, 1905 (Crustacea: Caligoida). *Pacific Science* 17, 195–242.

Lewis, A.G. (1969) A discussion of the maxillae of the 'Caligoidea' (Copepoda). *Crustaceana* 16, 65–77.

Lewis, R.M. and Hetler, W.F. (1968) Effect of temperature and salinity on the survival of young Atlantic menhaden, *Brevoortia tyrannus*. *Transactions of the American Fisheries Society* 97, 344–349.

Lin, C.L. (1989) A new species of *Caligus* parasitic on milkfish. *Crustaceana* 57, 225–246.

Lin, C.L. and Ho, J.-S. (1993) Life history of *Caligus epidemicus* Hewitt parasitic on tilapia (*Oreochromis mossambicus*) cultured in brackish water. In: Boxshall, G.A. and Defaye, D. (eds) *Pathogens of Wild and Farmed Fish: Sea Lice*. Ellis Horwood, New York, pp. 5–15.

Lindsay, J.A. and Moran, R.L. (1976) Relationships of parasitic isopods, *Lironeca ovalis* and *Olencira praegustator* to marine fish hosts in Delaware Bay. *Transactions of the American Fisheries Society* 327–332.

Löpmann, A. (1940) Uber die quantitative bestimmung des ergasilusbefalles an schleien (*Tinca vulgaris*). *Zeitschrift für Parasitenkunde* 11, 474–483.

MacKenzie, K. and Morrison, J.A. (1989) An unusually heavy infestation of herring (*Clupea harengus* L.) with the parasitic copepod *Caligus elongatus* Nordmann, 1832. *Bulletin of the European Association of Fish Pathologists* 9, 12–13.

MacKinnon, B.M. (1992) Egg production in sea lice, *Caligus elongatus*. *Bulletin of the Canadian Society for Zoology* 23, 79.

Malta, J.C.O. (1982) The argulids (Crustacea: Branchiura) of Amazonia, Brazil. 2. Aspects of the ecology of *Dolops geayi* Bouvier, 1897, and *Argulus juparanaensis* Castro, 1950. *Acta Amazonia* 12, 701–705.

Malta, J.C.O. and Silva, E.N.S. (1986) *Argulus amazonicus* n.sp. a crustacean parasite of fishes from the Brazilian Amazon (Branchiura: Argulidae). *Amazoniana* 9, 485–492.

Mann, H. (1952) *Lernaeocera branchialis* (Copepoda parasitica) und seine Schadwirkung bei einigen Gadiden. *Archiv für Fischereiwissenschaft* 4, 133–144.

Mann, H. (1970) Copepoda and Isopoda as parasites of marine fishes. In: Snieszko, S. (ed.) *A Symposium on Diseases of Fishes and Shellfishes*. American Fisheries Society, Special Publication 5, Washington, pp. 177–189.

Margolis, L., Kabata, Z. and Parker, R.R. (1975) Catalogue and synopsis of *Caligus*. *Bulletin of the Fisheries Research Board of Canada* 192, 1–117.

Martin, M. (1932) On the morphology and classification of *Argulus* (Crustacea). *Proceedings of the Zoological Society of London* X, 771–806.

Mattson, N.S., Egidius, E. and Solbakken, J.E. (1988) Uptake and elimination of (meth-14C)-trichlorfon in blue mussel (*Mytilus edulis*) and European oyster (*Ostrea edulis*) – impact of Neguvon disposal on mollusc farming. *Aquaculture*. 71, 9–14.

Maxwell, J.G.H. (1982) Infestation of the jack mackerel, *Trachurus declivis* (Jenyns), with the cymothoid isopod, *Ceratothoa imbricatus* in southeastern Australian waters. *Journal of Fish Biology* 20, 341–350.

Mayer, P. (1879) Carcinologische Mittheilungen. VI Ueber den Hermaphroditismus bei einigen Isopoden. *Mittheilungen Zoologische Station zu Neapel* 1, 165–179.

McGladdery, S.E. and Johnston, C.E. (1988) Egg development and control of the gill parasite, *Salmincola salmoneus*, on Atlantic salmon kelts (*Salmo salar*) exposed to four different regimes of temperature and photoperiod. *Aquaculture* 68, 193–202.

McGregor, E.A. (1963) *Publications on Fish Parasites and Diseases, 330 B.C. – A.D. 1923*. US Department of the Interior, Fish and Wildlife Service, Report-Fisheries No. 474, Washington.

McHenery, J.G., Turrell, W.R. and Munro, A.L.S. (1992) Control of the use of the insecticide dichlorvos in Atlantic salmon farming. In: Michel, C. and Alderman, D.J. (eds) *Chemotherapy in Aquaculture: from Theory to Reality*. Office International des Epizooties, Paris, pp. 179–186.

McLean, P.H., Smith, G.W. and Wilson, M.J. (1990) Residence time of the sea louse, *Lepeophtheirus salmonis* K., on Atlantic salmon, *Salmo salar* L., after immersion in fresh water. *Journal of Fish Biology* 37, 311–314.

McNeil, P.L., Jr. (1961) The use of benzene hexachloride as a copepodicide and some observations on lernaean parasites in trout rearing units. *Progressive Fish-Culturalist* 23, 127–133.

Meeüs, T., Renaud, F. and Gabrion, C. (1990) A model for studying isolation mechanisms in parasite populations: the genus *Lepeophtheirus* (Copepoda, Caligidae). *Journal of Experimental Zoology* 254, 207–214.

Meeüs, T., Raibaut, A. and Renaud, F. (1993a) Comparative life history of two species of sea lice. In: Boxshall, G.A. and Defaye, D. (eds) *Pathogens of Wild and Farmed Fish: Sea Lice*. Ellis Horwood, New York, pp. 143–150.

Meeus, T., Raibaut, A. and Renaud, F. (1993b) Speciation and specificity in parasitic copepods: caligids of the genus *Lepeophtheirus*, parasites of flatfish in the Mediterranean. In: Boxshall, G.A. and Defaye, D. (eds) *Pathogens of Wild and Farmed Fish: Sea Lice*. Ellis Horwood, New York, pp. 143–150.

Menezes, J., Ramos, M.A., Pereira, T.G. and Moreira da Silva, A. (1990) Rainbow trout culture failure in a small lake as a result of massive parasitosis related to careless fish introductions. *Aquaculture* 89, 123–126.

Menzies, R.J., Bowman, T.E. and Alverson, F.G. (1955) Studies of the biology of the fish parasite *Lironeca convexa* Richardson (Crustacea, Isopoda, Cymothoidae). *Wasmann Journal of Biology* 13, 277–295.

Minchin, E.A. (1909) Observations on the flagellates parasitic in the blood of freshwater fishes. *Proceedings of the Zoological Society of London* 1909, 2–30.

Möller, H. (1984) *Daten zur Biologie der Elbfische*. Möller, Kiel.

Möller, H. and Anders, K. (1986) *Diseases and Parasites of Marine Fishes*. Möller, Kiel.

Monod, T. (1926) Les Gnathiidae. *Mémoires de la Société des Sciences naturelles du Maroc* 13, 1–667.

Montalenti, G. (1948) Note sulla sistematica e la biologia di alcuni Cimotoidi del Golfo di Napoli. *Archivio di Oceanografia e Limnologia, Venezia* 5, 25–81.

Moravec, F. (1978) First record of *Molnaria erythrophthalmi* larvae in the intermediate host in Czechoslovakia. *Folia Parasitologia* 25, 141–142.

Morton, B. (1974) Host specificity and position on the host in *Nerocila phaeopleura* Bleeker (Isopoda, Cymothoidae). *Crustaceana* 26, 143–148.

Moser, M., Haldorson, L. and Field, L.J. (1985) The taxonomic status of *Sarcotaces komaii* and *Sarcotaces verrucosus* (Copepoda: Phylichthyidae) and host-parasite relationships between *Sarcotaces arcticus* and *Sebastes* spp. (Pisces). *Journal of Parasitology* 71, 472–480.

Moser, M. and Sakanari, J. (1985) Aspects of host location in the juvenile isopod *Lironeca vulgaris* (Stimpson, 1857). *Journal of Parasitology* 71, 464–468.

Moser, M. and Taylor, S. (1978) Effects of the copepod *Cardiodectes medusaeus* on the lanternfish *Stenobrachius leucopsarus* with notes on hypercastration by the hydroid *Hydrichthys* sp. *Canadian Journal of Zoology* 56, 2372–2376.

Mugridge, R.E.R. and Stallybrass, H.G. (1983) A mortality of eels, *Anguilla anguilla* L., attributed to Gnathiidae. *Journal of Fish Diseases* 6, 81–82.

Muroga, K., Kawatow, K. and Ichizono, H. (1981) Infestation by *Alella macrotrachelus* (Copepoda) of cultured black sea-bream. *Fish Pathology* 16, 139–144.

Nagasawa, K. (1987) Prevalence and abundance of *Lepeophtheirus salmonis* (Copepoda: Caligidae) on high-seas salmon and trout in the North Pacific Ocean. *Nippon Suisan Gakkaishi* 53, 2151–2156.

Nagasawa, K. and Maruyama, S. (1987) Occurrence and effects of *Haemobaphes diceraus* (Copepoda: Pennellidae) on brown sole *Limanda herzensteini* off the Okhotsk coast of Hokkaido. *Bulletin of the Japanese Society of Scientific Fisheries* 53, 991–994.

Nagasawa, K., Imai, Y. and Ishida, K. (1985) Distribution, abundance, and effects of *Pennella* sp. (Copepoda: Pennellidae), parasitic on the saury, *Cololabis saira* (Brevoort), in the western North Pacific Ocean and adjacent seas, 1984. *Bulletin of the Biogeographical Society of Japan* 40, 35–42.

Nagasawa, K., Imai, Y. and Ishida, K. (1988) Longterm changes in the population size and geographical distribution of *Pennella* sp. (Copepoda) on the saury, *Cololabis saira*, in the western North Pacific Ocean and adjacent seas. *Hydrobiologia* 167/168, 571–577.

Nagasawa, K., Ishida, I., Ogura, M., Tadokoro, K. and Hiramatsu, K. (1993) The abundance and distribution of *Lepeophtheirus salmonis* (Copepoda: Caligidae) on six species of Pacific salmon in offshore waters of the north Pacific Ocean and Bearing Sea. In: Boxshall, G.A. and Defaye, D. (eds) *Pathogens of Wild and Farmed Fish: Sea Lice*. Ellis Horwood, New York, pp. 166–178.

Nair, G.A. and Nair, N.B. (1982) Effect of certain organophosphate biocides on the juvenile of the isopod *Alitropus typus* M. Edwards (Crustacea: Flabellifera: Aegidae). *Journal of Animal Morphology and Physiology* 29, 265–271.

Nair, G.A. and Nair, N.B. (1983) Effect of infestation with the isopod, *Alitropus typus* M. Edwards (Crustacea: Flabellifera: Aegidae) on the haematological parameters of the host fish, *Channa striatus* (Bloch). *Aquaculture* 30, 11–19.

Nakajima, K., Izawa, S. and Egusa, S. (1974) Parasitic copepode, *Pseudergasilus zacconis* Yamaguti, found on the gills on cultured ayu, *Plecoglossus altivelis –* II. *Fish Pathology* 9, 95–99.

Natarajan, P. and Nair, N.B. (1973) Observations on the nature of attack of *Lernaeenicus hemirhamphi* Kirtisinghe on *Hemirhamphus xanthopterus* (Val.). *Journal of Animal Morphology and Physiology* 20, 56–63.

Natarajan, P. and Nair, N.B. (1976) Effects of infestation by *Lernaeenicus hemirhamphi* on the biochemical composition of the host fish *Hemirhamphus xanthopterus*.

Journal of Animal Morphology and Physiology 23, 25–31.

Neilson, J.D., Perry, R.I., Scott, J.S. and Valerie, P. (1987) Interactions of caligid ectoparasites and juvenile gadids on Georges Bank. *Marine Ecology (Progress Series)* 39, 221–232.

Neubert, J. (1984) Investigations on the application of trichlorfon in control of parasites and food organisms in inland fisheries. *Zeitschrift für die Binnenfischerei der DDR* 31, 334–336.

Nierkerk, J.P. and Kok, D.J. (1989) *Chonopeltis australis* (Branchiura): structural, developmental and functional aspects of the trophic appendages. *Crustaceana* 57, 51–56.

Nierkerk, J.P. and Van As, J.G. (1986) Ultrastructure of mouthparts to illustrate the feeding mechanism of *Chonopeltis australis* Boxshall, 1976 (Crustacea: Branchiura). *ICOPA VI Handbook* Abstract No. 682, 249.

Nigrelli, R.F. and Firth, F.E. (1939) On *Sphyrion lumpi* (Kroyer), a copepod parasite on the redfish, *Sebastes marinus* (Linnaeus), with special reference to the host-parasite relationships. *Zoologica* 24, 1–9.

Noble, E.R., King, R.E. and Jacobs, B.L. (1963) Ecology of the gill parasites of *Gillichthys mirabilis* Cooper. *Ecology* 44, 295–305.

Noga, E.J. (1986) The importance of *Lernaea cruciata* (Le Sueur) in the initiation of skin lesions in largemouth bass, *Micropterus salmoides* (Lacépède), in the Chowan River, North Carolina, USA. *Journal of Fish Diseases* 9, 295–302.

Noga, E.J., Mitchell, C.G., Groman, D.B. and Johnston, J.A.A. (1991) Dermatological diseases affecting fishes of the Tar-Pamlico Estuary, North Carolina. *Diseases of Aquatic Organisms* 10, 87–92.

Novotny, A.J. and Mahnken, C.W. (1971) Predation on juvenile salmon by a marine isopod, *Rocinela belliceps pugetensis. Fishery Bulletin* 69, 699–701.

Nylund, A., Bjorknes, B. and Wallace, C. (1991) *Lepeophtheirus salmonis* – a possible vector in the spread of diseases on salmonids. *Bulletin of the European Association of Fish Pathologists* 11, 213–216.

Nylund, A., Okland, S. and Bjorknes, B. (1992) Anatomy and ultrstructure of the alimentary canal in *Lepeophtheirus salmonis* (Copepoda: Siphonostomatoida). *Journal of Crustacean Biology* 12, 423–437.

Nylund, A., Wallace, C. and Hovland, T. (1993) The possible role of *Lepeophtheirus salmonis* (Kroyer) in the transmission of infectious salmon anaemia. In: Boxshall, G.A. and Defaye, D. (eds) *Pathogens of Wild and Farmed Fish: Sea Lice.* Ellis Horwood, New York, pp. 367–373.

Oldewage, W.H. and Van As, J.G. (1987) Observations on the attachment of a piscine gill parasitic ergasilid (Crustacea: Copepoda). *South African Journal of Zoology* 22, 313–317.

Olsson, P. (1872) Om *Sarcotaces* och *Acrobothrium*, tva nya parasitslagten fran fiskar. *Ofversigt af Kongl. Vetenskaps-Academiens Forhandlingarr* 29, 37–44.

Overstreet, R.M., Dykova, I. and Hawkins, W.E. (1992) Branchiura. In: *Microscopic Anatomy of Invertebrates Vol. 9: Crustacea.* Wiley-Liss, New York, pp. 385–413.

Palmer, R., Rodger, H., Drinan, E., Dwyer, C. and Smith, P.R. (1987) Preliminary trials of the efficacy of ivermectin against parasitic copepods of Atlantic salmon. *Bulletin of the European Association of Fish Pathologists* 7, 47–54.

Panikkar, N.K. and Sproston, N.G. (1941) Osmotic relations of some metazoan parasites. *Parasitology* 33, 214–223.

Paperna, I. (1975) Parasites and diseases of the grey mullet (Mugilidae) with special reference to the seas of the Near East. *Aquaculture* 5, 65–80.

Paperna, I. (1977) Copepod infections in fish in euryhaline environments. *Wiadomosci Parazytologiczne* 23, 183–187.

Paperna, I. (1991) Diseases caused by parasites in the aquaculture of warm water fish. *Annual Review of Fish Diseases* 1, 155–194.

Paperna, I. and Lahav, M. (1974) Mortality amoung gray mullets in a seawater pond due to caligiid parasitic copepod epizootic. *Bamidgeh* 26, 12–15.

Paperna, I. and Overstreet, R.M. (1981) Parasites and diseases of mullets (Mugilidae). In: Oren, O.H. (ed.) *Aquaculture of Grey Mullets.* University Press, Cambridge, pp. 411–493.

Paperna, I. and Por, F.D. (1977) Preliminary data on the Gnathiidae (Isopoda) of the northern Red Sea, the Bitter Lakes and the eastern Mediterranean and the biology of *Gnathia piscivora* n.sp. *Rapports et Proces-Verbeaux des Reunions Commission Internationale pour l'Exploration Scientifique de la Mer Mediterranee Monaco* 24(4), 195–197.

Paperna, I. and Zwerner, D.E. (1976a) Studies on *Ergasilus labracis* Kroyer (Cyclopidea: Ergasilidae) parasitic on striped bass, *Morone saxatilis*, from the lower Chesapeake Bay. I. Distribution, life cycle, and seasonal abundance. *Canadian Journal of Zoology* 54, 449–462.

Paperna, I. and Zwerner, D.E. (1976b) Parasites and diseases of striped bass, *Morone saxatilis* (Walbaum), from the lower Chesapeake Bay. *Journal of Fish Biology* 9, 267–287.

Paperna, I. and Zwerner, D.E. (1981) Host–parasite relationship of *Ergasilus labracis* Kroyer (Cyclopidea, Ergasilidae) and the striped bass, *Morone saxatilis* (Walbaum) from the lower Chesapeake Bay. *Annales de Parasitologie* 57, 393–405.

Parker, R.R. (1969) Validity of the binomen *Caligus elongatus* for a common parasitic copepod formerly misidentified with *Caligus rapax. Journal of the Fisheries Research Board of Canada* 26, 1013–1035.

Parker, R.R. and Margolis, L. (1964) A new species of parasitic copepod, *Caligus clemensi* sp. nov. (Caligoida, Caligidae) from pelagic fishes in the coastal waters of British Columbia. *Journal of the Fisheries Research Board of Canada* 21, 873–889.

Payne, A.I.L. (1986) Observations on some conspicuous parasites of the southern African kingklip *Genypterus capensis. South African Journal of Marine Science* 4, 163–168.

Perkins, P.S. (1983) The life history of *Cardiodectes medusaeus* (Wilson), a copepod parasite of lanternfishes (Myctophidae). *Journal of Crustacean Biology* 3, 70–87.

Perkins, P.S. (1985) Iron crystals in the attachment organ of the erythrophagous copepod *Cardiodectes medusaeus* (Pennellidae). *Journal of Crustacean Biology* 5, 591–605.

Petrushevsky, G.K. and Shulman, S.S. (1961) The parasitic diseases of fish in the natural waters of the USSR. In: Dogiel, V.A. *et al.* (eds) *Parasitology of Fishes.* Oliver & Boyd, Edinburgh, pp. 299–319.

Piasecki, W. (1989) Life cycle of *Tracheliastes maculatus* Koller, 1835 (Copepoda, Siphonostomatoida, Lernaeopodidae). *Wiadomosci Parazytologiczne* 35, 187–245.

Pike, A.W. (1989) Sea lice – major pathogens of farmed Atlantic salmon. *Parasitology Today* 5, 291–297.

Pike, A.W., Mackenzie, K. and Rowland, A. (1993) Ultrastructure of the frontal filament in chalimus larvae of *Caligus elongatus* and *Lepeophtheirus salmonis* from Atlantic salmon, *Salmo salar.* In: Boxshall, G.A. and Defaye, D. (eds) *Pathogens of Wild and Farmed Fish: Sea Lice.* Ellis Horwood, New York, pp. 99–113.

Poulin, R., Conley, D.C. and Curtis, M.A. (1990a) Effects of temperature fluctuations and photoperiod on hatching in the parasitic copepod *Salmincola edwardsii*. *Canadian Journal of Zoology* 68, 1330–1332.

Poulin, R., Curtis, M.A. and Rau, M.E. (1990b) Responses of the fish ectoparasite *Salmincola edwardsii* (Copepoda) to stimulation, and their implication for host-finding. *Parasitology* 100, 417–421.

Poulin, R. and FitzGerald, G.J. (1987) The potential of parasitism in the structuring of a salt marsh stickleback community. *Canadian Journal of Zoology* 65, 2793–2798.

Poulin, R. and FitzGerald, G.J. (1989a) Risk of parasitism and microhabitat selection in juvenile sticklebacks. *Canadian Journal of Zoology* 67, 14–18.

Poulin, R. and FitzGerald, G.J. (1989b) Shoaling as an anti-ectoparasite mechanism in juvenile sticklebacks (*Gasterosteus* spp.). *Behavioural Ecology and Sociobiology* 24, 251–255.

Poulin, R. and FitzGerald, G.J. (1989c) A possible explanation for the aggregated distribution of *Argulus canadensis* Wilson, 1916 (Crustacea: Branchiura) on juvenile sticklebacks (Gasterosteidae). *Journal of Parasitology* 75, 58–60.

Priebe, K. (1986) Some fish species of the northern Atlantic attacked by copepods of the genus *Sphyrion* and the pathological alterations in redfish fillets caused by *Sphyrion*-infestation. *Wiadomosci Parazytologiczne* 32, 501–504.

Priebe, K. (1989) An investigation about the attack of the parasitic copepod *Sphyrion lumpi* on redfish in the northern Atlantic with special consideration on the changes in fillets caused by this parasite. In: *Proceedings of the Workshop on* Sphyrion lumpi. Padogogische Hochschule, Gustrow, pp. 52–64.

Puffer, H.W. and Beal, M.L. (1981) Control of parasitic infestations in killifish (*Fundulus parvipinnis*). *Laboratory Animal Science* 31, 200–201.

Radhakrishnan, S. (1977) Description of a new species of *Peniculisa* including its immature stages. *Hydrobiologia* 52, 251–255.

Radhakrishnan, S. and Nair, N.B. (1981a) Histopathology of the infection of *Trichiurus savala* Cuvier by *Caligus uruguayenses* Thomsen (Copepoda: Caligidae). *Fisch und Umwelt* 10, 147–152.

Radhakrishnan, S. and Nair, N.B. (1981b) Nature of *Peniculisa wilsoni* Radhakrishnan (Copepoda: Lernaeoceridae) infestation of *Diodon hystrix* Lennaeus (Pisces: Diodontidae). *Journal of Animal Morphology and Physiology* 28, 73–81.

Radhakrishnan, S. and Nair, N.B. (1981c) Histopathology of the infestation of *Diodon hystrix* L. by *Peniculisa wilsoni* Radhakrishnan (Copepoda: Lernaeoceridae). *Journal of Fish Diseases* 4, 83–87.

Raibaut, A. (1985) Les cycles evolutifs des copepodes parasites et les modalites de l'infestation. *Annales de Biologie* 24, 233–274.

Raibaut, A., Ben Hassine, O.K. and Prunus, G. (1975) Etude de l'infestation de *Mugil (Mugil) cephalus* Linne, 1758 (Poissons, Teleosteens, Mugilides) par le copepode *Ergasilus nanus* van Beneden, 1870 dans le Lac Ischkeul (Tunisia). *Bulletin de la Société Zoologique de France* 100, 427–437.

Rand, T.G. (1986) The histopathology of infestation of *Paranthias furcifer* (L.) (Osteichthyes: Serranidae) by *Nerocila acuminata* (Schioedte and Meinert) (Crustacea: Isopoda: Cymothoidea). *Journal of Fish Diseases* 9, 143–146.

Rawson, M.V., Jr. (1977) Population biology of parasites of striped mullet, *Mugil cephalus* L. Crustacea. *Journal of Fish Biology* 10, 441–451.

Razmashin, D.A. and Shirshov, V.Y. (1981) Argulosis and its prevention in young coregonids reared in south-western Siberia fish farms. (*Fish Diseases and Aquatic Toxicology*), *Bolezni Rvb.I.Vodnava.Toksikologiya* 32.

Reichenbach-Klinke, H.H. and Landolt, M. (1973) *Fish Pathology*. TFH, Jersey City.

Reilly, P. and Mulcahy, M.F. (1993) Humoral antibody response in Atlantic salmon (*Salmo salar* L.) immunised with extracts derived from the ectoparasitic caligid copepods, *Caligus elongatus* (Nordmann, 1832) and *Lepeophtheirus salmonis* (Kroyer, 1838). *Fish and Shellfish Immunology* 3, 59–70.

Reimer, L.W. (1989) Preface. In: *Proceedings of the Workshop on* Sphyrion lumpi. Pedagogische Hochschule, Gustrow, pp. 2–3.

Rigby, D. and Tunnell, N. (1971) Internal anatomy and histology of female *Pseudocharopinus dentatus* (Copepoda, Lernaeopodidae). *Transactions of the American Microscopical Society* 90, 61–71.

Ritchie, G., Mordue, A.J., Pike, A.W. and Rae, G.H. (1993) The reproductive output of *Lepeophtheirus salmonis* adult females in relation to seasonal variability of temperature and photoperiod. In: Boxshall, G.A. and Defaye, D. (eds) *Pathogens of Wild and Farmed Fish: Sea Lice.* Ellis Horwood, New York, pp. 153–165.

Roberts, L.S. (1970) *Ergasilus* (Copepoda, Cyclopoida): revision and key to species in North America. *Transactions of the American Microscopical Society* 89, 134–161.

Robinson, G.R. (1981) Otter trawl sampling bias of the gill parasite *Lironeca vulgaris* from sand dab hosts, *Citharichthys* spp. *Fishery Bulletin* 80, 907–909.

Rodgers, L.J. and Burke, J.B. (1988) Aetiology of 'red spot' disease (vibriosis) with special reference to the ectoparasitic digenean *Prototransversotrema steri* (Angel) and the sea mullet, *Mugil cephalus* (Linnaeus). *Journal of Fish Biology* 32, 655–663.

Rogers, W.A. (1969) *Ergasilus cyprinaceus* sp.n. (Copepoda: Cyclopoida) from cyprinid fishes of Alabama, with notes on its biology and pathology. *Journal of Parasitology* 55, 443–446.

Romero, R.C. and Kuroki, H.B. (1986) Pre-metamorphosis stages of two pennellids (Copepoda, Siphonostomatoida) from their definitive hosts. *Crustaceana* 50, 166–175.

Romero, R.C. and Kuroki, H.B. (1989) Lamelliform structures on the proboscis of *Peniculus* and *Metapeniculus* (Copepoda: Pennellidae). *Proceedings of the Biological Society of Washington* 102, 912–915.

Romestand, B. (1979) Étude écophysiologique des parasitoses à Cymothoadiens. *Annales de Parasitologie* 54, 423–448.

Romestand, B. and Trilles, J.-P. (1975) Les relations immunologiques 'hôte-parasite' chez les Cymothoidae (Isopoda, Flabellifera). *Compte Rendu de l'Académie des Sciences, Paris*, Ser. D 280, 2171–2173.

Romestand, B. and Trilles, J.-P. (1976) Production d'une substance anticoagulante par les glandes exocrines cephalothoraciques des Isopodes Cymothoidae *Meinertia oestroides* (Risso, 1826) et *Anilocra physodes* (L., 1758) (Isopoda, Flabellifera, Cymothoidae). *Compte Rendu de l'Académie des Sciences, Paris* 282, 663–665.

Romestand, B. and Trilles, J.-P. (1977a) Dégénérescence de la langue des Bogues [(*Boops boops* L., 1758) (Téléostéens, Sparidae)] parasitées par *meinertia oestroides* (Risso, 1826) (Isopoda, Flabellifera, Cymothoidae). *Zeitschrift für Parasitenkunde* 54, 47–53.

Romestand, B. and Trilles, J.-P. (1977b) Influence des Cymothoadiens (Crustacea, Isopoda, Flabellifera) sur certaines constantes hématologiques des poissons hôtes. *Zeitschrift für Parasitenkunde* 52, 91–95.

Romestand, B. and Trilles, J.-P. (1979) Influence des Cymothoadiens *Meinertia oestroides, Meinertia parallela* et *Anilocra physodes* (Crustacés, Isopodes; parasites de poissons) sur la croissance des poissons hôtes *Boops boops* et *Pagellus erythrinus* (Sparidés). *Zeitschrift für Parasitenkunde* 59, 195–202.

Romestand, B., Janicot, M. and Trilles, J.-P. (1977) Modifications tissulaires et réactions de défense chez quelques Téléostéens parasités par les Cymothoidae (Crustacés, Isopodés, Hematophagés). *Annales de Parasitologie* 52, 171-180.

Romestand, B., Thuet, P. and Trilles, J.-P. (1982) Quelques aspects des mécanismes nutritionnels chez l'isopode Cymothoidae: *Ceratothoa oestroides* (Risso, 1826). *Annales de Parasitologie* 57, 79-89.

Rose, M. and Hamon, M. (1953) A propos de *Pennella varians* Steenstrup et Lutken, 1861, parasite des branchies de Cephalopodes. *Bulletin de la Société d'Histoire Naturelle de l'Afrique du Nord* 44, 172-183.

Roth, M. (1988) Morphology and development of the egg case in the parasitic copepod *Haemobaphes intermedius* Kabata, 1967 (Copepoda: Pennellidae). *Canadian Journal of Zoology* 66, 2573-2577.

Roth, M. and Richards, R.H. (1992) Trials on the efficacy of azamethiphos and its safety to salmon for the control of sea lice. In: Michel, C. and Alderman, D.J. (eds) *Chemotherapy in Aquaculture: from Theory to Reality*. Office International des Epizooties, Paris, pp. 212-218.

Roth, M., Richards, R.H. and Sommerville, C. (1993) Current practices in the chemotherapeutic control of sea lice infestations in aquaculture: a review. *Journal of Fish Diseases* 16, 1-26.

Roubal, F.R. (1981) The taxonomy and site specificity of the metazoan ectoparasites on the black bream, *Acanthopagrus australis* (Günther), in northern New South Wales. *Australian Journal of Zoology, Supplementary Series* 84, 1-100.

Roubal, F.R. (1986a) Studies on monogeneans and copepods parasitizing the gills of a sparid (*Acanthopagrus australis* (Günther)) in northern New South Wales. *Canadian Journal of Zoology* 64, 841-849.

Roubal, F.R. (1986b) The histopathology of the copepod, *Ergasilus lizae* Kroyer, on the pseudobranchs of the teleost, *Acanthopagrus australis* (Günther) (family Sparidae). *Zoologischer Anzeiger* 217, 65-74.

Roubal, F.R. (1987) Comparison of ectoparasite pathology on gills of yellowfin bream, *Acanthopagrus australis* (Günther) (Pisces: Sparidae): a surface area approach. *Australian Journal of Zoology* 35, 93-100.

Roubal, F.R. (1989a) Comparative pathology of some monogenean and copepod ectoparasites on the gills of *Acanthopagrus australis* (family Sparidae). *Journal of Fish Biology* 34, 503-514.

Roubal, F.R. (1989b) Pathological changes in the gill filaments of *Acanthopagrus australis* (family Sparidae) associated with the post-settlement growth of a lernaeopodid copepod, *Alella macrotrachelus. Journal of Fish Biology* 34, 333-342.

Roubal, F.R. (1990) Seasonal changes in ectoparasite infection of juvenile yellowfin bream, *Acanthopagrus australis* (Günther) (Pisces: Sparidae), from a small estuary in northern New South Wales. *Australian Journal of Marine and Freshwater Research* 41, 411-427.

Roubal, F.R. (1994) Histopathology caused by *Caligus epidemicus* Hewitt (Copepoda: Caligidae) on captive *Acanthopagrus australis* (Günther) (Pisces: Sparidae). *Journal of Fish Diseases* 17, 631-640.

Rousset, V. and Raibaut, A. (1989) Peculiar cases of intracardiac parasitism in the pilchard, *Sardina pilchardus* (Walbaum), by a pennellid copepod belonging to the genus *Lernaeenicus. Journal of Fish Diseases* 12, 263-268.

Ruangpan, L. and Kabata, Z. (1984) An invertebrate host for *Caligus* (Copepoda, Caligidae)? *Crustaceana* 47, 219-220.

Russell, F.S. (1933) On the occurrence of young stages of Caligidae on pelagic young fish in the Plymouth area. *Journal of the Marine Biological Association, UK* 18, 551-553.

Sadowsky, V. and Soares Moreira, P. (1981) Occurrence of *Squalus cubensis* Rivero, 1936, in the Western South Atlantic Ocean, and incidence of its parasitic copepod *Lironeca splendida* sp.n. *Studies on Neotropical Fauna and Environment* 16, 137–150.

Sadzikowski, M.R. and Wallace, D.C. (1974) The incidence of *Lironeca ovalis* (Say) (Crustacea, Isopoda) and its effects on the growth of white perch, *Morone americana* (Gmelin), in the Delaware River near Artificial Island. *Chesapeake Science* 15, 163–165.

Salte, R., Syversten, C., Kjoennoy, M. and Fonnum, F. (1987) Fatal acetylcholinesterase inhibition in salmonids subjected to a routine organophosphate treatment. *Aquaculture* 61, 173–179.

Samaras, W.F. and Durham, F.E. (1985) Feeding relationship of two species of epizoic amphipods and the grey whale, *Eschrichtius robustus*. *Bulletin of the Southern California Academy of Science* 84, 113–126.

Samuelsen, O.B. (1987) Aeration rate, pH and temperature effects on the degradation of trichlorfon to DDVP and the half-lives of trichlorfon and DDVP in sea water. *Aquaculture* 66, 373–380.

Sandifer, P.A. and Kerby, J.H. (1983) Early life history and biology of the common fish parasite, *Lironeca ovalis* (Say) (Isopoda, Cymothoidae). *Estuaries* 6, 420–425.

Sarig, S. (1971) *Diseases of Fishes, Book 3: The Prevention and Treatment of Diseases of Warmwater Fishes*. TFH, Neptune City.

Schafer, J.W., Enriquez, R. and Monras, M. (1989) Preliminary results of two years ichthyopathological service (1987–1989) in the south of Chile. In: *IV EAFP Conference*, EAFP, Santiago. Abstract, p. 6.

Schäperclaus, W., Kulow, H. and Schreckenbach, K. (1991) *Fish Diseases*, 5th edn. Oxion Press, New Delhi.

Schlüter, U. (1978) Beobachtungen zum Befall des Wirtes durch die Karpfenlaus *Argulus foliaceus* L. (Crustacea, Branchiura). *Zoologischer Anzeiger* 200, 85–91.

Schlüter, U. (1979) Uber die Temperaturabhangigkeit des Wachstums und des Hautungszyklus von *Argulus foliaceus* (L.) (Branchiura). *Crustaceana* 37, 100–106.

Schram, T.A. (1979) The life history of the eye-maggot of the sprat, *Lernaeenicus sprattae* (Sowberby) (Copepoda, Lernaeoceridae). *Sarsia* 64, 279–316.

Schram, T.A. (1993) Supplementary descriptions of the developmental stages of *Lepeophtheirus salmonis* (Kroyer, 1837) (Copepoda: Caligidae). In: Boxshall, G.A. and Defaye, D. (eds) *Pathogens of Wild and Farmed Fish: Sea Lice*. Ellis Horwood, New York, pp. 30–47.

Schram, T.A. and Anstensrud, M. (1985) *Lernaeenicus sprattae* (Sowerby) larvae in the Oslofjord plankton and some laboratory experiments with the nauplius and copepodid (Copepoda: Penellidae). *Sarsia* 70, 127–134.

Schram, T.A. and Haug, T. (1988) Ectoparasites on the Atlantic halibut, *Hippoglossus hippoglossus* (L.), from northern Norway – potential pests in halibut aquaculture. *Sarsia* 73, 213–227.

Schultz, D.A. (1969) *How to Know the Marine Isopod Crustaceans*. Brown, Dubuque.

Scott, P.W. and Fogel, B. (1983) Treatment of ornamental carp with anchorworm. *Veterinary Record* 113, 421.

Segal, E. (1987) Behaviour of juvenile *Nerocila acuminata* (Isopoda, Cymothoidae) during attack, attachment and feeding on fish prey. *Bulletin of Marine Science* 41, 351–360.

Shafir, A. and Van As, J.G. (1986) Laying, development and hatching of eggs of the fish ectoparasite *Argulus japonicus* (Crustacea: Branchiura). *Journal of Zoology* 210, 401–414.

Shariff, M. (1981) The histopathology of the eye of big head carp, *Aristichthys nobilis* (Richardson), infested with *Lernaea piscinae* Harding, 1950. *Journal of Fish Diseases* 4, 161–168.

Shariff, M. and Roberts, R.J. (1989) The experimental histopathology of *Lernaea polymorpha* Yu, 1938 infection in naive *Aristichthys nobilis* (Richardson) and a comparison in naturally infected clinically resistant fish. *Journal of Fish Diseases* 12, 405–414.

Shariff, M. and Sommerville, C. (1986a) Identification and distribution of *Lernaea* spp. in peninsular Malaysia. In: Maclean, J.L., Dizon, L.B. and Hosillos, L.V. (eds) *First Asian Fisheries Forum*. Asian Fisheries Society, Manila, pp. 269–272.

Shariff, M. and Sommerville, C. (1986b) The life cycles of *Lernaea polymorpha* and *L. cyprinacea*. In: Maclean, J.L., Dizon, L.B. and Hosillos, L.V. (eds) *First Asian Fisheries Forum*. Asian Fisheries Society, Manila, pp. 273–278.

Shariff, M. and Sommerville, C. (1986c) Effects of *Lernaea polymorpha* on the growth of big head carp, *Aristichthys nobilis*. *ICOPA VI Handbook*, Abstract no. 598, 227.

Shariff, M. and Sommerville, C. (1990) Comparative morphology of adult *Lernaea polymorpha* Yu and *Lernaea cyprinacea* Linnaeus. In: Hirano, R. and Hanyu, I. (eds), *Second Asian Fisheries Forum*. Asian Fisheries Society, Manila, pp. 717–720.

Shariff, M., Kabata, Z. and Sommerville, C. (1986) Host susceptibility to *Lernaea cyprincaea* L. and its treatment in a large aquarium system. *Journal of Fish Diseases* 9, 393–401.

Sherman, K. and Wise, J.P. (1961) Incidence of the cod parasite *Lernaeocera branchialis* L. in the New England area and its possible use as an indicator of cod populations. *Limnology and Oceanography* 6, 61–67.

Shields, R.J. (1978) Procedures for the laboratory rearing of *Lernaea cyprinacea* L. (Copepoda). *Crustaceana* 35, 259–264.

Shields, R.J. and Goode, R.P. (1978) Host rejection of *Lernaea cyprinacea* L. (Copepoda). *Crustaceana* 35, 301–307.

Shields, R.J. and Sperber, R.G. (1974) Osmotic relationships of *Lernaea cyprinacea* L. (Copepoda). *Crustaceana* 26, 157–171.

Shields, R.J. and Tidd, W.M. (1968) Effect of temperature on the development of larval and transformed females of *Lernaea cyprinacea* L. (Lernaeidae). *Crustaceana* Supplement 1, 87–95.

Shields, R.J. and Tidd, W.M. (1974) Site selection on hosts by copepodids of *Lernaea cyprinacea* L. (Copepoda). *Crustaceana* 27, 225–230.

Shiino, S.M. (1958) Copepods parasitic on Japanese fishes. 17. Lernaeidae. *Report of the Faculty of Fisheries, Prefectural University of Mie* 3, 75–100.

Shimura, S. (1981) The larval development of *Argulus coregoni* Thorell (Crustacea: Branchiura). *Journal of Natural History* 15, 331–348.

Shimura, S. (1983a) Seasonal occurrence, sex ratio and site preference of *Argulus coregoni* Thorell (Crustacea: Branchiura) parasitic on cultured freshwater salmonids in Japan. *Parasitology* 86, 537–552.

Shimura, S. (1983b) SEM observation on the mouth tube and preoral sting of *Argulus coregoni* Thorell and *Argulus japonicus* Thiele (Crustacea: Branchiura). *Fish Pathology, Tokyo* 18, 151–156.

Shimura, S. and Asai, M. (1984) *Argulus americanus* (Crustacea: Branchiura) parasitic on the bowfin, *Amia calva*, imported from North America. *Fish Pathology, Tokyo* 18, 199–213.

Shimura, S. and Inoue, K. (1984) Toxic effects of extract from the mouth-parts of *Argulus coregoni* Thorell (Crustacea: Branchiura). *Bulletin of the Japanese Society of Scientific Fisheries* 50, 729.

Shimura, S., Inoue, K., Kasai, K. and Saito, H. (1983a) Hematological changes of *Oncorhynchus masou* (Salmonidae) caused by the infection of *Argulus coregoni* (Crustacea: Branchiura). *Fish Pathology, Tokyo* 18, 157–162.

Shimura, S., Inoue, K., Kudo, M. and Egusa, S. (1983b) Studies on effects of parasitism of *Argulus coregoni* (Crustacea: Branchiura) on furunculosis of *Oncorhynchus masou* (Salmonidae). *Fish Pathology, Tokyo* 18, 37–40.

Shotter, R.A. (1971) The biology of *Clavella uncinata* (Muller) (Crustacea: Copepoda). *Parasitology* 63, 419–430.

Shotter, R.A. (1973) A comparison of the parasite fauna of young whiting, *Odontogadus merlangus* (L.) (Gadidae) from an inshore and an offshore location off the Isle of Man. *Journal of Fish Biology* 5, 185–195.

Shotter, R.A. (1976) The distribution of some helminth and copepod parasites in tissues of whiting, *Merlangius merlangus* L., from Manx waters. *Journal of Fish Biology* 8, 101–117.

Singhal, R.N., Jeet, S. and Davies, R.W. (1986) Chemotherapy of six ectoparasitic diseases of cultured fish. *Aquaculture and Fisheries Management* 54, 165–171.

Slinn, D.J. (1970) An infestation of adult *Lernaeocera* (Copepoda) on wild sole, *Solea solea*, kept under hatchery conditions. *Journal of the Marine Biological Association, UK* 50, 787–800.

Smith, J.A. and Whitfield, P.J. (1988) Ultrastructural studies on the early cuticular metamorphosis of adult female *Lernaeocera branchialis* (L.) (Copepoda, Pennellidae). *Hydrobiologia* 167/168, 607–616.

Sproston, N.G. (1942) The developmental stages of *Lernaeocera branchialis* (Linn.). *Journal of the Marine Biological Association, UK* 25, 441–466.

Sproston, N.G. and Hartley, P.H.T. (1941a) The ecology of some parasitic copepods of gadoids and other fishes. *Journal of the Marine Biological Association, UK* 25, 361–392.

Sproston, N.G. and Hartley, P.H.T. (1941b) Observations on the bionomics and physiology of *Trebius caudatus* and *Lernaeocera branchialis* (Copepoda). *Journal of the Marine Biological Association, UK* 25, 393–417.

Squires, H.J. (1966) Reproduction in *Sphyrion lumpi*, a copepod parasitic on redfish (*Sebastes* spp.). *Journal of the Fisheries Research Board of Canada* 23, 521–526.

Stammer, J. (1959) Beiträge zur Morphologie, Biologie und Bekämpfung der Karpfenläuse. *Zeitschrift für Parasitenkunde* 19, 135–208.

Stekhoven, J.H.S. (1936) Beobachtungen zur Morphologie und Physiologie der *Lernaeocera branchialis* L. und der *Lernaeocera lusci* Bassett-Smith (Crustacea parasitica). *Zeitschrift für Parasitenkunde* 8, 659–697.

Stekhoven, J.H.S. and Punt, A. (1937) Weitere Beiträge zur Morphologie und Physiologie der *Lernaeocera branchialis* L. *Zeitschrift für Parasitenkunde* 9, 648–668.

Stephenson, A.B. (1976) Gill damage to fish produced by buccal parasites. *Records of the Aukland Institute and Museum* 13, 167–173.

Stephenson, A.B. (1987) Additional notes on *Lironeca neocyttus* (Isopoda: Cymothoidae). *Records of the Aukland Institute and Museum* 24, 135–142.

Stoll, C. (1962) Cycle evolutif de *Paragnathia formica* (Hesse) (Isopode-Gnathiidae). *Cahiers de Biologie Marine* 3, 401–416.

Stuart, R. (1990) Sea lice, a maritime perspective. *Bulletin of the Aquaculture Association of Canada* 90, 18–24.

Sutherland, D.R. and Wittrock, D.D. (1985) The effects of *Salmincola californiensis* (Copepoda: Lernaeopodidae) on the gills of farm-raised rainbow trout, *Salmo gairdneri*. *Canadian Journal of Zoology* 63, 2893–2901.

Sutherland, D.R. and Wittrock, D.D. (1986) Surface topography of the branchiuran *Argulus appendiculosus* Wilson, 1907 as revealed by scanning electron microscopy. *Zeitschrift für Parasitenkunde* 72, 405-415.

Szidat, L. (1966) Untersuchungen über den entwicklungszyklus von *Meinertia gaudichaudii* (Milne Edwards, 1840) Stebbing, 1886 (Isopoda, Cymothoidae) und die entstehung eines sekundären sexualdimorphismus bei parasitischen asseln der familie Cymothoidae Schioedte u. Meinert, 1881. *Zeitschrift für Parasitenkunde* 27, 1-24.

Templeman, W., Hodder, V.M. and Fleming, A.M. (1976) Infection of lumpfish (*Cyclopterus lumpus*) with larvae and of Atlantic cod (*Gadus morhua*) with adults of the copepod, *Lernaeocera branchialis*, in and adjacent to the Newfoundland area, and inferences therefrom on inshore-offshore migrations of cod. *Journal of the Fisheries Research Board of Canada* 33, 711-731.

Thatcher, V.E. (1988) *Asotana magnifica* n.sp. (Isopoda, Cymothoidae) an unusual parasite (commensal?) of the buccal cavities of piranhas (*Serrasalmus* sp.) from Roraima, Brazil. *Amazoniana* 10, 239-248.

Thatcher, V.E. and Boeger, W.A. (1983) The parasitic crustaceans of fishes from the Brazilian Amazon. 4. *Ergasilus colomesus* n.sp. (Copepoda: Cyclopoida) from an ornamental fish, *Colomesus asellus* (Tetraodontidae) and aspects of its pathogenicity. *Transactions of the American Microscopical Society* 102, 371-379.

Thatcher, V.E. and Carvalho, M.L. (1988) *Artystone minima* n.sp. (Isopoda, Cymothoidae) a body cavity parasite of the pencil fish (*Nannostomus beckfordi* Guenther) from the Brazilian Amazon. *Amazoniana* 10, 255-265.

Thomassen, J.M. (1993) Hydrogen peroxide as a delousing agent for Atlantic salmon. In: Boxshall, G.A. and Defaye, D. (eds) *Pathogens of Wild and Farmed Fish: Sea Lice*. Ellis Horwood, New York, pp. 290-295.

Thoney, D.A. and Burreson, E.M. (1988) Lack of a specific humoral antibody response in *Leiostomus xanthurus* (Pisces: Sciaenidae) to parasitic copepods and monogeneans. *Journal of Parasitology* 74, 191-193.

Thurston, J.P. (1969) The biology of *Lernaea barnimiana* (Crustacea: Copepoda) from Lake George, Uganda. *Revue de Zoologie et de Botanique Africaines* 80, 15-33.

Tidd, W.M. and Shields, R.J. (1963) Tissue damage inflicted by *Lernaea cyprinacea* Linnaeus, a copepod parasitic on tadpoles. *Journal of Parasitology* 49, 693-696.

Tikhomirova, V.A. (1983) Carp lice – intermediate hosts of Skrjabillanidae nematodes. *Vliyanie antropogennykh faktorov na strukturu i funktsionirovanie ekosystem, Kali* 104-109.

Treasurer, J.W. (1993) Management of sea lice (Caligidae) with wrasse (Labridae) on Atlantic salmon (*Salmo salar* L.) farms. In: Boxshall, G.A. and Defaye, D. (eds) *Pathogens of Wild and Farmed Fish: Sea Lice*. Ellis Horwood, New York, pp. 335-345.

Trilles, J.-P. (1964a) A propos d'un fait particulier d'éthologie parasitaire chez les isopodes Cymothoidae: la relation de taille entre parasites et poissons. Note préliminaire. *Vie et Milieu* 15, 365-369.

Trilles, J.-P. (1964b) Variations morphologiques du crâne chez les Téléostéens Sparidae et Centracanthidae en rapport avec l'existence sur ces poissons de certains Cymothoidae parasites. *Annales de Parasitologie* 39, 627-630.

Trilles, J.-P. (1975) Les Cymothoidae (Isopoda, Flabellifera) des côtes francaises. II. Les Anilocridae Schioedte et Meinert, 1881. Genres *Anilocra* Leach, 1818, et *Nerocila* Leach, 1818. *Bulletin du Museum National d'Histoire Naturelle, Paris. Section A.* 200, 303-346.

Tully, O. (1989) The succession of generations and growth of the caligoid copepods

Caligus elongatus and *Lepeophtheirus salmonis* parasitising farmed Atlantic salmon smolts (*Salmo salar* L.). *Journal of the Marine Biological Association, UK* 69, 279–287.

Tully, O. and Whelan, K.F. (1992) The impact of sea lice (*Lepeophtheirus salmonis*) infestation of sea trout (*Salmo trutta* L.) along the west coast of Ireland, 1989–1991. *Pathological Conditions of Wild Salmonids, SOAFD Marine Laboratory* (Abstract).

Tully, O., Poole, W.R., Whelan, K.F. and Merigoux, S. (1993) Parameters and possible causes of epizootics of *Lepeophtheirus salmonis* (Kroyer) infesting sea trout (*Salmo trutta*) off the west coast of Ireand. In: Boxshall, G.A. and Defaye, D. (eds) *Pathogens of Wild and Farmed Fish: Sea Lice*. Ellis Horwood, New York, pp. 202–213.

Turner, W.R. and Roe, R.B. (1967) Occurrence of the parasitic isopod *Olencira praegustator* in the yellowfin menhaden, *Brevoortia smithi*. *Transactions of the American Fisheries Society* 96, 357–359.

Tuuha, H., Valtonen, E.T. and Taskinen, J. (1992) Ergasilid copepods as parasites of perch *Perca fluviatilis* and roach *Rutilus rutilus* in central Finland: seasonality, maturity and environmental influence. *Journal of Zoology* 228, 405–422.

Uehara, J.K., Sholz, A.T., Lang, B.Z. and Anderson, E. (1984) Prevalence of the ectoparasitic copepod *Lernaea cyprinacea* L. on four species of fish in Medical Lake, Spokane County, Washington. *Journal of Parasitology* 70, 183–184.

Ueki, N. and Sugiyama, T. (1979) Mass mortality of cultured juvenile black sea bream *Mylio macrocephalus* in cold water season – 1. Influence of the gill-parasitic copepod *Clavellodes macrotrachelus*. *Bulletin of the Fisheries Experimental Station, Okayama, Japan* 1979, 197–201.

Urawa, S. and Kato, T. (1991) Heavy infections of *Caligus orientalis* (Copepoda: Caligidae) on caged rainbow trout *Oncorhynchus mykiss* in brackish water. *Fish Pathology* 26, 161–162.

Urawa, S., Muroga, K. and Izawa, K. (1979) *Caligus orientalis* Gussev (Copepoda) parasitic on Akame (*Lize akame*). *Fish Pathology* 13, 139–146.

Urawa, S., Muroga, K. and Kasahara, S. (1980a) Naupliar development of *Neoergasilus japonicus* (Copepoda: Ergasilidae). *Bulletin of the Japanese Society of Scientific Fisheries* 46, 941–947.

Urawa, S., Muroga, K. and Kasahara, S. (1980b) Studies on *Neoergasilus japonicus* (Copepoda: Ergasilidae), a parasite of freshwater fishes – II development in copepodid stage. *Journal of the Faculty of Applied Biological Science, Hiroshinma University* 19, 21–38.

Uzmann, J.R. and Rayner, H.J. (1958) Record of the parasitic copepod *Lernaea cyprinacea* L. in Oregon and Washington fishes. *Journal of Parasitology* 44, 452–453.

Valtonen, E.T., Koskivaara, M. and Brummer-Korvenkontio, H. (1987) Parasites of fishes in central Finland in relation to environmental stress. In: *Lake Paeijaenne Symposium, 1987*, Biol. Res. Rep. 10 Univ. Jyvaeskylae, pp. 129–130.

Van As, J.G. and Viljoen, S. (1984) A taxonomic study of sessile peritrichs (Ciliophora: Peritricha) associated with crustacean fish ectoparasites in South Africa. *South African Journal of Zoology* 19, 275–279.

Vaughan, G.E. and. Coble, D.W. (1975) Sublethal effects of three ectoparasites on fish. *Journal of Fish Biology* 7, 283–294.

Viljoen, S. and Van As, J.G. (1985) Cases of aquatic parasitism and hyperparasitism. *South African Journal of Science* 81, 701.

Voth, D.R. (1972) Life history of the caligid copepod *Lepeophtheirus hospitalis* Fraser, 1920 (Crustacea: Caligoida), PhD Thesis, Oregon State University.

Vu-Tân-Tuê, K. (1963) Sur la présence de dents vomériennes et ptérygoidïennes chez *Boops boops* (L.) (Pisces, Sparidae), en rapport avec l'Isopode phorétique intra-buccal *Meinertia*. *Vie et Milieu* 14, 225–232.

Wägele, J.-W. (1987) Description of the postembryonal stages of the Antarctic fish parasite *Gnathia calva* Vanhoffen (Crustacea: Isopoda) and synonymy with *Heterognathia* Amar & Roman. *Polar Biology* 7, 77–92.

Wägele, J.-W. (1988) Aspects of the life-cycle of the Antarctic fish parasite *Gnathia calva* Vanhoffen (Crustacea: Isopoda). *Polar Biology* 8, 287–291.

Wägele, J.-W. (1990) Growth in captivity and aspects of reproductive biology of the Antarctic fish parasite *Aega antarctica* (Crustacea, Isopoda). *Polar Biology* 10, 521–527.

Watanabe, Y., Kosaka, S., Tanino, Y. and Takahashi, S. (1985) Occurrence of parasitic copepod *Pennella* sp. on the Pacific saury *Cololabis saira* in 1983. *Bulletin of the Tohoku Regional Fisheries Research Laboratory* 47, 37–46.

Waugh, D.N., Bennett, T. and Dugoni, T.L. (1989) The incidence of the cymothoid isopod *Lironeca californica* on fishes in Campbell Cove, Sonoma County, California. *Bulletin of the Southern California Academy of Science* 88, 33–39.

Weinstein, M.P. and Heck, K.L. (1977) Biology and host-parasite relationships of *Cymothoa excisa* (Isopoda, Cymothoidae) with three species of snappers (Lutjanidae) on the Caribbean coast of Panama. *Fishery Bulletin* 75, 875–877.

White, H.C. (1940) 'Sea lice' (*Lepeophtheirus*) and death of salmon. *Canadian Journal of Fisheries and Aquatic Sciences* 5, 172–175.

White, H.C. (1942) Life history of *Lepeophtheirus salmonis*. *Journal of the Fisheries Research Board of Canada* 6, 24–29.

Whitfield, P.J., Pilcher, M.W., Grant, H.J. and Riley, J. (1988) Experimental studies on the development of *Lernaeocera branchialis* (Copepoda: Pennellidae): population processes from egg production to maturation on the flatfish host. *Hydrobiologia* 167/168, 579–586.

Wijeyaratne, M.J.S. and Gunawardene, R.S. (1988) Chemotherapy of ectoparasite, *Ergasilus ceylonensis* of Asian cichlid, *Etroplus suratensis*. *Journal of Applied Ichthyology* 4, 97–100.

Williams, E.H. and Williams, L.B. (1982) *Mothocya bohlkeorum*, new species (Isopoda: Cymothoidae) from West Indian cardinalfishes (Apogonidae). *Journal of Crustacean Biology* 2, 570–577.

Williams, E.H. and Williams, L.B. (1985a) *Cuna insularis*, n.gen. and n.sp. (Isopoda: Cymothoidae) from the gill chamber of the sergeant major, *Abudefduf saxatilis* (Linnaeus), (Osteichthyes) in the West Indies. *Journal of Parasitology* 71, 209–214.

Williams, E.H. and Williams, L.B. (1986) The first *Anilocra* and *Pleopodias* isopods (Crustacea: Cymothoidae) parasitic on Japanese fishes, with three new species. *Proceedings of the Biological Society of Washington* 99, 647–657.

Williams, E.H. and Williams, L.B. (1992) *Renocila loriae* and *R. richardsonae* (Crustacea: Isopoda: Cymothoidae), external parasites of coral reef fishes from New Guinea and the Philippines. *Proceedings of the Biological Society of Washington* 105, 299–309.

Williams, E.H., Williams, L.B., Waldner, R.E. and Kimmel, J.J. (1982) Predisposition of a pomacentrid fish, *Chromis multilineatus* (Guichenot) to parasitism by a cymothoid isopod, *Anilocra chromis* Williams and Williams. *Journal of Parasitology* 68, 942–945.

Williams, I.C. (1963) The infestation of the redfish *Sebastes marinus* (L.) and *S. mentella* Travin (Scleroparei: Scorpaenidae) by the copepods *Peniculus clavatus*

Muller), *Sphyrion lumpi* (Kroyer) and *Chondracanthopsis nodosus* (Muller) in the eastern North Atlantic. *Parasitology* 53, 501–525.

Williams, L.B. and Williams, E.H. (1979) The ability of various West Indian cleaners to remove parasitic isopod juveniles of the genus *Anilocra* – a preliminary report. *Proceedings of the Association of Island Marine Laboratories of the Caribbean* 14, 28.

Williams, L.B. and Williams, E.H. (1981) Nine new species of *Anilocra* (Crustacea: Isopoda: Cymothoidae), external parasites of West Indian coral reef fishes. *Proceedings of the Biological Society of Washington* 94, 1005–1047.

Williams, L.B. and Williams, E.H. (1985b) Brood pouch release of *Anilocra chromis* Williams and Williams (Isopoda, Cymothoidae) a parasite of brown chromis, *Chromis multilineatus* (Guichenot) in the Caribbean. *Crustaceana* 49, 92–95.

Wilson, C.B. (1902) North American parasitic copepods of the family Argulidae, with a bibliography of the group and a systematic review of all known species. *Proceedings of the US National Museum* 25, 635–742.

Wilson, C.B. (1905a) North American parasitic copepods belonging to the family Caligidae. Pt. 1. The Caliginae. *Proceedings of the US National Museum* 28, 479–672.

Wilson, C.B. (1905b) Habits and life-history of parasitic copepods. *Biological Bulletin of the Marine Biological Laboratory* 8, 236–237.

Wilson, C.B. (1911) North American parasitic copepods. Part 9. The Learnaeopodidae. *Proceedings of the US National Museum* 39, 189–226.

Wilson, C.B. (1915) North American parasitic copepods belonging to the family Lernaeopodidae, with a revision of the entire family. *Proceedings of the US National Museum* 47, 565–729.

Wilson, C.B. (1917) North American parasitic copepods belonging to the Lernaeidae with a revision of the entire family. *Proceedings of the US National Museum* 53, 1–150.

Wilson, C.B. (1919) North Americal parasitic copepods belonging to the new family Sphyriidae. *Proceedings of the US National Museum* 55, 549–604.

Woo, P.T.K. and Shariff, M. (1990) *Lernaea cyprinacea* L. (Copepoda: Caligidae) in *Helostoma temmincki* Cuvier and Valenciennes: the dynamics of resistance in recovered and naive fish. *Journal of Fish Diseases* 13, 485–494.

Wootten, R., Smith, J.W. and Needham, E.A. (1977) Studies on the salmon louse, *Lepeophtheirus*. *Bulletin de l'Office International des Epizooties* 87, 521–522.

Wootten, R., Smith, J.W. and Needham, E.A. (1982) Aspects of the biology of the parasitic copepods *Lepeophtheirus salmonis* and *Caligus elongatus* on farmed salmonids, and their treatment. *Proceedings of the Royal Society of Edinburgh* 81B, 185–197.

Yamaguti, S. (1963) *Parasitic Copepoda and Branchiura of Fishes*. Interscience, New York.

Zeddam, J.L., Berrebi, P., Renaud, F., Raibaut, A. and Gabrion, C. (1988) Characterization of two species of *Lepeophtheirus* (Copepoda, Caligidae) from flatfishes. Description of *Lepeophtheirus europaensis* sp.nov. *Parasitology* 96, 129–144.

Zmerzlaya, E.I. (1972) *Ergasilus sieboldi* Nordmann, 1832, its development, biology and epizootic significance. *Izvestiya Gosudarstvennogo Nauchno-issledovateleskogo Instituta Ozernogo i Rechnogo Rybonog Zhozuaistva* 80, 132–177.

14

Phylum Annelida: Hirudinea as Vectors and Disease Agents*

E.M. Burreson

School of Marine Science, Virginia Institute of Marine Science, College of William and Mary, Gloucester Point, Virginia 23062, USA.

INTRODUCTION

Leeches are clitellate annelids and are the only important fish pathogens in the phylum. Aquatic leeches, both freshwater and marine, occur in a diversity of habitats worldwide. They are most abundant in temperate lakes, ponds and streams and in polar to temperate seas. Leeches can potentially affect the health of fishes in a variety of ways, but most involve blood-feeding activities. Their most important role is as vectors of potentially pathogenic organisms. Both freshwater and marine leeches are known to transmit haemoflagellates of the genera *Trypanosoma* and *Cryptobia (= Trypanoplasma)* and the intracellular haemogregarines and piroplasmas to fish. There is accumulating evidence that leeches can also transmit viruses and bacteria. In addition, leeches may affect the host by the sheer amount of blood withdrawn during feeding. The feeding or attachment wounds caused by leeches may also serve as sites for secondary pathogenic invaders.

Aquatic leeches also serve as second intermediate hosts for Digenea, harbouring metacercariae of a number of different species. Most of these worms are adults in waterfowl, but some mature in freshwater fishes. However, none of these digeneans has been implicated in pathology of fishes so they are not considered further in this review. Additional information on the role of leeches as intermediate hosts is in Sawyer (1986) or papers by McCarthy (1990), Spelling and Young (1986a,b,c) and Vojtek *et al.* (1967).

Extensive reviews on the general biology of leeches have been published (Mann, 1961; Sawyer, 1986). Pathologic implications of marine leeches have been reviewed by Rohde (1984) and the role of leeches in pathology of fish cultured in the tropics has been reviewed by Kabata (1985).

---

*Virginia Institute of Marine Science contribution number 1931.

TAXONOMY AND GENERAL BIOLOGY

Leeches are generally classified in the class Hirudinea within the phylum Annelida, subphylum Clitellata. The class is separated into two subclasses, Acanthobdellidea and Euhirudinea. Members of the subclass Acanthobdellidea possess setae in cephalic segments and are thought to be the most primitive leeches. The subclass was generally considered to contain a single family, Acanthobdellidae, and a single genus, *Acanthobdella*, with two species, *A. peledina* Grube and *A. livanowi* Epshtein, both parasitic on freshwater fishes. Recently Epshtein (1987) proposed the family Paracanthobdellidae and the genus *Paracanthobdella* for *A. livanowi*. The subclass Euhirudinea contains two groups – those that possess a protrusible proboscis for feeding (order Rhynchobdellida) and those that possess a pharynx, with or without jaws (order Arhynchobdellida) (Sawyer, 1986). Leeches that are important in fish pathology are in the order Rhynchobdellida, families Glossiphoniidae and Piscicolidae. Body shapes and generalized internal anatomy of representative members of these two families are in Fig. 14.1.

Identification keys to freshwater leeches of the world and to marine leeches of the North Atlantic are in Sawyer (1986). Earlier keys to European and North American freshwater leeches and to the marine leech genera of the world are in Mann (1961). A key to genera of the family Piscicolidae can be found in Soós (1965). Keys to North American freshwater leeches are also in Klemm (1982) and Davies (1991), and keys to freshwater leeches of the USSR are in Lukin (1976) and Epshtein (1987, Piscicolidae only). A key to marine leeches of the Indian Ocean can be found in Sanjeeva Raj (1974).

Leeches that feed on the blood of fishes may be either temporary or semi-permanent parasites. Temporary parasites usually leave the host soon after a blood meal although it is not uncommon for them to remain on the host for a few more days. They usually seek a sheltered location among vegetation or under stones to digest the blood meal. However, some estuarine and marine leeches that inhabit soft substrate areas utilize crustaceans for attachment after leaving the fish host. Semi-permanent parasites remain on the fish and take successive blood meals; they leave the host only to deposit cocoons. Leeches are hermaphroditic and mating may involve copulation through the female gonopore or hypodermic implantation of spermatophores anywhere on the body. Mating may occur on or off the fish host, but cocoons are never deposited on the fish. The cocoons of piscicolid leeches, which are usually adhered to hard substrates such as vegetation, rocks, shells or even the carapace of live crustaceans, typically contain a single egg. However, some species of *Malmiana* may deposit up to five eggs in each cocoon. Cocoons are left unattended and newly hatched piscicolid leeches must find a fish host on their own; young leeches can usually survive for a week or more before their first blood meal. Hatching time is temperature dependent, but some species may aestivate in the cocoon until appropriate environmental conditions reoccur. Unlike the piscicolids, glossiphoniid leeches brood their eggs in fragile cocoons which, depending on the species, may be attached either to the substrate or to the ventral body surface of the leech. In either case, young leeches attach to the

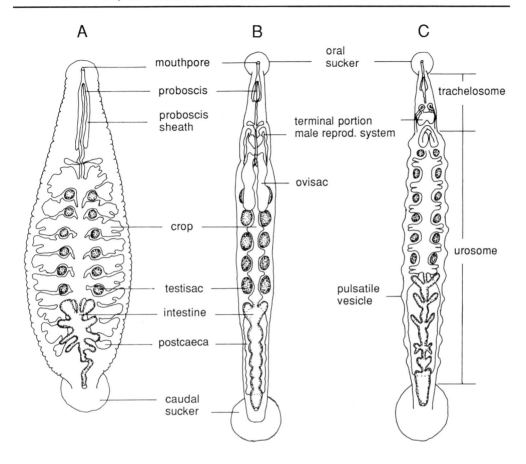

Fig. 14.1. Semi-diagrammatic representation of some rhynchobdellid leeches illustrating the main morphological features. A, family Glossiphoniidae, B, family Piscicolidae, subfamily Platybdellinae. C, family Piscicolidae, subfamily Piscicolinae.

parental ventral surface and are carried to a host for their first blood meal. Leeches usually die after cocoon deposition or brooding, but some species may produce successive broods.

Leeches can be difficult to collect and to identify. They often leave the host after feeding and, therefore, may go undetected even when abundance is high. They are usually sufficiently large to be detected by the naked eye and occur on the body surface and fins or in the gill cavity or mouth. If possible, leeches should be collected by gently dislodging the caudal sucker with forceps and placing them in a dish containing water. For proper identification it is important to observe leeches alive and to note as many external characters as possible. It may be necessary to relax leeches in weak alcohol or other narcotizing agent prior to examination. Careful observation should be made of pigmentation colour and pattern, number and arrangement of eyes on the oral sucker, ocelli on the body and caudal sucker, number of lateral pulsatile vesicles if present, and arrangements of papillae, tubercles or other obvious external characters. Leeches that have been fixed unrelaxed are almost impossible to identify because they usually contract strongly or curl into a tight ball making observation of important characters difficult. Leeches relaxed prior to fixation in formalin will usually retain their pigmentation and eyes for long periods; however, pigmentation fades rapidly after transfer to alcohol. Leeches that have been fixed and then preserved in alcohol are often difficult to identify, especially to species. Generic determination of many leeches, especially piscicolids, may depend on internal anatomy that can only be determined with serial sections. All these difficulties combine to make identification of leeches, even for experts, problematic.

LEECHES AS PATHOGENS

Leeches alone are generally not considered important fish pathogens. Effects are usually localized and restricted to attachment and/or feeding sites. The muscular caudal sucker used for attachment usually causes little damage; however, leeches that are semi-permanent parasites may elicit a substantial host tissue response at the attachment site. Rhynchobdellid leeches feed on host blood or tissue fluid by means of a protrusible proboscis that is inserted into host tissue and this feeding activity may produce localized petechial haemorrhage (Sloan *et al.*, 1984; Jones and Woo, 1990b). Even though pathology is usually localized, heavy infestations can result in severe epidermal erosion and even mortality because of large amounts of blood loss or secondary effects of multiple feeding wounds. Thus, leeches in aquaculture facilities should always be cause for concern. Fish can tolerate a high burden of leeches with little apparent effect, but pathology may depend on the relative size of the leech compared to the fish. Kabata (1985), without providing details, reported that the presence of 100 leeches on a single fish in Africa resulted in no serious harm to the host. Other reports, however, suggest that leeches may be pathogenic or provide portals of entry through their feeding or attachment wounds for secondary pathogens. For example, general statements by Bauer (1961a) Bauer *et al.* (1973) and Markevich (1963) suggest that *Piscicola geometra* can severely affect carp and other species in rearing ponds in eastern Europe by causing severe emaciation, small bleeding ulcers and secondary invasion by bacteria and fungi.

One of the earliest accounts of a leech causing mortality is the report by Badham (1916) involving the leech *Austrobdella translucens* and the sand whiting, *Sillago ciliata*, in Australia. A few dozen whiting were periodically stocked with other species in a salt water pond over a number of years and on each occasion the whiting were killed by the leeches. The fish developed large ulcerated patches on the skin and badly infested fish harboured as many as 100 leeches on the fins and body surface. *Austrobdella translucens* was specific for sand whiting and other species of fish in the pond apparently survived well, suggesting that mortality of whiting was not the result of unfavourable water quality.

There have been a number of reports of leeches causing pathology in feral freshwater and estuarine fishes in North America. An epizootic of *Piscicola punctata* caused pathology in the bigmouth buffalo, *Ictiobus cyprinellus*, in the Rock River near Rockford, Illinois, USA (Thompson, 1927). During February and March of 1926, almost every fish was heavily infested with the leech; intensity ranged up to 50 leeches per fish taken from the river channel, and over 100 leeches per fish collected from backwaters and sloughs. It is interesting that the leech had not been observed during the previous two winters even though sampling had been intensive, and fishermen reported no similar infestation in the last 50 years. Leeches were removed by fishermen prior to delivery to the retail market, but the raw and bleeding scars around the attachment sites made the fish difficult to sell. There is no mention of host mortality in the report. Leeches left the fish by the end of March when water temperature increased above 0°C. In late April fungal infections were observed on wounds apparently made by leeches. This leech outbreak was unusual, but there were no obvious, unusual environmental conditions known to be present that would have caused such an occurrence. Undoubtedly, a number of factors contributed to the greater than normal reproductive success and hence the high abundance.

Another unusual occurrence of fish mortality caused by leeches involved adult brook trout, *Salvelinus fontinalis*, and freshwater leeches in a small, shallow lake in Maine (Rupp and Meyer, 1954). During periods of hot weather, water temperature in the lake can become critically high and trout congregate around cooler underwater springs in shallow water. The congregating trout encouraged poaching and bird predation, so brush was placed in the spring to provide cover for the fish. Unfortunately, the brush provided an ideal habitat for two hirudinid leeches, *Macrobdella decora* and *Haemopis grandis*. These leeches are not normally associated with fish, but on one occasion 50 to 60 large trout were observed 'being fiercely attacked by hordes' of leeches from the brush. Despite continual harassment by the leeches, the trout remained congregated in the spring. On one occasion over 20 dead trout were reported during a 3-day period. One captured, dying trout had six leeches attached to the gill arches, isthmus and fin bases; one *M. decora* had rasped an opening through the body wall and into the ventral aorta. It appeared certain to the investigators that the fish was dying from loss of blood. This unusual occurrence was possible only because of high concentrations of fish and leeches in close proximity for prolonged periods.

Freshwater leeches have also caused mortality in fish hatcheries. A heavy infestation of *Piscicola salmositica* caused mortality as high as 25% in sac fry of pink salmon, *Oncorhynchus gorbuscha*, in the state of Washington (Earp and Schwab, 1954). Leeches invaded the hatchery through the freshwater supply from a nearby stream and attacked fry as they hatched in trays. Once attached to the fry, leeches quickly became gorged with blood and the fry invariably died, apparently from blood loss. Adult salmon migrants in the supply stream had heavy infestations of *P. salmositica* and indications were that some fish died before spawning.

Leeches that are semi-permanent parasites on fishes, such as *Acanthobdella peledina* or the freshwater and estuarine leech *Myzobdella lugubris*, tend to remain attached at a single site. They often cause very localized histopathological changes including cellular infiltration, erosion of the integument under the attachment site and hyperplasia of the epidermis around the caudal sucker. Localized subcutaneous haemorrhages often occur at leech feeding sites. Paperna and Zwerner (1974) reported removing over 500 *M. lugubris* from a single moribund white catfish, *Ictalurus catus*, in the York River estuary in Virginia. Leeches were in the mouth and under the operculum and externally on the skin fold behind the lower jaw and at the bases of the fins (Fig. 14.2). Extensive histopathological changes caused by the leech included inflammation, displacement and erosion of the dermis and hyperplasia of the epithelium. All pathological changes were attributed to leeches and it was concluded that leeches were at least a major contributing factor to the distressed condition of the fish. The same leech was recently implicated in an epidemic of oral ulcerations in adult largemouth bass, *Micropterus salmoides*, from Currituck Sound, North Carolina (Noga *et al.*, 1990). A systematic survey was not conducted, but sport fishermen reports suggested that 90% of legal size fish might have been affected. These fish had large ulcerations on the tongue and buccal cavity often extending to underlying musculature; leeches were always present in or near the wounds (Fig. 14.3). Localized pathology from *M. lugubris* also has been reported in logperch, *Percina caprodes*, and brown bullhead, *Ictalurus nebulosus*, in Ontario, Canada (Appy and Cone, 1982) and in mullet, *Mugil* sp. (Paperna and Overstreet, 1981). Heavy infestations of *M. lugubris* were reported in intensive striped bass (*Morone saxatilis*) culture facilities on the Chesapeake Bay, Maryland, USA, but there was no mention of pathology (Woods *et al.*, 1990).

Similar pathological effects caused by *Austrobdella bilobata* have been reported in yellowfin bream, *Acanthopagrus australis*, in New South Wales, Australia (Roubal, 1986) and in flounder *Rhombosolea tapirina*, in Tasmania (Sanjeeva Raj, 1974). Khan (1982) described subcutaneous lesions caused by *Johanssonia arctica* in Atlantic cod, *Gadus morhua*.

Damage from semi-permanent parasites can be extensive in heavily infested fish. *Acanthobdella peledina* usually attaches to the base of the dorsal fin in salmonids and feeds on fin tissue as well as blood (Andersson, 1988). Deep attachment wounds typically occur and the fin is often eroded; in heavy infestations leeches can cause complete destruction of the fin (Bauer, 1961b), but there is no evidence of host mortality.

It is clear that leeches alone have only rarely been implicated as serious pathogens of fishes. Their effects are usually localized to the feeding or attachment sites and only become serious when infestation is high. Documented fish mortality caused by leeches is extremely rare and appears to occur only when there is close, prolonged contact between large numbers of leeches and fish. Although a single leech may withdraw only a small amount of blood, it nevertheless weakens the host (Meyer, 1946a). The subtle effect of blood loss was elegantly demonstrated by Mace and Davis (1972) who studied the energetics of parasitism by the leech *Malmiana brunnea* on the shorthorn sculpin, *Myoxocephalus scorpius*, in Newfoundland. Energy budgets were developed

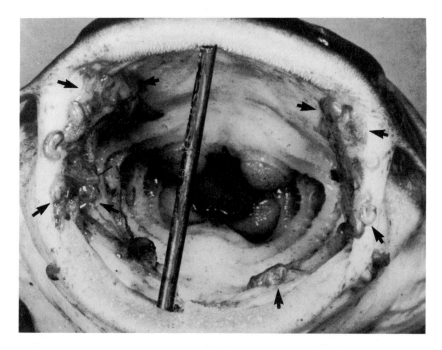

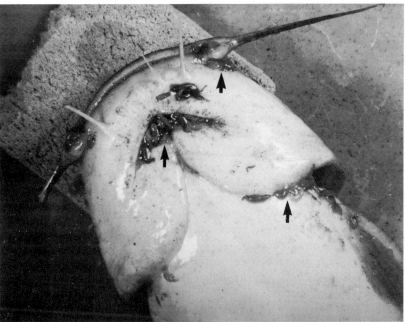

Fig. 14.2 Heavy infestation of *Myzobdella lugubris* on a white catfish, *Ictalurus catus*. Arrows identify concentrations of leeches. From Paperna and Zwerner (1974); courtesy of the *Proceedings of the Helminthological Society of Washington*.

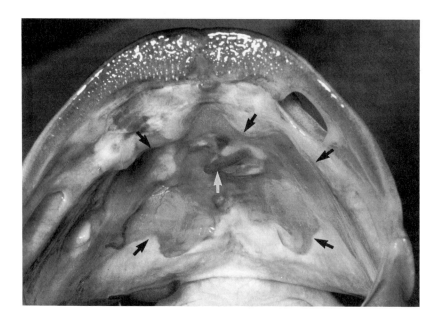

Fig. 14.3 Large ulceration of the hard palate of a largemouth bass, *Micropterus salmoides*, infected with *Myzobdella lugubris*. Black arrows delineate extent of the lesion; white arrow identifies a single leech. From Noga *et al.* (1990); courtesy of the *Journal of Wildlife Diseases*.

for leeches and fish, and growth was measured in two groups of unparasitized fish for five weeks. Growth of a parasitized group was significantly lower than the expected growth and the difference was attributed to the energy require-ments of the leeches. It was concluded that the additional energy consumption of the host because of the leeches was approximately $750\,\mathrm{cal}\,\mathrm{g}^{-1}$ of leech per week. According to Mace and Davis (1972), a well adapted host–parasite relationship should exert a metabolic demand that is little more than the energy requirements of the parasite. However, tissue damage, harmful meta-bolites and/or hormone leakage cause increased energy loss. They concluded that energy loss in sculpins was not entirely because of energy requirements of the leech, but probably also because of increased physiological strain from leech saliva secretions or undetected mechanical irritation of feeding leeches.

Subtle effects of leech infestation have also been demonstrated for *Acanthobdella peledina* on grayling, *Thymallus thymallus* L., in Sweden (Andersson, 1988). Mean weight of infested fish was reduced 5 to 6% com-pared to uninfested hosts when leech biomass exceeded 0.1% of fish weight and it was reduced 9 to 20% when leech biomass exceeded 1.0% of fish weight. In addition, haemoglobin values were reduced in heavily infested fish and gills were often pale.

LEECHES AS VECTORS OF PATHOGENS

Leeches are much more important as vectors of fish pathogens than they are as pathogens themselves. Haematozoa of fishes are all transmitted by rhynchobdellid leeches and, although less well documented, leeches may transmit viruses and bacteria as well. As vectors for other organisms, leeches must feed on at least two different host individuals; they must acquire the organisms by feeding on an infected host and they must then leave that host and attach to and feed on a different host individual. Thus, leeches that are semi-permanent parasites tend to be less efficient vectors. A good example is *Myzobdella lugubris*, which is common in fresh water in the United States and in southeastern estuaries. This leech occurs on a wide variety of hosts, many of which harbour haematozoa (unpublished data), but it has never been implicated as a vector. Another example is *Oceanobdella pallida*, which occurs in the mouth of English sole, *Parophrys vetulus*, in the northeastern Pacific. English sole are commonly infected with *Trypanosoma pacifica* off Oregon (Burreson and Pratt, 1972) and developmental stages have been observed in the crop of the leech, but never in the proboscis sheath. All attempts to transmit *T. pacifica* to uninfected sole with *O. pallida* failed because leeches removed from infected hosts could never be induced to reattach to other hosts (Burreson, 1975). The leech appears to be a semi-permanent parasite and is probably not the vector for *T. pacifica*. These results illustrate that the presence of developmental stages of haematozoa in a leech does not necessarily indicate that the leech is a vector.

Vectors for fish pathogens are restricted to the rhynchobdellid families Glossiphoniidae and Piscicolidae; however, within each family, behaviour of individual species and not taxonomic grouping determines its role as a vector. Thus, vectors are distributed throughout most subfamilies. Nevertheless, there appears to be a general trend of semi-permanent parasitism within the Piscicolid subfamily Platybdellinae and a general trend of leaving the host after the blood meal in the subfamily Piscicolinae and possibly also the subfamily Pontobdellinae. Thus, one expects more vectors in the subfamily Piscicolinae than in the Platybdellinae, and this seems to be true although it must be noted that vectors have been identified for only a small number of fish haematozoa. Of the 21 genera in the subfamily Platybdellinae (Sawyer, 1986), only one genus, *Malmiana*, is known to have species that are vectors (Burreson, 1982; Siddall and Desser, 1993). However, Sloan *et al.* (1984) reported that 71.0% of the *Notostomum cyclostomum* (Johansson) (subfamily Platybdellinae) examined in northern British Columbia, Canada harboured heavy infections of *Cryptobia* in the proboscis sheath and one leech had trypanosomes in the proboscis sheath. This leech is undoubtedly a vector, but the fish host of the flagellates is unknown. Of the 15 genera in the subfamily Piscicolinae, five genera, *Calliobdella, Johanssonia, Cystobranchus, Piscicola* and *Orientobdella* have been documented as vectors and *Calliobdella* and *Piscicola* have more than one species known to be vectors. Of the four genera in the subfamily Pontobdellinae (Sawyer, 1986) only one genus and species, *Pontobdella muricata*, has been shown to be a vector.

Leeches as vectors of viruses

There have only been two reports implicating leeches as vectors of fish viruses. Ahne (1985) demonstrated that *Piscicola geometra* mechanically transmitted spring viraemia of carp virus (SVCV, *Rhabdovirus carpio*) to carp, *Cyprinus carpio*. Leeches acquired the virus from infected carp during the first feeding and transmitted the virus to uninfected carp during two successive feedings. However, leeches eventually lost the virus and carp used for the fourth feeding did not become infected. Ahne concluded that SVC virus did not replicate in *P. geometra* and that the leech served only as a mechanical vector.

Mulcahy *et al.* (1990), on the other hand, present evidence that infectious hematopoietic necrosis (IHN) virus may replicate in the leech *Piscicola salmositica*. They isolated IHN virus from *P. salmositica* collected on sockeye salmon, *Oncorhynchus nerka*, and from the streambed of the Cedar River, Washington. Leeches from salmon were isolated in laboratory containers and over a 16-day-period the virus titre increased from 4.0×10^1 plaque forming units per gram to $6.5 \times 10^3 \, \mathrm{pfu \, g^{-1}}$. This increase suggests that the virus may be replicating in the leeches; however viral replication was not demonstrated with certainty. Prevalence of IHN virus in leeches from the streambed decreased from 57.0% to 20.0% over 72 days suggesting that the leeches were gradually losing the virus. Although no experiments were conducted using infected leeches to transmit the virus to uninfected fish, the authors concluded that the leeches probably increase the infection rate of the virus in spawning sockeye salmon. The virus is apparently not transmitted vertically in the leeches as the virus was not isolated from leech cocoons.

Leeches as vectors of bacteria

Although there has been speculation that leeches are capable of transmitting pathogenic bacteria to fishes, there have been few cases of transmission actually documented (Cusack and Cone, 1986). Negele (1975) reported that *Piscicola geometra* transmitted *Aeromonas hydrophila* to fish during the feeding process, but gave no details. Bragg *et al.* (1989) isolated a bacterium from the leech *Batrachobdelloides tricarinata* that was biochemically and serologically identical to *Streptococcus* sp. pathogenic to rainbow trout. The authors proposed that the leech was a possible reservoir for the bacterium, but no transmission experiments were conducted. Dombrowski (1953) reported that *P. geometra* transmitted the pathogenic *Pseudomonas punctata* to carp.

Leeches, including the important fish parasites *Piscicola geometra* and *Hemiclepsis marginata*, harbour endosymbiotic bacteria in specialized anterior crop diverticula called mycetomes or oesophageal diverticula (Jennings and Van Der Lande, 1967; Sawyer, 1986). These bacteria aid in digestion of blood by providing normally deficient digestive enzymes; however, they may be mistaken for bacteria that are pathogenic to fish.

Leeches as vectors of haematozoic protozoa

According to Laveran and Mesnil (1912), Leydig first observed flagellates in the leeches *Piscicola* and *Pontobdella* in 1857 and Doflein in 1901 first suggested that leeches were vectors of fish haemoflagellates. This speculation was confirmed by a number of separate studies in the early 1900s (Lom, 1979). The leeches that have been ascertained to be vectors of fish haematozoa through experimental transmission studies are listed in Table 14.1. Although a large number of haemoflagellates have been described from fishes (Lom, 1979; Woo, 1987) very few vectors have been identified. Even less is known about leeches as vectors of intraerythrocytic parasites of fish. Although leeches have long been suspected as vectors of these parasites in fishes, the first life cycles were not confirmed until the early 1980s (Khan, 1980; Lainson, 1981). Much more research remains to be done, but it is apparent, especially from the work by Khan in Newfoundland, that a very few species of leeches in a given area transmit many different haematozoa.

Of the leech vectors listed in Table 14.1, five of them are especially important, either because they have a wide geographic and host range or because they parasitize commercially important hosts. These are *Hemiclepsis marginata*, *Piscicola geometra*, *Johanssonia arctica*, *Calliobdella vivida*, and *Piscicola salmositica*; each will be discussed in detail in the following sections.

Hemiclepsis marginata (O.F. Müller)

Hemiclepsis marginata transmits *Cryptobia borreli*, a known pathogen in European freshwater fish (Dyková and Lom, 1979). The life cycle of *C. borreli* and vector role of *H. marginata* were elucidated by Robertson (1911). This leech also transmits *Cryptobia ompoki* and *Trypanosoma punctati* in India (Shanavas *et al.*, 1989; Shanavas, 1991) and at least three trypanosomes in Europe, *Trypanosoma cobitis* in the stone loach, *Nemacheilus barbatulus*, and other hosts (Letch, 1979, 1980), *T. tincae* in the tench, *Tinca tinca* (see Needham, 1969) and *T. granulosum* in eels *Anguilla anguilla* (see Brumpt, 1906a,b). In addition, developmental stages of *T. danilewskyi* were found in the proboscis sheath of *H. marginata*; however, experimental transmission to the fish was not attempted (Qadri, 1962).

Hemiclepsis marginata is widely distributed from the United Kingdom to most of Europe (Harding, 1910; Wilkialis, 1970; Elliott and Mann, 1979) and Asia (Shulman, 1961; Khalifa, 1985; Sawyer, 1986; Singhal *et al.*, 1986; Shanavas *et al.*, 1989). It is abundant in small, hard-water ponds or slow moving streams that contain abundant vegetation, especially large sheath-bearing plants; it does not occur in rapidly flowing water (Mann, 1955; Sawyer, 1986). Although apparently widespread, both Robertson (1911) and Needham (1969) had difficulty collecting *H. marginata*. From late spring to early autumn *H. marginata* is primarily in the leaf bases of large marginal reeds; it is rarely found on fish. During winter *H. marginata* is among the plant rhizomes (Needham, 1969).

Long known as a fish parasite, *H. marginata* has been shown to feed on a wide variety of freshwater fishes under laboratory conditions (Robertson,

Table 14.1. Leech vectors of fish haematozoa.

Leech species	Distribution and hosts	Haematozoa transmitted
Freshwater **Glossiphoniidae** *Hemiclepsis marginata* (O.F. Müller)	Western Europe to northeast Asia, India (Sawyer, 1986). Not host specific (Robertson, 1911; Letch and Ball, 1979)	*Trypanosoma punctati* (Shanavas, 1991); *T. cobitis, T. granulosum* and probably other trypanosomes (Brumpt, 1906a,b; Needham, 1969; Letch, 1980); *Cryptobia borreli* (Robertson, 1911), *C. ompoki* (Shanavas *et al.*, 1989)
Haementeria 'lutzi' Pinto (According to Sawyer, 1986, this leech is *species inquirendae*)	Brazil on *Synbranchus marmoratus* (Lainson, 1981)	*Trypanosoma bourouli, Cyrilia gomesi* (Lainson, 1981)
Desserobdella phalera (Graf)	United States and Canada (Klemm, 1982) on bowfin, *Amia calva,* and largemouth bass, *Micropterus salmoides,* and other fishes (Jones and Woo, 1990a)	*Trypanosoma phaleri* (Jones and Woo, 1990b)
Piscicolidae *Piscicola geometra* L.	Eurasia, North America (?). Not host specific (Sawyer, 1986; Epshtein, 1987; Madill, 1988)	*Cryptobia borreli, Trypanosoma danilewskyi* and possibly other trypanosomes (Brumpt, 1906b; Khaibulaev and Guseinov, 1982; Kruse *et al.*, 1989)
Piscicola salmositica Meyer	Northwestern North America, Primarily on salmonids (*Oncorhynchus* spp. and sculpins (*Cottus* spp.) (Becker and Katz, 1965b)	*Cryptobia salmositica* (Becker and Katz, 1965a)
Cystobranchus virginicus Hoffman	Eastern North America (range undetermined) primarily on cyprinids (Hoffman, 1964; Putz, 1972a)	*Cryptobia cataractae* (Putz, 1972a,b)

Marine
Piscicolidae

Pontobdella muricata L.	Northeast Atlantic Ocean, Mediterranean Sea, on skates (*Raja* spp.) (Sawyer, 1986)	*Trypanosoma rajae* (Brumpt, 1906a; Robertson, 1907, 1909; Neumann, 1909)
Johanssonia arctica (Johansson)	Arctic seas, North Atlantic, northeast Pacific to California. Not host specific (Khan, 1982; Meyer and Khan, 1979; Madill, 1988; Burreson, unpublished data)	*Trypanosoma murmanensis, Haemohormidium terraenovae, H. beckeri, Haemogregarina uncinata* (Khan, 1976, 1978a, 1980, 1984)
Calliobdella punctata Beneden and Hesse	Northeast Atlantic Ocean. Distribution and hosts (*Enophrys bubalis, Blennius pholis*) poorly known (Sawyer, 1986)	*Trypanosoma cotti* (Brumpt, 1906a; Khan, 1978b)
Calliobdella vivida (Verrill)	East and Gulf coasts of North America, primarily in estuaries. Not host specific (Sawyer *et al.*, 1975; Appy and Dadswell, 1981; Burreson and Zwerner, 1982; Madill, 1988)	*Cryptobia bullocki* (Burreson, 1982)
Malmiana diminuta Burreson	Northeast Pacific. Not host specific (Burreson, 1977)	*Cryptobia* (?) *beckeri* in cabezon, *Scorpaenichthys marmoratus* (Burreson, 1979)
Malmiana scorpii (Malm)	North Atlantic on *Myoxocephalus octodecemspinosus* and *M. scorpius* (Khan and Meyer, 1976)	*Haemogregarina myoxocephali* (Siddall and Desser, 1993)

1911). Letch and Ball (1979) suggested that, because of habitat similarities and the inactive nature of the fish, *H. marginata* feeds primarily on the stone loach, *Nemacheilus barbatulus*, bullhead, *Cottus gobio*, and gudgeon, *Gobio gobio*, under natural conditions in England. Fish hosts in culture ponds in India are listed in Singhal *et al.* (1986). After feeding, the leech leaves the host and seeks a concealed area among plant leaves to digest its blood meal.

Hemiclepsis marginata is in the family Glossiphoniidae, subfamily Glossiphoniinae. Mature adults are up to 30 mm long by 7 mm wide; colour is green to pale yellow with seven longitudinal rows of yellow spots (Mann, 1961). There are usually two pairs of eyes on the head region, but the anterior pair may be coalesced. This species can easily be separated from other glossiphoniids by the expanded head and oral sucker region (Fig. 14.4) which, at rest, is wider than the body segments immediately posterior to it (Mann, 1961; Sawyer, 1986). A key to the currently recognized European glossiphoniids is in Sawyer (1986).

An annual life cycle for *H. marginata* has been proposed by Needham (1969). Eggs are laid in late spring and early summer when water temperature rises to 15°C. Cocoons are deposited on a substrate, usually the protective shelter of leaf bases of reeds, and are brooded by the adult. When eggs hatch, young leeches attach to the venter of the adult and are eventually carried to a fish host. On the host the young leeches leave the adult and take their first blood meal. The adult leech soon dies, and the young leave the fish host and return to the shelter of the reeds to digest their blood meal. A period of subsequent feeding and growth occurs through summer and early autumn. During winter, feeding is reduced or may cease, but large leeches can survive 10 months without feeding (Sawyer, 1986). As the water temperature increases in early spring, feeding activity increases prior to the breeding cycle. *Hemiclepsis marginata* is apparently relatively easy to maintain in the laboratory if vegetation is provided for cover.

Piscicola geometra L.

On the basis of flagellate developmental stages in the crop, *P. geometra* has long been implicated as a vector for fish trypanosomes and cryptobiids (Léger, 1904; Keysselitz, 1906; Lom, 1979). However, none of the early researchers clearly documented transmission of either trypanosomes or cryptobiids to uninfected fishes through the feeding activity of *P. geometra*. For example, Brumpt (1905) discussed development of cryptobiids in *P. geometra* and *Hemiclepsis marginata* and stated that successful transmission was achieved using leeches, but he did not specifically state that transmission was successful using *P. geometra*. Keysselitz (1906) did not observe developmental stages of cryptobiids in the proboscis sheath of *P. geometra* and was unable to actually transmit the parasite to uninfected fish via leeches because of a lack of uninfected hosts. Brumpt (1906a) stated that *Trypanosoma granulosum* developed in *P. geometra* for about 15 days, but all flagellates eventually died. Lom (1979) reported that *T. danilewskyi* developed in *P. geometra*, but made no mention of successful transmission. Khaibulaev (1970) reported development of both trypanosomes and cryptobiids in *P. geometra*, but his paper is

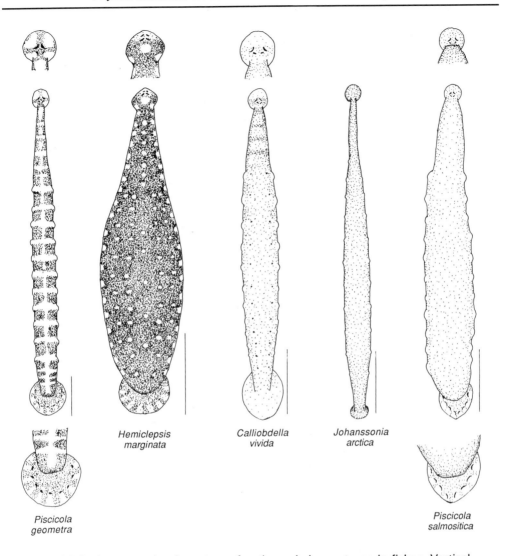

Fig. 14.4 Important leech vectors of pathogenic haematozoa in fishes. Vertical line beside each leech is 5.0 mm. See text for identification characters. *Hemiclepsis marginata* redrawn from Mann (1961), courtesy of Pergamon Press; *Piscicola salmositica* redrawn from Klemm (1982); courtesy of US Environment Protection Agency.

unaccompanied by figures and should be treated with scepticism. Needham (1969) reviewed the photographs in Khaibulaev's doctoral dissertation and reported that the flagellates observed in *P. geometra* appeared to be leech spermatozoa. In any case, Khaibulaev did not report transmission of flagellates to uninfected fish through the feeding activities of *P. geometra*. The only documented experimental transmission of *Cryptobia borreli* to uninfected fish through the feeding activity of *P. geometra* is the report by Kruse *et al*. (1989). *Cryptobia* infections in carp, *Cyprinus carpio*, were detected 6 days after feeding by infected leeches. They also confirmed the observation of Keysselitz (1906) that flagellates did not invade the proboscis sheath. The only evidence that *P. geometra* transmits trypanosomes to fish is the report by Khaibulaev and Guseinov (1982). They were apparently able to transmit *T. danilewskyi* to *Lebistes reticulatus, T. percae* to *Cobitis taenia* and *T. scardinii* to *Pseudorasbora parva* through the feeding activity of *P. geometra*.

Piscicola geometra is a serious pest in commercial fish ponds (Bauer, 1961a; Bielecki, 1988), but its vector relationships, especially for trypanosomes, remain unclear. Flagellates develop much higher abundance and persist for much longer in *H. marginata* than in *P. geometra* (Lom, 1979); in addition, *C. borreli* invades the proboscis sheath of *H. marginata*, but not *P. geometra*. These observations, all of which need confirmation, suggest that *H. marginata* is the natural vector for European freshwater haemoflagellates and that *P. geometra* has only secondarily acquired the ability to transmit flagellates and is less efficient.

Piscicola geometra occurs throughout most of Eurasia (Elliot and Mann, 1979: Sawyer, 1986; Shulman, 1961; Epshtein, 1987) and may be circumpolar. There has been speculation (Klemm, 1982) that *Piscicola milneri* (Verrill), which is widely distributed in northern North America, is identical to *P. geometra*, and Sawyer (1986) states that *P. milneri* is inadequately distinguished from *P. geometra*. Nevertheless, Madill (1988) and Davies (1991) list both species in North America. *Piscicola geometra* appears to be tolerant of diverse environmental conditions and has been collected in a wide range of habitats, including estuaries of the Caspian and Baltic Seas (Epshtein, 1987). However, *P. geometra* requires cool water that is relatively highly oxygenated and it is most abundant in rapidly flowing streams and large lakes where fish and vegetation are abundant (Mann, 1961: Sawyer, 1986). *Piscicola geometra* is not found in stagnant ponds or slow moving streams, the preferred habitat of *Hemiclepsis marginata*.

Over 30 species of fish have been recorded as hosts for *P. geometra*. Leeches feed anywhere on the body including skin, fins, gills and mouth cavity and they may remain on the host for successive blood meals. However, most leeches leave the host within seven days after feeding (Sawyer, 1986).

Piscicola geometra is in the family Piscicolidae, subfamily Piscicolinae. It is a cylindrical, elongate leech up to 50 mm in length, but usually 15–30 mm long with a caudal sucker diameter 1.5–2.0 times the maximum body width. The urosome diameter is nearly constant throughout its length. The body colour may be brownish olive, greenish or grey with segmental, unpigmented transverse bands and usually a longitudinal, mid-dorsal unpigmented stripe. Two pairs of eyes are present on the oral sucker (Fig. 14.4) and 12 to 14 ocelli usually occur on the caudal sucker alternating with radiating pigment bands. Recent keys that include *P. geometra* are in Sawyer (1986) and Epshtein (1987).

The life cycle of *P. geometra* was studied extensively by Malecha (1984) in northern France. The life span of an individual leech is about seven to nine months and three to four generations are produced each year. The overwintering generation deposits cocoons during March and adults die between

April and June. Cocoons are deposited on any hard substrate including vegetation. Cocoons hatch in late April and early May and leeches reach sexual maturity during June. Adults reproduce more or less continuously through August, producing two or three generations. The winter generation emerges from cocoons in September and individuals feed on fish for much of the winter. No reproduction occurs during this period and leeches may become quite large compared with the summer generations. Reproduction eventually begins in late February and March when cocoons that will become the next summer generation are deposited.

Johanssonia arctica (Johansson)

Johanssonia arctica is important as a vector for a variety of haematozoa in northern seas. It transmits *Trypanosoma murmanensis* to a number of commercially important hosts including Atlantic cod, *Gadus morhua*, American plaice, *Hippoglossoides platessoides*, and yellowtail flounder, *Limanda ferruginea* in the northwest Atlantic (Khan, 1976, 1977, 1991). It also transmits the intraerythrocyte parasites *Haemohormidium beckeri* in Perciform fishes (Khan, 1980) and *H. terraenovae* in pleuronectiform fishes (Khan, 1984). In addition, *J. arctica* may transmit *Haemogregarina uncinata*; developmental stages were observed in the leech crop, but transmission was not demonstrated because of a lack of uninfected hosts (Khan, 1978a). Khan (1982) has speculated that *J. arctica* also transmits an undescribed cryptobiid. *Johanssonia arctica* is the only known vector for haematozoa in the Newfoundland region even though it has a rich leech fauna (Khan and Meyer, 1976; Meyer and Khan, 1979; Appy and Dadswell, 1981). Early developmental stages of *H. beckeri* were observed (Khan, 1980) in another leech, *Platybdella olriki*, but presence of developmental stages alone is not sufficient evidence of a vector role. Recent studies by Siddall and Desser (1992a, 1993) on the developmental sequence of *Haemogregarina myoxocephali* in *Malmiana scorpii* strongly suggest that this leech is the vector for the parasite; however, actual transmission to fish has not been attempted.

Johanssonia arctica is commonly encountered in Arctic seas (Epshtein, 1961, 1962), in the northwestern Atlantic Ocean (Meyer and Khan, 1979; Khan, 1982) and in deep waters in the northeastern Pacific Ocean as far south as California (unpublished data). In Newfoundland, leeches have been collected in nature from Atlantic cod, *Gadus morhua*, polka-dot seasnail, *Liparis cyclostigma*, and American plaice, *Hippoglossoides platessoides*. In laboratory studies, 19 additional hosts were identified, but leeches seemed to prefer American plaice and Atlantic cod (Khan, 1982). Preferred feeding sites are the head of Atlantic cod and the dorsal and ventral fins of American plaice, although leeches also feed on other regions of the body.

Johanssonia arctica is in the family Piscicolidae, subfamily Piscicolinae. The species was originally described by Johansson (1899) as *Oxytonostoma arctica* from Greenland. Epshtein (1968) transferred the species to *Johanssonia* Selensky. Individuals are subcylindrical, up to 29 mm total length, with a narrow body and an especially narrow neck region. Unengorged leeches can appear almost threadlike. The body is devoid of pigmentation and lacks eyes on the oral sucker and ocelli on the body or caudal sucker; the integument is thin and transparent. The body has 11 pairs of very small, lateral, pulsatile vesicles and 12 longitudinal rows of small papillae; both vesicles and papillae may be difficult to discern in preserved specimens. The oral sucker, attached strongly eccentrically, is wider than the maximum body width of unengorged individuals, an unusual character. The caudal sucker is relatively small and approximately the same size as the oral sucker (Fig. 14.4). Recent keys that include *J. arctica* are in Appy and Dadswell (1981) and in Sawyer (1986). An excellent photograph of *J. arctica* appears in Sawyer (1986); the drawings of the suckers in Appy and Dadswell (1981) are more representative of the actual leech than those in Meyer and Khan (1979).

The life cycle of *J. arctica* in Newfoundland waters has been well studied (Khan, 1982). The leech needs a stable physical environment, with relatively constant temperatures of −1 to 2°C, and its life cycle involves little seasonality. Laboratory experiments show that leeches cannot tolerate water temperature greater than 5°C. Leeches mature after four blood meals and mating and cocoon deposition take place on the spider crab *Chionoecetes opilio* and occasionally on the toad crab, *Hyas coarctatus*. In the northeastern Pacific, cocoons of *J. arctica* have been observed on the tanner crab, *Chionoecetes tanneri* (unpublished data). In the laboratory, engorged leeches, both mature and immature, rapidly leave the fish host and attach to crabs, which serve as hard substrates for attachment in an otherwise muddy environment. Cocoons are deposited on the ventral femoral surfaces of the first and second walking legs. One naturally infested crab harboured 1246 cocoons (Khan, 1982). At ambient temperatures of −1 to 2°C, cocoons hatched in 176 to 253 days. In laboratory experiments newly hatched leeches held at 0 to 1°C began depositing cocoons after the fourth blood meal and deposited from 2 to 26 cocoons after each feeding through eight blood meals at which time leeches were two years old. Average clutch size was 16 cocoons and a mean of 62 cocoons per leech was produced over the two year period. In nature, young leeches were observed throughout the year, but were most abundant from May to October.

Calliobdella vivida (Verrill)

Calliobdella vivida transmits *Cryptobia bullocki*, which has been shown to be pathogenic to summer flounder, *Paralichthys dentatus*, a species of commercial and recreational importance along the east coast of the United States (Burreson, 1982; Burreson and Zwerner, 1984). A trypanosome in American eels, *Anguilla rostrata*, is also transmitted by *C. vivida* in Chesapeake Bay (unpublished data) and it is likely that *C. vivida* is also the vector for *Haemogregarina platessae* although no experimental transmission studies have been conducted on this parasite.

Calliobdella vivida occurs during winter in estuaries and is known from Louisiana to western Florida in the Gulf of Mexico and along the east coast of North America from Georgia to Newfoundland (Sawyer *et al.*, 1975; Appy and Dadswell, 1981; Madill, 1988). Although extensive collections have not been conducted in all localities, the leech appears to be most abundant in the large estuaries of Virginia, North Carolina and South Carolina. Leeches are most commonly collected at salinities between 10 and 22 ppt, but individuals can withstand oceanic salinity and pond water for long periods (Sawyer and Hammond, 1973; Burreson and Zwerner, 1982).

The host range of *C. vivida* is extremely wide. In Chesapeake Bay, *C. vivida* was collected from nine different hosts and fed on another five host species in the laboratory (Burreson and Zwerner, 1982). No host was found to which *C. vivida* would not attach and feed. Common hosts in Chesapeake Bay include summer flounder, *Paralichthys dentatus*, hogchoker, *Trinectes maculatus*, menhaden, *Brevoortia tyrannus*, and oyster toadfish, *Opsanus tau*. On the basis of studies in other areas and on the host distribution of *Cryptobia bullocki, C. vivida* seems to prefer pleuronectiform fishes and menhaden (Laird and Bullock, 1969; Becker and Overstreet, 1979). Leeches feed anywhere on the body including the fins.

Calliobdella vivida is in the family Piscicolidae, subfamily Piscicolinae. The species was originally described, inadequately, by Verrill (1872) as *Cystobranchus vividus*, but no type material has been found. Sawyer and Chamberlain (1972) described *Calliobdella carolinensis* from South Carolina estuaries, but later, on the basis of further collections, Sawyer *et al.* (1975) considered *C. carolinensis* to be a junior synonym of *C. vividus*. However, they transferred the species to the genus *Calliobdella* as *Calliobdella vivida*. The species was thoroughly described by Sawyer and Chamberlain (1972). The subcylindrical body is not distinctly divided into trachelosome and urosome; total length can be up to 40 mm, but most mature individuals are approximately 18–30 mm long. The oral sucker is well developed and has two pairs of concentric eyes (Fig. 14.4). The caudal sucker is slightly wider than the maximum body width, attached strongly eccentrically, and lacks ocelli. Pigmentation varies from none to faint, segmental transverse bands, usually reddish brown in colour, especially obvious on the trachelosome. Paired, segmental, punctiform ocelli are often visible dorsolaterally and ventrolaterally on the body. Eleven pairs of pulsatile vesicles are usually obvious on the lateral margins of the urosome. Recent keys that include *C. vivida* are in Appy and Dadswell (1981) and Sawyer (1986).

The life cycle of *C. vivida* has been studied through field collections and laboratory experiments by Sawyer and Hammond (1973) in South Carolina and by Burreson and Zwerner (1982) in Chesapeake Bay. In Chesapeake Bay, leeches are abundant from December through March when water temperature averages 4–8°C. Leeches begin to deposit cocoons as early as February, but most cocoon deposition occurs in late April and early May when water temperature is about 15–17°C. In the laboratory, a single large leech deposited 51 cocoons over a six day period (Burreson and Zwerner, 1982). Leeches die after depositing cocoons and no leech has been collected in nature after June. Eggs oversummer in the cocoon and begin hatching in the autumn when water temperature decreases to between 15 and 18°C, usually late October to early November in Chesapeake Bay, but as late as December in South Carolina. Unfed *C. vivida* are strong swimmers and are routinely collected in plankton nets. They invade fish tanks through an unfiltered flowing seawater system each winter at the Virginia Institute of Marine Science on Chesapeake Bay (unpublished data).

The marked seasonality of *C. vivida* has implications for its role as a vector. Most fish species in the Chesapeake Bay are migratory and leave the Bay during the autumn to spawn on the continental shelf or overwinter south of Cape Hatteras, North Carolina; they return to the Bay in late spring. Thus, most fish are migrating out of the Bay at the time *C. vivida* is hatching from the cocoon and are re-entering the Bay as leeches are dying after cocoon deposition. Thus, most fish species are exposed to *C. vivida* for only short periods and prevalence of *Cryptobia bullocki* is low in migratory fishes. Resident fish, such as hogchoker, oyster toadfish and juvenile summer flounder are exposed to leeches throughout the winter and harbour higher prevalences of *C. bullocki*.

Piscicola salmositica Meyer

Piscicola salmositica is the vector for *Cryptobia salmositica*, an important pathogen in Pacific salmon of the northwest coast of North America (Wood, 1979; Bower and Margolis, 1984) and in rainbow/steelhead trout (Wales and Wolf, 1955; Woo, 1979; Mundie and Traber, 1983). The elucidation of the vector role of *P. salmositica* by Becker and Katz (1965a) was the first documentation of a freshwater leech transmiting haematozoa to fish in North America.

Piscicola salmositica is restricted to the northern Pacific coast of North America where it occurs in moderate to rapidly flowing streams with low water temperatures, high dissolved oxygen content and gravelled beds. It has been collected from northern California to central British Columbia (Becker and Katz, 1965b). In California it is known from the Eel River and from Fall Creek, a tributary of the Klamath River. The coastal Alsea and Nehalem Rivers and Eagle Creek, a tributary of the Willamette River, have yielded specimens in Oregon and leeches have been collected from most of the coastal rivers in Washington and from the Columbia River system. Leeches have been collected from streams on both the east and west coast of Vancouver Island and from tributaries in the Fraser River system in British Columbia, Canada. The report of *P. salmositica* on cutthroat trout in Wyoming by Cope (1958) needs confirmation.

Hosts for *P. salmositica* include most of the Pacific salmon, *Oncorhynchus* spp., including *O. kisutch, O. nerka, O. tshawytscha, O. gorbuscha*, and *O. keta*, rainbow/steelhead trout, *O. mykiss (Salmo gairdneri)*, and the sculpins *Cottus rhotheus* and *C. gulosus*. However, on the basis of the host range of *Cryptobia salmositica*, the host range for *P. salmositica* is probably much greater than indicated by actual collections (Becker and Katz, 1965c; Becker, 1980; Bower and Margolis, 1984). Leeches usually feed in the axillae of the pectoral or pelvic fins or on gill lamellae.

Piscicola salmositica is in the family Piscicolidae, subfamily Piscicolinae; it was thoroughly described by Meyer (1946b). The body is subcylindrical and the general form is not as elongate as most other members of the genus (Fig. 14.2). The urosome is rather broad throughout and contains 11 pairs of lateral pulsatile vesicles; total length is up to about 40 mm. Pigmentation is uniformly dark grey. The oral sucker has two pairs of eyes, a large, crescentiform anterior pair and a smaller dash-like posterior pair. The caudal sucker is relatively small, not as wide as the maximum body width, and harbours eight to ten strongly crescentiform ocelli. The posterior margin of the caudal sucker may be angular rather than discoidal (Fig. 14.2). Recent keys that include *P. salmositica* are Klemm (1982) and Sawyer (1986).

The biology of *P. salmositica* was studied by Becker and Katz (1965b) in the vicinity of the Green River salmon hatchery located on Soos Creek, a tributary of the Green River near Auburn, Washington. The leech exhibits a definite seasonal cycle tied to spawning migrations of the host salmon. Leeches, small and few in number, first appear in late September attached to spawning chinook salmon. The leech becomes more abundant and individual leeches are larger during the coho salmon migration in November and December. Mean number of leeches per host for ten hosts was as high as 40 in December, 1961 and the total leech population in Soos Creek was estimated at over 1 million (Becker and Katz, 1965b). As they mature, leeches attach to the underside of stones, mate, deposit cocoons and die. The salmon spawning migrations terminate in February and leeches disappear by March. On the basis of field studies it was concluded that cocoons hatched in seven days, an unusually short time period, especially at ambient temperatures below 10°C. This short hatching time is contrary to results in laboratory experiments in which leeches took 300 days to hatch (Bower and Thompson, 1987). Becker and Katz (1965b) concluded that only small leeches survived the summer, although none was ever collected, and it was these leeches that attached to the first returning salmon each autumn. On the basis of more recent research with other species that demonstrate marked seasonality (Burreson and Zwerner, 1982) it is more likely that only eggs survive the summer period and hatch in the autumn when water temperature decreases. This view was supported by Bower and Margolis (1984) who did not find leeches during the summer, but more recent research (Bower and Thompson, 1987), in which leeches hatched during the autumn even when held at constant temperature, suggests that hatching may not be environmentally controlled.

Newly hatched leeches may feed on the resident torrent sculpin, *C. rhotheus*, known to serve as a reservoir for *Cryptobia salmositica*, prior to arrival of the salmon. In this manner leeches may already be infected with *C. salmositica* when the first migrating salmon enter the streams. Leeches may also periodically feed on sculpins during the winter. Thus, the wide host range of *P. salmositica* facilitates the transfer of *C. salmositica* to salmon.

CONTROL OF FISH LEECHES

Much of the research on control of leeches has been oriented toward *P. geometra* in eastern European fish ponds and has been summarized by Bauer (1961a), Bauer *et al.* (1973), Negele (1975), Kabata (1985) and Bielecki (1988).

Therapeutic agents listed in these papers and summarized below may not be approved for food fish in all countries.

Prevention is the best control, and, according to Bauer *et al.* (1973), the presence of leeches in a culture pond is a sign of neglect, primarily the failure to remove aquatic vegetation that provides ideal substrate for cocoon deposition. Sand or gravel filters can prevent introduction of leeches with the water supply. In carp culture in eastern Europe leeches must be removed from fish before they are transferred to the wintering ponds. The following baths have been shown to be effective against *P. geometra*:

1. 2.5% sodium chloride for one hour. Leeches leave the fish host, but are not killed; care must be taken not to reintroduce them to the ponds (Kabata, 1985).
2. 200 g quicklime (calcium oxide) per 100 l of water for 5 seconds. Concentration and time are critical, too high a concentration or too long a time can cause death of the fish. The quicklime must be fresh and free of impurities (Negele, 1975).
3. 0.2% solution of lysol or 0.4% solution of priasol (creosote derivative) for 5-15 seconds (Conroy and Herman, 1970).
4. 0.005% solution of cupric chloride for 15 minutes. The solution should be changed after each group of 15-20 kg of fish (Bauer *et al.*, 1973).
5. The proprietary varieties of Dylox are successful leech therapeutic agents. For example, 2.5% Masoten for 5 minutes is effective, but old baths are toxic to fish (Negele, 1975).

Varieties of Dylox have also been utilized to treat ponds infested with leeches. Neguvon at 1 ppm for 5 days is effective against adult leeches, but this concentration does not kill embryos in cocoons. A concentration of 10 ppt is necessary to kill embryos, but this concentration is also toxic to fish (Prost *et al.*, 1974). Masoten is effective in ponds at a concentration of 1 g per 4 m^3, but it is apparently harmful to zooplankton as well (Negele, 1975). Chlorofos has been used at 2.0 ppm for 4 days (Kabata, 1985). Correct dosage may be difficult to calculate in ponds where the water volume is not known.

Massive infestations of *P. geometra* require that the pond be drained and disinfected. Although desiccation for 24 hours kills embryos in cocoons (Prost *et al.*, 1974) it is very difficult to effectively dry a pond. Thus, disinfection with quicklime (calcium oxide) is a common prophylactic measure. Normal concentrations are 2000-3000 kg ha^{-1}, but much less can be used if the calcium oxide purity is at least 93%. It is critical that the calcium oxide concentration is high enough to yield a pH of at least 8.6 to ensure death of both adults and embryos in cocoons (Prost *et al.*, 1974). The general procedure is to drain the pond, spread the lime evenly over the pond bottom, and then flood the pond for 7-10 days (Kabata, 1985).

Hemiclepsis marginata has also been the subject of chemotherapeutic studies in fish culture ponds in India. A number of different treatments are effective (Singhal *et al.*, 1986). A concentration of 0.001 mg l^{-1} glacial acetic acid followed by 10 mg l^{-1} potassium permanganate with an immersion time of 5 minutes each reduced *H. marginata* infection prevalence in treatment aquaria by 80% compared to untreated control fish. A single treatment of 5 mg l^{-1} gammexane (1-6-hexachlorocyclohexane) sprayed over the surface of the culture ponds removed most of the leeches from the fish, but it was not reported whether the leeches actually died from the treatment. A 15 minute immersion in 30 mg l^{-1} sodium chloride solution caused the leeches to detach from the fish hosts and was found to be the most effective treatment.

Because of its importance as a vector and parasite in salmonids in hatcheries, *P. salmositica* has been the subject of a number of studies on control. In an outbreak of *P. salmositica* on pink salmon fry, none of the chemicals commonly employed to control bacterial and other parasitic diseases (formalin, roccal, and pyridylmercuric acetate) had any effect on the leeches (Earp and Schwab, 1954). However, the leeches were sensitive to undiluted sea water (salinity of at least 28 ppt); one-hour exposures killed the leeches, but did not harm the salmon eggs or fry. Meyer (1969) conducted a study to examine the efficacy of three organophosphate insecticides, Dylox, Baygon and Baytex, on leeches held in small beakers. The effects of Baygon and Baytex were only temporary, but Dylox caused permanent damage and a concentration of 0.5 ppm killed *P. salmositica* in 96 hours. Another fish leech, *Myzobdella lugubris*, was more sensitive to Dylox; a concentration of 0.125 ppm killed it in 96 hours.

In more recent studies, Bower and co-workers have investigated the use of chlorine for controlling *P. salmositica*. They found a wide range of chlorine concentrations and exposure times that would kill juvenile leeches, but adult leeches were more resistant (Bower *et al.*, 1985). In a subsequent paper, Bower and Thompson (1987) found cocoons of *P. salmositica* to be surprisingly resistant to various chemical treatments. High concentrations of an iodophore disinfectant and of chlorine (2 to 13 times higher than concentrations lethal to juvenile leeches) had little effect on cocoons. A chlorine concentration that was lethal to leech embryos was also potentially harmful to fish. Hatching was prevented by freezing cocoons, in or out of water, or by desiccation for at least 7 days; even desiccation for 24 hours reduced hatching by 50%.

The freshwater and estuarine leech *Myzobdella lugubris* has been the subject of chemotherapeutic studies in striped bass (*Morone saxatilis*) intensive culture facilities located on the Chesapeake Bay, Maryland, USA (Woods *et al.*, 1990). Dylox at a concentration of 40 ppm caused detachment of all leeches from fish after three hour static baths at 11.5°C. Potassium permanganate at 10 ppm was not effective at that temperature and Dylox was not effective at temperatures below 10°C.

Chemotherapeutic control of marine leeches has not been attempted, but with increasing culture of marine fish species they may become important pests in the future. Freshwater baths would probably be as effective for truly marine leeches as salt water baths are for freshwater leeches.

SUMMARY AND CONCLUSIONS

Leeches can be directly pathogenic to fishes, especially if conditions allow close, continuous contact between large numbers of leeches and their hosts. Thus, the presence of leeches in aquaculture facilities should always be cause for concern. However, their most important role is as vectors of pathogenic organisms. The study of leeches as vectors of pathogenic viruses and bacteria is still in its infancy and should be a productive area of research in the future. Even though the role of leeches as vectors for the various haematozoa groups is now well documented, vectors have been identified for only a very few of the myriad species known to inhabit the blood of fishes, especially for the intraerythrocytic protozoa. Current knowledge suggests that, in a given area, only one or two leech species transmit a wide variety of blood parasites, but vectors are known for so few haematozoa that this may not be a valid conclusion.

Verification of transmission is difficult because laboratory-raised fish hosts, known to be uninfected, are rarely available and results of transmission trials using hosts from the study area must be interpreted with caution. A false negative result may be obtained if the host had lost a previous infection and was now immune, and a false positive result may be obtained if the host already carried a light infection that was not detected prior to the experiment. However, if the biology of both host and putative vector are well known, it may be possible to find life history stages of the host (for example, juveniles) that do not co-occur with the vector spatially and/or temporally. These hosts can often be used with confidence in transmission experiments if a sample suggests that they are not infected with the blood parasite. The false positive problem can also be eliminated with a sufficient number of control hosts that are not exposed to infected leeches. If none, or only a few, of the control hosts become infected, but all of the hosts exposed to leeches become infected, the investigator can be confident that infections in the experimental host were initiated by leeches. This approach requires an abundance of leeches and fish hosts, which may not be available.

One of the intriguing questions remaining to be answered is the vector specificity of fish haematozoa. It has been generally assumed that vector specificity is high and that a haematozoan species utilizes only a single leech species as a vector. However, this assumption is contradicted by *Cryptobia borreli*, which appears to utilize two vectors that belong to different families, *H. marginata*, a glossiphoniid and *P. geometra*, a piscicolid. Rigorous comparative studies are needed to confirm the vector relationships of both *P. geometra* and *H. marginata* and to ascertain that they are transmitting the same species of flagellates. Recent experimental studies with frog and turtle trypanosomes (Siddall and Desser, 1992b) identify multiple leech vectors, although in both cases one of the leech species appears to be a more important vector than the other species. These results raise many interesting questions about the evolution of the vector–parasite relationships.

There are regions with an abundant leech and haematozoa fauna where important research could be conducted to determine if fish haematozoa species can be transmitted by more than one leech. For example, developmental stages of various haematozoa were reported from many different leech species in Newfoundland (Khan *et al.*, 1991), but experiments are necessary to demonstrate whether other leeches can actually transmit haematozoa known to utilize *Johanssonia arctica* as a vector. Similar experiments should be conducted to determine whether *Myzobdella lugubris* can also transmit haematozoa known to utilize *Calliobdella vivida* in estuaries of the southeastern United States.

A final plea is for leech taxonomists. The leech fauna in many parts of the world is still poorly known. Our knowledge is especially inadequate for the marine fauna of the southern hemisphere, but even in areas where the fauna is relatively well known taxonomic confusion persists (Barta and Sawyer, 1990). Unfortunately, there are very few active leech taxonomists. Leeches are a difficult group to work with and special training is required to learn the intricacies of internal anatomy necessary for generic assignment. Current employment and funding opportunities in taxonomy make it difficult to attract students. The end result will be that fewer studies will be conducted on the vector role of leeches because of identification difficulties.

REFERENCES

Ahne, W. (1985) *Argulus foliaceus* L. and *Piscicola geometra* L. as mechanical vectors of spring viraemia of carp virus (SVCV). *Journal of Fish Diseases* 8, 241–242.

Andersson, E. (1988) The biology of the fish leech *Acanthobdella peledina* Grube. *Zoologische Beiträge Neue Folge* 32, 31–50.

Appy, R.G. and Cone, D.K. (1982) Attachment of *Myzobdella lugubris* (Hirudinea: Piscicolidae) to logperch, *Percina caprodes*, and brown bullhead, *Ictalurus nebulosus. Transactions of the American Microscopical Society* 101, 135–141.

Appy, R.G. and Dadswell, M.J. (1981) Marine and estuarine piscicolid leeches (Hirudinea) of the Bay of Fundy and adjacent waters with a key to species. *Canadian Journal of Zoology* 59, 183–192.

Badham, C. (1916) On an ichthyobdellid parasitic on the Australian sand whiting (*Sillago ciliata*). *Quarterly Journal of Microscopical Science* (new series) 62, 1–41.

Barta, J.R. and Sawyer, R.T. (1990) Definition of a new genus of glossiphoniid leech and a redescription of the type species, *Clepsine picta* Verrill, 1872. *Canadian Journal of Zoology* 68, 1942–1950.

Bauer, O.N. (1961a) Parasitic diseases of cultured fishes and methods of their prevention and treatment. In: Dogiel, V.A., Petrushevski, G.K. and Polyenski, Yu.I. (eds), *Parasitology of Fishes*. Oliver and Boyd, Edinburgh, pp. 265–298.

Bauer, O.N. (1961b) Relationships between host fishes and their parasites. In: Dogiel, V.A., Petrushevski, G.K. and Polyanski, Yu.I. (eds) *Parasitology of Fishes*. Oliver and Boyd, Edinburgh, pp. 84–103.

Bauer, O.N., Musselius, V.A. and Strelkov, Yu.A. (1973) *Diseases of Pond Fishes*. Israel Program for Scientific Translations, Jerusalem.

Becker, C.D. (1980) Haematozoa from resident and anadromous fishes of the central Columbia River: a survey. *Canadian Journal of Zoology* 8, 356–362.

Becker, C.D. and Katz, M. (1965a) Transmission of the hemoflagellate, *Cryptobia salmositica* Katz, 1951, by a rhynchobdellid vector. *Journal of Parasitology* 51, 95–99.

Becker, C.D. and Katz, M. (1965b) Distribution, ecology, and biology of the salmonid leech, *Piscicola salmositica* (Rhynchobdellae: Piscicolidae). *Journal of the Fisheries Research Board of Canada* 22, 1175–1195.

Becker, C.D. and Katz, M. (1965c) Infections of the hemoflagellate, *Cryptobia salmositica* Katz, 1951, in freshwater teleosts of the Pacific coast. *Transactions of the American Fisheries Society* 94, 327–333.

Becker, C.D. and Overstreet, R.M. (1979) Haematozoa of marine fishes from the northern Gulf of Mexico. *Journal of Fish Diseases* 2, 469–479.

Bielecki, A. (1988). Leeches (Hirudinea) – parasites of fish. *Wiadomosci Parazytologiczne* 34, 3–10 (in Polish.)

Bower, S.M. and Margolis, L. (1984) Distribution of *Cryptobia salmositica*, a haemoflagellate of fishes, in British Columbia and the seasonal pattern of infection in a coastal river. *Canadian Journal of Zoology* 62, 2512–2518.

Bower, S.M. and Thompson, A.B. (1987) Hatching of the Pacific salmon leech (*Piscicola salmositica*) from cocoons exposed to various treatments. *Aquaculture* 66, 1–8.

Bower, S.M., Margolis, L. and MacKay, R.J. (1985) Potential usefulness of chlorine for controlling Pacific salmon leeches, *Piscicola salmositica*, in hatcheries. *Canadian Journal of Fisheries and Aquatic Science* 42, 1986–1993.

Bragg, R.R., Oosthuizen, J.H. and Lordan, S.M. (1989) The leech *Batrachobdelloides tricarinata* Blanchard, 1987 (Hirudinea: Glossiphoniidae) as a possible reservoir of the rainbow trout pathogenic *Streptococcus* spp. *Onderstepoort Journal of Veterinary Research* 56, 203–204.

Brumpt, M.E. (1905) Trypanosomes et trypanosomoses. *Revue Scientifique* 4, 321–332.

Brumpt, M.E. (1906a) Mode de transmission et évolution des trypanosomes des poissons – Description de quelques espéces de trypanoplasmes des poissons d'eau douce – Trypanosome d'un crapaud African. *Comptes Rendus des Séances de la Société de Biologie* 60, 162–164.

Brumpt, M.E. (1906b) Expériences relatives au mode de transmission des trypanosomes et des trypanoplasmes par les hirudinées. *Comptes Rendus des Séances de la Société de Biologie* 61, 77–79.

Burreson, E.M. (1975) Biological studies on the hemoflagellates of Oregon marine fishes and their potential leech vectors. Unpublished PhD thesis, Oregon State University.

Burreson, E.M. (1977) Two new species of *Malmiana* (Hirudinea: Piscicolidae) from Oregon coastal waters. *Journal of Parasitology* 63, 130–136.

Burreson, E.M. (1979) Structure and life cycle of *Trypanoplasma beckeri* sp. n. (Kinetoplastida), a parasite of the cabezon, *Scropaenichthys marmoratus*, in Oregon coastal waters. *Journal of Protozoology* 26, 343–347.

Burreson, E.M. (1982) The life cycle of *Trypanoplasma bullocki* (Strout) (Zoomastigophorea: Kinetoplastida). *Journal of Protozoology* 29, 72–77.

Burreson, E.M. and Pratt, I. (1972) *Trypanosoma pacifica* sp. n. from the English sole *Parophrys vetulus* Girard from Oregon. *Journal of Protozoology* 9, 555–556.

Burreson, E.M. and Zwerner, D.E. (1982) The role of host biology, vector biology, and temperature in the distribution of *Trypanoplasma bullocki* infections in the lower Chesapeake Bay. *Journal of Parasitology* 68, 306–313.

Burreson, E.M. and Zwerner, D.E. (1984) Juvenile summer flounder, *Paralichthys dentatus*, mortalities in the western Atlantic Ocean caused by the hemoflagellate *Trypanoplasma bullocki*: evidence from field and experimental studies. *Helgoländer Meeresuntersuchungen* 37, 343–352.

Conroy, D.A. and Herman, R.L. (1970) *Erwin Amlacher Textbook of Fish Diseases.* TFH Publications, Jersey City.

Cope, O.B. (1958) Incidence of external parasites on cutthroat trout in Yellowstone Lake. *Proceedings of the Utah Academy of Science Arts and Letters* 35, 95–100.

Cusack, R. and Cone, D.K. (1986) A review of parasites as vectors of viral and bacterial diseases of fish. *Journal of Fish Diseases* 9, 169–171.

Davies, R.W. (1991) Annelida: leeches, polychaetes, and acanthobdellids. In: Thorp, J.H. and Covich, A.P. (eds) *Ecology and Classification of North American Freshwater Invertebrates.* Academic Press, San Diego, pp. 437–479.

Dombrowski, H. (1953) Die Nahrungsmenge des Fischegels *Piscicola geometra* L. (Zugleich ein Beitrage zur Physiologie des Blutes des Karpfens *Cyprinus carpio* L.) *Biologische Zentralblatt, Leipzig* 72, 311–314.

Dyková, I. and Lom, J. (1979) Histopathological changes in *Trypanosoma danilewskyi* Laveran & Mesnil, 1904 and *Trypanoplasma borelli* Laveran & Mesnil, 1902 infections of goldfish, *Carassius aurata* (L.). *Journal of Fish Diseases* 2, 381–390.

Earp, B.J. and Schwab, R.L. (1954) An infestation of leeches on salmon fry and eggs. *Progressive Fish Culturist* 16, 122–124.

Elliott, J.M. and Mann, K.H. (1979) *A Key to the British Freshwater Leeches with Notes on their Life Cycle and Ecology.* Freshwater Biological Association. Scientific Publication No. 40.

Epshtein, V.M. (1961) A review of the fish leeches (Hirudinea, Piscicolidae) from the northern seas of the SSSR. *Doklady Akademii Nauk SSSR* 141, 1121–1124.

Epshtein, V.M. (1962) A survey of fish leeches (Hirudinea, Piscicolidae) from and Okhotsk Seas and from the Sea of Japan. *Doklady Akademii Nauk SSSR* 141, 648–651.

Epshtein, V.M. (1968) Revision of the genera *Oxytonostoma* and *Johanssonia* (Hirudinea: Piscicolidae). *Zoologicheskii Zhurnal* 47, 1011–1021 (in Russian). (translated by P.G. Rossbacher, Oregon State University).

Epshtein, V.M. (1987) Phylum Annelida. In: Bauer, O.N. (ed.) *Opredelitel Parazitov Presnovodnykh Ryb Fauny SSSR, Volume 3. Paraziticheskie Mnogokletochnye (Part 2).* Nauka, Leningrad, pp. 340–372. (Translated from Russian by the Multi-lingual Translation Directorate, Canada.)

Harding, W.A. (1910) A revision of the British leeches. *Parasitology* 3, 130–201.

Hoffman, R.L. (1964) A new species of *Cystobranchus* from southwestern Virginia (Hirudinea: Piscicolidae). *American Midland Naturalist* 72, 390–395.

Jennings, J.B. and Van Der Lande, V.M. (1967) Histochemical and bacteriological studies on digestion in nine species of leeches. (Annelida: Hirudinea). *Biological Bulletin* 133, 166–183.

Johansson, L. (1899) Die Ichthyobdelliden im Zoologischen Reichsmuseum in Stockholm. *Öfversigt Af Kungliga Vetenskapsakademiens Förhandlingar* 55, 665–687.

Jones, S.R.M. and Woo, P.T.K. (1990a) Redescription of the leech *Desserobdella phalera* (Graf, 1899) n. comb. (Rhynchobdellida: Glossiphoniidae), with notes on its biology and occurrence on fishes. *Canadian Journal of Zoology* 68, 1951–1955.

Jones, S.R.M. and Woo, P.T.K. (1990b) The biology of *Trypanosoma phaleri* n.sp. from bowfin, *Amia calva* L., in Canada and the United States. *Canadian Journal of Zoology* 68, 1956–1961.

Kabata, Z. (1985) *Parasites and Diseases of Fish Cultured in the Tropics.* Taylor & Francis, London.

Keysselitz, G. (1906) Generations- und Wirtswechsel von *Trypanoplasma borreli* Laveran et Mesnil. *Archiv für Protistenkunde* 7, 1–74.

Khaibulaev, K.Kh. (1970) The role of leeches in the life cycle of blood parasites of fishes. *Parazitologiya* 4, 13–17. (in Russian).

Khaibulaev, K.Kh. and Guseinov, M.A. (1982) Experimental study of the biology of some flagellates from the genera *Trypanosoma* Grudy, 1841 (Trypanosomidae Doflein, 1911) and *Cryptobia* Leidy, 1846 (Bodonidae Stenn, 1878). *Izvestiya Akademii Nauk Azerbaidzhanskoi SSR, Biologicheskie Nauki* 2, 87–91.

Khalifa, K.A. (1985) Leeches on freshwater farmed fishes in Iraq. *Journal of Wildlife Diseases* 21, 312–313.

Khan, R.A. (1976) The life cycle of *Trypanosoma murmanensis* Nikitin. *Canadian Journal of Zoology* 54, 1840–1849.

Khan, R.A. (1977) Susceptibility of marine fish to trypanosomes. *Canadian Journal of Zoology* 55, 1235–1241.

Khan, R.A. (1978a) A new hemogregarine from marine fishes. *Journal of Parasitology* 64, 35–44.

Khan, R.A. (1978b) A redescription of *Trypanosoma cotti* Brumpt and Lebailly, 1904 and its development in the leech, *Calliobdella punctata*. *Annales de Parasitologie* 53, 461–466.

Khan, R.A. (1980) The leech as a vector of a fish piroplasm. *Canadian Journal of Zoology* 58, 1631–1637.

Khan, R.A. (1982) Biology of the marine piscicolid leech *Johanssonia arctica* (Johansson) from Newfoundland. *Proceedings of the Helminthological Society of Washington* 49, 266–278.

Khan, R.A. (1984) Simultaneous transmission of a piscine piroplasm and trypanosome by a marine leech. *Journal of Wildlife Diseases* 20, 339–341.

Khan, R.A. (1991) Trypanosome occurrence and prevalence in the marine leech *Johanssonia arctica* and its host preferences in the northwestern Atlantic Ocean. *Canadian Journal of Zoology* 69, 2374–2380.

Khan, R.A. and Meyer, M.C. (1976) Taxonomy and biology of some Newfoundland marine leeches (Rhynchobdellae: Piscicolidae). *Journal of the Fisheries Research Board of Canada* 33, 1699–1714.

Khan, R.A., Lee, E.M. and Whitty, W.S. (1991) Blood protozoans of fish from the Davis Strait in the northwestern Atlantic Ocean. *Canadian Journal of Zoology* 69, 410–413.

Klemm, D.J. (1982) *Leeches (Annelida: Hirudinea) of North America.* EPA-600/3-82-025, US Environmental Protection Agency, Cincinnati.

Kruse, P., Steinhagen, D. and Körting, W. (1989) Development of *Trypanoplasma borreli* (Mastigophora: Kinetoplastida) in the leech vector *Piscicola geometra* and its infectivity for the common carp, *Cyprinus carpio*. *Journal of Parasitology* 75, 527–530.

Lainson, R. (1981) On *Cyrilia gomesi* (Neiva & Pinto, 1926) gen. nov. (Haemogregarinidae) and *Trypanosoma bourouli* Neiva & Pinto, in the fish *Synbranchus marmoratus*: simultaneous transmission by the leech *Haementeria lutzi*. In: Canning, E.U. (ed.), *Parasitological Topics*. Society of Protozoologists, pp. 150–158.

Laird, M. and Bullock, W.L. (1969) Marine fish haematozoa from New Brunswick and New England. *Journal of the Fisheries Research Board of Canada* 26, 1075–1102.

Laveran, A. and Mesnil, F. (1912) *Trypanosomes et Trypanosomiases*, 2nd edn. Masson et Cie, Paris.

Léger, L. (1904) Sur les hémoflagellés des *Cobitis barbatula* L., 1. *Trypanosoma barbatulae*. *Comptes Rendus des Séances de la Société de Biologie* 57, 344–345.

Letch, C.A. (1979) Host restriction, morphology and isoenzymes among trypanosomes of some British freshwater fishes. *Parasitology* 79, 107–117.

Letch, C.A. (1980) The life-cycle of *Trypanosoma cobitis* Mitrophanow 1883. *Parasitology* 80, 163–169.

Letch, C.A. and Ball, S.J. (1979) Prevalence of *Trypanosoma cobitis* Mitrophanow, 1883 in fishes from the River Lee. *Parasitology* 79, 119–124.

Lom, J. (1979) Biology of the trypanosomes and trypanoplasms of fish. In: Lumsden, W.H.R. and Evans, D.A. (eds) *Biology of the Kinetoplastida*. Vol. 2. Academic Press, London, pp. 269–237.

Lukin, E.I. (1976) Leeches of fresh and brackish water bodies. In: *Fauna of the USSR*, Vol. 1. Academy of Science of the USSR, Leningrad (in Russian).

McCarthy, A.M. (1990) Experimental observations on the specificity of *Apatemon minor* Yamaguti, 1933 (Digenea: Strigeidae) toward leech (Hirudinea) second intermediate hosts. *Journal of Helminthology* 64, 161–167.

Mace, T.F. and Davis, C.C. (1972) Energetics of a host–parasite relationship as illustrated by the leech *Malmiana nuda*, and the shorthorn sculpin *Myoxocephalus scorpius*. *Oikos* 23, 336–343.

Madill, J. (1988) New Canadian records of leeches (Annelida: Hirudinea) parasitic on fish. *Canadian Field Naturalist* 102, 685–688.

Malecha, J. (1984) Cycle biologique de l'hirudinée rhynchobdelle *Piscicola geometra* L. *Hydrobiologia* 118, 237–243.

Mann, K.H. (1955) The ecology of the British freshwater leeches. *Journal of Animal Ecology* 24, 98–119.

Mann, K.H. (1961) *Leeches (Hirudinea). Their Structure, Physiology, Ecology and Embryology*. Pergamon Press, New York.

Markevich, A.P. (1963) *Parasitic Fauna of Freshwater Fish of the Ukrainia SSR*. Israel Program for Scientific Translations, Jerusalem.

Meyer, F.P. (1969) A potential control for leeches. *Progressive Fish Culturist* 31, 160–163.

Meyer, M.C. (1946a) Further notes on the leeches (Piscicolidae) living on fresh-water fishes of North America. *Transactions of the American Microscopical Society* 65, 237–249.

Meyer, M.C. (1946b) A new leech, *Piscicola salmositica* n.sp. (Piscicolidae), from steelhead trout (*Salmo gairdneri gairdneri* Richardson, 1838). *Journal of Parasitology* 32, 467–476.

Meyer, M.C. and Khan, R.A. (1979) Taxonomy, biology, and occurrence of some marine leeches in Newfoundland waters. *Proceedings of the Helminthological Society of Washington* 46, 254–264.

Mulcahy, D., Klaybor, D. and Batts, W.N. (1990) Isolation of infectious hematopoietic necrosis virus from a leech (*Piscicola salmositica*) and a copepod (*Salmincola* sp.), ectoparasites of sockeye salmon *Oncorhynchus nerka*. *Diseases of Aquatic Organisms* 8, 29–34.

Mundie, J.H. and Traber, R.E. (1983) Carrying capacity of an enhanced side-channel for rearing salmonids. *Canadian Journal of Fisheries and Aquatic Science* 40, 1320–1322.

Needham, E.A. (1969) Protozoa parasitic in fish. Unpublished PhD thesis. University of London.

Negele, R.-D. (1975) Fish leeches as pests and vectors of disease. *Fish und Umwelt* 1, 123–126. (Canadian translation of fisheries and aquatic sciences No. 4812.)

Neumann, R.O. (1909) Studien über protozoische Parasiten im Blut von Meeresfischen. *Zeitschrift für Hygiene und Infektionskrankheiten* 64, 1–112.

Noga, E.J., Bullis, R.A. and Miller, G.C. (1990) Epidemic oral ulceration in large-mouth bass (*Micropterus salmoides*) associated with the leech *Myzobdella lugubris*. *Journal of Wildlife Diseases* 26, 132–134.

Paperna, I. and Overstreet, R.M. (1981) Parasites and diseases of mullets (Mugilidae). In: Oren, O.H. (ed.), *Aquaculture of Grey Mullets*. Cambridge University Press, Cambridge.

Paperna, I. and Zwerner, D.E. (1974) Massive leech infestation on a white catfish (*Ictalurus catus*): a histopathological consideration. *Proceedings of the Helminthological Society of Washington* 41, 64–67.

Prost, M., Studnicka, M. and Niezgoda, J. (1974) Efficacy of some methods of controlling leeches in water. *Aquaculture* 3, 287–294.

Putz, R.E. (1972a) *Cryptobia cataractae* sp. n. (Kinetoplastida: Cryptobiidae), a hemoflagellate of some cyprinid fishes of West Virginia. *Proceedings of the Helminthological Society of Washington* 39, 18–22.

Putz, R.E. (1972b) Biological studies on the hemoflagellates *Cryptobia cataractae* and *Cryfobia salmositica*. *US Sport Fisheries and Wildlife Technical Paper* 63, 3–25.

Qadri, S.S. (1962) An experimental study of the life cycle of *Trypanosoma danilewskyi* in the leech, *Hemiclepsis marginata*. *Journal of Protozoology* 9, 254–258.

Robertson, M. (1907) Studies on a trypanosome found in the alimentary canal of *Pontobdella muricata*. *Proceedings of the Royal Physical Society of Edinburgh* 17, 83–108.

Robertson, M. (1909) Further notes on a trypanosome found in the alimentary tract of *Pontobdella muricata*. *Quarterly Journal of Microscopical Science* 54, 119–139.

Robertson, M. (1911) Transmission of flagellates living in the blood of certain freshwater fishes. *Philosophical Transactions of the Royal Society of London*, Serial B 202, 29–50.

Rohde, K. (1984) Diseases caused by metazoans: helminths. In: Kinne, O. (ed.), *Diseases of Marine Animals*, Vol. 4, part 1. Biologische Anstalt Helgoland, Hamburg, pp. 193–320.

Roubal, F.R. (1986) Histopathology of leech, *Austrobdella bilobata* Ingram, infestation on the yellowfin bream, *Acanthopagrus australis* (Günther), in northern New South Wales. *Journal of Fish Diseases* 9, 213–223.

Rupp, R.S. and Meyer, M.C. (1954) Mortality among brook trout, *Salvelinus fontinalis*, resulting from attacks of freshwater leeches. *Copeia* 1954, 294–295.

Sanjeeva Raj, P.J. (1974) A review of the fish-leeches of the Indian Ocean. *Journal of the Marine Biological Association of India* 16, 381–397.

Sawyer, R.T. (1986) *Leech Biology and Behaviour*. Vol. 2. Feeding Biology, Ecology. and Systematics. Oxford Scientific Publications, Oxford.

Sawyer, R.T. and Chamberlain, N.A. (1972) A new species of marine leech (Annelida: Hirudinea) from South Carolina, parasitic on the Atlantic menhaden, *Brevoortia tyrannus*. *Biological Bulletin* 142, 470–479.

Sawyer, R.T. and Hammond, D.H. (1973) Observations on the marine leech *Calliobdella carolinensis* (Hirudinea: Piscicolidae), epizootic on the Atlantic menhaden. *Biological Bulletin* 145, 373–388.

Sawyer, R.T., Lawler, A.R. and Overstreet, R.M. (1975) Marine leeches of the eastern United States and the Gulf of Mexico with a key to the species. *Journal of Natural History* 9, 633–667.

Shanavas, K.R. (1991) *Trypanosoma punctati* Hasan and Qasim, 1962 from *Channa punctatus* Bloch in Kerala, India, with observations on its vector-phase development and transmission. *Archiv für Protistenkunde* 140, 201–208.

Shanavas, K.R., Ramachandran, P. and Janardanan, K.P. (1989) *Trypanoplasma ompoki* sp.n. from freshwater fishes in Kerala, India, with observations on its vector-phase development and transmission. *Acta Protozoologica* 28, 293–302.

Shulman, S.S. (1961) Zoogeography of parasites of USSR freshwater fishes. In: Dogiel, V.A., Petrushevski, G.K. and Polyanski, Yu.I. (eds) *Parasitology of Fishes*. Oliver and Boyd, Edinburgh, pp. 180–229.

Siddall, M.E. and Desser, S.S. (1992a) Ultrastructure of gametogenesis and sporogony of *Haemogregarina* (sensu lato) *myoxocephali* (Apicomplexa: Adeleina) in the marine leech *Malmiana scorpii*. *Journal of Protozoology* 39, 545–554.

Siddall, M.E. and Desser, S.S. (1992b) Alternative leech vectors for frog and turtle trypanosomes. *Journal of Parasitology* 78, 562–563.

Siddall, M.E. and Desser, S.S. (1993) Ultrastructure of merogonic development of *Haemogregarina* (sensu lato) *myoxocephali* (Apicomplexa: Adeleina) in the marine leech *Malmiana scorpii* and localization of infective stages in the salivary cells. *European Journal of Protistology* 29, 191–201.

Singhal, R.N., Jeet, S. and Davies, R.W. (1986) Chemotherapy of six ectoparasitic diseases of cultured fish. *Aquaculture* 54, 165–171.

Sloan, N.A., Bower, S.M. and Robinson, S.M.C. (1984) Cocoon deposition on three crab species and fish parasitism by the leech *Notostomum cyclostoma* from deep fjords in northern British Columbia. *Marine Ecology – Progress Series* 20, 51–58.

Soós, Á. (1965) Identification key to the leech (Hirudinoidea) genera of the world, with a catalogue of the species. I. Family: Piscicolidae. *Acta Zoologica Academiae Scientiarum Hungaricae* 11, 417–463.

Spelling, S.M. and Young, J.O. (1986a) Seasonal occurrence of metacercariae of the trematode *Cotylurus cornutus* Szidat in three species of lake-dwelling leeches. *Journal of Parasitology* 72, 837–845.

Spelling, S.M. and Young, J.O. (1986b) The population dynamics of metacercariae of *Apatemon gracilis* (Trematoda: Digenea) in three species of lake-dwelling leeches. *Parasitology* 93, 517–530.

Spelling, S.M. and Young, J.O. (1986c) The occurrence of metacercariae of the trematode *Cyathocotyle opaca* in three species of lake-dwelling leeches. *Freshwater Biology* 16, 609–614.

Thompson, D.H. (1927) An epidemic of leeches on fishes in Rock River. *Bulletin of the Illinois Natural History Survey* 17, 195–201.

Verrill, E.A. (1872) Descriptions of North American fresh water leeches. *American Journal of Science* 3, 126–139.

Vojtek, J., Opravilová, V. and Vojtková, L. (1967) The importance of leeches in the life cycle of the order Strigeidida (Trematoda). *Folia Parasitologica* 14, 107–119.

Wales, J.H. and Wolf, H. (1955) Three protozoan diseases of trout in California. *California Fish and Game* 41, 183–187.

Wilkialis, J. (1970) Investigations on the biology of leeches of the Glossiphoniidae family. *Zoologica Poloniae* 20, 29–54.

Woo, P.T.K. (1979) *Trypanoplasma salmositica*: experimental infections in rainbow trout, *Salmo gairdneri*. *Experimental Parasitology* 47, 36–48.

Woo, P.T.K. (1987) Cryptobia and cryptobiosis in fishes. *Advances in Parasitology* 26, 199–237.

Wood, J.W. (1979) *Diseases of Pacific Salmon: their Prevention and Treatment*, 3rd edn. State of Washington Department of Fisheries, Hatchery Division, Olympia.

Woods, L.C., III, McCarthy, M.A., Kraeuter, J.N. and Sager, D.R. (1990) Infestation of striped bass, *Morone saxatilis*, by the leech *Myzobdella lugubris*. In: Perkins, F.O. and Cheng, T.C. (eds) *Pathology in Marine Science*. Academic Press, San Diego, pp. 277–282.

15

Fish-borne Parasitic Zoonoses

R.C. Ko

Department of Zoology, University of Hong Kong, Hong Kong.

INTRODUCTION

A large number of marine and freshwater fishes can serve as a source of medically important parasitic zoonoses which include trematodiasis, cestodiasis and nematodiasis. Some of these infections are highly pathogenic. Fish protozoans are not known to be infective to man. Fish-borne diseases are mainly acquired through eating raw or undercooked fish. Previously, these infections were only prevalent in a few countries where the local population has a predilection for such eating habits. However, in recent years, with dramatic increase in popularity of 'sushi', 'sashimi' and similar food dishes, diseases transmitted by fish will probably become more widely distributed, and have greater economical and medical impacts. Compared to malaria, schistosomiasis, filariasis, trichinellosis etc., there are relatively few recent studies on fish-borne zoonoses. There is not any diagnostic test which can be used to screen fish on a large scale for infective stages of metazoan parasites.

Generally, fish can either be intermediate host of parasites involving man as the definitive host, or harbour larval parasites of other animals which can invade human tissues for a limited period, but can undergo no further development. The latter are considered incidental infections because the natural definitive hosts for these parasites are other animals, e.g. marine mammals or birds. However, the larval stages of a few species of parasites can mature both in animals and man.

The emphasis of this chapter will be on disease, i.e. pathogenicity, symptoms, methods of transmission, diagnosis, epidemiology and treatment of the major infections. The basic morphological features of the infective-stage and the adult parasite are only briefly mentioned.

TREMATODIASIS

Fish-borne trematodiasis is important in Southeast Asia and the Far East where many people are dependent on freshwater fish as the major source of protein. Among the fish-borne parasitic diseases, infections by both large and small digenetic trematodes are the most common. Although the diseases *per se* are seldom fatal, they can cause morbidity and serious complications. The route of infection is by ingesting the metacercariae of parasites which are in muscles, subcutaneous and other tissues of freshwater fishes.

Clonorchiasis

Clonorchiasis is caused by *Clonorchis sinensis* (Cobbold, 1875) Looss, 1907 which is located in the bile duct of man. It measures 8–15 mm by about 4 mm in width and is characterized by the presence of two highly branched testes and a prominent seminal receptacle (Fig. 15.1). The worm also matures in cats, dogs, pigs and rats. Although this trematode has also been referred to as *Opisthorchis* (see Muller, 1975), the old nomenclature is retained here because it has wider acceptance among the medical profession. Furthermore, in Asia, clonorchiasis is considered as a distinct disease from opisthorchiasis (which will be described in the next section).

Epidemiology

The prevalence of clonorchiasis remains high in Hong Kong, southern China, Taiwan and South Korea.

Duchastel (1984) reported that 13.4% of 25,095 Hong Kong residents who applied to emigrate to Canada during 1979–1981 had ova in stool samples. Most applicants were students under 25 years old. There is no natural transmission of the parasite in Hong Kong because the snail intermediate hosts cannot propagate in local fish ponds which have high salinities. This disease is due to the consumption of infected cyprinid fish imported from China where the infection occurs in 23 provinces. The infection rates in the Chinese population were reported to vary from 1–48% (Li, 1991).

In Taiwan, there were three endemic areas: Miao-li in the north, Sun-moon in the middle and Mei-nung in the south. However, more recent surveys have shown that the endemic areas have become more widespread due to an increase in consumption of raw fish. The infection seemed to be more common in Hakkanase and farmers than in other groups (Chen, 1991a).

In South Korea, stool samples of 13,373 people living within 6 km of some rivers and had an overall infection rate of 21.5% (Seo *et al.*, 1981). The rate was as high as 40% in the Nakdong River basins. A more recent study by Soh (1984) reported the rate was more than 10% in endemic areas. Rim (1986) extensively reviewed the epidemiology of the disease in South Korea and other countries.

The southern Chinese commonly acquire the infection by eating rice congee laden with pieces of raw or poorly cooked carp. Raw carp is also consumed during special festivals, e.g. Chinese New Year. The Japanese like to eat raw carp (particularly *Cyprinus carpio*, *Carassius carassius*) and Koreans also consume raw fish especially during drinking parties.

Biology of parasite

Adult worms in the bilary passage lay operculated ova which are passed into the external environment via the intestine. Each ovum (26–30 × 15–17 µm) is fully embryonated with a ciliated miracidium when laid (Fig. 15.2). The egg is characterized by a small protuberance on the posterior extremity. In water, the miracidium hatches and enters the first intermediate host (freshwater snails) either by direct penetration or ingestion. Inside the snail, the miracidium will develop into a sporocyst, then a redia, and finally the free-swimming cercaria which are released. These penetrate the skin of the second intermediate host (freshwater fish) to become the metacercaria. The infective metacercaria (135–145 × 90–100 µm) is enclosed by a thick cyst wall. It is located mainly in muscles and subcutaneous tissues but somtimes also on the scales, fins and gills of fish (Fig. 15.3).

The important first intermediate hosts are *Parafossarulus manchouricus*, *Bithynia* (= *Alocima*) *longicornis*, and *Bulimus fuchsiana*. Other susceptible species include *Semisulcospira libertina*, *Thiara granifera*, *Assiminea lutea*, and *Melanoides tuberculata*.

Many freshwater fishes, and in particular members of Cyprinidae, serve as second intermediate hosts. Yoshimura (1965) listed 81 species which are susceptible to *C. sinensis*, including 71 species in the Cyprinidae, two species each in the Ophiocephalidae and Eleotridae, and one species each in the Bagridae, Cyprinodontidae, Clupeidae, Osmeridae, Cichlidae and Gobiidae. In China, 43 species of freshwater fishes (Cyprinidae, Cyprinodontidae, Ophiocephalidae, Eleotridae and Gobidae) have been found infected. Among these, the most heavily and widely distributed species are *Pseudorasbora parva*, *Rhodeus ocellatus* and *Cultriculus kneri*. However, the major species which are commonly eaten raw are *Ctenophargyndon idellus*, *Hypothalmicthys molitrix*, *Mylopharyngodon aethiops* and *Culter aburnus*. In southern China, the traditional practice of building 'out-houses' above carp ponds and the use of night soil for pond enrichment help to maintain the infection in the cultured fish population (Fig. 15.4). In Taiwan, *Tilapia mossambica* and *Ophiocephalus maculatus* are two of the common species which are eaten raw. In South Korea, 36 species of fish belonging to Cyprinidae, Bargridae and Clupeidae can serve as the second intermediate host. The important species are, *P. parva*, *Sarcocheilichthys sinensis*, *Hemibarbus labeo*, *Acanthorhodeus gracilis*, *A. taenianalis*, *Puntungia herzi*, *Pseudogobio esocinus*, *Gnathopogon* spp., *Acheilognathus intermedia* and *C. kneri* (see Rim, 1986; Soh and Min, 1990). In Japan, Komiya and Suzuki (1964) reported that metacercariae were frequently found in *P. parva*, *Sarcocheilichthys variegatus*, *Acheilognathus lanceolata* and *Tribolodon hakonensis*. However, the importance of these fishes as important source of infection in recent years is not known.

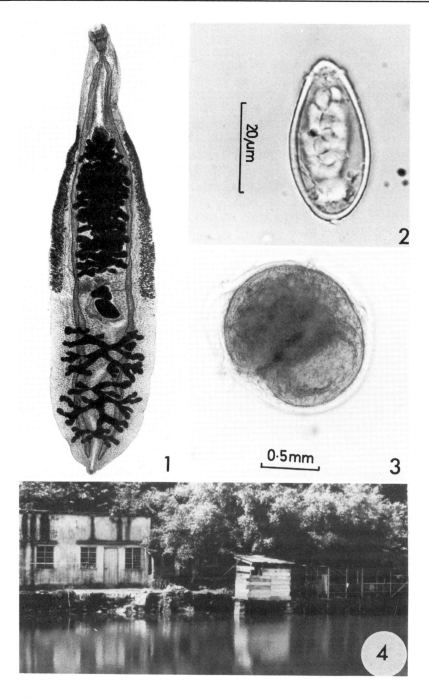

Fig. 15.1. *Clonorchis sinensis* whole worm (original, ×6). **Fig. 15.2.** Egg of *Clonorchis sinensis* recovered from faeces (original). **Fig. 15.3.** Metacercaria of *Clonorchis sinensis* (original). **Fig. 15.4.** A carp pond in Hong Kong, showing the location of 'out-house' over pond (original).

Pathogenesis

The pathology of clonorchiasis was studied extensively (Ong, 1962; Hou and Pang, 1964; Gibson and Sun, 1965; Mcfadzean and Yeung, 1966; Chan and Teoh, 1967). The infection has also been implicated in recurrent pyogenic cholangitis (Cook *et al.*, 1954; Ong, 1962), cholangiohepatitis (Fung, 1961) and cholangiocarcinoma (Hou, 1956; Belmaric, 1973). However, there are few recent reports on the pathogenesis of this parasite.

The histopathology of clonorchiasis was first described in detail by Hou (1955). After ingestion, the excysted metacercaria reaches the bile duct by direct migration from the intestine. The adult worms ingest blood in the bile ducts. Hepatic parenchymal damage and portal hypertension are usually absent in light and uncomplicated infections. The lesions are usually localized in the distal bilary passages, particularly those in the left lobe of liver. The extent of the lesions are dependent on worm burden and on the presence of a secondary bacterial infection. The latter complication commonly leads to biliary obstruction as a result of extensive adenomatous proliferation, calculi and cholangitis. During the acute stage of infection, the mucosa of the bile ducts is oedematous, with the desquamation of the biliary epithelium. Later, as a result of inflammatory responses, the bile ducts become thickened and crypt formation occurs. Metaplasia of the epithelial cells follows, resulting in the proliferation of glandular-like structures in the mucosa which are responsible for the production of mucin into the bile. During chronic and heavy infections, the continuous hyperplasia may lead to the formation of adenomatous tissue. This is probably caused by continuous mechanical and chemical stimulations (Flavell, 1981). Recurrent attacks of suppurative cholangitis may occur as a result of the biliary obstruction by the trematodes. Atrophy of the bilary epithelium and underlying hepatic cells may also develop.

Clonorchiasis may also cause pancreatitis (Hou, 1955). Hou reported that as many as 50 worms were recovered from pancreatic ducts. The main ducts are dilated. There is a limited degree of periductal fibrosis and adenomatous tissue formation but considerable squamous metaplasia. Mcfadzean and Yeung (1966) reported that in a series of 110 Chinese patients in Hong Kong who were suffering from acute pancreatitis of unknown aetiology, 83% were infected with *C. sinensis*.

Hou (1956) was first to note a close correlation between clonorchiasis and primary liver carcinoma. The close association between this trematode infection and cholangiocarcinoma were further elaborated by Belramic (1973), Purtillo (1976) and Schwartz (1980). Ona and Dytoc (1991) reported two cases with unusual manifestations. Kim (1984) suggested that exogenous carcinogenic promotors, nutritional, immunological and genetical factors, probably all induce goblet cell metaplasia and dysplasia of the bile duct epithelial cells, resulting in carcinoma.

Symptoms

In endemic areas, patients with an enlarged liver and a history of eating raw or undercooked freshwater fish should be examined for clonorchiasis. In light and acute infections, more than half of the patients are usually asymptomatic. However, some show general malaise, abdominal discomfort, occasional diarrhoea, slight fever, eosinophilia (5–20%), gall-bladder syndrome, paroxymal epigastric pain, mild jaundice and hepatomegaly. In heavy and chronic infections, loss of weight, lassitude, mental depression, anorexia, paresthesia, palpitation, tachycardia, headache, jaundice, tenderness of liver, tetanic cramps, tremors and toxaemia may occur (Belding, 1965; Kim *et al.*, 1982).

Laboratory diagnosis

The traditional laboratory diagnostic method is by stool examination. Rim (1988) reviewed some of the more recent methods. Although the infection can readily be detected by finding the charactersistic ova in faecal samples, serological tests are sometimes necessary when there is obstruction of biliary canals. Various serodiagnostic tests, which are based on detecting circulating antibodies, have also been used. However, the results are not satisfactory because a specific *Clonorchis* antigen (which will not cross-react with those of other trematodes) is still unavailable.

Three clinical stages are classified according to the pattern of antibody responses at various times postinfection. In experimentally infected cats, during the acute stage of infection, the major circulating antibodies are IgM, IgA and IgG. Peak antibody titres for the first two immunoglobulins occur at three and 11 weeks postinfection. In the subacute stage, only IgA and IgG predominate and in the chronic stage, only IgG is present (Chen *et al.*, 1989; en *et al.*, 1992).

The complement fixation test (CF) was first used by Ryoji (1922) and subsequently by Ito (1925), Chung *et al.* (1955), Wykoff (1959) and Sadun *et al.* (1959). Takei and Chun (1976) attempted to produce specific fractions from the crude extracts using gel filtration on Sephadex G-100 column, followed by disc electrophoresis. A fraction, KD1, containing protein and carbohydrate was produced and tested, using serum samples from 32 clonorchiasis patients. All the tests were positive. Negative reactions were observed in samples from uninfected persons and patients with schistosomiasis, ascariasis, filariasis, and hookworms. However, the specificity of this fraction has not been fully ascertained because only a limited cross-reactivity study was undertaken.

The precipation test, gel diffusion (DD), immunoelectrophoresis (IE), indirect fluorescent antibody test (IFA) and indirect haemagglutination test (IHA) have also been tried but the results were not satisfactory (Mori, 1957; Phillipson and Mcfadzean, 1962; Sun and Gibson, 1969; Lee, 1975; Pacheco *et al.*, 1960; Im, 1974; Cho and Soh, 1974, 1976; Kwon *et al.*, 1984).

Sawada *et al.* (1976) attempted to produce a specific antigen for IHA by using Biogel-A filtration. A fraction, K6-H, consisting of protein and carbohydrate, was tested on serum samples of the following patients: 38 clonorchiasis, three paragonimiasis, five schistosomiasis, ten ascariasis and five ancylostomiasis. Positive reactions were only observed in the samples from clonorchiasis patients. However, similar to the KD1 antigen, further studies should be carried on this fraction to evaluate its specificity.

Yen and Chen (1989) evaluated the counterimmunoelectrophoresis (CIE) method, on 70 clonorchiasis patients and seven cases of other parasitic infections. Antibody was detected in all clonorchiasis patients. Cross-reactions, however, were also observed in sera from patients with toxocariasis and angiostrongyliasis.

Using the charcoal granule agglutination test, Hu *et al.* (1989) observed positive reactions in 83.5% of 79 clonorchiasis sera and 19.5% of 41 sera from cases of other parasitic diseases.

The enzyme-linked immunosorbent assay (ELISA), has also been applied to diagnose clonorchiasis in man and animals (Jin *et al.*, 1983; Yang *et al.*, 1983, 1984; Hahm *et al.*, 1984; Han *et al.*, 1986 Sirisinha *et al.*, 1990; Yong *et al.*, 1991). Hahm *et al.* (1984) reported that ELISA detected 78% of 55 known clonorchiasis sera whereas IFA and CF detected 62% and 56% respectively. However, due to the non-specific nature of the antigens used, the ELISA studies reported in the literature commonly show significant cross-reactions. Future attempts must be directed towards the isolation of specific antigens using monoclonal antibodies, *in vitro* culture, chemical and electrical fractionation, and other immunochemical methods. Various ways of capturing antibodies or antigens should also be explored more extensively.

Immunoaffinity purification of whole worm extracts and characterization of egg and metacercaria antigens were carried out by Lee *et al.* (1988). Egg antigens vary from 15,000–200,000 M_r, metacercaria antigens from 15,000–100,000 M_r and adult crude antigens from 11,000–80,000 M_r. ELISA was more sensitive to the adult crude antigens than to those of metacercaria. Negative reactions were observed when egg antigens were used.

Significant cross-reactions were observed between *Clonorchis* and *Opisthorchis* in ELISA. Immunoblotting and radioimmunoprecipitation showed that a 89 kDa metabolic antigen which was predominant in *Opisthorchis viverrini*, reacted with antisera from both groups of patients. However, a 16 kDa antigen from *O. viverrini* appeared to be specific (Sirisinha *et al.*, 1990).

Recently, a monoclonal antibody, CsHyb 065–23, was produced and used in an ELISA-inhibition test. 78.5% of known clonorchiasis cases were tested positive. False positives were not found in sera from 16 paragonimiasis patients and 28 negative controls (Yong *et al.*, 1991). Nevertheless, the level of sensitivity of the test is too low for serodiagnosis.

ELISA has also been employed to monitor serum antibodies in patients after treatment with praziquantel. In five of ten patients, the titres returned to normal after one year (Chen *et al.*, 1987a,b).

Besides serodiagnosis, computed tomography scanning (CT) and ultrasound have also been used to diagnose clonorchiasis. In a study involving 17 patients with clonorchiasis alone and 25 patients with clonorchiasis and hepatobiliary malignancies, CT showed a diffuse, minimal or mild dilation of the intrahepatic bile ducts in 14 cases. However, a diffuse dilation of the ducts was observed in all the patients with both parasite and biliary malignancies (Choi *et al.*, 1989).

In a study of 947 clonorchiasis patients using ultrasound, gall stones were seen in 85 cases, common bile duct stones in three and hepatic stones in one

(Hon *et al.*, 1989). Lim *et al.* (1989) also presented sonographic findings in 59 proven cases.

In heavy infections, intraoperative cholangiography can show multiple filling defects in the common bile duct, common hepatic duct and left hepatic duct (Nishioka and Donnelly, 1990).

Treatment

Until praziquantel became available, there was no effective drug to treat clonorchiasis. Chen (1991b) reported a 96% cure rate among 356 patients with praziquantel ($60\,\text{mg}\,\text{kg}^{-1}$ of body weight, divided into three doses in one day).

Opisthorchiasis

Opisthorchiasis is a zoonosis which is mainly caused by *Opisthorchis viverrini* (Poirrier, 1886) Stiles and Hassall, 1896 and *O. felineus* (Rivolta, 1884) Blanchard, 1895. The natural definitive host for the former species is the civet cat (*Felis viverrini*). Cats, dogs, foxes and pigs are the common hosts for the latter. The pathogenesis and biology of these two trematodes are similar to those of *C. sinensis*.

Opisthorchis viverrini

Human infection is prevalent in Southeast Asia especially in northern Thailand where it is not only a medical problem, but also an economic impediment. The latter includes the high cost of mass treatment and the limitation of water resources development.

The adult worms (about $5.5–9.6\,\text{mm} \times 0.8–1.7\,\text{mm}$) live in the biliary passage. This species is distinguished from *C. sinensis* by having lobed testes and a different pattern of the flame cells in the cercaria and metacercaria.

EPIDEMIOLOGY

According to Harinasuta (1984), about 5–6 million people in northern Thailand are infected with this trematode. The infection rates vary from 10–90%. Males are more commonly infected than females and the prevalence is higher in adults. Kasuya *et al.* (1989) found that 7.5% of 491 primary school children in Chiangmai was infected. Maleewong *et al.* (1992c) reported that the prevalence was 66% in 2412 persons in northeast Thailand.

A pilot control programme for opisthorchiasis has recently been conducted in Khon Kaen Province in northeast Thailand. It is implemented by establishing a strong educational programme to change the eating habits, and by giving a free annual treatment of $40\,\text{mg}\,\text{kg}^{-1}$ dose of praziquantel (Sornmani, 1990).

The disease can be acquired by eating a local dish called 'Koi-pla' which contains infected raw fish. The high prevalence of this infection is mainly due to the eating and defaecation habits of the local inhabitants as well as the common occurrence of the snail intermediate hosts. Nevertheless, the infection is more commonly acquired at the end of the rainy season and the beginning of the dry (September to February) when the fish can be caught easily (Wykoff *et al.*, 1965).

BIOLOGY OF PARASITE

The adult worms in the biliary passage lay thousands of operculated ova per day. The ova (29 × 16μm) are discharged to the outside via the intestine. They are more elongated than those of *C. sinensis* and without a clear shoulder at the operculum.

Freshwater snails are the first intermediate hosts. The development of the parasite in snails is similar to that of *C. sinensis*. In Thailand, *Bithynia goniomphalus* is the important snail host in the northeast, *B. funiculata* in the north and *B. laevis* in the central region. The infection rate in snails in endemic areas is about 0.5% (Harinasuta, 1984).

The metacercariae (220 × 116μm) are located in the muscles of freshwater fish which are the second intermediate hosts. The important cyprinid fish hosts in northern Thailand are: *Cyclocheilicthys siaja*, *Hampala dispar* and *Puntius orphoides*. In highly endemic regions, the infection rates in fish are as high as 50–90%, each fish harbouring 20–50 metacercariae (Harinasuta, 1984).

PATHOGENESIS

Opisthorchiasis is a slow disabling disease (Riganti *et al.*, 1989). In highly endemic regions, recurrence of the infection is common and this often results in a heavy worm burden. Similar to clonorchiasis, the presence of a large number of worms in the bile duct and its canaliculi, the gall-bladder and hepatic ducts can elicit adenomatous hyperplasia of the bilary epithelium, resulting in the thickening of walls. In chronic infections, obstruction of the biliary system can occur. A heavy infection can be fatal because it leads to cirrhosis of the liver, cholangiocarcinoma or liver carcinoma and hepatic cyst formation. Using ultrasound, cholangiocarcinoma was shown to occur in eight of 87 patients with hepatomegaly or opisthorchiasis. This was confirmed by CT and endoscopic retrograde cholangiopancreatography (Vatanasapt *et al.*, 1990). Pungpak *et al.* (1990) suggested that the measurement of tumour markers, CA125 and CA19–9 may be useful in the early detection of opisthorchiasis associated cholangiocarcinoma.

Abnormalities have been observed in the high density lipoprotein content of the serum from opisthorchiasis patients. These are in form of low cholesterol and cholesteryl ester concentrations but high triglyceride. This indicates a disturbance of the synthesis of the high density liprotein or the removal of lipid content in lipoproteins (Changbrunrung *et al.*, 1988).

SYMPTOMS

Clinical manifestations cannot be observed in light infections (less than 1000 eggs g^{-1} of faeces). In moderate or heavy infections, there may be a slight fever, eosinophilia, diarrhoea, flatulent dyspnea, transient urticaria, weakness, anaemia, pain and 'hot sensation' over the liver which can be slightly enlarged, and jaundice. In chronic heavy infections, symptoms of gastroenteritis may occur together with cachexia, oedema and malnutrition (Belding, 1965; Harinasuta, 1984).

LABORATORY DIAGNOSIS

Although faecal examination is the traditional diagnostic method, as in clonorchiasis, serodiagnostic tests have also been used.

Aqueous somatic extracts of whole worm, Triton X-100 and dexoycholate extracts of surface and tegument as well as excretory/secretory antigens (ES) were analysed using radiolabelling, immunoprecipation, and polyacrylamide gel electrophoresis. A 89 kDa protein antigen was found to be predominant in the ES products (Wongratanacheewin and Sirisinha, 1987; Wongratana-cheewin et al., 1988; Amorpunt et al., 1991).

A 'sandwich' ELISA method has been used to detect parasite antigens in faecal samples (Sirisinha et al., 1992; Chaicumpa et al., 1992). An IgG monoclonal antibody, which was specific to the 89 kDa antigen, was first captured on a microtitre well by rabbit anti-mouse IgG. As little as 0.1 ng of parasite antigens could be detected. A specific DNA probe was constructed from a repetitive DNA containing 340 basepairs, which was used in a dot-blot hybridization of parasite DNA. The probe could detect DNA released from as few as five ova or even detect the presence of metacercariae in fish (Sirisinha et al., 1991). However, the disadvantage of the DNA probe is that this method cannot be used under field conditions.

The ribosomal RNA gene from this trematode has been cloned and characterized (Korbsrisate et al., 1992).

TREATMENT

Praziquantel, at a dosage of 25 mg kg^{-1} of body weight three times a day for one to two days, is effective. For mass treatment programmes, a single dose (40 mg kg^{-1}) is recommended (Harinasuta, 1984).

Opisthorchis felineus

Opisthorchiasis caused by O. felineus occurs in Poland, central and eastern Europe and Siberia. The highly endemic areas also include the Dnieper, Donetz and Desna Basins. The cat is the common host.

Adult worms are usually located in the distal and sometimes in the proximal bile ducts. This species morphologically resembles O. viverrini except its testes are less lobate and situated further from the ovary. The appearance and distribution of the vitellaria are also different. The two species are also distinguished by the pattern of flame cells in the cercaria and metacercaria (Belding, 1965).

EPIDEMIOLOGY

The disease usually occurs in villages and localities near rivers or reservoirs in Siberia and Central Europe where fish are eaten raw. In the Ukraine, most of the infections are acquired by eating fish on the first day of salting (see Muller, 1975). However, the prevalence of this infection in man is substantially lower than that of O. viverrini.

BIOLOGY OF PARASITE

The biology of this species is similar to that of *O. viverrini*. The first intermediate host is *Bithynia leachii* and cyprinids are the second intermediate
hosts. The important fish are *Cyprinus carpio*, *Idus melanotus*, *Abramis
brama*, *Barbus barbus*, *Tinca tinca*, *Blicca bjoerkna*, *Leuciscus rutilis* and
Scardinius erythrophthalmus (see Belding, 1965).

PATHOGENESIS, SYMPTOMS AND TREATMENT

Hepatic lesions, symptoms and treatment are similar to those of *O. viverrini*.

LABORATORY DIAGNOSIS

Faecal examination is the common method. The ovum is morphologically
similar to that of *O. viverrini*. Serodiagnosis has also been employed. Teplukhin
et al. (1987) used IHA and somatic or ES antigens to detect antibodies. They
concluded that somatic antigens were 1.5 times more sensitive than ES
antigens. Gitsu *et al.* (1987) reported that ELISA detected 31.8% of 323
patients passing less than 10 eggs g^{-1} of faeces and 86.3% with more than
1000 eggs g^{-1}.

SMALL TREMATODES

Besides *O. viverrini*, infections by numerous species of small trematodes
(0.5–3 mm) in length especially members of the Heterophyidae are also common in Southeast Asia and the Far East (northern Thailand, Laos, Cambodia,
Vietnam, Indonesia, the Philippines and South Korea). Infections have also
been reported in Hawaii, Egypt, Israel, Romania, Greece and eastern Siberia.
Some of the more important species are: *Heterophyes nocens*, *H. continua*,
H. heterophyes, *H. dispar*, *Heterophyopsis continua*, *Haplorchis taichui*,
H. pumilo, *Metagonimus yokogawai*, *M. takahashii*, *Pygidiopsis summa*, *Dior-
chitrema* (= *Stellantchasmus*) *falcatus*, *Stictodora fascatum*, *Centrocestus
armatus*, *Echinostoma hortense*, *E. cinetorchis*, *Echinochasmus japonicus*,
etc. (see Belding, 1965; Muller, 1975; Chai *et al.*, 1988; Chai and Lee, 1991).

Transmission is by eating raw or recently salted/pickled fish. *H. heterophyes* infection is common in Egypt where the pickled mullet (*Mugil cephalus*)
is traditionally eaten at the feast of Sham-al Nessim (Muller, 1975).

A more recent epidemiological study by Ahn *et al.* (1987) on *M. yokogawai*
in South Korea showed that 6.6% of faecal samples from 7357 inhabitants
of the eastern coast of Kanywon Province was positive for the infection. The
infection rate was higher in females than males. Metacercariae were found in
12 of 50 *Plecoglossus altivelis* caught in local streams.

Snails are the first intermediate host of these trematodes, e.g. *Tarebia
granifera* for *Haplorchis* spp.; *Pirenella conica*, *Cerithidea cingulata microp-
tera* for *Heterophyes* spp.; *Semisulcospira libertina*, *Thiara granifera* for
Metagonimus spp. Inside the snail, the development of the trematode usually
includes sporocyst, one to two generations of redia and cercaria.

The metacercariae are in various species of freshwater or brackish water fishes. Members of the Cyprinidae, Siluridae, and Coltidae commonly serve as the second intermediate host. Examples of the fish host are: for *Heterophyes* spp., *Mugil cephalus*, *M. japonicus*, *Tilapia nilotica*, *Acanthogobius*, *Sciaena aquilla*, *Solea vulgaris*; for *Metagonimus* spp., *Leuciscus hakuensis*, *Odontobutis obscurus*, *Plecoglossus altivelis*, *Salmo perryi*, *P. parva*, *Carassius carassius*; for *H. pumilo*, *Arius manilensis*.

Man acquires the infections by eating poorly cooked or raw fish and the metacercariae will develop into adults in the intestine. Concurrent infections of several species of the small sized trematodes are common. Therefore, the clinical manifestations of each species are not well documented. In general, the parasites only cause mild intestinal symptoms such as tenderness, chronic bloody diarrhoea, colicky pains, vomiting etc.

In heavy *Heterophyes* infections, superficial necrosis of the intestinal mucosa may occur. Occasionally, ova may enter the mesenteric circulation and produce granulomatous lesions in the myocardium and brain.

Diagnosis is by finding the characteristic ova of the various species in faecal samples. However, it is extremely difficult to differentiate them morphologically. An attempt was undertaken by Lee *et al.* (1993) to diagnose metagonimiasis using ELISA.

The treatment method is similar to that given for opisthorchiasis.

CESTODIASIS

There are relatively few cases of fish-borne cestode infections in man. The cestodes which mature in the small intestine of man are not very pathogenic and the diseases are never fatal. Diphyllobothriasis is the major cestodiasis transmitted by freshwater, marine and anadromous fishes.

Diphyllobothriasis

The disease is caused by pseudophyllid cestodes belonging to the genus *Diphyllobothrium*. The genus is characterized by a holdfast structure at the anterior extremity called bothrium (Fig. 15.5) and a highly coiled uterus with an uterine pore in mature proglottids. The adult worm grows to 3–10 m in length, comprising of about 4000 proglottids (each 2–4 mm × 10–12 mm). Most human infections are accidental as the natural definitive hosts of the genus include birds, bears, seals, dogs, walrus, etc. The method of infection is by ingesting raw or smoked fish which harbours the plerocercoid larva. However, there was one case which was presumably due to eating raw euphausiid crustaceans (Fukumoto *et al.*, 1988a).

The following are some of the major species which have been found in man: in North America and Europe, *D. latum* (Linnaeus, 1758) Luhe, 1910; in Greenland and Iceland, *D. cordatum* Leuckart, 1863; in the Pacific coast of South America, *D. pacifica* (Nybelin, 1931) Margolis, 1956; in Alaska, *D. dendriticum* Nitzsch, 1824, *D. alascense* Rausch and Williamson, 1958, *D.*

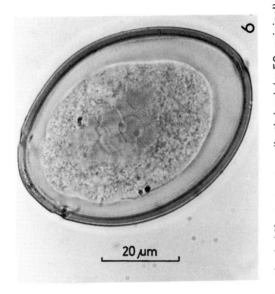

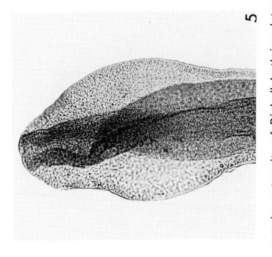

20 μm

Fig. 15.5. The anterior extremity of *Diphyllobothrium latum* showing the holdfast structure (bothrium) (× 50, original).
Fig. 15.6. Egg of *Diphyllobothrium latum* (original).

lanceolatum (Krabbe, 1865) Rausch and Hillard, 1970, *D. ursi* Rausch 1954, *D. dalliae* Rausch, 1956; in Japan, *D. nihonkaiense* Yamane *et al.* 1986, *D. hians* (Diesing, 1850) Markowski, 1952, *D. cameroni* Rausch, 1969, *D. yonagoense* Yamane *et al.*, 1981, *D. scoticum* (Rennie and Reid, 1912) Markowski, 1952.

Recently, several species have been reported in man in Japan; two of them were considered as new (Yamane *et al.*, 1981, 1986; Fukumoto *et al.*, 1988a; Kamo *et al.*, 1988a,b). Oshima (1984) suggested that the parasite from cherry salmon (*Oncorhynchus masou*) in Japan was probably not conspecific with *D. latum* described in North America and Europe because the life cycle of the former involves an anadromous host (salmon) whereas the latter has a freshwater life cycle. Also, the mature proglottid of the worms in Japan shows a horizontal cirrus sac and a small seminal vesicle whereas in *D. latum*, the cirrus sac is oblique. Clinically, megaloblastic anaemia is often concomitant in the European infections but not in the Japanese cases.

However, the taxonomic status of many species within this genus is confusing. The species are often distinguished on questionable morphological characters. Dick *et al.* (1991), Karlstedt *et al.* (1992) and Matsura *et al.* (1992) noted that it is extremely difficult to separate the different species by classical methods and they are investigating the possibility of using DNA probes for species identification.

Epidemiology

Sporadic cases of diphyllobothriasis are known to occur in many parts of the world and several cases were recently reported in Japan (Fukumoto *et al.*, 1988a; Hirai *et al.*, 1988; Kamo *et al.*, 1988a,b). According to Oshima (1984), the first scientific report on diphyllobothriasis in Japan was concerned with the self infection of Professor Iijima of Tokyo University. He swallowed several plerocercoids recovered from salmon at the Tone River (Iijima, 1889). The incidence of diphyllobothriasis started to decline from 1920 and, by the end of the 1950s, no new cases were reported. However, from the 1960s onwards, the disease reappeared along the coast of the Sea of Japan, from Hokkaido Island to Shimane Prefecture. About 100 cases are reported every year. Outbreaks usually occur at the beginning of April and subside by August, i.e. about one month after the fishing season for cherry salmon (Oshima and Waikai, 1983).

In Hokkaido, cherry salmon is the major source of the infection, with the humpback salmon playing a minor role. In a survey undertaken between 1977 and 1982, the infection of cherry salmon varied from 15.9–48.4% and the mean number of worms per infected fish was 2–3. Since 1955, the catches of cherry salmon have markedly increased, from several hundred tons before the Second World War, to about 4000 tons annually. With increasing affluence, more Japanese can afford to eat raw cherry salmon fillet as 'sushi' or 'sashimi'. This socioeconomical change mainly accounts for the resurgence of diphyllobothriasis in the country. However, there is evidence that the cherry salmon in Japan are not infected prior to their seaward migration. The majority of the migrating fish in the Sea of Japan are actually originated from the Siberian coastal area, e.g. Saghalin and Kamchatka Peninsula (Oshima and Waikai, 1983).

In Peru, 136 cases of diphyllobothriasis caused by *D. pacifica* were reported between 1962 and 1976 (Arambulo and Thakur, 1990). The infection is commonly acquired by eating a popular local dish called 'ceviche' which consists of raw marine fish marinated with lime juice.

Infections are also common in Baltic countries and the former USSR where raw or smoked fish is eaten. Slices of raw fish known as 'Strogonina' are the main source of infection. A study undertaken in the region near the Krasnoyark reservoir on River Enisel, Russia, between 1984 and 1985 showed that 7.7% of 679 villagers were passing *D. latum* ova as well as 27% of 34 dogs. 87–95% of *Esox lucius* and 11.4–17.5% of *Perca fluviatilis* harboured plerocercoids. This focus of infection occurred about ten years after the construction of the reservoir (Plyuscheva *et al.*, 1987a). Plyuscheva *et al.* (1987b) discussed the role of anthropogenic factors in the establishment of infection foci in reservoirs. They suggested that foci will not be established in heavily polluted industrial areas because the intermediate hosts cannot survive in such environments.

In the western coastal region of the Okhotsk Sea, 1% of 1846 human faecal samples taken in the Khabarorsk Territory was positive for *D. latum* ova. Plerocercoids were found in 27.3% of *O. keta*, 23.5% of *O. gorbuscha* and 30.7% *Salvelinus leucomaenis* (see Dovgalev, 1988).

More recent epidemiology data were given by Curtis and Bylund (1991), Muratov *et al.* (1992) and Revenga (1993).

Biology of parasite

Only the life cycle of *D. latum* is described briefly. The adult worm lives in the small intestine where it produces unembryonated ova. Each ovum, which is operculated and with a small knob at the opposing end, measures about 55–76 × 40–60μm in size (Fig. 15.6). In water, it embryonates into a ciliated larva (coracidium) which bears three pairs of hooks. After hatching, the free-swimming coracidium is ingested by copepods (e.g. *Cyclops* spp., *Diaptomus* spp.), the first intermediate host. The procercoid develops in the haemocoele. It has three pairs of hooks at an expanded posterior region called the cercomere.

In the second intermediate host (fish) the procercoid transforms into a worm-like, wrinkled plerocercoid stage (metacestode, 10–20 mm in length) in muscles, viscera and connective tissues. The plerocercoid is the infective stage and is found in plankton-feeding as well as carnivorous fishes. The latter act as paratenic hosts. Members of the Salmonidae, Percidae and various species which are common in lakes and reservoirs can be infected e.g. *Esox lucius*, *Stizostedion vitreum*, *S. canadense*, *Lota maculosa*, *L. lota*, *Perca flavescens*, *Oncorhynchus nerka*, *O. keta*, *O. gorbuscha*, *Salmo gairdneri*, *Salvelinus leucomaenis*, *S. malma*, *S. namaycush*, etc.

Pathogenesis

The adult worm attaches to the mucosa of the ileum and occasionally, the jejunum. Although a catarrhal condition may be produced in the mucosa, the parasite generally elicits little pathogenicity. For unknown reasons, the worm depletes vitamin B12 from its definitive host for its own extensive growth and development.

Symptoms

Most infections are asymptomatic but sometimes, epigastric pains, cramp-like abdominal pains, diarrhoea or constipation, nausea, vomiting, loss of weight, anorexin, and eosinophilia may occur. The major clinical manifestation in some patients is pernicious anaemia due to vitamin B12 deficiency, and there is central nervous system involvement. Typically, this includes paresthesia, numbness, impaired vibration sense and weakness. Folate absorption may also be decreased.

A 13–20-fold increase in lysozyme activity has been found in urine of patients (Gosteva *et al.*, 1991).

Laboratory diagnosis

A brief review on diagnosis was given by Kamo (1988). The traditional method is by stool examination. The operculated ova are relatively large and can easily be seen.

Serodiagnosis is seldom used. Antigenic cross-reactivities between the various species have not been fully established. Fukumoto *et al.* (1988b) compared the antigenic structures between *D. latum* and *D. nihonkaiense* by immunoelectrophoresis. Some differences were observed. Using heterologous antiserum, *D. latum* and *D. nihonkaiense* were found to share only six bands. But the number of characteristic bands for the former was 21 whereas for the latter, it was 14.

Treatment

The drugs of choice are niclosamide (Yomesan), bithionol (Bitin) and para-momycin sulphate. The dosages for the first two drugs are: Yomesan, 1 g given early in the morning, then 1 g after one hour; bithionol, 50–60 mg kg^{-1} of body weight in a single dose.

Hayashi and Kamo (1983) found that the administration of paramomycin sulphate would instantly immobilize the worm and then destroy the tegument especially in the anterior region. This would loosen the holdfast structure. A single dose of 50 mg kg^{-1} of body weight can result in 100% efficacy.

NEMATODIASIS

Fish-borne nematodiases are generally caused by the incidental infection of man with nematodes whose natural definitive hosts are marine mammals, birds, pigs or other animals. Freshwater, brackish or marine fishes are the second intermediate host. In most infections, the worms can only survive for a limited period after the initial invasion of the gastrointestinal tract. The method of infection is by ingesting the infective larvae which are located in the muscles, intestine or viscera of fish. Unlike cestodiasis, some nematode infections can be fatal.

Capillariasis

Capillariasis is caused by *Capillaria philippinensis* Chitwood *et al.*, 1968. This species is most unusual because it can mature either in man or in experimentally infected monkeys and birds. Moreover, the worm can either produce eggs or larvae. The disease is relatively new because it was known to occur in man only in 1963 when the nematode was first found at the autopsy of a Filipino in Manila (Chitwood *et al.*, 1964). *C. philippinensis* has since been recognized as a medically important parasite.

A brief update on the disease was given by Cross (1990). *C. philippinensis* is a trichuroid nematode with a narrow and filiform anterior region. The male worms measure 1.5–3.9 mm in length; the females are 2.5–5.3 mm in length. A trichuroid nematode is also characterized by a glandular structure called the stichosome which is located posterior to the muscular oesophagus. Most known species have extraordinary life-cycles. However, *C. philippinensis* is the only known capillarid which is transmitted to man via fish. The life-cycle has only been established by experimental infections of monkeys, Mongolian gerbils, fish and birds (Cross *et al.*, 1972; Bhaibulaya and Indra-Ngarm, 1979). Presumably the natural cycle is similar (Cross and Basaca-Sevilla, 1989).

Epidemiology

Capillariasis was originally presumed to be an indigeneous disease of the Philippines where an epidemic was first recorded in 1967. However, subsequently, the disease was also found in Thailand, Japan, Taiwan, Indonesia, Korea, Iran, Egypt and India (Nawa *et al.*, 1988; Cross and Basaca-Servilla, 1991; Chunlertrith *et al.*, 1992; Chichino *et al.*, 1992; Lee *et al.*, 1993; Kang *et al.*, 1994).

In the Philippines, capillariasis has been reported from several islands, i.e. Luzon, Bohol, Mindanao and Leyte. In 1967, the number of confirmed new cases was 1037, with 77 deaths but by 1989, the incidence decreased to nine, with no death. This is probably due to the improved recognition of the disease and the establishment of an effective therapeutic regimen using mebendazole. More cases were reported during the rainy season (from May to November). The infection rate was significantly higher in males than females. Of 1849 cases, 70% were male and the prevalence was highest in the 20–29 age group. This is probably attributed to the fact that, in the endemic areas, most men are farmers who occasionally fish in lagoons, lakes and rivers. They frequently eat part of their catch raw as lunch. Young men also like to gather at night in their village to drink 'basi' (Philippine gin) and eat raw food (Cross and Basaca-Sevilla, 1989).

Only small fish (6–7 cm in length) are eaten raw. They are known locally as bagsit (*Hypseleotris bipartita*), bagsan (*Ambassis miops*) and bacto (*Eleotris melanosoma*). Larger fish are cooked before eating.

In Thailand, more than 100 cases have been reported. All of them involved fishermen and farmers in the central, northern and northeastern parts of the country. The inhabitants in these regions enjoy eating a special dish which consists of chopped raw freshwater fish, seasoned with lime juice, red peppers and juice from fermented fish (Cross and Bhaibulaya, 1983).

The few cases reported elsewhere all had a history of eating raw fish.

Biology of parasite

Several species of monkeys (*Macaca cyclopis*, *M. fascicularis* and *M. mulatta*) and Mongolian gerbils (*Meriones unguiculatus*) were successfully infected in the laboratory. Small wild rats (*Rattus* sp.) are also susceptible but the patent period is short (Cross and Bhaibulaya, 1983).

Fish-eating birds, however, are presumed to be the natural definitive host. *Amauronis phoenicuris*, *Ardeiola bacchus*, *Bulbulcus ibis*, *Nyticorax nyticorax* and *Ixobrychus sinensis* were successfully infected under laboratory conditions (Bhaibulaya and Indra-Ngarm, 1979; Cross *et al.*, 1979). A natural infection was observed in yellow bittern in the Philippines (Cross and Basaca-Sevilla, 1989).

Adult worms are located in the lumen of the small intestine. The female worms produce first-stage larvae as early as 13–14 days postinfection. The first generation larvae are retained in the intestine where they develop into a second generation of males and females in about 10 days. Most females in the second generation produce eggs (36–45 × 20 µm) which are peanut-shaped, with a thin shell and flattened bioperculated plugs (Fig. 15.7). The eggs are passed out with the faeces. But some females in the population are ovoviviparous, producing larvae and this results in autoinfection. Cross *et al.* (1978) suggested that autoinfection is an integral part of the life cycle, both initially and in maintaining the infections. However, the genetic and molecular mechanisms which control such an unique developmental process have not been studied.

Embryonation of the eggs occurs in water and, upon ingestion by freshwater or brackish water fish, first-stage larvae hatch in the intestine. The worms grow to maximal length (250–300 µm) without moulting, in about three weeks. The fully grown larvae are infective to the definitive host.

The following freshwater and brackish water fishes were infected in the laboratory: *Hypseleotris bipartita*, *Ambassis miops*, *Eleotris melanosoma*, *Scyopterus* sp., *Poecilia reticulata*, *Cyprinus carpio*, *Puntius gonionotus*, *Aplocheilus panchax*, *Gambusia holbrooki*, *Rasbora borapetenis* and *Trichopsis vittatus* (see Bhaibulaya *et al.*, 1979; Cross and Bhaibulaya, 1983).

Pathogenesis

Although capillariasis is basically an intestinal disease, untreated infections can be fatal. This is in part due to autoinfection which is an inherent feature of the life cycle. Hence the parasite population can be increased dramatically within a short period. At the necropsy of two gerbils (each given two larvae) 46 and 47 days postinfection, 2520 and 5353 worms were recovered (Cross and Bhaibulaya, 1983). The presence of a large number of parasites will invariably disturb the normal gastrointestinal functions, resulting in extensive pathophysiological lesions.

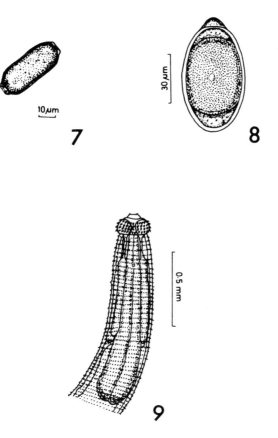

Fig. 15.7. Egg of *Capillaria philippenensis* (original). **Fig. 15.8.** Egg of *Gnathostoma spinigerum* (original). **Fig. 15.9.** Anterior region of third-stage larva of *Gnathostoma spinigerum* from catfish (*Clarias fuscus*) (original).

The lesions are usually more prominent in the jejunum. Atrophy of the crypts of Lieberkuhn occurs and the villi are denuded and flattened. The lamina propria is mainly infiltrated with mononuclear leukocytes. Ultrastructural studies have shown a complete loss of adhesion specialization and widespread separation of the jejunal epithelial cells, with microulceration and compressive degeneration (Sun *et al.*, 1974).

The pathophysiological process during the acute phase of infection seems to be caused by alterations in the gastrointestinal functions. Toxins or other secretions released by the worms may be responsible for the diarrhoea and deficiencies in digestion and absorption, as well as the increased permeability of the intestinal mucosa to protein loss from the plasma. Since an increase in IgE has been observed (Rosenberg *et al.*, 1970), changes in the gastrointestinal functions may be due to an anaphylactic response (Watten *et al.*, 1972).

Symptoms

The onset is marked by borborygmi and diffuse abdominal pain. This is followed by intermittent diarrhoea, two to five times a day especially after every meal. The stool is watery, greenish yellow in colour with a foul odour and the volume may reach to two litres per day in severely ill patients. A marked protein-losing enteropathy and steatorrhoea generally results in the loss of large quantities of protein, fat and minerals in the stool. With the onset of diarrhoea, rapid weight loss occurs. Muscle wasting, weakness, emaciation, abdominal distension and oedema are prominent. Distant heart sounds, hypotension and other cardiac abnormalities may be present in some patients. Death due to cardiac failure may occur two weeks to two months after the initial onset of symptoms.

Laboratory studies have shown a decrease in the excretion of xylose, hypokalaemia, hypocalcaemia and hypoproteinaemia. An increase in IgE is common (Watten et al., 1972; Cross and Basaca-Sevilla, 1989).

Laboratory diagnosis

The standard laboratory diagnostic method is by stool examination to search for the characteristic eggs, larvae and adults. Several stool samples may sometimes be required to detect the infection. Duodenal aspiration can be used to recover the worms.

IHA and double diffusion tests using antigens prepared from a chicken capillarid (C. obsignata) have been evaluated for serodiagnosis but the test lacks both specificity and sensitivity (Banzon et al., 1975). An ELISA test using somatic antigens of adult C. philippinensis was also found to be unsatisfactory (Cross and Chi, 1978).

Treatment

Mebendazole is the current drug of choice. The dosage is $400\,\mathrm{mg\ day^{-1}}$ (in divided doses) for 20–30 days. Few relapses occur with this regimen. Albendazole is also effective at the same regimen as mebendazole (Cross and Basaca-Sevilla, 1987). However, treatment with thiabendazole may result in relapses.

GNATHOSTOMIASIS

Gnathostomiasis is caused by members of the genus Gnathostoma which undergo visceral larval migration. Four species have been found in man, i.e. G. hispidum Fedchenko, 1872. G. spinigerum Owen, 1836, G. doloresi Tubangui, 1925 and G. nipponicum Yamaguti, 1941. Gnathostomatid nematodes are characterized by a headbulb and a fluid-filled cephalic–cervical system. The wall of the headbulb is lined with circular musculature (Ko et al., 1980). Both the headbulb and body cuticle bear rows of spines.

The definitive hosts are pigs, cats (and other felids), wild boars and weasels. They serve as the source for human infections. The adult worm is located in a tumour-like structure in the stomach or oesophagus; the head of the worm is embedded inside the wall. The infective third-stage larva undergoes extensive tissue migration if ingested by man. The infection may be fatal.

Epidemiology

The disease occurs in Southeast Asia, China, Japan, Korea, the Indian subcontinent and Middle-East. However, due to the nature of the infection and the lack of a specific seroepidemiological method, large scale epidemiological studies have seldom been carried out.

In Thailand, gnathostomiasis, which is caused by *G. spinigerum*, is endemic in several provinces in the central region where local people eat raw fish dishes known as 'Hu-Sae', 'Som-Fak' and 'Pla-Som'. During 1985–1988, more than 800 suspected cases were encountered annually at two major hospitals in Bangkok. Also during 1987–1989, *G. spinigerum* larvae were found in about 80% of freshwater eels (*Fluta alba*) purchased from a local market (Setasuban *et al.*, 1991). Daengsvang *et al.* (1964) examined seven species of freshwater fishes and found the infection rates varied from 2.3 to 16.7%. The rates (33–83%) appeared to be markedly higher in four species of snakes.

A survey undertaken in 1965–1970 showed that 4% of cats and 1.1% of dogs in Bangkok were infected (Harinasuta, 1984). Most of the human cases were encountered during the rainy season (May to September). About 96% of the patients drank untreated water from ponds or canals which were also suspected to be possible sources of infection (Punyagupta and Bunnag, 1991).

In Hong Kong, although the occurrence of gnathostomiasis has not been officially documented, two (4%) of the 49 parasitic cases referred to the author's laboratory (1987–1990) were due to this infection. One of the patients probably acquired the infection in Thailand. The second patient had eaten raw fish in a local restaurant. Of the 28 freshwater catfish (*Clarias fuscus*) which was purchased from local markets and examined by the pepsin digestion method, 11% was found infected with *G. spinigerum*; 106 third-stage larvae were recovered. Larvae were also found in *C. batrachus*, *Ophiocephalus* spp., frogs and snakes (Ko, unpublished data).

In Japan, gnathostomiasis was rare and it was caused by *G. spinigerum*. However, since the 1980s, sporadic cases of *G. hispidum* infection have been reported in urban areas. The Japanese are fond of eating a freshwater loach (*Misgurnus anguillicaudatus*) and sometimes the fish is swallowed live. Leaches have disappeared from the rice paddies in Japan due to extensive applications of agricultural chemicals. As a result, the fish has to be imported from China, Taiwan and South Korea where it is commonly infected with *G. hispidum*. About 10% of the loaches was infected, each fish with 6–7 larvae (Oshima, 1984).

Two cases of *G. doloresi* infection were also found recently in Japan. Both patients had a history of eating raw fish (Ogata *et al.*, 1988).

A survey in Bangladesh (1968–1970) showed that six species of fish, one species of frog and one species of snake were infected with *G. spinigerum* larvae. The rates of infection in fish varied from 15–60% (Bashirullah, 1972).

Biology of parasite

Only the biology of *G. spinigerum* and *G. hispidum* will be briefly described.

The adult worms are located in the stomach wall of the definitive host where unembryonated ova are deposited and passed to the outside via the intestine. The ovum (62–79 × 36–42 µm) is characterized by a polar thickening (Fig. 15.8). Embryonation occurs in fresh water and, upon ingestion by copepods, the hatched first-stage larva penetrates the digestive tract and reaches the haemocoele where it develops into the second-stage larva. When the infected copepod is ingested by the second intermediate host (a freshwater fish or frog), the second-stage larva enters the muscle and moults into the third-stage worm which is eventually encysted. A fully grown third-stage larva (4–5 mm) is infective to the definitive host (Fig. 15.9). Other fish, amphibians, snakes, mice, rats or chickens can serve as the paratenic host where third-stage larvae are encysted in muscles.

Daengsvang (1968) fed experimentally infected copepods to frogs, lizards, chicks, rats, mice, tree-shrews and pigs and found advanced third-stage larvae in the liver, muscle and other organs. He also recovered larvae from two new-born mice after 22 pregnant females were fed infected copepods.

The parasite has a low host specificity. Many species of copepods, amphibians, reptiles, freshwater fish and birds can serve as the intermediate and paratenic hosts. The following copepods have been commonly used in experimental studies: *Cyclops strenus*, *C. vicinus*, *Eucyclops serrulatus* and *Mesocyclops leuckarti*.

The common fish hosts are: *Ophicephalus argus*, *O.(= Channa) striata*, *O. micropeltes*, *O. maculata*, *Mogurnda obscura*, *Misgurnus anguillicaudatus*, *Clarias fuscus*, *C. batrachus*, *Anguilla japonica*, *Parasilurus asotus*, *Cyprinus auratus*, *C. carpio*, *Acanthogobius flavimanus*, *Heteropneustes fossilis* and *Fluta alba*.

Pathogenesis

If man ingests the third-stage larva from fish, frogs or snakes, the worm undergoes visceral migration. A similar migration will also occur upon ingestion of infected copepods (Daengsvang, 1968). The degree of pathogenesis, however, is dependent on the extent of tissue invasion by the larvae (Sagara, 1953; Nishibuko, 1961).

Takakura (1988) showed that, in experimentally infected rats, *G. hispidum* larvae will enter the abdominal cavity by penetrating the intestinal wall within 12 hours postinfection. The worms then invade the liver and about three weeks postinfection, they enter the muscles of the inner abdominal wall.

Some of the pathology of gnathostomiasis is probably caused by toxins secreted by the migrating larvae. These include acetylcholine, proteolytic enzymes, a haemolytic compound and hyaluronidase-like substances (see Muller, 1975).

The migrating larvae in the subcutaneous tissues cause creeping eruptions along their route. Intermittent swellings as a result of erythematous oedema are present. Invasion of the liver can disturb the normal function of this organ and the invasion of the eye, bladder or uterus can cause permanent impairment of vision, iritis, haemorrhages, orbital cellulitis, haematuria and leukorrhea (Daengsvang, 1949). If the larvae enter the brain and spinal cord, eosinophilic encephalomyelitis will occur (Chitanondh and Rosen, 1967). Abscesses of the breast, mastoid region and subcutaneous tissues have also been reported.

Typical lesions of the infection usually show areas of degeneration, necrosis and perivascular oedema, with heavy cellular infiltrations of polymorphs. Haemorrhagic exudates, mononuclear leukocytes and fibroplasia are also present.

Symptoms

The disease is characterized by eosinophilia (5–90%) and the presence of migratory swellings of different sizes, which can be located on the chest, abodomen, head, eyelid and other parts of the body. The swelling is sometimes itchy, irritating and tender but shows no pitting on pressure. Involvement of the eyes may show iritis, ecchymosis, subconjunctival haemorrhages and orbital cellulitis. Involvement of the lungs will lead to mild coughing and difficulty in breathing. Involvement of the central nervous system will result in meningeal irritation, headache, convulsion, hypoesthesia, paresis, paraplegia, coma and sometimes death. Only occasionally there is a slight fever.

When there is central nervous system involvement, the symptoms of gnathostomiasis can be distinguished from those of angiostrongyliasis (a neurotropic nematodiasis also endemic in Southeast Asia) by the fact that the former disease is usually in the form of eosinophilic myeloencephalitis, characterized by transverse or ascending myelitis following the symptoms of nerve root pain. The cerebral spinal fluid (CSF) is bloody or xanthochromic (Punyagupta, 1979). Also, due to the visceral migration of the larvae, there are signs and symptoms referable to other organs besides the central nervous system which have not been seen in angiostrongyliasis (Chitanondh and Rosen, 1967). Angiostrongyliasis is in the form of eosinophilic meningoencephalitis and the CSF is clear or turbid (Ko *et al.*, 1987).

Laboratory diagnosis

Although third-stage larvae can be surgically removed from the superficial swellings, serodiagnosis should serve as the key diagnostic method. However, in ELISA and IFAT, there are significant cross-reactions between the crude antigens and heterologous antiserum of *G. spinigerum*, *G. hispidum* and *A. cantonensis* (Ko *et al.*, 1987). Therefore, before serological methods can be used routinely, it is essential to isolate specific antigens from *Gnathostoma* spp.

Mimori *et al.* (1987) used ELISA and crude *G. doloresi* antigens to test the sera samples of known gnathostomiasis cases in Ecuador. Sera from 15 of 16 patients and 5 of 18 controls were found to be positive. Tada *et al.* (1987) reported a close correlation between the results obtained by *G. doloresi* and *G. hispidum* antigens in ELISA when they were used to screen sera samples from 30 patients. However, *G. doloresi* antigens have been reported to be less specific than those of *G. spinigerum* (Anantaphruti, 1989a, 1989b).

According to Chaicumpa *et al.* (1991) and Nopparatana *et al.* (1991), a specific 24 kDa protein antigen has been isolated from the third-stage larvae of *G. spinigerum* by gel filtration, chromatofocusing and anion exchange chromatography. The antigen, which has a pI of 8.5, was used in the indirect ELISA to screen sera samples from four parasitologically confirmed patients, 15 clinically diagnosed patients, 136 patients with other parasitic diseases and 25 controls. The sensitivity, specificity and predictive values of the assay were 100%.

Using immunoblotting, Akao *et al.* (1989) studied sera samples from nine patients who were infected with *G. hispidum*. An epitope having a M_r of 31,500 was prominent in all samples.

Attempts have also been undertaken to detect circulating antigens. Tuntipopipat *et al.* (1989) successfully detected antigens in the CSF using a sandwich ELISA (using antibodies from rabbits immunized with ES antigens from third-stage larvae). With a biotin streptavidin procedure, antigens as low as $2\,\text{ng}\,\text{ml}^{-1}$ could be detected. However, of the 11 patients showing symptoms of cerebral involvement, antigens were only detected in the CSF of one patient, whereas IgG antibodies were detected in nine patients. One other patient had immune complexes in the CSF.

A two-site ELISA was developed to detect circulating antigens in mice (Maleewong *et al.*, 1992b).

Treatment

The only treatment available is surgical excision of worms. There is no effective chemotherapy. Thiabendazole, at a dosage of 2 g daily for 20 days may help to alleviate some of the symptoms (Harinasuta, 1984). Albendazole has also been tried but the results were unsatisfactory (Suntharasamai *et al.*, 1990; Maleewong *et al.*, 1992a).

Anisakiasis

Anisakiasis refers to infection by larval ascaridoid nematodes whose normal definitive hosts are marine mammals. The genera involved are *Anisakis*, *Pseudoterranova* and *Contracaecum*. Larvae (located in squids and marine fish) can invade the gastrointestinal tract of man, causing an eosinophilic granuloma syndrome which, in Europe, has also been referred to as the 'herring worm' disease. The first documented human infection was in the Netherlands in 1955 and then later in Japan (van Thiel *et al.*, 1960; Asami *et al.*, 1965). However, anisakiasis has almost disappeared in the Netherlands, due to a legislation against eating raw herring and freezing of fish prior to sale. Freezing fish for 24 hours or heating the processed fish to 65°C kill the larvae. Also the gutting of fish soon after they are caught supposedly prevents the migration of larvae to muscles.

More than 2000 cases have been noted in Japan and sporadic cases are also known to occur in North America and other parts of the world (Chitwood, 1970, 1975; Little and MacPhail, 1972; Suzuki *et al.*, 1972; Little and Most, 1973; Kates *et al.*, 1973; Lichtenfels and Brancato, 1976; Oshima, 1984; Huber *et al.*, 1989; Kowalowska-Grochouska *et al.*, 1989).

The disease was reviewed extensively by Oshima (1972) and an update on the bibliography was given by Huang and Bussieras (1988). The taxnononomy and morphogenesis of the anisakid nematodes are confusing. This is probably due to the fact that the life cycles of these nematodes have not been elucidated under experimental conditions. The identification of the adults and the larvae found in fish and squid is presently based on the classical morphometric method which has great limitations. However, the larvae of the following species are considered major aetiological agents of anisakiasis: *A. simplex* (Rudolphi, 1809), *A. typica* (Diesing, 1861), *A. physeteris* (Baylis, 1923), *P. decipiens* (Krabbe, 1878) and *C. osculatum* (Rudolphi, 1802). The worms are characterized by three lips and a tooth at the anterior extremity.

There are three types of *Anisakis* larvae, i.e. Type I, Type II and Type III. Type I larva, which measures 19–36 mm in length, and has a short tail (0.08–0.16 mm). The body length of Type II is similar (19.0–25.5 mm) but the tail is longer (0.18–0.32 mm in length). Type III larva, which measures 23.8–38.4 mm in length, has the widest body (0.65–0.97 mm in width). The larva of *Pseudoterranova* is referred to as Type A and that of *Contracaecum* as Type B (Oshima, 1972).

Attempts have been undertaken to differentiate the various larvae using molecular biology methods. An analysis of restriction fragment length polymorphisms (RFLPs) was applied to distinguish Type I, Type II and Type B larvae. Different patterns of RFLPs of genomic DNA were observed. The patterns of the two different paratenic host-derived DNA were the same in hybridized fragments generated by endonucleases (Sugane *et al.*, 1989).

Epidemiology

More than 100 species of fish which the Japanese like to eat raw can harbour anisakid larvae. In Japan, eating raw fish is a traditional way of life. Hence, anisakiasis still occurs regularly.

The following Japanese dishes include raw sea food: 'sashimi' is raw fish or squid fillet eaten with shoyu and wasabi; 'sunomono' is pickled fillet of fish or squid with vinegar; 'isushi' is pickled rice with raw fillet of chum salmon, masu, cod or cod roes.

Anisakiasis due to Type I larvae has been reported throughout Japan, from Hokkaido Island to Okinawa. The most commonly infected fishes are the common mackerel, Pacific pollack, cod, various salmon, herring and tuna. Slightly salted roes of cod and herring are also sources of infection. The majority of worms removed from the stomach wall of patients by gastrofibrescopy were Type I larvae (Oshima, 1984).

Extensive surveys on the prevalence of anisakid larvae in fishes, squids and marine mammals have been undertaken in Japan (see Oshima, 1972). A study was carried out by Kobayashi *et al.* (1966) and Koyama *et al.* (1969) to determine the prevalence of infection in 15 species of fishes and five species of squids which were commonly available at the Tokyo Metropolitan Fish Market. The prevalence rates of Type I larva were 100% in *Theragra chalcogramma* (Pacific pollack) and *Oncorhynchus masou* (masu salmon), 83% in *Pneumatophorus japonicus japonicus* (common mackerel) and 50% in *T. pacificus* (squid). These are the species which are most commonly eaten raw by the Japanese.

In Hokkaido, the incidence of regional enteritis has been observed to increase from October to March and decrease during summer (Ishikura, 1968). This seems to coincide with the fishing of Pacific pollack. However, in other parts of Japan, seasonal variations in the incidence of anisakiasis has not been observed.

Seasonal variations have been observed in the infection of squids (*Todarodes pacificus*) with Type I and Type II larvae. Squids collected in June usually have the lowest prevalence and intensity of infection (Yamaguchi *et al.*, 1968; Kagei, 1969; Kosugi *et al.*, 1970). This was attributed to the seasonal migration of squids along the coast of Japan.

In Japan, anisakiasis appears to be more common in males than females. Of 95 patients seen by Yokogawa and Yoshimura (1967), 65 were male. This is probably because men frequently consume raw fish at drinking parties.

In the USA, a recent study has shown that all the 171 mature herring (*Clupea harengus pallasi*) from the Puget Sound, Washington, were infected with *Anisakis* and *Contracaecum* larvae. The average worm burden was 27 per fish and 36% of fish harboured worms in muscles. The prevalence in the Pacific herring is markedly higher than those recorded previously. The increase is probably related to an increase in number of marine mammals in Puget Sound and along the coastline, after the enactment of Marine Mammal Protection Act of 1972 (Adams *et al.*, 1990).

Despite the common occurrence of anisakid larvae in fish in USA, only a handful of anisakiasis cases have been reported because most Americans do not eat raw fish or squid.

Biology of parasite

The life cycle of anisakid nematodes has not been convincingly elucidated. It is not possible to infect marine mammals experimentally. The following brief account is based on the reports by Oshima (1972) and Smith (1983).

Adult worms live in the stomach of the definitive host. Eggs which are ellipsoidal in shape have a thick shell and measure about 46–58 × 41–53 μm. They are passed to the outside with the faeces. Embryonation of the egg occurs in the sea. The egg hatches to release the second-stage larva which is ingested by an euphausiid crustacean. In the haemocoele of the euphausiid, the second-stage larva moults to the third-stage worm which can grow to a length of about 30 mm. Upon ingestion by a fish or squid (paratenic host), the third-stage larva penetrates the gastrointestinal tract and enters the viscera or muscles. The odontocete marine mammals acquire the infection by consuming infected fish or squids. The worm attaches to the stomach wall and moults within 3–5 days to become the fourth-stage larva. After a subsequent moult, the worm develops to the adult stage.

Oshima (1972) provided a detailed list of fish in which anisakid larvae have been found. One hundred and twenty-three species were listed for Type I larvae, 25 for Type II, two for Type III, eight for Type A and four for Type B. These include *Oncorhynchus* spp., *Theragra chalcogramma*, *Gadus macrocephalus*, *Pneumatophorus japonicus*, *Clupea pallasi*, *Sarda orientalis*,

Epinephelus spp., *Sebastes* spp., *Fugu* spp., *Osmerus dentex, Hexagrammos otakii, Sebasticus albofasciatus,* etc.

The common cetacean hosts are *Stenella caeruleo-alba, Phocaena phocoena, Phocoenoides dalli* and the pinniped hosts are *Phoca vitulina, Callorhinus ursinus* and *Eumetopias jubata* (Oshima, 1972).

The common euphausiid hosts are *Thysanoessa raschii, T. longipes* and *Euphausia pacifica.*

Pathogenesis

The pathology of anisakiasis in man and experimental animals has been studied extensively (Meyers, 1963; Ashby *et al.*, 1964; Kuiper, 1964; Yokogawa *et al.*, 1965; Usutani, 1966; Yoshimura, 1966; Oyangi, 1967; Young and Lowe, 1969: Gibson, 1970; Ruitenberg *et al.*, 1971; Kikuchi *et al.*, 1972; Oshima, 1972; Jones *et al.*, 1990).

Anisakid larvae can penetrate into the stomach wall of rats within 1–4 hours postinfection. Meyers (1963) recovered Type I larvae from the mesentery, pancreas, liver and other tissues of experimentally infected guinea-pigs. The degeneration of larvae has been observed in the submucosa of rabbits and dogs ten days postinfection (Kuiper, 1964: Usutani, 1966; Oyangi, 1967; Ruitenberg *et al.*, 1971).

The human cases diagnosed in Europe mainly involved the intestine whereas in Japan, both gastric and intestinal infections are equally common. Lesions of human anisakiasis are classified into five types (Oshima, 1972). The foreign body response type is characterized by cellular infiltrations, consisting mainly of neutrophils, with a few eosinophils and giant cells. This type is caused by a primary anisakid infection without any presentation.

The phlegmonous type is characterized by extensive oedematous thickening in the submucosa, with heavy infiltrations of eosinophils, lymphocytes, monocytes, neutrophils and plasma cells. Haemorrhages and fibrin exudation are also present. Such lesions are common in acute intestinal cases within one week of infection.

The abscess type is marked by the presence of numerous eosinophils, histiocytes and lymphocytes around the worms, surrounded by a granulomatous zone. Necrosis, haemorrhages, fibrin exudation or fibrinoid degeneration occur near the inner region of the surrounding granuloma. Such lesions are found in chronic cases.

The abscess-granulomatous type shows a reduction of the central abscess around highly degenerated worms which is circumscribed by conspicuous granulation tissue with slight collagenization. Eosinophilic infiltration is less extensive than the abscess type. This type occurs in infections of more than six months.

In the granulomatous type, the abscess is completely replaced by granulomatous tissue with eosinophilic infiltration. Occasionally, remnants of degenerated worm tissues can be seen. This type of lesion is rare.

Symptoms

Acute gastric infections are manifested by gastric pain, nausea and vomiting about 4–6 hours after ingestion of raw infected sea food. During the chronic phase, vague epigastric pain, nausea and vomiting may last from several weeks to two years. In more than 50% of the recorded cases, eosinophils varied from 4–41%. Occult blood may occur in the gastric juice and stools.

The onset of intestinal anisakiasis occurs within seven days after ingestion of infected raw sea food. The symptoms are severe pain in the lower abdomen, with nausea and vomiting. The pressure point, unlike appendicitis, is vague and there is no muscular rigidity. Marked leukocytosis is common but eosinophilia is rare. Straw-coloured ascites may be present.

Laboratory diagnosis

Fibre-optic gastroscopy is the most useful method to diagnose gastric anisakiasis. The diagnosis of intestinal cases, however, relies mainly on clinical symptoms and a history of eating raw sea food. X-rays, after administration of a barium meal, may show segemented movement of the barium and jagged stricture of regional intestine with proximal enlargement and a radiolucent area (Oshima, 1972; Higashi *et al.*, 1987). Ultrasound was also used to diagnose intestinal anisakiasis (Shirahama *et al.*, 1992).

There is still no satisfactory serodiagnostic method although the skin test, CFT and IFAT have been studied (Yoshimura, 1966; Suzuki *et al.*, 1970; Kikuchi *et al.*, 1970; Ruitenberg, 1970; Boczon, 1988). The major problem is the failure to isolate specific antigens.

Takahasi *et al.* (1986) developed two hybridomas, An1 and An2 from Type I larvae. Using SDS-PAGE and western blotting, An1 precipitated a band of 34 kDa and An2, bands of 40 and 42 kDa respectively. An2 was also detected in the ES antigens using ELISA. Yagihasi *et al.* (1990) reported that serum from patients reacted strongly to An2 in micro-ELISA. Kennedy *et al.* (1988) reported extensive antigenic relationships between the somatic and ES antigens of *A. simplex* and those of other nematodes such as *Ascaris suum*, *A. lumbricoides* and *Toxocara canis*.

Although a reliable serological method still has not been developed to screen fish routinely for anisakid worms, Huber *et al.* (1989) briefly reported on the use of ELISA to detect antibodies against *A. simplex* in the saithe fish, *Pollachius virens*. A good correlation was observed between antibody titres against ES antigens and the number of anisakid larvae in fish. However, it is not known whether a cross-reactive study has been carried by these authors and whether the fish used were also naturally infected with other species of parasites.

ES antigens were used in immunoblots to diagnose human anisakiasis (Akao *et al.*, 1990; Petithory *et al.*, 1991; Iglesias *et al.*, 1993). Specific IgE was also found in patients (Kasuya and Koga, 1992).

Treatment

Removal of the worms by a gastrofibrescope is the method of choice in acute gastric infection. For chronic infections, a partial resection of a small area of the lesion has been recommended.

For intestinal anisakiasis, only conservative treatment can be used. However, recently, colonoscopic removal of worms was successful (Minamoto *et al.*, 1991).

FUTURE PERSPECTIVES

Since the biology and pathogenesis of many fish-borne parasites still remain unknown, it is possible that new zoonoses may be discovered if the clinicians and the public health workers are better informed about parasitic diseases.

Two fish parasites are known to be potentially infective to man. Ko (1976) sucessfully infected monkeys, dogs and kittens with the larvae of a gnathostomatid nematode, *Echinocephalus sinensis* Ko, 1975. This worm is a parasite of eagle ray, *Aebates flagellum*. The larvae, which were recovered from oysters (*Crassostrea gigas*), could produce lesions in the experimentally infected animals as in gnathostomiasis and anisakiasis. Ko (1977) also demonstrated that the number of successful infections produced and the number of worms in tissues of kittens were directly dependent on the temperature of acclimation of the oyster host. The study has demonstrated for the first time that the abnormal infection of a homeothermic host by a larva of a fish parasite can be influenced by ambient temperatures. Therefore, Ko suggested that, in the tropics, where the ambient temperature is always high, some other parasites can also infect mammals (including man).

Norris and Overstreet (1976) reported that a *Hysterothylacium* nematode larva from cutlass fish (*Trichiurus lepturus*) produced gastrointesinal lesions in mice. The authors suggested that, in sensitized animals with existing lesions, even dead larvae or their products would elicit an allergic reaction similar to that in anisakiasis. Since the worm was also shown to be infective to monkeys (Overstreet and Meyers, 1981), this further reinforces the belief that it can also invade the gastrointestinal tract of man. Recently, Vidal-Martinez *et al.* (1994) also succeeded to infect kittens with larvae of an avian *Contracaecum* recovered from euryhaline cichlid which is frequently eaten by Mexicans.

Theoretically, fish-borne parasitic zoonoses can easily be prevented by refraining from eating raw sea food. However, in some parts of the world, such an eating habit represents an established way of life and cannot be easily changed, even by the implementation of a strong education programme or the passage of a legislation. Therefore, these diseases will remain as public health problems and there is a need to undertake regular epidemiological studies. These studies, however, cannot be carried out effectively without the development of more sensitive and specific diagnostic methods which can be used in large scale screenings of fish and man.

REFERENCES

Adams, A.M., Berry, M., Wekell, M.M. and Deardorff, T.L. (1990) Juvenile anisakids in Pacific herring. *Abstracts of 33rd Southeast Asian Medical Education Organization Tropical Medicine Regional Seminar, Chiangmai, Thailand*, p. 42.

Ahn, Y.K., Chung, P.R., Lee, K.T. and Soh, C.T. (1987) Epidemiological survey of *Metagonimus yokogawai* infection in the Eastern coast area of Kangwon Province, Korea. *Korean Journal of Parasitology* 25, 59–68 (in Korean).

Akao, N., Ohyama, T., Kondo, K. and Takakura, Y. (1989) Immunoblot analysis of human gnathostomiasis. *Annals of Tropical Medicine and Parasiology* 83, 635–637.

Akao, N., Ohyama, T. and Kondo, K. (1990) Immunoblot analysis of serum IgG, IgA and IgE response against larval excretory-secretory antigens of *Anisakis simplex* in patients with gastric anisakiasis. *Journal of Helminthology* 64, 310–318.

Amornpunt, S., Sarasombath, S., Sirisinha, S. (1991). Production and characterization of monoclonal antibodies against the excretory-secretory antigens of the liver fluke (*Opisthorchis viverrini*). *International Journal for Parasitology* 21, 421–428.

Anantaphruti, M.T. (1989a) ELISA for diagnosis of gnathostomiasis using antigens from *Gnathostoma doloresi* and *G. spinigerum*. *Southeast Asian Journal of Tropical Medicine and Public Health* 20, 297–304.

Anantaphruti, M.T. (1989b) Demonstration of species specific antigens of *Gnathostoma spinigerum,* a preliminary report. *Southeast Asian Journal of Tropical Medicine and Public Health* 20, 305–312.

Arambulo, P.V. and Thakur, A. (1990) Current status of food-borne parasitic zoonoses in Latin America and the Carribean. *Abstracts of 33rd Southeast Asian Medical Organization Tropical Medicine Regional Seminar, Chiangmai, Thailand*, p. 117.

Asami, K., Watanuki, T., Sakai, H., Imano, H. and Okamoto, R. (1965) Two cases of stomach granuloma caused by *Anisakis*-like larvae nematodes in Japan. *American Journal of Tropical Medicine and Hygiene* 14, 119–123.

Ashby, B.S., Appleton, P. and Dawson, I. (1964) Eosinophilic granuloma of gastrointestinal tract caused by herring parasite, *Eustoma rotundatum. British Medical Journal* 1, 1141–1145.

Banzon, T.C., Lewer, R.M. and Yagore, M.G. (1975) Serology of *Capillaria philippinensis* infection: reactivity of human sera to antigens prepared from *Capillaria obsignata* and other helminths. *American Journal of Tropical Medicine and Hygiene* 24, 256–263.

Bashirullah, A.K.M. (1972) Occurence of *Gnathostoma spinigerum* Owen, 1836, in Dacca, Bangladesh. *Journal of Parasitology* 58, 187–188.

Belamaric, J. (1973) Intrahepatic bile duct carcinoma and *Clonorchis sinensis* in Hong Kong. *Cancer* 31, 468–473.

Belding, D.L. (1965) *Text Book of Parasitology*. Appleton-Century-Crofts, New York.

Bhaibulaya, M. and Indra-Ngarm, S. (1979) *Amaurornis phoenicurus* and *Ardeiola bacchus* as experimental definitive hosts for *Capillaria philippinensis* in Thailand. *International Journal for Parasitology* 9, 321–322.

Bhaibulaya, M., Indra-Ngarm, S. and Anathapruit, M. (1979) Freshwater fishes of Thailand as experimental intermediate host for *Capillaria philippinensis. International Journal for Parasitology* 9, 105–108.

Boczon, K. (1988) Diagnosis of anisakis infection. *Wiadomosci Parazytologiczne* 34, 11–17 (in Polish).

Chai, J.Y. and Lee, S.H. (1991) Intestinal trematodes infecting humans in Korea. *Southeast Asian Journal of Tropical Medicine and Public Health* 22 (suppl.), 163–170.

Chai, J.Y., Hong, S.J., Lee, S.H. and Seo, B.S. (1988) *Stictodora* sp. (Trematoda: Heterophyidae) recovered from a man in Korea. *Korean Journal of Parasitology* 26, 127–132.

Chaicumpa, W., Ruangkunaporn, V., Nopparatana, C., Chongsa, N.M., Tapachaisri,

P. and Sebasuban, P. (1991) Monoclonal antibody to a diagnostic Mr 24,000 antigen, of *Gnathostoma spinigerum. International Journal for Parasitology* 21, 735–738.

Chaicumpa, W., Ybanez, L., Kitikoon, V., Pungpak, S., Ruangkunaporn, Y., Chongsanguan, M. and Sornmani, S. (1992) Detection of *Opisthorchis viverrini* antigens in stools using specific monoclonal antibody. *International Journal for Parasitology* 22, 527–531.

Chan, P.H. and Teoh, T.B. (1967) The pathology of *Clonorchis sinensis* infestation of the pancreas. *Journal of Pathology and Bacteriology* 93, 185–189.

Changbrunrung, S., Ratarasaru, S., Hongtong, K., Miagasena, P., Vutikes, S. and Migasen, S. (1988) Lipid composition of serum lipoprotein in opisthorchiasis. *Annals of Tropical Medicine and Parasitology* 82, 263–269.

Chen, C.Y., Hsieh, W.C., Shih, H.H. and Chen, S.N. (1987a) Detection of serum antibody to *Clonorchis sinensis* by enzyme-linked immunosorbent assay. *Journal of the Formosan Medical Association* 86, 706–711.

Chen, C.Y., Hsieh, W.C., Shih, H.H. and Chen, S.N. (1987b) Evaluation of enzyme-linked immunosorbent assay for immunodiagnosis of clonorchiasis. *Chinese Journal of Microbiology and Immunology* 20, 241–246.

Chen, C.Y., Shin, J.W., Chen, S.N. and Hsieh, W.C. (1989) A preliminary study of clinical staging in clonorchiasis. *Chinese Journal of Microbiology and Immunology* 22, 193–200.

Chen, E.R. (1991a) Food-borne parasitic zoonoses in Taiwan. *Southeast Asian Journal of Tropical Medicine and Public Health* 22 (suppl.), 62–63.

Chen, E.R. (1991b) Clonorchiasis in Taiwan. *Southeast Asian Journal of Tropical Medicine and Public Health* 22 (suppl.), 184–185.

Chichino, G., Bernuzzi, A.M., Bruno, A., Cevini, C., Atzori, C. Malfitano, A. and Scaglia, M. (1992) intestinal capillariasis (*Capillaria philippinensis*) acquired in Indonesia: a case report. *American Journal of Tropical Medicine and Hygiene* 47, 10–12.

Chitonandh, H. and Rosen, L. (1967) Fatal encephalomyelitis caused by the nematode, *Gnathostoma spinigerum. American Journal of Tropical Medicine and Hygiene* 16, 638–645.

Chitwood, M.B. (1970) Nematodes of medical significance found in fish market. *American Journal of Tropical Medicine and Hygiene* 19, 599–602.

Chitwood, M.B. (1975) *Phocanema*-type larval nematode coughed up by a boy in California. *American Journal of Tropical Medicine and Hygiene* 24, 710–711.

Chitwood, M.B., Velasquez, C. and Salazar, N.G. (1964) Physiological changes in a species of *Capillaria* (Trichuroidea) causing a fatal case of human intestinal capillariasis. *Proceedings of the First International Congress of Parasitology* (Rome, Italy) 2, 797.

Chitwood, M.B., Velasquez, C. and Salazar, N.G. (1968) *Capillaria philippinensis* sp.n. (Nematoda: Trichuroidea) from intestine of man in the Philippines. *Journal of Parasitology* 54, 368–371.

Cho, K.M. and Soh, C.T. (1974) Evaluation of the indirect fluorescent antibody test with adult worm antigen for the immunodiagnosis of clonorchiasis. *Yonsei Reports of Tropical Medicine* 5, 45–56.

Cho, K.M. and Soh, C.T. (1976) Indirect fluorescent antibody test with adult worm antigen for the immunodiagnosis of clonorchiasis. *Yonsei Reports of Tropical Medicine* 7, 26–39.

Choi, B.I., Kim, H.J., Han, M.C., Do, Y.S., Han, M.H. and Lee, S.H. (1989) CT findings of clonorchiasis. *American Journal of Roentgenology* 152, 281–284.

Chung, H.L., Weng, H.C., Hou, T.C. and Ho, L.Y. (1955) Cross intradermal reactions of patients with paragonimiasis, clonorchiasis and schistosomiasis to different trematode antigen and their clinical significance. *Chinese Medical Journal* 73, 368–378.

Chunlertrith, K., Mairiang, P. and Sukeepaisarnjaroen, W. (1992) Intestinal capillariasis: a cause of chronic diarrhea and hypoalbuminea. *Southeast Asian Journal of Tropical Medicine and Public Health* 23, 433–436.

Cook, J., Hou, P.C., Ho, H.C. and Mcfadzean, A.J.S. (1954) Recurrent pyogenic cholangitis. *British Journal of Surgery* 42, 188–203.

Cross, J.H. (1990) Intestinal capillariasis. *Parasitology Today* 6, 26–28.

Cross, J.H. and Basaca-Sevilla, V. (1987) Albendazole in the treatment of intestinal capillariasis. *Southeast Asian Journal of Tropical Medicine and Public Health* 18, 507–510.

Cross, J.H. and Basaca-Sevilla, V. (1989) Intestinal capillariasis. *Progress in Clinical Parasitology* 1, 105–119.

Cross, J.H. and Basaca-Sevilla (1991) *Capillaria phillipinensis*: a fish-borne parasitic zoonosis. *Southeast Asian Journal of Tropical Medicine and Public Health* 22 (suppl.), 153–157.

Cross, J.H. and Bhaibulaya, M. (1983) Intestinal capillariasis in the Phillipines and Thailand. In: Croll, N. and Cross, J.H. (eds), *Human Ecology and Infectious Diseases*. Academic Press, Orlando, pp. 103–136.

Cross, J.H. and Chi, J.C.H. (1978) The ELISA test in the detection of antibodies to some parasitic diseases. *Proceedings of Southeast Asia Medical Education Organization Tropical Medicine 18th Seminar*, Kuala Lumpur, Malaysia, pp. 178–182.

Cross, J.H., Banzon, T.C., Clarke, M.D., Basaca-Sevilla, V., Watten, R.H. and Dizon, J.J. (1972) Studies on the experimental transmission of *Capillaria philippinensis* in monkeys. *Transactions of the Royal Society of Tropical Medicine and Hygiene* 66, 819–827.

Cross, J.H., Banzon, T.C. and Singson, C.N. (1978) Further studies on *Capillaria phillipinensis*: development of the parasite in the Mongolian gerbil. *Journal of Parasitology* 64, 208–213.

Cross, J.H., Singson, C.N., Battad, S. and Basaca-Sevilla, V. (1979) Intestinal capillariasis: epidemiology, parasitology and treatment. In: *Health Policies in Developing Countries. Royal Society of Medicine International Congress Serial* 24, 82–87.

Curtis, M.A. and Bylund, G. (1991) Diphyllobothriasis: fish tapeworm disease in the circumpolar north. *Arctic Medical Research* 50, 18–24.

Daengsvang, S. (1949) Human gnathostomiasis in Siam with reference to the method of prevention. *Journal of Parasitology* 35, 116–121.

Daengsvang, S. (1968) Further observations on the experimental transmission of *Gnathostoma spinigerum*. *Annals of Tropical Medicine and Parasitology* 62, 88–94.

Daengsvang, S., Chulalerk, U., Papasarathorn, T. and Tongkoom, B. (1964) Epidemiological observations on *Gnathostoma spinigerum* in Thailand. *Journal of Tropical Medicine and Hygiene* 67, 144–147.

Dick, T.A., Dixon, B. and Choudhury (1991) *Diphyllobothrium, Anisakis* and other fish-borne parasitic zoonoses. *Southeast Asian Journal of Tropical Medicine and Public Health* 22 (suppl.), 150–152.

Dovgalev, A.S. (1988) Diphyllobothriasis in the western coastal region of the Okhotsk Sea. *Meditsinskaya Parazitologigya i Parazitarnyl Bolezni* 4, 67–71 (in Russian).

Duchastel, P. (1984) Prevalence of parasites in stools of Hong Kong residents and Indochinese refugees applying for emigration to Canada: retrospective study over two year period (1979–81). In: Ko, R.C. (ed.) *Current Perspectives in Parasitic Diseases*. Departments of Zoology and Medicine, University of Hong Kong, pp. 53–54.

Flavell, D.J. (1981) Liver-fluke infection as an aetiological factor in bile-duct carcinoma of man. *Transactions of the Royal Society of Tropical Medicine and Hygiene* 75, 814–824.

Fukumoto, S., Yazaki, S., Maejima, J., Kamo, H., Takao, Y. and Tsutsumi, H. (1988a) The first report of human infection with *Diphyllobothrium scoticum* (Rennie et Reid, 1912). *Japanese Journal of Parasitology* 37, 84–90.

Fukumoto, S., Yazak, S., Kamo, H., Yamane, Y. and Tsuji, M. (1988b) Distinction between *Diphyllobothrium nihonkaiense and Diphyllobothrium latum* by immunoelectrophoresis. *Japanese Journal of Parasitology* 37, 91–95.

Fung, J. (1961) Liver fluke infestation and cholangio-hepatitis. *British Journal of Surgery* 48, 404–451.

Gibson, D.I. (1970) Aspects of the development of herring worm (*Anisakis* sp. larvae) in experimentally infected rats. *Norwegian Journal of Zoology* 18, 175–187.

Gibson, J.B. and Sun, T. (1965) Chinese liver fluke – *Clonorchis sinensis*. Its occurrence in Hong Kong. *International Pathology* 6, 93–98.

Gitsu, G.A., Ballard, N.E. and Zavoikin, V.D. (1987) The efficacy of the enzyme immunoassay in the diagnosis of opisthorchiasis. *Meditsinskaya Parazitologiya i Parazitarnye Bolezni* 6, 17–20 (in Russian).

Gosteva, L.A., Sergeeva, V.T. and Makarevichn, N.I. (1991) Lyzome activity in diphyllobothriasis patients. *Meditsinskaya Parazitologiya Moskow* 4, 45–46 (in Russian).

Hahm, J.H., Lee, J.S. and Rim, H.J. (1984) Comparative study on the indirect immunofluorescent antibody test, complement fixation test and ELISA in diagnosis of human clonorchiasis. *Korea University Medical Journal* 21, 177–184 (in Korean).

Han, J.H., Eom, K.S., and Rim, H.J. (1986) Comparative studies on the immuno-diagnosis of clonorchiasis by means of micro-ELISA using sera and blood collected in filter paper. *Korea University Medical Journal* 23, 13–25 (in Korean).

Harinasuta, C. (1984) Parasitic diseases of public health importance in Southeast Asia – epidemiology, treatment and control. In: Ko, R.C. (ed.) *Current Perspectives in Parasitic Diseases*. Departments of Zoology and Medicine, University of Hong Kong, pp. 1–28.

Hayashi, S. and Kamo, H. (1983) Studies on the effect and mode of action of paramomycin sulfate against tapeworm. *Japanese Journal of Antibiotics* 34, 552–565 (in Japanese).

Higashi, M., Tanaka, K., Kitada, T., Nakatake, K. and Tsuji, M. (1987) Anisakiasis confirmed by radiography of the large intestine. *Gastrointestinal Radiology* 13, 85–86.

Hirai, K., Torii, M., Suzuki, N. and Kamo, H. (1988) Occurrence of human cases infected with *Diphyllobothrium yonagoense* in Shikoku Island, Japan. *Japanese Journal of Parasitology* 37, 13–19 (in Japanese).

Hon, M.F., Ker, C.G., Sheen, P.C. and Chen, E.R. (1989) The ultrasound survey of gallstone diseases of patients infected with *Clonorchis sinensis* in southern Taiwan. *Journal of Tropical Medicine and Hygiene* 92, 108–111.

Hou, P.C. (1955) The pathology of *Clonorchis sinensis* infestation of the liver. *Journal of Pathology and Bacteriology* 70, 53–64.

Hou, P.C. (1956) Relationship between primary carcinoma of liver and infestation with *Clonorchis sinensis*. *Journal of Pathology and Bacteriology* 72, 239–246.

Hou, P.C. and Pang, S.C.L. (1964) *Clonorchis sinensis* infestation in man in Hong Kong. *Journal of Pathology and Bacteriology* 87, 245–250.

Hu, Y.X., Cao, W.J., Tan, W., Hu, R.Y. and Qi, Z.Q. (1989) Preliminary study on the diagnosis of clonorchiasis by charcoal granule agglutination test. *Chinese Journal of Parasitic Diseases Control* 2, 279–281 (in Chinese).

Huang, W. and Bussieras, J. (1988) Anisakides et anisakidoses humaines. Premier partie: Donnes bibliographiques. *Annales de Parasitologie Humaine et Comparée* 63, 119–132.

Huber, C., Martlbauer, E., Priebe, K. and Terplan, G. (1989) Entwicklung und anwendung eines ELISA zum nachweis von antikorpern gegen *Anisakis simplex* (Nematoda) beim seelachs *Pollachius virens*. *Deutsche Veterinarmedizinische Gesellschaft* 28, 272–275 (in German).

Hubert, B., Bacou, J. and Belveze, H. (1989) Epidemiology of human anisakiasis: incidence and sources in France. *American Journal of Tropical Medicine and Hygiene* 40, 301–303.

Iglesias, R., Leiro, J., Ubeira, F.M., Santamarina, M.T. and Sanmartin, M.L. (1993) *Anisakis simplex*: antigen recognition and antibody production in experimentally infected mice. *Parasite Immunology* 15, 243–250.

Iijima, T. (1889) The source of *Bothriocephalus latus* in Japan. *Journal of College of Science, Imperial University, Tokyo* 12, 49–56.

Im, K.I. (1974) Indirect fluorescent antibody test for the diagnosis of clonorchiasis in rabbit and human. *Yonsei Journal of Medical Science* 7, 194–205 (in Korean).

Ishikura, H. (1968) On the anisakiasis. *Hokkaido Igaku Zasshi* 43, 83–99 (in Japanese).

Ito, K. (1925) On the complement fixation reaction on experimental clonorchiasis in animals. *Aichi Igakkai Zasshi* 32, 900–914 (in Japanese).

Jin, S.W., Lee, J.S. and Rim, H.J. (1983) Comparative studies of ELISA test by use of the immune animal sera in clonorchiasis and paragonimiasis. *Korea University Medical Journal* 20, 191–199 (in Korean).

Jones, R.E., Deardorff, T.L. and Kayes, S.G. (1990) *Anisakis simplex*: histopathological changes in experimentally infected CBA/J mice. *Experimental Parasitology* 70, 305–313.

Kagei, N. (1969) Life cycle of the genus *Anisakis*. *Saishin Igaku* 24, 389–400 (in Japanese).

Kamo, H. (1988) Diphyllobothriasis. In: Balows, A., Hausler, W.J.Jr., Ohaski, M. and Turano, A.J. (eds) *Laboratory Diagnosis of Infectious Diseases. Vol I. Bacterial, Mycotic and Parasitic Diseases*. Springer-Verlag, New York, pp. 821–830.

Kamo, H., Maejima, J., Yazaki, S., Fukumoto, S. and Yammishi, Y. (1988a) Human infection with *Diphyllobothrium yonagoense* in Kinki-Tokai Districts. *Japanese Journal of Parasitology* 37, 62–66.

Kamo, H., Yazaki, S., Fukumoto, S., Fujino, T., Koga, M., Ishii, Y. and Matsuo, E. (1988b) The first human case infected with *Diphyllobothrium hians* (Diesing, 1850). *Japanese Journal of Parasitology* 37, 29–35.

Kang, G., Mathan, M., Ramakrishna, B.S., Mathai, E. and Sarada, V. (1994) Human intestinal capillariasis: first report from India. *Transactions of the Royal Society of Tropical Medicine and Hygiene* 88, 204.

Karlstedt, K.A., Paatero, G.I., Makela, J.H. and Wikgren, B.J. (1992) A hidden break in the 28.0S rRNA from *Diphyllobothrium dendriticum*. *Journal of Helminthology* 66, 193–197.

Kasuya, S. and Koga, K. (1992) Significance of detection of specific IgE in *Anisakis*-related diseases. *Areugi* 41, 106–110 (in Japanese).

Kasuya, S., Khambooruang, C., Amano, K., Murase, T., Araki, H., Kato, Y., Kumada, Y., Kajama, A., Higuchi, M., Naklamura, J., Tomida, K. and Makina, S. (1989) Intestinal parasitic infections among school children in Chiang Mai, Northern Thailand: an analysis of the present situation. *Journal of Tropical Medicine and Hygiene* 92, 360–364.

Kates, S., Wright, K.A. and Wright, R. (1973) A case of human infection with the cod nematode *Phocanema* sp. *American Journal of Tropical Medicine and Hygiene* 22, 606–608.

Kennedy, M.W., Tierney, J., Ye, P., McMonagle, F.A., McIntosh, A., Mclaughlin, D. and Smith, J.W. (1988) The secreted and somatic antigens of the third-stage larva of *Anisakis simplex* and antigenic relationship with *Ascaris suum, Ascaris lumbricoides* and *Toxocara canis. Molecular and Biochemical Parasitology* 31, 35–46.

Kikuchi, K., Toyokawa, O., Nakamura, K., Ishiyama, H., Yokota, H., Sato, H., Natori, T., Ishikura, H. and Aziawa, M. (1970) Immunopathology of experimental anisakiasis. *Minophagen Medical Review* 15, 54–58 (in Japanese).

Kikuchi, S.H., Kosugi, H., Satoh, H. and Hayashi, S. (1972) Studies on the pathogenicity of the larvae of a species of *Terranova* (Anisakinae, Nematoda) to experimental animals. *Yokohama Igaku* 22, 297–304 (in Japanese).

Kim, M.S., Lee, J.S. and Rim, H.J. (1982) Studies on the clinical aspects of clonorchiasis in Korea. *Korea University Medical Journal* 19, 107–121 (in Korean).

Kim, Y.L. (1984) Liver carcinoma and liver fluke infection. *Arzneimittel-Forschung* 34, 1121–1126.

Ko, R.C. (1976) Experimental infection of mammals with larval *Echinocephalus sinensis* (Nematoda: Gnathostomatidae) from oysters (*Crassostrea gigas*). *Canadian Journal of Zoology* 54, 597–609.

Ko, R.C. (1977) Effects of temperature acclimation on infection of *Echinocephalus sinensis* (Nematoda: Gnathostomatidae) from oysters to kittens. *Canadian Journal of Zoology* 55, 1129–1132.

Ko, R.C., Ling, J. and Adal, M.N. (1980) Cephalic anatomy of a gnathostomatid nematode, *Echinocephalus sinensis*, parasite of oysters and rays. *Journal of Morphology* 165, 301–317.

Ko, R.C., Chan, S.W., Lam, K., Farrington, M., Wong, H.W. and Yuen, P. (1987) Four documented cases of eosinophilic meningoencephalitis due to *Angiostrongylus cantonensis* in Hong Kong. *Transactions of the Royal Society of Tropical Medicine and Hygiene* 81, 807–810.

Kobayashi, A., Koyama, T., Kumada, M., Komiya, Y., Oshima, T., Kagei, N., Ishii, T. and Machida, M. (1966) A survey of marine fishes and squids for the presence of anisakinae larvae. *Japanese Journal of Parasitology* 15, 348–349 (in Japanese).

Komiya, Y. and Suzuki, N. (1964) Biology of *Clonorchis sinensis. Progress of Medical Parasitology in Japan* 1, 551–600.

Korbsrisate, S., Mongkolsuk, S., Haynes, J.R., Wong, Q. and Sirisinha, S. (1992) Cloning and characterization of ribosomal RNA genes from *Opisthorchis viverrini. Parasitology* 104, 323–329.

Kosugi, K., Kikuchi, S., Hirabayashi, H. and Hayashi, S. (1970) Seasonal occurrence of the larvae of *Anisakis* and related nematodes in fishes from Sagami Bay, the results of two years observation, 1968 to 1969. *Japanese Journal of Parasitology* 19, 106–107 (in Japanese).

Kowalowska-Grochouska, K., Quinn, J., Perry, I. and Sherbanink, R. (1989) A case of anisakiasis-Alberta. *Canadian Diseases Weekly Report* 15, 221–223.

Koyama, T., Kobayashi, A., Kumada, M., Komiya, A., Oshima, T., Kagei, N., Ishii, T. and Machida, M. (1969) Morphological and taxonomical studies on anisakidae larvae found in marine fishes and squids. *Japanese Journal of Parasitology* 18, 466–487 (in Japanese).

Kuiper, F.C. (1964) Eosinophilic phlegmonous inflammation of the alimentary canal caused by a parasite from the herring. *Pathologia et Microbiologia* 27, 925–930.

Kwon, K.H., Lee, J.S. and Rim, H.J. (1984) The use of IFAT in the diagnosis of human clonorchiasis. *Korea University Medical Journal* 21, 91–100.

Lee, J.S. (1975) Immunoelectrophoretic studies of *Clonorchis sinensis*. *Korea University Medical Journal* 12, 1–7 (in Korean).

Lee, O.R., Chung, P.R. and Nam, H.S. (1988) Studies on the immunodiagnosis of rabbit clonorchiasis. 2. Immunoaffinity purification of whole worm antigen and characterization of egg, metacercariae and adult antigens of *Clonorchis sinensis*. *Korean Journal of Parasitology* 26, 73–86.

Lee, S.C., Chung, Y.B., Kong, K.Y. and Cho, S.Y. (1993) Antigenic protein fractions of *Metagonimus yokogawi* reacting with patient sera. *Kisaengchunghak-Chapchi* 31, 43–48.

Lee, S.H., Hong, S.T., Chai, J.Y., Kim, W.H., Kim, Y.T., Song, I.S., Kim, S.W., Chi, B.I. and Cross, J.H. (1993) A case of intestinal capillariasis in the Republic of Korea. *American Journal of Tropical Medicine and Hygene* 48, 542–546.

Li, X. (1991) Food-borne parasitic zoonoses in the People's Republic of China. *Southeast Asian Journal of Tropical Medicine and Public Health* 22 (suppl.), 31–34.

Lichtenfels, J.R. and Brancato, F.P. (1976) Anisakid larva from the throat of an Alaskan Eskimo. *American Journal of Tropical Medicine and Hygiene* 25, 691–693.

Lim, J.H., Ko, Y.T., Lee, D.H. and Kim, S.Y. (1989) Clonorchiasis: sonographic findings in 59 proven cases. *American Journal of Roentgenology* 152, 761–764.

Little, M.D. and MacPhail, J.C. (1972) Large nematode larva from the abdominal cavity of a man in Massachusetts. *American Journal of Tropical Medicine and Hygiene* 21, 948–950.

Little, M.D. and Most, H. (1973) Anisakid larva from the throat of a woman in New York. *American Journal of Tropical Medicine and Hygiene* 22, 609–612.

Maleewong, W., Loahabhan, P., Wongkham, C., Intapan, P., Morakote, N. and Khamboonruang, C. (1992a) Effects of albendazole on *Gnathostoma spinigerum* in mice. *Journal of Parasitology* 78, 125–126.

Maleewong, W., Wongkham, C., Intapan, P., Mahaisavariya, Danseegaew, W., Pipitgool, V. and Morakote, N. (1992b) Detection of circulating parasite antigens in murine gnathostomiasis by a two-site enzyme-linked immunosorbent assay. *American Journal of Tropical Medicine and Hygiene* 46, 80–84.

Maleewong, W., Intapan, P., Wongwajana, S., Sitthithaworn, P., Pipitogool, V., Wongkham, C. and Daenseegaew, W. (1992c) Prevalence and intensity of *Opisthorchis viverrini* in rural community near the Mekong River on the Thai–Laos border in northeast Thailand. *Journal of the Medical Association of Thailand* 75, 231–235.

Matsura, T., Bylund, G. and Sugane, K. (1992) Comparison of restriction fragment length polymorphisms of ribosomal DNA between *Diphyllobothrium nihonkaiense* and *D. latum*. *Journal of Helminthology* 66, 261–266.

Mcfadzean, A.J.S. and Yeung, R.T.T. (1966) Acute pancreatitis due to *Clonorchis sinensis*. *Transactions of the Royal Society of Tropical Medicine and Hygiene* 60, 466–470.

Meyers, B. (1963) The migration of *Anisakis*-type larvae in experimental animals. *Canadian Journal of Zoology* 41, 147–148.

Mimori, T., Tada, I., Kawabat, M., Ollague, L.W., Calero, H.G. and Chong, Y.F. (1987) Immunodiagnosis of human gnathostomiasis in Ecuador by skin test and ELISA using *Gnathostoma doloresi* antigens. *Japanese Journal of Tropical Medicine and Hygiene* 15, 191–196.

Minamoto, T., Sawaguchi, K., Ogino, T. and Mai, M. (1991) Anisakiasis of the colon: report of two cases with emphasis on the diagnostic and therapeutic value of colonoscopy. *Endoscopy* 23, 50–52.

Mori, H. (1957) Studies on the liver fluke. *Acta Scholae Medica Gifu* 5, 601–603 (in Japanese).

Muller, R. (1975) *Worms and Disease – A Manual of Medical Helminthology*. William Heinemann Medical Books Ltd, London.

Muratov, I.V., Posokhov, P.S., Romanenko, N.A., Zimin, A.S. and Glazzyrina, G.F. (1992) The epidemiological characteristics of diphyllobothriasis caused by *Diphyllobothrium klebanovskii* in the Amur River basin. *Meditsinskaya Parazitologiya Moskow* 3, 46–47 (in Russian).

Nawa, Y., Imai, J.I., Abe, T., Kisanuki, H. and Tsuda, K. (1988) A case report of intestinal capillariasis – the second case found in Japan. *Japanese Journal of Parasitology* 37, 113–118.

Nishibuko, K. (1961) Studies on experimental gnathostomiasis with special reference to host–parasite relationship in *Gnathostoma spinigerum*. I. Experimental feeding of albino rat with larval *Gnathostoma spinigerum* obtained from *Ophicephalus argus*. *Endemic Disease Bulletin Nagasaki University* 5, 199–207 (in Japanese).

Nishioka, N.S. and Donnelly, S.S. (1990) A 72-year-old Chinese woman with recent abdominal pain and a right-sided abdominal mass. *New England Journal of Medicine* 323, 467–475.

Nopparatana, C., Setasuban, P., Chaicumpa, W. and Tapchaisri, P. (1991) Purification of *Gnathostoma spinigerum* specific antigen and immunodiagnosis of human gnathostomiasis. *International Journal for Parasitology* 21, 677–687.

Norris, D.E. and Overstreet, R.M. (1976) The public health implications of larval *Thynnascaris* nematodes from shellfish. *Journal of Milk Food Technology* 39, 47–54.

Ogata, K., Imai, J.I. and Yukifumi, N. (1988) Three confirmed and five suspected cases of *Gnathostoma doloresi* infection found in Miyazaki Prefecture, Kyushu. *Japanese Journal of Parasitology* 37, 358–364.

Ona, F.V. and Dytoc, J.N. (1991) *Clonorchis* associated cholangiocarcinoma: a report of two cases with unusual manifestations. *Gasteroenterology* 101, 831–839.

Ong, G.B. (1962) A study of recurrent pyogenic cholangitis. *Archives of Surgery* 84, 192–225.

Oyangi, T. (1967) Experimental studies on the visceral migrans of gastro-intestinal walls due to *Anisakis* larvae. *Japanese Journal of Parasitology* 16, 470–493 (in Japanese).

Oshima, T. (1972) *Anisakis* and anisakiasis in Japan and adjacent area. *Progress of Medical Parasitology in Japan* 4, 301–393.

Oshima, T. (1984) Anisakiasis, diphyllobothriasis and creeping disease – changing pattern of parasitic diseases in Japan. In: Ko, R.C. (ed.), *Current Perspectives in Parasitic Diseases*. Departments of Zoology and Medicine, University of Hong Kong, pp. 93–102.

Oshima, T. and Wakai, R. (1983) Epidemiology of *Diphyllobothrium latum* infection among Japanese people, especially on the infection of cherry salmon with *D. latum* plerocercoid. *Japanese Journal of Antibiotics* 36, 566–572 (in Japanese).

Overstreet, R.M. and Meyers, G.W. (1981) Hemorrhagic lesions in stomach of rhesus monkeys caused by a piscine ascaridoid nematode. *Journal of Parasitology* 67, 226–235.

Pacheco, G., Wykoff, D.E. and Jung, R.C. (1960) Trial of an indirect haemagglutination test for the diagnosis of infections with *Clonorchis sinensis*. *American Journal of Tropical Medicine and Hygiene* 9, 367–370.

Petithory, J.C., Rousseau, M. and Siodlak, F. (1991) Seroepidemiological data on anisakiasis: prophylactic consequences in fish products. *Bulletin de l'Académie Nationale de Médecine* 175, 273–277 (in French).

Phillipson, R.F. and Mcfadzean, J.A. (1962) *Clonorchis, Opisthorchis* and *Paragonimus* gel diffusion studies. *Transactions of the Royal Society of Tropical Medicine and Hygiene* 56, 13.

Plyuscheva, G.L., Romanenko, N.A., Gerasimov, I.V., Suleimanov, N.T., Stepanov, L.G., Akulova, L.M., Volodin, Yu, F., Vorobeva, N.P. and Khrolenko, E.T. (1987a) Establishment of diphyllobothriasis foci in the Krasnoyarsk reservoir. *Meditsinskaya Parazitologiya i Parazitarnye Bolezni* 1, 64–67 (in Russian).

Plyuscheva, G.L., Romanenko, N.A. and Gerasimov, I.V. (1987b) The role of anthropogenic factors in the establishment of foci of diphyllobothriasis in reservoirs. *Meditsinskaya Parazitologiya i Parazitarnye Bolezni* 6, 74–78 (in Russian).

Pungpak, S., Akai, P.S., Longenecker, B.M., Ho, M., Befus, A.D. and Bunnag, D. (1990) Tumour markers in the detection of opisthorchiasis associated cholangiocarcinoma. *Abstracts of 33rd Southeast Asian Medical Organization Tropical Medicine Regional Seminar, Chiangmai, Thailand*, p. 55.

Punyagupta, S. (1979) Angiostrongliasis: clinical features and human pathology. In: Cross, J. (ed)., *Studies on Angiostrongyliasis in Eastern Asia and Australia. Special Publication of U.S. Naval Medical Research Unit No. 2*, pp. 138–150.

Punyagupta, S. and Bunnag, T. (1990) Eosinophilic myeloencephalitis: invasion of the central nervous system by *Gnathostoma spinigerum*. *Abstracts of 33rd Southeast Asian Medical Education Organization-Tropical Medicine Regional Seminar*, Chiangmai, Thailand, p. 65.

Purtilo, D.T. (1976) Clonorchiasis and hepatic neoplasms. *Tropical and Geographical Medicine* 28, 21–27.

Revenga, J.E. (1993) *Diphyllobothrium dendriticum* and *Diphyllobothrium latum* in fishes from southern Argentina. *Journal of Parasitology* 79, 379–383.

Riganti, M., Pungpak, S., Punpoowong, B., Bunnag, D. and Harinasuta, T. (1989) Human pathology of *Opisthorchis viverrini* infection: a comparison of adults and children. *Southeast Asian Journal of Tropical Medicine and Public Health* 20, 95–100.

Rim, H.S. (1986) The current pathobiology and chemotherapy of *Clonorchiasis*. *Korean Journal of Parasitology* 24, Supplement, Monographic Series No. 3, Korean Society for Parasitology, Seoul, South Korea.

Rim, H.S. (1988) Clonorchiasis. In: Balows, A., Hausler, W.J.Jr, Ohashi, M. and Turano, A. (eds) *Laboratory Diagnosis of Infectious Diseases. Vol. 1. Bacterial, Mycotic and Parasitic Diseases*. Springer-Verlag, New York, pp. 801–810.

Rosenberg, E.B., Whalen, G.E., Bennich H., Johansson, S.G.O. (1970) Increased circulating IgE in a new parasitic disease – human intestinal capillariasis. *New England Journal of Medicine* 283, 1148–1149.

Ruitenberg, E.J. (1970) Anisakiasis – pathogenesis, serodiagnosis and prevention. PhD Thesis in Rijiks University, Utrecht, the Netherlands.

Ruitenberg, E.J., Berkvens, J.M. and Duyzings, M.J. (1971) Experimental *Anisakis marina* infections in rabbits. *Journal of Comparative Pathology* 81, 157–163.

Ryoji, S. (1922) Diagnostic value of complement fixation test for clonorchiasis. *Okayama Igakkai Zasshi* 384, 1–12 (in Japanese).

Sadun, E.H., Walton, B.C., Buck, A.A. and Lee, B.K. (1959) The use of purified antigen in the diagnosis of *Clonorchiasis sinensis* by means of intradermal and complement fixation test. *Journal of Parasitology* 45, 129–134.

Sagara, I. (1953) Studies on *Gnathostoma*. Part II. Migration route of larvae of *Gnathostoma spinigerum* in the rat's body and histopathological changes caused along the route. *Igaku Kenkyuu* 23, 822–836 (in Japanese).

Sawada, T., Kazutoshi, T. and Chun, S.K. (1976) Studies on the purification of antigens for hemagglutination test on clonorchiasis. *Japanese Journal of Experimental Medicine* 46, 337–342.

Schwartz, D.A. (1980) Review: Helminths in the induction of cancer: *Opisthorchis viverrini, Clonorchis sinensis* and cholangiocarcinoma. *Tropical and Geographical Medicine* 32, 95–100.

Seo, B.S., Lee, S.H., Cho, S.Y., Chai, J.Y., Hong, S.T., Han, I.S., Sohn, J.S., Cho, B.H., Ahn, S.R., Lee, S.K., Chung, S.C., Kang, K.S., Shin, H.S. and Hwang, I.S. (1981) An epidemiologic study on clonorchiasis and metagonimiasis in riverside area in Korea. *Korean Journal of Parasitology* 19, 137–150.

Setasuban, P., Jewjhangarnwanit, S., Rojanakittikoon, V., Yaemput, S., Dekumyoy, P., Akabane, H. and Kojima, S. (1991) Gnathostomiasis in Thailand: a survey on intermediate hosts of *Gnathostoma* spp. with special reference to a new type of larvae found in *Fluta alba. Southeast Asian Journal of Tropical Medicine and Public Health* 22 (suppl.), 220–224.

Shirahama, M., Koga, T., Ishibashi, H., Uchida, S., Ohta, Y. and Shimoda, Y. (1992) Intestinal anisakiasis: US in diagnosis. *Radiology* 185, 789–793.

Sirisinha, S., Sahassananda, D., Bunnag, D. and Rim, H.J. (1990) Immunological analysis of *Opisthorchis and Clonorchis* antigens. *Journal of Helminthology* 64, 133–138.

Sirisinha, S., Chaewengkirttikul, R. and Sermswan, R. (1991) Immunodiagnosis of opisthorchiasis. *Southeast Asian Journal of Tropical Medicine and Public Health.* 22 (suppl.), 179–183.

Sirisinha, S., Chaewengkirkttikul, R., Tayapiwatana, C., Naiyanetr, C., Waikagul, J., Radomyos, P. and Podoprigora, G.I. (1992) Specific and cross-reactive monoclonal antibodies to the 89 kDa antigen of *Opisthorchis viverrini. Southeast Asian Journal of Tropical Medicine and Public Health* 23, 489–490.

Smith, J.W. (1983) *Anisakis simplex* (Rudolphi, 1809, det. Krabbe, 1878) (Nematoda: Ascaridoidea): morphology and morphometry of larvae from euphausiids and fish and a review of the life history and ecology. *Journal of Helminthology* 57, 205–224.

Soh, C.T. (1984) The current status of human parasitic infections in Korea. In: Ko, R.C. (ed.), *Current Perspectives in Parasitic Diseases.* Department of Zoology and Medicine, University of Hong Kong, pp. 83–92.

Soh, C.T. and Min D.Y. (1990) Field operational research on the of clonorchiasis in Korea. *Abstracts of 33rd Southeast Asian Medical Organization Tropical Medicine Regional Seminar, Chiangmai, Thailand*, p. 52.

Sornmani, S. (1990) Opisthorchiasis and the control programme in Thailand. *Abstracts of 33rd Southeast Asian Medical Organization Tropical Medicine Regional Seminar, Chiangmai, Thailand*, p. 53.

Sugane, K., Liu, Q. and Matsuura, T. (1989) Restriction fragment length polymorphism of anisakinae larvae. *Journal of Helminthology* 63, 269–274.

Sun, S.C., Cross, J.H., Berg, H.S., Kau, S.L., Singson, C.N., Banzon, T.C. and Watten, R.H. (1974) Ultrastructural studies of intestinal capillariasis *Capillaria philippinensis* in human and gerbil hosts. *Southeast Asian Journal of Tropical Medicine and Public Health* 5, 524–533.

Sun, T. and Gibson, J.B. (1969) Antigens of *Clonorchis sinensis* in experimental and human infections. An analysis by gel diffusion technique. *American Journal of Tropical Medicine and Hygiene* 18, 241–252.

Suntharasamai, P., Riganti, M., Chittamas, S. and Desakorn, V. (1990) Treatment of gnathostomiasis with albendazole: a randomized double blind placebo controlled trial. *Abstracts of 33rd Southeast Asian Medical Education Organization Tropical Medicine Regional Seminar, Chiangmai, Thailand*, p. 69.

Suzuki, H., Ohnuma, H., Karasawa, Y., Ohbayashi, M., Koyama, T., Kumada, M. and Yokogawa, M. (1972) *Terranova* (Nematoda: Anisakidae) infection in man. 1. Clinical features of five cases of *Terranova* larva infection. *Japanese Journal of Parasitology* 21, 252–256.

Suzuki, T., Shiraki, T., Seikino, S., Otsuru, M. and Ishikura, H. (1970) Studies on the immunodiagnosis of anisakiasis. III. Intradermal test with purified antigen. *Japanese Journal of Parasitology*, 19, 1–9 (in Japanese).

Tada, I, Araki, T., Matsuda, H., Araki, K., Akakane, H. and Mimori, T. (1987) A study on immunodiagnosis of gnathostomiasis by ELISA and double diffusion with special reference to the antigenicity of *Gnathostoma doloresi*. *Southeast Asian Journal of Tropical Medicine and Public Health* 18, 444–448.

Takahasi, S., Sato, N. and Ishikura, H. (1986) Establishment of monoclonal antibodies that discriminate the antigen distribution specifically found in *Anisakis* larva (type I). *Journal of Parasitology* 72, 960–962.

Takakura, Y. (1988) Experimental studies on *Gnathostoma hispidum* Fedchenko, 1872: migration and development of the larvae in the rats and piglets. *Japanese Journal of Parasitology* 37, 67–75.

Takei, K. and Chun, S.K. (1976) Purification of antigens from *Clonorchis sinensis* worm for complement fixation test. *Japanese Journal of Experimental Medicine* 46, 399–403.

Teplukhin, Yu. V., Karal'Nik, B.V., Gorbunova, L.A., Slemnev, V.F. and Nikityak, G.V. (1987) Detection of antibodies in patients with chronic opisthorchiasis. *Meditsinskaya Parazitologiya i Parazitarnye Bolezni* 6, 21–24 (in Russian).

Tuntipopipat, S., Chawengkiattikul, R., Witoon-Panich, R., Chemichanya, S. and Sirisinha, S. (1989) Antigens, antibodies and immune complexes in cerebral fluid of patients with cerebral gnathostomiasis. *Southeast Asian Journal of Tropical Medicine and Public Health* 20, 439–446.

Usutani, T. (1966) Histological studies on experimental animals administered with *Anisakis*-like larvae from marine fish. *Shikoku Acta Medica* 22, 486–503 (in Japanese).

Van Thiel, P.H.F., Kuipers, F.C. and Roskam, R. (1960) A nematode parasite of herring, causing acute abodominal syndromes in man. *Tropical Geographical Medicine* 2, 97–113.

Vatanasapt, V., Uttaravichien, T., Mairiang, E.O., Pavioijkul, C., Chartbanchanchai, W. and Haswell-Elkins, M. (1990) Cholangiocarcinoma in north-east Thailand. *Lancet* 335, 116–117.

Vidal-Martinez, V.M., Osorio-Sarabia, D. and Overstreet, R.M. (1994) Experimental infection of *Contracaeum multipapillatum* (Nematoda: Aniskinae) from Mexico in the domestic cat. *Journal of Parasitology* 80, 576–579.

Watten, R.H., Becker, W.M., Cross, J.H., Gunning, J.J. and Jarimillo, J. (1972) Clinical studies of capillariasis philippinensis. *Transactions of the Royal Society of Tropical Medicine and Hygiene* 66, 828–834.

Wongratanacheewin, S. and Sirisinha, S. (1987) Analysis of *Opisthorchis viverrini* antigens: physiochemical characterization and antigen localization. *Southeast*

Asian Journal of Tropical Medicine and Public Health 18, 511–520.

Wongratanacheewin, S., Chawengkirttikul, R., Bunnag, D. and Sirisinha, S. (1988) Analysis of *Opisthorchis viverrini* antigens by immunoprecipation and poly-acrylamide gel electrophoresis. *Parasitology* 96, 119–128.

Wykoff, D.E. (1959) Studies on *Clonorchis sinensis*. II. Development of an antigen for complement fixation and studies on the antibody response in infected rabbits. *Experimental Parasitology* 8, 51–57.

Wykoff, D.E., Harinasuta, C., Juttijudata, P. and Winn, M.M. (1965) *Opisthorchis viverrini* in Thailand – the life cylce and comparison with *O. felineus*. *Journal of Parasitology* 51, 207–214.

Yagihashi, A., Sato, N., Takahashi, S., Ishikura, H. and Kikuch, K. (1990) A serodiagnostic assay by microenzyme-linked immunosorbent assay for human anisakiasis using a monoclonal antibody specific for *Anisakis* larvae antigen. *Journal of Infectious Diseases* 161, 995–998.

Yamaguchi, T., Kudo, N., Kawadam S., Nakade, Y. and Takada, N. (1968) Studies on larva migrans (24). The incidence of infection of *Anisakis* larvae in marine fishes. *Japanese Journal of Parasitology* 17, 262.

Yamane, Y., Kamo, H., Yazaki, S., Fukumoto, S. and Maejima, J. (1981) On a new marine species of the genus *Diphyllobothrium* (Cestoda: Pseudophyllidea) found from a man in Japan. *Japanese Journal of Parasitology* 30, 101–111.

Yamane, Y., Kamo, H., Bylund, G. and Wikgren, B.J.P. (1986) *Diphyllobothrium nihonkaisiense* sp. nov. (Cestoda: Diphyllobothridae) – revised identification of Japanese broad tapeworm. *Shimane Journal of Medical Science* 10, 29–48.

Yang, W.Y., Lee, J.S. and Rim, H.J. (1983) The use of ELISA in the diagnosis of human clonorchiasis. *Korea University Medical Journal* 20, 201–210 (in Korean).

Yang, W.Y., Lee, J.S. and Rim, H.J. (1984) Studies on the changing patterns of specific IgG antibody in the sera of rabbits infected with *Clonorchis sinensis*. *Korea University Medical Journal* 21, 81–88 (in Korean).

Yen, C.M. and Chen, E.R. (1989) Counterimmunoelectrophoresis test on human *Clonorchis sinensis* infection. *Southeast Asian Journal of Tropical Medicine and Public Health* 20, 433–438.

Yen, C.M., Chen, E.R., Hou, M.F. and Chang, J.H. (1992) Antibodies of different immunoglobulin isotypes in serum and bile of patients with clonorchiasis. *Annals of Tropical Medicine and Parasitology* 86, 263–269.

Yokogawa, M. and Yoshimura, H. (1967) Clinico-pathologic studies on larval anisa-kiasis in Japan. *American Journal of Tropical Medicine and Hygiene* 16, 723–728.

Yokogawa, M., Yoshimura, H. and Tsuji, M. (1965) Experimental studies on *Anisakis*-like larva infection. I. Inoculation to a small laboratory animals and immunological reaction. *Japanese Journal of Parasitology* 14, 606–607 (in Japanese).

Yong, T.S. Im, K.I. and Chung, P.R. (1991) Diagnosis of clonorchiasis by ELISA-inhibition test using *Clonorchis sinensis* specific monoclonal antibody. *Southeast Asian Journal of Tropical Medicine and Public Health* 22 (suppl.), 186–188.

Yoshimura, H. (1965) The life cycle of *Clonorchis sinensis*: a comment on the presenta-tion in the seventh edition of Craig and Faust's *Clinical Parasitology*. *Journal of Parasitology* 61, 961–966.

Yoshimura, H. (1966) Eosinophilic granuloma due to *Anisakis* larva penetrating the gas-trontestinal tract of man. *Minophagen Medical Review* 11, 105–114 (in Japanese).

Young, P.C. and Lowe, D. (1969) Larval nematodes from fish of the subfamily Anisakinae and gastrointestinal lesions in mammals. *Journal of Comparative Pathology* 79, 301–313.

16

Parasitic Diseases of Shellfish

S.M. Bower

Department of Fisheries and Oceans, Biological Sciences Branch, Pacific Biological Station, Nanaimo, British Columbia, Canada V9R 5K6.

INTRODUCTION

Numerous species of parasites have been described from various shellfish, especially representatives of the Mollusca and Crustacea (see Lauckner, 1983; Sparks, 1985; Sindermann and Lightner, 1988; Sindermann, 1990). Surprisingly few have proven to cause diseases of significance to host populations. However, some parasites may have the potential if conditions are in their favour. The following chapter is confined to parasites that cause disease in economically important shellfish that are utilized for either aquaculture or commercial harvest. Parasites with the potential of causing significant disease are briefly mentioned. The pathogenic parasites are grouped taxonomically. With noted exceptions, the protozoans are classified according to Levine *et al.* (1980).

PHYLUM SARCOMASTIGOPHORA

Four orders within the Sarcomastigophora (Amoebida, class Lobosea; Dinoflagellida and Euglenida, class Phytomastigophorea; and Diplomonadida, class Zoomastigophorea) contain species that are significant pathogens of shellfish. *Paramoeba perniciosa* (order Amoebida) is the best known and causes the greatest economic impact. It will be considered in detail while the other species will be briefly mentioned.

Order Amoebida

Taxonomy
Paramoeba perniciosa (suborder Conopodina, family Paramoebidae) is the cause of 'grey crab disease' or paramoebiasis in the blue crab (*Callinectes sapidus*) (see Sprague *et al.*, 1969).

Distribution and economic importance

Paramoeba perniciosa has been reported in blue crabs along the east coast of the United States from Connecticut to Florida including the high salinity areas of Chincoteague Bay and Chesapeake Bay (Sparks, 1985; Johnson, 1988). It has also been reported from the rock crab *Cancer irroratus*, the lobster *Homarus americanus*, and the green crab *Carcinus maenas* (Messick and Sindermann, 1992).

Paramoeba perniciosa caused many mass mortalities of blue crabs along the southeastern Atlantic coast (Couch, 1983; Sparks, 1985). Epizootics with high mortalities (about 17%) were reported from Chincoteague Bay in early summer and mortalities (20–30%) were observed in shedding tanks (for production of newly moulted softshell crab) (Johnson, 1988). Since 1967, the blue crab fishery has experienced periodic high losses and ongoing low level mortalities due to paramoebiasis (Couch, 1983).

Morphology and life cycle

Distinctive characteristics are its parasitic habit, its failure to survive in common culture media, its relatively small size, and its linguiform lobopodia (Sprague *et al.*, 1969). The amoebae are round to elongate and can be differentiated into small (3–12 µm) and large (15–35 µm) forms (Johnson, 1977; Couch, 1983). Each amoeba contains a vesicular nucleus with a large central endosome and a 'second nucleus', 'Nebenkörper', or elongate parasome (1–4 µm) with a Feulgen-positive middle bar and two opposing basophilic polar caps. Due to the unusual ultrastructure of the parasome, which is characteristic for *Paramoeba* spp., Perkins and Castagna (1971) proposed that the parasome may be a discrete organism of unknown taxonomic affinities and not an organelle of the amoeba.

Paramoeba perniciosa is a parasite of the connective tissues and haemal spaces, invading the circulating blood only when the infection is terminal (Johnson, 1977). In light infections, the parasite is in connective tissues along the midgut, antennal gland, and Y organ. Haemal spaces in gills are usually invaded in medium and heavy infections and eventually the infection becomes systemic (Sparks, 1985). However, peripheral epithelial tissues of the gut, hepatopancreas, hypodermis, and gonad are not invaded even in advanced infections (Couch, 1983). In heavy infections, pathological changes caused by large numbers of amoebae include: tissue displacement; probable lysis of some types of tissue including haemocytes; and significant decreases in protein, haemocyanin (the oxygen-binding and -transport molecule of crustaceans), and glucose (Pauley *et al.*, 1975; Johnson, 1977). Sparks (1985) suggested that the probable cause of death was a combination of anoxia and nutrient deficiency. Terminal infections are usually observed during the late spring to early autumn but infected blue crabs are found throughout the year (Johnson, 1988). The mode of transmission in the field has not been fully elucidated.

Host–parasite relationships

Most infected blue crabs demonstrate a defence response which is usually manifested as phagocytosis of amoebae by hyaline haemocytes and infrequently as encapsulation of amoebae by haemocytes, but destruction of amoebae by humoral factors also occurred (Johnson, 1977; Messick and Sindermann, 1992). Occasionally a blue crab would overcome the infection. Some attempts to infect blue crabs by injecting amoebae or feeding infected crab tissues failed (Newman and Ward, 1973; Couch, 1983). However, Johnson (1977) observed the disease in two blue crabs 34 and 39 days post inoculation with infected haemolymph and Sparks (1985) claimed that the disease was transmitted by consumption of moribund or dead infected blue crab.

Propagation

Paramoeba perniciosa has not been cultured in medium that supported continuous growth of *Paramoeba eilhardi*, a free-living species originally isolated from algal material. Other attempts to maintain *P. perniciosa* in various media were also unsuccessful. However, *P. perniciosa* survived for about 2 weeks in 10% calf serum agar overlaid with sterile sea water and incubated at 18°C (Sprague *et al.*, 1969).

Clinical signs and diagnosis of infection

Signs of infection include a greyish discoloration of the ventral exoskeleton, general sluggishness, reduced or absence of clotting of the haemolymph, and poor survival subsequent to handling or holding in tanks (Sparks, 1985). Infection is easily diagnosed only in the terminal phase when numerous amoebae and virtually no haemocytes are present in circulating blood. Amoebae in blood can be observed with phase contrast either live or fixed in 5% formalin seawater and stained with dilute methylene blue. Permanent smears are fixed in Bouin's, Davidson's, Hollande's, or 10% formalin solutions and stained with iron haematoxylin or Giemsa's stain. Before amoebae appear in the circulation, they may be observed in squashes of subepithelial connective tissue examined with phase contrast if the infection is sufficiently advanced (Johnson, 1988).

Related pathogen

In the early 1980s mass mortalities of the sea urchin, *Strongylocentrotus droebachiensis* along the Atlantic coast of Nova Scotia were attributed to a previously undescribed species, *Paramoeba invadens* (Scheibling and Stephenson, 1984; Jones, 1985; Jones and Scheibling, 1985; Jones *et al.*, 1985). From 1980 to 1983 sea urchin mortalities were estimated to be at least 245,000 tons (Miller, 1985). No mortalities were observed in echinoderms including other echinoids, asteroids, and ophiuroids from the same area (Scheibling and Stephenson, 1984). Although *S. droebachiensis* is not currently commercially harvested, extensive reduction of this dominant herbivore resulted in a rapid recolonization of the rocky subtidal area by fleshy macrophytes. The transformation of echinoid-dominated 'barren grounds' into kelp beds provided increased areas for American lobster (*Homarus americanus*) recruitment and thus increased lobster productivity (Wharton and Mann, 1981).

Paramoeba invadens is similar in size (20–35 μm in length and 8–15 μm in width) to *P. perniciosa*. However, *P. invadens* is more elongated in shape with a length: width ratio of about 2 and has digitiform pseudopodia. Also, the parasome (2–3 µm in size) has Feulgen-positive poles but no Feulgen positive central band and it lies adjacent to the nucleus. Characteristic signs of paramoebiasis in sea urchins included muscle necrosis, general infiltration of coelomocytes, reddish-brown discoloration, and high mortalities (Jones, 1985; Jellett *et al.*, 1988). Transmission was direct and the infection was water-borne (Jones, 1985). This amoeba was easily cultured on malt yeast and on non-nutrient agar with marine bacteria as a food source although there was some loss of virulence after 15 weeks in monoxenic culture and 58 weeks in polyxenic culture (Jones and Scheibling, 1985; Jellett and Scheibling, 1988).

Order Dinoflagellida

Hematodinium perezi (family Syndinidae) originally described from the hae-molymph of crabs (*Carcinus maenas* and *Portunus depurator*) from European waters (Chatton and Poisson, 1930) was reported from crabs (*Callinectes sapidus, Cancer irroratus, Cancer borealis*, and *Ovalipes ocellatus*) along the east coast of the United States from New Jersey to the western coast of Florida (MacLean and Ruddell, 1978; Couch, 1983). Prevalence of infection was low in all hosts except for the blue crab (*C. sapidus*) with up to 30% infected in some localities (Newman and Johnson, 1975). Infected blue crabs were found only in areas above 11 ppt salinity and in all seasons except late winter and early spring (Messick and Sindermann, 1992).

In blue crabs, the most abundant form of *H. perezi* is a round cell (about 6 µm in diameter) with a single dinokaryon nucleus which is characteristic of the dinoflagellates. Binucleate cells and multinucleate plasmodia (8–64 µm in diameter or length) are also observed in the haemal sinuses (Couch, 1983). However, the flagellated dinospore is usually not found in the blue crab.

The haemolymph of heavily infected crabs is opalescent or milky, slow to clot, devoid of haemocytes, and filled with non-motile dinoflagellates. Also, there may be total lysis of hepatopancreatic tubules and partial destruc-tion of muscle fibres (Couch, 1983). Although insufficient data is available to assess the annual impact on blue crab populations and fishery success, Newman and Johnson (1975) suggested that this parasite may cause high mortalities among blue crabs in enzootic areas along the east coast of North America.

Recently, a *Hematodinium*-like dinoflagellate was described as causing an astringent after-taste (bitter crab syndrome) and mortalities in tanner crabs (*Chionoecetes bairdi* and *Chionoecetes opilio*) in southeast Alaska and the Bering Sea (Meyers *et al.*, 1987, 1990; Eaton *et al.*, 1991). Infected tanner crabs apparently express the disease about 12–18 months following exposure. Numerous non-motile vegetative stages are observed in the haemal sinuses of all tissues, causing opaque haemolymph and milky emaciated musculature. During July through October, when the prevalence and intensity were highest, two types of motile dinospores were observed in tissues of some crabs.

Sporulation is the last stage in the infection; this results in death of the tanner crab with peak mortalities occurring during August and September (Love *et al.*, 1993). Reduction in prevalence of the disease in crab populations occurs in the winter. Laboratory transmission by injection indicated that both types of dinospores are infectious (Eaton *et al.*, 1991). However, as for *Hematodinium perezi*, the complete life cycle of this parasite in the field is poorly understood.

Meyers *et al.* (1987) suggested that the bitter flavour in cooked infected tanner crabs is the result of either the dinoflagellate itself or its metabolite(s). Meyers *et al.* (1990) conservatively estimated that the total economic loss to fisherman due to rejected diseased crabs was about 5% of the catch for the 1988/1989 season. In addition, data from the commercial tanner crab fishery suggested that there is an increase in prevalence and spread of the disease to new areas. The management of bitter crab syndrome may be possible by harvesting tanner crabs in the winter when fewer crabs are severely parasitized and meats are more marketable. Also, proper disposal of infected tanner crabs is essential in controlling dissemination of the parasite (Meyers *et al.*, 1990).

Order Euglenida

An *Isonema*-like flagellate was reported to cause significant mortalities in cultured larval geoduck clams (*Panope abrupta*). The flagellates penetrate the mantle and proliferate within the coelom resulting in the death of heavily infected geoduck larvae (Kent *et al.*, 1987). There have been no other reports of invasive, pathogenic *Isonema* sp. in shellfish and Kent *et al.* (1987) suggested that conditions within the culture system may have predisposed the geoduck larvae to infection and resulting mortalities.

Order Diplomonadida

Members of the genus *Hexamita*, including *H. inflata* and *H. nelsoni*, have been reported as cosmopolitan inhabitants of oysters and clams (Lauckner, 1983; Vivarès *et al.*, 1986; López-Ochoterena *et al.*, 1988). Several workers have attributed oyster mortalities to these flagellates (Mackin *et al.*, 1952; Stein *et al.*, 1959; Laird, 1961; Feng and Stauber, 1968; Sparks, 1985). Nevertheless, the pathogenic nature of the flagellates has been questioned because, in most instances, oysters experiencing mortalities were held under poor environmental conditions (Scheltema, 1962; Lauckner, 1983; Vivarès *et al.*, 1987). In British Columbia, *Hexamita* sp. has been observed in the mantle cavity and intestinal tract of up to 45% of the Pacific oysters (*Crassostrea gigas*) with no associated mortalities. Only in one instance were high numbers of *Hexamita* sp. observed in the tissues of about 75% of the moribund and decomposing small Pacific oysters (about 1 cm in shell length) from one location and these Pacific oysters may have been dying from other unknown

but unrelated cause(s) (S.M. Bower, unpublished data). The amenability of
Hexamita sp. to axenic cultivation (Khouw *et al.*, 1968) provides a means of
establishing the identification and number of species of *Hexamita* in oysters
and their relative pathogenicity under various conditions.

PHYLUM LABYRINTHOMORPHA

Although several molluscan pathogens have been reported from the Labyrinthomorpha (i.e. in lesser octopus (*Eledone cirrhosa*) (Polglase, 1980), in squid
(*Illex illecebrosus*) (Jones and O'Dor, 1983), and in nudibranch (*Tritonia
diomedea*) (McLean and Porter, 1987)), only one species in this phylum has
been documented as a pathogen of economically important shellfish (i.e.
cultured abalone).

Order Labyrinthulida

Taxonomy
Labyrinthuloides haliotidis is pathogenic to small juvenile northern abalone
(*Haliotis kamtschatkana*) and small juvenile red abalone (*Haliotis rufescens*)
(Bower, 1987a). This parasite is closely related to the thraustochytrids (achlo-
rophyllous, eukaryotic protists) which have often been grouped with the lower
fungi but were included in the subkingdom Protozoa by Levine *et al.* (1980).

Distribution and economic importance
To date, this parasite has only been observed in small abalone (less than 1 cm
in shell length) from the only abalone culture facility in British Columbia.
Within 2 weeks of first being detected in a raceway, over 90% of the 100,000
small abalone in a raceway succumbed to the infection and the disease quickly
spread between raceways. The high mortalities caused by *L. haliotidis* was one
of the reasons that this particular abalone culture facility is no longer in
operation.

Small abalone that are susceptible to infection have not been found in the
field. Thus, the geographical distribution of this parasite and its effect on wild
stocks is not known.

Morphology and life cycle
The vegetative stage of *L. haliotidis* was spheroid (5–9 µm in diameter) and
had a unique organelle of this phylum called the sagenogenetosome. Several
sagenogenetosomes were scattered on the surface of each organism forming
openings in the thin laminated cell wall and each produced the ectoplasmic
membrane net upon which the parasite moved ($0.3 \pm 0.16 \,\mu m \, min^{-1}$) and
obtained nutrients. Following removal from sources of nutrients (i.e. place-
ment in sterile sea water), the vegetative cell underwent synchronous multiple
fission to form a zoosporoblast (6–10 µm in diameter) containing about ten
developing zoospores. The uninucleate, slightly oval, motile biflagellated

zoospores (4–6 µm long and 2.5–4.5 µm wide) escaped through a rupture in the zoosporoblast wall. The two flagella were laterally attached with a brush of mastigonemes along one side (about one third of the circumference) of the anterior flagellum (9–15 µm in length) and with a tapered tip on the naked posterior flagellum (5–10 µm in length) (Bower, 1987a). The flagella were shed when the zoospore contacted a hard surface or after about 24 hours of active swimming in sea water. The resulting cell was morphologically similar to the vegetative stage and survived in sterile sea water for at least 2 years (Bower, 1987b).

Vegetative stages that developed from zoospores were infective to small abalone. Within 4 hours of contacting the host, sagenogenetosomes were produced and extracellular lytic activity disrupted the plasmalemma layer of the host epithelial cells adjacent to the parasite eventually lysing the host cell. By 24 hours post exposure, the ectoplasmic net was well developed, allowing the parasite to move into and within the head and foot tissues of the abalone and dividing forms of the parasite were observed (Bower *et al.*, 1989a). Within 10 days after exposure to about 10^4 parasites in 20 ml of sea water, about 90% of the abalone (less than 4.0 mm in shell length and 140 days of age) died with numerous parasites throughout the head and foot (Bower, 1987c). As dead abalone decomposed, vegetative stages released from the tissues developed into zoosporoblasts that produced zoospores within about 24 to 72 hours. Parasites released from infected abalone were infective to other abalone on contact. Alternate hosts have not been described, but *L. haliotidis* can utilize diverse sources of nutrients although it seems incapable of coexisting with bacteria.

Host–parasite relationships

The tissues of heavily infected abalone were slightly swollen with a loss of integrity. Prevalence and intensity decreased, and time to death increased as the abalone increased in age and size. Abalone greater than 15 mm in shell length could not be infected even when about 1.5×10^4 *L. haliotidis* were injected intramuscularly. The mechanism of defence against this parasite is not known. There was no indication of an inflammatory response in young susceptible abalone. Possibly the resistance of older abalone corresponded to the development of cellular or humoral defence mechanisms as the abalone matured (Bower, 1987c).

Small juvenile Japanese scallops (*Patinopecten yessoensis*) and juvenile Pacific oysters (*Crassostrea gigas*) both less than 8 months of age were resistant to infection. However, two of the oysters with badly cracked shells became infected suggesting that *L. haliotidis* was capable of utilizing oyster tissue as nutrients for growth and multiplication if it was able to gain access to the soft tissues of the oyster (Bower, 1987c).

Propagation

Axenic cultures of *L. haliotidis* grew well on several different liquid media and agar based solid media. An *in vitro* life cycle could be produced by alternately placing the parasite in minimum essential medium with 10% fetal calf serum (where rapid production of vegetative stages occurred through binary fission) and sterile sea water (where the vegetative stages transformed into zoosporo-blasts and zoospores were produced) (Bower, 1987d). Cultured *L. haliotidis* were infective to small abalone (Bower, 1987c; Bower *et al.*, 1989a). The vegetative stage also grew on pine pollen (*Pinus contorta*) in sea water but failed to produce zoosporoblasts and zoospores (Bower, 1987d).

Diagnosis of infection

The spheroid parasite is readily observed with light microscopy (100 × magni-fication) of the head and foot of the small abalone squashed in sea water between a glass slide and coverslip. *Labyrinthuloides haliotidis* is also evident in histological sections prepared using routine procedures. However, due to the morphological similarities of *L. haliotidis* to other thraustochytrids, identification of the parasite outside its host is almost impossible. A direct fluorescent antibody technique showed promise in facilitating the detection of this parasite (Bower *et al.*, 1989b). However, this technique has not been fully tested to verify its specificity.

Prevention and control

The source of infection in the abalone hatchery was not established. Trans-mission of the disease between raceways could be prevented by employing sanitary techniques. The parasite was destroyed in 20 minutes when exposed to $25 \, \text{mg} \, \text{l}^{-1}$ of chlorine in sea water. The fungicide cyclohexamide at 1–$2 \, \text{mg} \, \text{l}^{-1}$ for 23 hours per day on 5 consecutive days cured infected abalone. However, this treatment had the disadvantages of: (i) being detrimental to diatoms upon which the abalone fed; (ii) being ineffective against non-growing but infective zoospores such that reinfection occurred within 2 to 3 weeks following treatment; and (iii) inducing resistant forms (as few as three succes-sive treatments resulted in the production of forms twice as resistant to cyclohexamide) (Bower, 1989). Ozone treatment of incoming water may only be efficacious if ozone exposure is greater than $0.97 \, \text{mg} \, \text{ozone} \, \text{l}^{-1}$ for 25 minutes (Bower *et al.*, 1989c).

PHYLUM APICOMPLEXA

Three orders within the Apicomplexa (Perkinsorida in the class Perkinsasida (nomenclature modified by Levine (1988)), and Eugregarinida and Eucoc-cidiida in the class Sporozoea) contain species that are significant pathogens of shellfish. Of these species, *Perkinsus marinus* is the best known and will be considered in detail. Other pathogenic *Perkinsus* spp. and other species of lesser significance in the other two orders will be briefly mentioned.

Order Perkinsorida

Taxonomy

Perkinsus marinus, a pathogen of eastern oysters (*Crassostrea virginica*), was the first species recognized in this order (Levine, 1978). This parasite was originally described as *Dermocystidium marinum* (Mackin *et al.*, 1950) and later changed to *Labyrinthomyxa marina* by Mackin and Ray (1966) due to the presence of some stages with superficial similarities to the Labyrinthulales (originally grouped with the fungi but are now considered protozoa in the phylum Labyrinthomorpha). The apical complex in the zoospore was used to justify placing this parasite within the phylum Apicomplexa (Perkins, 1976a; Levine, 1978). Despite the taxonomic changes, the disease caused by *P. marinus* is popularly known as 'Dermo' and fungal terminology is still occasionally used for various stages in the life cycle.

Distribution and economic importance

Perkinsus marinus has been reported from the east coast of the United States from Massachusetts to Florida, along the Gulf of Mexico to Venezuela and in Puerto Rico, Cuba, and Brazil (Kern, 1985). However, Delaware Bay is periodically free of the disease owing to: (i) poor propagation of the parasite due to cool temperatures; and (ii) an embargo placed on importation of eastern oysters from more southern areas (Andrews, 1988a). The life cycle of *P. marinus* has a statistically significant positive correlation with temperature and salinity (Crosby and Roberts, 1990; Ford, 1992). The parasite is most virulent in eastern oysters at salinities above 15 ppt during periods of elevated water temperatures (above 20°C for at least 1 month) (Kern, 1985; Chu and Greene, 1989). Thus, the disease is prominent for about half the year in Chesapeake Bay and active for most of the year in the Gulf of Mexico (Lauckner, 1983). Powell *et al.* (1992) proposed that *P. marinus* prevalence and intensity may eventually be predictable from climatic models especially in the Gulf of Mexico.

Extensive losses of eastern oysters were attributed to this disease. Although the disease is chronic in cooler waters, its severity increases southward to the Gulf of Mexico where mortalities of adult eastern oysters may reach 50% each year. In addition to mortalities, meat yields are drastically reduced by high levels of infection and infections may reach 100% in eastern oysters exposed to two consecutive summers of *P. marinus* activity (Andrews and Ray, 1988; Crosby and Roberts, 1990). Andrews (1988a) indicated that it is the most destructive oyster parasite in the Gulf of Mexico and in some areas along the Atlantic coast of the United States (i.e. high salinity areas of Chesapeake Bay). Mass mortalities of eastern oysters (30–34 million oysters or 90–99% of the stock) imported into Pearl Harbour, Hawaii were attributed to this pathogen (Kern *et al.*, 1973).

Morphology and life cycle

In the eastern oyster, the vegetative stage or trophozoite (aplanospore or meront) ranges in size from 2 to 20 µm with the smaller stages often observed within the phagosome of a haemocyte. At maturity, the trophozoite (10–20 µm in diameter) has an eccentric vacuole (often containing a refringent vacuoplast) that may comprise 90% of the cell volume. The peripheral location of the nucleus produces the characteristic signet ring configuration. The mature trophozoite undergoes schizogony by successive bipartitioning of the protoplast (alternating karyokinesis and cytokinesis) to form a sporangium (ranging from 15 to 100 µm but usually less than 25 µm in diameter) containing from 2 to 64 (usually 8–32) trophozoites (coccoid or cuneiform and 2–4 µm in the longest axis) which are released when the sporangium ruptures (Perkins, 1969a, 1976b, 1988).

At death and decomposition of an infected host or when infected tissue is placed in anaerobic fluid thioglycollate medium (Ray, 1966a), the trophozoites transform into prezoosporangia (hypnospores) that enlarge to sizes between 20 and 100 µm (Perkins, 1976b). Upon release into sea water (aerobic conditions), the prezoosporangia (from either dead oysters or thioglycollate medium) differentiate into flask-shaped zoosporangia within which biflagellated zoospores develop in about 4 days. Perkins and Menzel (1967) estimated that a zoosporangium 85 µm in diameter contained from 1000 to 2000 zoospores. However, calculations from data presented by Chu and Greene (1989) suggests that on the average each zoosporangium can produce about 350,000 zoospores. The biflagellated zoospores (ovoid body 4–6 µm by 2–3 µm) have a row of long filamentous mastigonemes (tinsels) along the length of the anterior flagellum and a naked posterior flagellum. It also has an apical complex consisting of a conoid, polar ring, up to 39 subplasmalemnal microtubules, rhoptries, and micronemes (Perkins, 1976a, 1988). Zoospores escape from the zoosporangium via the discharge tube and may initiate infection in the epithelium of gill, mantle, or gut where they become trophozoites (Perkins and Menzel, 1967; Perkins, 1976a, 1988). Apparently trophozoites contained within haemocytes that underwent diapedesis are infective and represent an alternate means by which the infection is transmitted.

Hoese (1964) speculated that dissemination of *P. marinus* might be achieved through scavengers because he detected prezoosporangia in the faeces of fishes, oyster drills, and crabs that fed on dead or moribund infected oysters. He was able to infect eastern oysters with this material. In addition to a direct life cycle, laboratory studies indicated that *P. marinus* might also be transmitted by an ectoparasitic pyramidellid snail, *Boonea impressa* (White *et al.*, 1987). Although the method of transmission was not identified, White *et al.* (1987) speculated that either *P. marinus* was directly injected into the oyster mantle by an infected snail during feeding or *P. marinus* was transferred through the water between the snail and oyster, perhaps entering the feeding wound made by the snail.

Clinical signs and host–parasite relationships

Gross signs of 'Dermo' are severe emaciation, gaping, pale appearance of digestive gland, shrinkage of mantle away from the outer edge of the shell, inhibition of gonadal development, retarded growth and occasionally presence of pus-like pockets (Lauckner, 1983; Sindermann, 1990). In the early stages of infection, many *P. marinus* trophozoites are engulfed by haemocytes and the infection spreads systemically throughout the oyster. Although the extent of the inflammatory response was variable, encapsulation of trophozoites by several layers of haemocytes has been reported. Also, host cell destruction seemed to be limited to the immediate vicinity of the pathogen (Perkins, 1976b). Advanced infections seem to be characterized by haemocyte activation and recruitment, with concomitant exuberant production of haemocyte-derived oxygen intermediates (oxyradicals) that may participate in the pathogenesis of the disease (Anderson *et al.*, 1992).

Foci of infection or abscesses containing thousands of *P. marinus* and host debris may attain several hundred microns in diameter during later stages of infection. In addition, the pathogen often occludes haemolymph sinuses. Although the epithelium and adductor muscle are invaded, they do not appear to be damaged until late in the infection. By the time the eastern oyster becomes moribund, large numbers of *P. marinus* have accumulated in all tissues (Perkins, 1976b). Paynter and Burreson (1991) have indicated that in Chesapeake Bay, groups of eastern oysters, which incurred high prevalences and intensities of infection, exhibited low mortalities during their first year but suffered high mortalities during the following year.

In addition to the developmental stages of *P. marinus* in the ectoparasitic snail *Boonea impressa*, *Perkinsus* sp. have been reported in numerous other Mollusca (see below). Thus, a complete list of species that serve as suitable hosts for *P. marinus* has not been developed. Another complicating factor in identifying the host specificity of *P. marinus* is that the *Perkinsus* in eastern oysters may represent more than one species due to ultrastructural differences observed in parasites from eastern oysters in Virginia and Florida as indicated by Lauckner (1983).

Propagation

In vitro propagation of the trophozoite was recently described by LaPeyre *et al.* (1993), Gauthier and Vasta (1993) and Kleinschuster and Swink (1993). In addition to having biological characteristics similar to the histozoic stages of *P. marinus* (i.e. morphology, antigenicity, biochemistry, and development in thioglycollate medium as described by Ray (1966a)), some cultured isolates were infective to eastern oysters.

The transformation of the trophozoites into prezoosporangia in thioglycollate medium (Ray, 1966a) is frequently referred to as a culture technique. However, prezoosporangia survive only about 10 days in fluid thioglycolate medium (Perkins, 1988). The subsequent transformation of the prezoosporangia into zoospore-producing zoosporangia can then be achieved within 2 to 4 days by transferring the prezoosporangia from the thioglycollate medium to sea water.

Diagnosis of infection

The thioglycollate culture technique as reviewed by Ray (1966a) is routinely used in the diagnosis of *Perkinsus* infections. Prezoosporangia readily stand out at 30–40 × magnification as dark brown to blue-black spheres in tissue that had been incubated in thioglycollate medium for at least one week and subsequently stained with dilute Lugol's iodine solution. A semi-quantitative estimate of disease intensity is determined by the apparent percentage of squashed mantle or rectal tissue that contained *P. marinus* prezoosporangia (Andrews, 1988a; Choi *et al.*, 1989). Gauthier and Fisher (1990) demonstrated that haemolymph could be assayed to produce a sensitive, reliable, and completely quantitative method of estimating the intensity of infection. However, Bushek (1993) indicated that this method is inadequate for detecting light infections but is advantageous in that it does not require that the oyster be sacrificed. The thioglycollate culture technique is not species specific, consequently it has been utilized to detect other species of *Perkinsus* in various Mollusca (see below).

Monoclonal and polyclonal antibodies against the prezoosporangia, have been produced and can be used in ELISA or immunofluorescent assays for identification and quantification of *P. marinus* (Choi *et al.*, 1991; Dungan and Roberson, 1993). As expected, the various antibodies show differences in cross-reactivity with other life stages of *P. marinus* and for other species of *Perkinsus* (Dungan and Roberson, 1993).

Prevention and control

Continuous bath treatment with low levels of cyclohexamide ($1 \mu g\,ml^{-1}$ per week for 45 days) prolonged the life of laboratory stocks of eastern oysters infected with *P. marinus* (Ray, 1966b). However, chemical treatment is impractical in the field. Andrews (1988a), Andrews and Ray (1988), and Sindermann (1990) indicated that control of the disease caused by *P. marinus* depends on isolation and manipulation of seed stock and recommended the following procedures: (i) avoid use of infected seed stocks; (ii) plant oysters thinly on beds; (iii) isolate newly planted beds (0.4 km) from infected eastern oysters; (iv) continually monitor eastern oysters (25 oysters at 2 years of age or older in the late summer or early autumn) for the disease using the thioglycollate culture technique; (v) harvest early if beds become infected; and (vi) fallow beds after harvest to allow all infected oysters to die before replanting. Goggin *et al.* (1990) further recommended that the spread of *Perkinsus* sp. from shellfish processing plants could be prevented by not returning untreated mollusc tissues to the sea.

Although *P. marinus* persisted in eastern oysters held at low salinities (6 ppt), it was less virulent at salinities below 9 ppt (Ragone and Burreson, 1993). The occurrence of disease only at higher salinities has been used in management practices (Paynter and Burreson, 1991). In Chesapeake Bay, uninfected eastern oyster seed are acquired from areas of low salinity, which are not suitable for oyster culture because a reduction in oyster growth and condition is mitigated by low salinity. In the Gulf of Mexico where warmer temperatures allow the infection to remain active year round, fresh water diversions into high salinity bays have been proposed in order to revive or enhance areas that are marginally productive for eastern oysters (Andrews and Ray, 1988). The possibility of breeding eastern oysters that are resistant to *P. marinus* is under investigation (Burreson, 1991). Also, the introduction of a non-endemic species, the Pacific oyster that is more tolerant of *P. marinus* (Meyers *et al.*, 1991), is being considered as a method for the recovery of stable oyster production in areas of Chesapeake Bay where native eastern oysters have been eliminated (Mann *et al.*, 1991).

Related pathogens

Perkins (1988) noted that, after treatment in thioglycollate medium, prezoo-sporangia of *Perkinsus* sp. were observed in at least 34 species of bivalve molluscs from temperate to tropical waters of the Atlantic and Pacific oceans and Mediterranean Sea. On the Great Barrier Reef, 30 of 84 species of bivalves (160 of 644 individuals) were infected with *Perkinsus* spp. (Goggin and Lester, 1987; Lester *et al.*, 1990). Although *Perkinsus* sp. were associated with giant clam (*Tridacna gigas*) mortalities (Alder and Braley, 1989; Anderson, 1990) and lesions in the tissues of pearl oysters (*Pinctada maxima*) (Norton *et al.*, 1993), many *Perkinsus* sp. infections seem to have no detectable adverse affects on their hosts (Goggin *et al.*, 1990).

In addition to *P. marinus*, at least four species of *Perkinsus* have been identified based on differences in host, morphology, and geographical location. One species in Baltic macoma clams (*Macoma balthica*) from the York River area in Chesapeake Bay (Perkins, 1968) remains unnamed. Based on ultrastructural differences, Perkins (1988) proposed that this *Perkinsus* sp. belongs to a group that is distinct from *P. marinus*.

Perkinsus karlssoni caused tissue lesions and mortalities in brood stock of bay scallops (*Argopecten irradians*) being conditioned for spawning in warmed sea water (20°C) in Atlantic Canada (McGladdery, 1990; McGladdery *et al.*, 1991; Whyte *et al.*, 1994). Since the bay scallop is a non-indigenous species in Canada and no similar parasites have been detected in native bivalves, *P. karlssoni* was probably imported with the bay scallop. A similar parasite occurred in native bay scallops from the Rhode Island area of the eastern United States (McGladdery *et al.*, 1991). The transmission of *P. karlssoni* from adult bay scallops to eggs at the time of spawning and the post-spawning mortality that is characteristic of bay scallops may explain why *P. karlssoni* escaped detection during at least four generations that the bay scallops spent in quarantine prior to release in Atlantic Canada (McGladdery *et al.*, 1993).

Perkinsus olseni causes soft yellow pustules in the muscle and haemo-lymph of the blacklip abalone (*Haliotis ruber*) from South Australia and has been suspected of causing appreciable mortalities in greenlip abalone (*Haliotis laevigata*) from the Gulf of St. Vincent, South Australia (Lester and Davis, 1981; Lester, 1986; Lester *et al.*, 1990). *Perkinsus olseni* differs from *P. marinus* in having a larger trophozoite (14–18 µm in diameter), a weakly eosinophilic vacuoplast when present, and a short discharge tube on the zoosporangia.

The fourth species, *Perkinsus atlanticus*, causes high mortalities in the European littleneck clams (*Tapes* (= *Ruditapes*) (= *Venerupis*) *decussatus*) from Portugal where it occurs most frequently in milky white cysts on the gills but also in the foot and mantle of heavily infected clams (Azevedo, 1989). In comparison to *P. marinus, P. atlanticus* has larger trophozoites (30–40 μm in diameter) and the zoospores are more uniform in size and structure and have some minor ultrastructural differences (Azevedo *et al.*, 1990). The relationship between *P. atlanticus* and the *Perkinsus* sp. causing mortalities in clams (European littleneck clam and *Venerupis aurea*) in the Mediterranean basin (DaRos and Canzonier, 1985), in the European littleneck clam (= carpet-shell clam) in Galicia, Spain (Figueras *et al.*, 1992) and in the Baltic macoma clam on the east coast of the United States needs to be investigated.

To differentiate the various species of *Perkinsus*, host specificity studies as well as examinations for differences in ultrastructure, genetics (DNA analysis), cytochemistry, and immunogenic components are essential. Host specificity trials that have been conducted in the United States and Australia are contradictory. *Perkinsus* spp. from the east and south coasts of the United States apparently had fairly rigid host specificity while those from the coast of Queensland, Australia were easily cross transmitted between several species of bivalves and gastropods (Goggin *et al.*, 1989; Lester *et al.*, 1990). Sorting out the identity, host specificity, and respective pathogenicity of this group of parasites is a prerequisite to controlling the disease.

Order Eugregarinida

Gregarines (family Porosporidae) are digenetic and usually alternate between arthropod and bivalve hosts. In most species, the reproductive stages (including trophozoites that undergo syzygy to give rise to gametocysts and gametes) occurs in the gut of the arthropod. Gymnospores produced by sporulation of the zygote are infective to bivalves. In bivalves, the gymnospores are intracellular (usually within haemocytes) where they mature into sporocysts. Each sporocyst contains one or more uninucleate sporozoites which are infective to arthropods (Lauckner, 1983). Species of the genera *Porospora* and *Nematopsis* are found in numerous species of crustaceans and bivalves worldwide. In most cases there is little or no evidence of disease in either host. However, there are a few inconclusive reports concerning pathogenicity. For example: *Nematopsis* sporocysts in the mantle tissues of the blue mussel (*Mytilus edulis*) in Europe has been associated with the production of pearls or calcareous deposits on the inner surface of the valves (Götting, 1979); heavy infections of *Nematopsis ostrearum* in eastern oysters on the east coast of the United States may cause mechanical interference with host physiology; and heavy infections of *Nematopsis penaeus* may cause extensive damage to the intestinal epithelium of brown shrimp (*Penaeus aztecus*) from the Gulf of Mexico (Sindermann, 1990).

Order Eucoccidiida

Coccidians belonging to *Pseudoklossia, Klossia*, and other related genera have been observed in the epithelial cells of the kidney of many molluscs. All recognized stages of each species have only been reported from one host and life cycles are thought to be direct (Lauckner, 1983; Morado *et al.*, 1984). Hosts include: sea scallops (*Pecten maximus*), flat oysters (*Ostrea edulis*), and clams from Europe (Lauckner, 1983); bay scallops from the east coast of Canada and adjacent United States (Leibovitz *et al.*, 1984; McGladdery, 1990; Whyte *et al.*, 1994); blue mussels from the east coast of the United States (Farley, 1988) and the west coast of Canada (Bower, 1992); Pacific little-neck clams (*Protothaca staminea*) from the State of Washington and adjacent British Columbia (Morado *et al.*, 1984); and black abalone (*Haliotis crache-rodii*) and other species of abalone from California (Steinbeck *et al.*, 1989; Friedman *et al.*, 1993). Although the pathology caused by this group of parasites seems to be focal and often confined to hypertrophy of the infected host cell, there is potential for these coccidians to adversely impact on their host, especially molluscs under stressful conditions.

PHYLUM MICROSPORA

Many diverse species of microsporidians, all from the order Microsporida (genera *Agmasoma, Ameson, Microsporidium* (generic group), *Nosema, Pleistophora,* and *Thelohania*), have been described from shrimps, crabs, and freshwater crayfish worldwide (Overstreet, 1973; Sparks, 1985; Sindermann, 1990). The majority of these parasites occur in low prevalences (<1%) in wild populations (Butler, 1980; Olson and Lannan, 1984; Parsons and Khan, 1986; Owens and Hall-Mendelin, 1990). Thus, the economic impact of most species of Microsporida on crustacean fisheries is unknown. However, several species were perceived to have adverse economic impacts.

Agmasoma (= *Thelohania*) *penaei* infects at least two species of shrimp, white shrimp (*Penaeus setiferus*) and pink shrimp (*Penaeus duorarum*), in the Gulf of Mexico. It was presumed to cause high prevalence of parasitic castration in white shrimp (Sparks, 1985).

Ameson (= *Nosema*) *nelsoni* infects at least six species of shrimp throughout the Gulf of Mexico and north along the Atlantic coast of the United States to Georgia (Sparks, 1985). It is a common disease and mortality in bait and food shrimps results in significant financial losses to the industry (Sindermann, 1990). Microsporidosis in captive-wild *Penaeus* brood stock (infections not apparent at time of collection) resulted in losses of up to 20% (Lightner, 1988). Also, prevalences (16% and 15% respectively) of *A. nelsoni*, in pond reared brown shrimp from Texas, and in white shrimp in a net enclosed bay in Florida suggest a potential threat to shrimps reared in extensive culture (Lightner, 1975).

Ameson (= *Nosema*) *michaelis* is widely distributed at low prevalences in blue crab on the Gulf and Atlantic coasts of the United States (Sparks, 1985). Diseased blue crab often inhabit sheltered areas near shore and experience high mortalities when stressed (Overstreet, 1988). However, unlike *Ameson* in shrimps, the transmission of *A. michaelis* is direct, i.e. by ingestion of infected tissue (Sparks, 1985; Overstreet, 1988). Some authors indicated that this parasite was a significant factor in blue crab mortality and thus a potential threat to the industry. However, more information is needed on pathogenicity, geographic distribution, and prevalence in various populations before its economic significance can be established (Sparks, 1985).

Unidentified Microsporida have been presumed to cause high mortalities in freshwater crayfish in western Europe and England (Pixell Goodrich, 1956). Anderson (1990) noted that up to 8% of the freshwater crayfish (*Cherax quadricarinatus*) from some Australian rivers are infected with *Thelohania* sp. and that microsporidians may be the only enzootic parasite to pose problems for culturing freshwater crayfish. Recently, a Microsporida in the hepatopan-creatocytes of tiger shrimp (*Penaeus monodon*) was associated with low production, slow growth rates, and occasional mortalities in brackishwater pond culture in Malaysia (Anderson *et al.*, 1989). The impact that Microsporida will have on the rapidly developing culture of crustaceans has yet to be determined.

Each species of Microsporida is characterized by the number of spores per sporont, the spore size, the tissues infected, and to some extent by the host species (Couch, 1978; Sparks, 1985). Infected tissue, especially muscle, is even-tually replaced by spores giving it an opaque appearance. Due to this white discoloration, heavy infections are apparent and justify the common names of 'cotton', 'milk', or 'cooked' shrimp and crabs. The fluorescent screening technique for Microsporida in histological sections described by Weir and Sullivan (1989) may be useful for detecting light infections.

The only known method of prevention is removal and destruction (freezing may not destroy spores) of infected individuals (Lightner, 1988; Overstreet, 1988). In penaeid culture, it would be prudent to exclude from culture systems and water supplies the intermediate hosts (finfish) of the parasite (i.e. *A. penaei* became infective for pink shrimp following passage through the gut of a shrimp predator the spotted sea trout (*Cynoscion nebulosus*) (Iversen and Kelly, 1976; Lightner, 1988)). A single treatment of buquinolate (used to treat coccidiosis in broiler chickens) prevented microsporidosis caused by *A. michaelis* in most exposed blue crab (Overstreet, 1975). Lightner (1988) suggested that Fumidil B (an antibiotic used to control microsporidosis in honeybees) and Benomyl (a systemic fungicide used to control microsporidosis in the alfalfa weevil) may be suitable treatments for this disease in penaeid shrimps. However, Overstreet (1975) found that Fumidil B seemed to exacer-bate *A. michaelis* infection in blue crabs and benomyl was not as effective as buquinolate and it apparently killed some crabs.

PHYLUM ASCETOSPORA

The Ascetospora contain three distinct groups of bivalve pathogens. One group consists of members of the order Blanosporida (class Stellatosporea). The second group was initially included in the same order (Perkins, 1976c) and later transferred to the order Occlusosporida (class Stellatosporea) (Sprague, 1979; Levine *et al.*, 1980). Desportes (1984) moved them to the class Paramyxea and order Marteiliida and Desportes and Perkins (1990) suggested that the class Paramyxea be raised to the rank of phylum. The third group has not yet been placed within an order and are referred to as microcells in the present review. Unlike the other Ascetospora, spore stages of microcells have not been observed and direct transmission occurs. Thus, microcells may be transferred to another phylum at some future date. The parasite with the greatest economic impact in each of the three groups is considered in detail. Other related species are briefly mentioned.

Order Balanosporida

Taxonomy

In this order the most significant pathogens occur in oysters. They have undergone several changes in classification since the parasites were first recognized in the early 1960s. The order was changed from Haplosporida to Balanosporida in the reclassification of Levine *et al*. (1980) and, more recently, the members of this group were considered to constitute the phylum Haplosporidia (Perkins, 1990). Also, the genus was changed from *Haplosporidium* to *Minchinia* and back again. The first named species parasitic to marine bivalves was *Haplosporidium costale* in eastern oysters (Wood and Andrews, 1962). Sprague (1963) transferred *H. costale* along with other species to the genus *Minchinia*. A few years later, Haskin *et al*. (1966) described a related eastern oyster parasite *Minchinia nelsoni* that surpassed *H. costale* in pathogenicity and economic impact. Subsequently, all species within *Minchinia* were retransferred to *Haplosporidium* and the genus *Minchinia* was considered a nomen nudum (Lauckner, 1983). Recently, Perkins (1988) recognized the genus *Minchinia* which he placed in the phylum Haplosporidia and differentiated it from *Haplosporidium* by the presence of prominent extensions (tails) on the spore wall (visible by light microscopy). Using this criterion, the haplosporidians from eastern oysters belong in the genus *Haplosporidium* while a recently described species from flat oysters (*Ostrea edulis*) is a *Minchinia*. The following review concentrates on *Haplosporidium nelsoni* (commonly known as MSX, Multinucleate Sphere X) with *H. costale* (commonly known as SSO, SeaSide Organism) and other species mentioned under the heading 'Related pathogens'.

Distribution and economic importance

The catastrophic epizootics caused by *Haplosporidium nelsoni* were first recognized in the late 1950s in eastern oysters from Delaware Bay. This parasite currently occurs along the eastern coast of the United States from North Carolina to Massachusetts. Enzootic areas are limited to Delaware Bay with occasional epizootics in Chesapeake Bay, Long Island area, and Cape Cod (Andrews, 1988b; Haskin and Andrews, 1988). The disease is restricted to salinities over 15 ppt, rapid and high mortalities occur at 20 ppt, and the disease may be limited by salinities between 30 and 35 ppt (Andrews, 1988b).

In 1957, 85% mortality (with 50% dead within 6 weeks) occurred among eastern oysters planted in Delaware Bay. The high mortality represented a loss in production from 7.5 million pounds of shucked meats prior to the enzootic to about 100,000 pounds production in 1960 and production has not significantly recovered (Lauckner, 1983; Sindermann, 1990). Since the onset, average mortalities in enzootic areas have been estimated at 50–60% in the first year with a 50% further loss in the second year of oyster grow-out (Andrews, 1988b). Also, eastern oyster culture in the lower Chesapeake Bay (late-summer with salinity >20 ppt) was abandoned for 25 years due to this disease (Andrews, 1988b).

Morphology and life cycle

Despite more than 30 years of intensive research, the complete life cycle, mode of infection, and several aspects of the general biology remain obscure (Haskin and Andrews, 1988; Ford, 1992). Spherical plasmodia (4–30 μm in diameter) are usually multinucleate (up to about 60 nuclei per plasmodium with each nucleus from 1.5 to 7.5 μm in diameter). Smaller plasmodia are formed by cytoplasmic cleavage of the larger ones. They are first observed in the gills, palps, and suprabranchial chambers but subsequently occur in the vesicular connective tissues adjacent to the digestive tract, and gonad alveoli. Eventually the infection becomes systemic (Lauckner, 1983). Plasmodia and prespore stages are most frequently observed while sporocysts containing mature spores are rare in adult oysters (<0.01%). However, sporulation occurs in at least 75–85% of the infected young oysters (<1 year) in Delaware Bay (Barber et al., 1991b). Sporulation, when present, occurs in the epithelial cells of the digestive gland tubules. Spores are operculate and measure 7.5 × 5.4 μm unfixed (Couch et al., 1966; Rosenfield et al., 1969). Ford and Haskin (1982) and Perkins (1993) noted that the parasite could not be transmitted in the laboratory with either infected tissues or spore suspensions and that foci of infection persisted in areas where eastern oysters were sparse. No host other than eastern oysters has been found (Haskin and Andrews, 1988). However, there is speculation that an intermediate host is required for the completion of the life cycle of H. nelsoni as well as for all other haplosporidians (Burreson, 1988).

The development of sexual stages prior to spore formation was proposed from light microscopic studies by Farley (1967). However, no evidence of sexuality was observed during ultrastructural examination of a closely related species H. costale (Perkins, 1969b). Farley (1967) also described encysted or 'cystoid' stages which occur in the oyster in unfavourable conditions. The ultrastructure of these 'cystoid' plasmodia has not been described.

Extensive epizootiological data indicates that infections acquired in the early summer become patent in July and mortalities begin in early August and peak in September with a subsequent decline to low levels by November (Andrews, 1982). A few mortalities occur in late winter followed by increased mortalities in June and July of the second year resulting from infections acquired during the late summer and autumn of the previous year (Haskin and Andrews, 1988). Essentially the disease is regulated by temperature with both parasite and host being inactive below 5°C. Between 5 and 20°C, the parasite multiplies faster than the host can contain it. Above 20°C resistant oysters can inhibit parasite multiplication or undergo remission (Sindermann, 1990). MSX levels have fluctuated in a cyclic pattern with peaks in prevalence every 6–8 years and reduced parasite activity following very cold winters (Ford and Haskin, 1982).

Host–parasite relationships

Haplosporidium nelsoni is highly virulent for eastern oysters and the occurrence of moribund oysters with relatively light infections suggests a toxic effect (Andrews, 1988b). The course of the infection seemed dependent on the history of exposure in the eastern oyster stocks (Farley, 1975). In susceptible populations such as those in Chesapeake Bay, the prevalence of infection can reach 100% with mortalities ranging between 40 and 80% (Andrews, 1988b). However in enzootic areas such as Delaware Bay, natural selection has increased the proportion of disease resistant eastern oysters and mortalities were about half (Ford and Haskin, 1982).

During their second year, eastern oysters that survived the infection were able to suppress or rid themselves of the parasite in the late spring as temperatures approached 20°C (Ford and Haskin, 1982). Remission was characterized by diminution of infection and localization of parasites to external epithelium with diapedesis resulting in the deposition of moribund parasites and necrotic tissues against the shell followed by external conchiolinous encapsulation (Farley, 1968). Ford and Haskin (1982) demonstrated that resistance to mortalities was not correlated with an ability to prevent infection but with restriction of parasites to localized nonlethal lesions. Chintala and Fisher (1991) indicated that lectins in the haemolymph could be related to disease resistance or affected by *H. nelsoni* infection.

Parasitism was associated with reduced meat yield, impaired gonadal development and lower fecundity (Barber *et al.*, 1988). The greatest effect on reproduction occurred when gametes were in the formative stage rather than after they matured (Ford *et al.*, 1990a). There was also a three-fold increase in the proportion of females among infected oysters which Ford *et al.* (1990a) suggested was due to the inhibition in development of male more than female gametes. However, infected oysters that underwent temperature associated remission during the summer developed mature gonads and spawned before new or recurrent infections proliferated in autumn (Ford and Figueras, 1988).

Propagation

Haplosporidium nelsoni has not been cultured and controlled transmission has not been achieved (Haskin and Andrews, 1988). Even transplantation of infected tissues was unsuccessful (Lauckner, 1983). Andrews (1979) suggested that long prepatent periods may account for the failure of some laboratory transmission studies. Enriched suspensions of *H. nelsoni* can be obtained using the 'panning' technique described by Ford *et al.* (1990b).

Clinical signs and diagnosis of infection

The only specific but rare sign of this disease is a whitish discoloration of the digestive gland tubules due to the presence of mature spores. Other nonspecific signs are: emaciation, mantle recession, failure of shell growth, retracted mantle, and rarely brown patches of periostracum opposite lesions on the mantle surface (Lauckner, 1983; Andrews, 1988b). Histological examination is used to confirm the presence of infection. However, heavy infections can be diagnosed by microscopic examination of stained haemolymph smears (Andrews, 1988b; Burreson *et al.*, 1988; Ford and Kanaley, 1988). When mature spores are present, the sporoplasm specifically stains bright red with a modified Ziehl-Neelsen carbol fuchsin technique (Farley, 1965). Barrow and Taylor (1966) and Burreson (1988) illustrated the potential use of immuno-assays for detecting infection and possibly for identifying alternate hosts.

Prevention and control

Reduced salinities (<10 ppt) adversely affected the pathogenicity and survival of the parasite in oysters (Haskin and Ford, 1982; Haskin and Andrews, 1988). Thus, management strategies rely, in large measure, on avoiding the disease by culturing oysters in areas of low salinity and/or altering the time at which oysters are moved to enzootic areas of high salinity to take advantage of better growth. Continuous monitoring and early diagnosis of infections are impor-tant because they allow mortality to be predicted so growers and managers can make informed decisions on when or whether to plant and harvest (Ford and Haskin, 1988).

Excellent survival has been achieved in enzootic areas using eastern oysters that were experimentally selected for disease resistance (Haskin and Andrews, 1988; Ford *et al.*, 1990a). Barber *et al.* (1991a) indicated that resistance in the selected strain may be the result of physiological responses that inhibit parasite development and basic metabolic adjustments to parasitism. Much of the physiological response may be derived from an increased number of haemo-cytes that plug lesions, remove debris, and repair tissue thereby helping resis-tant oysters to survive infection (Ford *et al.*, 1993). Although resistant oysters have not yet been produced in commercial quantities because of economic limitations (Andrews, 1988b), triploid eastern oysters seem more resistant to the disease than diploid cohorts (Matthiessen and Davis, 1992). The increased resistance in triploids may provide a viable alternative for the eastern oyster culture industry in areas where the disease occurs.

Related pathogens

Unidentified species of *Haplosporidium* or *Minchinia* were reported from Pacific oysters in Japan, Korea, California, and France, from flat oysters and gaper clams (*Tresus capax*) in Oregon, USA (Lauckner, 1983; Sparks, 1985; Andrews, 1988a; Sindermann, 1990; Comps and Pichot, 1991), from Australian flat oysters (*Ostrea angasi*) imported into France from Australia (Bachère *et al.*, 1987), from flat oysters in Europe (Bonami *et al.*, 1985; Pichot, 1986), from blue mussels on the coast of Maine (Sherburne and Bean, 1986; Figueras *et al.*, 1991), from European littleneck clams (= carpet-shell clams) in Galicia, Spain (Figueras *et al.*, 1992), from blue crab in Virginia and North Carolina (Sparks, 1985), and from cultured shrimp (*Penaeus vannamei*) imported into Cuba from Nicaragua (Dyková *et al.*, 1988). A few haplosporidian-infected spat and adult Pacific oysters from Matsushima Bay, Japan contained advanced sporogonic infections in which the mature operculate spores within the digestive tubule epithelium closely resembled *H. nelsoni* (Friedman *et al.*, 1991). In addition, four named species occur in bivalves of economic importance.

Haplosporidium costale (SSO, seaside organism) causes disease in eastern oysters from Maine to Virginia but is important only in high salinity (>25 ppt) areas from Delaware to Virginia (Andrews, 1988c). It can be differentiated from *H. nelsoni* by: (i) the smaller spore (3.1 × 2.6 μm); (ii) antigenic differences; (iii) occurrence of sporulation throughout all connective tissue and not in the epithelium of the digestive gland; (iv) being confined to high salinity (>25 ppt) waters; and (v) a regular and clearly defined life cycle (the period of disease and sporulation is seasonal (May and June), short (4–6 weeks), and concurrent with mortalities followed by a 8–10 month prepatent period in newly exposed oysters) (Couch *et al.*, 1966; Andrews, 1982). *Haplosporidium costale* is not as serious a pathogen as *H. nelsoni* and losses can be minimized by harvesting oysters at 18–24 months of age (Andrews, 1988c).

Minchinia armoricana causes brown meat disease in flat oysters in Brittany (France) and in the Netherlands among flat oysters imported from Brittany (van Banning, 1985a). The enormous number of operculate spores (5.0–5.5 μm × 4.0–4.5 μm) with two long projections (70–100 μm) in groups of 100–150 within sporocysts (35–50 μm in diameter) throughout the connective tissue gave the brownish discoloration to heavily infected flat oysters. Although the disease is fatal, the prevalence of infection to date has been low (<1%) with insignificant impact on the flat oyster culture industry in Europe (van Banning, 1979; Lauckner, 1983).

Haplosporidium tumefacientis was associated with 'tumefaction' (swellings) in the digestive gland and kidney of 2.1% of California mussels (*Mytilus californianus*) from California (Taylor, 1966). The low prevalence and lack of additional information suggests that this disease is not economically significant.

The final species, *Haplosporidium* (= *Minchinia*) *tapetis*, was described from European littleneck clams (*Tapes* (= *Ruditapes*) *decussatus*) in Portugal and France (Lauckner, 1983; Chagot *et al.*, 1987). Slightly ovoid spores (4–6 μm in diameter) were observed in the connective tissue below the digestive gland tubules and gills. Reported prevalences of infection were low (4%) and pathogenicity was minimal.

Order Marteiliida

Taxonomy

Marteilia refringens, a morphologically unique protozoan, presumed to cause recurrent serious mortalities (from 1967 to about 1977) in flat oysters (*Ostrea edulis*) in Europe is the aetological of Aber disease or digestive gland disease (Grizel *et al.*, 1974). Ultrastructural features led Perkins (1976c) to suggest that this parasite belongs with the haplosporidans (phylum Ascetospora, class Haplosporea). Sprague (1979) provisionally placed *M. refringens* in the class Stellatosporea and order Occlusosporida and this was supported by Levine *et al.* (1980). More recently, *M. refringens* was reclassified into the class Paramyxea and order Marteiliida because of the unique characteristics of sporulation where spores (propagules) are formed from cells enclosed inside each other by a process of internal cleavage or endogenous budding (Desportes, 1984; Perkins, 1988). Desportes (1984) and Perkins (1993) suggested that the class Paramyxea may be raised to the rank of phylum.

Distribution and economic importance

Mortalities exceeding 90% were first noted in the summer of 1967 from the Aber Wrach on the northwestern coast of Brittany, France (Alderman, 1979). The disease spread and peaked in the early 1970s then began to decline in 1977 (Sindermann, 1990). To date, *M. refringens* has been reported from the Netherlands to the Atlantic coast of Spain (Comps, 1985a). In addition to flat oysters, *M. refringens* was reported in Europe from blue mussels, imported Pacific oysters, and experimentally infective to dredge oysters (*Tiostrea* (= *Ostrea*) *chilensis* (= *lutaria*)) and Australian flat oysters (Cahour, 1979; Grizel *et al.*, 1983; Bougrier *et al.*, 1986).

Alderman (1979) indicated that the decline (to about 47%) of the original flat oyster production in France over 6 years (from 18,000 metric tons in 1969 to 8400 metric tons in 1975) was a direct result of the spread of Aber disease. However, even during periods when the disease was most severe, high mortalities did not occur in all areas where the parasite was found and there were several river estuaries where flat oysters were thought to be free of infection (Alderman, 1979; Balouet, 1979; Grizel, 1979). As Aber disease seemed to be subsiding in the late 1970s the flat oyster industry in Europe was struck by another devastating disease caused by the microcell *Bonamia ostreae* (see below).

Morphology and life cycle

Early stages of infection, presumably initiated by a plasmodium (primary cell or stem cell, 5–8 μm in diameter), are believed to occur in the epithelial cells of the gut or gills (Grizel *et al.*, 1974). The plasmodia contain striated inclusions (1–5 μm in length and thought to consist of protein) within the cytoplasm (Perkins, 1976c). Perkins (1988) suggested that the earliest stage may be a primary uninucleate cell which already has a secondary uninucleate cell in a vacuole within its cytoplasm. The process of cell multiplication within the host prior to sporulation has not been described (Perkins, 1993). The infection proceeds to the digestive gland epithelium where sporulation occurs (Fig. 16.1). Unique to this group of protozoa, the plasmodium undergoes a series of internal cleavages to produce cells within cells during sporulation.

At the initiation of sporulation, uninucleate segments become delimited within the plasmodial cytoplasm to form the sporangial primordia (secondary cells). Sporulation progresses by the formation of about eight sporangial primordia (each about 12 μm in diameter at maturity) within the original plasmodium which has enlarged from about 15 to 30 μm in diameter and retains its nucleus. The sporangial primordia form an outer wall followed

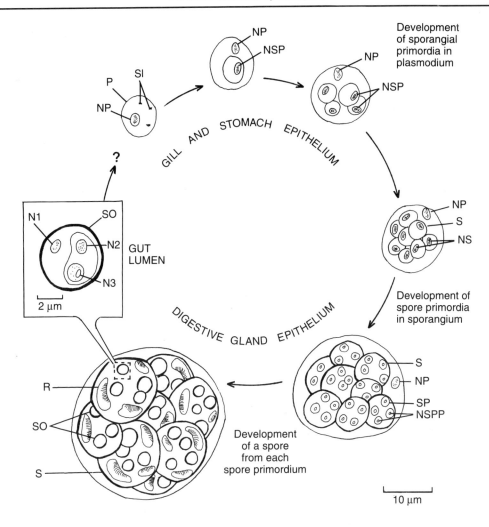

Fig. 16.1. Schematic drawing, to scale, of the developmental cycle of *Marteilia refringens* based on descriptions by Grizel *et al.* (1974) and Perkins (1976c, 1988). P = plasmodium (or primary cell), NP = nucleus of plasmodium, SI = striated inclusions, NSP = nucleus of sporangial primordium, S = sporangium (or secondary cell), NS = nucleus of sporangium, SP = spore primordium (or tertiary cell), NSPP = nucleus of spore primordium, SO = spore, R = refringent bodies, N1 = nucleus of outermost sporoplasm, N2 = nucleus of middle sporoplasm, N3 = nucleus of innermost sporoplasm.

by nuclear multiplication and cytoplasmic cleavage resulting in the formation of three or four spore primordia (tertiary cells) that mature into spores. Each spore contains three uninucleate sporoplasms of graded sizes, with each of the smaller sporoplasms within the cytoplasm of the next largest one (i.e. consecutive internal cleavage of two sporoplasms within the spore primordium, Fig. 16.1) (Perkins, 1976c). A continuous spore wall with no operculum occurs around each spheroid mature spore that measures 3.5–4.5 µm in diameter. As the spores mature, light refractile inclusion bodies, as large or larger than the spores, appear in the sporangial cytoplasm surrounding the spores. The specific name of this parasite was derived from these refringent inclusion bodies. From light microscopic studies, Franc (1980) proposed that *M. refringens* may have a sexual phase in the oyster.

The mechanism of infection has not been determined. Experimental attempts to transmit the disease to flat oysters in the laboratory met with failure although field exposures were successful (Balouet *et al.*, 1979). As with *Haplosporidium* spp. and *Minchinia* spp., an intermediate host is suspected (Perkins, 1993).

The development of *M. refringens* in flat oysters was directly related to water temperature with the disease being most severe towards the end of summer. However, mortalities usually commenced in May and persisted until December (Lauckner, 1983; Comps, 1985a; Grizel, 1986). Results of transplantation experiments suggested that new infections were acquired from early May to early September (Grizel, 1979). Long-term observations indicated that mature spores and refringent bodies within sporangia were present from about May to January, whereas young plasmodia persisted during the winter and re-initiated new clinical infections in the following May (Lauckner, 1983).

Clinical signs and host–parasite relationships

Characteristic signs of disease in flat oysters include a poor condition index with glycogen loss (emaciation), discoloration of the digestive gland, cessation of growth, tissue necrosis, and mortalities (Comps, 1985a; Sindermann, 1990). However, the pathogenesis of *M. refringens* remains obscure due to the lack of consistent correlation between the degree of infection and mortality (Lauckner, 1983). Some flat oysters kept in high-prevalence areas for extended periods showed characteristic signs of disease without notable numbers of parasites, while other flat oysters heavily infected with young plasmodia and mature sporangia exhibited virtually no histological alterations. To explain these inconsistencies, van Banning (1979) and Balouet (1979) suggested that: (i) the parasite inconsistently produced toxins; (ii) the parasite required the synergistic affect of another as yet unidentified pathogen; (iii) an intermediate host was required to amplify parasite abundance; and/or (iv) unfavourable environmental conditions (i.e. physicochemical factors in seawater) played prominent roles in determining the apparent pathogenicity of *M. refringens*. Also, *M. refringens* showed little or no pathogenicity to other susceptible bivalves such as Pacific oysters and European cockles (*Cardium edule*) (Cahour, 1979).

Propagation

Marteilia refringens has not been propagated or transmitted between shellfish in the laboratory. Extensive attempts to transmit the parasite in the field and under laboratory conditions and through potential intermediate hosts (shrimps (*Crangon crangon*), shore crabs (*Carcinus maenas*), and gammarid amphipods (*Marinogammarus marinus*)) were unsuccessful (van Banning, 1979).

Diagnosis of infection

Since there are no specific clinical signs, the disease can best be detected by histological examination (Grizel, 1979). A diagnostic feature is the presence of *Marteilia* in histological sections of the digestive gland tubule epithelium and occasionally in the gills and palps (Sindermann, 1990). Gutiérrez (1977) described a modified staining technique for enhancing the detection of the parasite in paraffin embedded histological sections.

Prevention and control

Since transmission of infection usually occurred in July and August, planting of flat oyster seed during that period was curtailed to reduce the risk of infecting seed oysters (Sindermann, 1990). Also, high salinities (35–37 ppt) limited the development of *M. refringens* (Grizel, 1986). Alternately, Pacific oysters, that seem to be resistant to the disease, were cultured in most infected zones (Comps, 1985a). Therapeutic treatments with some chemicals used in aquaculture (methylene blue, malachite green, and Furanace) were ineffective (Grizel, 1979). A microsporean hyperparasite (*Nosema ormieresi*) that caused necrotic changes in the sporogenic stages of *M. refringens* and reduced the number of spores may have potential as a biological control agent (Comps *et al.*, 1979).

Related pathogens

To date, four other species of *Marteilia* have been described. *Marteilia sydneyi* was described from the digestive gland of Sydney rock oysters (*Saccostrea* (= *Crassostrea*) *commercialis*) and possibly also from black lip oysters (*Crassostrea echinata*) in Australia. It can be differentiated from *M. refringens* by: (i) the lack of striated inclusions in the plasmodia; (ii) the formation of 8–16 sporangial primordia in each plasmodium instead of eight; (iii) each sporangium contains two infrequently three rather than four spores; and (iv) the heavy layer of concentric membranes surrounding mature spores in comparison to the lack of such a covering around *M. refringens* spores (Perkins and Wolf, 1976). Infections of *M. sydneyi* were associated with high mortalities (often exceeding 80%) in subtropical and tropical regions of southern Queensland and northern New South Wales (Wolf, 1979a; Anderson, 1990). As with *M. refringens*, the life cycle is not direct (Roubal *et al.*, 1989). The incubation period of the parasite in the host is less than 60 days from early infection to death of the oyster (Wolf, 1979a). Disease control is attempted by altering culture techniques (Wolf, 1979b; Anderson, 1990). Roubal *et al.* (1989) have developed an indirect antibody fluorescent test for detecting most stages of *M. sydneyi*.

    Marteilia lengehi was described from the epithelium of the stomach and digestive gland of rock oysters (*Saccostrea* (= *Crassostrea*) *cucullata*) from the Persian Gulf (Comps, 1976). Spores were not observed and features that differentiate this species from *M. refringens* were not given.

    Marteilia maurini was described from epithelium of the stomach and digestive gland of Mediterranean mussels (*Mytilus galloprovincialis*) and blue mussels from the Atlantic coast of Spain and France and the Persian Gulf

(Comps *et al.*, 1981; Auffret and Poder, 1983; González *et al.*, 1987). Because the morphology and developmental sequence of *M. maurini* closely resemble those of *M. refringens*, there is controversy over the validity of the former species. However, the availability of technology to obtain purified isolates of *Marteilia* from both mussels and oysters should facilitate the ultrastructural, biochemical, and physiological characterization of *Marteilia* from mussels and oysters in order to establish their taxonomic affinities (Mialhe *et al.*, 1985). Nevertheless, properly controlled cross infection studies are essential to establish the relationship between the two species. Disease and mortality associated with the *Marteilia* sp. from mussels is poorly understood (Lauckner, 1983). High prevalences (37–70%) in blue mussels from the north coast of Brittany (Auffret and Poder, 1983) and 46% cumulative mortalities (over 19 months) with significant inhibition of gametogenesis in mussels from various localities in northern Spain (Villalba *et al.*, 1990) suggest that *Marteilia* sp. represents a risk for mussel culture.

The final species *Marteilia christenseni* was described from the digestive gland epithelium of a pelecypod (*Scrobicularia piperata*) at Ronce-les-Bains, France (Comps, 1985b). The differentiation of the cytoplasmic contents of the sporangia and the morphological characteristics of the spore distinguish this parasite from the other species of *Marteilia* (Comps, 1985b).

Marteilia sp. has also been reported from various organs of an amphipod (*Orchestia gammarellus*) along the coast of France (Ginsburger-Vogel and Desportes, 1979). The identity of the *Marteilia* sp. from imported Pacific oysters and European cockles in Europe is in question (Cahour, 1979) and the relationship of these parasites with the other *Marteilia* has yet to be determined.

Microcells

Taxonomy

Tiny intracellular protozoan parasites (2–6 µm in diameter) of oysters, commonly referred to as microcells, have been affiliated with the haplosporidians due to the presence of organelles called haplosporosomes (Lauckner, 1983). However, because no spores have been found, Balouet *et al.*, (1983) have questioned the taxonomic position of these parasites.

Microcells are currently divided into two genera, *Bonamia* and *Mikrocytos* (Farley *et al.*, 1988). *Bonamia ostreae*, the first described species in this group (Pichot *et al.*, 1980), has had the greatest economic impact on the oyster culture industry. Thus, the following section will concentrate on *B. ostreae*. Other species of microcells will be described under the heading 'Related pathogens'.

Distribution and economic importance

Bonamia ostreae parasitizes flat oysters along the coast of Europe from Spain to Denmark and England and Ireland where it causes bonamiasis (Comps, 1985c; McArdle *et al.*, 1991). However, flat oysters on the Mediterranean coast have remained free of bonamiasis despite several inadvertent transfers of infected oysters to this area (Grizel *et al.*, 1988; Montes, 1991). *Bonamia ostreae* also occurs in some introduced flat oyster populations in California and Washington on the west coast of the United States (Elston *et al.*, 1986; Friedman *et al.*, 1989). Elston *et al.* (1986) hypothesized that the epizootic bonamiasis in Europe might have been due to importation of infected flat oysters from a California estuary to the French coast. Microcells were observed in moribund flat oysters imported into California from Milford, Connecticut as early as 1965 (Katkansky *et al.*, 1969). However, the early history and movements of introduced flat oysters in the United States is unclear and the original source of the *B. ostreae* in the flat oysters in California is open to speculation (Elston *et al.*, 1987).

This parasite was first associated with mortalities in Brittany, France in 1979 and the disease quickly spread throughout the major flat oyster culture areas in Europe. Average losses were about 80% or higher (Comps, 1985c; Grizel, 1986; Montes and Melendez, 1987; Montes, 1991; van Banning, 1991; Hudson and Hill, 1991). In conjunction with the protozoa *M. refringens*, *B. ostreae* reduced the flat oyster production in Brittany by 1.8 billion francs (1983 value) between 1974 and 1982 (an average of about 200 million francs per year) with a further loss of 1.3 billion francs in associated industries (Meuriot and Grizel, 1984; Grizel *et al.*, 1986). The persistence of *B. ostreae*, despite attempts to eradicate it, has lead to further economic strife by forcing commercial oyster culture firms to grow other bivalves with less profitable returns (Grizel, 1986; van Banning, 1987).

Morphology and life cycle

Two morphological forms of *B. ostreae* have been identified (Lauckner 1983; Grizel *et al.*, 1988). The most frequently observed 'dense forms' (2–3 µm in diameter) have basophilic, dense cytoplasm with a pale halo around the nucleus. This form is usually free (extracellular) in tissues altered by the disease and may represent the transmission stage. The slightly larger and less dense 'clear forms' (2.5–5 µm in diameter) may be the vegetative, schizogonic stage and typically occurs within the cytoplasm of haemocytes and in branchial epithelial cells (Montes *et al.*, 1994). The usual mode of multiplication in the oyster is by simple binary fission. However, Brehélin *et al.* (1982) described a true plasmodial form with three to five nuclei and about 6 µm in diameter. This multinucleate stage was only observed in dying and dead oysters and may represent a rare event or may be typical of the terminal stage of the disease.

Unlike the other haplosporidians, *B. ostreae* can be directly transmitted between flat oysters and lethal infections usually develop within 3–6 months after exposure (Grizel *et al.*, 1988; Sindermann, 1990). Transmission occurs year round with the highest prevalence of infection found during the summer. Van Banning (1990) suggested that an infectious phase may occur in the ovarian tissue of flat oysters.

Host–parasite relationships

Bonamiasis is usually systemic because *B. ostreae* normally resides within haemocytes. Early stages of infection are often accompanied by dense, focal haemocyte infiltration into the connective tissue of the gill and mantle, and around the gut. Many of the infiltrating haemocytes contain several microcells which are often in cytoplasmic vacuoles. As the infection progresses, infected haemocytes occur in the vascular sinuses and microcells may be free in necrotic tissues. Free microcells were probably released by the lysis of haemocytes (Balouet *et al.*, 1983). Infections of *B. ostreae* may result in lesions in the connective tissues of the gills, mantle, and digestive gland (Bucke and Feist, 1985; Comps, 1985c).

Bonamia ostreae was also infective to dredge oysters (Grizel *et al.*, 1983) and Australian flat oysters (Bougrier *et al.*, 1986), and reported from Olympia oysters (*Ostrea conchaphila* (= *lurida*)) (Farley *et al.*, 1988). However, Pacific oysters were not infected by cohabitation or by injection. Other economically important bivalves (blue mussels, European littleneck clams, Manila clams (*Tapes philippinarum*), and European cockles) from *Bonamia*-contaminated habitats were not infected (Grizel *et al.*, 1988). Laboratory experiments that examine the interaction of *B. ostreae* with haemocytes from susceptible and resistant oysters (i.e. flat oysters and Pacific oysters respectively) could lead to a better understanding of host-parasite interactions (Fisher, 1988; Grizel *et al.*, 1988).

Propagation

Bonamia ostreae is readily propagated *in vivo* by injection of infected haemocytes or purified parasite suspensions and by cohabitation of diseased and uninfected oysters (Hervio *et al.*, 1994). Comps (1983) reported *in vitro* proliferation of *B. ostreae* in the presence of flat oyster cells after 48 hours of incubation but the viability of the cultures over longer periods was not indicated.

Diagnosis of infection

The disease is characterized by the presence of microcells as observed histologically (Bucke and Feist, 1985; Grizel *et al.*, 1988). Although many infected oysters appear normal, others may have yellow discoloration and/or extensive lesions (i.e. perforated ulcers) on the gills and mantle. The isolation and purification of *B. ostreae* from infected flat oysters (Mialhe *et al.*, 1988a,b) has lead to the production of monoclonal antibodies (Rogier *et al.*, 1991), the development of an indirect fluorescent antibody assay (Boulo *et al.*, 1989), and an ELISA diagnostic technique with 90% reliability in comparison to standard histopathologic light microscopic examinations (Cochennec *et al.*, 1992). The monoclonal antibodies may be commercially available from SANOFI Santé Nutrition Animale (La Ballastière, B.P. 126, 33501 Libourne Cédex, France). Because classical histological techniques and immunoassays (ELISA) are unreliable for detecting light infections during the early stages, DNA probes are being developed. It is hoped that the DNA probes (polymerase chain reaction) will be a more sensitive diagnostic method to detect low numbers of *B. ostreae* (Hervio *et al.*, 1990).

Prevention and control

Following the recognition of bonamiasis in Europe, measures such as the destruction of infected stocks and restricting movement of flat oysters were implemented (van Banning, 1985b; Grizel *et al.*, 1986; Hudson and Hill, 1991). In many instances these measures were employed too late to prevent the spread of the pathogen. Studies in the Netherlands indicated that *B. ostreae* persisted in low levels for at least 6 years in areas where flat oysters were virtually eradicated (van Banning, 1987). Sensitive diagnostic tests can be employed to monitor the presence of *B. ostreae* (Mialhe *et al.*, 1988b).

Alternate species such as Pacific oysters are now being cultured in areas where flat oyster populations were devastated by bonamiasis. However, flat oyster production has marginally persisted in a few areas of France in which the seeding of young oysters was reduced from 5 to between 1 and 2 metric tons per hectare, and by the use of 'deep water' where the parasite apparently is not transmitted (Grizel *et al.*, 1986). Also, the absence of infection in juveniles has allowed the use of oyster seed produced in areas where *B. ostreae* occurs (Grizel *et al.*, 1988). The possibility of breeding disease resistant flat oysters is also being investigated (Elston *et al.*, 1987; Grizel *et al.*, 1988; Hervio *et al.*, 1994). *In vitro* studies on the interactions between flat oyster haemocytes and *B. ostreae* could broaden the current means to combat bonamiasis and may lead to new therapeutic techniques (Grizel *et al.*, 1988).

Related pathogens

Disease and mortalities caused by other microcells have been reported in three other species of oysters. The first microcell to be recognized was *Mikrocytos mackini* in connective tissue cells of Pacific oysters from Denman Island, British Columbia (Farley *et al.*, 1988). This microcell causes Denman Island disease and can be differentiated from *B. ostreae* by the eccentric location of the nucleolus (as apposed to a peripheral location in *B. ostreae*), the apparent lack of mitochondria, and the focal nature of the infection with vesicular connective tissue cells being the main host cell rather than haemocytes. This parasite has been observed in Pacific oysters from 13 locations in British Columbia. Laboratory studies indicated that the parasite can be directly transmitted but only develops in Pacific oysters held below about 10°C for at least 3 months (Bower, unpublished data). This requirement for cool temperatures and the long prepatent period may explain why the disease only occurs during the spring and seems to be confined to oysters cultured in more northerly locations. Mortalities caused by Denman Island disease can be circumvented by well-timed plantings and harvests of Pacific oysters in relation to season and tide levels (Bower, 1988).

The microcell that causes winter mortalities in Sydney rock oysters (*Saccostrea commercialis*) in New South Wales, Australia was described as *Mikrocytos roughleyi* (Farley *et al.*, 1988). The morphological features that separate *M. roughleyi* from *M. mackini* and *B. ostreae* have not been identified. Like *B. ostreae*, *M. roughleyi* also occurs within haemocytes and is associated with focal abscess-type lesions in the gill, connective, and gonadal tissues and alimentary tract. However, the disease caused by *M. roughleyi* is associated with low temperatures and high salinities (30–35 ppt). It can kill up to 70% of mature Sydney rock oysters in their third winter before marketing, and mortalities seem to be highest in autumns and winters with low rainfall (Wolf, 1979a). The high mortalities can be reduced by harvesting large oysters before the winter and by overwintering smaller oysters on up-river leases where lower salinities and higher racks protect them from the disease (Anderson, 1990).

Finally, an unidentified species of *Bonamia* sp. has devastated dredge oyster (*Tiostrea chilensis* (= *T. lutaria*)) populations in New Zealand (Dinamani *et al.*, 1987; Hine, 1991a). Like *B. ostreae*, *Bonamia* sp. from dredge oysters resides in haemocytes, is small in size (2–7 µm), and has light and dense forms which vary in prevalence seasonally (Hine, 1991a,b). However, *Bonamia* sp. can be differentiated from *B. ostreae* by the morphology of the mitochondria, smaller nucleus to cytoplasm ratio (Dinamani *et al.*, 1987) and antigenic differences (Mialhe *et al.*, 1988c). Ultrastructural examination of cytoplasmic structures in *Bonamia* sp. by Hine and Wesney (1992) led to the suggestion that the haplosporosome-like bodies (the only feature that affiliates this group of protozoa with the haplosporidians) may be a sign of disease within the microcells. In part this suggestion was derived because of the similarities between haplosporosomes and virus-like particles, and between haplosporogenesis and virus production within host cells. This suggestion and further speculation by Hine and Wesney (1992) that an underlying disease in *Bonamia* sp. (and possibly in other oyster pathogens such as *Haplosporidium nelsoni*) may relate to the sudden appearance and pathogenicity of infection is worthy of further investigation.

PHYLUM CILIOPHORA

Two closely related genera of holotrich ciliates (class Oligohymenophorea, order Scuticociliatida) (Small and Lynn, 1985) periodically cause high mortalities among crustacens in captivity. *Mesanophrys* (= *Paranophrys*, = *Anophrys*) sp. has been observed in high numbers in the haemolymph of Dungeness crabs (*Cancer magister*) on the west coast of North America, and in edible crabs (*Cancer pagurus*) and green crabs (*Carcinus maenas*) in Europe, and *Mugardia* (= *Paranophrys*, = *Anophrys*) sp. occurs in lobsters (*Homarus americanus*) on the east coast of North America (Sparks *et al.*, 1982; Sparks, 1985; Sindermann, 1988a,b, 1990; Armetta, 1990; Sherburne and Bean, 1991). Although the ciliates are rare in the haemolymph of most wild caught crustaceans, they were observed in stained hepatopancrease smears from all 89 lobsters from ten locations along the coast of Maine in November 1990 (Sherburne and Bean, 1991). The ciliate is infectious and lethal for animals held in artificial enclosures where the disease is usually detected and can be responsible for high mortalities. Infected individuals are very lethargic. Presumptive diagnosis is made by observing numerous ciliates of typical elongate form and reduced numbers of haemocytes in the haemolymph. Ciliates can also be observed in histological sections of the soft tissues, especially the heart and gills and may be associated with tissue destruction especially of the intestine (Sherburne and Bean, 1991). Protargol stained preparations are required for specific identification (Armstrong *et al.*, 1981). Because most reports of infection pertained to injured crabs and lobsters being held in enclosures, lowering densities (i.e. less stress of crowding) and reducing mechanical damage during holding may be beneficial (Sindermann, 1988a).

PHYLUM ANNELIDA

The cosmopolitan spionid polychaetes include several species (most in the genus *Polydora*) that burrow into the shells of living molluscs. Due to the overall low economic significance of this group of parasites, the taxonomic problems as indicated by Lauckner (1983) will not be reiterated here. Instead, instances where *Polydora* sp. caused disease among mussels, oysters, and scallops in various parts of the world will be mentioned.

In European waters, mortalities and loss of market quality of blue mussels were caused by *Polydora ciliata* (Lauckner, 1983). The burrows excavated by *P. ciliata* in blue mussel shells not only caused unsightly blisters containing compacted mud but also resulted in significant reductions in shell strength, thereby increasing susceptibility to predation by birds and shore crabs (Kent, 1981). Nacreous blisters produced by blue mussels in response to *P. ciliata* may result in atrophy and detachment of the adductor muscle and possibly interference with gamete production when the calcarious ridges occur adjacent to these organs (Lauckner, 1983).

On the east and south coasts of North America, *Polydora websteri* may cause unsightly mud blisters in the shell and yellowish abscesses in the adductor muscle (when the burrow comes in contact with the muscle tissue) of eastern oysters (Lauckner, 1983). Prevalence and intensity vary considerably with local ecological conditions, but there is a general tendency for infection to be more severe on the south and southeast coasts. Infection rarely causes mortalities and infected oysters can be marketed. However, mud blisters may interfere with shucking and reduce the commercial value of oysters to be served on the half-shell. *Polydora websteri* was also considered to be a contributing factor to unusually high mortalities of bay scallop in Massachusetts (Lauckner, 1983).

In British Columbia, stunting and high mortalities caused by high intensities of *P. websteri* (burrows too numerous and interwoven to count in shells of dead scallops) has precluded the culture of introduced Japanese scallops in a few localities (Bower, 1990). However, *P. websteri* only occurred in low intensities (less than ten per shell) and had no apparent affect on Pacific oysters and giant rock scallops (*Crassedoma giganteum*) cultured in the same localities (Bower, unpublished data).

In Japan, four species of *Polydora* (*P. conveaz, P. concharum, P. variegata,* and *P. websteri*) were observed in cultured and natural populations of Japanese scallops (Sato-Okoshi and Nomura, 1990). Although prevalence was 100% in some areas, associated pathology was not indicated by Sato-

Okoshi and Nomura (1990) but Mori *et al.* (1985) suggested that *Polydora* may have an adverse affect on the growth of Japanese scallops in some areas.

In southern Australia, five species of polydorid polychaetes (*Polydora haswelli, Polydora hoplura, P. websteri, Boccardia chilensis*, and *Boccardia polybranchia*) were observed in up to 95% of blue mussels. Although the intensity of infection was generally low, about 15% of the blue mussels from two localities had serious shell damage attributed to polydorids. The most heavily infested blue mussels were from bottom samples (Pregenzer, 1983). Also, the spread of *P. websteri* along the east coast of Australia changed Sydney rock oyster production from bottom culture to an intertidal stick and tray culture system (Anderson, 1990).

PHYLUM PLATYHELMINTHES

Class Digenea

Numerous species of digenean trematodes have been described from various shellfish worldwide. In general, the trematodes that cause the greatest economic impact are species in the families Bucephalidae and Fellodistomidae that utilize bivalves as primary hosts. In such instances, miracidia are infective to bivalves and the larval trematode life stages of sporocysts and development of cercariae occur within the tissues of the bivalve. Five cases in which trematodes from other families were reported to cause pathology are noted. In addition, there is a lung fluke disease in humans caused by metacercariae of *Paragonimus westermani* from edible freshwater crabs in Asia.

Family Bucephalidae

Taxonomy

Numerous species of Bucephalidae (suborder Gasterostomata) have been described from marine and freshwater fishes and the larval forms have been reported from bivalves worldwide. However, few experimental life cycle studies have been conducted. Thus, the taxonomy of many larval Bucephalidae in bivalves remains obscure (Lauckner, 1983). For simplicity, the Bucephalidae will be considered as a group with examples of certain species presented where appropriate.

Distribution and economic importance

Larval bucephalids infecting commercially important scallops, oysters, and mussels are possibly the most deleterious metazoan parasites of marine bivalves (Lauckner, 1983). Examples have been reported from various locations. Scallops (*Pecten alba*) from Bass Strait, Australia, parasitized by *Bucephalus* sp. (prevalence of 31%) were castrated and had significant adductor muscle (the only part of the scallop that is marketed) atrophy (Sanders and Lester, 1981). *Bucephalus longicornutus* caused castration and significant mortalities of infected dredge oyster (*Tiostrea chilensis* (= *Ostrea lutaria*)), under laboratory conditions with suspected impact on wild stocks in New Zealand (Millar, 1963; Howell, 1967). Weakness and gaping caused by *Prosorhynchus squamatus* in blue mussels from northwestern Europe, can reduce product value during shipping and marketing (Coustau *et al.*, 1990). Surprisingly, parasitic castration of blue mussels caused by *P. squamatus* (Coustau *et al.*, 1993) was once thought to be beneficial for blue mussel culture, especially in southern waters where blue mussels spawn more heavily. Parasitized blue mussels remained in good condition all summer. However, consumption of trematode-infested molluscs may be hazardous to humans due to accumulation of toxic metabolites (butyric and other short-chain fatty acids) resulting from degeneration of the host's neutral fats by parasite-secreted enzymes (Cheng, 1967; Lauckner, 1983).

Morphology and life cycle

Bucephalids have fairly uniform life-cycle patterns. Sporocysts and cercariae invariably occur in bivalves with remarkable specificity. Metacercariae occur in various parts of the central and peripheral nervous systems or in internal organs and musculature of teleost fish, and adults inhabit the alimentary tract of piscivorous fish (Lauckner, 1983). In bivalves, the large dichotomously branching sporocyst forms a dense interwoven network that infiltrates practically every organ except the foot. Usually the gonad is the most heavily infected resulting in partial or complete castration. Infection is terminal following growth into and occlusion of the haemal sinuses, and gradual destruction and replacement of molluscan tissue by the sporocyst. Prevalence of infection usually increases with age (Matthews, 1974). Cercariae, often several hundred at a time, are forcibly discharged through the bivalve's exhalant siphon. Although they are not active swimmers, transmission to the intermediate host is aided by the long, extendable and retractible furcae (Matthews, 1974; Lauckner, 1983).

Host–parasite relationships

Bucephalid sporocysts and cercariae cause castration of infected bivalves, tissue necrosis, and debilitation expressed as significant reduction to the host's overall resistance to environmental stress (Lauckner, 1983). Despite the severe pathology associated with *Bucephalus* sp. infection in eastern oysters, there is usually little host response to the parasite but massive biochemical alterations have been observed (Lauckner, 1983).

Prevention and control

Haplosporidian hyperparasites have been described from *Bucephalus* sp. parasitic in eastern oysters (Lauckner, 1983) and from *Bucephalus longicornutus* parasitic in dredge oysters (Howell, 1967). Although both hyperparasites are pathogenic for the bucephalids, Howell (1967) concluded that ecological conditions as well as the difficulty of collecting large numbers of infective spores precluded the effective use of the hyperparasites as biological controls.

Coustau *et al.* (1990) showed that blue mussels are more susceptible to *P. squamatus* than hybrids of blue mussels and Mediterranean mussels and suggested that it may be possible to select for a mussel stock that is resistant to this parasite.

Family Fellodistomidae

Although numerous species of this family parasitize many marine pelecypods as primary hosts and secondary hosts worldwide (Lauckner, 1983; Wolf *et al.*, 1987), *Proctoeces maculatus*, which infects blue mussels as well as other molluscs, has the greatest economic impact. Thus, this section is confined to *P. maculatus*.

Proctoeces maculatus from shellfish appear in the literature under a variety of synonyms (*Cercaria tenuans*, *Cercaria milfordensis*, *Cercaria brachidontis*, *Proctoeces subtenuis*, *Proctoeces scrobiculariae*, and *Proctoeces buccini*). For other synonyms from fish see Bray (1983). Life stages have been described from a wide variety of bivalves and gastropods (Bray, 1983). Meta-cercariae were in various Mollusca including species of Amphineura, Gastropoda, Cephalopoda, and Lamellibranchiata, in Polychaeta (Annelida), and in Echinoidea (Echinodermata). Adults have been reported in mollusc-eating fishes (mainly labrids and sparids) in tropical and subtropical areas as well as in some Gastropoda, Lamellibranchiata, and Polychaeta. However, sporocysts have only been reported from blue mussels, Mediterranean mussels, and hooked mussels (*Ischadium recurvum*). The wide host tolerance, global distribution in tropical and temperate marine waters, and morphometric variability (Wardle, 1980; Bray, 1983) lead Lauckner (1983) to speculate that more than one species of trematode may have been included in *P. maculatus*. Thus, this species (group) requires further study using biochemical and DNA analysis as well as life cycle studies in the laboratory.

Procteces maculatus in mussels was reported from up to 46% of blue mussels and Mediterranean mussels on both sides of the North Atlantic Ocean and in the Mediterranean and Black Seas (Lauckner, 1983). In Mediterranean mussels from the Black Sea, up to 28,000 sporocysts per mussel, comprising 20% of the biomass of the soft tissues, were observed (Machkevski, 1985). In Italy, extensive mortalities in cultured mussels were attributed to this parasite which was thought to have been introduced via a depuration plant located nearby (Munford *et al.*, 1981).

In mussels, sporocysts of *P. maculatus* usually occur in the vascular system of the mantle (Lauckner, 1983). Infection causes an alteration in haemolymph components, a sharp decrease in energy stores, a reduction in growth rate, and weakness with respect to valve closure and attachment to the substrate (Mulvey and Feng, 1981; Shchepkina, 1982; Machkevski, 1985, 1988). In heavily infected mussels, sporocysts developing in the mantle can seriously reduce the glycogen content of the tissues and efficiency of the circulatory system. This results in disturbances to gametogenesis and possibly castration and death (Mulvey and Feng, 1981; Machkevski and Shchepkina, 1985; Pascual *et al.*, 1987; Feng, 1988). Mussels may also serve as a final host for *P. maculatus* (Lauckner, 1983). Progenetic development (Fig. 16.2)

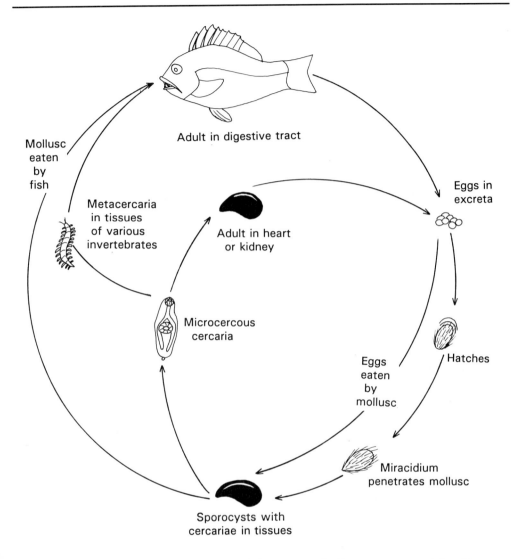

Fig. 16.2. Life cycle alternatives of the trematode *Proctoeces maculatus*. The occurrence of a progenetic cycle (adult stages of *P. maculatus* in the heart or kidney of the blue mussel (*Mytilus edulis*)) eliminates the requirement of a final (definitive) fish host for the completion of the life cycle.

represents a mechanism by which *P. maculatus* could become established in new localities as a result of moving infected stocks. *Proctoeces maculatus* probably represents a threat to mussel culture worldwide. However, Lauckner (1983) indicated that the hazard would be minimal due to the supposed narrow range of ecological conditions under which *P. maculatus* is capable of disseminating.

Other pathogenic Digenea

In addition to the Bucephalidae and *P. maculatus*, several other digenetic trematodes have been reported as pathogens of shellfish. On the French Atlantic and Italian Mediterranean coasts, *Cercaria pectinata* (family Fellodistomidae) was responsible for large scale fluctuations in clam (*Tapes* spp. and *Donax vittatus*) populations (Lauckner, 1983). In the North Sea and adjacent areas, reduced byssal production and impaired shell cleaning was reported in young blue mussels infected with metacercariae of the bird trematode *Himasthla elongata* (family Echinostomatidae) (Lauckner, 1984). Also, *Himasthla elongata* and *Renicola roscovita* were thought to control European cockle populations on the German North Sea coast (Lauckner, 1983). On the west coast of Sweden, high intensities of *Cercaria cerastodermae* (family Monorchiidae) in about 20% of the cockles (*Cardium (= Cerastoderma) edule*) led to severe tissue damage, impairment of burrowing, and eventual mortalities (Jonsson and André, 1992). Several species of gymnophallid metacercariae that occur between the mantle and shell of various lamellibranchs on the European coasts and the American North Atlantic coast are reported to cause soft tissue pathology, induction of pearl formation, and shell deformities (Lauckner, 1983). Unidentified metacercariae (possibly belonging to the family Microphallidae) were found encysted in the nervous system in 4% of Dungeness crabs (*Cancer magister*) from Puget Sound in Washington State with one of the eight infected crabs showing marked ataxia (Sparks and Hibbits, 1981). Also, unidentified metacercariae and associated histopathology were reported from the muscles, gonad, and digestive gland of 8 of 13 cultured freshwater prawns (*Macrobrachium rosenbergii*) from Thailand (Nash, 1989).

Apart from the trematodes that are pathogenic to shellfish, metacercariae of *Paragonimus westermani*, from various species of edible freshwater crabs in Asia, cause lung fluke disease or paragonimiasis in humans (Chapter 15).

As aquaculture operations expand and diversify, diseases caused by various trematodes will probably be encountered. However, the requirement of at least two different hosts for completing the life cycle in most species renders these parasites vulnerable to control once the life cycles have been identified. Aquaculture practices alone may be sufficient to create an unfavourable environment for the completion of a trematode life cycle as illustrated by the reduced prevalence of *Renicola roscovita* in farmed (4% to 12%) as opposed to natural populations (96% to 100%) of blue mussels from the west coast of Sweden (Svärdh and Thulin, 1985).

PHYLUM PLATYHELMINTHES

Class Cestoidea

Metacestodes (larval cestodes) have been reported from a wide variety of aquatic invertebrates. Among marketed shellfish, metacestode infections are economically insignificant. Nevertheless, there are a few isolated instances of high prevalences and intensities of metacestodes in bivalves and crustaceans from various subtropical and tropical areas of the world (Overstreet, 1973; Lauckner, 1983; Sparks, 1985; Sindermann, 1990). Metacestodes of *Echeneibothrium* spp. were associated with unusual behaviour of Pacific littleneck clams (*Protothaca* (= *Venerupis*) *staminea*) and fringed littleneck clams (*Protothaca laciniata*) in California (Sparks and Chew, 1966; Warner and Katkansky, 1969) and caused histopathology and gonad atrophy in Atlantic calico scallops (*Argopecten gibbus*) in North Carolina (Singhas *et al.*, 1993). In most cases, the final hosts of the cestodes are fishes, mainly elasmobranchs.

PHYLUM NEMATODA

Nematodes are uncommon as parasites of shellfish (Lauckner, 1983; Sindermann, 1990). The exceptions are all larval stages. Various species of the gnathostomid genus *Echinocephalus* are found in oysters, scallops, and abalone from tropical and subtropical marine waters. Although the pathology in the bivalve hosts is minimal, there is concern that at least some species may have public health significance as potential invaders of the human digestive tract (see Chapter 15). The species (*Echinocephalus pseudouncinatus*) in pink abalone (*Haliotis corrugata*) from California causes blisters and weakens the foot as a holdfast organ in heavily infected specimens (Sindermann, 1990). An ascaridoid *Sulcascaris sulcata* is widespread in warm seas and has a considerable host range including scallops and clams (Lauckner, 1983; Sindermann, 1990). Although *S. sulcata* is a minor pathogen for its hosts, significant economic impact occurred on the east coast of North America where a haplosporidian hyperparasite (*Urosporidium spisuli*) caused the usually white to yellowish coloured worm to become dark brown. The epizootic spread of the hyperparasite in *S. sulcata* parasitizing Atlantic surfclams (*Spisula solidissima*) in the mid-seventies caused considerable economic concern for aesthetic reasons (Payne *et al.*, 1980). *Angiostrongylus cantonensis*, the rat lungworm that causes human eosinophilic meningoencephalitis in parts of Asia can utilize eastern oysters and quahogs (*Mercenaria mercenaria*) as aberrant intermediate hosts under experimental conditions (Sparks, 1985). These findings could be significant for some of the Pacific Islands where the rat lung worm occurs and oysters and clams may be eaten raw or poorly cooked (Lauckner, 1983). The 'codworm' *Phocanema decipiens* in the North Atlantic has been observed in blue mussels and softshell clams (*Mya arenaria*) which may serve as paratenic hosts for this parasite (Lauckner, 1983).

PHYLUM ARTHROPODA

The pathogenic arthropods all belong to the class Crustacea (subclass Copepoda, the majority of which are in the order Cyclopoida and subclass Malacostraca, order Isopods). Since the economic significance of all species is either disputable or confined to small local areas, these pathogens will only be briefly mentioned.

Subclass Copepoda

Most of the parasitic copepods of shellfish are within the order Cyclopoida (Lauckner, 1983). The cyclopoid copepods presumed to cause the most significant mortalities among shellfish belong to the genus *Mytilicola*. These copepods have a direct life cycle and reside in the intestinal tract of a wide variety of bivalves (Dare, 1982; Gee and Davey, 1986). Prevalence and intensity of *Mytilicola intestinalis* in mussels in Europe can be high. For example, in Cornwall, England, the prevalence in mussels from some localities only fell below 90% during the early summer months and intensity of infection often exceeded 30 copepods per mussel (Davey, 1989). Several workers concluded that some of the periodic mass mortalities in cultured mussels in Europe were attributable to *M. intestinalis* (Sparks, 1985; Blateau *et al.*, 1990). However, these conclusion were: (i) not substantiated by statistical analysis; (ii) not supported by experimental evidence; and (ii) did not rule out the possibility that microscopic pathogens were responsible for the mortalities (Lauckner, 1983). From the results of a 10 year study conducted in Cornwall, England, Davey (1989) concluded that *M. intestinalis* is not a harmful parasite. Nevertheless, more work is required before the pest status of *M. intestinalis* can be fully appreciated, especially with respect to its synergistic relations with other pathogens and/or pollutants (Davey and Gee, 1988).

A closely related species, *Mytilicola orientalis*, was thought to originate in mussels (*Mytilus crassitesta*) from the Inland Sea of Japan. This parasite has now spread along the Pacific coast of North America where it is found in up to 58% of blue mussels and 65% of California mussels as well as in other bivalves (Chew *et al.*, 1964; Bernard, 1969). *Mytilicola orientalis* was recently introduced into France with imported Pacific oysters and is now found in blue mussels and Mediterranean mussels in Europe where dual infections with *M. intestinalis* occur (Lauckner, 1983). Apparently, *M. orientalis* is not pathogenic to mussels or other bivalves. However, synergistic interactions with other parasites including *M. intestinalis* should be closely monitored.

A recently described parasitic copepod of unknown taxonomic affinity, *Pectenophilus ornatus*, was originally thought to be a species of rhizocephalan in the subclass Cirripedia (Nagasawa *et al.*, 1988). The bright yellowish or orange female (up to 8 mm wide), attached to the gills of Japanese scallops and other commercially valuable scallops (*Chlamys* spp.) in Japan, consists mainly of a brood pouch and lacks appendages (Nagasawa *et al.*, 1991). The parasite feeds on the blood of its host, and heavy intensities of infection (greater than 20 *P. ornatus* per scallop) have detrimental affects on the condition of cultured Japanese scallops and reduces market acceptability. This parasite is considered a serious pest of commercial scallop production in Japan (Nagasawa and Nagata, 1992).

Subclass Malacostraca

Members of the Family Bopyridae within the order Isopoda are common parasites of the branchial chamber of many species of shrimp worldwide.

Infected shrimp are conspicuous due to the protruding lump on the lateral aspect of the carapace of the cephalothorax caused by the presence of the bopyrid (Sparks, 1985). Although the prevalence of bopyrids is usually low (<5%), a few instances of high prevalences and associated pathology have been noted. Japanese red prawns (*Penaeopsis akayebi*) were frequently infected (up to 70%) with *Epipenaeon japonicus* with associated gonad reduction or castration in some male prawns (Sindermann, 1990). In the Gulf of Carpentaria, Australia, the bopyrid *Epipenaeon ingens* infects up to 25% of the grooved tiger prawns (*Penaeus semisulcatus*) which it not only castrates but also alters the growth and geographical distribution in comparison to uninfected prawns (Somers and Kirkwood, 1991). In conjunction with other parasites on various species of prawns, bopyrids cause an estimated $5 million (Australian) loss of production to prawn fisheries in the Gulf of Carpentaria through reductions to prawn reproduction (Owens, 1990). In Australia, the liriopsid hyperparasite *Cabriops ordionei* (Isopods: Epicaridea), which sterilizes its bopyrid hosts is being considered as a potential biological control agent for the bopyrids (Owens, 1990).

CONCLUSIONS

A wide variety of parasites have been identified as causing significant economic losses in shellfish production worldwide. An even greater number of parasites have been identified as having the potential of causing significant losses either in endemic areas or if they inadvertently become established in other areas. In the past, the movement of commercial species of shellfish has been notorious for the accidental introduction of associated parasites (Sindermann, 1990, 1993). In order to avoid future disasters, all movements of shellfish must be conducted with caution. It is also essential that information on agents of disease including parasites be amassed so that risk involved with past and impending movements and aquaculture practices can be accurately assessed. This information should also prove useful for treating or controlling a disease in the advent that an accidental introduction occurs.

REFERENCES

Alder, J. and Braley, R. (1989) Serious mortality in populations of giant clams on reefs surrounding Lizard Island, Great Barrier Reef. *Australian Journal of Marine and Freshwater Research* 40, 205–213.

Alderman, D.J. (1979) Epizootiology of *Marteilia refringens* in Europe. *Marine Fisheries Review* 41, 67–69.

Anderson, I.G. (1990) Diseases in Australian invertebrate aquaculture. In: *Proceedings, Fifth International Colloquium on Invertebrate Pathology and Microbial Control*, Society for Invertebrate Pathology, 20–24 Aug. 1990, Adelaide, Australia, pp. 38–48.

Anderson, I.G., Shariff, M. and Nash, G. (1989) A hepatopancreatic microsporidian in pond-reared tiger shrimp, *Penaeus monodon*, from Malaysia. *Journal of Invertebrate Pathology* 53, 278–280.

Anderson, R.S., Paynter, K.T. and Burreson, E.M. (1992) Increased reactive oxygen intermediate production by hemocytes withdrawn from *Crassostrea virginica* infected with *Perkinsus marinus*. *Biological Bulletin* 183, 476–481.

Andrews, J.D. (1979) Oyster disease in Chesapeake Bay. *Marine Fisheries Review* 41, 45–53.

Andrews, J.D. (1982) Epizootiology of late summer and fall infections of oysters by *Haplosporidium nelsoni*, and comparison to annual life cycle of *Haplosporidium costalis*, a typical haplosporidan. *Journal of Shellfish Research* 2, 15–23.

Andrews, J.D. (1988a) Epizootiology of the disease caused by the oyster pathogen *Perkinsus marinus* and its effects on the oyster industry. *American Fisheries Society Special Publication* 18, 47–63.

Andrews, J.D. (1988b) *Haplosporidium nelsoni* disease of oysters. In: Sindermann, C.J. and Lightner, D.V. (eds) *Disease Diagnosis and Control in North American Marine Aquaculture*. Elsevier, Amsterdam, pp. 291–295.

Andrews, J.D. (1988c) *Haplosporidium costale* disease of oysters. In: Sindermann, C.J. and Lightner, D.V. (eds) *Disease Diagnosis and Control in North American Marine Aquaculture*. Elsevier, Amsterdam, pp. 296–299.

Andrews, J.D. and Ray, S.M. (1988) Management strategies to control the disease caused by *Perkinsus marinus*. *American Fisheries Society Special Publication* 18, 257–264.

Armetta, T.M. (1990) Hemolymph changes in Dungeness crab caused by infection with the ciliated protozoan parasite *Mesanophrys* sp. MSc Thesis, University of Washington, 58 pp.

Armstrong, D.A., Burreson, E.M. and Sparks, A.K. (1981) A ciliate infection (*Paranophrys* sp.) in laboratory-held Dungeness crabs, *Cancer magister*. *Journal of Invertebrate Pathology* 37, 201–209.

Auffret, M. and Poder, M. (1983) Studies on *Marteilia maurini*, parasite of *Mytilus edulis* from the north coasts of Brittany. *Revue des Travaux de l'Institut des Pêches Maritimes* 47, 105–109 (in French, with English abstract).

Azevedo, C. (1989) Fine structure of *Perkinsus atlanticus* n.sp. (Apicomplexa, Perkinsea) parasite of the clam *Ruditapes decussatus* from Portugal. *Journal of Parasitology* 75, 627–635.

Azevedo, C., Corral, L. and Cachola, R. (1990) Fine structure of zoosporulation in *Perkinsus atlanticus* (Apicomplexa: Perkinsea). *Parasitology* 100, 351–358.

Bachère, E., Chagot, D., Tigé, G. and Grizel, H. (1987) Study of a haplosporidian (Ascetospora), parasitizing the Australian flat oyster *Ostrea angasi*. *Aquaculture* 67, 266–268.

Balouet, G. (1979) *Marteilia refringens* – considerations of the life cycle and development of Abers disease in *Ostrea edulis*. *Marine Fisheries Review* 41, 64–66.

Balouet, G., Cahour, A. and Chastel, C. (1979) Epidemiology of flat oyster digestive tract disease: hypothesis on *Marteilia refringens* cycle. *Haliotis* 8, 323–326 (in French, with English abstract).

Balouet, G., Poder, M. and Cahour, A. (1983) Haemocytic parasitosis: morphology and pathology of lesions in the French flat oyster, *Ostrea edulis* L. *Aquaculture* 34, 1–14.

Barber, B.J., Ford, S.E. and Haskin, H.H. (1988) Effects of the parasite MSX (*Haplosporidium nelsoni*) on oyster (*Crassostrea virginica*) energy metabolism. I. Condition index and relative fecundity. *Journal of Shellfish Research* 7, 25–31.

Barber, B.J., Ford, S.E. and Littlewood, D.T.J. (1991a) A physiological comparison of resistant and susceptible oysters *Crassostrea virginica* (Gmelin) exposed to the endoparasite *Haplosporidium nelsoni* (Haskin, Stauber & Mackin). *Journal of Experimental Marine Biology Ecology* 146, 101–112.

Barber, R.D., Kanaley, S.A. and Ford, S.E. (1991b). Evidence for regular sporulation by *Haplosporidium nelsoni* (MSX) (Ascetospora; Haplosporidiidae) in spat of the American oyster, *Crassostrea virginica*. *Journal of Protozoology* 38, 305–306.

Barrow, J.H. and Taylor, B.C. (1966) Fluorescent-antibody studies of haplosporidian parasites of oysters in Chesapeake and Delaware bays. *Science* 153, 1531–1533.

Bernard, F.R. (1969) The parasitic copepod *Mytilicola orientalis* in British Columbia bivalves. *Journal of the Fisheries Research Board of Canada* 26, 190–191.

Blateau, D., LeCoguic, Y., Mialhe, E., Grizel, H. and Flamion, G. (1990) Treatment of mussels (*M. edulis*) against the copepode *Mytilicola intestinalis*. In: Figueras, A. (ed.) *Abstracts, Fourth International Colloquium on Pathology in Marine Aquaculture* 17–21 September 1990, Vigo (Pontevedra), Spain, pp. 97–98 (in French).

Bonami, J.R., Vivarès, C.P. and Brehélin, M. (1985) Study of a new haplosporidian parasite of the flat oyster *Ostrea edulis* L.: morphology and cytology of different stages. *Protistologica* 21, 161–173 (in French, with English summary).

Bougrier, S., Tigé, G., Bachère, E. and Grizel, H. (1986) *Ostrea angasi* acclimatization to French coasts. *Aquaculture* 58, 151–154.

Boulo, V., Mialhe, E., Rogier, H., Paolucci, F. and Grizel, H. (1989) Immuno-diagnosis of *Bonamia ostreae* (Ascetospora) infection of *Ostrea edulis* L. and subcellular identification of epitopes by monoclonal antibodies. *Journal of Fish Diseases* 12, 257–262.

Bower, S.M. (1987a) *Labyrinthuloides haliotidis* n.sp. (Protozoa: Labyrinthomorpha), a pathogenic parasite of small juvenile abalone in a British Columbia mariculture facility. *Canadian Journal of Zoology* 65, 1996–2007.

Bower, S.M. (1987b) The life cycle and ultrastructure of a new species of thrausto-chytrid (Protozoa: Labyrinthomorpha) pathogenic to small abalone. *Aquaculture* 67, 269–272.

Bower, S.M. (1987c) Pathogenicity and host specificity of *Labyrinthuloides haliotidis* (Protozoa: Labyrinthomorpha), a parasite of juvenile abalone. *Canadian Journal of Zoology* 65, 2008–2012.

Bower, S.M. (1987d) Artificial culture of *Labyrinthuloides haliotidis* (Protozoa: Labyrinthomorpha), a pathogenic parasite of abalone. *Canadian Journal of Zoology* 65, 2013–2020.

Bower, S.M. (1988) Circumvention of mortalities caused by Denman Island oyster disease during mariculture of Pacific oysters. *American Fisheries Society Special Publication* 18, 246–248.

Bower, S.M. (1989) Disinfectants and therapeutic agents for controlling *Labyrinthuloides haliotidis* (Protozoa: Labyrinthomorpha), an abalone pathogen. *Aquaculture* 78, 207–215.

Bower, S.M. (1990) Shellfish diseases on the west coast of Canada. *Bulletin of the Aquaculture Association of Canada* 90-3, 19–22.

Bower, S.M. (1992) Diseases and parasites of mussels. In: Gosling, E. (ed.) *The Mussel Mytilus, Ecology, Physiology, Genetics and Culture*. Elsevier Amsterdam, pp. 543–563.

Bower, S.M., McLean, N. and Whitaker, D.J. (1989a) Mechanism of infection by *Labyrinthuloides haliotidis* (Protozoa: Labyrinthomorpha), a parasite of abalone (*Haliotis kamtschatkana*) (Mollusca: Gastropoda). *Journal of Invertebrate Pathology* 53, 401–409.

Bower, S.M., Whitaker, D.J. and Elston, R.A. (1989b) Detection of the abalone parasite *Labyrinthuloides haliotidis* by a direct fluorescent antibody technique. *Journal of Invertebrate Pathology* 53, 281–283.

Bower, S.M., Whitaker, D.J. and Voltolina, D. (1989c) Resistance to ozone of zoospores of the thraustochytrid abalone parasite, *Labyrinthuloides haliotidis* (Protozoa: Labyrinthomorpha). *Aquaculture* 78, 147–152.

Bray, R.A. (1983) On the fellodistomid genus *Proctoeces* Odhner, 1911 (Digenea), with brief comments on two other fellodistomid genera. *Journal of Natural History* 17, 321–339.

Brehélin, M., Bonami, J.R., Cousserans, F. and Vivarès, C.P. (1982) True plasmodial forms exist in *Bonamia ostreae*, a pathogen of the European flat oyster *Ostrea edulis*. *Compte Rendu Hebdomadaire des Séanaces de l'Académie des Sciences, Paris. Série III* 295, 45–48 (in French, with English abstract).

Bucke, D. and Feist, S. (1985) Bonamiasis in the flat oyster *Ostrea edulis*, with comments on histological techniques. In: Ellis, A.E. (ed.) *Fish and Shellfish Pathology*. Academic Press, London, pp. 387–392.

Burreson, E.M. (1988) Use of immunoassays in haplosporidan life cycle studies. *American Fisheries Society Special Publication* 18, 298–303.

Burreson, E.M. (1991) Effects of *Perkinsus marinus* infection in the eastern oyster, *Crassostrea virginica*: I. susceptibility of native and MSX-resistant stocks. *Journal of Shellfish Research* 10, 417–423.

Burreson, E.M., Robinson, M.E. and Villalba, A. (1988) A comparison of paraffin histology and hemolymph analysis for the diagnosis of *Haplosporidium nelsoni* (MSX) in *Crassostrea virginica* (Gmelin). *Journal of Shellfish Research* 7, 19–23.

Bushek, D. (1993) Evaluation of *Perkinsus marinus* quantification techniques using fluid thioglycollate media. *Journal of Shellfish Research* 12, 126. (Abstract.)

Butler, T.H. (1980) Shrimps of the Pacific coast of Canada. *Canadian Bulletin of Fisheries and Aquatic Sciences* 202, 280 pp.

Cahour, A. (1979) *Marteilia refringens* and *Crassostrea gigas*. *Marine Fisheries Review* 41, 19–20.

Chagot, D., Bachère, E., Ruano, F., Comps, M. and Grizel, H. (1987) Ultrastructural study of sporulated instars of a haplosporidian parasitizing the clam *Ruditapes decussatus*. *Aquaculture* 67, 262–263.

Chatton, E. and Poisson, R. (1930) On the existence of dinoflagellate parasites *Hematodinium perezi* n.g., n.sp. (Syndinidae) in the blood of crabs. *Comptes Rendus des Seances de la Société de Biologie et de ses Filiales* 105, 553–557 (in French).

Cheng, T.C. (1967) Marine molluscs as hosts for symbioses with a review of known parasites of commercially important species. In: Russell, F.S. (ed.) *Advances in Marine Biology, Volume 5*. Academic Press, New York, pp. 1–424.

Chew, K.K., Sparks, A.K. and Katkansky, S.C. (1964) First record of *Mytilicola orientalis* in the California mussel *Mytilus californianus* Conrad. *Journal of the Fisheries Research Board of Canada* 21, 205–207.

Chintala, M.M. and Fisher W.S. (1991) Disease incidence and potential mechanisms of defense for MSX-resistant and -susceptible eastern oysters held in Chesapeake Bay. *Journal of Shellfish Research* 10, 439–443.

Choi, K., Lewis, D.H., Powell, E.N., Frelier, P.F. and Ray, S.M. (1991) A polyclonal antibody developed from *Perkinsus marinus* hypnospores fails to cross react with other life stages of *P. marinus* in oyster (*Crassostrea virginica*) tissues. *Journal of Shellfish Research* 10, 411–415.

Choi, K.S., Wilson, E.A., Lewis, D.H., Powell, E.N. and Ray, S.M. (1989) The energetic cost of *Perkinsus marinus* parasitism in oysters: quantification of the thioglycollate method. *Journal of Shellfish Research* 8, 125–131.

Chu, F.E. and Greene, K.H. (1989) Effect of temperature and salinity on *in vitro* culture of the oyster pathogen, *Perkinsus marinus* (Apicomplexa: Perkinsea). *Journal of Invertebrate Pathology* 53, 260–268.

Cochennec, N., Hervio, D., Panatier, B., Boulo, V., Mialhe, E., Rogier, H., Grizel, H. and Paolucci, F. (1992) A direct monoclonal antibody sandwich immunoassay for detection of *Bonamia ostreae* (Ascetospora) in hemolymph samples of the flat oyster *Ostrea edulis* (Mollusca: Bivalvia). *Diseases of Aquatic Organisms* 12, 129–134.

Comps, M. (1976) *Marteilia lengehi* n.sp., parasite of the oyster *Crassostrea cucullata* Born. *Revue des Travaux de l'Institut des Pêches Maritimes* 40, 347–349 (in French, with English summary).

Comps, M. (1983) Culture *in vitro* of *Bonamia ostreae* hemocytic parasite of the flat oyster *Ostrea edulis* L. *Compte Rendu Hebdomadaire des Séances de l'Académie des Sciences, Paris. Série III* 296, 931–933 (in French, with English abstract).

Comps, M. (1985a) Digestive gland disease of the flat oyster. In: Sindermann, C.J. (ed.) *Fiches d'Identification des Maladies et Parasites des Poissons, Crustacés et Mollusques*. Conseil International pour l'Exploration de la Mer, Copenhagen, 19, 1–5.

Comps, M. (1985b) Morphological study of *Marteilia christenseni* sp.n. parasite of *Scrobicularia piperata* P. (Mollusc Pelecypod). *Revue des Travaux de l'Institut des Pêches Maritimes* 47, 99–104 (in French, with English abstract).

Comps, M. (1985c) Haemocytic disease of the flat oyster. In: Sindermann, C.J. (ed.) *Fiches d'Identification des Maladies et Parasites des Poissons, Crustacés et Mollusques*. Conseil International pour l'Exploration de la Mer, Copenhagen, 18, 1–5.

Comps, M. and Pichot, Y. (1991) Fine spore structure of a haplosporidan parasitizing *Crassostrea gigas*: taxonomic implications. *Diseases of Aquatic Organisms* 11, 73–77.

Comps, M., Pichot, Y. and Deltreil, J.P. (1979) Evidence for a microsporidan parasite of *Marteilia refringens* the agent of digestive gland disease in *Ostrea edulis* L. *Revue des Travaux de l'Institut des Pêches Maritimes* 43, 409–412 (in French).

Comps, M., Pichot, Y. and Papagianni, P. (1981) Research on *Marteilia maurini* n.sp. a parasite of the mussel *Mytilus galloprovincialis* Lmk. *Revue des Travaux de l'Institut des Pêches Maritimes* 45, 211–214 (in French, with English abstract).

Couch, J.A. (1978) Diseases, parasites, and toxic responses of commercial penaeid shrimps of the Gulf of Mexico and South Atlantic coasts of North America. *Fishery Bulletin* 76, 1–44.

Couch, J.A. (1983) Diseases caused by protozoa. In: Provenzano, A.J. (ed.) *The Biology of Crustacea. Volume 6, Pathobiology*. Academic Press, New York, pp. 79–111.

Couch, J.A., Farley, C.A. and Rosenfield, A. (1966) Sporulation of *Minchinia nelsoni* (Haplosporida, Haplosporidiidae) in *Crassostrea virginica* (Gmelin). *Science* 153, 1529–1531.

Coustau, C., Combes, C., Maillard, C., Renaud, F. and Delay, B. (1990) *Prosorhynchus squamatus* (Trematoda) parasitosis in the *Mytilus edulis*-Mytilus galloprovincialis complex: specificity and host–parasite relationships. In: Perkins, F.O. and Cheng, T.C. (eds) *Pathology in Marine Science*. Academic Press, New York, pp. 291–298.

Coustau, C., Robbins, I., Delay, B., Renaud F. and Mathieu, M. (1993) The parasitic castration of the mussel *Mytilus edulis* by the trematode parasite *Prosorhynchus squamatus*: specificity and partial characterization of endogenous and parasite-induced anti-mitotic activities. *Comparative Biochemistry and Physiology* 104A, 229–233.

Crosby, M.P. and Roberts, C.F. (1990) Seasonal infection intensity cycle of the parasite *Perkinsus marinus* (and an absence of *Haplosporidium* spp.) in oysters from a South Carolina salt marsh. *Diseases of Aquatic Organisms* 9, 149–155.

Dare, P.J. (1982) The susceptibility of seed oysters of *Ostrea edulis* L. and *Crassostrea gigas* Thunberg to natural infestation by the copepod *Mytilicola intestinalis* Steuer. *Aquaculture* 26, 201–211.

DaRos, L. and Canzonier, W.J. (1985) *Perkinsus*, a protistan threat to bivalve culture in the Mediterranean basin. *Bulletin of the European Association of Fish Pathologists* 5, 23–25.

Davey, J.T. (1989) *Mytilicola intestinalis* (Copepoda: Cyclopoida): A ten year survey of infested mussels in a Cornish estuary, 1978–1988. *Journal of the Marine Biological Association of the United Kingdom* 69, 823–836.

Davey, J.T. and Gee, J.M. (1988) *Mytilicola intestinalis*, a copepod parasite of blue mussels. *American Fisheries Society Special Publication* 18, 64–73.

Desportes, I. (1984) The Paramyxea Levine 1979: an original example of evolution towards multicellularity. *Origins of Life* 13, 343–352.

Desportes, I. and Perkins, F.O. (1990) Phylum Paramyxea. In: Margulis, L., Corliss, J.O., Melkonian, M. and Chapman, D.J. (eds) *Handbook of Protoctista*. Jones and Bartlett Publishers, Boston, pp. 30–35.

Dinamani, P., Hine, P.M. and Jones, J.B. (1987) Occurrence and characteristics of the haemocyte parasite *Bonamia* sp. in the New Zealand dredge oyster *Tiostrea lutaria*. *Diseases of Aquatic Organisms* 3, 37–44.

Dungan, C.F. and Roberson, B.S. (1993) Binding specificities of mono- and polyclonal antibodies to the protozoan oyster pathogen *Perkinsus marinus*. *Diseases of Aquatic Organisms* 15, 9–22.

Dyková, I., Lom, J. and Fajer, E. (1988) A new haplosporean infecting the hepatopancreas in the penaeid shrimp, *Penaeus vannamei*. *Journal of Fish Diseases* 11, 15–22.

Eaton, W.D., Love, D.C., Botelho, C., Meyers, T.R., Imamura, K. and Koeneman, T. (1991) Preliminary results on the seasonality and life cycle of the parasitic dinoflagellate causing bitter crab disease in Alaskan Tanner crabs (*Chionoecetes bairdi*). *Journal of Invertebrate Pathology* 57, 426–434.

Elston, R.A., Farley, C.A. and Kent, M.L. (1986) Occurrence and significance of bonamiasis in European flat oysters *Ostrea edulis* in North America. *Diseases of Aquatic Organisms* 2, 49–54.

Elston, R.A., Kent, M.L. and Wilkinson, M.T. (1987) Resistance of *Ostrea edulis* to *Bonamia ostreae* infection. *Aquaculture* 64, 237–242.

Farley, C.A. (1965) Acid-fast staining of haplosporidian spores in relation to oyster pathology. *Journal of Invertebrate Pathology* 7, 144–147.

Farley, C.A. (1967) A proposed life cycle of *Minchinia nelsoni* (Haplosporida, Haplosporidiidae) in the American oyster *Crassostrea virginica*. *Journal of Protozoology* 14, 616–625.

Farley, C.A. (1968) *Minchinia nelsoni* (Haplosporida) disease syndrome in the American oyster *Crassostrea virginica*. *Journal of Protozoology* 15, 585–599.

Farley, C.A. (1975) Epizootic and enzootic aspects of *Minchinia nelsoni* (Haplosporida) disease in Maryland oysters. *Journal of Protozoology* 22, 418–427.

Farley, C.A. (1988) A computerized coding system for organs, tissues, lesions, and parasites of bivalve mollusks and its application in pollution monitoring with *Mytilus edulis*. *Marine Environmental Research* 24, 243–249.

Farley, C.A., Wolf, P.H. and Elston, R.A. (1988) A long-term study of 'microcell' disease in oysters with a description of a new genus, *Mikrocytos* (g.n.) and two new species, *Mikrocytos mackini* (sp.n.) and *Mikrocytos roughleyi* (sp.n.). *Fishery Bulletin* 86, 581-593.

Feng, S.L. (1988) Host response to *Proctoeces maculatus* infection in the blue mussel, *Mytilus edulis* L. *Journal of Shellfish Research* 7, 118. (Abstract.)

Feng, S.Y. and Stauber, L.A. (1968) Experimental hexamitiasis in the oyster *Crassostrea virginica*. *Journal of Invertebrate Pathology* 10, 94-110.

Figueras, A.J., Jardon, C.F. and Caldas, J.R. (1991) Diseases and parasites of mussels (*Mytilus edulis*, Linneaus, 1758) from two sites on the east coast of the United States. *Journal of Shellfish Research* 10, 89-94.

Figueras, A., Robledo, J.A.F. and Novoa, B. (1992) Occurrence of haplosporidian and *Perkinsus*-like infections in carpet-shell clams, *Ruditapes decussatus* (Linnaeus, 1758), of the Ria de Vigo (Galicia, NW Spain). *Journal of Shellfish Research* 11, 377-382.

Fisher, W.S. (1988) *In vitro* binding of parasites (*Bonamia ostreae*) and latex particles by hemocytes of susceptible and insusceptible oysters. *Developmental and Comparative Immunology* 12, 43-53.

Ford, S.E. (1992) Avoiding the transmission of disease in commercial culture of molluscs, with special reference to *Perkinsus marinus* (Dermo) and *Haplosporidium nelsoni* (MSX). *Journal of Shellfish Research* 11, 539-546.

Ford, S.E. and Figueras, A.J. (1988) Effects of sublethal infection by the parasite *Haplosporidium nelsoni* (MSX) on gametogenesis, spawning, and sex ratios of oysters in Delaware Bay, USA. *Diseases of Aquatic Organisms* 4, 121-133.

Ford, S.E. and Haskin, H.H. (1982) History and epizootiology of *Haplosporidium nelsoni* (MSX), an oyster pathogen in Delaware Bay, 1957-1980. *Journal of Invertebrate Pathology* 40, 118-141.

Ford, S.E. and Haskin, H.H. (1988) Management strategies for MSX (*Haplosporidium nelsoni*) disease in eastern oysters. *American Fisheries Society Special Publication* 18, 249-256.

Ford, S.E. and Kanaley, S.A. (1988) An evaluation of hemolymph diagnosis for detection of the oyster parasite *Haplosporidium nelsoni* (MSX). *Journal of Shellfish Research* 7, 11-18.

Ford, S.E., Figueras, A.J. and Haskin, H.H. (1990a) Influence of selective breeding, geographic origin, and disease on gametogenesis and sex ratios of oysters, *Crassostrea virginica*, exposed to the parasite *Haplosporidium nelsoni* (MSX). *Aquaculture* 88, 285-301.

Ford, S.E., Kanaley, S.A., Ferris, M. and Ashton-Alcox, K.A. (1990b) 'Panning', a technique for enrichment of the oyster parasite *Haplosporidium nelsoni* (MSX). *Journal of Invertebrate Pathology* 56, 347-352.

Ford, S.E., Kanaley, S.A. and Littlewood, D.T.J. (1993) Cellular responses of oysters infected with *Haplosporidium nelsoni*: changes in circulating and tissue-infiltrating hemocytes. *Journal of Invertebrate Pathology* 61, 49-57.

Franc, A. (1980) On some unpublished aspects in the cycle of *Marteilia refringens* Grizel *et al.*, 1974, parasite of the flat oyster *Ostrea edulis* L. *Cahiers de Biologie Marine* 21, 99-106 (in French, with English summary).

Friedman, C.S., Cloney, D.F., Manzer, D. and Hedrick, R.P. (1991) Haplosporidiosis of the Pacific oyster, *Crassostrea gigas*. *Journal of Invertebrate Pathology* 58, 367-372.

Friedman, C.S., McDowell, T., Groff, J.M., Hollibaugh, J.T., Manzer, D. and Hedrick, R.P. (1989) Presence of *Bonamia ostreae* among populations of the European flat oyster, *Ostrea edulis* Linné in California, USA. *Journal of Shellfish Research* 8, 133-137.

Friedman, C.S., Roberts, W., Kismohandaka, G. and Hedrick, R.P. (1993) Trans-missibility of a coccidian parasite of abalone, *Haliotis* spp. *Journal of Shellfish Research* 12, 201–205.

Gauthier, J.D. and Fisher, W.S. (1990) Hemolymph assay for diagnosis of *Perkinsus marinus* in oysters *Crassostrea virginica* (Gmelin, 1791). *Journal of Shellfish Research* 9, 367–371.

Gauthier, J.D. and Vasta, G.R. (1993) Continuous *in vitro* culture of the eastern oyster parasite *Perkinsus marinus*. *Journal of Invertebrate Pathology* 62, 321–323.

Gee, J.M. and Davey, J.T. (1986) Stages in the life cycle of *Mytilicola intestinalis* Steuer, a copepod parasite of *Mytilus edulis* (L.), and the effect of temperature on their rates of development. *Journal du Conseil International pour l'Exploration de la Mer* 42, 254–264.

Ginsburger-Vogel, T. and Desportes, I. (1979) Structure and biology of *Marteilia* sp. in the amphipod, *Orchestra gammarellus*. *Marine Fisheries Review* 41, 3–7.

Goggin, C.L. and Lester, R.J.G. (1987) Occurrence of *Perkinsus* species (Protozoa, Apicomplexa) in bivalves from the Great Barrier Reef. *Diseases of Aquatic Organisms* 3, 113–117.

Goggin, C.L., Sewell, K.B. and Lester, R.J.G. (1989) Cross-infection experiments with Australian *Perkinsus* species. *Diseases of Aquatic Organisms* 7, 55–59.

Goggin, C.L., Sewell, K.B. and Lester, R.J.G. (1990) Tolerances of *Perkinsus* spp. (Protozoa, Apicomplexa) to temperature, chlorine and salinity. *Journal of Shellfish Research* 9, 145–148.

González, P., Pascual, C., Quintana, R. and Morales, J. (1987) Parasites of cultivated Galician Spain mussels: 1. Protozoa with reference to *Marteilia maurini* and *Steinhausia mytilovum*. *Alimentaria* 24, 37–44 (in Spanish with English abstract).

Götting, K.J. (1979) Pearl formation in *Mytilus edulis* L. (Bivalvia) through parasitic induction. *Malacologia* 18, 563–567 (in German with English abstract).

Grizel, H. (1979) *Marteilia refringens* and oyster disease – recent observations. *Marine Fisheries Review* 41, 38–39.

Grizel, H. (1986) Epidemiology of bivalve molluscs: analysis of present data and perspectives of development. *European Aquaculture Society, Special Publications* 9, 1–22 (in French, with English abstract).

Grizel, H., Comps, M., Bonami, J.R., Cousserans, F., Duthoit, J.L. and LePennec, M.A. (1974) Research on the agent of digestive gland disease of *Ostrea edulis* Linne. *Science et Pêche Bulletin d'Information et de Documentation de l'Institut Scientifique et Technique des Pêches Maritimes* 240, 7–30 (in French).

Grizel, H., Comps, M., Raguenes, D., Leborgne, Y., Tigé, G. and Martin, A.G. (1983) Results of acclimatization trials of *Ostrea chilensis* on the coast of Brittany. *Revue des Travaux de l'Institut des Pêches Maritimes* 46, 209–225 (in French, with English abstract).

Grizel, H., Bachère, E., Mialhe, E. and Tigé, G. (1986) Solving parasite-related pro-blems in cultured molluscs. In: Howell, M.J. (ed.) *Parasitology – Quo Vadit? Pro-ceedings of the Sixth International Congress of Parasitology*. Australian Academy of Science, Canberra, pp. 301–308.

Grizel, H., Mialhe, E., Chagot, D., Boulo, V. and Bachère, E. (1988) Bonamiasis: a model study of diseases in marine molluscs. *American Fisheries Society Special Publication* 18, 1–4.

Gutiérrez, M. (1977) Technique for staining the agent of digestive gland disease of flat oysters, *Ostrea edulis* L. *Investigacion Pesquera (Barcelona)* 41, 643–645 (in Spanish, with English summary).

Haskin, H.H. and Andrews, J.D. (1988) Uncertainties and speculations about the life cycle of the eastern oyster pathogen *Haplosporidium nelsoni* (MSX). *American Fisheries Society Special Publication* 18, 5–22.

Haskin, H.H. and Ford, S.E. (1982) *Haplosporidium nelsoni* (MSX) on Delaware Bay seed oyster beds: a host-parasite relationship along a salinity gradient. *Journal of Invertebrate Pathology* 40, 388–405.

Haskin, H.H., Stauber, L.A. and Mackin, J.A. (1966) *Minchinia nelsoni* n.sp. (Haplosporida, Haplosporidiidae): causative agent of the Delaware Bay oyster epizoötic. *Science* 153, 1414–1416.

Hervio, D., Lubat, V., Mialhe, E. and Grizel, H. (1990) Nucleic acid probes in the diagnosis and study of *Bonamia ostreae* (Ascetospora), intrahemocytic parasite of the flat oyster *Ostrea edulis*. In: Figueras, A. (ed.) *Abstracts, Fourth International Colloquium on Pathology in Marine Aquaculture*, 17–21 Sept. 1990, Vigo (Pontevedra), Spain, pp. 71.

Hervio, D., Bachère, E., Boulo, V., Cochennec, N., Vuillemin, V., LeCoguic, Y., Cailletaux, G., Mazurie, J. and Mialhe, E. (1995) Establishment of an experimental infection protocol for the flat oyster *Ostrea edulis* with the intrahaemocytic protozoan parasite *Bonamia ostreae*: application in the selection of parasite resistant oysters. *Aquaculture* (in press).

Hine, P.M. (1991a) The annual pattern of infection by *Bonamia* sp. in New Zealand flat oysters, *Tiostrea chilensis*. *Aquaculture* 93, 241–251.

Hine, P.M. (1991b) Ultrastructural observations on the annual infection pattern of *Bonamia* sp. in flat oysters *Tiostrea chilensis*. *Diseases of Aquatic Organisms* 11, 163–171.

Hine, P.M. and Wesney, B. (1992) Interrelationships of cytoplasmic structures in *Bonamia* sp. (Haplosporidia) infecting oysters *Tiostrea chilensis*: an interpretation. *Diseases of Aquatic Organisms* 14, 59–68.

Hoese, H.D. (1964) Studies on oyster scavengers and their relation to the fungus *Dermocystidium marinum*. *Proceedings of the National Shellfisheries Association* 53, 161–174.

Howell, M. (1967) The trematode, *Bucephalus longicornutus* (Manter, 1954) in the New Zealand mud-oyster, *Ostrea lutaria*. *Transactions of the Royal Society of New Zealand, Zoology* 8, 221–237.

Hudson, E.B. and Hill, B.J. (1991) Impact and spread of bonamiasis in the UK. *Aquaculture* 93, 279–285.

Iversen, E.S. and Kelly, J.F. (1976) Microsporidiosis successfully transmitted experimentally in pink shrimp. *Journal of Invertebrate Pathology* 27, 407–408.

Jellett, J.F. and Scheibling, R.E. (1988) Virulence of *Paramoeba invadens* Jones (Amoebida, Paramoebidae) from monoxenic and polyxenic culture. *Journal of Protozoology* 35, 422–424.

Jellett, J.F., Wardlaw, A.C. and Scheibling, R.E. (1988) Experimental infection of the echinoid *Strongylocentrotus droebachiensis* with *Paramoeba invadens*: quantitative changes in the coelomic fluid. *Diseases of Aquatic Organisms* 4, 149–157.

Johnson, P.T. (1977) Paramoebiasis in the blue crab, *Callinectes sapidus*. *Journal of Invertebrate Pathology* 29, 308–320.

Johnson, P.T. (1988) Paramoebiasis of blue crabs. In: Sindermann, C.J. and Lightner, D.V. (eds), *Disease Diagnosis and Control in North American Marine Aquaculture*. Elsevier, Amsterdam, pp. 204–207.

Jones, G.M. (1985) *Paramoeba invadens* n.sp. (Amoebida, Paramoebidae), a pathogenic amoeba from the sea urchin, *Strongylocentrotus droebachiensis*, in eastern Canada. *Journal of Protozoology* 32, 564–569.

Jones, G.M. and O'Dor, R.K. (1983) Ultrastructural observations on a thraustochytrid fungus parasitic in the gills of squid (*Illex illecebrosus* Lesueur). *Journal of Parasitology* 69, 903–911.

Jones, G.M. and Scheibling, R.E. (1985) *Paramoeba* sp. (Amoebida, Paramoebidae) as the possible causative agent of sea urchin mass mortality in Nova Scotia. *Journal of Parasitology* 71, 559–565.

Jones, G.M., Hebda, A.J., Scheibling, R.E. and Miller, R.J. (1985) Histopathology of the disease causing mass mortalities of sea urchins (*Strongylocentrotus droebachiensis*) in Nova Scotia. *Journal of Invertebrate Pathology* 45, 260–271.

Jonsson, P.R. and André, C. (1992) Mass mortality of the bivalve *Cerastoderma edule* on the Swedish west coast caused by infestation with the digenean trematode *Cercaria cerastodermae* I. *Ophelia* 36, 151–157.

Katkansky, S.C., Dahlstrom, W.A. and Warner, R.W. (1969) Observations on survival and growth of the European flat oyster, *Ostrea edulis*, in California. *California Fish and Game* 55, 69–74.

Kent, M.L., Elston, R.A., Nerad, T.A. and Sawyer, T.K. (1987) An *Isonema*-like flagellate (Protozoa: Mastigophora) infection in larval geoduck clams, *Panope abrupta*. *Journal of Invertebrate Pathology* 50, 221–229.

Kent, R.M.L. (1981) The effect of *Polydora ciliata* on the shell strength of *Mytilus edulis*. *Journal du Conseil International pour l'Exploration de la Mer* 39, 252–255.

Kern, F.G. (1985) *Perkinsus marinus* parasitism, a sporozoan disease of oyster. In: Sindermann, C.J. (ed.) *Fiches d'Identification des Maladies et Parasites des Poissons, Crustacés et Mollusques/Identification Leaflets for Diseases and Parasites of Fish and Shellfish*. International Council for the Exploration of the Sea, Copenhagen, 30, 1–4.

Kern, F.G., Sullivan, L.C. and Takata, M. (1973) *Labyrinthomyxa*-like organisms associated with mass mortalities of oysters *Crassostrea virginica*, from Hawaii. *Proceedings of the National Shellfisheries Association* 63, 43–46.

Khouw, B.T., McCurdy, H.D. and Drinnan, R.E. (1968) The axenic cultivation of *Hexamita inflata* from *Crassostrea virginica*. *Canadian Journal of Microbiology* 14, 184–185.

Kleinschuster, S.J. and Swink, S.L. (1993) A simple method for the *in vitro* culture of *Perkinsus marinus*. *Journal of Shellfish Research* 12, 110. (Abstract.)

Laird, M. (1961) Microecological factors in oyster epizootics. *Canadian Journal of Zoology* 39, 449–485.

LaPeyre, J.F., Faisal, M. and Burreson, E.M. (1993) *In vitro* propagation of the protozoan *Perkinsus marinus*, a pathogen of the eastern oyster, *Crassostrea virginica*. *Journal of Eukaryotic Microbiology* 40, 304–310.

Lauckner, G. (1983) Diseases of Mollusca: Bivalvia. In: Kinne, O. (ed.) *Diseases of Marine Animals. Volume II: Introduction, Bivalvia to Scaphopoda*. Biologische Anstalt Helgoland, Hamburg, pp. 477–961.

Lauckner, G. (1984) Impact of trematode parasitism on the fauna of a North Sea tidal flat. *Helgoländer Meeresuntersuchungen* 37, 185–199.

Leibovitz, L., Schott, E.F. and Karney, R.C. (1984) Diseases of wild, captive and cultured scallops. *Journal of the World Mariculture Society* 15, 269–283.

Lester, R.J.G. (1986) Abalone die-back caused by protozoan infection? *Australian Fisheries* 45, 26–27.

Lester, R.J.G. and Davis, G.H.G. (1981) A new *Perkinsus* species (Apicomplexa, Perkinsea) from the abalone *Haliotis ruber*. *Journal of Invertebrate Pathology* 37, 181–187.

Lester, R.J.G., Goggin, C.L. and Sewell, K.B. (1990) *Perkinsus* in Australia. In:

Perkins, F.O. and Cheng, T.C. (eds) *Pathology in Marine Science*. Academic Press, San Diego, pp. 189–199.

Levine, N.D. (1978) *Perkinsus* gen. n. and other new taxa in the protozoan phylum Apicomplexa. *Journal of Parasitology* 64, 549.

Levine, N.D. (1988) *The Protozoan Phylum Apicomplexa. Vol. 1*. CRC Press, Inc., Boca Raton, Florida.

Levine, N.D., Corliss, J.O., Cox, F.E.G., Deroux, G., Grain, J., Honigberg, B.M., Leedale, G.F., Loeblich, A.R., Lom, J., Lynn, D., Merinfeld, E.G., Page, F.C., Poljansky, G., Sprague, V., Vavra, J. and Wallace, F.G. (1980) A newly revised classification of the Protozoa. *Journal of Protozoology* 27, 37–58.

Lightner, D.V. (1975) Some potentially serious disease problems in the culture of penaeid shrimp in North America. In: *Proceedings of the Third US - Japan Meeting on Aquaculture*, 15–16 Oct. 1974, Tokyo, Japan, pp. 75–97.

Lightner, D.V. (1988) Cotton shrimp disease of penaeid shrimp. In: Sindermann, C.J. and Lightner, D.V. (eds) *Disease Diagnosis and Control in North American Marine Aquaculture*. Elsevier, Amsterdam, pp. 70–75.

López-Ochoterena, E., Madrazo-Garibay, M. and Pérez-Reyes, R. (1988) *Hexamita nelsoni* (Sarcomastigophora, Diplomonadida) and its association with diverse species of molluscs of the Laguna de Términos, Campeche and the fluvial-lagoon system Atasta-Pom. *Anales del Instituto de Ciencias del Mar y Limnología Universidad Nacional Autónoma de México* 15, 259–264 (in Spanish, with English abstract).

Love, D.C., Rice, S.D., Moles, D.A. and Eaton, W.D. (1993) Seasonal prevalence and intensity of bitter crab dinoflagellate infection and host mortality in Alaskan Tanner crabs *Chionoecetes bairdi* from Auke Bay, Alaska, USA. *Diseases of Aquatic Organisms* 15, 1–7.

Machkevski, V.K. (1985) Some aspects of the biology of the trematode, *Proctoeces maculatus*, in connection with the development of mussel farms on the Black Sea. In: Hargis, J.W.J. (ed.) *Parasitology and Pathology of Marine Organisms of the World Ocean*. United States Department of Commerce, Seattle (NOAA technical report NMFS 25) pp. 109–110.

Machkevski, V.K. (1988) Effect of *Proctoeces maculatus* parasitization on the growth of *Mytilus galloprovincialis*. *Parazitologiya* 22, 341–344 (in Russian, with English summary).

Machkevski, V.K. and Shchepkina, A.M. (1985) Infection of the Black Sea mussels with larval *Proctoeces maculatus* and its effects on the glycogen content in the tissues of the host. *Ehkologiya Morya* 20, 69–73 (in Russian, with English summary).

Mackin, J.G. and Ray, S.M. (1966) The taxonomic relationship of *Dermocystidium marinum* Mackin, Owen, and Collier. *Journal of Invertebrate Pathology* 8, 544–545.

Mackin, J.G., Owen, H.M. and Collier, A. (1950) Preliminary note on the occurrence of a new protistan parasite, *Dermocystidium marinum* n.sp. in *Crassostrea virginica* (Gmelin). *Science* 111, 328–329.

Mackin, J.G., Korringa, P. and Hopkins, S.H. (1952) Hexamitiasis of *Ostrea edulis* L. and *Crassostrea virginica* (Gmelin). *Bulletin of Marine Science of the Gulf and Caribbean* 1, 266–277.

MacLean, S.A. and Ruddell, C.L. (1978) Three new crustacean hosts for the parasitic dinoflagellate *Hematodinium perezi* (Dinoflagellata: Syndinidae). *Journal of Parasitology* 64, 158–160.

Mann, R., Burreson, E.M. and Baker, P.K. (1991) The decline of the Virginia oyster fishery in Chesapeake Bay: considerations for introduction of a non-endemic species, *Crassostrea gigas* (Thunberg, 1793). *Journal of Shellfish Research* 10, 379–388.

Matthews, R.A. (1974) The life-cycle of *Bucephaloides gracilescens* (Rudolphi, 1819) Hopkins, 1954 (Digenea: Gasterostomata). *Parasitology* 68, 1–12.

Matthiessen, G.C. and Davis, J.P. (1992) Observations on growth rate and resistance to MSX (*Haplosporidium nelsoni*) among diploid and triploid eastern oysters (*Crassostrea virginica* (Gmelin, 1791)) in New England. *Journal of Shellfish Research* 11, 449–454.

McArdle, J.F., McKiernan, F., Foley, H. and Jones, D.H. (1991) The current status of *Bonamia* disease in Ireland. *Aquaculture* 93, 273–278.

McGladdery, S. (1990) Shellfish parasites and diseases on the east coast of Canada. *Bulletin of the Aquaculture Association of Canada* 90-3, 14–18.

McGladdery, S.E., Cawthorn, R.J. and Bradford, B.C. (1991) *Perkinsus karlssoni* n.sp. (Apicomplexa) in bay scallops *Argopecten irradians*. *Diseases of Aquatic Organisms* 10, 127–137.

McGladdery, S.E., Bradford, B.C. and Scarratt, D.J. (1993) Investigations into the transmission of parasites of the bay scallop, *Argopecten irradians* (Lamarck, 1819), during quarantine introduction to Canadian waters. *Journal of Shellfish Research* 12, 49–58.

McLean, N. and Porter, D. (1987) Lesions produced by a thraustochytrid in *Tritonia diomedea* (Mollusca: Gastropoda: Nudibranchia). *Journal of Invertebrate Pathology* 49, 223–225.

Meuriot, E. and Grizel, H. (1984) Note on the economic impact of diseases of the flat oyster in Bretagne. *Rapports Techniques de l'Institut Scientifique et Technique des Pêches Maritimes* 12, 1–20 (in French).

Messick, G.A. and Sindermann, C.J. (1992) Synopsis of principal diseases of the blue crab, *Callinectes sapidus*. *NOAA Technical Memorandum NMFS-F/NEC* 88, 1–24.

Meyers, J.A., Burreson, E.M., Barber, B.J. and Mann, R. (1991) Susceptibility of diploid and triploid Pacific oysters, *Crassostrea gigas* (Thunberg, 1793) and eastern oysters, *Crassostrea virginica* (Gmelin, 1791), to *Perkinsus marinus*. *Journal of Shellfish Research* 10, 433–437.

Meyers, T.R., Koeneman, T.M., Botelho, C. and Short, S. (1987) Bitter crab disease: a fatal dinoflagellate infection and marketing problem for Alaskan Tanner crabs *Chionoecetes bairdi*. *Diseases of Aquatic Organisms* 3, 195–216.

Meyers, T.R., Botelho, C., Koeneman, T.M., Short, S. and Imamura, K. (1990) Distribution of bitter crab dinoflagellate syndrome in southeast Alaskan Tanner crabs *Chionoecetes bairdi*. *Diseases of Aquatic Organisms* 9, 37–43.

Mialhe, E., Bachère, E., LeBec, C. and Grizel, H. (1985) Isolation and purification of *Marteilia* (Protozoa: Ascetospora) parasites of marine Bivalvia: ultrastructural study of pansporoblasts. *Compte Rendu Hebdomadaire des Séances de l'Académie des Sciences, Paris. Série III* 301, 137–142 (in French, with English abstract).

Mialhe, E., Bachére, E., Chagot, D. and Grizel, H. (1988a) Isolation and purification of the protozoan *Bonamia ostreae* (Pichot *et al.* 1980), a parasite affecting the flat oyster *Ostrea edulis* L. *Aquaculture* 71, 293–299.

Mialhe, E., Boulo, V., Grizel, H., Rogier, H. and Paolucci, F. (1988b) Monoclonal antibodies: a tool for molluscan pathology. *American Fisheries Society Special Publication* 18, 304–310.

Mialhe, E., Boulo, V., Elston, R., Hill, B., Hine, M., Montes, J., van Banning, P. and Grizel, H. (1988c) Serological analysis of *Bonamia* in *Ostrea edulis* and

Tiostrea lutaria using polyclonal and monoclonal antibodies. *Aquatic Living Resources* 1, 67–69.

Millar, R.H. (1963) Oysters killed by trematode parasites. *Nature* 197, 616.

Miller, R.J. (1985) Succession in sea urchin and seaweed abundance in Nova Scotia, Canada. *Marine Biology* 84, 275–286.

Montes, J. (1991) Lag time for the infestation of flat oyster (*Ostrea edulis* L.) by *Bonamia ostreae* in estuaries of Galicia (NW Spain). *Aquaculture* 93, 235–239.

Montes, J. and Meléndez, I. (1987) Data on the parasitism of *Bonamia ostreae* in the flat oyster of Galicia, north-west coast of Spain. *Aquaculture* 67, 195–198 (in French, with English abstract).

Montes, J., Anadón, R. and Azevedo, C. (1994) A possible life cycle for *Bonamia ostreae* on the basis of electron microscopy studies. *Journal of Invertebrate Pathology* 63, 1–6.

Morado, J.F., Sparks, A.K. and Reed, S.K. (1984) A coccidian infection of the kidney of the native littleneck clam *Protothaca staminea*. *Journal of Invertebrate Pathology* 43, 207–217.

Mori, K., Sato, W., Nomura, T. and Imajima, M. (1985) Infestation of the Japanese scallop *Patinopecten yessoensis* by the boring polychaetes, *Polydora*, on the Okhotsk Sea coast of Hokkaido, especially in Abashiri waters. *Bulletin of the Japanese Society of Scientific Fisheries* 51, 371–380 (in Japanese, with English abstract).

Mulvey, M. and Feng, S.Y. (1981) Hemolymph constituents of normal and *Proctoeces maculatus* infected *Mytilus edulis*. *Comparative Biochemistry and Physiology* 70A, 119–125.

Munford, J.G., DaRos, L. and Strada, R. (1981) A study on the mass mortality of mussels in the Laguna Veneta. *Journal of the World Mariculture Society* 12, 186–199.

Nagasawa, K. and Nagata, M. (1992) Effects of *Pectenophilus ornatus* (Copepoda) on the biomass of cultured Japanese scallop *Patinopecten yessoensis*. *Journal of Parasitology* 78, 552–554.

Nagasawa, K., Bresciani, J. and Lützen, J. (1988) Morphology of *Pectenophilus ornatus*, new genus, new species, a copepod parasite of the Japanese scallop *Patinopecten yessoensis*. *Journal of Crustacean Biology* 8, 31–42.

Nagasawa, K., Takahashi, K., Tanaka, S. and Nagata, M. (1991) Ecology of *Pectenophilus ornatus*, a copepod parasite of the Japanese scallop *Patinopecten yessoensis*. *Bulletin of the Plankton Society of Japan, Special Volume – Proceedings of the Fourth International Conference on Copepoda* 1991, 495–502.

Nash, G. (1989) Trematode metacercarial infection of cultured giant freshwater prawns, *Macrobrachium rosenbergii*. *Journal of Invertebrate Pathology* 53, 124–127.

Newman, M.W. and Johnson, C.A. (1975) A disease of blue crabs (*Callinectes sapidus*) caused by a parasitic dinoflagellate, *Hematodinium* sp. *Journal of Parasitology* 61, 554–557.

Newman, M.W. and Ward, G.E. (1973) An epizootic of blue crab, *Callinectes sapidus*, caused by *Paramoeba perniciosa*. *Journal of Invertebrate Pathology* 22, 329–334.

Norton, J.H., Shepherd, M.A., Perkins, F.P. and Prior, H.C. (1993) *Perkinsus*-like infection in farmed golden-lipped pearl oyster *Pinctada maxima* from the Torres Strait, Australia. *Journal of Invertebrate Pathology* 62, 105–106.

Olson, R.E. and Lannan, C.N. (1984) Prevalence of microsporidian infection in commercially caught pink shrimp, *Pandalus jordani*. *Journal of Invertebrate Pathology* 43, 407–413.

Overstreet, R.M. (1973) Parasites of some penaeid shrimps with emphasis on reared hosts. *Aquaculture* 2, 105–140.

Overstreet, R.M. (1975) Buquinolate as a preventive drug to control microsporidosis in the blue crab. *Journal of Invertebrate Pathology* 26, 213–216.

Overstreet, R.M. (1988) Microsporosis of blue crabs. In: Sindermann, C.J. and Lightner, D.V. (eds) *Disease Diagnosis and Control in North American Marine Aquaculture*. Elsevier, Amsterdam, pp. 200–203.

Owens, L. (1990) Biology of *Cabriops orbionei* (Epicaridea; Liriopsidae), biological control agent for bopyrids in northern Australia. In: *Proceedings, Fifth International Colloquium on Invertebrate Pathology and Microbial Control*, Society for Invertebrate Pathology, 20–24 August 1990, Adelaide, Australia, pp. 406. (Abstract).

Owens, L. and Hall-Mendelin, S. (1990) Diseases relevant to penaeid mariculture in tropical Australia. In: Perkins, F.O. and Cheng, T.C. (eds) *Pathology in Marine Science*. Academic Press, San Diego, pp. 421–432.

Parsons, D.G. and Khan, R.A. (1986) Microsporidiosis in the northern shrimp, *Pandalus borealis*. *Journal of Invertebrate Pathology* 47, 74–81.

Pascual, C., Quintana, R., Molares, J. and González, P. (1987) Parasites of cultivated Galician mussels: 2. Metazoans with special reference to trematodes and copepods. *Alimentaria* 24, 31–36. (in Spanish, with English abstract).

Pauley, G.B., Newman, M.W. and Gould, E. (1975) Serum changes in the blue crab, *Callinectes sapidus*, associated with *Paramoeba perniciosa*, the causative agent of grey crab disease. *Marine Fisheries Review* 37, 34–38.

Payne, W.L., Gerding, T.A., Dent, R.G., Bier, J.W. and Jackson, G.J. (1980) Survey of the U.S. Atlantic coast surf clam, *Spisula solidissima*, and clam products for anisakine nematodes and hyperparasitic protozoa. *Journal of Parasitology* 66, 150–153.

Paynter, K.T. and Burreson, E.M. (1991) Effects of *Perkinsus marinus* infection in the eastern oyster, *Crassostrea virginica*: II. disease development and impact on growth rate at different salinities. *Journal of Shellfish Research* 10, 425–431.

Perkins, F.O. (1968) Fine structure of zoospores from *Labyrinthomyxa* sp. parasitizing the clam *Macoma balthica*. *Chesapeake Science* 9, 198–202.

Perkins, F.O. (1969a) Ultrastructure of vegetative stages in *Labyrinthomyxa marina* (= *Dermocystidium marinum*), a commercially significant oyster pathogen. *Journal of Invertebrate Pathology* 13, 199–222.

Perkins, F.O. (1969b) Electron microscope studies of sporulation in the oyster pathogen, *Minchinia costalis* (Sporozoa: Haplosporida). *Journal of Parasitology* 55, 897–920.

Perkins, F.O. (1976a) Zoospores of the oyster pathogen, *Dermocystidium marinum*. I. Fine structure of the conoid and other sporozoan-like organelles. *Journal of Parasitology* 62, 959–974.

Perkins, F.O. (1976b) *Dermocystidium marinum* infection in oysters. *Marine Fisheries Review* 38, 19–21.

Perkins, F.O. (1976c) Ultrastructure of sporulation in the European flat oyster pathogen, *Marteilia refringens* – taxonomic implications. *Journal of Protozoology* 23, 64–74.

Perkins, F.O. (1988) Structure of protistan parasites found in bivalve molluscs. *American Fisheries Society Special Publication* 18, 93–111.

Perkins, F.O. (1990) Phylum Haplosporidia. In: Margulis, L., Corliss, J.O., Melkonian, M. and Chapman, D.J. (eds) *Handbook of Protoctista*. Jones and Bartlett Publishers, Boston, pp. 19–29.

Perkins, F.O. (1993) Infectious diseases of molluscs. In: Couch, J.A. and Fournie, J.W. (eds) *Advances in Fisheries Science: Pathobiology of Marine and Estuarine Organisms.* CRC Press, Boca Raton, pp. 255–287.

Perkins, F.O. and Castagna, M. (1971) Ultrastructure of the Nebenkörper or 'secondary nucleus' of the parasitic amoeba *Paramoeba perniciosa* (Amoebida, Paramoebidae). *Journal of Invertebrate Pathology* 17, 186–193.

Perkins, F.O. and Menzel, R.W. (1967) Ultrastructure of sporulation in the oyster pathogen *Dermocystidium marinum. Journal of Invertebrate Pathology* 9, 205–229.

Perkins, F.O. and Wolf, P.H. (1976) Fine structure of *Marteilia sydneyi* sp.n. – haplosporidan pathogen of Australian oysters. *Journal of Parasitology* 62, 528–538.

Pichot, Y. (1986) Sporulation of *Haplosporidium* sp. (Haplosporida, Haplosporidiidae) in the flat oyster *Ostrea edulis* L. in the Bay of Arcachon (France). *European Aquaculture Society, Special Publications* 9, 119–126 (in French, with English abstract).

Pichot, Y., Comps, M., Tigé, G., Grizel, H. and Rabouin, M.A. (1980) Research on *Bonamia ostreae* gen. n., sp.n., a new parasite of the flat oyster *Ostrea edulis* L. *Revue des Travaux de l'Institut des Pêches Maritimes.* 43, 131–140 (in French).

Pixell Goodrich, H. (1956) Crayfish epidemics. *Parasitology* 46, 480–483.

Polglase, J.L. (1980) A preliminary report on the thraustochytrid(s) and labyrinthulid(s) associated with a pathological condition in the lesser octopus *Eledone cirrhosa. Botanica Marina* 23, 699–706.

Powell, E.N., Gauthier, J.D., Wilson, E.A., Nelson, A., Fay, R.R. and Brooks, J.M. (1992) Oyster disease and climate change. Are yearly changes in *Perkinsus marinus* parasitism in oysters (*Crassostrea virginica*) controlled by climatic cycles in the Gulf of Mexico. *Pubblicazioni della Stazione Zoologica de Napoli I: Marine Ecology* 13, 243–270.

Pregenzer, C. (1983) Survey of metazoan symbionts of *Mytilus edulis* (Mollusca: Pelecypoda) in Southern Australia. *Australian Journal of Marine and Freshwater Research* 34, 387–396.

Ragone, L.M. and Burreson, E.M. (1993) Effect of salinity on infection progression and pathogenicity of *Perkinsus marinus* in the eastern oyster, *Crassostrea virginica* (Gmelin). *Journal of Shellfish Research* 12, 1–7.

Ray, S.M. (1966a) A review of the culture method for detecting *Dermocystidium marinum*, with suggested modifications and precautions. *Proceedings of the National Shellfisheries Association* 54, 55–69.

Ray, S.M. (1966b) Cycloheximide: inhibition of *Dermocystidium marinum* in laboratory stocks of oysters. *Proceedings of the National Shellfisheries Association* 56, 31–36.

Rogier, H., Hervio, D., Boulo, V., Clavies, C., Hervaud, E., Bachère, E., Mialhe, E., Grizel, H., Pau, B. and Paolucci, F. (1991) Monoclonal antibodies against *Bonamia ostreae* (Protozoa: Ascetospora), an intrahaemocytic parasite of flat oyster *Ostrea edulis* (Mollusca: Bivalvia). *Diseases of Aquatic Organisms* 11, 135–142.

Rosenfield, A., Buchanan, L. and Chapman, G.B. (1969) Comparison of the fine structure of spores of three species of *Minchinia* (Haplosporida, Haplosporidiidae). *Journal of Parasitology* 55, 921–941.

Roubal, F.R., Masel, J. and Lester, R.J.G. (1989) Studies on *Marteilia sydneyi*, agent of QX disease in the Sydney rock oyster, *Saccostrea commercialis*, with implications for its life cycle. *Australian Journal of Marine and Freshwater Research* 40, 155–167.

Sanders, M.J. and Lester, R.J.G. (1981) Further observations on a bucephalid trematode infection in scallops (*Pecten alba*) in Port Phillip Bay, Victoria. *Australian Journal of Marine and Freshwater Research* 32, 475–478.

Sato-Okoshi, W. and Nomura, T. (1990) Infestation of the Japanese scallop *Patinopecten yessoensis* by the boring polychaetes *Polydora* on the coast of Hokkaido and Tohoku district. *Nippon Suisan Gakkaishi* 56, 1593–1598 (in Japanese, with English summary).

Scheibling, R.E. and Stephenson, R.L. (1984) Mass mortality of *Strongylocentrotus droebachiensis* (Echinodermata: Echinoidea) off Nova Scotia, Canada. *Marine Biology* 78, 153–164.

Scheltema, R.S. (1962) The relationship between the flagellate protozoon *Hexamita* and the oyster *Crassostrea virginica*. *Journal of Parasitology* 48, 137–141.

Shchepkina, A.M. (1982) On pathogenic influence of larval trematodes on *Mytilus galloprovincialis*. In: Burukovskii, R.N. (ed.) *Problemy Ratsional'nogo Ispol'zovaniya Promyslovykh Bespozvonochnykh. Tezisy Dokladov III Vsesoyuznoi Konferentsii*, 12–16 October 1982. AtlantNIRO, Kaliningrad, pp. 229–231 (in Russian).

Sherburne, S.W. and Bean, L.L. (1986) A synopsis of the most serious diseases occurring in maine shellfish. *Fish Health Section, American Fisheries Society Newsletter* 14, 5.

Sherburne, S.W. and Bean, L.L. (1991) Mortalities of impounded and feral Maine lobsters, *Homarus americanus* H. Milne-Edwards, 1837, caused by the protozoan ciliate *Mugardia* (formerly *Anophrys* = *Paranophrys*), with initial prevalence data from ten locations along the Maine coast and one offshore area. *Journal of Shellfish Research* 10, 315–326.

Sindermann, C.J. (1988a) Ciliate (*Paranophrys*) disease of Dungeness crabs. In: Sindermann, C.J. and Lightner, D.V. (eds) *Disease Diagnosis and Control in North American Marine Aquaculture*. Elsevier, Amsterdam, pp. 220–221.

Sindermann, C.J. (1988b) Ciliate (*Paranophrys*) disease of lobsters. In: Sindermann, C.J. and Lightner, D.V. (eds) *Disease Diagnosis and Control in North American Marine Aquaculture*. Elsevier, Amsterdam, pp. 258–259.

Sindermann, C.J. (1990) *Principal Diseases of Marine Fish and Shellfish. Volume 2, Diseases of Marine Shellfish*, 2nd edn. Academic Press, San Diego.

Sindermann, C.J. (1993) Disease risks associated with importation of nonindigenous marine animals. *Marine Fisheries Review* 54, 1–10.

Sindermann, C.J. and Lightner, D.V. (eds) (1988) *Disease Diagnosis and Control in North American Marine Aquaculture*. Elsevier, Amsterdam.

Singhas, L.S., West, T.L. and Ambrose, W.G. (1993) Occurrence of *Echeneibothrium* (Platyhelminthes, Cestoda) in the calico scallop *Argopecten gibbus* from North Carolina. *Fishery Bulletin* 91, 179–181.

Small, E.B. and Lynn, D.H. (1985) Phylum Ciliophora Doflein, 1901. In: Lee, J.J., Hutner, S.H. and Bovee, E.C. (eds) *Illustrated Guide to the Protozoa*. Society of Protozoologists, Lawrence, Kansas, pp. 393–575.

Somers, I.F. and Kirkwood, G.P. (1991) Population ecology of the grooved tiger prawn, *Penaeus semisulcatus*, in north-western Gulf of Carpentaria, Australia: growth, movement, age structure and infestation by the bopyrid parasite *Epipenaeon ingens*. *Australian Journal of Marine and Freshwater Research* 42, 349–367.

Sparks, A.K. (1985) *Synopsis of Invertebrate Pathology Exclvsive of Insects*. Elsevier, Amsterdam.

Sparks, A.K. and Chew, K.K. (1966) Gross infestation of the littleneck clam, *Venerupis staminea*, with a larval cestode (*Echeneibothrium* sp.). *Journal of Invertebrate Pathology* 8, 413–416.

Sparks, A.K. and Hibbits, J. (1981) A trematode metacercaria encysted in the nerves of the Dungeness crab, *Cancer magister*. *Journal of Invertebrate Pathology* 38, 88–93.

Sparks, A.K., Hibbits, J. and Fegley, J.C. (1982) Observations on the histopathology of a systemic ciliate (*Paranophrys* sp.?) disease in the Dungeness crab, *Cancer magister*. *Journal of Invertebrate Pathology* 39, 219–228.

Sprague, V. (1963) Revision of genus *Haplosporidium* and restoration of genus *Minchinia* (Haplosporidia, Haplosporidiidae). *Journal of Protozoology* 10, 263–266.

Sprague, V. (1979) Classification of the Haplosporidia. *Marine Fisheries Review* 41, 40–44.

Sprague, V., Beckett, R.L. and Sawyer, T.K. (1969) A new species of *Paramoeba* (Amoebida, Paramoebidae) parasitic in the crab *Callinectes sapidus*. *Journal of Invertebrate Pathology* 14, 167–174.

Stein, J.E., Denison, J.G. and Mackin, J.G. (1959) *Hexamita* sp. and an infectious disease in the commercial oyster *Ostrea lurida*. *Proceedings of the National Shellfisheries Association* 50, 67–81.

Steinbeck, J.R.S., Groft, J.M., Friedman, C.S., McDowelly, T. and Hedrick, R.P. (1989) A coccidian-like protozoan from the California black abalone *Haliotis creacherodii*. In: *First International Symposium on Abalone Biology, Fisheries, and Culture*, 21–25 November 1989, La Paz, Mexico, pp. 23. (Abstract.)

Svärdh, L. and Thulin, J. (1985) The parasite fauna of natural and farmed *Mytilus edulis* from the west coast of Sweden, with special reference to *Renicola roscovita*. *Meddelanden från Havsfiskelaboratoriet Lysekil* 312, 1–16.

Taylor, R.L. (1966) *Haplosporidium tumefacientis* sp.n., the etiologic agent of a disease of the California sea mussel, *Mytilus californianus* Conrad. *Journal of Invertebrate Pathology* 8, 109–121.

van Banning, P. (1979) Haplosporidian diseases of imported oysters, *Ostrea edulis*, in Dutch estuaries. *Marine Fisheries Review* 41, 8–18.

van Banning, P. (1985a) *Minchinia armoricana* disease of the flat oyster. In: Sindermann, C.J. (ed.), *Fiches d'Identification des Maladies et Parasites des Poissons, Crustacés et Mollusques*. Conseil International pour l'Exploration de la Mer, Copenhagen, 17, 1–4.

van Banning, P. (1985b) Control of *Bonamia* in Dutch oyster culture. In: Ellis, A.E. (ed.), *Fish and Shellfish Pathology*. Academic Press, London, pp. 393–396.

van Banning, P. (1987) Further results of the *Bonamia ostreae* challenge tests in Dutch oyster culture. *Aquaculture* 67, 191–194.

van Banning, P. (1990) The life cycle of the oyster pathogen *Bonamia ostreae* with a presumptive phase in the ovarian tissue of the European flat oyster, *Ostrea edulis*. *Aquaculture* 84, 189–192.

van Banning, P. (1991) Observations on bonamiasis in the stock of the European flat oyster, *Ostrea edulis*, in the Netherlands, with special reference to the recent developments in Lake Grevelingen. *Aquaculture* 93, 205–211.

Villalba, A., López, M.C., Mourelle, S.G. and Carballal, M.J. (1990) Assessment of the effects of the infection by *Marteilia* sp. on the reproduction of mussel, *Mytilus galloprovincialis* LMK., in estuaries of Galicia (NW of Spain). In: Figueras, A. (ed.) *Abstracts, Fourth International Colloquium on Pathology in Marine Aquaculture*, 17–21 Sept. 1990, Vigo (Pontevedra), Spain, pp. 76.

Vivarès, C.P., Papayianni, P. and Boemare, N. (1986) Ecopathological study of host-parasite relationships between *Hexamita nelsoni* Schlicht and Mackin, 1968 and Mediterranean oysters. *European Aquaculture Society, Special Publications* 9, 43–58. (In French, with English abstract).

Vivarès, C.P., Papayanni, P. and Quiot, J.M. (1987) *In vivo* and *in vitro* study of the pathogenic effect of *Hexamita nelsoni* Schlicht and Mackin 1968, on oysters. *Aquaculture* 67, 165–170 (in French with English abstract).

Wardle, W.J. (1980) On the life cycle stages of *Proctoeces maculatus* (Digenea: Fellodistomidae) in mussels and fishes from Galveston Bay, Texas. *Bulletin of Marine Science* 30, 737–743.

Warner, R.W. and Katkansky, S.C. (1969) Infestation of the clam *Protothaca staminea* by two species of tetraphyllidian cestodes (*Echeneibothrium* spp.). *Journal of Invertebrate Pathology* 13, 129–133.

Weir, G.O. and Sullivan, J.T. (1989) A fluorescence screening technique for microsporida in histological sections. *Transactions of the American Microscopical Society* 108, 208–210.

Wharton, W.G. and Mann, K.H. (1981) Relationship between destructive grazing by the sea urchin, *Strongylocentrotus droebachiensis*, and the abundance of American lobster, *Homarus americanus*, on the Atlantic coast of Nova Scotia. *Canadian Journal of Fisheries and Aquatic Sciences* 38, 1339–1349.

White, M.E., Powell, E.N., Ray, S.M. and Wilson, E.A. (1987) Host-to-host transmission of *Perkinsus marinus* in oyster (*Crassostrea virginica*) populations by the ectoparasitic snail *Boonea impressa* (Pyramidellidae). *Journal of Shellfish Research* 6, 1–5.

Whyte, S.K., Cawthorn, R.J. and McGladdery, S.E. (1994) Co-infection of bay scallops *Argopecten irradians* with *Perkinsus karlssoni* (Apicomplexa, Perkinsea) and an unidentified coccidian parasite. *Diseases of Aquatic Organisms* 18, 53–62.

Wolf, P.H. (1979a) Diseases and parasites in Australian commercial shellfish. *Haliotis* 8, 75–83.

Wolf, P.H. (1979b) Life cycle and ecology of *Marteilia sydneyi* in the Australian oyster, *Crassostrea commercialis*. *Marine Fisheries Review* 41, 70–72.

Wolf, P.H., Winstead, J.T. and Couch, J.A. (1987) *Proctoeces* sp. (Trematode: Digenea) in Australian oysters, *Saccostrea commercialis* and *Crassostrea amasa*. *Transactions of the American Microscopical Society* 106, 379–380.

Wood, J.L. and Andrews, J.D. (1962) *Haplosporidium costale* (Sporozoa) associated with a disease of Virginia oysters. *Science* 136, 710–711.

17

The Piscine Immune System: Innate and Acquired Immunity

W.B. van Muiswinkel

*Department of Experimental Animal Morphology and Cell Biology,
Wageningen Agricultural University, PO Box 338, 6700 AH Wageningen,
The Netherlands.*

INTRODUCTION

Vertebrates are distinguished from invertebrates by an internal skeleton of cartilage or bone. The subphylum Vertebrata includes the jawless fish (Agnatha) such as hagfish and lamprey, the Placodermi which is the earliest group of the jawed fish, the cartilaginous fish (Chondrichthyes) such as sharks and rays, the bony fish (Osteichthyes) such as sturgeon, trout, carp and perch, the amphibians, the reptiles, the birds and the mammals. Fish are the oldest animal group with an immune system showing clear similarities with the defence systems of mammals and birds. The cellular and humoral responses have the expected characteristics of specificity and memory. Moreover, the appearance of immunoglobulin (Ig) as well as other members of the 'Ig superfamily' is observed for the first time in this animal group. Classic immunoglobulin molecules are lacking in the invertebrate phyla. This chapter provides a general overview of the defence mechanisms in fish. Most data are derived from bony fish, but some interesting differences with cartilaginous or jawless fish will be discussed. It is known from fossil remains that the earliest fish appeared some 350–400 million years ago. This implies that unique specializations in the defence systems of certain fish species may have developed over the years.

NON-LYMPHOID DEFENCE MECHANISMS

Epithelial barriers

The first line of defence includes structures which form stable physical and/or chemical barriers against invading microorganisms. The epithelial surfaces

729

(e.g. skin, gills and gut) are examples of these barriers. It is of prime importance for the fish to maintain the integrity of covering epithelia because they are important in defence and for osmoregulation. Hence wound healing is a remarkably rapid process in fish. Normal epithelia are covered by a mucous layer which is secreted by goblet cells. The most important function of mucus is to prevent the attachment of bacteria, fungi or parasites to epithelial surfaces (Pickering and Richards, 1980).

Transferrin

In addition to mucosal factors, several other serum factors have been detected and have been shown to be protective in fish. One of them is the iron binding protein, transferrin. This glycoprotein limits the amount of free iron in the bloodstream, thus making it unavailable for bacteria during infection. The various transferrin genotypes in coho salmon (*Oncorhynchus kisutch*) were correlated with degrees of resistance to bacterial kidney disease. The difference in disease resistance was suggested to lie in the avidity of the transferrin genotypes for iron (Suzumoto *et al.*, 1977). However, Houghton *et al.* (1991) did not observe any influence of the transferrin genotype on the resistance to *Aeromonas salmonicida* in carp (*Cyprinus carpio*).

Lectins

Lectins (or natural agglutinins) in fish can be detected as natural precipitins or agglutinins. They are usually cross-linking carbohydrate moieties on the surface of xenogeneic erythrocytes or microorganisms. They are presumed to be important in neutralizing bacterial components (e.g. exotoxins) or in immobilizing microorganisms and hence will facilitate phagocytosis (Fletcher, 1982). Fish lectins are not structurally related to Ig, but resemble plant or invertebrate agglutinins. Some lectins have also been found in fish eggs (Wiens, personal communication).

C-reactive protein

In teleost fish, C-reactive protein (CRP) is a serum component which increases rapidly upon exposure to bacterial endotoxins (Ingram, 1980) or experimental infection with bacterial pathogens (Murai *et al.*, 1990). CRP reacts with phosphorylcholine molecules at the cell surface of microorganisms. It has lectin-like properties and can act as an opsonin to enhance phagocytosis or to activate the complement system in a manner similar to Ig (White *et al.*, 1981). CRP from rainbow trout was recently isolated and characterized as a 66 kDa glycoprotein which contains two protein subunits (Murai *et al.*, 1990). The CRP from channel catfish may have five subunits (Szalai, personal communication).

Interferon

Fish interferon (IFN) is a protein which is produced during viral infections. It increases the resistance of cells to different viruses (De Kinkelin and Dorson, 1973). IFN in teleosts is species specific, e.g. IFN produced by rainbow trout does not protect cyprinid cells *in vitro*. *In vivo* synthesis of IFN during a viral infection peaks after 2–3 days and usually precedes the virus neutralizing effects of circulating antibodies, which appear 1 or 2 weeks later (De Kinkelin *et al.*, 1977). It is interesting to mention that type 1 and type 2 IFN can be distinguished in rainbow trout based upon acid stability (pH 2) and relative temperature resistance (60°C) (Secombes, 1991).

Complement

The complement system consists of a group of protein and non-protein components which are involved in both innate defence mechanisms and in specific adaptive immunity. The complement system can be activated along two major routes: (i) the classical pathway, which is stimulated by antigen-antibody immune complexes; (ii) the alternative pathway, which is started by contact with certain microbial cell wall polysaccharides. In both cases the activation results in the opsonization and/or lysis of foreign cells. Day *et al.* (1970) demonstrated that most classes of fishes including Agnatha possess a lytic complement system. More recently, Nanoka *et al.* (1981, 1984) isolated C3 and C5 from rainbow trout plasma and C3 from lamprey plasma. Yano *et al.* (1986) have shown that C1-C4 are present in carp plasma. These studies suggest that almost all the mammalian complement factors (C1-C9) are present in fish blood and that both pathways are operating in fish. Woo (1992) has shown that the alternative pathway is the protective mechanism against haemoflagellate parasites (*Cryptobia*) in naive fish. The classical pathway turned out to be important in acquired immunity after survival of parasitic infections. It is tempting to speculate which pathway predominates in fish. Yano *et al.* (1988) did suggest that the alternative pathway in fish is more active than in mammals (Fig. 17.1). However, this may also depend on the ambient temperature (season), age or condition of the animals.

Inflammation

Inflammation is a local reaction brought about by high numbers of granulocytes and macrophages. It confers some degree of protection by 'walling off' an infected area from the rest of the body. Histopathological studies in fish provide evidence for inflammatory responses in bacterial, viral, fungal, protozoan and metazoan parasitic infections (Roberts, 1978; van Muiswinkel and Jagt, 1984) . Acute inflammation responses in bony fish are comparable with those in mammals (Finn and Nielsen, 1971). Granulocyte infiltration appears 12–24 hours after injection of bacteria or Freund's complete adjuvant in

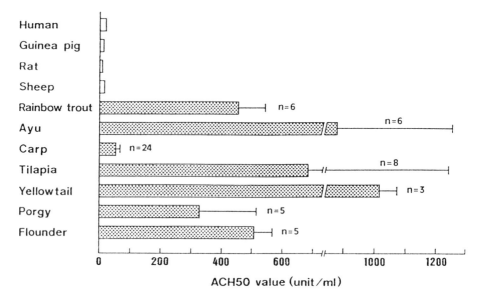

Fig. 17.1. Alternative complement pathway activity (ACH50 value) in mammals and bony fish. (From Yano *et al.*, 1988, with permission.)

rainbow trout. The infiltrating cells (granulocytes and macrophages) increase in numbers until day 2–4. Necrotic tissue is later replaced with fibrous tissue.

Phagocytic cells

Phagocytosis of antigenic material by macrophages is not only an activity of the non-specific innate defence system but can also be an initial step in the specific adaptive immune response. As in mammals, we are probably dealing with subpopulations of mononuclear phagocytes which differ in function. In this respect it is interesting to mention that macrophages from immune fish are more active in phagocytosis than those from control animals. This is probably due to opsonization of the antigen by antibodies or due to metabolic activation of the macrophages (Griffin, 1983). Sakai (1984) has even suggested that salmonid macrophages have Fc and C3 receptors on their surface facilitating the binding and subsequent phagocytosis of opsonized material. Most macrophages from the hindgut of carp are binding purified immunoglobulin which is an indication for Fc receptor on these cells (Koumans-Van Diepen and Rombout, personal communication).

Non-specific cytotoxic cells

Studies in channel catfish reveal the presence of non-specific cytotoxic cells (NCC) in these bony fish (Graves *et al.* 1984). The monocyte-like NCC show

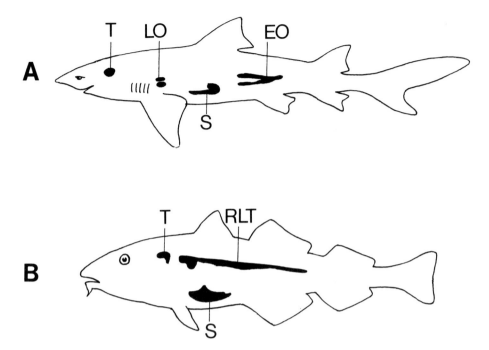

Fig. 17.2. The main lymphoid organs in cartilaginous fish (A) and bony fish (B). EO = epigonal organ; LO = Leydig organ; RLT = renal lymphoid tissue (pronephros and mesonephros); S = spleen; T = thymus. (From Fänge, 1982, slightly modified.)

a clear lytic activity against certain transformed mammalian cell lines *in vitro*. So far, the exact biological function of these cells remains unclear. They may be involved in killing protozoan parasites.

LYMPHOID CELLS AND ORGANS

Lymphocyte heterogeneity (T- and B-cells) has been demonstrated in hapten-carrier studies (Stolen and Mäkela, 1975), by using monoclonal antibodies (Secombes et al., 1983; DeLuca *et al.*, 1983) and by functional tests for cell cooperation (Miller *et al.* 1985, 1987). The main lymphoid organs in cartilaginous fish (Fig. 17.2) are thymus, Leydig organ, epigonal organ, kidney and spleen (Fänge, 1982). There are also indications for substantial gut-associated tissue in these animals (Tomonaga *et al.*, 1986). Teleosts do not have a Leydig organ nor epigonal tissue. However, thymus, head kidney (pronephros), trunk kidney (mesonephros), spleen and intestine contain high numbers of leucocytes (Fänge, 1982; Rombout *et al.*, 1986). Some leucocytes are also found in skin and gills. Bone marrow, bursa of Fabricius, Peyer's patches and lymph nodes, which are present in birds and/or mammals are not found in fish. Most observations indicate that the spleen of bony fish is

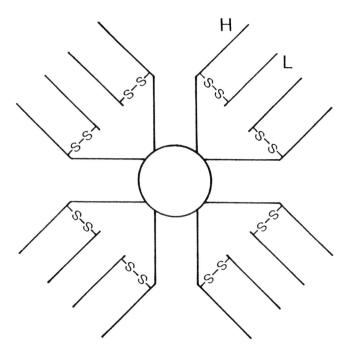

Fig. 17.3. The tetrameric serum immunoglobulin of bony fish. H = heavy chain; L = light chain; S-S = disulphide bridges between H and L.

an erythropoietic and secondary lymphoid organ, whereas the thymus is a primary lymphoid organ. The kidney is probably analogous to mammalian bone marrow (Zapata, 1979; Lamers, 1985). Therefore, it may function as a primary organ (B-lymphocyte differentiation) but also as a secondary organ (memory cell and plasma cell development).

ANTIGEN RECOGNITION AND PRESENTATION

Ig structure

The major Ig in bony fish consists of heavy (H) chains and light (L) chains and hence is similar to that in other vertebrates. Figure 17.3 shows the native Ig molecule which is a tetramer with four structural units (H_2L_2) 4. It contains 4×2 antigen binding sites and has a molecular mass of approximately 700 kDa (Table 17.1). The molecule is usually called IgM because of its high molecular weight and polymeric structure. Moreover, the amino acid sequence of the four constant (C) domains in the H chain shows a 24% homology with the mouse μ chain (Ghaffari and Lobb, 1989a). Interesting enough, the VH genes of channel catfish (Ghaffari and Lobb, 1989b) or rainbow trout (Matsunaga *et al.*, 1990) show much higher amino acid sequence identity

Table 17.1. The immunoglobulin molecule in cartilaginous and bony fish.

Animal group	Molecular mass (kDa)			
	Native	H-chain	L-chain	Formula
Cartilaginous fish	850–908	69–77	20–23	$(H_2L_2)_5$
	160–198	69–77	20–23	H_2L_2
Bony fish	608–900	60–77	22–26	$(H_2L_2)_4$

(45–60%) with mammals than the C domain genes. In other words the antigen binding part of the Ig molecule is better conserved in evolution than the 'so-called' constant part of the same molecule.

Ig isotypes

Most authors agree that cartilaginous fishes (e.g. sharks) have both pentameric and monomeric serum Ig (Clem and Small, 1967; Frommel *et al.*, 1971). In both Ig types the H chain corresponds with the mammalian μ chain. No real proof for the existence of other isotypes homologous to mammalian γ, α, δ or ε have been found in fish. On the other hand, convincing evidence for the existence of IgM isotypes were found in molecular studies on the C genes of sharks (Kokubu *et al.*, 1987) or in biochemical studies on the Ig H chains in teleosts (Lobb and Olson, 1988). Moreover, the existence of a separate mucosal Ig (sub)class in bile and mucus of carp can not be excluded when the results of Rombout *et al.* (1993) are taken into account. They developed monoclonal antibodies specific for mucosal Ig H chains which are staining skin epithelium and bile capillaries in histological sections. At least two distinct L chain types (F and G) were found in channel catfish. This was based upon differences in molecular weight, antigenic structure and peptide mapping (Lobb *et al.*, 1984).

Antibody repertoire

The mechanism by which antibody diversity is generated in mammals is well known and involves recombination of various Ig gene segments (V, D, J and C) during differentiation from haemopoietic stem cell to B lymphocyte (Tonegawa, 1983). Studies in sharks show a remarkable Ig gene organization (Hinds and Litman, 1986). In contrast to mammals, we see high numbers ($\geq$200) of closely linked clusters of V, D, J and C segments in genomic DNA of these marine fish (Fig. 17.4). Inter-cluster rearrangements are not thought to occur during B-cell development, which will limit antibody diversity in these 'primitive' fish. In bony fish the organization of V, D, J and C segments is almost identical to those in mammals (Fig 17.4). There are at least 150 V, two or more D, at least six J and a single C_μ region (Ghaffari and Lobb, 1991;

A. CARTILAGINOUS FISH (SHARKS)

$$-V_1 D_1 D_1 J_1 C_\mu\text{———}V_2 D_2 D_2 J_2 C_\mu\text{———}V_{\geqslant 200} D_{200} D_{200} J_{200} C_\mu\text{———}$$

B. BONY FISH

$$-V_1 V_2 \cdots V_{\geqslant 150}\text{———}D_1 D_2 \cdots D_?\text{———}J_1 J_2 J_3 J_4 J_5 J_6\text{———}C_\mu\text{———}$$

C. MAMMALS

$$-V_1 V_2 \cdots V_{300}\text{———}D_1 D_2 \cdots D_{12}\text{———}J_1 J_2 J_3 J_4\text{———}C_\mu C_\gamma C_\delta C_\epsilon C_\alpha\text{———}$$

Fig. 17.4. Schematic presentation of Ig heavy chain loci in germline DNA of cartilaginous fish, bony fish and mammals. V = variable gene segment; D = diversity gene segment; J = joining gene segment; C = constant gene segment. The V, D, J and C gene segments are recombined during B cell development in bony fish and mammals. In cartilaginous fish this process has taken place already at the early germline stage.

Matsunaga and Tömänen, 1990) which means that there are numerous possibilities for recombination during B cell development in bony fish. However, an Ig class switch (change of H chain) has not been observed. This may explain why somatic hypermutation and the subsequent selection of high affinity B-cell clones is restricted in bony fish (Du Pasquier, 1982).

T cell receptor

We know from molecular studies in mammals that Ig and T cell receptors (TCR) are related protein molecules which are characterized by an extreme variation in antigen binding sites based upon rearrangements of V, J, C and sometimes D region gene segments in the genome of early B or T cells (Hood *et al.*, 1985). The information on TCR in fish is very limited at this time. Marchalonis and Schluter (1989) isolated three DNA clones from a goldfish genomic library by using a modified human TCRβ probe. Amino acid sequencing of the clones showed a 40% homology with the corresponding Vβ region of man and some homology with mammalian IgL chains. It is expected that more information on the TCR genes and the protein structure will become available in the near future.

Major histocompatibility complex

Studies in mammals and birds have shown that the gene products of the major histocompatibility complex (MHC) play a key role in the regulation of the immune response (Klein, 1986). The MHC incorporates a group of closely linked genes which show a high degree of polymorphism. They code for membrane glycoproteins, which can be divided into class I and II. Class I molecules are present on all nucleated cells, whereas class II molecules are more or less restricted to cells of the immune system, i.e. lymphocytes and antigen-presenting macrophages. These MHC molecules also play an important role in the development of the T cell repertoire (self tolerance) and in antigen presentation (Fig. 17.5). Until recently there was only indirect evidence for the presence of a MHC in fish. This idea was based upon the existence of an integrated immune system characterized by the cooperation of different cell types during humoral and cellular immune responses. More precise information on the MHC is now available based on studies on genes and gene products.

Research at the gene product or protein level

One of the prerequisites for this approach is the availability of antisera recognizing only MHC molecules. These antisera are produced by immunizing an animal with the cells of another, which differs at the MHC. Recent development of homozygous fish lines is very valuable in this respect because specific combinations of donor and recipient animals can be chosen based upon MHC differences. Using this approach Kaastrup *et al.* (1989) were able to show MHC class I-like molecules on the surface of erythrocytes and lymphocytes of carp. These allo-antibodies can also be used for purification and analysis of fish MHC molecules or for isolation of DNA clones from cDNA libraries.

Research at the gene or DNA level

Using the polymerase chain reaction (PCR) Hashimoto *et al.* (1990) were able to demonstrate that the genome of carp contains nucleotide sequences which show considerable homology with MHC class I and II sequences in man and mice. The new DNA probes, which became available from these elegant studies, can be used for MHC typing of cyprinid fish populations and may allow further studies on the MHC of fish and other vertebrates. Restriction fragment length polymorphisms (RFLPs) have been identified recently in the MHC of carp using class I and class II specific DNA probes (Stet *et al.*, 1993). It is expected that more details on the structure and function of fish MHC molecules will be published in the near future (for review see also Stet and Egberts, 1991).

THE HUMORAL AND CELLULAR RESPONSE

Cell cooperation

The specific response in fish shows the expected characteristics of specificity and memory. At the start of the humoral (antibody) response it takes some time before the first specific antibodies appear in the circulation. This lag phase is needed for antigen processing and cell cooperation between distinct leucocyte populations (accessory cells, B and T cells). Accessory cells (monocytes and macrophages) process different antigens and present the processed

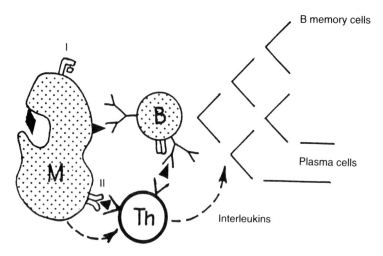

Fig. 17.5. Schematic diagram of cell interactions during the humoral response in vertebrates. Specialized macrophages (M) are able to trap and process foreign molecules or particles, i.e. antigen (♦). These macrophages will present relative small antigenic determinants (▶) associated with MHC class II molecules to lymphoid cells. Subsequently specific T helper cells (Th) are activated by interaction with the antigenic determinant and factors secreted by the macrophages (interleukin-1). The activated Th cells stimulate the differentiation and proliferation of effector cells as B lymphocytes (B) and cytotoxic T cells (not shown in this scheme) by secretion of different factors (e.g. interleukin-2). Depending on the circumstances, B cells will develop into long-lived B memory cells or short-lived plasma cells. These plasma cells secrete huge amounts of specific antibodies (immunoglobulins) which will bind or kill invading microorganisms showing the corresponding determinant. < = proliferation: ---> = interleukins: ⌐⌐ = MHC class I molecule: ⋎ = B cell receptor (immunoglobulin): ⋔ = MHC class II molecule: ⋎ = T cell receptor.

antigenic determinants in association with MHC class II molecules to lymphocytes (Fig. 17.5). We know from mammalian studies that the TCR on the cell membrane of a T helper (Th) cell is important for the recognition of the antigenic determinant. Also, other molecules such as CD3 and CD4 are essential as co-receptors. Homologues of these molecules await identification in fish. Activated macrophages will secrete interleukin-1 (IL-1) which is essential for the induction of the response by activating Th cells. Th cells regulate the proliferation and differentiation of B cells into antibody secreting plasma cells by producing IL-2 and other Th substances (Fig. 17.5) . Most of these B, Th and accessory cell functions have been verified by using monoclonal antibodies and functional *in vitro* tests for channel catfish (Miller *et al.*, 1985, 1987) or carp (Caspi and Avtalion, 1984; Grondel and Harmsen, 1984). However, except for the presence of Ig molecules on B cells, the precise nature of other membrane molecules on fish leucocytes, such as TCR, MHC class II, CD3 and CD4 needs further clarification.

Cooper, 1963). Avtalion and coworkers (1981) studied the effects of temperature on the antibody production in carp and *Tilapia* against bovine serum albumin and hapten-carriers. They showed that synthesis and release of antibody could take place at low temperatures ($\leq 12^\circ$C) if fish were kept at high temperatures (25°C) during the early phase of the response. It was suggested that antigen processing and subsequent cooperation between macrophages, Th and B cells is a temperature sensitive event which lasts 3–4 days in these warm-water fish. The temperature sensitivity of T cells was confirmed later by Miller and Clem (1984) who showed that low temperatures inhibit the generation of putative carrier-specific memory Th cells from virgin Th cells. The normal function of fish lymphocytes at different temperatures is highly dependent on homoviscous adaptation of membrane lipids (Abruzzini *et al.*, 1982). It is most likely that the fatty acid composition (unsaturated versus saturated) is a factor in determining the fluidity and permeability of membranes as well as the activity of membrane-associated receptors and enzymes. Sheldon and Blazer (1991) working with channel catfish at optimal (28°C) and suboptimal (19°C) temperatures observed a positive correlation between the bactericidal activity of macrophages and the level of highly unsaturated fatty acids in the diet. This opens new perspectives for the improvement of disease resistance of fish at lower temperatures.

Stress

It is known that stress conditions will influence the health of fish. Crowded blue gourami are releasing pheromone-like crowding factors which suppress the immune response to infectious pancreatic necrosis virus (Perlmutter *et al.*, 1973). Carp under crowded conditions show a high susceptibility to *Aeromonas* infections (Avtalion *et al.*, 1973). Numerous other studies suggest that prolonged stress causes lymphocyte depletion in peripheral blood and lymphoid organs (see for review Zapata *et al.*, 1992). High corticosteroid levels are usually present in the blood of stressed fish (Maule *et al.*, 1987). The correlation between corticosteroid levels and immune suppression is also found in other studies. In juvenile carp the application of corticosteroids supresses the Ab response to the ectoparasite *Ichthyophthirius multifiliis* (Houghton and Matthews, 1990). Injection of corticosteroids in immunized rainbow trout reduces the humoral response to a *Yersinia ruckeri* bacterin (Anderson *et al.*, 1982). The addition of cortisol or aldosterone to cultures of Coho salmon leucocytes reduces the mitogenic response to lipopolysaccharide or the Ab response to a hapten-carrier, TNP-LPS (Tripp *et al.* 1987). Recent studies in mammals demonstrate that the immunosuppressive effects of glucocorticosteroids are an example of the highly complicated interaction between the immune system and the endocrine system (Jankovic *et al.*, 1987). It is tempting to speculate that the neuroendocrine system in fish is as important in the regulation of the immune system as in mammals. This hypothesis is supported by elegant work showing that endogenous opioids are in part responsible for the immunesuppression in submissive *Tilapia* after confrontation with aggressive fish (Faisal *et al.*, 1989).

GENETIC ASPECTS AND DISEASE RESISTANCE

The identification of genes involved in the regulation of defence mechanisms might be important for understanding and perhaps improving disease resistance in fish. Studies with several mammalian species have shown that the products of the MHC genes play a key role in the regulation of the immune response (Klein, 1986). Moreover, an association has been established between certain MHC alleles and the susceptibility for specific diseases in birds and mammals (De Vries *et al.*, 1979; Svejgaard *et al.*, 1982). The increasing knowledge of the MHC in fish will certainly be important for the formulation of our ideas about regulation of the immune response in fish (see also pp. 736–738). Several examples of genetic differences in disease resistance in fish have been described (Chevassus and Dorson, 1990; Houghton *et al.*, 1991; Wiegertjes *et al.*, 1993), but well defined genetic markers have not been identified. It is tempting to speculate that typing of the MHC will provide us with some of these markers.

CONCLUSIONS

During the last 10–20 years considerable progress has been made in describing and understanding the immune system of fish. Antigenic stimulation in fish evokes responses which are comparable with those in warm-blooded vertebrates. For example, the humoral and cellular responses show the expected characteristics of specificity and memory. There are clear influences of environmental factors such as temperature, diet and stress conditions. Our knowledge of the immune system of fish can be used for evaluation of the health status of fish under different conditions, but can also be used for vaccination and breeding for disease resistance.

ACKNOWLEDGEMENTS

Part of this review paper was written during a sabbatical leave at the Immunology laboratory (Dr S.L. Kaattari), department of Microbiology, Oregon State University, Corvallis supported by a Fulbright grant from the Netherlands America Commission for Educational Exchange. The valuable comments from Drs G. Wiens and P.T.K. Woo are also gratefully acknowledged.

REFERENCES

Abruzzini, A.F., Ingram, L.O. and Clem, L.W. (1982) Temperature-mediated processes in teleost immunity: Homeoviscous adaptation in teleost lymphocytes. *Proceedings of the Society for Experimental Biology and Medicine* 169, 12–18.
Anderson, D.P., Roberson, B.S. and Dixon, O.W. (1982) Immunosuppression induced

by a corticosteroid or an alkylating agent in rainbow trout (*Salmo gairdneri*) administered a *Yersinia ruckeri* bacterin. *Developmental and Comparative Immunology*, Suppl. 2. 197–204.

Arkoosh, M.R. and Kaattari, S.L. (1991) Development of immnological memory in rainbow trout (*Oncorhynchus mykiss*). I. An immunochemical and cellular analysis of the B cell response. *Developmental and Comparative Immunology* 15, 279–293.

Avtalion, R.R. (1981) Environmental control of the immune response in fish. *Critical Reviews on Environmental Control* 11, 163–188.

Avtalion, R.R., Wojdani, A., Malik, Z., Sharabani, R. and Duczyminer, M. (1973) Influence of environmental temperature on the immune response in fish. *Current Topics in Microbiology and Immunology* 61, 1–35.

Bisset, K.A. (1948) The effect of temperature upon antibody production in cold-blooded vertebrates. *Journal of Pathology and Bacteriology* 60, 87–92.

Borysenko, M. and Hildemann, W.H. (1970) Reactions to skin allografts in the horned shark, *Heterodontus francisci. Transplantation* 10, 545–557.

Caspi, R.R. and Avtalion, R.R. (1984) Evidence for the existence of an IL-2 like lymphocyte growth promoting factor in bony fish, *Cyprinus carpio. Developmental and Comparative Immunology* 8, 51–66.

Chevassus, B. and Dorson, M. (1990) Genetics of resistance to disease in fishes. *Aquaculture* 85, 83–107.

Chiller, J.M., Hodgins, H.O. and Weiser, R.S. (1969) Antibody response in rainbow trout (*Salmo gairdneri*) II. Studies on the kinetics of development of antibody-producing cells and on complement and natural hemolysin. *Journal of Immunology* 102, 1202–1207.

Clem, L.W. and Small, P.A. (1967) Phylogeny of immunoglobulin structure and function I. Immunoglobulins of the lemon shark. *Journal of Experimental Medicine* 125, 893–920.

Day, N.K.B., Gewurz, H., Johannsen, R., Finstad, J. and Good, R.A. (1970) Complement and complement-like activity in lower vertebrates and invertebrates. *Journal of Experimental Medicine* 132, 941–950.

De Kinkelin, P. and Dorson, M. (1973) Interferon production in rainbow trout (*Salmo gairdneri*) experimentally infected with Egtved virus. *Journal of General Virology* 19, 125–127.

De Kinkelin, P., Baudouy, A.M. and Le Berre, M. (1977) Réaction de la truite fario (*Salmo trutta*, L. 1766) et arc-en-ciel (*Salmo gairdneri* Richardson, 1836) à l'infection par un nouveau rhabdovirus. *Comptes Rendus de l'Academie des Sciences Paris* 248, 401–404.

DeLuca, D., Wilson, M. and Warr, G.W. (1983) Lymphocyte heterogeneity in the trout, *Salmo gairdneri*, defined with monoclonal antibodies to IgM. *European Journal of Immunology* 13, 546–551.

De Vries, R.R.P., Meera Khan, P., Bernini, L.F., Van Loghem E. and Van Rood, J.J. (1979) Genetic control of survival to epidemics? *Journal of Immunogenetics* 6, 271–287.

Du Pasquier, L. (1982) Antibody diversity in lower vertebrates – Why is it so restricted? *Nature* 296, 311–313.

Ellis, A.E. (ed.) (1988) *Fish Vaccination*. Academic Press, London.

Faisal, M., Chiappelli, F., Ahmed, I.I., Cooper, E.L. and Weiner, H. (1989) Social confrontation 'stress' in aggressive fish is associated with an endogenous opioid-mediated suppression of proliferative response to mitogens and nonspecific cytotoxicity. *Brain, Behavior and Immunity* 3, 223–233.

Fänge, R. (1982) A comparative study of lymphomyeloid tissue in fish. *Developmental and Comparative Immunology* 6 (Suppl. 2), 23–33.

Finn, J.P. and Nielsen, N.O. (1971) The inflammatory response in rainbow trout. *Journal of Fish Biology* 3, 463–478.

Fletcher, T.C. (1982) Non-specific defence mechanisms of fish. *Developmental and Comparative Immunology* (Suppl. 2), 123–132.

Frommel, D., Litman, G.W., Finstad, J. and Good, R.A. (1971) The evolution of the immune response XI. The immunoglobulins of the horned shark, *Heterodontus francisci*: purification, characterization and structural requirements for antibody activity. *Journal of Immunology* 106, 1234–1243.

Ghaffari, S.H. and Lobb, C.J. (1989a) Cloning and sequence analysis of channel catfish heavy chain cDNA indicate phylogenetic diversity within the IgM immunoglobulin family. *Journal of Immunology* 142, 1356–1365.

Ghaffari, S.H. and Lobb, C.J. (1989b) Nucleotide sequence of channel catfish heavy chain cDNA and genomic blot analysis. Implications for the phylogeny of Ig heavy chains. *Journal of Immunology* 143, 2730–2739.

Ghaffari, S.H. and Lobb, C.J. (1991) Heavy chain variable region gene families evolved early in phylogeny: immunoglobulin complexity in fish. *Journal of Immunology* 146, 1037–1046.

Graves, S.S., Evans, D.L., Cobb, D. and Dawe, D.L. (1984) Non-specific cytotoxic cells in fish (*Ictalurus punctatus*) I. Optimum requirements for target cell lysis. *Developmental and Comparative Immunology* 8, 293–302.

Griffin, B.R. (1983) Opsonic effect of rainbow trout (*Salmo gairdneri*) antibody on phagocytosis of *Yersinia ruckeri* by trout leukocytes. *Developmental and Comparative Immunology* 7, 253–260.

Grondel, J.L. and Harmsen, E.G.M. (1984) Phylogeny of interleukins: growth factors produced by leukocytes of the cyprinid fish, *Cyprinus carpio* L. *Immunology* 52, 477–482.

Harrell, L.W. (1979) Immunization of fish in world mariculture: A review. *Proceedings of the World Mariculture Society* 10, 534–544.

Harrell, L.W., Etlinger, H.M. and Hodgins, H.O. (1975) Humoral factors important in resistance of salmonid fish to bacterial disease I. Serum antibody protection of rainbow trout (*Salmo gairdneri*) against vibriosis. *Aquaculture* 6, 211–219.

Hashimoto, K., Nakanishi, T. and Kurosawa, Y. (1990) Isolation of carp genes encoding major histocompatibility antigens. *Proceedings of the National Academy of Sciences, USA* 87, 6863–6867.

Hildemann, W.H. and Cooper, E.L. (1963) Immunogenesis of homograft reactions in fishes. *Federation Proceedings* 22, 1145–1151.

Hinds, K.R. and Litman, G.W. (1986) Major reorganization of immunoglobulin VH segmental elements during vertebrate evolution. *Nature* 320, 546–549.

Hood, L., Kronenberg, M. and Hunkapiller, T. (1985) T cell antigen receptors and the immunoglobulin supergene family. *Cell* 40, 225–229.

Houghton, G. and Matthews, R.A. (1990) Immunosuppression in juvenile carp, *Cyprinus carpio* L.: the effect of the corticosteroids triamcinolone acetonide and hydrocortisone 21-hemisuccinate (cortisol) on acquired immunity and the humoral antibody response to *Ichthyophthirius multifiliis* Fouquet. *Journal of Fish Diseases* 13, 269–280.

Houghton, G., Wiegertjes, G.F., Groeneveld, A. and Van Muiswinkel, W.B. (1991) Differences in resistance of carp, *Cyprinus carpio* L., to atypical *Aeromonas salmonicida*. *Journal of Fish Diseases* 14, 333–341.

Ingram, G. (1980) Substances involved in the natural resistance of fish to infection. A review. *Journal of Fish Biology* 16, 23–60.

Jankovic, B.D., Markovic, B.M. and Spector N.H. (eds) (1987) Neuroimmune interactions: proceedings of the second international workshop on neuroimmunomodulation. *Annals of the New York Academy of Sciences* Vol. 496.

Kaastrup, P., Stet, R.J.M., Tigchelaar, A.J., Egberts, E. and Van Muiswinkel, W.B. (1989) A major histocompatibility locus in fish: serological identification and segregation of transplantation antigens in the common carp (*Cyprinus carpio* L.). *Immunogenetics* 30, 284–290.

Klein, J. (1986) *Natural History of the Major Histocompatibility Complex*. John Wiley and Sons, New York.

Kokubu, F.K., Hinds, K., Litman, R., Shamblott, M.J. and Litman, G.W. (1987) Extensive families of constant region genes in a phylogenetically primitive vertebrate indicate an additional level of immunoglobulin complexity. *Proceedings of the National Academy of Sciences, USA* 84, 5868–5872.

Lamers, C.H.J. (1985) The reaction of the immune system of fish to vaccination. PhD Thesis, Agricultural University, Wageningen, The Netherlands.

Lamers, C.H.J., De Haas, M.J.H. and Van Muiswinkel, W.B. (1985) Humoral response and memory formation in carp after injection of *Aeromonas hydrophila* bacterin. *Developmental and Comparative Immunology* 9, 65–75.

Lobb, C.J. and Olson, M.O.J. (1988) Immunoglobulin heavy chain isotypes in a teleost fish. *Journal of Immunology* 141, 1236–1245.

Lobb, C.J., Olson, M.O. and Clem, L.W. (1984) Immunoglobulin light chain classes in a teleost fish. *Journal of Immunology* 132, 1917–1923.

Marchalonis, J.J. and Schluter, S.F. (1989) Immunoproteins in evolution. *Developmental and Comparative Immunology* 13, 285–301.

Matsunaga, T., Chen, T. and Törmänen, V. (1990) Characterization of a complete immunoglobulin heavy-chain variable region germline gene of rainbow trout. *Proceedings of the National Academy of Sciences, USA* 87, 7767–7771.

Matsunaga, T. and Törmänen V. (1990) Evolution of antibody and T-cell receptor V genes – the antibody repertoire might have evolved abruptly. *Developmental and Comparative Immunology* 14, 1–8.

Maule, A.G., Schreck, C.B. and Kaattari, S.L. (1987) Changes in the immune system of Coho salmon (*Oncorhynchus kisutch*) during parr-to-smolt transformation and after implantation of cortisol. *Canadian Journal of Fisheries and Aquatic Sciences* 44, 161–166.

Miller, N.W. and Clem, L.W. (1984) Temperature-mediated processes in teleost immunity: differential effects of temperature on catfish *in vitro* antibody responses to thymus-dependent and thymus-independent antigens. *Journal of Immunology* 133, 2356–2359.

Miller, N.W., Sizemore, R.C. and Clem, L.W. (1985) Phylogeny of lymphocyte heterogeneity: the cellular requirements for *in vitro* antibody responses of channel catfish leukocytes. *Journal of Immunology* 134, 2884–2888.

Miller, N.W., Bly, J.E., Van Ginkel, F.W. and Clem, L.W. (1987) Phylogeny of lymphocyte heterogeneity: identification and separation of functionally distinct subpopulations of channel catfish lymphocytes with monoclonal antibodies. *Developmental and Comparative Immunology* 11, 739–747.

Murai, T., Kodama, H., Naiki, M., Mikami, T. and Isawa, H. (1990) Isolation and characterization of rainbow trout C-reactive protein. *Developmental and Comparative Immunology* 14, 49–58.

Nanoka, M., Natsume-Sakai, S. and Takahashi, M. (1981) The complement system of rainbow trout (*Salmo gairdneri*) II. Purification and characterization of the fifth component (C5). *Journal of Immunology* 126, 1495–1498.

Nanoka, M., Iwaki, M., Nakai, C., Nozaki, M., Kaidoh, T., Nonaka, M., Natsuume-Sakai, S. and Takahashi, M. (1984) Purification of a major serum protein of rainbow trout (*Salmo gairdneri*) homologous to the third component of mammalian complement. *Journal of Biological Chemistry* 259, 6327–6333.

Pickering, A.D. and Richards, R.H. (1980) Factors influencing the structure, function and biota of the salmonid epidermis. *Proceedings of the Royal Society, Edinburgh* 79B, 93–104.

Perey, D.Y.E., Finstad, J., Pollara, B. and Good, R.A. (1968) Evolution of the immune response VI. First and second set homograft rejections in primitive fishes. *Laboratory Investigation* 19, 591–598.

Perlmutter, A., Sarot, D.A., Yu, M.L., Filazzola, R.J. and Seeley, R.J. (1973) The effect of crowding on the immune response of the blue gourami, *Trichogaster trichopterus*, to infectious pancreatic necrosis (IPN) virus. *Life Sciences* 13, 363–375.

Rijkers, G.T. (1980) The immune system of cyprinid fish. PhD thesis, Wageningen Agricultural University, The Netherlands.

Rijkers, G.T. (1982) Kinetics of humoral and cellular immune reactions in fish. *Developmental and Comparative Immunology* Suppl. 2, 93–100.

Rijkers, G.T. and Van Muiswinkel, W.B. (1977) The immune system of cyprinid fish. The development of cellular and humoral responsiveness in the rosy barb (*Barbus conchonius*). In: Solomon, J.B. and Horton, J.D. (eds) *Developmental Immunobiology*. Elsevier/North-Holland, Amsterdam, pp. 233–240.

Rijkers, G.T., Frederix-Wolters, E.M.H. and van Muiswinkel, W.B. (1980a) The immune system of cyprinid fish. Kinetics and temperature dependence of antibody producing cells in carp (*Cyprinus carpio*). *Immunology* 41, 91–97.

Rijkers, G.T., Frederix-Wolters, E.M.H. and van Muiswinkel, W.B. (1980b) The immune system of cyprinid fish. The effect of antigen dose and route of administration on the development of immunological memory in carp (*Cyprinus carpio*) In: Manning, M.J. (ed.) *Phylogeny of Immunological Memory*. Elsevier/North-Holland, Amsterdam, pp. 93–102.

Roberts, R.J. (ed.) (1978) *Fish Pathology*. Baillière Tindall, London.

Rombout, J.H.W.M., Blok, L.J., Lamers, C.H.J. and Egberts, E. (1986) Immunization of carp (*Cyprinus carpio*) with a *Vibrio anguillarum* bacterin: indications for a common mucosal immune system. *Developmental and Comparative Immunology* 10, 341–351.

Rombout, J.H.W.M. and Van den Berg, A.A. (1989a) Immunological importance of the second gut segment of carp. I. Uptake and processing of antigens by epithelial cells and macrophages. *Journal of Fish Biology* 35, 13–22.

Rombout, J.H.W.M., Bot, H.E. and Taverne-Thiele, J.J. (1989b) Immunological importance of the second gut segment of carp. III. Systemic and/or mucosal immune responses after immunization with soluble or particulate antigen. *Journal of Fish Biology* 35, 179–186.

Rombout, J.H.W.M., Taverne, N., Van de Kamp, M. and Taverne-Thiele, A.J. (1993) Differences in mucus and serum immunoglobulin of carp (*Cyprinus carpio* L.). *Developmental and Comparative Immunology* 17, 309–317.

Sakai, D.K. (1984) Opsonization by fish antibody and complement in immune phagocytosis by peritoneal exudate cells isolated from salmonid fish. *Journal of Fish Diseases* 7, 29–38.

Sailendri, K. and Muthukkaruppan, Vr. (1975) The immune response of the teleost, *Tilapia mossambica* to soluble and cellular antigens. *Journal of Experimental Zoology* 191, 371–381.

Secombes, C.J. (1991) The phylogeny of cytokines. In: Thomson, A.W. (ed.) *The Cytokine Handbook*. Academic Press, London, pp. 387–412.

Secombes, C.J., Van Groningen, J.J.M. and Egberts, E. (1983) Separation of lymphocyte subpopulations in carp, *Cyprinus carpio* L. by monoclonal antibodies: immunohistochemical studies. *Immunology* 48, 165–175.

Sheldon, W.M. and Blazer, V.S. (1991) Influence of dietary lipid and temperature on bactericidal activity of channel catfish macrophages. *Journal of Aquatic Animal Health* 3, 87–93.

Stet, R.J.M. and Egberts, E. (1991) The histocompatibility system in teleostean fishes: from multiple histocompatibility loci to a major histocompatibility complex. *Fish and Shellfish Immunology* 1, 1–16.

Stet, R.J.M., Van Erp, S.H.M., Hermsen, T., Sultmann, H.A. and Egberts, E. (1993) Polymorphism and estimation of the number of *MhcCyca* Class I and Class II genes in laboratory strains of the common carp (*Cyprinus carpio* L.) *Developmental and Comparative Immunology* 17, 141–156.

Stolen, J.S. and Mäkela, O. (1975) Carrier preimmunization in the anti-hapten response of a marine fish. *Nature* 254, 718–719.

Stroband, H.W.J. and Van der Veen, F.H. (1981) Localization of protein absorption during transport of food in the intestine of the grasscarp, *Ctenopharyngodon idella* (Val.). *Journal of Experimental Zoology* 218, 149–156.

Suzumoto, B.K., Schreck, C.B. and McIntyre, J.D. (1977) Relative resistances of three transferrin genotypes of Coho salmon (*Oncorhynchus kisutch*) and their hematological responses to bacterial kidney disease. *Journal of the Fisheries Research Board of Canada* 34, 1–8.

Svejgaard, A., Platz, P. and Ryder, L.P. (1982) HLA and disease: 1982 – a review. *Immunological Reviews* 70, 193–218.

Sypek, J.P. and Burreson, M. (1983) Influence of temperature on the immune response of juvenile summer flounder, *Paralichthys dentatus*, and its role in the elimination of *Trypanoplasma bullocki* infections. *Developmental and Comparative Immunology* 7, 277–286.

Tomonaga, S., Kobayashi, K., Hagiwara, K., Yamaguchi, K. and Awaya, K. (1986) Gut associated lymphoid tissue in elasmobranchs. *Zoological Science* 3, 23–29.

Tonegawa, S. (1983) Somatic generation of antibody diversity. *Nature* 302, 575–581.

Tripp, R.A., Maule, A.G., Schreck, C.B. and Kaattari, S.L. (1987) Cortisol mediated suppression of salmonid lymphocyte responses *in vitro*. *Developmental and Comparative Immunology* 11, 565–576.

van Muiswinkel, W.B. and Jagt, L.P. (1984) Host-parasite relationships in fish and other ectothermic vertebrates. *Developmental and Comparative Immunology* Suppl. 3, 205–208.

White, A., Fletcher, T.C., Pepys, M.B. and Baldo, B.A. (1981) The effect of inflammatory agents on C-reactive protein and serum amyloid P-component levels in plaice (*Pleuronectes platessa* L.) serum. *Comparative Biochemistry and Physiology* 69C, 325–329.

Wiegertjes, G.F., Daly, J.G. and Van Muiswinkel, W.B. (1993) Disease resistance of carp, *Cyprinus carpio* L.: identification of individual genetic differences by bath challenge with atypical *Aeromonas salmonicida*. *Journal of Fish Diseases* 16, 569–576.

Woo, P.T.K. (1992) Immunological responses of fish to parasitic organisms. *Annual Review of Fish Diseases* 2, 339-366.

Yano, T., Matsuyama, H. and Nakao, M. (1986) An intermediate complex in immune hemolysis by carp complement homologous to mammalian EAC1, 4. *Bulletin of the Japanese Society of Scientific Fisheries* 52, 281-286.

Yano, T., Hatayama, Y., Matsuyama, H. and Nakao, M. (1988) Titration of the alternative complement pathway activity of representative cultured fishes. *Nippon Suisan Gakkaishi* 54, 1049-1054.

Zapata, A. (1979) Ultrastructural study of the teleost fish kidney. *Developmental and Comparative Immunology* 3, 55-65.

Zapata, A.G., Varas, A. and Torroba, M. (1992) Seasonal variations in the immune system of lower vertebrates. *Immunology Today* 13, 142-147.

18

Immunological Approaches and Techniques

P.T. Thomas[1] and P.T.K. Woo[2]

[1]*Department of Zoology, U.C. College PO, Aluva 683 102, Kerala, India;*
[2]*Department of Zoology, University of Guelph, Guelph, Ontario, Canada N1G 2W1.*

INTRODUCTION

Parasitic and microbial diseases sometimes exact tremendous tolls on fish production especially when fish are under intensive culture. In recent years there have been concerted efforts made to better understand the piscine immune system especially as it relates to protection against infectious agents. In general, the piscine immune system is relatively well understood and it is in many respects similar to that in mammals (e.g. Woo and Jones, 1989; Faisal and Hetrick, 1992; Chapter 17, this volume). Some of the recent advances in the study of fish diseases include the prospects of using immunological strategies against a limited number of parasitic infections and the adaptation of immunological techniques to study the disease process (Woo, 1987, 1992).

The biology of pathogenic piscine parasites (e.g. taxonomy, systematics, development, transmission) has been relatively well studied (see earlier chapters); however, in many parasitic groups (e.g. Microsporidia, Monogenea, Nematoda, Arthropoda) we know little about immunological responses of fish to these parasites. This may partly be because parasitologists with more classical training do not feel comfortable about using immunological approaches and techniques. However, these techniques can help us to better understand the host–parasite relationship and to resolve phylogenetic relationships between species or groups of parasites. They may also be useful for serodiagnosis of infections, and for evaluating protection in recovered and immunized fish. We hope the present review will encourage more parasitologists to adapt immunological techniques to study their parasite systems.

We have summarized approaches that have been employed successfully to study immunity (innate and acquired) against parasites in fishes. Also, simple parasite isolation techniques are included because pure antigen(s) are often necessary for immunological approaches (e.g. used in serodiagnosis of infections, in parasite taxonomy and systematics).

INNATE IMMUNITY

Innate immunity refers to natural protection against parasitic organisms and this type of resistance does not require prior exposure to the parasite. Very little is known about innate immunity to parasites (Woo, 1992). The protection consists of humoral and/or cell-mediated components and techniques used are discussed in the following section.

Humoral response

Alternative pathway of complement activation

Freshly collected plasma of naturally resistant fish (resistant to a specific parasite) when incubated with the parasite often cause significant visible damage (e.g. lysis of protozoans and/or cytotoxic effects on metazoans) to the parasite. The damage may be due to activation of factor D (or properdir factor D) in the blood of resistant fish, by certain components (e.g. polysaccharides) on the parasite cell membrane. The activated properdin in combination with plasma factor B (properdin factor B) activates C3b to form the unstable C3bB complex. Properdin (P), another natural plasma component, reacts with C3bB to form the stable C3bBP. The C3bBP then cleaves C3 in the plasma into C3a and C3b fragments and the C3b fragments form more C3bB complexes which start the complement cascade (C5 to C9) with eventual lysis of cells containing the initiating polysaccharide (Saki, 1992). This is the alternative pathway of complement activation and it protects fish from some parasites.

The *in vitro* plasma incubation technique was described and used by Bower and Woo (1977) to study innate immunity due to the alternative pathway of complement activation. The haemoflagellate, *Cryptobia catostomi*, from infected white suckers, *Catostomus commersoni* was incubated with freshly collected fish plasma. The parasite was lysed in plasma of resistant fishes (e.g. goldfish, rainbow trout) while it was alive and active in plasma from susceptible fish. This *in vitro* technique confirmed the *in vivo* study where parasites were experimentally inoculated into fishes. The *in vitro* plasma incubation technique was also used to study innate immunity and to confirm the *in vivo* susceptibility or resistance of fishes to the pathogenic haemoflagellate, *Cryptobia salmositica*, a parasite in salmonids (Wehnert and Woo, 1980; Forward *et al.* 1995).

According to Wood and Matthews (1987), the plasma of uninfected mullet, *Chelon labrosus*, had cytotoxic effects on cercariae of a digenean, *Cryptocotyle lingua*. They concluded that the cytotoxic effects were caused by activation of complement via the alternative pathway. Similarly, Harvey and Meade (1969) found that sera from uninfected fish affected the survival of cercariae but not the metacercariae of *Posthodiplostomum minimum*.

Cell-mediated response

Nonspecific cytotoxicity

Nonspecific cytotoxic cells (or 'natural killer' cells) are also involved in innate resistance. The cells are normally collected from peripheral blood, the anterior kidney or the thymus using density gradient centrifugation in Percoll or Ficoll-Paque. The cells are then incubated with target cells, which are labelled with radioactive chromium (e.g.^{51}Cr-Na$_2$CrO$_4$). After incubation, the supernatant fluid is assayed for radioactivity using a gamma counter and radioactivity in the supernatant fluid indicates lysis of the labelled target cells.

Antiprotozoan activity of nonspecific cytotoxic cells was detected by Graves *et al.* (1984) in channel catfish, *Ictalurus punctatus*. The cells were from the anterior kidney and they caused up to 60% mortality of deciliated *Tetrahymena pyriformis*. Faisal *et al.* (1990) also used this technique on cytotoxic cells in arrowtooth flounder (*Atheresthes stomias*) infected with the parasitic copepod, *Phrixocephalus cincinnatus*.

Nonspecific lymphocyte proliferation

T-cell mitogens (e.g. phytohaemagglutinin and concanavalin A) and B-cell mitogens (e.g. lipopolysaccharide) induce *in vitro* proliferation of respective cell-lines. This results in increased DNA synthesis which is measured by the incorporation of radio-labelled DNA precursors (e.g. tritiated thymidine) in the cells. The cells are cultured in a suitable culture medium with the mitogen for 3–5 days. Radiolabelled DNA precursor is added to the cultures 24 hours before harvesting. The DNA of the harvested cells is precipitated and the radioactivity is measured in a liquid scintillation counter.

This technique was employed by Faisal *et al.* (1990) to study the responsiveness of lymphocytes in *A. stomias* infected with *Phrixocephalus cincinnatus*.

Respiratory burst activity of phagocytes

The phagocyte cell membrane is stimulated during phagocytosis. The stimulation results in respiratory burst activities which cause the production of highly reactive oxygen species such as O_2^- and H_2O_2. The reduction of the added ferricytochrome C in the system is used to demonstrate the production of O_2^- while horseradish peroxidase-dependent oxidation in phenol red is used to detect H_2O_2 production.

Whyte *et al.* (1989, 1990) used this technique to assay the activity of macrophages from rainbow trout against diplostomules of the eye fluke, *Diplostomum spathaceum*. Briefly, head kidney macrophage monolayers in microtitre plates were incubated with ferricytochrome C and parasite antigens (either whole or sonicated diplostomules). Superoxide dismutase was then added to convert the O_2^- to H_2O_2 and the optical densities were determined spectrophotometrically at 550 nm.

For detection of H_2O_2 production, head kidney macrophages were incubated in microtitre plates with phenol red containing horseradish peroxidase and parasite antigens (either whole or sonicated diplostomules). The reaction was stopped by the addition of NaOH and the optical densities were measured using a spectrophotometer at 620 nm.

ACQUIRED IMMUNITY

The exposure of fish to parasites usually provokes a detectable acquired response (humoral and/or cell-mediated) a few days after infection. The immune response is influenced by numerous factors and these include water temperature, type of and amount of antigen, route of exposure, age and species of fish.

Humoral response (polyclonal antibodies)

Antibodies produced in animals to infections are polyclonal because they react with most determinants of the parasite antigen(s). However, monoclonal antibodies are more specific and they react with only one antigenic determinant (see pp. 763–764). Numerous techniques are used to detect and to quantify specific polyclonal antibodies in antisera from animals. Detection of antibodies, which are secreted into blood and/or tissue fluids, is based on their reaction(s) with the parasite antigen(s).

Antigen–antibody reactions may result in: (i) clumping or agglutination of particulate antigens; (ii) precipitation of soluble antigens as antigen–antibody complexes; (iii) lysis of intact antigenic cells with complement-fixing antibodies; and/or (iv) an antigen-antibody reaction which is detected/measured indirectly. In the 'indirect' techniques, a fluorescent dye, or a radioactive chemical or an enzyme is first conjugated to the antibody. The degree of fluorescence or radiation or colour reaction (developed by adding the specific substrate to the enzyme-conjugated antibody) in the antigen-antibody complex is detected and measured using specialized equipment.

Agglutination tests

Parasite agglutination involves the incubation of intact parasites with serial dilutions of a specific antiserum and noting the highest antiserum dilution with agglutination. Various modifications of this basic technique (e.g. absorption of soluble parasite antigens on various carriers) are used routinely for diagnostic and/or screening purposes in mammalian parasitic diseases (Houba, 1980).

Clark *et al.* (1987) demonstrated agglutinating antibodies in channel catfish, *Ictalurus nebulosus*, that had recovered from infection with the ciliate, *Ichthyophthirius multifiliis*. They used the infective tomites to titrate the antisera that came from immune fish. Agglutination of intact organisms (usually with protozoans) shows interaction between antibodies and surface antigen(s) of the parasite and it is used with either living or killed parasites. Killed parasites fixed in glutaraldehyde (2.5% in phosphate-buffered saline) ensure that excretory/secretory products from live parasites do not interfere with the reaction.

The indirect (passive) haemagglutination test is carried out after adsorption of soluble parasite antigens on to a carrier (e.g. red blood cells) and their subsequent agglutination by specific antibodies (Barrett, 1983). Indirect haemagglutination is also used as a diagnostic technique in numerous mammalian diseases (Houba, 1980; Walls and Schantz, 1986). This is a very sensitive serological procedure and can easily be performed either in test tubes or in microtitre plates; however, adequate controls (e.g. carrier with antigen but no antibody, carrier without antigen but with antibody) have to be included to ensure that the agglutination is specific to the parasite antigen. The technique has rarely been used in studying fish parasites and is one that should be considered in future studies.

Indirect haemagglutination was used to detect and measure antibody production in rainbow trout infected with *Cryptobia salmositica* (Jones and Woo, 1987). Thomas and Woo (1990a) also used this technique to follow antibody production in *C. salmositica*-infected trout which were on diets with no pantothenic acid or on low protein. McArthur (1978) demonstrated antibodies against the digenean *Telogaster opisthorchis* in the serum of infected New Zealand eel, *Anguilla australis schmidtii*, and in the gut mucus of *A. dieffenbachii*. Similarly, Wood and Matthews (1987) detected antibodies in the thick-lipped grey mullet, *Chelon labrosus* infected with *Cryptocotyle lingua*.

Direct haemagglutination of red cells (without adsorption of antigen on red cells) is used to assess the effects a parasitic infection has on the humoral response in the immune system of the host (e.g. immunodepression). Erythrocytes from sheep, horse or man (O-group) are normally used as the immunizing antigen. The infected animal is inoculated with a suspension of washed erythrocytes and after one or more booster doses, the antiserum from the host is collected and titrated against the homologous erythrocytes. The agglutinating titre is compared with its preimmunization titre and also those from uninfected control animals (same immunization protocol as infected animals). The difference between the two groups indicate changes in the ability to produce antibodies to a second antigen in infected animals.

Laudan *et al.* (1986a,b) used this approach to show the depression of antibody production in winter and summer flounders, *Pseudopleuronectes americanus* and *Paralichthys dentatus*, infected with the microsporidian, *Glugea stephani*. Similarly, Jones *et al.* (1986) demonstrated immunodepression in rainbow trout infected with *C. salmositica* while Thomas and Woo (1992a) used it to show that anorexia contributed to the immunodepression during salmonid cryptobiosis.

Neutralization test

Neutralization of the infectivity of parasites after their exposure to antisera is used to determine protective immunity due to antibodies. Briefly, parasites incubated in heat-inactivated antiserum are inoculated into a susceptible host. The receipient host is later examined for parasites. The test works well with parasites (e.g. blood flagellates) that multiply readily in the susceptible host. It is not generally suitable for metazoan parasites.

Although the test is normally used to detect protection due to antibodies it has also been used as a serodiagnostic test in some diseases (e.g. Fife, 1976). The neutralization activity of the antiserum is quantified with the serial dilutions of the antiserum.

The pathogenic *Trypanosoma danilewskyi* after incubation in heat inactivated plasma of recovered goldfish, *Carassius auratus*, was not infective when inoculated into goldfish (Woo, 1981). This confirmed the *in vivo* study which showed that the recovered goldfish were refractive to infection. In another study, Jones and Woo (1987) showed that infectivity of *C. salmositica* from the blood of infected rainbow trout was not affected by antisera from immune fish while the infectivity of culture forms was severely reduced after incubation in the same antisera. They suggested the difference was due to blockage of antigenic sites on the cell membrane of blood forms with trout serum components during infections; hence neutralizing antibodies were unable to attach to their antigenic sites on blood forms.

In vitro immune lysis test

The *in vitro* immune lysis test is a direct approach to detect protective antibodies (complement-fixing) in the blood of infected or immunized animals. The heat-inactivated antiserum is incubated, usually in microtitre wells, with living protozoa and complement. Lysis of the protozoan usually occurs after an hour and this can be determined when the microtitre plate is examined using an inverted microscope. The technique can also be used on metazoan parasites; however, a longer incubation time may be required before mortality and/or significant damage to the metazoan parasite is evident. As in other serological tests, it is used with serial dilutions of the antiserum.

Jones and Woo (1987) used this technique to demonstrate complement-fixing antibodies in the blood of *C. salmositica*-infected or recovered rainbow trout. The technique was employed to elucidate the mechanism of protection during cryptobiosis and the anamnestic response in fish vaccinated with a live attenuated *C. salmositica* vaccine (Li and Woo, 1994) and the effects of dietary ascorbic acid deficiency in protection against cryptobiosis (Li *et al.*, 1993).

Immunodiffusion tests

In immunodiffusion tests the precipitation of antibodies by soluble antigens is usually carried out in purified agarose gels. The assays include single and double diffusions in one dimension and double diffusions in two dimensions. Antigens and antibodies are either loaded in wells punched into gels or incorporated into gels. Precipitates are formed where there are optimal specific antigen-antibody concentrations. The double diffusion (Ouchterlony method) is particularly useful in that several different antigens and antisera are tested at the same time and information on non-identity, identity or partial identity of antigens determined. The immunoprecipitates can also be stained for protein, lipid or special enzyme activities (Barrett, 1983; Stites and Rodgers, 1991).

The single radial immunodiffusion can be used to quantify total immunoglobulin in infected animals. Briefly, antibodies against the immunoglobulin of the animal species is produced (see p. 760) and incorporated into agar or agarose in a petri dish (gel). Several dilutions of the antigen (serum from the infected animal) are placed into wells in the gel. The antigen diffuses out of each well to form a gradient. A precipitation ring forms at a distance where the antigen concentration is equivalence to the antibody in the gel. The area of the ring is directly proportional to the initial antigen concentration.

This technique can be used to determine immunoglobulin levels in blood during an infection. Laudan *et al.* (1986a,b, 1989) used it to show declines in IgM in both adult and juvenile winter flounders, *P. americanus* infected or injected with the microsporidian, *Glugea stephani*.

The sensitivity of this technique is increased by lowering the concentration of antiserum in the gel. This gives a wider ring because the antigen must reach a lower concentration to be at equivalence with the antiserum (Berzofsky *et al.*, 1989).

Cottrell (1977) demonstrated antibody response in plaice (*Pleuronectes platessa*) against the metacercariae of two digeneans, *Cryptocotyle lingua* and *Rhipidocotyle johnstonei* using the Ouchterlony method. Similarly, McArthur (1978) detected antibody present in the serum of the New Zealand eel, *Anguilla australis schmidtii* and in the gut mucus of *A. dieffenbachii* against *Telogaster opisthorchis* using this technique.

The immunodiffusion technique can and has been used to determine the origin (host animal) of a blood meal isolated from the gut of a blood-sucking invertebrate (Boreham and Gill, 1973). Khan (1991) used this approach to determine the host preference of the marine leech, *Johanssonia arctica* in the northwestern Atlantic Ocean.

Immunoassays using conjugated antibodies

Immunoassays are used for both qualitative and quantitative evaluations. In a qualitative assay, labelled antibody or antigen is used to visualize the immune reaction and the label is usually a fluorescein dye (fluorescent antibody techniques) or an enzyme (enzyme-linked immunosorbent assay – ELISA). It is used either as a direct or an indirect technique. In the direct technique, the specific antibody against the antigen under investigation is tagged with fluorescein or an enzyme and added to the suspected antigen which is on an immobilizing surface such as a microscope slide or microtitre plate or nitrocellulose membrane. The antigen–antibody reaction is then visualized using an appropriate microscope (for detecting fluorescence), or using an ELISA reader (when the assay is done in microtitre plates) or the naked eye (when the antigen is immobilized on nitrocellulose) to detect colour development after addition of the substrate specific to the conjugated enzyme.

In the indirect technique, the antigen is reacted with unlabelled specific antibody (first antibody). The second antibody, which is against the first antibody, is conjugated with fluorescein or enzyme, is then added to the antigen–first antibody complex and is examined for fluorescence or colour development after reaction with the appropriate substrate. The indirect method has the advantage over the direct method in that it is more sensitive, since the unlabelled antibody, while serving as antigen to the labelled antibody, provides more combining sites than the original antigen. Also, it significantly decreases the number of labelled antibodies that must be prepared. However, the indirect procedure has an obvious disadvantage because it requires more time and more reagents.

Fluorescent antibody techniques have been used as a screening and/or diagnostic tool in numerous mammalian parasitic diseases (Houba, 1980; Garcia and Bruckner, 1988). In fish diseases, Wolf and Markiw (1985) used the direct fluorescent antibody technique for the identification of *Myxosoma cerebralis*, the aetiological agent of whirling disease in salmonids. Markiw (1989) also used the technique to study the antigenic relatedness of two stages in the life cycle of the parasite – spores from fish cartilage and the triactinomyxon stage from the gut of the aquatic tubificid, *Tubifex tubifex*. The technique was also used for the detection of *Labyrinthuloides haliotidis*, a protozoan parasite of abalone, *Haliotis kamtschatkana* (see Bower *et al.*, 1989).

The indirect fluorescent antibody technique was used to detect prespore stages and spores of *M. cerebralis* within sections of rainbow trout cartilage (Hamilton and Canning, 1988). This technique was also employed by Bartholomew *et al.* (1989a) to determine host immune response against the myxosporean parasite, *Ceratomyxa shasta*. Dickerson *et al.* (1989) used it to localize membrane-associated immobilization antigens on *Ichthyophthirius multifiliis*, and Woo (1990) used it to detect antibody response in rainbow trout to *Cryptobia salmositica*. Similarly, Whyte *et al.* (1987) used it to detect specific antibodies in rainbow trout against cercariae of *Diplostomum spathaceum*, and Boulo *et al.* (1989) used it as a diagostic technique for *Bonamia ostreae*, a pathogenic protozoan in flat oysters, *Ostrea edulis*.

Stables and Chappell (1986) adapted the immunoperoxidase technique (modified ELISA) of Secombes *et al.* (1983) for detction of antibodies in rainbow trout against cercariae of *D. spathaceum*.

Bortz *et al.* (1984) studied the primary and secondary antibody response in rainbow trout against metacercariae of *D. spathaceum* using the ELISA. Specific humoral antibodies against cercariae and diplostomules of *D. spathaceum* were found in antisera of infected rainbow trout (Whyte *et al.*, 1987). The technique was to screen supernatants of hybridomas for monoclonal antibodies against *Cryptobia salmositica* (Verity and Woo, 1993). ELISA was also used recently to follow immune responses to *Diphyllobothrium dendriticum*, to *Amyloodinium ocellatum*, to *Cryptobia* (as *Trypanoplasma*) *borreli*, and to *C. salmositica* in freshwater fishes (Sharp *et al.*, 1992; Smith *et al.*, 1992; Jones *et al.*, 1993; Sitja-Bobadilla and Woo, 1994). It was also used as a very reliable diagnostic technique for the detection of *B. ostreae* in oysters (Cochennec *et al.*, 1992).

Woo (1990) described and used MISET (microscopic immuno-substrate-enzyme technique) as a serodiagnostic technique in rainbow trout infected with *C. salmositica* (Fig. 18.1). He detected antibodies in about two weeks after infection. MISET is similar to the peroxidase technique (Stables and Chappell, 1986); it is however different in that Woo used a phosphatase-labelled goat anti-trout immunoglobulin to replace the second (rabbit anti-trout immunoglobulin) and third (conjugated goat anti-rabbit immunoglobulin) antibodies.

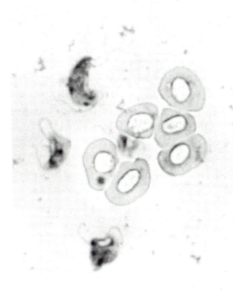

Fig. 18.1. Positive MISET reaction on *Cryptobia salmositica*; note that the flagella, nucleus, kinetoplast and parasite membrane are well stained. (From Woo, 1990; courtesy of *Journal of Parasitology*.)

Jones and Woo (1992) also found MISET to be a useful technique to show that the antigenic differences between three species of piscine trypanosomes were on the body surface. MISET is less expensive to use and seems to be more sensitive than IFAT (Woo, 1990). It can also be used to assist identification of parasites (e.g. metazoan parasites in tissue sections or protozoan parasites in vectors) when a specific first antibody (usually a monoclonal antibody) is available.

The dot-ELISA may be used to titrate either antigen or antibody. In this assay, serially diluted parasite antigen (antigen titration) or a suitable concentration of the antigen (antibody titration) are applied to nitrocellulose membranes or filters as dots. The dots are treated with the first and second antibodies (in antigen titration, a fixed dilution of first antibody and in antibody titration serially diluted antibody are used) and the reaction is visualized by adding the chromogenic substrate specific to the enzyme conjugated to the second antibody. The dot-ELISA is a very sensitive assay for detecting or quantifying antigen or antibody. Further, it needs less expensive equipment, is time saving and also provides a permanent record of the assay (Pappas, 1988). This technique has been used in the diagnosis of numerous parasitic diseases in mammals. Clark *et al.* (1988) used it to demonstrate antibody response to ciliary antigens of *Ichthyophthirius multifiliis* in fish. In this case the antigen was blotted onto a nitrocellulose filter. Specific anti-ciliary antibodies were quantified photometrically.

SDS-PAGE and Western immunoblot

A combination of SDS-PAGE (sodium dodecyl sulphate polyacrylamide gel electrophoresis) and Western immunoblot is a most useful approach to study the antibody production against a complex antigen (e.g. a parasite). The antigen is first separated into polypeptides during SDS-PAGE and the molecular masses of these can be determined based on simultaneous electrophoresis of protein standards of known molecular masses. The separated polypeptides are then blotted onto a nitrocellulose membrane and treated in the same way as for a dot-ELISA.

Dickerson *et al.* (1989) used this approach to resolve the ciliary membrane proteins of *Ichthyophthirius multifiliis*. Bartholomew *et al.* (1989a) employed the Western immunoblot to determine the host immune response against *Ceratomyxa shasta*.

This is potentially a very useful approach in parasite taxonomy especially in distinguishing morphologically similar parasite species from different hosts and from different geographical locations. Woo and Thomas (1991) found the polypeptide (using SDS-PAGE) and antigenic (Western immunoblot) profiles of three species of *Cryptobia* to be quite distinct. Similarly, Jones and Woo (1992) showed that this approach was also useful in distinguishing species of piscine trypanosomes.

Production of rabbit anti-fish immunoglobulin

Thomas and Woo (1988) produced rabbit anti-trout immunoglobulin in their study on the pathobiology of *Cryptobia salmositica* in rainbow trout. The procedure was adapted from Hodgins *et al.* (1965). Briefly, rainbow trout were inoculated intraperitoneally with 5% washed rabbit red blood cells (Rrbc) in Alserver's solution. A booster was given 40 days later, fish were bled for antisera 25 days after the booster and agglutinating antibody titre against Rrbc was determined using the direct haemagglutination test. After that the trout antisera were incubated with washed Rrbc for 2.5 hours, washed twice in Alserver's solution and Rrbc injected subcutaneously into a rabbit. A similar injection was given 35 days later; 14 days after the booster, the rabbit was bled and the antiserum collected.

The specificity of the rabbit antiserum to rainbow trout immunoblobulin was determined using a precipitation test (Aiyedun and Amodu, 1973). Briefly, antiserum from a rabbit was drawn into haematocrit capillary tubes containing antisera from Rrbc immunized fish or sera from non-immunized fish. The controls include pre-immunized serum from the same rabbit with either antisera from Rrbc immunized fish or sera from non-immunized fish. Tubes were supported vertically and precipitation was observed after 30 minutes only in tubes which contained both rabbit antiserum and immunized fish antiserum.

Anti-globulin test (Coombs' test)

A positive anti-globulin test indicates immune complexes and/or antibody on the surface of red cells. In a positive reaction, red cells agglutinate (Fig. 18.2) when they are mixed with specific antibodies against the immunoglobulin of the host species.

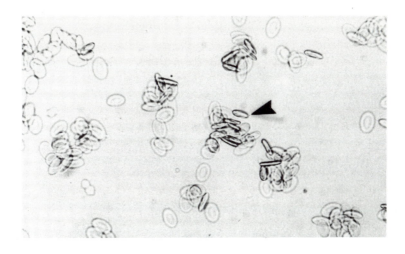

Fig. 18.2. Positive Coombs' reaction. Red blood cells with immune complexes on their surface are agglutinated (arrow) upon addition of anti-immunoglobulin antibody.

Using rabbit anti-trout immunoglobulin antiserum, Thomas and Woo (1988) showed that red cells from rainbow trout infected with *Cryptobia salmositica* were Coombs' positive (Fig. 18.2) 4 weeks after infection. The rabbit sera (preimmune serum and antiserum) were diluted (1 : 5) to avoid agglutination of trout red cells due to natural agglutinins in rabbit serum. Washed red cells (either from infected or uninfected fish) were incubated with the diluted rabbit serum (either preimmune serum or antiserum) for about 5 minutes at 20°C. A drop of the suspension was then examined under a microscope for red cell agglutination. Diluted preimmune rabbit serum should not agglutinate red cells (either from infected or uninfected fish) while rabbit antiserum would agglutinate cells if trout immunoglobulins were on their surface.

Haemolysis occurred when Coombs' positive red cells from infected fish were incubated in trout complement (Thomas and Woo, 1988). This confirmed that the red cells had immune complexes on their surface.

Complement level

Assays of complement involve either evaluation of the total haemolytic activity or that of individual components (Ruddy, 1986; Stites and Rodgers, 1991). Complement levels in many mammalian diseases, e.g. trypanosomosis, leishmaniosis, amoebosis, cryptosporidiosis, malaria, babesiosis, theileriosis, toxoplasmosis, schistosomosis, cysticercosis/hydatidosis, nippostrongylosis, filarosis, sarcoptidosis (Leid, 1988), are usually lowered.

Thomas and Woo (1989) showed depressed haemolytic activity of complement in rainbow trout infected with *Cryptobia salmositica* based on the activation of rainbow trout complement by rabbit erythrocytes.

Cell-mediated response

Cell-mediated immune responses in fish against parasites is little explored despite the well-documented occurrence of functional T lymphocytes and a number of lymphokines including the macrophage migration inhibition factor, mitogenic factors, chemotactic factors and interleukin 1 and 2 (Trust, 1986; Jurd, 1987; Pasquier, 1989).

Macrophage migration inhibition

The macrophage migration inhibition test is one that assays cell-mediated immune response *in vitro*. In this technique, a suspension of macrophages and lymphocytes is drawn into capillary tubes and the cells are packed by centrifugation. The tubes are then cut at the cell-fluid interface and the stubs with the packed cells are fixed to the bottom of tissue culture wells using vaseline. The antigen suspension is then added to the wells and incubated. Macrophages from an animal sensitized with the antigen do not migrate from the tube because of the release of a lymphokine, a macrophage migration inhibition factor, from the T-like lymphocytes. However, unsensitized macrophages migrate out of the tube as the antigen does not elicit the release of the lymphokine. No inhibition of migration occurs in sensitized macrophages when no antigen is added to the culture well. The macrophage migration inhibition test has been used in amoebosis and filarosis (Houba, 1980).

Thomas and Woo (1990b) used this technique to demonstrate cell-mediated immune response in rainbow trout against *Cryptobia salmositica* and they also showed depressive effects of deficient diets (in pantothenic acid or low protein) on cell-mediated immune response to cryptobiosis. These were confirmed using the *in vivo* delayed hypersensitivity reaction (see below).

Antibody-dependent cell-mediated cytotoxicity

The cytotoxic activity exhibited by immune effector cells as a result of their binding to specific antibody is termed antibody dependent cell-mediated cytotoxicity. The effector cells include macrophages and granulocytes.

Whyte *et al.* (1989) showed increased *in vitro* killing of diplostomules (metacarcariae) of the eye fluke, *Diplostomum spathaceum* by incubating them in microtitre plates with rainbow trout anti-diplostomule antiserum and activated macrophages in comparison with diplostomules opsonized with naive trout serum. This was confirmed in an *in vivo* study by inoculating fish with cercariae opsonized with either anti-cercarial or anti-diplostomule antisera. Fish inoculated with cercariae opsonized with naive trout serum (control) had significantly higher numbers of diplostomules (Whyte *et al.*, 1990).

Under *in vitro* conditions, Li and Woo (1994) showed antibody dependent and antibody independent cytotoxicity in trout immunized against cryptobiosis with a live attenuated vaccine.

Delayed type hypersensitivity reaction

The delayed type hypersensitivity reaction is an *in vivo* manifestation of cell-mediated immunity. Antigen is injected intradermally into the animal and

Collection of 'pure' helminth parasites is comparatively easier because of their large size and their usual location in lumens of organs. Similarly, ectoparasites (e.g. copepods) are also relatively easier to collect. Whyte *et al.* (1988) collected eggs, cercariae and diplostomules of *Diplostomum spathaceum* from the faeces of herring-gull, snails and rainbow trout respectively. Parasite eggs were purified by sieving the faeces. Snails infected in the laboratory with miracidia from the eggs were exposed to bright light to release cercariae. These were concentrated using a vacuum apparatus with a filter. Pieces of rainbow trout skin with the scales and subcutaneous muscle removed were kept in a two-chambered tube and cercariae in artificial pond water applied to the epidermal side of the skin in the upper chamber. The lower chamber had culture medium containing antibiotics. Cercariae penetrated the skin, become transformed to diplostomules and these were collected in the lower chamber. The diplostomules were washed with sterile medium and cultured *in vitro*. Cercariae and diplostomules were sonicated before using them in ELISA or other immunological assays as parasite antigen (Whyte *et al.*, 1987, 1989, 1990).

CONCLUSIONS

It is clear that the study of piscine immune responses against parasites is still in its infancy (Woo, 1992; present review) when compared to the wealth of knowledge about the immune responses against bacteria and viruses (Hennessen, 1981; Schill *et al.*, 1989; Hetrick, 1989). The involvement of complement as well as cell-mediated immune responses are, in particular, two areas that have been almost neglected. With the increased awarenesss of the similarities between the piscine immune system and that of higher vertebrates, a number of techniques (e.g. evaluation of various complement components, mitogen responses, graft versus host reaction, effect of thymectomy) may be profitably used to further our knowledge.

Similarly, there is ample scope for biochemical studies on fish parasites. Little is known about the metabolic pathways, and studies along this line will be rewarding. It is hoped that this article will stimulate and challenge workers to address many of these areas of study.

REFERENCES

Aiyedun, B.A. and Amodu, A.A. (1973) The capillary-tube agglutination (CA) test: a proposed aid for the diagnosis of trypanosomiasis in the field. *Zeitschrift für Tropenmedizin und Parasitologie* 24, 67–71.

Barrett, J.T. (1983) *Textbook of Immunology. An Introduction to Immunochemistry and Immunobiology.* 4th edn. The C.V. Mosby Company, St Louis, 520 pp.

Bartholomew, J.L., Smith, C.E., Rohovec, J.S. and Fryer, J.L. (1989a) Characterization of a host response to the myxosporean parasite, *Ceratomyxa shasta* (Noble) by histology, scanning electron microscopy and immunological techniques. *Journal of Fish Diseases* 12, 509–522.

Bartholomew, J.L., Rohovec, J.S. and Fryer, J.L. (1989b) Development, characterization and use of monoclonal and polyclonal antibodies against the myxosporean, *Ceratomyxa shasta*. *Journal of Protozoology* 36, 397–401.

Bartholomew, J.L., Yamamoto, T., Rohovec, J.S. and Fryer, J.L. (1990) Immunohistochemical characterization of monoclonal antibody against *Ceratomyxa shasta*. *Journal of Aquatic Animal Health* 2, 68–71.

Berzofsky, J.A., Epstein, S.L. and Berkower, I.J. (1989) Antigen-antibody interactions and monoclonal antibodies. In: Paul, W.E. (ed.) *Fundamental Immunology*. 2nd edn, Raven Press, New York, pp. 315–356.

Boreham, P.F.L. and Gill, G.S. (1973) Serological identification of reptile feeds of *Glossina*. *Acta Tropica* 30, 356–365.

Bortz, B.M., Kenny, G.E., Pauley, G.B., Garcia-Ortigoza, E. and Anderson, D.P. (1984) The immune response in immunized and naturally infected rainbow trout (*Salmo gairdneri*) to *Diplostomum spathaceum* as described by the enzyme-linked immunosorbent assay (ELISA). *Developmental and Comparative Immunology* 8, 813–822.

Boulo, V., Mialhe, E., Rogier, H., Paolucci, F. and Grizel, H. (1989) Immunodiagnosis of *Bonamia ostreae* (Ascetospora) infection of *Ostrea edulis* L. and subcellular identification of epitopes by monoclonal antibodies. *Journal of Fish Diseases* 12, 257–262.

Bower, S. and Woo, P.T.K. (1977) *Cryptobia catostomi*: incubation in plasma of susceptible and refractory fishes. *Experimental Parasitology* 43, 63–68.

Bower, S.M., Whitaker, D.J. and Elston, R.A. (1989) Detection of the abalone parasite *Labyrinthuloides haliotidis* by a direct fluorescent antibody technique. *Journal of Invertebrate Pathology* 53, 281–283.

Clark, T.G., Dickerson, H.W., Gratzek, J.B. and Findly, R.C. (1987) *In vitro* response of *Ichthyophthirius multifiliis* to sera from immune channel catfish. *Journal of Fish Biology* 31 (Suppl. A), 203–208..

Clark, T.G., Dickerson, H.W. and Findly, R.C. (1988) Immune response of channel catfish to ciliary antigens of *Ichthyophthirius multifiliis*. *Developmental and Comparative Immunology* 12, 581–594.

Cochennec, N., Hervio, D., Panatier, B., Boulo, V., Mialhe, E., Rogier, H., Grizel, H. and Paolucci, F. (1992) A direct monoclonal antibody sandwich immunoassay for detection of *Bonamia ostreae* (Ascetospora) in hemolymph samples of the flat oyster *Ostrea edulis* (Mollusca: Bivalvia). *Diseases of Aquatic Organisms* 12, 129–134.

Cottrell, B. (1977) The immune response of plaice *Pleuronectes platessa* L.) to the metacercariae of *Cryptocotyle lingua* and *Rhipidocotyle johnstonei*. *Parasitology* 74, 93–107.

Dickerson, H.W., Clark, T.G. and Findly, R.C. (1989) *Ichthyophthirius multifiliis* has membrane-associated immobilization antigens. *Journal of Protozoology* 36, 159–164.

Dungan, C.F. and Roberson, B.S. (1993) Binding specificities of mono- and polyclonal antibodies to the protozoan oyster pathogen *Perkinsus marinus*. *Diseases of Aquatic Organisms* 15, 9–22.

Ekejindu, G.O.C., Maikaje, D.B. and Ekejindu, I.M. (1986) The combined procedures of Ficoll-Paque centrifugation and anion exchange separation for the recovery of trypanosomes from blood and other tissues. *Transactions of the Royal Society of Tropical Medicine and Hygiene* 80, 879–882.

Faisal, M. and Hetrick, F.M. (eds) (1992) *Annual Review of Fish Diseases*. Vol. 2. Pergamon Press, New York, 403 pp.

Faisal, M., Perkins, P.S. and Cooper, E.L. (1990) Infestation by the pennellid copepod *Phrixocephalus cincinnatus* modulates cell-mediated immune responses in the Pacific arrowtooth flounder, *Atheresthes stomias*. In: Perkins, F.O. and Cheng, T.C. (eds) *Pathology in Marine Science*. Academic Press, San Diego, pp. 471–478.

Fife, E.H. (1976) Serodiagnosis of trypanosomiasis. In: Cohen, S. and Sadun, E. (eds) *Immunology of Parasitic Infections*. Blackwell Scientific Publications, Oxford. pp. 77–93.

Forward, G.M., Ferguson, M.M. and Woo, P.T.K. (1993) Susceptibility of brook charr, *Salvelinus fontinalis* to *Cryptobia salmositica* and the mechanism of innate immunity. *Bulletin, Canadian Society of Zoologists* 24(2), S1-SS (Abstract).

Galfre, G. and Milstein, C. (1981) Preparation of monoclonal antibodies: strategy and procedures. *Methods in Enzymology* 73, 1–46.

Garcia, L.S. and Bruckner, D.A. (1988) *Diagnostic Parasitology*. Elsevier, New York, 500 pp.

Goding, J.W. (1986) *Monoclonal Antibodies: Principles and Practice*, 2nd edn. Academic Press, London. 315 pp.

Graves, S.S., Evans, D.L., Cobb, D. and Dawe, D.L. (1984) Nonspecific cytotoxic cells in fish (*Ictalurus punctatus*) I. Optimal requirement for target cell lysis. *Developmental and Comparative Immunology* 7, 253–259.

Hamilton, A.J. and Canning, E.U. (1988) The production of mouse anti-*Myxosoma cerebralis* antiserum from Percoll-purified spores and its use in immunofluorescent labelling of Historesin-embedded cartilage derived from infected rainbow trout, *Salmo gairdneri* Richardson. *Journal of Fish Diseases* 11, 185–190.

Harvey, J.S., Jr. and Meade, T.G. (1969) Observations on the effects of fish serum on cercarial and metacercarial stages of *Posthodiplostomum minimum* (Trema toda: Diplostomidae). *Proceedings of the Helminthological Society of Washington* 36, 211–214.

Hennessen, W. (acting ed.) (1981) *Developments in Biological Standardization*. Vol. 49. S. Karger, Basel, 489 pp.

Hetrick, F.M. (1989) Fish viruses. In: Austin, B. and Austin, D.A. (eds) *Methods for the Microbiological Examination of Fish and Shellfish*. Ellis Horwood, England, pp. 216–239.

Hodgins, H.O., Ridgway, G.J. and Utter, F.M. (1965) Electrophoretic mobility of an immune globulin from rainbow trout serum. *Nature* 208, 1106–1107.

Houba, V. (ed.) (1980) *Immunological Investigation of Tropical Parasitic Diseases*. Churchill Livingstone, Edinburgh, 170 pp.

Humphryes, K.C. (1969) The separation of *brucei* subgroup trypanosomes from rat blood. *East African Trypanosomiasis Research Organization Annual Report*, pp. 32–36.

Jones, S.R.M. and Woo, P.T.K. (1987) The immune response of rainbow trout, *Salmo gairdneri* Richardson, to the haemoflagellate, *Cryptobia salmositica* Katz, 1951. *Journal of Fish Diseases* 10, 395–402.

Jones, S.R.M. and Woo, P.T.K. (1991) Culture characteristics of *Trypanosoma catostomi* and *Trypanosoma phaleri* from North American freshwater fishes. *Parasitology* 103, 237–243.

Jones, S.R.M. and Woo, P.T.K. (1992) Antigenic characterization of cultured trypanosomes isolated from three species of fishes. *Systematic Parasitology* 23, 43–50.

Jones, S.R.M., Woo, P.T.K. and Stevenson, R.M.W. (1986) Immunosuppresion in rainbow trout, *Salmo gairdneri* Richardon, caused by the haemoflagellate *Cryptobia salmositica* Katz, 1951. *Journal of Fish Diseases* 9, 431–438.

Jones, S.R.M., Palmen, M. and van Muiswinkel, W.B. (1993) Effects of inoculum route and dose on the immune response of common carp, *Cyprinus carpio* to the blood parasite, *Trypanoplasma borreli*. *Veterinary Immunology and Immunopathology* 36, 369–378.

Jurd, R.D. (1987) Hypersensitivity in fishes: a review. *Journal of Fish Biology* 31 (Suppl. A), 1–7.

Khan, R.A. (1991) Trypanosome occurrence and prevalence in the marine leech *Johanssonia arctica* and its host preferences in the northwestern Atlantic Ocean. *Canadian Journal of Zoology* 69, 2374–2380.

Lanham, S.M. and Godfrey, D.G. (1970) Isolation of salivarian trypanosomes from man and other mammals using DEAE-cellulose. *Experimental Parasitology* 28, 521–534.

Laudan, R., Stolen, J.S. and Cali, A. (1986a) The immune response of a marine teleost, *Pseudopleuronectes americanus*, (winter flounder) to the protozoan parasite, *Glugea stephani*. *Veterinary Immunology and Immunopathology* 12, 403–412.

Laudan, R., Stolen, J.S. and Cali, A. (1986b) Immunoglobulin levels of the winter flounder (*Pseudopleuronectes americanus*) and the summer flounder (*Paralichthys dentatus*) injected with the microsporidian parasite *Glugea stephani*. *Developmental and Comparative Immunology* 10, 331–340.

Laudan, R., Stolen, J.S. and Cali, A. (1989) The effect of the microsporida *Glugea stephani* on the immunoglobulin levels of juvenile and adult winter flounder (*Pseudopleuronectes americanus*). *Developmental and Comparative Immunology* 13, 35–41.

Leid, R.W. (1988) Parasites and complement. *Advances in Parasitology* 27, 131–168.

Li, S. and Woo, P.T.K. (1994) Efficacy of a live *Cryptobia salmositica* vaccine, and the mechanism of protection in vaccinated *Oncorhynchus mykiss* against cryptobiosis. *Veterinary Immunology and Immunopathology* (in press).

Li, S., Cowey, C. and Woo, P.T.K. (1993) The effects of dietary ascorbic acid deficiency on *Cryptobia salmositica* infection and on vaccination against cryptobiosis in *Oncorhynchus mykiss*. *Bulletin, Canadian Society of Zoologists* 24(2), 74–75 (Abstract).

Li, S., Verity, C. and Woo, P.T.K. (1994) Monoclonal and DNA probes in salmonid cryptobiosis. *VIII International Congress of Parasitology* 1,5 (Abstract).

McArthur, C.P. (1978) Humoral antibody production by New Zealand eels against the intestinal trematode *Telogaster opisthorchis* Macfarlane, 1945. *Journal of Fish Diseases* 1, 377–387.

Markiw, M.E. (1989) Salmonid whirling disease: myxosporean and actinosporean stages cross-react in direct fluorescent antibody test. *Journal of Fish Diseases* 12, 137–141.

Mialhe, E., Boulo, V., Elston, R., Hill, B., Hine, M., Montes, J., Banning, P. and Erizel, H. (1988) Serological analysis of *Bonamia* in *Ostrea edulis* and *Tiostrea lutaria* using polyclonal and monoclonal antibodies. *Aquatic Living Resources* 1, 67–69.

Pappas, M.G. (1988) Dot enzyme-linked immunosorbent assays. In: Collins, W.P. (ed.), *Complementary Immunoassays*. John Wiley and Sons, Chichester, pp. 113–134.

Pasquier, L.D. (1989) Evolution of the immune system. In: Paul, W.E. (ed.) *Fundamental Immunology*. Raven Press, New York, pp. 139–165.

Rogier, H., Hervio, D., Boulo, V., Clavies, C., Hervaud, E., Bachere, E., Mialhe, E., Grizel, H., Pau, B. and Paolucci, F. (1991) Monoclonal antibodies against *Bonamia ostreae* (Protozoa: Ascetospora), an intrahaemocytic parasite of flat oyster *Ostrea edulis* (Mollusca: Bivalvia). *Diseases of Aquatic Organisms* 11, 135–142.

Ruddy, S. (1986) Complement. In: Rose, N.R., Friedman, H. and Fahey, J.L. (eds) *Manual of Clinical Laboratory Immunology*. 3rd edn. American Society for Microbiology, Washington, D.C, pp. 175–184.

Sakai, D.K. (1992) Repertoire of complement in immunological defense mechanisms of fish. *Annual Review of Fish Diseases* 2, 223–247.

Schill, W.B., Bullock, G.L. and Anderson, D.P. (1989) Serology. In: Austin, B. and Austin, D.A. (eds) *Methods for the Microbiological Examination of Fish and Shellfish*. Ellis Horwood, Chichester, pp. 98–140.

Secombes, C.J., van Gronigen, J.J. and Egberts, E. (1983) Separation of lymphocyte subpopulations in carp *Cyprinis carpio* L. by monoclonal antibodies: immunohistochemical studies. *Immunology* 48, 165–175.

Sharp, G.J.E., Pike, A.W. and Secombes, C.J. (1992) Sequential development of the immune response in rainbow trout, [*Oncorhynchus mykiss* (Walbaum, 1792)] to experimental plerocercoid infections of *Diphyllobothrium dendriticum* (Nitzsch, 1824). *Parasitology* 104, 169–178.

Sitja-Bobadilla, A. and Woo, P.T.K. (1994) An enzyme-linked immunosorbent assay (ELISA) for the detection of antibodies against the pathogenic haemoflagellate, *Cryptobia salmositica* Katz and protection against cryptobiosis in juvenile *Oncorhynchus mykiss* (Walbaum) inoculated with a live *Cryptobia* vaccine. *Journal of Fish Diseases* 17, 399–408.

Smith, S.A., Levy, M.G. and Noga, E.J. (1992) Development of an enzyme-linked immunosorbent assay (ELISA) for the detection of antibody to the parasitic dinoflagellate *Amyloodinium ocellatum* in *Oreochromis aureus*. *Veterinary Parasitology* 42, 145–155.

Stables, J.N. and Chappell, L.H. (1986) Putative immune response of rainbow trout, *Salmo gairdneri*, to *Diplostomum spathaceum* infections. *Journal of Fish Biology* 29, 115–122.

Stites, D.P. and Rodgers, R.P.C. (1991) Clinical laboratory methods for detection of antigens and antibodies. In: Stites, D.P. and Terr, A.I. (eds) *Basic and Clinical Immunology*. 7th edn. Appleton and Lange, Norwalk, pp. 217–262.

Thomas, P.T. and Woo, P.T.K. (1988) *Cryptobia salmositica*: an *in vitro* and *in vivo* study on the mechanism of anaemia in infected rainbow trout, *Salmo gairdneri* Richardson. *Journal of Fish Diseases* 11, 425–431.

Thomas, P.T. and Woo, P.T.K. (1989) Complement activity in *Salmo gairdneri* Richardson infected with *Cryptobia salmositica* Sarcomastigophora: Kinetoplastida) and its relationship to the anaemia in cryptobiosis. *Journal of Fish Diseases* 12, 395–397.

Thomas, P.T. and Woo, P.T.K. (1990a) Dietary modulation of humoral immune response and anaemia in rainbow trout, *Oncorhynchus mykiss* (Walbaum), infected with *Cryptobia salmositica* Katz, 1951. *Journal of Fish Diseases* 13, 435–446.

Thomas, P.T. and Woo, P.T.K. (1990b) *In vivo* and *in vitro* cell-mediated immune responses of rainbow trout, *Oncorhynchus mykiss* (Walbaum), against *Cryptobia salmositica* Katz, 1951 (Sarcomastigophora: Kinetoplastida). *Journal of Fish Diseases* 13, 423–433.

Thomas, P.T. and Woo, P.T.K. (1992a) Anorexia in rainbow trout, *Oncorhynchus mykiss* (Walbaum), infected with *Cryptobia salmositica* (Sarcomastigophora: Kinetoplastida): its onset and contribution to the immunodepression. *Journal of Fish Diseases* 15, 443–447.

Thomas, P.T. and Woo, P.T.K. (1992b) *In vitro* culture and multiplication of *Cryptobia catostomi* and experimental infection of white sucker (*Catostomus commersoni*). *Canadian Journal of Zoology* 70, 201–204.

Thomas, P.T., Ballantyne, J.S. and Woo, P.T.K. (1992) *In vitro* oxygen consumption and motility of *Cryptobia salmositica, C. bullocki* and *C. catostomi* (Sarcomastigophora: Kinetoplastida). *Journal of Parasitology* 78 747–749.

Trust, T.J. (1986) Pathogenesis of infectious diseases of fish. *Annual Review of Microbiology* 40, 479–502.

Verity, C. and Woo, P.T.K. (1993) Antigenic characterization of *Cryptobia salmositica* using monoclonal antibodies and Western immunoblot. *Bulletin, Canadian Society of Zoologists* 24(2) 112–113.

Verity, C. and Woo, P.T.K. (1994) Characterization of a monoclonal antibody against *Cryptobia salmositica*: use of the antibody in an antigen-ELISA for diagnosis of cryptobiosis. *Bulletin, Canadian Society of Zoologists* 25(2), 106 (Abstract).

Walls, K.W. and Schantz, P.M. (eds) (1986) *Immunodiagnosis of Parasitic Diseases.* Vol. 1, Academic Press, Orlando, 301 pp.

Wehnert, S.D. and Woo, P.T.K. (1980) *In vivo* and *in vitro* studies on the host specificity of *Trypanoplasma salmositica. Journal of Wildlife Diseases* 16, 183–187.

Whyte, S.K., Allan, J.C., Secombes, C.J. and Chappell, L.H. (1987) Cercariae and diplostomules of *Diplostomum spathaceum* (Digenea) elicit an immune response in rainbow trout, *Salmo gairdneri* Richardson. *Journal of Fish Biology* 31 (Suppl. A), 185–190.

Whyte, S.K., Chappell, L.H. and Secombes, C.J. (1988) *In vitro* transformation of *Diplostomum spathaceum* (Digenea) cercariae and short-term maintenance of post-penetration larve *in vitro. Journal of Helminthology* 62, 293–302.

Whyte, S.K., Chappell, L.H. and Secombes, C.J. (1989) Cytotoxic reactions of rainbow trout, *Salmo gairdneri* Richardson, macrophages for larvae of the eye fluke *Diplostomum spathaceum* (Digenea). *Journal of Fish Biology* 35, 333–345.

Whyte, S.K., Chappell, L.H. and Secombes, C.J. (1990) Protection of rainbow trout, *Oncorhynchus mykiss* Richardson, against *Diplostomum spathaceum* (Digenea): the role of specific antibody and activated macrophages. *Journal of Fish Diseases* 13, 281–291.

Wolf, K. and Markiw, M.E. (1985) *Salmonid Whirling Disease.* Fish Disease Leaflet 69. US Fish and Wildlife Service. Cited in Amos, K.H. (ed.) 1985. *Procedures for the Detection and Identification of Certain Fish Pathogens.* 3rd edn. Fish Health Section, American Fisheries Society, Oregon.

Woo, P.T.K. (1979) *Trypanoplasma salmositica*: experimental infections in rainbow trout *Salmo gairdneri. Experimental Parasitology* 47, 36–48.

Woo, P.T.K. (1981) Acquired immunity against *Trypanosoma danilewskyi* in goldfish, *Carassius auratus. Parasitology* 83, 343–346.

Woo, P.T.K. (1987) Immune response of fish to protozoan infections. *Parasitology Today* 3, 186–188.

Woo, P.T.K. (1990) MISET: an immunological technique for the serodiagnosis of *Cryptobia salmositica* (Sarcomastigophora: Kinetoplastida) infection in *Oncorhynchus mykiss. Journal of Parasitology* 76, 752–755.

Woo, P.T.K. (1992) Immunological responses of fish to parasitic organisms. *Annual Review of Fish Diseases* 2, 339–366.

Woo, P.T.K. and Jones, S.R.M. (1989) The piscine immune system and the effects of parasitic protozoans on the immune response. In: Ko, R.C. (ed.) *Current Concepts in Parasitology.* Hong Kong University Press, Hong Kong, pp. 47–64.

Woo, P.T.K. and Thomas, P.T. (1991) Polypeptide and antigen profiles of *Cryptobia salmositica, C. bullocki* and *C. catostomi* (Kinetoplastida: Sarcomastigophora) isolated from fishes. *Diseases of Aquatic Organisms* 11, 201–205.

Woo, P.T.K., Leatherland, J.F. and Lee, M.S. (1987) *Cryptobia salmositica*: cortisol increases the susceptibility of *Salmo gairdneri* Richardson to experimental cryptobiosis. *Journal of Fish Biology* 10, 75–83.

Wood, B.P. and Matthews, R.A. (1987) The immune response of the thick-lipped grey mullet, *Chelon labrosus* (Risso, 1862), to metacercarial infections of *Cryptocotyle lingua* (Creplin, 1825). *Journal of Fish Biology* 31 (Suppl. A), 175–183.

Glossary

Abdomen: The part of the crustacean body that lies behind the genital segments; the last genital segment normally corresponds to the last thoracic segment.

Abiotic factors: Physical factors which affect the development/survival of an organism.

Abscess: A localized collection of pus in a cavity formed by disintegration of tissues.

Adenoma: A tumour which consists of glandular tissues.

Acanthella: An acanthocephalan larva that lacks a proboscis and develops from an acanthor in the intermediate host.

Acanthor: An acanthocephalan larva that develops from a zygote and is infective to the intermediate host.

Accessory filaments: Fibrillar structures, such as striated lamella, that may form part of the support in cells; accessory filaments are distinct from microtubules.

Aclid organ: A structure with large, tissue-cutting spines which is located anteriorly on the acanthor.

Acquired immunity: A series of host defences against a parasite which are characterized by extreme specificity and immunological memory; the defences are mediated by antibody and/or T cells.

Adhesion: A pathological condition which results in connective tissue proliferation within and around an organ.

Affinity: The strength of interaction or binding between an antibody binding site and an antigenic determinant.

Agglutinate: The aggregation of organisms or particles into clumps; this may be due to the reaction of specific antibodies against surface antigenic determinants on the organism.

Agglutinin: Any substance, not necessarily antibody, which is capable of forming bridges between antigenic determinants on contiguous cells to form visible clumps.

Alae: Lateral cuticular expansions along the body of some nematodes.

Alternative complement pathway: The activation of complement through involvement of properdin factor D, properdin factor B, C3b, C3 and then progressing as in the classical pathway.

Amastigote: The ovoid developmental stage (usually intracellular) of a trypanosome in which there is no emergent flagellum.

Anadromous: Fishes that migrate from the sea to fresh water to spawn.

Anaemia: A deficiency in blood or red blood cells.

Anchor: An outgrowth from the head of some female copepods; it is used for attachment to the host.

Anchoring disc: A laminar structure at the anterior end of a microsporidia spore.

Anorexia: The loss of appetite for food.

Anoxia: The absence of oxygen in tissues.

Antibody (Ab): A protein with the molecular properties of an immunoglobulin and capable of combination with an antigen; it is produced by lymphoid cells, especially plasma cells, in response to antigenic stimulation.

Antibody titre: A measure of the amount of antibody, usually in the blood of an animal.

Antigen: A substance or cell that elicits a specific immune response when introduced into an animal.

Antiglobulin test (Coombs' test): A technique to detect cell-bound immunoglobulin.

Antisera: Sera containing antibodies induced in animals against known antigens.

Aplanospore: The trophozoite or vegetative stage of the protozoa *Perkinsus* spp. and Labyrinthomorpha.

Appendix: A finger-like posterior projection of the ventriculus, common in some ascaridoid nematodes.

Ascites: The presence of fluid in the abdominal cavity.

Atrophy: A decrease in the amount of tissue or the size of an organ after normal growth has been achieved.

Attachment plate: A region of the trophont of a dinoflagellate which is modified for attachment to the host.

Autogamy: A kind of sexual reproduction in which the zygote is formed by fusion of two haploid nuclei from one individual.

Axial structure: A supporting structure aligned along the long axis of a cell.

Anoneme: The core of a cilium or flagellum which comprises the microtubule skeleton.

Axostyle: A rod-like supporting organelle composed of microtubules arranged in sheets or ribbons, e.g. in the Trichomonadida.

Basement membrane: The extracellular supporting layer of mucopolysaccharides and proteins underlying the epithelium.

Benthos: The bottom of the sea.

Binary fission: Asexual reproduction by an organism which results in two progenies.

Biomass: Mass of biological tissue commonly expressed in units of weight.

B lymphocyte (also B cells): A subpopulation of lymphocytes that develops into antibody producing cells after contact with an antigen; their development is independent of the thymus.

Borborygmus: A rumbling noise caused by gas in the intestine.

Boring tooth: A cuticular 'tooth'-like projection at the cephalic end of a larval nematode.

Branchiae: Leaf-like or finger-like lateral appendages on the urosome of some leeches.

Buccal: Pertaining to the oral cavity.

Buccal cavity: A pouch in ciliates equipped with compound ciliary organelles (*see* Cilium) and it precedes the cytostome.

Buccal ciliary apparatus: The specialized ciliature inside and around the buccal cavity in ciliates for food uptake.

Cachexia: The wasting of body tissues due to extreme malnutrition.

Caecum: An anterior projection of the intestine from the region of the oesophageal–intestine junction, common in some ascaridoids and spirurids.

Calcification: The deposition of calcium compounds occasionally associated with capsule formation.

Carcinoma: An epithelial tumour which tends to invade the lymph spaces of surrounding connective tissues.

Catadromous: Fishes that migrate from inland waters to the sea to spawn.

Cataract: The loss of transparency in the eye lens.

Catarrhal: The discharge of fluid due to inflammation of the mucous membrane.

Caudal cilium: An unusually long somatic cilium at the posterior pole of a ciliate.

cDNA: Copy DNA, made from isolated messenger RNA.

C domain: The constant homology region of an immunoglobulin molecule.

Cell-mediated immunity (also cellular immunity): Specific immune response which is dependent on T lymphocytes.

Cercariae: Free swimming larvae of digeneans; these stages are usually released from the gastropod intermediate host.

Chalimus: The modified copepodid stage of a copepod; the parasite attaches to its host by a filament.

Chemotherapy: The use of chemicals to control an infection.

Cholangitis: The inflammation of the bile duct.

Cholangiocarcinoma: Cancer of the bile duct.

Chondroblasts: Cartillage forming cells.

Chorioallantois: The vascular membrane which encloses the chick embryo.

Chromophores: Dermal cells which contribute to skin colour.

Cilium: A locomotory organelle rooted in a kinetosome, composed of a microtubular axoneme and covered by the outer cell membrane. The axoneme consists of nine peripheral doublets and a central pair of microtubules. Cilia may be grouped together in functionally specialized **compound ciliary organelles** (e.g. membranelles).

Cirrus: Coarse cilia which are longer than normal.

Classical complement pathway: The activation of complement components (in a sequential enzymatic cascade; C1, 4, 2, 3, 5, 6, 7, 8, 9) as a result of an antigen–antibody reaction which ultimately leads to cell lysis.

Clitellum: The portion of a leech which is located between the trachelosome and urosome; it secretes the cocoon.

Clone: A population which is derived from a single individual.

Club cells: Large, roundish or club-shaped cells in the epidermis of fishes. In carps and crucian carps they are the Leydig cells, in channel catfish they are the alarm substance cells and in eel and loach they are the clavate cells.

Coccoid: A globular shape structure that resembles a coccus bacterium.

Cocoon: A brooding capsule for leech eggs.

Coelozoic parasites: Parasites which occur in organ cavities.

Collagen: The principal proteinaceous substance in connective fibres.

Complement: A group of serum proteins that is the primary humoral mediator of antigen–antibody reactions and as opsonins in cellular immunity to enable leucocytes to phagocytose foreign particles.

Complement fixation: A serologic assay used for the detection of an antigen–antibody reaction in which complement is fixed as a result of the formation of an immune complex.

Conchiolin: A nitrogenous albuminoid substance which is usually dark brown in colour; it is the organic base of many molluscan shells.

Congested: An excessive or abnormal accumulation of blood in an organ or tissues.

Conjugation: A sexual process in ciliates, in which two individuals fuse; their micronuclei undergo meiotic division followed by exchange and fusion of the haploid nuclei in the conjugants.

Contractile vacuole (or water expulsion vesicle): The pulsating vacuole in a protozoan that has osmoregulatory and excretory functions.

Copepods: Small planktonic or bottom dwelling crustaceans that are important components of the food chain of aquatic animals.

Copepodid: The larval stage of a copepod; it is between the nauplius and the pre-adult stages.

Cornified: Containing fleshy fibrous tissue.

Corticosteroid: Any of the adrenal-cortex steroids.

Cortisol: The major natural glucocorticoid synthesized by the adrenal gland.

C-reactive protein (also **acute phase protein**): A serum protein which increases during acute inflammation; it has calcium-dependent phosphorylcholine binding properties.

C3 receptor: The cell surface receptor which is capable of selectively combining with the third component of complement.

Cyst: A non-motile, resistant, dormant stage.

Cystacanth: A juvenile acanthocephalan that develops from an acanthella; it possesses a proboscis, and is infective to the definitive host.

Cystogenous glands: Glands which produce encystment substances.

Cytopharynx: A tube (**cytopharyngeal tube**) which leads from the cytostome into the endoplasm of a protozoan.

Cytostome: An oral opening or aperture for ingestion of food particles.

Definitive host: The host in which the parasite usually undergoes sexual reproduction.

Delayed hypersensitivity: A cell-mediated immune response producing a cellular infiltrate and oedema that are maximal about 48 hours after antigen challenge.

Dermatitis: The inflammation of the skin.

Diadromous: The life history of fish with both freshwater and marine phases.

Diapedesis: The movement of haemocytes through intact epithelium (i.e. through the gut or mantle) and thus, are voided from the body.

Digitiform: Finger-like.

Dimorphic: Existing in two morphologically different forms.

Dinospore: A unicellular, haploid, motile, stage of dinoflagellate parasites that resembles a free-living dinoflagellate.

Diplokaryon: Two closely apposed nuclei with their membranes adhering to each other in a binucleated cell.

Diplomonad: A flagellate with one or two karyomastigonts, each with one to four flagella and accessory organelles; diplozoic individuals have bilateral symmetry.

Diplozoic: Having two karyomastigonts.

Disporic: A small sporogonic plasmodium or pseudoplasmodium which produce only one spore.

Disporoblastic sporogony: A process of spore production in which two sporoblasts arise from one sporont.

DNA probes: Segments of DNA from an organism which are used to identify homologous segments of DNA from another organism. These probes are usually used for rapid identification of parasites.

Ductus ejaculatorius: The functional penis in flatworms.

Dyplasia: The regressive change in adult cells.

Dysnea: Difficulty in breathing.

Ecchymosis: A small haemorrhagic spot, larger than a petechia, in the skin or mucous membrane.

Ecdysis: The shedding of the nematode cuticle following moulting.

Electrophoresis: The separation of proteins across an electrical gradient according to their mass and charge.

ELISA (Enzyme-linked immunosorbent assay): An immunological test to detect minute quantities of an antigen (antigen capture ELISA) or an antibody (antibody ELISA).

Emaciation: The deficiency of muscle and fat in an organism due to undernourishment.

Encapsulation: The covering of a parasite by the host with materials, mostly, if not entirely, of host origin.

Encephalomyelitis: The inflammation of the brain and the spinal cord.

Encystment: The covering of a parasite with materials of parasite origin.

Endocytosis: The uptake by a cell of material from the environment by invagination of its plasma membrane.

Endoplasmic reticulum: The network of fine tubules in the cytoplasm of cells for structural framework and circulating pathway.

Endosome: The central dark staining body in the nucleus of an amoeba that probably corresponds to the nucleolus.

Endospore: The thick inner chitinous layer of the wall in a microsporidia spore.

Enteritis: The inflammation of the intestine.

Eosinophilia: An increase in the number of eosinophils in circulation or in tissues.

Eosinophilic granulocytes: Leucocytes which contain acid staining granules.

Eosinophilic meningoencephalitis: Meningitis characterized by the presence of numerous eosinophils.

Epimastogote: The elongated developmental stage of a trypanosome in which the flagellum arises anterior to the nucleus and emerges laterally to form a short undulating membrane along the body to the anterior end.

Epistomial disc: The vaulted apical cell surface in peritrichous ciliates, it is encircled by the haplo- and polykinety and these two rows together form the **adoral zone.**

Epithelioid cells: Mononuclear cells which are involved in the inflammatory reaction and they are common in granulomatous lesions.

Epithelioid cells: Cells which are formed from the transformation of macrophages during encapsulation (common in granulomas).

Epitope: The structural component of an antigen against which an immune response is directed and to which a specific antibody binds.

Epizootic: An outbreak of a disease or an unusually large increase in prevalence and/or intensity of a parasite in an animal population.

Epizootiology: The study of disease in animal populations.

Eukaryote: An organism whose cell(s) has a double nuclear envelope.

Eutropic: An aquatic habitat with abundant organic matter.

Exfoliation: The sloughing of cells from the tissue surface.

Exocytosis: The discharge from a cell of particles that are too large to diffuse through the cell membrane.

Exogenous budding: Reproduction by outgrowth of a protrusion (one or more buds) that is smaller than the parental cell.

Exophthalmus: Protrusion of the eye.

Exospore: The proteinaceous outer layer of the wall in a microsporidia spore.

Extracellular: Situated or occurring outside a cell.

Extrasporogonic: A phase of the developmental cycle which occurs parallel to the sporogonic phase.

Extravascular: Situated or occurring outside a vessel.

Extrusion apparatus: An elaborate apparatus that serves for injection of the sporoplasm of a microsporidia into the host cell.

Fc receptor: The receptor on a cell surface which selectively combines with the crystallizable fragment of an immunoglobulin.

Fibroblastic response: The response characterized by intensive hyperplasia of fibroblasts.

Fibroblasts: A type of connective tissue cell which is responsible for synthesis of collagen.

Fibrosis: The proliferation of connective tissues which consists of a high proportion of fibroblasts.

Fingerling: A young salmonid fish which is older than a fry, and is approximately finger size.

Flagellar pocket: An elongate tubular intucking of the body surface from which emerges a flagellum.

Flame cells: Cilia bearing cells in flatworms that have excretory and/or osmoregulatory functions.

Flashing: An erratic swimming behaviour in fishes.

Formalin: A 37% solution of formaldehyde.

Freund's complete adjuvant: A water-in-oil emulsion with killed mycobacteria; addition of an antigen to the emulsion significantly increases the production of antibody against the antigen.

Friable: Easily pulverized.

Funis: One of three bands of microtubular ribbons in *Hexamita* and *Spironucleus*; the ribbon (M_3) originates at kinetosome R_1 and passes from a nucleus posteriorly along the flagellar pocket.

Gamogony: Part of sexual reproduction; it generally precedes fertilization of macrogametes by microgametes.

Gastritis: The inflammation of the stomach.

Gastroderm: The lining of the gut.

Glomerulitis: The inflammation of the glomeruli of the kidney, with proliferative or necrotizing changes of the endothelial or epithelial cells or thickening of the basement membrane.

Glycocalyx: The glycoprotein and polysaccharide covering that surrounds many cells.

Golgi complex: The internal reticular apparatus in the cell cytoplasm which is involved in the secretory process.

Gonopore: The external opening of the reproductive system of a crustacean.

Graft: A piece of transplanted tissue.

Granulation tissues: Newly formed tissues (proliferating fibroblasts, endothelial cells, histiocytes and macrophages) involved in the healing of various types of lesions and in the reparative stage of inflammation.

Granulocytes: Granular leucocytes (eosinophils, basophils and neutrophils).

Granuloma: A lesion resembling a tumour which results from chronic inflammation and consisting primarily of aggregation of macrophages, epithelioid cells and some connective tissue elements.

Granulomatous inflammation: The type of inflammation characterized by the exudate composed of macrophages, epithelioid and giant cells and by connective tissue formation.

Haematophagous: Feeding on blood or lymph.

Haematuria: The presence of blood in the urine.

Haemodilution: The increase of fluid content in the blood with resulting decrease in concentration of its red blood cells.

Haemorrhage: Internal bleeding and subsequent clotting caused by the rupture of blood vessels.

Hamuli: Large sclerites that occur in pairs which are located centrally in the haptor of monogeneans.

Haplokinety: Paroral membrane of exceptional length which is around the epistomial disc in peritrichous ciliates.

Hapten: A substance that cannot initiate an immune response in an animal unless it is bound to a carrier because of its small size.

Haptor: The posterior and principal organ of attachment used by monogeneans.

Hepatomegaly: The enlargement of the liver.

Hepatopancreas: A part of the digestive tract in crustacea that extends from the midgut and is composed of numerous ducts and blind tubules.

Hermaphrodites: Organisms with both male and female genital systems.

Heteroxenous parasite: A parasite that requires more than one host to complete its cycle.

Histiocyte: A type of cell with phagocytic 'macrophage'-like properties; it is found in connective tissues.

Histolysis: The breakdown of host tissues.

Histophagous: Feeding on tissues of animals.

Holarctic: Of the entire Arctic region.

Holdfast: The attachment organ of some parasites (e.g. copepods, cestodes).

Holotrichous: Ciliates with cilia which cover the entire body surface.

Humoral immunity: Specific immunity mediated by antibodies.

Hyalinization: Describes an abnormally smooth and glassy appearance in tissues under the light microscopic.

Hyaluronidase: An enzyme that catalyses the hydrolysis of hyaluronic acid, the cement substance of tissues.

Hydropic: Pertaining to or affected with dropsy.

Hyperaemia: The increase in blood flow during acute inflammation.

Hyperbiotic: Pathological processes which have in common an increase in cell metabolism resulting in excessive growth.

Hyperosmotic: The increase in concentration of osmotically active components in a solution.

Hyperplasia: The increase in size of a tissue or an organ by the formation and growth of new cells.

Hyperplastic: Pertaining to abnormal multiplication or increase in the number of normal cells in normal arrangement in a tissue.

Hypertrophic: Pertaining to enlargement or overgrowth of an organ.

Hypertrophy: An increase in size of a tissue or an organ due to an increase in size of individual cells.

Hypnospores: The prezoosporangium of *Perkinsus* spp. that is produced by incubating the parasite from hosts in fluid thioglycollate medium for several days.

Hypobiotic: Pathological processes which have in common the decrease in cell metabolism resulting in deficient growth or abnormal patterns of growth.

Hypochromic: A condition in which the amount of haemoglobin in red blood cells is lower than normal.

Hypoesthesia: Abnormal decreased sensitivity to stimulation.

Hypoxia: The reduction of oxygen supply to tissues below physiological levels despite adequate perfusion of the tissue by blood.

Immunization: The protection of an organism against disease by exposing the animal to parasite antigens so that its immune system recognizes and produces a rapid response against the same antigens on subsequent exposures.

Immunodepression: The decrease in response of the immune system to antigenic materials because of an established infection or exposure to some chemical agents.

Immunoglobulin (Ig): A family of proteins which is made up of light and heavy chains linked together by disulphide bonds; it is usually produced in response to an antigenic stimulation.

Indirect fluorescent antibody technique (IFAT): A technique in which unlabelled antibody is incubated with the antigen and then overlaid with a fluorescent conjugated anti-immunoglobulin to form a sandwich.

Indirect (passive) agglutination test: The agglutination by specific antibodies of particles or red blood cells to which antigens have been chemically adsorbed on their surfaces.

Inflammation: The initial response in vertebrates to tissue injury; this is characterized by the release of amines which cause vasodilation, infilatration of blood cells and proteins, and redness.

Infranuclear microtubules: One of three bands of microtubular ribbons in *Hexamita* and *Spironucleus*; the ribbon (M_2) originates at kinetosome R_1, passes below one nucleus, crosses to the other, and then passes posteriorly along the flagellar pocket.

Infundibulum: The funnel-shaped buccal cavity of peritrichous ciliates; it is equipped with buccal ciliary apparatus.

Innate immunity: Various host defences against a parasite which are not dependent on prior exposure to the parasite nor are they dependent on immunological memory.

Integument: The tissue which covers an organism.

Intensity: The number of parasites per infected host.

Intercellular: Situated or occurring between cells in a structure.

Interleukin (IL): Glycoproteins derived from macrophages or T cells which exert regulatory effects on other cells.

Intermediate host: A host in which there is development of the asexual or immature stages of a parasite.

Interspecific: Between two different species.

Intracellular: Situated or occurring inside a cell.

Intraperitoneal: Within the peritoneal cavity.

Intraspecific: Among individuals of a species.

Isoelectric focusing (IEF): A method that separates isozymes based on their PI (isoelectric point) on a gel.

Isozymes: Different molecular forms of an enzyme with different physical properties.

Karyomastigont: The unit of mastigonts and associated nucleus.

Kilodaltons: The expression of the molecular weight (size) of a compound and is expressed as thousand atomic mass.

Kinetoplast: The part of the mitochondrion which contains DNA and is conspicuous after staining; it is located near the base of the flagellum.

Kinetosomal complexes: Clusters of kinetosomes, in *Hexamita* and *Spironucleus*, each complex comprises one anterior pair and one posterior pair of kinetosomes.

Kinetosomal pocket: A depression in the nucleus that accommodates the kinetosome.

Kinetostome (or basal body): A subpellicular cylinder of nine skewed, peripheral structures each composed of three microtubules. It produces a cilium or flagellum at its distal end.

Kinety: Rows of cilia; structurally and functionally integrated entity of kinetosomes and their cilia but not all kinetosomes are necessarily ciliferous.

Lacunar system: A system of fluid-filled sinuses in the tegument of acanthocephalans that may act to distribute nutrients.

Lacustrine: Of the lake habitat.

Lanceolate: Tapering at both ends.

Lemniscus: One of a pair of tegumental projections into the anterior portion of the acanthocephalan pseudocoelom.

Lesions: Abnormal changes in tissues or body functions.

Leucocyte (or leukocyte): A white blood cell.

Leucocytosis: The increase in the number of white blood cells.

Leucorrhoea: A whitish discharge from the vagina.

Linguiform lobopodium: Tongue-shaped, lobular pseudopodium of an amoeba that often has a hyaline cap at the advancing tip; used both in locomotion and feeding.

Liquefaction: The result of necrobiosis; conversion of tissue into a semisolid or fluid mass.

Littoral: The coastal zone.

Lymphocyte: A mononuclear, nongranular leucocyte with a dark staining nucleus and pale blue cytoplasm; this class of leucocyte includes cells responsible for producing antibodies and are in circulation and in organs such as lymph nodes, spleen, thymus, tonsils.

Lysin: An antibody which causes cell lysis (e.g. haemolysin) or a bacterial toxin which lyses cells.

Macrogamogenesis: The formation of macrogamonts during sexual reproduction.

Macronucleus: A large, usually highly polyploid vegetative nucleus in ciliates which controls trophic functions.

Macrophages: Leucocytes which are important in cellular defence and they have phagocytic activity.

Macrospores: Large spores in microsporidia characterized by spore di- or polymorphism.

Manca: The larval stage of a parasitic isopod; it has six pairs of legs and usually has good swimming ability.

Manubrium: The stalk of the bulla in a lernaeopodid copepod.

Marsupium: The chamber in which eggs are brooded in an isopod and it is formed by oostegites.

Mastigonemes: Fine filamentous structures arranged in one or more rows or bands and typically appearing at right angles to the flagellar shaft of some protozoa that have flagellated stages.

Mastigont: A complex of flagella-associated organelles in flagellated protozoa, including basal bodies (also called kinetosomes) and projecting and trailing flagella.

Mehlis gland: A gland which evacuates into the ootype in the digenean female system.

Melanomacrophages: Pigment bearing macrophages which aggregate in the haemopoietic tissues to form melanomacrophage centres. They contain melanin, haemosiderin and lipid pigments.

Melanophores: Dermal cells containing melanin.

Membranelle: A ciliary organelle in the buccal cavity of ciliates, it is composed of three short rows of densely set kinetosomes.

Memory (immunological): The ability of the immune system to recall a previous exposure to an antigen and to respond with an earlier and/or higher response.

Merogony: The sequence of asexual multiplication (syn. schizogony) to produce a great number of vegetative parasite stages.

Meront: An asexual developmental stage to produce merozoites.

Merozoite: A crescentic stage formed by a meront during asexual reproduction; a stage which may initiate another merogony or gamogony.

Mesangial cells: Irregularly shaped, phagocytic cells which support capillary loops in the renal corpuscles.

Mesotrophic: Aquatic habitat with intermediate organic load.

Metacercariae: Encysted cercariae of digeneans.

Metaplasia: A change in tissue type.

Metatrypanosome: The trypomastigote stage of a trypanosome that is formed in the vector after cyclopropagative development.

Microcytic: A condition in which red blood cells are smaller than normal.

Microfilaments: Very fine filaments, such as those occurring in longitudinal rows of tufts on the surface of *Spironucleus elegans*.

Microgamete: A male gamete which is usually motile.

Microgamogenesis: The formation of microgamonts and microgametes.

Microgamont: A cell which produces microgametes.

Micronucleus: A usually very small, diploid generative nucleus in ciliates which carries the genetic information.

Micropyle: An opening of the oocyst wall.

Microspores: Small spores in microsporidia which are di- or polymorphic.

Microtubular ribbons: Multiple linked microtubules in a flat, ribbon-like arrangement; they are also called microtubular bands, or fibres.

Microtubule: A hollow rigid cylindrical structure of 20–25 nm in diameter; microtubules may be linked to others to form a ribbon or band; they comprise the '9 + 2' structure in flagella.

Mictosporic: A myxosporean species whose plasmodia forms various types of spores.

Miracidia: The larval stage that hatches from a digenean egg.

Mitogen: Any agent which induces mitosis in cells.

Monoclonal antibody: Identical antibody (immunoglobulin) molecules produced by a cloned antibody producing cell.

Monocytes: Non-granular, usually phagocytic leucocytes.

Monopisthocotyleans: Tissue feeding monogeneans which have sclerotized hamuli in the haptor at some point in their life cycle.

Monosporic: A plasmodium or pseudoplasmodium which produce a single spore.

Monoxenous: A life cycle which involves one host.

Monozoic: Flagellates that have a single karyomastigont or cestodes that have a single set of reproductive organs.

Moulting: The shedding of an old cuticle during the development of a nematode.

mRNA: Messenger RNA.

Mucocyst: A membranous and/or fibrous structure located beneath the surface membrane of a ciliate or a flagellate; it is discharged through an opening in the pellicle as a mucus-like mass.

Mucopolysaccarids: Mucin substances which are composed of protein–sugar complexes.

Mucron: A cuticular spine at the posterior end in a nematode.

Mucus: A viscid secretion that is usually rich in mucin and is produced by epithelial goblet cells.

Multiple fission: Cell division in which the parent cell undergoes a series of nuclear divisions before the cytoplasm divides to produce daughter cells.

Muscularis externa: The external muscular layer of the vertebrate gut.

Myelitis: The inflammation of the spinal cord.

Myofibrils: Microfibrillar strands or bundles with contractile function in protozoan cells.

Nacre: The iridescent inner layer (mother-of-pearl) of many mollusc shells.

Nauplius: The early larval stage of copepods and barnacles; it has three pairs of swimming appendages.

Necrosis: The alteration of tissue which results in cell death and formation of exudate.

Necrotic: Pertaining to cell death and enzymatic degradation in a limited area in an organism.

Nematocyst: A specialized organelle in coelenterates which contain a capsule with extrusible polar filament. Serves mostly for capture of prey.

Nematodesma: A bundle of microtubules in paracrystalline arrangement, associated with kinetosomes, reinforcing the walls of the cytopharyngeal tube in ciliates.

Neutralization: The process by which an antibody neutralizes the infectivity of a parasite.

Neutrophils: Leucocytes with polymorphic nuclei, and neutral staining granules; they have phagocytic properties and are present during inflammation.

Northern blot: A technique in which RNA is transferred from an agarose gel to a membrane and hybridized with the complementary DNA.

Obligate parasite: An organism that can only survive as a parasite on or in another animal.

Ocelli: Punctiform or crescentiform pigment markings, usually arranged segmentally on the urosome of leeches or radially around the caudal sucker.

Oedema: The accumulation of fluid in tissues.

Oligohaline: Water that is less salty than sea water, i.e. brackish, usually refers to the area where fresh water and sea water mix such as in an estuary, bay or areas adjacent to the shoreline.

Oligonucleotide primers: Short single strands of DNA used to initiate the DNA polymerase chain reaction on the template for the synthesis of the complementary strand.

Omnivorous: An animal with diverse diet.

Oncomiracidium: The free-swimming or crawling infection stage of monogeneans.

Oocyst: A one- to three-layered wall or membrane bound sack surrounding a sporont or sporocysts.

Oostegite: Thoracid plates that extend under a female isopod to form a marsupium in which the eggs are carried.

Ootype: A central chamber in the digenean female system which connects the oviduct, the seminal receptacle, the vaginal and the vitelline ducts.

Operculum: A cup-like opening of the egg of flatworms.

Opsonin: A factor which binds to particles or parasites and increases their susceptibility to phagocytosis.

Oral kinety: Any kinety associated with the oral region in ciliates.

Oral vaccination: The immunization of an animal in which the vaccine is mixed with food.

Osmolarity: The concentration of osmotically active particles in solution.

Osmoregulation: The adjustment of internal osmotic pressure in an organism in relation to changes in the external osmotic environment.

Oviparous: An organism that lays eggs.

Palintomy: Rapid binary fissions typically within a cyst and essentially without intervening growth.

Pansporoblast: A spore-producing formation within a polysporic plasmodium. It originates by the union of two generative cells (the pericyte and the sporogonic cell). The pericyte gives rise to the pansporoblast envelope while the sporogonic cell divides to produce the sporoblast cells.

Papillae: Small conical projections on the body surface.

Paraoral membrane: A ciliary organelle at the right border of the buccal cavity in ciliates. It consists of two rows of kinetosomes, of which only the outer one is ciliated.

Paratenic host: A transport host in which the larval stage of a parasite undergoes no development and its only function is to transfer the parasite to the next host.

Paraxial rod: An electron-dense ribbon or rod along the axoneme in the flagellar membrane.

Paresthesia: Morbid or abnormal sensation such as burning, prickling.

PAS reaction: The Periodic-Acid-Schiff (PAS) reaction used for specific staining of glycogen, neutral mucin, basement membranes and fungi.

Pathognomonic: Specifically distinctive or characteristic of a disease or pathological condition; a sign or indicant on which a diagnosis can be made.

Pectinelle: One of an equatorial band of short oblique rows of up to about six closely apposed cilia. It is used for locomotion in telotrochs of peritrichous ciliates; in adult ciliates the cilia are usually resorbed.

Pellicle: The cortex of the ciliate cell; it is composed of an outer cell membrane subtended with flat pellicular alveoli with underlying fibrous layer, the epiplasm.

Peniculus: A modified membranelle, with kinetosomes slightly more apart, forming sometimes more than three rows (up to seven) across.

Pentamer: A molecule consisting of five (sub)units.

Perforatorium: A cortical structure at the anterior of theronts and trophonts; it is used for penetration of epithelium and feeding.

Perichondrium: Cartilage forming tissue in an animal.

Pericyte: See pansporoblast.

Periopod: A walking leg of a crustacean.

Peristomial lip: A pellicular fold in peritrichous ciliates, it skirts the groove in which the cilia encircling the epistornial disc are inserted.

Petechial haemorrhage: A tiny haemorrhage about the size of a pin head.

Phagocytosis: The engulfment of microorganisms or other particulate material by cells.

Phagosome: A closed membrane-bound intracellular vesicle in a phagocyte formed by invagination of the cell membrane and the phagocytized material.

Pharynx: The muscular bulb at the anterior end of the gut in flatworms.

Plankton: Small organisms which float in water.

Plasma membrane: The trilaminated cell membrane.

Plasmodium: A multinucleate mass of protoplasm which is generally produced from a uninucleate stage.

Plasmotomy: The division of a multinucleate cell by cleavage in two or more cells.

Pleopod: An abdominal appendage of a crustacean.

Plerocercoid: The larval stage of cestodes which are adapted to aquatic transmission; the larval stage is usually in the fish intermediate host.

Poikilotherms: Vertebrates (fish, amphibians and reptiles) whose body temperature fluctuates with the ambient temperature.

Polar cap: A body stained by PAS (Periodic-Acid-Schiff) in the apex of a microsporidian spore; it is essentially identical with the anchoring disc.

Polaroplast: An organelle in the anterior part of a microsporidian spore which consists of an anterior membraneous and posterior vesicular part; it is instrumental in the extrusion of the polar tube.

Polar capsule: A thick-walled vesicle in myxozoan spores with an inverted polar filament; it forms a coil and on stimulation everts inside out.

Polar filament: *See* Polar capsule.

Polar tube: A tubular coiled organelle inside a microsporidian spore; it is used to inject the sporoplasm into a host cell.

Polykinety: An extremely elongated membranelle in peritrichous ciliates which encircles the epistomial disc.

Polymorphic: The ability of biomolecules, such as some enzymes, to exist in more than one molecular form.

Polyopisthocotyleans: These are blood feeding monogeneans, which attach to fish gills by means of sclerotized clamps.

Polypeptide: A peptide which on hydrolysis yields more than two amino acids.

Polysporic: A plasmodium which produces several spores.

Postoral kinety: A kinety which runs from the buccal cavity on the ventral side of a ciliate.

Praniza: The larval stage of a gnathid isopod.

Precipitin: An antibody or other substance which reacts with a soluble antigen to form a precipitate.

Premunition: The resistance to infection against the same or closely related pathogen after an acute infection has become chronic. The protection lasts as long as the infection persists.

Prepatent period: The period after infection but before the causative agent can be found using the usual diagnostic techniques.

Presporogonic: A sequence of development which precedes sporogony.

Prevalence: The percentage of animals in a population which are infected at any one time by particular organism.

Primary lymphoid organ: An organ which serves as the primary site where lymphocytes mature.

Proboscis: A muscular, protrusible feeding organ in rhynchobdellid leeches.

Proboscis sheath: The space surrounding the proboscis when it is retracted.

Procercoid: The first larval stage of many cestodes which develop inside the body cavity of the invertebrate (first) intermediate host.

Productive inflammation: *See* Proliferative inflammation.

Progenesis: Larval stages which attain sexual maturity and produce eggs.

Prohaptor: A non-sclerotized anterior organ of attachment used by monogeneans.

Proliferative inflammation: An inflammation which is characterized by a pronounced multiplication of fibroblasts; histiocytes and endothelial cells.

Pronephros: The primary kidney which is incorporated into the head kidney in adult fish.

Prostate gland: A gland which provides the supporting medium for male gametes.

Protandry: A type of hermaphroditism in which the individual first functions as a male and then as a female.

Protease: A general term for proteolytic enzymes.

Protista: Single-celled (or very few-celled and, therefore, still microscopic) eukaryotic microorganisms which diversified to give rise to animal, plant and fungal evolutionary lines.

Pseudocyst: A cyst-like structure surrounded by a dense fibrous capsule. It does not possess an inner lining and the formation of pseudocyst follows necrotic changes.

Pseudopodium: A temporary cytoplasmic projection of an organism which is used for feeding and/or for locomotion.

Pulsatile vesicles: Blister-like lateral appendages on the urosome of some piscicolid leeches; they are part of the coelomic system, and they contract/expand to circulate coelomic fluid.

Pyknotic: The degeneration of a cell in which the nucleus shrinks and the chromatin condenses to a solid amorphous mass.

Pyriform: Pear-shaped.

Raceway: An elongated earth pond or cement lined container with constant water flow.

Recurrent: Directed towards the posterior end.

Redia: The third larval stage of digeneans.

Regressive changes: A pathological process which is related to metabolic disorders and it includes dystrophic (degenerative) changes and necrosis.

Repair: A process to reestablish anatomical and functional integrity of tissues after an injury or inflammation.

Restriction fragment length polymorphism (RFLP): DNA fragments of different lengths after the DNA has been cleaved by restriction endonucleases at specific sites.

Rete system: A series of canals associated with the musculature in the body wall in an acanthocephala.

Retina: The sensory lining on the back of the eye.

Rhizocyst: A nail-like organelle which projects from the attachment disc of some parasitic dinoflagellates, e.g. *Piscinoodinium*; the organelle penetrates the host cells to provide anchorage for the parasite.

Rhizoids: Modified cytoplasmic projections which arise from the basal end of some parasitic dinoflagellates, e.g. *Amyloodinium*, *Crepidoodinium*; these projections attach to or penetrate host cells to provide anchorage.

Ribosomes: Intracytoplasmic granules which are rich in RNA and function in protein synthesis.

Rootlet fibril: *See* Striated lamella.

Sagenogenetosomes: Complex pit-like structures in the cell surface of the Labyrinthomorpha; they are presumed responsible for production of the ectoplasmic net which is involved in movement and nutrient acquisition.

Sanguiniferous: Blood feeders.

Schizogony: *See* Merogony.

Schizont: *See* Meront.

Scolex: *See* Holdfast.

Scolex pleuronectis: A scientific name which is used widely to describe tetraphyllidean plerocercoids whose adult stages are unknown.

Scopula: A small area of the pellicle at the aboral body end of sessiline peritrich ciliates. It consists of very short, imperfect and immobile cilia and is used for attachment to the substrate, either directly or by means of a secreted stalk.

Secondary lymphoid organ: An organ which contains the effector cells (activated lymphocytes and plasma cells) after an antigenic stimulation.

SEM (scanning electron microscopy): A technique which utilizes electrons to study the surface structures of an organism; this method provides increased resolutions at high magnifications compared to what is achieved with light microscopy.

Senescent: The depression of cellular functions in an animal as part of the aging process.

Serosanguineous: Pertaining to or containing both serum and blood.

Shell valve: One of the parts of the myxosporean spore wall.

Shucking: A procedure in which a mollusc is removed from its shell.

Smolt: A young salmon or sea trout that is about two years old.

Somatic: Pertaining to the body.

Somatic cilia: All cilia on the body surface which are outside the oral region.

Sonicated: Disruption of an organism by ultrasonic waves.

Spermatophore: A package of sperm, usually deposited by a male crustacean on to a female.

Sphaeromastigote: The round developmental stage of a trypanosome in which the flagellum is visible but the anterior end of the organism cannot be identified.

Splenomegaly: The enlargement of the spleen.

Spongiosis: Intercellular oedema of the spongy layer (malpighian layer) of the skin.

Sporangium: The developmental stage of some protozoa that is sac-shaped and is destined to produce multiple spores or trophozoites as in *Perkinsus* sp.

Spore: The infective stage of an organism that is usually protected from the environment by one or more protective membranes.

Sporoblast: A cell which develops into a spore; it is produced generally by the division of a sporont.

Sporocyst: A membrane bound sac which contains sporozoites of Apicomplexa or the second larval stage of digeneans in the gastropod intermediate host.

Sporogonial plasmodium: A multinucleate body that divides directly, or via sporoblast mother cells, into sporoblasts (undergoes sporogony).

Sporogonic cell: A cell which is produced by division of the sporoblast cell.

Sporogonic plasmodium: A multinuclear cell with many generative cells engaged in sporogony.

Sporogony: A phase in the development of an Apicomplexa in which the zygote initiates asexual reproduction and results in production of infective sporozoites.

Sporont: A developmental stage that gives rise to one or many sporoblasts.

Sporophorocyst: A thick envelope which is produced by early stages of heterosporis, and later is enlarged to contain all ensuing stages of merogony and sporogony.

Sporophorous vesicle: A solid envelope of parasite origin which encloses sporogonic stages and mature spores when they are formed.

Sporoplasm: The infectious component in spores.

Sporoplasmosome: An electron-dense inclusion in the sporoplasm.

Sporozoite: A nucleated infective stage formed by division of the sporont.

Sporulation: The formation of sporocysts and sporozoites within an oocyst.

Stigma: A pigmented red spot in dinoflagellates, it may also be present in the dinospore and other stages.

Stomopode: A membrane-bound tubular projection which arises proximal to the sulcus in the trophont stage of *Amyloodinium*.

Stress: An adverse stimulus which results in disruption of homeostasis in an organism.

Striated lamella: The fibrillar structure which extends from a kinetosome (R in Hexamitinae), into the deeper cytoplasm of the cell; typically it is cross-banded (striated).

Sublabia: Lip-like structures which are located below true lips, and they open into the oral cavity in some nematodes.

Suctorial tentacle: A rod-like organelle which extends from the body of suctorian ciliates. It is retractile and contains a microtubular cylinder, and has a capitate end. This end has a battery of haptocysts whose extrusomes contain toxic and lytic substances to capture prey and the prey cytoplasm is ingested through the tentacle.

Supranuclear microtubules: One of three bands of microtubular ribbons in *Hexamita* and *Spironucleus*; the ribbon (M_1) originates at kinetosome K_1, and passes from the top of one nucleus across to the top of the other nucleus.

Surfactant: A surface-active agent such as a soap or a detergent.

Suture line: A line of convergence of somatic kineties from different areas of the surface of a ciliate. There may be preoral sutures (in front of) and postoral sutures (behind the oral area).

Symmetrogenic division: A type of binary fission (generally longitudinal) in which the progenies are mirror images of each other wish respect to principal structures.

Syzygy: Side-by-side or end-to-end association of gamonts after the formation of gametocysts and gametes.

T cell: A thymus derived lymphocyte (*see* T lymphocyte).

Tegument: The nonciliated body wall of platyhelminths.

Tetramer: A molecule which consists of four (sub)units.

Theca/rhizotheca: A multilayered, membranous sheath which covers most of the surface of dinoflagellates.

Theront: The free-swimming infective form of *Ichthyophthirius and Cryptocaryon*.

Thigmotactic ciliature: Specialized somatic cilia which are presumed to have sensory-tactile or adhering function.

Thrombus (thrombi, *pl.*): Blood clot.

Thyroxine: A hormone secreted by the thyroid; its main function is to increase the rate of cell metabolism.

T lymphocyte: A subpopulation of lymphocyte whose development occurs in the thymus; these are effector cells in cell-mediated immune response and regulatory cells of the immune system.

Tomites: Cells within the tomont which result from serial binary division.

Tomont: A cyst-like structure formed by the trophont following detachment from the host; it supports binary division of the tomites and their maturation.

Torus (tori, *pl.*): Rounded swelling.

Trachelosome: The neck region of a piscicolid leech that includes clitellar and preclitellar segments.

Trophont: The feeding and growing stage of a dinoflagellate or a ciliate which differentiates into the reproductive tomont following detachment.

Trophozoite: The feeding, vegetative, and non-dividing stage in parasitic protozoa; it is also called a trophont.

Trunk: The 'body' of many parasitic copepods; it is formed from fused and enlarged genital segments and may also incorporate other thoracic segments.

Trypomastigote: The elongated developmental stage of a trypanosome in which the flagellum arises posterior to the nucleus and emerges laterally to form a long undulating membrane along the body to the anterior end.

Ulcer: The excavation of the surface of an organ or tissue and is produced by sloughing of necrotic inflammatory tissue.

Ulceration: A pathological condition which is usually associated with the skin or mucosal surfaces; it involves lesions with erosion of surface epithelia, and inflammation with infiltration of leucocytes.

Urosome: The main body region of a piscicolid leech between the clitellum and caudal sucker.

Urticaria: An allergic condition marked by red wheals.

Vaccine: An antigen preparation which consists of either whole cells or extracts of cells and is used to immunize animals.

Vacuolated: Containing spaces or cavities in the cytoplasm of a cell.

Valvogenic cell: One of the cells in a sporoblast in *Myxosporea* which differentiates into the shell valve.

Vascularization: The increase in supply of blood to a tissue; this is either by increasing blood volume (dilation of blood vessels) or the development of new blood vessels.

Vector: Any agent that transmits an infectious organism.

Ventriculus: The posterior glandular part of the oesophagus, and this structure is characteristic of many ascaridoid nematodes.

Vesicle: A small circumscribed elevation on the epidermis which contains serous fluid.

Vestibulum: A depression of body surface which leads to the cytostome and is equipped with ciliature of somatic origin; if it is in the shape of a groove, it is called a **vestibular groove**.

VH genes: Genes which code for the variable region of the heavy chain in immunoglobulins.

Virulence: The capacity of a parasite to cause disease in an animal; the damage may be modified by the defence mechanism of the host.

Vitellaria: Yolk glands.

Vitreous humour: The posterior chamber of the eye.

Viviparous: The bearing of live young.

Western immunoblotting: A technique to identify specific antigens in a mixture by separation on polyacrylamide gels, blotting onto nitrocellulose, and labelling with radiolabelled or enzyme-labelled antibodies.

Xanthochromic: Yellow in colour.

Xenoma: A symbiotic complex formed by a hypertrophic host cell and an intracellular parasite which proliferates in its cytoplasm.

Yearling: A one-year-old salmonid fish.

Zoonosis: A disease shared in nature by man and other vertebrates.

Index